TEACHER'S EDITION & PLANNIN

Algebra Essentials and Applications

by Joseph C. Power and Marie Petranic Power

HOLT, RINEHART AND WINSTON

A Harcourt Classroom Education Company

Austin · New York · Orlando · Atlanta · San Francisco · Boston · Dallas · Toronto · London

Project Editors
Andrew Roberts, *Executive Editor*
Higinio Dominguez, *Associate Editor*
Eileen Shihadeh, *Associate Editor*
June Turner, *Associate Editor*
April Warn, *Associate Editor*

Editorial Staff
Gary Standafer, *Associate Director*
Marty Sopher, *Managing Editor*

Book Design
Diane Motz, *Senior Design Director*
Lisa Woods, *Designer*
Tim Hovde, *Designer*
Charlie Taliaferro, *Design Associate*

Manufacturing
Jevara Jackson, *Manufacturing Coordinator*

Production
Gene Rumann, *Production Manager*
Rose Degollado, *Senior Production Coordinator*

Cover Design
Pronk & Associates

Research and Curriculum
Kathy McKee
Barbara Ryan
Guadalupe Solis
Jennifer Swift
Mike Tracy

Requests for permission to make copies of any part of the work should be mailed to the following address: Permissions Department, Holt, Rinehart and Winston, 1120 South Capital of Texas Highway, Austin, Texas 78746-6487

Photo Credits: Page T2(b), Yoav Levy/HRW Photo; T6(tr), Michelle Bridwell/HRW Photo; T9(tl), Randal Alhadeff/HRW Photo; T10(tr), Scott Van Osdol/HRW Photo.

Printed in the United States of America
ISBN: 0-03-064613-8
1 2 3 4 5 6   030   05 04 03 02 01 00

Algebra Essentials and Applications

A Comprehensive Algebra 1 Intervention Program

Contents of the Teacher's Edition and Planning Guide

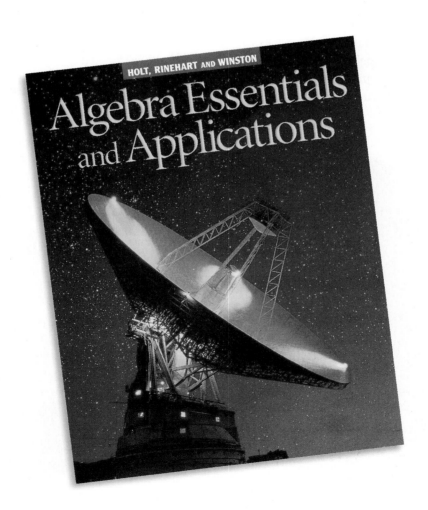

Algebra Essentials and Applications
Program Overview

A Complete Instructional Package *Algebra Essentials and Applications* is a comprehensive set of intervention materials designed to help first-year algebra students reach the levels of proficiency required by the mathematics content standards of most local and state boards of education. The creation of these materials is a conscientious effort to respond to the need for intervention materials for use in school districts that require all students to complete a course in algebra 1 prior to high school graduation.

Algebra Essentials and Applications is designed to be used with Holt, Rinehart and Winston's *Algebra 1,* to help students who experience difficulties in understanding algebra 1 concepts. Teachers who use Holt, Rinehart and Winston's *Algebra 1* as their primary algebra 1 text have the option of using *Algebra Essentials and Applications* whenever intervention is needed.

This combination provides teachers with more alternatives so they can more effectively meet the needs of all first-year algebra students.

Since *Algebra Essentials and Applications* directly addresses the core content of most algebra 1 courses of study, it can also be used as a stand-alone program. With detailed teaching guidance for each lesson as well as assessment support, this is a viable option for teachers who prefer to use a direct approach in the teaching of mathematics.

This program consists of five components. The primary component for students, *Algebra Essentials and Applications,* is designed to assist them in reaching proficiency of the core content of any algebra 1 curriculum. A number sense booklet, *Making Sense of Numbers: A Resource for Parents and Students,* is structured to provide

parents with a valuable tool to help their children succeed in operations with fractions, decimals, and percents.

Teacher support for classroom presentation of the content is provided in the *Algebra Essentials and Applications Teacher's Edition and Planning Guide*. Professional assistance is provided in *Teaching Algebra in the Middle Grades: A Professional Development Guide*. Various assessment options are included in the *Algebra*

Essentials and Applications Assessment Resources. Together, these five components help students attain algebra 1 proficiency by encouraging a collaborative effort among students, teachers, and administrators.

The diagram below shows how *Algebra Essentials and Applications* supports Holt, Rinehart and Winston's *Algebra 1* for those who use the program for intervention and how it can be used as a stand-alone program.

Algebra Essentials and Applications Pupil's Edition

Direct Instruction *Algebra Essentials and Applications Pupil's Edition* is the core component of the transitional package. This textbook provides students with a complete intervention course in algebra 1. Tailored to encompass local and state algebra 1 content standards, it provides direct instruction in a clearly designed and easy-to-follow format.

Each of the ten chapters begins with a two-page chapter opener that previews the chapter content, introduces new vocabulary, and provides a chart that illustrates how mathematical reasoning and skills interrelate with the chapter content. Each chapter contains six to nine lessons with each lesson providing direct instruction of one to three key algebra

skills. Each skill is presented in a two-page format with the first page providing instruction of a key skill or application and the second page providing practice. A continuous mixed review is included as an integral part of every practice page. Each chapter ends with a two-page chapter assessment.

Altogether, the student transitional component contains 77 lessons and 191 skills and applications. All skills and applications are targeted at student mastery of algebra 1 content as well as key pre-algebra concepts that are prerequisite to mastery of algebra 1.

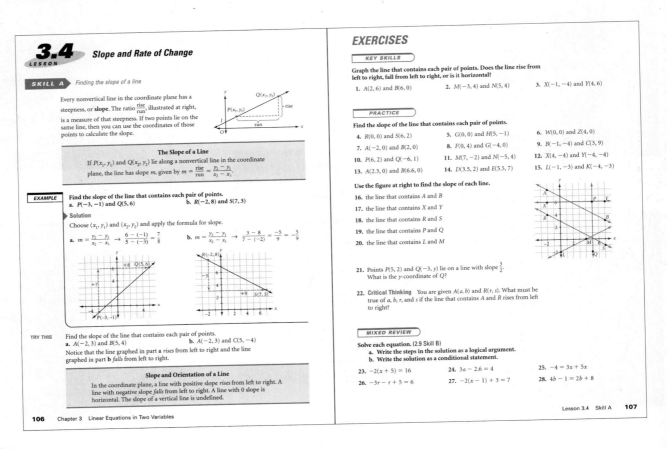

Algebraic and Linguistic Transitions for ESL Students

Transition As Change   Any transitional process involves change. When change is designed to occur in a carefully planned manner, transition results in progression and growth. For English as a Second Language (ESL) students, transition is a twofold process because these students need to develop both their academic and linguistic skills simultaneously.

Implementing Transition   Transition must capitalize on the knowledge students bring from their diverse backgrounds. To provide a point of departure that supports the learning needs of ESL students, *Algebra Essentials and Applications* includes lessons and practice material designed to accelerate the development of algebraic and linguistic skills. The book's direct instruction approach has a highly structured format that allows students to predict how the next algebraic skill will be presented. The use of skills with performance-type objectives tells students what learning behavior is expected of them as they progress through each lesson.

Lessons are structured with step-by-step examples followed by "Try This" exercises to provide immediate assessment of student understanding. Language used is simple, yet precise; accessible to all students, yet mathematically rigorous. This linguistic adaptation is based on important findings in the field of second language acquisition. According to these findings, certain English structures (e.g., two-word verbs, regular past, third person -*s*, articles, and irregular past) are more difficult to acquire than others (e.g., progressive -*ing*, auxiliary *be*, and plural -*s*).

> *"The success of implementing any vision of school algebra ultimately lies in creating conditions, policies, assessment, curriculum materials, and support that enables teachers to provide the kind of algebra experience that is essential for all students."*
>
> A Promising Practice, Algebra Working Group–1998

Maintaining Transition   It is important to recognize that ESL students do not become proficient in algebra or English in a short period of time. Their academic and linguistic transition requires continued support. An important distinction between *basic interpersonal communication skills and cognitive academic language proficiency* points to the fact that while some ESL students may possess fairly developed conversational skills, they may not have had opportunities or the time required to develop academic language, the language used in the learning, discussion, and communication of academic subject matter. A balanced combination of textbook reading, classroom discussion, and group interaction contributes significantly to accelerate ESL students' academic language learning.

Algebra Essentials and Applications emphasizes mathematics vocabulary through repetition of terms within the lesson material. New vocabulary terms are listed in the chapter openers. When a new vocabulary term is used, it appears in boldface type and is defined at point of use.

Exercises are carefully sequenced in increasing order of difficulty. Each exercise set begins with exercises that cover the key skills found in the instructional part of that particular skill's section. A mixed review of important math content from previous lessons is integrated into every exercise set.

Specific features found throughout the program that provide additional ESL support include the following.

- The use of italics to present key mathematical terms.
- The use of red annotations to explain the steps used in examples.
- The use of red call-out boxes to provide clues, tips, and reminders.
- The use of samples with solutions within the exercise sets to help students understand the exercise direction lines.
- The inclusion of a glossary and index at the end of the book to facilitate vocabulary comprehension.

Building a Community of Learners   To help ESL students develop their mathematical and linguistic proficiency, a connection between content, teaching, and learning must be established. Learning the content of algebra includes learning the symbolic system (the ability to interpret information encoded in mathematical symbols, such as $\frac{-b \pm \sqrt{b^2 - 4ac}}{2a}$), the structural system (the ability to perform traditional mathematical operations on algebraic expressions, such as $2b + 5b = 7b$), and the procedural system (the ability to follow procedures correctly, such as $2[a + b] = 2a + 2b$).

"Until a student is able to conceive of an algebraic expression as a mathematical object rather than as a process, algebraic manipulation can be a source of conflict."

Kieren, C. (1992). The Handbook of Research on Mathematics Teaching and Learning

---

In order to obtain the greatest benefit from our algebra transitional materials, meaningful home-school partnerships must be established. Such partnerships represent a source of invaluable help for ESL students in particular, and for all students in general, who can find additional support for learning by working closely with interested members of these partnerships.

Three additional resources have been prepared to enhance expertise of teachers and parents. *Algebra Essentials and Applications, Teacher's Edition and Planning Guide* and *Teaching Algebra in the Middle Grades: A Professional Development Guide* offer teachers ideas about how to present the theoretical background of key algebra topics to be taught in the classroom and the procedures required to solve problems. *Making Sense of Numbers* is a practical guide for parents to learn the meaning of numbers and operations with numbers in order to assist their children in understanding the conceptual and procedural aspects of algebra.

Holt, Rinehart and Winston's transitional materials are important resources designed to help students learn algebra. The program encourages a collaborative effort that involves continued participation of students, teachers, parents, and administrators.

Algebra Essentials and Applications
Assessment Resources

The assessment component of the transitional materials contains a variety of assessment resources.

- diagnostic screening tests
- quick warm-ups
- lesson quizzes
- chapter assessments
- mid-semester assessments
- semester assessments
- end-of-year assessments

Assessment Options   The lesson quizzes provide the teacher with a lesson-by-lesson assessment of the standards throughout the school year. These quizzes, along with the various chapter assessments and semester assessments, provide the teacher with continuous assessment of student proficiency with respect to the standards.

Diagnostic Screening Test   The diagnostic screening test is a tool that enables the teacher to determine a student's strengths and weaknesses in pre-algebra skills and concepts. If administered at the beginning of the school year, the teacher can pinpoint student deficiencies and provide individualized teaching and reinforcement using the *Making Sense of Numbers* student workbook. In combination with the diagnostic test, other measures can assist the teacher in placing students in one of the following programs.

Regular program
- *Algebra 1*

Transitional program
- *Algebra Essentials and Applications*
- *Making Sense of Numbers*

NAME _____ CLASS_____ DATE _____

3.4 Quick Warm-Up
LESSON   Slope and Rate of Change

Find each difference.

1. $9 - (-3)$ _____   2. $-8 - (-8)$ _____   3. $5 - 11$ _____

Solve. Assume that y varies directly as x.

4. When $y = 8$, $x = 2$. Find y when $x = 16$. _____

5. When $y = 3$, $x = 4$. Find y when $x = 12$. _____

3.4 Lesson Quiz
LESSON   Slope and Rate of Change

Find the slope of the line that contains each pair of points.

1. $E(5, 1)$ and $F(8, 6)$ _____   2. $P(-2, -1)$ and $Q(2, -5)$ _____

3. $A(-3, -9)$ and $B(-3, -1)$ _____   4. $S(-4, 7)$ and $T(10, 7)$ _____

5. A line with slope $-\frac{3}{4}$ contains $R(0, -2)$.

   a. Graph the line on the coordinate plane at right.

   b. Find the coordinates of a second point on the line.

6. Are $G(-3, 0)$, $H(3, -2)$, and $J(15, -6)$ collinear? Explain.

   _____

   _____

Calculate each rate of change.

7. In the year 1990, the population of Midville was 12,000. In the year 2000, the population of Midville was 11,400. _____

8. After 3 minutes of flight, a plane's altitude was 5000 feet. After 7 minutes of flight, the plane's altitude was 12,000 feet. _____

NAME _____ CLASS_____ DATE _____

3 Chapter Assessment, *page 1*
CHAPTER

Write the letter of the best answer.

_____ 1. Which point do you reach when you carry out these instructions in the coordinate plane?
   Start at the origin. Move 9 units to the left and then move 5 units up. Stop.
   a. $A(4, 0)$ b. $B(9, -5)$ c. $C(-9, 5)$ d. $D(-9, -5)$

_____ 2. Which ordered pair is *not* a solution to $y = -2x^2$?
   a. $(0, 0)$ b. $(-2, 1)$ c. $(-1, -2)$ d. $(1, -2)$

_____ 3. Which of the following sets of ordered pairs represent functions?
   I. $\{(-1, 1), (-1, 2)\}$ II. $\{(1, -1), (2, -1)\}$ III. $\{(-1, -1), (-2, -2)\}$
   a. I   b. I and III   c. II and III   d. I, II, and III

_____ 4. What is the range of the function at right?
   a. $\{1\}$ b. $\{-3, -2, -1, 0, 1, 2, 3\}$
   c. all integers   d. all real numbers

_____ 5. The equation $v = 10n$ expresses the value, v, of a collection of dimes as a function of the number of dimes, n. What is the domain of this function?
   a. all whole-number multiples of 10   b. all whole numbers
   c. all integers   d. all real numbers

_____ 6. Which equation gives the sale price, s, of an item being sold at 60% of its original price, t?
   a. $s = t - 40$ b. $s = t - 60$ c. $s = 60t$ d. $s = 0.6t$

_____ 7. If y varies directly as x, and $y = 1.2$ when $x = 4$, find y when $x = 6$.
   a. $y = 28.8$ b. $y = 20$ c. $y = 1.8$ d. $y = 0.8$

_____ 8. In four hours, the temperature rose from 23°F to 51°F. What was the rate of change?
   a. 4°F per hour   b. 7°F per hour   c. 24°F per hour   d. 28°F per hour

_____ 9. Find the slope of the line that contains the points $R(1, -9)$ and $S(-4, 1)$.
   a. -2 b. -0.5 c. 0 d. no slope

_____ 10. Which point is *not* collinear with $L(-6, 4)$ and $M(3, -2)$?
   a. $N(12, -8)$ b. $O(0, 0)$ c. $P(-15, 10)$ d. $Q(-9, -6)$

_____ 11. Which statement is true of the graph of $3x - 3y = 6$?
   a. Its y-intercept is -2.   b. Its y-intercept is 2.
   c. Its y-intercept equals its x-intercept.   d. It has no y-intercept.

T7

Algebra Essentials and Applications
Teacher's Edition and Planning Guide

The teacher's primary component of the transitional package, *Algebra Essentials and Applications Teacher's Edition and Planning Guide*, provides teaching support for every example and every skill of each student lesson in the *Algebra Essentials and Applications, Pupil's Edition*. Presented in a scripted and directed format, this important teaching tool follows a research-based three-phase instructional model.

Three-Phase Instructional Model

Phase 1 of the three-phase instructional model provides guidance for the teacher in presenting each example. Teachers are given alternate formats for presenting the examples as well as key questions for student discussion. Suggestions

to actively involve the student, such as visual strategies, are also provided. Through the use of various problem-solving strategies such as "guess and check," students learn to appreciate the power of algebraic methods. During the introductory stage of new material, students are not passive listeners. They are asked to examine relationships, describe algebraic processes and concepts, discuss answers, and work in groups or pairs.

Phase 2 provides guidance for the "Try This" activity that follows each example. Designed to take the student to the independent application of a new concept, the "Try This" activity provides students with a problem or question that requires students to work on their own while the teacher monitors performance. *The Teacher's*

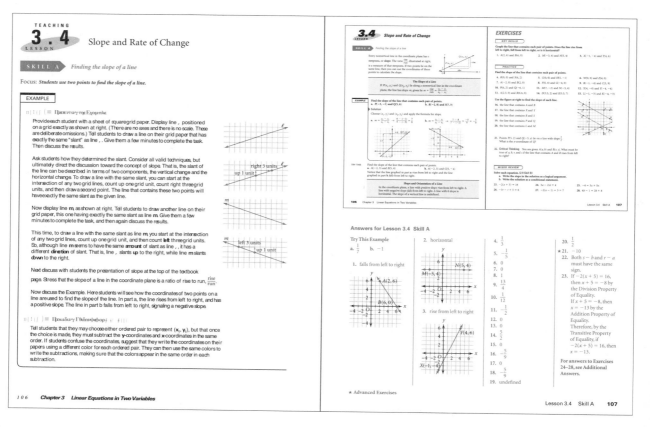

Edition and Planning Guide prepares teachers to recognize how students may interpret new concepts and then provides helpful hints to prompt students toward better understanding. It alerts teachers to mistakes students are likely to make when they are first asked to work on their own. Teachers are also cautioned of situations when students are likely to become confused during the learning process. Suggestions are then given for the appropriate instruction that best helps the student understand the concept.

In Phase 3, assessment and closure, the teacher asks the student to discuss, compare, summarize, consider alternatives, write, and otherwise test their understanding of the concepts and skills taught in the lesson. Students are also shown how the material is a prerequisite to the lessons that follow. For example, Phase 3 in Lesson 3.7, "Finding the Equation of a Line Given Two Points," states that in subsequent lessons, students will use this type of equation to model sequences of numbers and to examine relationships between lines on a coordinate plane. The closure portion of this phase of the instructional model reviews the goals and objectives of the lesson. Phase 3 also makes reference to the frequent quizzes and tests found in the *Assessment Resources* component of the transitional materials package.

In addition to the directed three-phase instructional model, the *Teacher's Edition and Planning Guide* contains a reduced reproduction of every student page with an on-page answer key to all of the student exercises. A diagram showing how the five components of the transition package are linked to Holt's regular *Algebra 1* program can be found on page T3 in this *Teacher's Edition and Planning Guide.*

Features

Focus   A Focus statement immediately follows the Skill statement at the beginning of each lesson in the *Teacher's Edition and Planning Guide.* This statement provides the teacher with a more detailed articulation of the lesson objective and thus sets the stage for the three-phase instructional model.

Examples   The Examples in the *Teacher's Edition and Planning Guide* provide the teacher with guidance for teaching the lesson concepts for each corresponding example in the *Pupil's Edition.* This guidance includes teaching suggestions with specific mathematical models for classroom presentation. Useful mathematical content information is also provided to assist teachers who may need to hone their own knowledge base of algebra topics. Within each of these examples, teachers can find the guidance needed to cover Phase 1 and Phase 2 of the three-phase instructional model.

Assessment and Closure   This section of the *Teacher's Edition and Planning Guide* brings closure to each lesson based on the objectives of all lesson skills. This section also connects the lesson content with the mathematical topics that follow. Reference is also made to the corresponding assessment material found in the *Assessment Resources.*

Teaching Algebra in the Middle Grades: A Professional Development Guide

Instructional materials play an important role not only for instruction of pupils, but also for instruction of teachers. *Teaching Algebra In the Middle Grades: A Professional Development Guide* is designed for use by middle school teachers whose teacher training is not in secondary mathematics. *Teaching Algebra in the Middle Grades* provides teachers with background knowledge of key algebra 1 level concepts. The goal of this booklet is to provide elementary or middle school teachers the knowledge they need to become confident algebra teachers. The content is organized into nine chapters.

1. The Role of the Mathematics Teacher in the Middle Grades
2. Linear Equations and Inequalities in One Variable
3. Linear Equations and Inequalities in Two Variables
4. Systems of Linear Equations and Inequalities
5. Polynomials and Factoring
6. Quadratic Equations and Functions
7. Rational Expressions and Functions
8. Functions
9. Mathematical Structure and Reasoning

The material is presented in an accessible writing style, with emphasis on relating algebraic concepts to algebraic processes. Chapter 1 addresses the many issues that middle school teachers face both in terms of where they fit into the instructional structure and in terms of what goals they are expected to reach. This chapter describes the transition that students must make as they move from a focus on numerical skills to the acquisition of intellectual and abstract concepts of algebra.

Teaching Algebra in the Middle Grades integrates procedural knowledge (calculation, measurement, formulas, data, and algorithms) and conceptual knowledge (algebraic functions, multiple representations, problem-solving strategies, abstractions, reasoning, and number sense). Throughout the book, teachers are invited to assign equal importance to both types of knowledge. Suggestions to achieve this balance are offered, such as discussions of algebraic methods and interpretations of answers, whiteboard demonstrations, guided and independent practice, and step-by-step solutions to problems.

The presentation of content is expressly written with a focus on the middle grades teacher as the audience. The purpose of the writing style is not to challenge the reader with abstract mathematical content. Instead, algebraic concepts are explained in an accessible writing style that relates the mathematical concepts and topics to the type of situations with which the middle school teacher is most familiar.

Chapter 2, "Linear Equations and Inequalities in One Variable," begins with a sample problem that illustrates making the transition from arithmetic to algebra. This chapter then moves to a discussion involving open sentences before delving into equation solving. The teacher is then instructed on how to chain various properties together to make equivalent equations until a solution of a complicated equation is found. From linear equations, the chapter makes the key transition to linear inequalities in one variable, thus emphasizing the importance of transferring knowledge between different but related skills.

Chapter 3 explores linear equations and inequalities in two variables. Here the coordinate plane is introduced before solving application problems that require the use of linear equations in two variables. The various aspects of slope are then discussed before the reader is required to find an equation for a line and later the solutions of linear inequalities in two variables.

Chapters 4, 5, and 6 focus on the traditional but somewhat more rigorous algebra 1 topics of systems of linear equations and inequalities, polynomials, factoring, and quadratic equations. Chapter 7 explains various skills that students must learn in order to work with rational expressions. In this chapter, teachers can review and preview adding, subtracting, multiplying, and dividing fractions and mixed numbers. As a result, teachers reinforce what they already know and lay the foundation for new algebraic work. The reader learns the meaning of a rational function, performs operations on rational functions, and uses them to solve real-world problems.

The concept of function is the topic of Chapter 8. The objective of this chapter is to provide the teacher with knowledge about this important mathematical concept so that they can teach students to 1) identify functions and relations represented as symbolic expressions, a table of ordered pairs, or a graph and 2) identify the domain and range of a given function. Specific types of functions studied in this chapter include linear functions, step functions, and quadratic functions.

Since mathematical competence depends not only on getting correct answers, but also on learning how to reason correctly according to rules of logic, the book ends with a chapter that discusses mathematical structure and reasoning. Chapter 9 brings the content of *Teaching Algebra in the Middle Grades* into a resourceful mathematical perspective for the middle school teacher. The first part of this chapter covers the mathematical structure of number sets and the second part covers mathematical reasoning. From deductive reasoning to indirect proof, the teacher is shown how the student takes logically correct steps to arrive at a correct solution to a mathematical problem.

Making Sense of Numbers: A Resource for Parents and Students

Parental Involvement Designed to facilitate parental involvement by way of its simple, direct, step-by-step presentation, *Making Sense of Numbers: A Resource for Parents and Students* provides comprehensive instruction and practice on topics such as fractions, decimals, ratios, proportions, percentages, integers, and rational numbers. These skills are often seen by teachers as a barrier to student success in algebra 1, and many students come to the algebra 1 class lacking these key prerequisite skills. State and local mathematics curricula require students in grades 6 and 7 to solve problems involving basic number sense skills. Specifically, these standards require students to use ratios to describe proportional relationships and to add, subtract, multiply, and divide decimals and fractions. Students must also be proficient in converting between fractions, decimals, and percents, and to calculate given percentages of quantities when solving problems involving interest earned, tips, and discounts on sales. Written in a format similar to the *Algebra Essentials and Applications Pupil's Edition,* this component provides instruction and practice for over 200 number sense skills.

A *Teacher's Manual and Answer Key* accompanies the student and parent workbook. This booklet provides the teacher with suggestions on how the workbook can be used in the classroom and at home. Also included are answers to all the exercises in the workbook as well as to the diagnostic screening tests. The diagnostic test allows the teacher to pinpoint student deficiencies in pre-algebra skills and directs the teacher to the appropriate practice pages in the *Making Sense of Numbers* workbook.

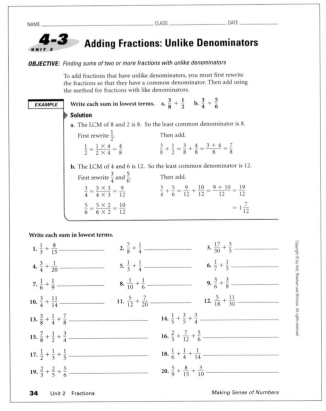

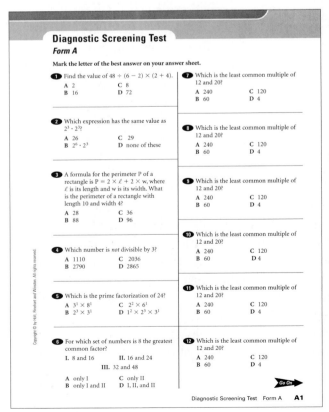

Algebra Essentials and Applications
A Complete Package of Transitional Materials

Algebra Essentials and Applications from Holt, Rinehart and Winston is a comprehensive package of transitional materials whose goal is to support teachers, students, and parents in bringing first-year algebra students to a level of proficiency. This is accomplished through the use of:

- a direct and simple approach that facilitates learning of algebra content

- continual assessment that focuses on learning of key algebra concepts while providing immediate feedback of student understanding,

- integrated mixed review that ensures retention of key algebra skills and concepts,

- immediate practice of each skill in the form of "Try This" exercises and exercise sets that provides immediate feedback of student comprehension,

- a format that is supportive of the learning needs of English language learners,

- a no-frills design that ensures focus on the important steps to learning the skills, and

- an arithmetic booklet, *Making Sense of Numbers,* that provides presentation and practice of all number sense skills for fractions, decimals, and percents.

Applications

Algebra Essentials and Applications provides a wide variety of problem-solving experiences for students. Approximately 20% of *Algebra Essentials and Applications* skills are designated "Applications Skills" because of their emphasis on word problems in realistic and meaningful student contexts.

Applications Skills are placed directly after skills that instruct and practice the conceptual and procedural skills the students need to solve the applications. Applications Skills are integrated into every chapter, appearing whenever they are appropriate. A list of applications and their respective page numbers are provided for reference below.

Applications

Problem-Solving Strategies

Problem solving is a process that involves concepts, skills, and an understanding of problem-solving strategies. *Algebra Essentials and Applications* instructs students in the following problem-solving plan, found in Lesson 2.7 Skill C on page 74.

1. Understand the problem.
2. Choose a strategy.
3. Solve the problem.
4. Check the results.

The second step in the problem-solving plan involves selecting and implementing one or more problem-solving strategies. This step is considered a key step in the problem-solving plan. Because of its importance, problem-solving strategies are utilized and labeled throughout the text.

A list of strategies and respective page numbers follows. By examining some of these pages, you can see how these strategies can be applied for different problems and in different ways.

Problem Solving Strategies

Make a diagram.	74, 174, 180, 260, 316	Look for key words.	4, 6, 26, 30, 34, 54, 142, 152, 158, 214
Make a graph.	100, 394		
Make a table.	14, 94, 116, 134, 136, 168, 200, 260, 286, 288, 316, 334, 336, 380	Use a formula.	168, 194, 208, 238, 244, 336, 366, 370, 394, 414, 418
Make an organized list.	16, 208, 270, 272	Write an equation.	30, 34, 74, 80, 136, 194, 226, 228, 316, 334, 336, 376, 378
Guess and check.	16, 86		
Look for a pattern.	330	Write an inequality.	152, 158, 162, 174, 180, 214

Logical Reasoning

Algebra Essentials and Applications is carefully structured and organized to promote logical reasoning. Skills provide exposure to logical reasoning in the form of step-by-step solutions to examples with justifications provided for each step. Exercise sets regularly ask students to justify their answers and Critical Thinking exercises are provided in most exercise sets.

Algebra Essentials and Applications also contains skills that explicitly instruct students in logical reasoning. These skills begin in Chapter 1 when students determine whether a statement is true or false by finding counterexamples. In Lesson 2.9, students are introduced to deductive and indirect reasoning, and in Lesson 3.9 students learn about inductive reasoning.

Algebra Essentials and Applications introduces the topic of mathematical proof in an accessible way for students. Students are instructed on how to write two-column proofs, and numerous proofs are provided for students as examples. For example, Lesson 8.4 Skill C on pages 322–323 is devoted to an understanding of the proof of the quadratic formula.

Planning and Pacing Guide

Lesson		Regular (days)	Block (days)
Chapter 1			
1.1	Skill A	1	
1.1	Skill B	1	1
1.2	Skill A	2	1
1.2	Skill B	1	
1.2	Skill C	1	1
1.3	Skill A	1	
1.3	Skill B	1	1
1.3	Skill C	1	
1.4	Skill A	1	1
1.5	Skill A	1	
1.6	Skill A	0.5	1
1.6	Skill B	0.5	
1.7	Skill A	1	
1.8	Skill A	1	1
1.9	Skill A	1	
1.9	Skill B	1	1
	Total	**16**	**8**

Lesson		Regular (days)	Block (days)
Chapter 2			
2.1	Skill A	1	
2.1	Skill B	1	1
2.1	Skill C	1	
2.2	Skill A	1	1
2.2	Skill B	1	
2.3	Skill A	0.5	1
2.3	Skill B	0.5	
2.3	Skill C	1	
2.4	Skill A	0.5	1
2.4	Skill B	0.5	
2.4	Skill C	1	
2.5	Skill A	1	1
2.5	Skill B	1	
2.6	Skill A	0.5	1
2.6	Skill B	0.5	

Lesson		Regular	Block
2.7	Skill A	1	
2.7	Skill B	0.5	1
2.7	Skill C	0.5	
2.8	Skill A	1	1
2.8	Skill B	1	
2.8	Skill C	0.5	
2.9	Skill A	0.5	1
2.9	Skill B	0.5	
2.9	Skill C	0.5	
	Total	**18**	**9**

Lesson		Regular	Block
Chapter 3			
3.1	Skill A	1	
3.1	Skill B	1	1
3.2	Skill A	1	
3.2	Skill B	1	1
3.2	Skill C	1	
3.3	Skill A	1	1
3.3	Skill B	1	
3.4	Skill A	1	
3.4	Skill B	0.5	1
3.4	Skill C	0.5	
3.5	Skill A	1	
3.5	Skill B	0.5	1
3.5	Skill C	0.5	
3.6	Skill A	1	
3.6	Skill B	1	1
3.7	Skill A	1	
3.7	Skill B	1	1
3.8	Skill A	1	
3.8	Skill B	0.5	1
3.8	Skill C	0.5	
3.9	Skill A	1	
3.9	Skill B	0.5	1
3.9	Skill C	0.5	
	Total	**19**	**9**

Lesson		Regular	Block
Chapter 4			
4.1	Skill A	1	
4.1	Skill B	1	1
4.1	Skill C	1	
4.2	Skill A	1	1
4.2	Skill B	1	
4.2	Skill C	0.5	1
4.3	Skill A	1	
4.3	Skill B	1	1
4.3	Skill C	0.5	
4.4	Skill A	1	1
4.4	Skill B	1	
4.5	Skill A	1	
4.5	Skill B	0.5	1
4.5	Skill C	0.5	
4.6	Skill A	1	1
4.6	Skill B	1	
4.6	Skill C	0.5	
4.7	Skill A	1	1
4.7	Skill B	0.5	
4.7	Skill C	0.5	
4.8	Skill A	0.5	1
	Total	**17**	**9**

Lesson		Regular	Block
Chapter 5			
5.1	Skill A	1	1
5.2	Skill A	1	
5.2	Skill B	1	1
5.2	Skill C	1	
5.3	Skill A	1	
5.3	Skill B	0.5	1
5.3	Skill C	0.5	
5.4	Skill A	1	1
5.4	Skill B	1	
5.5	Skill A	0.5	1
5.5	Skill B	0.5	
5.6	Skill A	1	
5.6	Skill B	0.5	1
5.6	Skill C	0.5	
	Total	**11**	**6**

Lesson		Regular	Block
Chapter 6			
6.1	Skill A	1	1
6.1	Skill B	1	
6.1	Skill C	0.5	
6.2	Skill A	1	1
6.2	Skill B	0.5	
6.3	Skill A	1	1
6.3	Skill B	1	
6.4	Skill A	1	
6.4	Skill B	0.5	1
6.4	Skill C	0.5	
6.5	Skill A	1	
6.5	Skill B	0.5	1
6.5	Skill C	0.5	
6.6	Skill A	1	
6.6	Skill B	0.5	1
6.6	Skill C	0.5	
6.7	Skill A	1	1
6.7	Skill B	1	
	Total	**14**	**7**

Lesson		Regular	Block
Chapter 7			
7.1	Skill A	1	1
7.1	Skill B	1	
7.1	Skill C	1	1
7.2	Skill A	1	
7.2	Skill B	0.5	
7.3	Skill A	1	1
7.3	Skill B	0.5	
7.4	Skill A	1	1
7.4	Skill B	1	
7.4	Skill C	1	1
7.5	Skill A	1	
7.5	Skill B	1	1
7.5	Skill C	1	
7.6	Skill A	1	1
7.6	Skill B	1	
7.7	Skill A	1	
7.7	Skill B	0.5	1
7.7	Skill C	0.5	
	Total	**16**	**8**

Chapter 8

Lesson		Regular	Block
8.1	Skill A	1	
8.1	Skill B	1	1
8.1	Skill C	1	
8.2	Skill A	1	1
8.2	Skill B	1	
8.2	Skill C	1	1
8.3	Skill A	1	
8.3	Skill B	0.5	1
8.3	Skill C	0.5	
8.4	Skill A	1	1
8.4	Skill B	1	
8.4	Skill C	1	
8.5	Skill A	0.5	1
8.5	Skill B	0.5	
8.6	Skill A	1	
8.6	Skill B	0.5	1
8.6	Skill C	0.5	
8.7	Skill A	1	1
8.7	Skill B	1	
Total		**16**	**8**

Chapter 9

Lesson		Regular	Block
9.1	Skill A	1	1
9.2	Skill A	1	
9.2	Skill B	1	1
9.2	Skill C	1	
9.3	Skill A	1	
9.3	Skill B	0.5	1
9.3	Skill C	0.5	
9.4	Skill A	1	1
9.4	Skill B	1	
9.4	Skill C	1	1
9.5	Skill A	1	
9.5	Skill B	1	1
9.5	Skill C	1	
9.6	Skill A	1	1
9.6	Skill B	1	
9.7	Skill A	1	
9.7	Skill B	0.5	1
9.7	Skill C	0.5	
9.8	Skill A	1	
9.8	Skill B	1	1
9.8	Skill C	0.5	
9.9	Skill A	0.5	
Total		**19**	**9**

Chapter 10

Lesson		Regular	Block
10.1	Skill A	1	1
10.2	Skill A	1	
10.2	Skill B	1	1
10.2	Skill C	1	
10.3	Skill A	1	1
10.3	Skill B	1	
10.3	Skill C	1	1
10.4	Skill A	1	
10.4	Skill B	1	1
10.4	Skill C	1	
10.5	Skill A	1	
10.5	Skill B	0.5	1
10.5	Skill C	0.5	
10.6	Skill A	1	
10.6	Skill B	0.5	1
10.6	Skill C	0.5	
Total		**14**	**7**

Total Teaching Days	**160**	**80**

Table of Contents

CHAPTER 1

CHAPTER 2 Operations With Real Numbers 37

CHAPTER 7

CHAPTER 8

CHAPTER 10

Chapter 1 Solving Simple Equations	State or Local Standards	Corresponding Lessons in *Algebra 1*, Schultz et al.
1.1 Relating Words and Mathematical Symbols		1.2
Skill A: Translating verbal phrases into mathematical expressions		
Skill B: APPLICATIONS Expressing real-world situations with mathematical symbols		
1.2 Numerical Expressions and Algebraic Expressions		1.2, 2.3, 2.5
Skill A: Identifying and using basic properties of numbers		
Skill B: Simplifying numerical expressions		
Skill C: Evaluating algebraic expressions and formulas		
1.3 Equations and Logical Reasoning		1.2, 2.5, 11.6
Skill A: Using replacement sets in solving equations		
Skill B: Determining whether a statement is true or false		
Skill C: Using basic properties of equality in proofs		
1.4 Solving Simple Equations Using Addition		3.1
Skill A: Using addition to solve an equation in one step		
1.5 Solving Simple Equations Using Subtraction		3.1
Skill A: Using subtraction to solve an equation in one step		
1.6 Solving Simple Equations by Choosing Addition or Subtraction		3.1
Skill A: Choosing addition or subtraction as the solution method		
Skill B: APPLICATIONS Choosing addition or subtraction in solving real-world problems		
1.7 Solving Simple Equations Using Multiplication		3.2
Skill A: Using multiplication to solve an equation in one step		
1.8 Solving Simple Equations Using Division		3.2
Skill A: Using division to solve an equation in one step		
1.9 Solving Simple Equations by Choosing Multiplication or Division		3.2
Skill A: Choosing multiplication or division as the solution method		
Skill B: APPLICATIONS Choosing multiplication or division in solving real-world problems		

1

Solving Simple Equations

▶ **What You Already Know**

For some years now, you have studied how to add, subtract, multiply, and divide many different kinds of numbers. For example, you have worked with whole numbers, fractions, decimals, and percents.

▶ **What You Will Learn**

In Chapter 1, you will learn to solve equations by using what you already know about addition, subtraction, multiplication and division.

In Chapter 1, you first learn how to translate verbal phrases into mathematical symbols. You will then reinforce your understanding of the order of operations by simplifying and evaluating various expressions.

Finally, you will examine different kinds of equations, some that are always true and others that are sometimes or never true.

VOCABULARY

Basic Properties of Equality
   reflexive   symmetric
   transitive
additive identity
algebraic expression
counterexample
equation
evaluate
expression
formula
identity
inverse operations
mathematical expression
mathematical proof
multiplicative identity
numerical expression

open sentence
opposite
order of operations
Properties of Real Numbers
   closure   commutative
   associative   identity
   inverse   distributive
Properties of Equality
   addition   subtraction
   multiplication division
reciprocal
replacement set
simplify
solution
Substitution Principle
variable

The diagram below shows how mathematical skills and mathematical reasoning are interrelated with the skills and concepts in Chapter 1. Notice that learning how to solve simple equations is a major focus of Chapter 1.

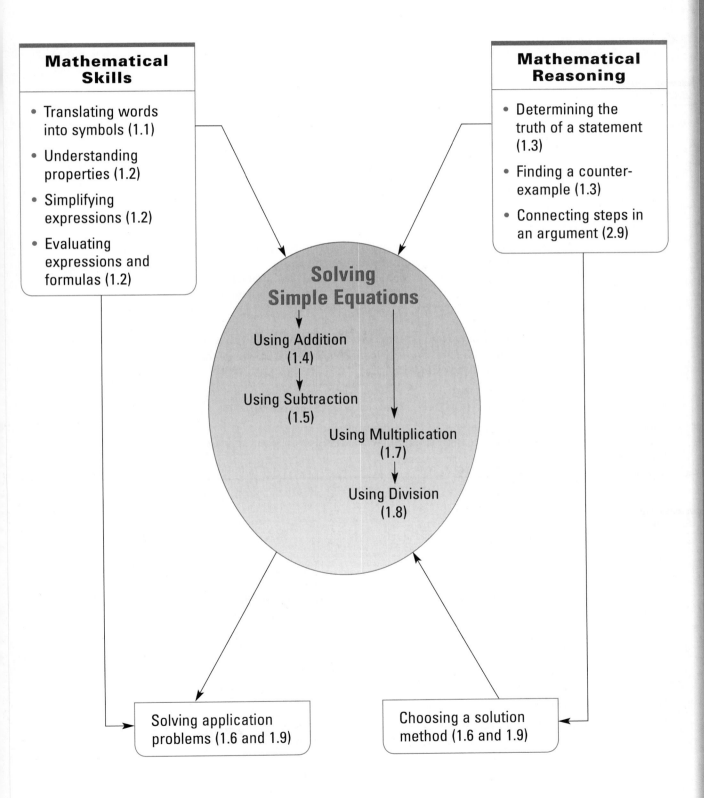

Mathematical Skills

- Translating words into symbols (1.1)
- Understanding properties (1.2)
- Simplifying expressions (1.2)
- Evaluating expressions and formulas (1.2)

Mathematical Reasoning

- Determining the truth of a statement (1.3)
- Finding a counter-example (1.3)
- Connecting steps in an argument (2.9)

Solving Simple Equations

Using Addition (1.4)

Using Subtraction (1.5)

Using Multiplication (1.7)

Using Division (1.8)

Solving application problems (1.6 and 1.9)

Choosing a solution method (1.6 and 1.9)

Relating Words and Mathematical Symbols

SKILL A ▶ *Translating verbal phrases into mathematical expressions*

Focus: *Students translate verbal phrases about addition, subtraction, multiplication, division, and grouping into mathematical symbols.*

EXAMPLE 1

PHASE 1: Presenting the Example

Display the traffic-sign shapes shown at right. Each sign indicates a driving "operation" to be performed. Ask students to identify the operations. [**top to bottom: stop; yield**]

Now display these "signs": $+$ $-$ $\times$ $\div$ $\bullet$ $/$ $(a)(b)$ $\dfrac{a}{b}$

Point out that, just as with the traffic signs, each mathematical sign indicates an operation to be performed. Review the indicated operations with students. [**left to right: addition; subtraction; multiplication; division; multiplication; division; multiplication; division**] Although variables are not formally introduced until later, be sure students understand that the letters a and b in the last two examples are standing in for unspecified numbers.

Now display these terms: addition, subtraction, multiplication, division. Work with students to develop a list of words and phrases associated with each. Elicit responses that include those in the box at the top of the textbook page.

PHASE 2: Providing Guidance for TRY THIS

Some students may write the difference as $3 - 12$. Commutativity is not discussed formally until a later lesson, but you can point out the differences in meaning using the language at right.

$12 - 3$: twelve less three
$3 - 12$: three less twelve

EXAMPLE 2

PHASE 1: Presenting the Example

After doing the example with the students, ask them why it is incorrect to write $(9 \times 3) - 4 + 2$. [**The interpretation would then be:** *the difference of the product of 9 and 3 minus 4, increased by 2.*] This indicates the importance of parentheses.

PHASE 2: Providing Guidance for TRY THIS

After doing the Try This exercise, have students work in groups of four. Each group member should write the symbols for a different one of the following phrases. Group members can then compare and contrast the results.

the product of 10 and 2, decreased by the sum of 3 and 4	$[(10 \times 2) - (3 + 4)]$
the sum of 10 and 2, decreased by the product of 3 and 4	$[(10 + 2) - (3 \times 4)]$
the sum of 10, 2, and 3, decreased by 4	$[(10 + 2 + 3) - 4]$
the sum of the product of 10 and 2 and the product of 3 and 4	$[(10 \times 2) + (3 \times 4)]$

1.1 LESSON

Relating Words and Mathematical Symbols

SKILL A ▸ *Translating verbal phrases into mathematical expressions*

You can use operation symbols, such as $+$, $-$, $\times$, or $\div$, to translate words and phrases into *mathematical expressions*. In algebra, letters can be used to represent numbers, such as the *a* and *b* in the table below.

Verbal Phrases	Expressions
a plus *b*, *a* increased by *b*, the sum of *a* and *b*, *b* added to *a*	$a + b$
b subtracted from *a*, *b* taken from *a*, *a* decreased by *b*, *a* less *b*, difference of *a* minus *b*	$a - b$
a multiplied by *b*, the product of *a* and *b*, *a* times *b*	$a \times b \quad a \cdot b \quad (a)(b) \quad ab$
a divided by *b*, the quotient of *a* and *b*, *a* per *b*	$a \div b \quad \frac{a}{b} \quad a/b$

EXAMPLE 1   Write each phrase as a mathematical expression.
  a. *the sum of 5 and 4*   b. *the difference of 5 minus 4*
  c. *the product of 5 and 4*   d. *the quotient of 5 divided by 4*

▸ **Solution**

Look for key words.

  a. the sum of 5 and 4   b. the difference of 5 minus 4
      $5 + 4$ $5 - 4$
  c. the product of 5 and 4   d. the quotient of 5 divided by 4
      5×4 $5 \div 4$

TRY THIS   Write each phrase as a mathematical expression.
  a. the sum of 12 and 3   b. 12 less 3
  c. 12 multiplied by 3   d. the quotient of 12 divided by 3

To indicate grouping in mathematics, you can use parentheses ( ).

EXAMPLE 2   Write the following phrase as a mathematical expression:
  the difference of the product of 9 and 3 minus the sum of 4 and 2.

▸ **Solution**

Look for key words.

  1. *Difference* indicates subtraction.   (product of 9 and 3) $-$ (sum of 4 and 2)
  2. Write the product and the sum 9×3 $4 + 2$
      with operation symbols.
  3. Use parentheses to indicate grouping. (9×3) $-$ $(4 + 2)$

TRY THIS   Write the following phrase as a mathematical expression:
  the difference of the sum of 4 and 3 minus the product of 10 and 2.

EXERCISES

KEY SKILLS

Write each mathematical expression in words.

1. 6.5×4.5 2. $6.5 - 4.5$ 3. $6.5 \div 4.5$
4. $6.5 + 4.5$ 5. $12 - (3.5 + 1)$ 6. $(3 + 4) - (3 + 1)$

PRACTICE

Write each phrase as a mathematical expression. Do not calculate.

7. the difference of 9 minus 9   8. the quotient of 9 divided by 9
9. the sum of 3 and 3   10. the product of 5 and 6
11. 12 less 4   12. 6 increased by 5
13. 18 divided by 6   14. the quotient of 6 divided by 18
15. 20 less than the sum of 6 and 2   16. 7 less than the sum of 5 and 10
17. the sum of 3 and 3, less the sum of 0 and 7
18. the sum of 7 and 12, decreased by the sum of 3 and 5
19. the sum of the product of 5 and 4 and the product of 3 and 6
20. the difference of 7 times 6 minus 5 times 4
21. the product of 3 and 4, plus the quotient of 20 divided by 4
22. the sum of the quotient of 18 divided by 9 and the quotient of 20 divided by 4

Critical Thinking   Write each phrase in as many ways as you can.

23. $\frac{1}{2}$ of *y*   24. *a* divided by $\frac{1}{2}$ 25. *b* added *a* times   26. *b* subtracted *a* times

MIXED REVIEW

Find each sum, difference, product, or quotient. Round your answers to two decimal places. (previous courses)

27. $100 + 1$ 28. $100 + 1.1$ 29. $100.1 + 1.1$ 30. $100.1 + 1.01$
31. $100.01 + 1.01$ 32. $100 - 1$ 33. $100 - 1.1$ 34. $100.1 - 1.1$
35. $100.1 - 1.01$ 36. $100.01 - 1.01$ 37. 100×1 38. 100×1.1
39. 100.1×1.1 40. 100.1×1.01 41. 100.01×1.01 42. $100 \div 1$
43. $100 \div 1.1$ 44. $100.1 \div 1.1$ 45. $100.1 \div 1.01$ 46. $100.01 \div 1.01$
47. 32×10 48. $32 \times 100,000$ 49. $32 \div 10$ 50. $32 \div 100,000$

Answers for Lesson 1.1   Skill A

Try This Example 1
a. $12 + 3$
b. $12 - 3$
c. 12×3
d. $12 \div 3$

Try This Example 2
$(4 + 3) - (10 \times 2)$

Answers for Exercises 1–6 may vary. Sample answers are given.
1. 6.5 times 4.5
2. 6.5 less 4.5
3. 6.5 divided by 4.5
4. the sum of 6.5 and 4.5
5. the difference when the sum of 3.5 and 1 is taken from 12
6. the sum of 3 and 4, less the sum of 3 and 1
7. $9 - 9$
8. $9 \div 9$
9. $3 + 3$
10. 5×6

11. $12 - 4$
12. $6 + 5$
13. $18 \div 6$
14. $6 \div 18$
15. $20 - (6 + 2)$
16. $(5 + 10) - 7$
17. $(3 + 3) - (0 + 7)$
18. $(7 + 12) - (3 + 5)$
19. $(5 \times 4) + (3 \times 6)$
20. $(7 \times 6) - (5 \times 4)$
21. $3 \times 4 + \dfrac{20}{4}$
22. $\dfrac{18}{9} + \dfrac{20}{4}$
23. Sample answer:
$\frac{1}{2}$ times y, $\frac{1}{2}$ multiplied by y, the product of $\frac{1}{2}$ and y,
$\frac{1}{2} \times y$, $\frac{1}{2} \cdot y$
$\frac{1}{2}y$, $\left(\frac{1}{2}\right)(y)$

24. Sample answer: the quotient of a and $\frac{1}{2}$
$a \div \frac{1}{2}$, $\dfrac{a}{\left(\frac{1}{2}\right)}$, $a / \left(\frac{1}{2}\right)$

25. Sample answer: the sum of a b's, b times a, a times b, $a \times b$, $a \cdot b$, $(a)(b)$, ab, $b \times a$, $b \cdot a$, $(b)(a)$, ba

26. Sample answer: the difference of a b's, $-b$ times a, a times $-b$, $-a \times b$, $-a \cdot b$, $(-a)(b)$, $-ab$, $-b \times a$, $-b \cdot a$, $(-b)(a)$, $-ba$, $-(a \times b)$, $-(a \cdot b)$, $-(a)(b)$, $-(ab)$

27. 101
28. 101.1
29. 101.2
30. 101.11
31. 101.02
32. 99
33. 98.9
34. 99
35. 99.09
36. 99
37. 100
38. 110
39. 110.11
40. 101.101
41. 101.0101
42. 100
43. 90.91
44. 91
45. 99.11
46. 99.02
47. 320
48. 3,200,000
49. 3.2
50. 0.00032

Focus: *Students translate real-world situations involving addition, subtraction, multiplication, division, and grouping into mathematical symbols.*

EXAMPLE 1

PHASE 1: Presenting the Example

As soon as students see numbers in a problem, they may immediately begin to calculate, giving little thought to their choice of operations. To help them focus on operations, first display the problem at right and have them write a brief response to the question. [**sample: Multiply the cost of each item by the number of items, and then add the results.**] Then show students how to translate their responses into the specifics of the example.

You buy some items at one cost and some at a different cost. How would you find the total cost?

EXAMPLE 2

PHASE 1: Presenting the Example

PHASE 2: Providing Guidance for **TRY THIS**

Some students may remember the rectangle perimeter formula in the form shown at right. Using this form, they can write the perimeter in the example as $(2 \times 5.5) + (2 \times 4.5)$ and the perimeter in Try This as $(2 \times 0.02) + (2 \times 0.01)$. In Lesson 1.2, students will learn why these expressions are equivalent to $2 \times (5.5 + 4.5)$ and $2 \times (0.02 + 0.01)$.

$P = 2l + 2w$

P: perimeter
l: length
w: width

EXAMPLE 3

PHASE 1: Presenting the Example

Have students first translate the problem into mathematical symbols using $\div$ as the division symbol. [$(87 + 90 + 78) \div 3 = (87 \div 3) + (90 \div 3) + (78 \div 3)$] Then show them how to use the fraction bar. Stress that the fraction bar is also a *grouping symbol*, making parentheses unnecessary in the numerator $87 + 90 + 78$ and in the quotients $\frac{87}{3}$, $\frac{90}{3}$, and $\frac{78}{3}$.

PHASE 3: ASSESSMENT AND CLOSURE for Lesson 1.1

- Have students write a verbal description of the operations in the expression at right. [**the product of 5 and the sum of 8 and 4**] Then have them write a real-world problem that the expression could represent. [**Answers will vary.**]

$5 \times (8 + 4)$

- Students should be aware that the skill of translating words into symbols is needed for solving all problems encountered in this course.

☞ *For a Lesson 1.1 Quiz, see* Assessment Resources *page 1.*

You can begin to solve a real-world problem by writing a statement that contains both words and mathematical symbols. The symbol "=" indicates a statement.

Verbal Statement	Mathematical Statement
a equals b, a is b, a is the same as b	$a = b$

EXAMPLE 1

The price for one CD is $14.95 and the price for one tape is $4.95. Write a statement that gives the total cost of three CDs and five tapes. Do not calculate.

Look for key words.

▶ **Solution**

The total cost equals the cost of three CDs plus the cost of five tapes.
cost of three CDs: $3 \times \$14.95$ cost of five tapes: $5 \times \$4.95$
total cost $= (3 \times \$14.95) + (5 \times \$4.95)$

TRY THIS The prices for one shirt and one pair of pants are $22.95 and $44.50, respectively. Write a statement that gives the total cost of two shirts and two pairs of pants.

EXAMPLE 2

The perimeter of a rectangle is found by multiplying the sum of its length and width by 2. The length of a rectangle is 5.5 feet and its width is 4.5 feet. Write a statement for the perimeter.

▶ **Solution**

The perimeter equals 2 times the sum of the length and width.
perimeter $= 2 \times (5.5 + 4.5)$
perimeter $= 2(5.5 + 4.5)$

TRY THIS Write a statement for the perimeter of a rectangle which has a length of 0.02 in. and a width of 0.01 in.

EXAMPLE 3

One way of finding the average of the test scores 87, 90, and 78 is to add the scores and divide this sum by 3. Another way is to divide each score by 3 and then add these quotients. Write a mathematical statement to show that the results of these two methods are the same.

▶ **Solution**

Adding, then dividing   Dividing, then adding
$(87 + 90 + 78) \div 3 = (87 \div 3) + (90 \div 3) + (78 \div 3)$
$\frac{87 + 90 + 78}{3} = \frac{87}{3} + \frac{90}{3} + \frac{78}{3}$

TRY THIS Rework Example 3 for the scores 80, 97, and 75.

EXERCISES

KEY SKILLS

Write each statement using both words and mathematical symbols. Do not calculate.
Sample: the total number of miles a car traveled after 5 trips of 15 miles each
Solution: total miles traveled = 5 trips × 15 miles= 5 × 15

1. the total cost of 12 tickets if each ticket costs $5

2. the total amount of water in 12 glasses that each holds 8 ounces

3. the amount of money in the bank after $35 is withdrawn from $1200

4. the number of cookies each of four students receives when 24 cookies are shared equally

PRACTICE

Write each statement using words and mathematical symbols as in Examples 1 and 2. Do not calculate.

5. the total cost of 7 CDs at $14.95 each plus 3 tapes at $4.50 each

6. the weight of 7 books each weighing 1.3 pounds added to the weight of 5 notebooks each weighing 0.3 pounds

7. the total cost of 7 cans of paint at $20.14 each, less 3 cans of paint costing $14.25 each

8. the perimeter of a rectangle whose length is 126 feet and whose width is 44.5 feet

9. the perimeter of a rectangle whose length and width are both 12.5 feet

Write each sentence as a mathematical statement, as in Example 3.

10. When 5 and 18 are added and the sum is then divided by 2, the result is the same as one half of 5 plus one half of 18.

11. When you add 3 to 5 and then add 6, you get the same result as adding 6 to 3 and then adding 5.

MIXED REVIEW

Solve each problem. (previous courses)
Sample: $60 \times \frac{3}{5}$ Solution: $\frac{\overset{12}{60}}{1} \times \frac{3}{\underset{1}{5}} = 36$

12. $42 \times \frac{1}{3}$ 13. $42 \times \frac{1}{6}$ 14. $42 \times \frac{2}{3}$ 15. $42 \times \frac{5}{6}$

Sample: $33 \div \frac{3}{4}$ Solution: $\frac{\overset{11}{35}}{1} \times \frac{4}{\underset{1}{3}} = 44$

16. $42 \div \frac{1}{3}$ 17. $42 \div \frac{1}{6}$ 18. $42 \div \frac{2}{3}$ 19. $42 \div \frac{6}{7}$

Answers for Lesson 1.1 Skill B

Try This Example 1
$(2 \times 22.95) + (2 \times 44.50)$

Try This Example 2
$2(0.02 + 0.1)$, or $(2 \times 0.02) + (2 \times 0.1)$

Try This Example 3
$\frac{80 + 97 + 75}{3} = \frac{80}{3} + \frac{97}{3} + \frac{75}{3}$

1. 12×5
2. 12×8
3. $1200 - 35$
4. $\frac{24}{4}$
5. $(7 \times 14.95) + (3 \times 4.50)$
6. $(7 \times 1.3) + (5 \times 0.3)$
7. $(7 \times 20.14) + (3 \times 14.25)$
8. $2(126 + 44.5)$
9. $2(12.5 + 12.5)$
★10. $\frac{5 + 18}{2} = \frac{5}{2} + \frac{18}{2}$, or $\frac{5 + 18}{2} = \frac{1}{2}(5) + \frac{1}{2}(18)$
★11. $(5 + 3) + 6 = (3 + 6) + 5$
12. 14

13. 7
14. 28
15. 35
16. 126
17. 252
18. 63
19. 49

★ Advanced Exercises

Numerical Expressions and Algebraic Expressions

SKILL A *Identifying and using basic properties of numbers*

Focus: *Students learn the fundamental properties of addition and multiplication: closure, commutativity, associativity, identities, inverses, and distributivity.*

EXAMPLE 1

PHASE 1: Presenting the Example

Be sure students understand that the letters *a*, *b*, and *c* represent any numbers. They will learn more about *variables* like these in Lesson 1.2 Skill C.

Divide students into six groups and assign each group a different property. Tell students to read the mathematical language for the property and write a verbal description in their own words. Here are some possible descriptions:

Closure	*The sum or product of two numbers is also a number.*
Commutative	*The order in which you add (multiply) two numbers does not change the sum (product).*
Associative	*When you add (multiply) three numbers, changing the way that you group the numbers does not change the sum (product).*
Identity	*The sum of any number and 0 is equal to the number.*
	The product of any number and 1 is equal to the number.
Inverse	*The sum of a number and its opposite is 0.*
	The product of a number and its reciprocal is 1.
Distributive	*The product when a number is multiplied by a sum of two other numbers is the same as the first number times the second plus the first number times the third.*

PHASE 2: Providing Guidance for TRY THIS

Be sure that students distinguish properties of *addition* from properties of *multiplication*. Note that the Distributive Property involves both operations.

EXAMPLE 2

PHASE 1: Presenting the Example

You may want to separate the steps in Example 2, as shown at right.

$$(1)(10) + 3\left(\frac{1}{3}\right) = 10 + 3\left(\frac{1}{3}\right)$$ ← *Identity Property of Multiplication*
$$= 10 + 1$$ ← *Inverse Property of Multiplication*

PHASE 2: Providing Guidance for TRY THIS

Some students will need to write the decimals as fractions in order to see the reciprocals. It may be helpful to rewrite the exercise as shown at right.

$$10(0.1) + 100(0.01)$$
$$\downarrow \qquad\qquad \downarrow$$
$$10\left(\frac{1}{10}\right) + 100\left(\frac{1}{100}\right)$$

1.2 Numerical Expressions and Algebraic Expressions

LESSON

SKILL A | *Identifying and using basic properties of numbers*

In arithmetic, you learned how to add and multiply numbers. The basic properties of real numbers that govern these operations are shown below.

Let a, b, and c represent real numbers.

Property	Addition	Multiplication
Closure	$a + b$ is a real number.	ab is a real number.
Commutative	$a + b = b + a$	$a \cdot b = b \cdot a$
Associative	$a + b + c = a + (b + c)$ $= (a + b) + c$	$a \cdot b \cdot c = a \cdot (b \cdot c)$ $= (a \cdot b) \cdot c$
Identity	Because $a + 0 = 0 + a = a$, 0 is called the **additive identity**.	Because $a \cdot 1 = 1 \cdot a = a$, 1 is called the **multiplicative identity**.
Inverse	For every real number a, there is an **opposite** real number, $-a$, and $a + (-a) = -a + a = 0$.	For every nonzero real number a, there is a **reciprocal**, $\frac{1}{a}$, and $a \cdot \frac{1}{a} = \frac{1}{a} \cdot a = 1$.

The Distributive Property links addition and multiplication.

> **Distributive Property**
> If a, b, and c are real numbers, then $a(b + c) = ab + ac$.

EXAMPLE 1
Which property is illustrated by this statement?
Adding two numbers and then adding a third number gives the same answer as adding the sum of the second two numbers to the first.

▶ **Solution**
The statement illustrates the Associative Property of Addition, $(a + b) + c = a + (b + c)$.

TRY THIS
In Example 1, which property is illustrated if *sum* is replaced by *product* and *adding* is replaced by *multiplying*?

EXAMPLE 2
Use properties to simplify $(1)(10) + 3\left(\frac{1}{3}\right)$.

▶ **Solution**

Identity Property of Multiplication → $(1)(10) + 3\left(\frac{1}{3}\right)$ ← Inverse Property of Multiplication
$10 + 1$
11

TRY THIS
Use properties to simplify $10(0.1) + 100(0.01)$.

EXERCISES

KEY SKILLS

Which property is illustrated by each statement?

1. $2 + 0 = 2$
2. $(1)2.5 = 2.5$
3. $2.4 + 7.1 = 7.1 + 2.4$
4. $1 + (2 + 5) = (1 + 2) + 5$
5. $5\left(\frac{1}{5}\right) = 1$
6. $4 \times 10 = 10 \times 4$

PRACTICE

Which property is illustrated by each statement?

7. Multiplying 5 by 9 gives the same answer as multiplying 9 by 5.
8. Adding 5 to 9 gives the same number as adding 9 to 5.
9. Add 0 to any number and you get the same number.
10. Multiply any number by 1 and you get the same number.
11. Multiply any nonzero number by its reciprocal and the result is 1.

Simplify each expression. Use mental math where possible. Identify the property or properties used.

12. $5 \times \frac{1}{5}$
13. $\frac{5}{3} \times \frac{3}{5}$
14. $1 \times \frac{1}{7}$
15. $\frac{1}{3} \times 1$
16. $0 + \frac{5}{2}$
17. $3 + 0 + \frac{5}{2}$
18. $3 + \frac{5}{2} + 0$
19. $1 \times 3 \times 5$
20. $4 \times 1 \times 7$
21. $\left(\frac{5}{2} \times \frac{2}{5}\right) + \left(3 \times \frac{1}{3}\right)$
22. $\left(1 \times \frac{2}{5}\right) + \left(1 \times \frac{3}{5}\right)$
23. $(1 \times 6) + \left(\frac{5}{3} \times \frac{3}{5}\right)$
24. $(1 \times 7) + \left(\frac{2}{9} \times \frac{9}{2}\right)$
25. $2(4.5 + 0)$
26. $3(12 + 0)$

27. **Critical Thinking** Show a convenient way to calculate a 15% tip on a meal that costs $12.00 by using the Distributive Property. Do not calculate.

MIXED REVIEW

Find the value of each expression. (previous courses)
(Recall that a^2 means $a \times a$.)

28. 2^2
29. 3^2
30. 7^2
31. 5^2
32. 10^2
33. 100^2
34. 1^2
35. 0^2
36. $\left(\frac{1}{2}\right)^2$
37. $\left(\frac{3}{4}\right)^2$
38. $\left(\frac{5}{2}\right)^2$
39. $\left(\frac{7}{4}\right)^2$
40. $(0.5)^2$
41. $(0.2)^2$
42. $(0.05)^2$
43. $(0.002)^2$

Answers for Lesson 1.2 Skill A

Try This Example 1
Associative Property of Multiplication

Try This Example 2
2

1. Additive Identity
2. Multiplicative Identity
3. Commutative Property of Addition
4. Associative Property of Addition
5. Multiplicative Inverse
6. Commutative Property of Multiplication
7. Commutative Property of Multiplication
8. Commutative Property of Addition
9. Additive Identity
10. Multiplicative Identity
11. Multiplicative Inverse
12. 1; Multiplicative Inverse

13. 1; Multiplicative Inverse
14. $\frac{1}{7}$; Multiplicative Identity
15. $\frac{1}{3}$; Multiplicative Identity
16. $\frac{5}{2}$; Additive Identity
17. $\frac{11}{2}$, or $5\frac{1}{2}$; Additive Identity
18. $\frac{11}{2}$, or $5\frac{1}{2}$; Additive Identity
19. 15; Multiplicative Identity
20. 28; Multiplicative Identity
21. 2; Multiplicative Inverse
22. 1; Multiplicative Identity
23. 7; Multiplicative Identity, Multiplicative Inverse
24. 8; Multiplicative Identity, Multiplicative Inverse
25. 9; Additive Identity
26. 36; Additive Identity

27. Sample answer: $12(1 + 0.15)$
28. 4
29. 9
30. 49
31. 25
32. 100
33. 10,000
34. 1
35. 0
36. $\frac{1}{4}$
37. $\frac{9}{16}$
38. $\frac{25}{4}$
39. $\frac{49}{16}$
40. 0.25
41. 0.04
42. 0.0025
43. 0.000004

SKILL B Simplifying numerical expressions

Focus: *Students use the order of operations to simplify numerical expressions.*

EXAMPLE 1

PHASE 1: Presenting the Example

Display the expression $2 + 3 \times 8$. Point out that it involves two operations, addition and multiplication. Tell half of the class to perform the addition first, and then the multiplication. Tell the other half to perform the multiplication first, and then the addition. Compare their answers. [**40, 26**]

Tell students that this activity demonstrates the need for a set of rules for expressions that involve more than one operation. For this reason, mathematicians accept the *order of operations* outlined on the textbook page. Ask students to find the value of $2 + 3 \times 8$ according to these rules. [**26**] Be sure they understand that an answer of 40 is incorrect.

Before presenting the Example, you may wish to review exponents. Remind students that an expression like 4^2 is read as "4 squared" or "4 to the second power," and it means "4 multiplied by itself," or 4×4.

PHASE 2: Providing Guidance for TRY THIS

Some students may experience difficulty with this expression because the fraction bar is both a grouping symbol and a division symbol. These students may find it helpful to rewrite such expressions in two steps, as shown below.

$$3 + \frac{(2 + 3)^2}{2(2 + 1) + 2}$$

Step 1
Group the numerator.
Group the denominator.

$$= 3 + \frac{[(2 + 3)^2]}{[2(2 + 1) + 2]}$$

Step 2
Rewrite the fraction using the division symbol $\div$.

$$= 3 + [(2 + 3)^2] \div [2(2 + 1) + 2]$$

EXAMPLE 2

PHASE 1: Presenting the Example

PHASE 2: Providing Guidance for TRY THIS

When they encounter expressions like those in Example 2, some students are confused by what they see as a maze of symbols. You may want to suggest that these students always work vertically, maintaining a very strict vertical alignment. A sample display of the expression in Example 2 is shown at right.

$$\begin{array}{l}
[2(5 - 1)^2 - 3(10 - 8)^2]^2 \\
[2(\ \ 4\ \ )^2 - 3(\ \ \ 2\ \ \ )^2]^2 \\
[2(\ 16\ ) - 3(\ \ \ 4\ \ \ )\,]^2 \\
(\ \ \ 32\ \ \ \ -\ \ \ \ 12\ \ \ )^2 \\
(\ \ \ \ \ \ \ \ \ \ 20\ \ \ \ \ \ \ \ \ \ )^2 \\
\ \ \ \ \ \ \ \ \ \ \ 400
\end{array}$$

A **numerical expression** contains operation symbols and numbers. To **simplify** a numerical expression, perform all the indicated operations using the following order of operations.

Order of Operations
1. Simplify expressions within grouping symbols.
2. Simplify expressions involving *exponents*.
3. Multiply and divide from left to right.
4. Add and subtract from left to right.

In addition to parentheses ( ), grouping symbols include square brackets [ ] and braces { }. The fraction bar is also a grouping symbol.

EXAMPLE 1 Simplify $\dfrac{(12-3)^2}{2(4+1)-2}$.

▸ **Solution**

$$\frac{(12-3)^2}{2(4+1)-2} = \frac{9^2}{2\times5-2} \quad \longleftarrow \text{ Work within parentheses first.}$$
$$12-3=9 \quad 4+1=5$$
$$= \frac{81}{2\times5-2}$$
$$= \frac{81}{10-2} \quad \longleftarrow \text{ Multiply before subtracting.}$$
$$= \frac{81}{8}, \text{ or } 10\frac{1}{8}$$

TRY THIS Simplify $3 + \dfrac{(2+3)^2}{2(2+1)+2}$.

In some expressions, you may need to work with groups within groups. Begin with the innermost grouping symbols and work outward.

EXAMPLE 2 Simplify $[2(5-1)^2 - 3(10-8)^2]^2$.

▸ **Solution**

$$[2(5-1)^2 - 3(10-8)^2]^2$$
$$[2\times(4)^2 - 3\times(2)^2]^2 \quad \longleftarrow \text{ Work in the innermost grouping symbols first.}$$
$$(2\times16 - 3\times4)^2 \quad \longleftarrow 4^2 = 16 \text{ and } 2^2 = 4$$
$$(32-12)^2 \quad \longleftarrow \text{ Multiply.}$$
$$20^2 \quad \longleftarrow \text{ Subtract.}$$
$$400 \quad \longleftarrow 20^2 = 20\times20$$

TRY THIS Simplify $4[3(10-8)^2 + 4(10-9)^2]$.

EXERCISES

Write the expression that results when you perform the operation(s) in parentheses. Do not continue after that step.

1. $4 + 3(10-3)$ 2. $3(12+3) - 5(10-1)$ 3. $(3+1)^2 + 5\times4$

4. $3\times7 - (5-1)^2$ 5. $\dfrac{(7+2)^2}{3\times2}$ 6. $\dfrac{(3+2)^2}{(10-8)^2}$

Use the order of operations to simplify each expression.

7. $2(3+4) + 3\times5$ 8. $3\times5 - 2(3+4)$ 9. $(3-1)(10+4)$

10. $(13-3)(13+3)$ 11. $\dfrac{(3+2)^2}{2(10-8)+1}$ 12. $\dfrac{(7+1)^2}{3(7-1)+18}$

13. $\dfrac{3+(5+2)^2}{7(7-1)-16}$ 14. $\dfrac{(5+2)^2-24}{6-(2-1)}$ 15. $\dfrac{(5+2)^2-(5+1)^2}{16-2(5+2)}$

16. $\dfrac{(3-2)^2+(1+1)^2}{2(5+3)-9}$ 17. $\dfrac{(3-2)^2+(1+1)^2}{(6-2)^2-(2+1)^2}$ 18. $\dfrac{(6-3)^2+(3+1)^2}{(6+4)^2-(4+1)^2}$

19. $\dfrac{4+4}{2(1+1)} + \dfrac{(3+1)^2}{4(3-1)}$ 20. $\dfrac{20+4}{3(3-1)} - \dfrac{(3+1)^2}{4(3-1)}$ 21. $[20(2+3) - 3(5+5)]^2$

22. $[2(2+1) + 3(2+2)]^2$ 23. $3[2(2+1)]^2$ 24. $7[2(6-1)]^2$

25. $[2(6-1)]^2 + 3(10-7)^2$ 26. $30(9-6)^2 - [2(6-1)]^2$

27. $\dfrac{[(3.2+0.8)^2 + (0.5+0.5)^2]}{[(6-1)^2 - (5-1)^2]^2}$ 28. $\dfrac{[(7.3+2.7)^2 - (5.5+0.5)^2]}{[(3-1)^2 + (3-1)]^2}$

Find each sum. (previous courses)

29. $\frac{1}{2} + \frac{1}{2}$ 30. $\frac{1}{3} + \frac{1}{3}$ 31. $\frac{2}{3} + \frac{2}{3}$ 32. $\frac{3}{4} + \frac{3}{4}$

Sample: $\frac{1}{4} + \frac{1}{2}$ Solution: $\left(\frac{2}{2}\right)\frac{1}{2} = \frac{2}{4} \rightarrow \frac{1}{4} + \frac{2}{4} = \frac{3}{4}$

33. $\frac{3}{4} + \frac{1}{2}$ 34. $\frac{1}{6} + \frac{1}{3}$ 35. $\frac{3}{10} + \frac{2}{5}$ 36. $\frac{1}{2} + \frac{1}{6}$

Sample: $\frac{1}{5} + \frac{1}{6}$ Solution: $\left(\frac{6}{6}\right)\left(\frac{1}{5}\right) + \left(\frac{5}{5}\right)\left(\frac{1}{6}\right) = \frac{6}{30} + \frac{5}{30} = \frac{11}{30}$

37. $\frac{1}{2} + \frac{1}{3}$ 38. $\frac{1}{2} + \frac{2}{3}$ 39. $\frac{1}{3} + \frac{3}{4}$ 40. $\frac{1}{3} + \frac{2}{5}$

41. $\frac{4}{5} + \frac{2}{3}$ 42. $\frac{2}{5} + \frac{3}{4}$ 43. $\frac{3}{5} + \frac{1}{6}$ 44. $\frac{3}{5} + \frac{5}{6}$

Answers for Lesson 1.2 Skill B

Try This Example 1
$6\frac{1}{8}$, or 6.125

Try This Example 2
64

1. $4 + 3(7)$
2. $3(15) - 5(9)$
3. $4^2 + 5\times4$
4. $3\times7 - (4)^2$
5. $\dfrac{9^2}{3\times2}$
6. $\dfrac{5^2}{2^2}$
7. 29
8. 1
9. 28
10. 160
11. 5
12. $\frac{16}{9}$, or $1\frac{7}{9}$
13. 2
14. 5

15. $\frac{13}{2}$, or $6\frac{1}{2}$
16. $\frac{5}{7}$
17. $\frac{5}{7}$
18. $\frac{1}{3}$
19. 4
20. 2
21. 4900
22. 324
23. 108
24. 700
25. 127
26. 170
27. $\frac{17}{81}$
28. $\frac{16}{9}$
29. 1
30. $\frac{2}{3}$

31. $\frac{4}{3}$, or $1\frac{1}{3}$
32. $\frac{3}{2}$, or $1\frac{1}{2}$
33. $\frac{5}{4}$, or $1\frac{1}{4}$
34. $\frac{1}{2}$
35. $\frac{7}{10}$
36. $\frac{2}{3}$
37. $\frac{5}{6}$
38. $\frac{7}{6}$, or $1\frac{1}{6}$
39. $\frac{13}{12}$, or $1\frac{1}{12}$
40. $\frac{11}{15}$

41. $\frac{22}{15}$, or $1\frac{7}{15}$
42. $\frac{23}{20}$, or $1\frac{3}{20}$
43. $\frac{23}{30}$
44. $\frac{43}{30}$, or $1\frac{13}{30}$

Focus: *Students evaluate algebraic expressions for given values of each variable.*

EXAMPLE 1

PHASE 1: Presenting the Example

Give students the "quiz" at right. They should have little difficulty with the first four expressions, but probably will be puzzled by the fifth. Note that this expression has an unknown quantity, represented by the letter *n*. Tell them that *n* is a *variable*, and $3 + n$ is an *algebraic expression*.

Before presenting Example 1, tell students that to *evaluate* an algebraic expression, you first substitute values for the variables; then you simplify the numerical expression that results. Then discuss Example 1.

> **Simplify.**
> 1. $3 + 9$ [12]
> 2. $3 + 0$ [3]
> 3. $3 + 0.65$ [3.65]
> 4. $3 + 8\frac{1}{2}$ $\left[11\frac{1}{2}\right]$
> 5. $3 + n$ [**cannot be simplified**]

PHASE 2: Providing Guidance for TRY THIS

To reinforce the concept of substitution, have students write $3(a + 6) - 9$ in pencil. Then tell them to erase *a* and replace it with the value 7.

EXAMPLE 2

PHASE 1: Presenting the Example

Give students the problem at right. They should quickly realize that there is not enough information to complete the task. Lead them to see that, in order to evaluate an expression like this one, it is necessary to identify substitutions for all of the variables involved. Give them the additional substitution $b = 2$, and discuss the solution outlined on the textbook page.

> *Evaluate*
> $$\frac{3a + 5b + 1}{4a - 11b}$$
> given $a = 7$.

PHASE 2: Providing Guidance for TRY THIS

Some students may confuse the values of *r* and *s*. Suggest that they copy the expression and write the variables in different colors, perhaps using red for *r* and blue for *s*, each time that they appear in the expression. Students should then use the same colors when substituting 10 for *r* and 5 for *s*.

PHASE 3: ASSESSMENT AND CLOSURE for Lesson 1.2

- Given the expression at right, have students replace each box with a different operation symbol to make three expressions. The value of one expression must be less than 5, the value of the second must be between 5 and 20, and the value of the third must be greater than 20.

- Students should be aware that most future topics in this course will require skill in working with numerical and algebraic expressions.

> $16 \ \square \ 8 \ \square \ 4 \ \square \ 2$
>
> [**samples:**
>
> $16 \div 8 + 4 - 2 = 4;$
> $16 + 8 \div 4 - 2 = 16;$
> $16 + 8 \times 4 - 2 = 46$]

 *For a Lesson 1.2 Quiz, see* Assessment Resources *page 2.*

A **variable** is a letter that is used to represent numbers. An **algebraic expression** is an expression that contains numbers, variables and operation symbols. To *evaluate* an algebraic expression, substitute numbers for the variables in the expression and then calculate. When you evaluate, you are using the Substitution Principle.

The Substitution Principle

For any numbers a and b, if $a = b$, then a and b may be substituted for each other.

EXAMPLE 1 Evaluate $2(x - 3) + 5$ given $x = 7$.

▶ **Solution**

$2(x - 3) + 5$
$2(7 - 3) + 5$ ⟵ Replace x with 7, then simplify.
$2(4) + 5$
13

TRY THIS   Evaluate $3(a + 6) - 9$ given $a = 7$.

An algebraic expression may contain more than one variable.

EXAMPLE 2 Evaluate $\dfrac{3a + 5b + 1}{4a - 11b}$ given $a = 7$ and $b = 2$.

▶ **Solution**

$\dfrac{3a + 5b + 1}{4a - 11b} \rightarrow \dfrac{3(7) + 5(2) + 1}{4(7) - 11(2)}$ ⟵ Replace a with 7 and b with 2.

$= \dfrac{21 + 10 + 1}{28 - 22}$ ⟵ Multiply.

$= \dfrac{32}{6}$, or $5\dfrac{1}{3}$

TRY THIS   Evaluate $\dfrac{4r - 5s + 5}{3r + s}$ given $r = 10$ and $s = 5$.

A **formula** is an equation that shows a mathematical relationship between two or more quantities.

For example, the surface area, S, of a rectangular box is given by the formula $S = 2lw + 2wh + 2lh$. You can find the surface area of the box at right by evaluating the formula.

$h = 5.0$
$l = 9.5$ $w = 2.0$

$$S = \overset{l\quad w}{2(9.5)(2.0)} + \overset{w\quad h}{2(2.0)(5.0)} + \overset{l\quad h}{2(9.5)(5.0)} = 153 \text{ units}^2$$

12   Chapter 1   Solving Simple Equations

EXERCISES

Write the numerical expression that results when you apply the Substitution Principle. Do not evaluate.

Sample: $x + 3$, given $x = 2$ Solution: $2 + 3$

1. $3x^2 - 5x + 4$, given $x = 5$

2. $3(a + 3) - 2(a - 1)$, given $a = 9$

3. $\dfrac{3m + 3n}{4n}$, given $m = 3$ and $n = 2$

4. $\dfrac{7k - 3p}{4p + 1} + \dfrac{k}{p}$, given $k = 3$ and $p = 2$

PRACTICE

Evaluate each expression.

5. $3d + 4(d - 1)$, given $d = 7$

6. $8s - 4(s + 1)$, given $s = 3$

7. $8b^2 - 4b$, given $b = 5$

8. $2k^2 + 7k$, given $k = 10$

9. $0.5h^2 - 0.25h$, given $h = 4$

10. $0.5h^2 + 0.6h$, given $h = 10$

11. $\dfrac{1}{2}(a - 1) + \dfrac{1}{3}$, given $a = 9$

12. $\dfrac{m + 4}{2} + \dfrac{m - 2}{5}$, given $m = 6$

13. $\dfrac{5(n + 3) - 1}{4} - \dfrac{3(n + 1)}{3}$, given $n = 1$

14. $\dfrac{5(z - 3) + 1}{4} + \dfrac{2(z + 1)}{3}$, given $z = 5$

15. $a + 2(1 + b) - 2b$, given $a = 3$ and $b = 1$

16. $7x - 2(1 + y) + 1$, given $x = 5$ and $y = 2$

17. $\dfrac{7(m - n)}{3(m + n)}$, given $m = 7$ and $n = 1$

18. $\dfrac{5(2m - 3n)}{4(m - n)}$, given $m = 10$ and $n = 6$

19. $\dfrac{7(m - n)}{3(m + n)}$, given $m = \dfrac{3}{4}$ and $n = \dfrac{1}{4}$

20. $\dfrac{5(2m - 3m)}{4(m - n)}$, given $m = \dfrac{1}{2}$ and $n = \dfrac{1}{3}$

21. The amount in a savings account is calculated using the formula $A = p(1 + r)t$, where A is the amount, p is the *principal*, r is the annual interest rate, and t is the time in years. What is the amount in an account after 1 year if the principal is $1200 and the interest rate is 0.05?

MIXED REVIEW

Write each fraction as a decimal and each decimal as a fraction. (previous courses)

Sample: $\dfrac{1}{8}$ Solution: $8\overline{)1.000}$ = 0.125   Sample: 0.125   Solution: $\dfrac{125}{1000} = \dfrac{1}{8}$

22. $\dfrac{1}{2}$ 23. $\dfrac{1}{4}$ 24. $\dfrac{1}{8}$ 25. $\dfrac{1}{20}$ 26. $\dfrac{1}{40}$ 27. $\dfrac{3}{4}$

28. $\dfrac{4}{5}$ 29. $\dfrac{3}{8}$ 30. $\dfrac{7}{8}$ 31. 0.3   32. 0.33   33. 3.3

34. 3.03   35. $\dfrac{1}{6}$ 36. $\dfrac{5}{6}$ 37. $\dfrac{2}{3}$ 38. $0.\overline{6}$ 39. 0.66

Answers for Lesson 1.2 Skill C

Try This Example 1
30

Try This Example 2
$\dfrac{4}{7}$

1. $3(5)^2 - 5(5) + 4$
2. $3(9 + 3) - 2(9 - 1)$
3. $\dfrac{3(3) + 3(2)}{4(2)}$
4. $\dfrac{7(3) - 3(2)}{4(2) + 1} + \dfrac{3}{2}$
5. 45
6. 8
7. 180
8. 270
9. 44
10. 56
11. $\dfrac{10}{3}$, or $3\dfrac{1}{3}$
12. $\dfrac{29}{5}$, or $5\dfrac{4}{5}$
13. $\dfrac{11}{4}$, or $2\dfrac{3}{4}$

14. $\dfrac{27}{4}$, or $6\dfrac{3}{4}$
15. 5
16. 30
17. $\dfrac{7}{4}$, or $1\dfrac{3}{4}$
18. $\dfrac{5}{8}$
19. $\dfrac{7}{6}$, or $1\dfrac{1}{6}$
20. 0
21. $1260
22. 0.5
23. 0.25
24. 0.125
25. 0.05
26. 0.025
27. 0.75
28. 0.8
29. 0.375
30. 0.875
31. $\dfrac{3}{10}$

32. $\dfrac{33}{100}$
33. $\dfrac{33}{10}$, or $3\dfrac{3}{10}$
34. $\dfrac{303}{100}$, or $3\dfrac{3}{100}$
35. $0.1\overline{6}$
36. $0.8\overline{3}$
37. $0.\overline{6}$
38. $\dfrac{2}{3}$
39. $\dfrac{33}{50}$

TEACHING
LESSON **1.3**

Equations and Logical Reasoning

SKILL A   *Using replacement sets in solving equations*

Focus: *Students find which, if any, numbers in a finite replacement set satisfy a given equation.*

EXAMPLE

PHASE 1: Presenting the Example

Display the equations at right. Ask students to write brief answers to the following questions:

1. $15 - 4 = 11$
2. $z - 8 = 12$

- How is **2** similar to **1**?

- How is **2** different from **1**?

Discuss their responses.

Among the similarities, students should notice that both items involve the equal sign, $=$. Point out that each is called an *equation*, and review the definition given in the text: An equation is a mathematical statement that two expressions are equal.

Among the differences, students should notice that equation **2** involves a variable, while equation **1** does not. This is the key difference. Since there are no unknowns in equation **1**, it is possible to identify the statement as true or false. Students should easily see that it is true.

In equation **2**, however, the value of the variable is unknown. Therefore, equation **2** by itself is neither true nor false. It is called an *open sentence*. That is, whether the equation is true or false is an open issue, depending on the value that is substituted for the variable. Ask students to give a value that makes a false statement when it is substituted for z. [**any number other than 20**] Then ask them to give a value that makes a true statement when it is substituted for z. [**20**] In fact, 20 is the only replacement for z that makes a true statement. Note that 20 is called a *solution* to $z - 8 = 12$, and review the definition of solution given in the text.

Now display the equation in the Example, $2x + 1 = 11$. Ask students to vote: Is this equation *true, false,* or *open*? [**open**] Then lead them through the Example, where they learn to search for a solution within a restricted *replacement set*. They will see that it is possible for an equation to have a solution in one replacement set while it has no solution in a different replacement set.

PHASE 2: Providing Guidance for **TRY THIS**

Check students' understanding of replacement sets by asking them to identify a replacement set other than $\{0, 2, 4, 6, 8\}$ in which $3x - 5 = 13$ has a solution. [**sample: $\{0, 3, 6, 9, 12, 15, 18\}$**] Then ask them to identify a replacement set other than $\{1, 3, 5, 7, 9\}$ in which $3x - 5 = 13$ has *no* solution. [**sample: $\{10, 20, 30, 40\}$**]

1.3 Equations and Logical Reasoning
LESSON

SKILL A *Using replacement sets in solving equations*

An **equation** is a mathematical statement that two expressions are equal
(=). An equation with at least one variable is called an **open sentence**.
Examples of equations that are open sentences follow.

$$x + 5 = 8 \qquad x = x + 9 \qquad x + 4 = 4 + x$$

A **solution** to an open sentence is any value of the variable that makes the
equation true. A solution is said to *satisfy* the equation.

- There is exactly one value of x, 3, that satisfies $x + 5 = 8$.
- No value of x satisfies $x = x + 9$. This equation has no solution.
- The equation $x + 4 = 4 + x$ is true for all values of x because of the
  Commutative Property of Addition. An equation that is true for all
  values of the variable is called an **identity**.

A set of possible replacements for a variable is called a **replacement set**,
and the set is enclosed with braces { }.

EXAMPLE Given each replacement set, find the solution to the equation $2x + 1 = 11$.
  a. {0, 2, 4, 6, 8} **b.** {1, 3, 5, 7, 9}

▶ **Solution**
Make a table.
  a. {0, 2, 4, 6, 8}

x	$2x + 1$	Solution?
0	$2(0) + 1 = 1$	No, $1 \neq 11$.
2	$2(2) + 1 = 5$	No, $5 \neq 11$.
4	$2(4) + 1 = 9$	No, $9 \neq 11$.
6	$2(6) + 1 = 13$	No, $13 \neq 11$.
8	$2(8) + 1 = 17$	No, $17 \neq 11$.

$2x + 1 = 11$ does not have a solution
in {0, 2, 4, 6, 8}.

  b. {1, 3, 5, 7, 9}

x	$2x + 1$	Solution?
1	$2(1) + 1 = 3$	No, $3 \neq 11$.
3	$2(3) + 1 = 7$	No, $7 \neq 11$.
5	$2(5) + 1 = 11$	Yes, $11 = 11$. ✓
7	$2(7) + 1 = 15$	No, $15 \neq 11$.
9	$2(9) + 1 = 19$	No, $19 \neq 11$.

$2x + 1 = 11$ has one solution, 5,
in {1, 3, 5, 7, 9}.

TRY THIS Given each replacement set, find the solution to the equation $3x - 5 = 13$.
  a. {0, 2, 4, 6, 8} **b.** {1, 3, 5, 7, 9}

The Example shows that an equation can have a solution in one
replacement set but no solution in another replacement set. When
no replacement set is specified, assume that the replacement set is
all real numbers.

EXERCISES

KEY SKILLS

State whether each item is an equation. Write *yes* or *no*.

  1. $2(x + 5) - 3(x + 1)$ **2.** $3(a + 5) = 4(a - 1) + 2$ **3.** $4x + 2y = 5$

State whether each item is an open sentence. Write *yes* or *no*.

  4. $5x = 75$ **5.** $5x = 7y$ **6.** $4(5) = 20$

PRACTICE

Find the values in the replacement set {0, 1, 2, 3, 4, 5, 6} that are solutions
to each equation. If the equation has no solution, write *none*.

  7. $x - 1 = 2$ **8.** $a - 1 = 4$ **9.** $w + 1 = 8$
  10. $t + 1 = 0$ **11.** $2y = 8$ **12.** $5c = 10$
  13. $2h + 1 = 7$ **14.** $2z - 1 = 11$ **15.** $3x - 2 = 7$
  16. $4d + 1 = 8$ **17.** $3(r + 3) = 12$ **18.** $5(k - 2) = 20$
  19. $2(n + 1) + 1 = 10$ **20.** $3(v - 1) + 2 = 11$ **21.** $5(u + 1) + 1 = 11$

Is each equation an identity? If not, explain.

  22. $x - 1 = x - 1$ **23.** $2(p - 1) = (p - 1)(2)$ **24.** $2g + 3 = 9$
  25. $2z - 1 = 5$ **26.** $2s + 8 = 8 + 2s$ **27.** $3(2w) = 2(3w)$
  28. $(2q)(3) = (2)(3)q$ **29.** $3x - 5 = 13$ **30.** $7 = 4y - 2$

  31. Critical Thinking Does the equation $2x + 1 = 6$ have a solution in
  the replacement set of:
    a. all even numbers? **b.** all odd numbers? **c.** all mixed numbers?

MIXED REVIEW

Find each product or quotient. (previous courses)

Sample: $\frac{5}{12} \times \frac{3}{4}$ Solution: $\frac{5}{\cancel{12}_4} \times \frac{\cancel{3}^1}{4} = \frac{5}{16}$ Sample: $\frac{3}{5} \div \frac{8}{10}$ Solution: $\frac{3}{8} \times \frac{\cancel{10}^2}{\cancel{8}_4} = \frac{6}{7}$

  32. $\frac{1}{3} \times \frac{1}{2}$ **33.** $\frac{1}{3} \times \frac{1}{5}$ **34.** $\frac{2}{3} \times \frac{1}{5}$ **35.** $\frac{2}{3} \times \frac{1}{2}$ **36.** $\frac{2}{3} \times \frac{3}{5}$
  37. $\frac{2}{3} \times \frac{4}{5}$ **38.** $\frac{3}{7} \times \frac{7}{3}$ **39.** $\frac{4}{7} \times \frac{7}{4}$ **40.** $\frac{3}{7} \times \frac{14}{3}$ **41.** $\frac{3}{7} \times \frac{21}{9}$
  42. $\frac{1}{3} \div \frac{1}{2}$ **43.** $\frac{1}{3} \div \frac{1}{5}$ **44.** $\frac{2}{3} \div \frac{1}{5}$ **45.** $\frac{2}{3} \div \frac{1}{2}$ **46.** $\frac{2}{3} \div \frac{3}{5}$
  47. $\frac{2}{3} \div \frac{4}{5}$ **48.** $\frac{3}{7} \div \frac{7}{3}$ **49.** $\frac{4}{7} \div \frac{7}{4}$ **50.** $\frac{3}{7} \div \frac{14}{3}$ **51.** $\frac{3}{7} \div \frac{21}{9}$

Answers for Lesson 1.3 Skill A

Try This Example
a. 6
b. none

1. no
2. yes
3. yes
4. yes
5. yes
6. no
7. 3
8. 5
9. none
10. none
11. 4
12. 2
13. 3
14. 6
15. 3
16. none
17. 1
18. 6
19. none
20. 4

21. 1
22. yes
23. yes
24. No; for example, 0 is not a solution.
25. No; for example, 0 is not a solution.
26. yes
27. yes
28. yes
29. No; for example, 0 is not a solution.
30. No; for example, 0 is not a solution.
31. **a.** no
    b. no
    c. yes
32. $\frac{1}{6}$
33. $\frac{1}{15}$
34. $\frac{2}{15}$

35. $\frac{1}{3}$
36. $\frac{2}{5}$
37. $\frac{8}{15}$
38. 1
39. 1
40. 2
41. 1
42. $\frac{2}{3}$
43. $\frac{5}{3}$, or $1\frac{2}{3}$
44. $\frac{10}{3}$, or $3\frac{1}{3}$
45. $\frac{4}{3}$, or $1\frac{1}{3}$
46. $\frac{10}{9}$, or $1\frac{1}{9}$
47. $\frac{5}{6}$
48. $\frac{9}{49}$

49. $\frac{16}{49}$
50. $\frac{9}{98}$
51. $\frac{9}{49}$

Focus: *Students use a counterexample to demonstrate that a statement is false, and they identify instances when a statement is true.*

EXAMPLE 1

PHASE 1: Presenting the Example

Display the statement *The sum of two even numbers is always an odd number.* Tell students to write on their papers one instance in which this statement is *false.*

Now ask just one student to read the response from his or her paper. [**sample: 2 + 4 = 6**] Point out that, while the other students' papers may show several different instances, this one *counterexample* is all that is needed to identify the given statement as false.

PHASE 2: Providing Guidance for **TRY THIS**

Some students may justify their responses with the statement *The sum of two odd numbers is always an even number.* This statement is true, of course, but it is important for students to realize that they presently cannot prove it. At this point, the proper technique for proving the given statement false is to find a single counterexample.

EXAMPLE 2

PHASE 1: Presenting the Example

PHASE 2: Providing Guidance for **TRY THIS**

Give students the multiplications shown below and ask them to find the products.

9 × 5	[**45**]
7 × 12	[**84**]
12 × 20	[**240**]
10 × 5.8	[**58**]
100 × 1.332	[**133.2**]

Ask if it is true that, in each of these instances, the product is greater than either of the factors. [**yes**] Then proceed with the solution outlined on the textbook page.

Tell students that this example illustrates a basic principle of algebraic reasoning: You cannot declare a statement to be true simply because you can show one or two true examples. It is not even sufficient to show a million true examples. This is in stark contrast to the process of declaring a statement *false*, where only one counterexample is required. Students will learn how to show that a statement is true in Skill C.

Be sure students understand the phrase *sometimes true.* A statement that is only sometimes true is considered, over all, to be false. However, knowing when the statement is true can be helpful. Ask students to use their knowledge about when the Example is true to make a guess about the Try This exercise. [**sample: Since the product is greater when *a* and *b* are greater than 1, the quotient will be greater when *a* or *b* is less than 1.**] Have students test their guess by creating an example and a counterexample for the Try This exercise.

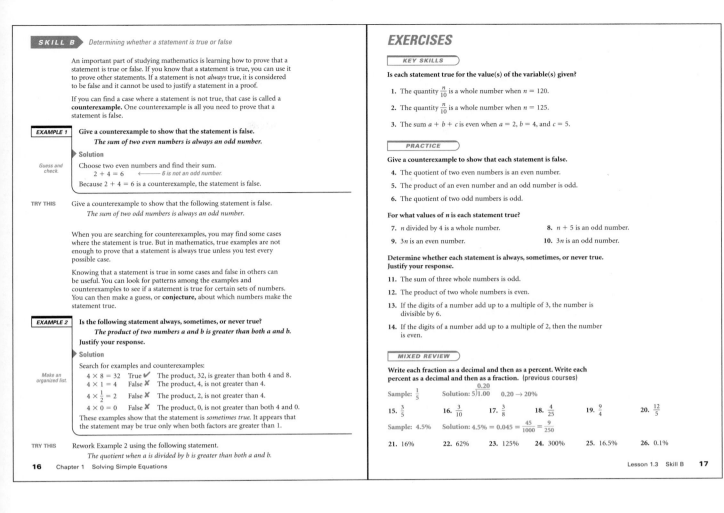

An important part of studying mathematics is learning how to prove that a statement is true or false. If you know that a statement is true, you can use it to prove other statements. If a statement is not *always* true, it is considered to be false and it cannot be used to justify a statement in a proof.

If you can find a case where a statement is not true, that case is called a **counterexample.** One counterexample is all you need to prove that a statement is false.

EXAMPLE 1

Give a counterexample to show that the statement is false.

The sum of two even numbers is always an odd number.

▶ **Solution**

Guess and check.

Choose two even numbers and find their sum.

$2 + 4 = 6$ ◀——— *6 is not an odd number.*

Because $2 + 4 = 6$ is a counterexample, the statement is false.

TRY THIS

Give a counterexample to show that the following statement is false.

The sum of two odd numbers is always an odd number.

When you are searching for counterexamples, you may find some cases where the statement is true. But in mathematics, true examples are not enough to prove that a statement is always true unless you test every possible case.

Knowing that a statement is true in some cases and false in others can be useful. You can look for patterns among the examples and counterexamples to see if a statement is true for certain sets of numbers. You can then make a guess, or **conjecture,** about which numbers make the statement true.

EXAMPLE 2

Is the following statement always, sometimes, or never true?

The product of two numbers a and b is greater than both a and b.

Justify your response.

▶ **Solution**

Make an organized list.

Search for examples and counterexamples:

$4 \times 8 = 32$ True ✔   The product, 32, is greater than both 4 and 8.
$4 \times 1 = 4$ False ✘   The product, 4, is not greater than 4.
$4 \times \frac{1}{2} = 2$ False ✘   The product, 2, is not greater than 4.
$4 \times 0 = 0$ False ✘   The product, 0, is not greater than both 4 and 0.

These examples show that the statement is *sometimes true.* It appears that the statement may be true only when both factors are greater than 1.

TRY THIS

Rework Example 2 using the following statement.

The quotient when a is divided by b is greater than both a and b.

EXERCISES

KEY SKILLS

Is each statement true for the value(s) of the variable(s) given?

1. The quantity $\frac{n}{10}$ is a whole number when $n = 120$.

2. The quantity $\frac{n}{10}$ is a whole number when $n = 125$.

3. The sum $a + b + c$ is even when $a = 2$, $b = 4$, and $c = 5$.

PRACTICE

Give a counterexample to show that each statement is false.

4. The quotient of two even numbers is an even number.

5. The product of an even number and an odd number is odd.

6. The quotient of two odd numbers is odd.

For what values of *n* is each statement true?

7. *n* divided by 4 is a whole number.   8. $n + 5$ is an odd number.

9. $3n$ is an even number.   10. $3n$ is an odd number.

Determine whether each statement is always, sometimes, or never true. Justify your response.

11. The sum of three whole numbers is odd.

12. The product of two whole numbers is even.

13. If the digits of a number add up to a multiple of 3, the number is divisible by 6.

14. If the digits of a number add up to a multiple of 2, then the number is even.

MIXED REVIEW

Write each fraction as a decimal and then as a percent. Write each percent as a decimal and then as a fraction. (previous courses)

Sample: $\frac{1}{5}$ Solution: $5\overline{)1.00}^{\,0.20}$ $0.20 \rightarrow 20\%$

15. $\frac{3}{5}$ 16. $\frac{3}{10}$ 17. $\frac{3}{8}$ 18. $\frac{4}{25}$ 19. $\frac{9}{4}$ 20. $\frac{12}{5}$

Sample: 4.5%   Solution: $4.5\% = 0.045 = \frac{45}{1000} = \frac{9}{250}$

21. 16%   22. 62%   23. 125%   24. 300%   25. 16.5%   26. 0.1%

Answers for Lesson 1.3   Skill B

Try This Example 1

Answers may vary. Sample answer: The numbers 1 and 3 are odd, but their sum, 4, is even.

Try This Example 2

Answers may vary. Sample answer:

$12 \div 4 = 3$ False
$4 \div 1 = 4$ False
$4 \div \frac{1}{2} = 8$ True
$\frac{1}{2} \div \frac{1}{4} = 2$ True

These examples show that the statement is sometimes true. It appears that the statement may be true only when *a* or *b* (or both) is less than 1.

1. yes

2. no

3. no

4. Answers may vary. Sample answer: The quotient when 2 is divided by 4 is not a whole number, so it cannot be classified as even or odd.

5. Answers may vary. Sample answer: Consider 3 and 4. The product is even, not odd.

6. Answers may vary. Sample answer: The quotient when 3 is divided by 5 is not a whole number, so it cannot be classified as even or odd.

7. 0, 4, 8, 12, 16, . . .

8. all even whole numbers

9. all even whole numbers

10. all odd whole numbers

11. sometimes true; $1 + 3 + 5 = 9$ is odd, but $2 + 3 + 5 = 10$ is even.

12. sometimes true; $2 \times 3 = 6$ is even, but $3 \times 3 = 9$ is odd.

★13. sometimes true; the digits of 12 add to a multiple of 3, and 12 is divisible by 6; but the digits of 15 add to a multiple of 3, and 15 is not divisible by 6.

★14. sometimes true; the digits of 26 add to 8, which is even, and 26 is even; but the digits of 17 also add to 8, and 17 is odd.

15. 0.6; 60%

16. 0.3; 30%

17. 0.375; 37.5%

18. 0.16; 16%

For answers to Exercises 19–26, see Additional Answers.

★ **Advanced Exercises**

Focus: *Students learn basic properties of equality and use them together with definitions and other properties to prove statements.*

EXAMPLES 1 and 2

PHASE 1: Presenting the Examples

Display the statement *For all numbers x, x + x = 2x.* Ask one half of the class to write on their papers one example in which this statement is true. Ask the other half to write one example in which the statement is false. Then ask several students to read the responses from their papers. [**Answers will vary for true examples, but students should not have been able to find false examples.**] As a class, have the students decide whether they think the statement is true or false.

Students will probably believe that the statement is true. Remind them, however, that you cannot declare a statement to be true simply because you have found a few true examples. What is needed is a way to show that the statement is true in *all* instances. Ask students if they think they can write an example for all instances. They should realize that this is an impossible task, requiring infinitely many examples.

So how do you show that a statement is true for all numbers?

Point out to students that the properties they have studied so far are true for all numbers. So if a given statement can be "backed up," or *justified*, by a property, then that statement must also be true for all numbers.

In Examples 1 and 2, several justified statements are linked together to show that a given statement must be true. The set of justified statements taken together forms a *mathematical proof* of the given statement.

PHASE 2: Providing Guidance for **TRY THIS**

Before they are able to write an entire proof independently, some students need practice in simply justifying the steps of a given proof. You can use Skill C Exercises 1 and 2 on the facing page for this purpose.

PHASE 3: ASSESSMENT AND CLOSURE for Lesson 1.3

- Have students write a false statement containing $k + 6$ and show that it is false. Then have them write a true statement containing $k + 6$ and show that it is true.

  [**false: For all real numbers k, $k + 6 = 6k$. A counterexample is $k = 1$.**

  true: For all real numbers k, $k + 6 = 6 + k$ by the Commutative Property of Addition.]

- This lesson introduced the concept of an *equation*. Much of this course will be devoted to studying types of equations and methods for solving them.

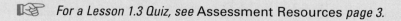 *For a Lesson 1.3 Quiz, see* Assessment Resources *page 3.*

In arithmetic, you may use some concepts that seem very obvious. But in algebra, it is important to understand, define, and name these concepts. These include the Properties of Numbers (1.2 Skill A) and the Basic Properties of Equality defined below.

Basic Properties of Equality

Reflexive Property: For all real numbers a, $a = a$.

Symmetric Property: For all real numbers a and b, if $a = b$, then $b = a$.

Transitive Property: For all real numbers a, b, and c, if $a = b$ and $b = c$, then $a = c$.

Properties of Numbers, Properties of Equality, and definitions are often used to write a **mathematical proof**, which is a convincing argument using *logic* to show that a statement is true.

EXAMPLE 1 Prove that $\frac{1}{2} = 50\%$.

▶ Solution

Statements	Reasons
1. $\frac{1}{2} = \frac{1}{2} \times \frac{50}{50}$	1. Identity Property of Multiplication $\left(1 = \frac{50}{50}\right)$
2. $\frac{1}{2} \times \frac{50}{50} = \frac{1 \times 50}{2 \times 50} = \frac{50}{100}$	2. Definition of multiplication of fractions
3. $\frac{50}{100} = 50\%$	3. Definition of percent
4. Thus, $\frac{1}{2} = 50\%$.	4. Transitive Property of Equality

TRY THIS Prove that $\frac{3}{4} = 75\%$.

You can use Properties of Numbers to show that a statement containing a variable is true.

EXAMPLE 2 Prove that $3a = 2a + a$.

▶ Solution

Statements	Reasons
1. $3a = a + a + a$	1. Definition of multiplication by 3
2. $= (a + a) + a$	2. Associative Property of Addition
3. $= 2a + a$	3. Definition of multiplication by 2

TRY THIS Prove that $4a = 2a + 2a$.

18 Chapter 1 Solving Simple Equations

EXERCISES

Justify each step below.

1. $2n + (2n + 1) = (2n + 2n) + 1$
$= (n + n + n + n) + 1$
$= 4n + 1$

2. $2n + 1 + 2n + 3 = 2n + 2n + 1 + 3$
$= (2n + 2n) + (1 + 3)$
$= (n + n + n + n) + 1 + 3$
$= 4n + 4$

Prove the following statements.

3. $\frac{3}{5} = 60\%$ 4. $\frac{1}{10} = 10\%$ 5. $\frac{4}{10} = 40\%$ 6. $\frac{3}{12} = 25\%$

7. $3(x + 3) + 4 = 3x + 13$ 8. $2(x + 1) + 5 = 2x + 7$

9. $5a = 3a + 2a$ 10. $6a = 5a + a$

11. $3 + 4(y + 2) = 4y + 11$ 12. $6 + 4(r + 3) = 4r + 18$

Critical Thinking If n is a whole number, then $2n$ is an even number and $2n + 1$ is an odd number. Does the given expression represent an even number or an odd number? Justify your answer.

13. $2n + 2n$ 14. $2n + (2n + 1)$

15. $(2n + 1) + (2n + 1)$ 16. $(2n + 1) + (2n + 3)$

Write each phrase as a mathematical expression. Do not evaluate. (1.1 Skill A)

17. 5 more than 3 18. the sum of 2.5 and 8

Write each sentence as an equation. (1.1 Skill B)

19. One plus twice ten is twenty-one. 20. The sum of 3 and 4 less 2 equals 5.

Use the order of operations to simplify each expression. (1.2 Skill B)

21. $\frac{(3 + 2)^2}{15 - 4 \times 3}$ 22. $3[4(6 - 5)]^2$

Evaluate each expression. (1.2 Skill C)

23. $3.5(x + 3) + 4.5(x + 7)$, given $x = 5$ 24. $4(rs) + 3(r + s)$, given $r = 3$ and $s = 9$

Find the values in the replacement set $\{0, 1, 2, 3, 4, 5, 6\}$ that are solutions to each equation. If the equation has no solution, write *none*. (1.3 Skill A)

25. $2x + 5 = 11$ 26. $3x + 5 = 13$ 27. $2(a - 1) - 5 = 11$ 28. $y + (y + 5) = 13$

Lesson 1.3 Skill C 19

Answers for Lesson 1.3 Skill C

Try This Example 1

$\frac{3}{4} = \frac{3}{4} \times \frac{25}{25}$ Identity Property of Multiplication

$\frac{3}{4} \times \frac{25}{25} = \frac{75}{100}$ Multiplication of fractions

$\frac{75}{100} = 75\%$ Definition of percent

$\frac{3}{4} = 75\%$ Transitive Property of Equality

Try This Example 2

$4a = a + a + a + a$ Definition of multiplication by 4

$= (a + a) + (a + a)$ Associative Property of Addition

$= 2a + 2a$ Definition of multiplication by 2

1. Associative Property of Addition, definition of multiplication by 2, definition of multiplication by 4
2. Commutative Property of Addition, Associative Property of Addition, definition of multiplication by 2, definition of multiplication by 4

Answers for Exercises 3–12 may vary. Sample answers are given.

3. $\frac{3}{5} = \frac{3}{5} \cdot \frac{20}{20}$ Identity Property of Multiplication

$\frac{3}{5} \cdot \frac{20}{20} = \frac{60}{100}$ Multiplication of fractions

$\frac{60}{100} = 60\%$ Definition of percent

$\frac{3}{5} = 60\%$ Transitive Property of Equality

4. $\frac{1}{10} = \frac{1}{10} \cdot \frac{10}{10}$ Identity Property of Multiplication

$\frac{1}{10} \cdot \frac{10}{10} = \frac{10}{100}$ Multiplication of fractions

$\frac{10}{100} = 10\%$ Definition of percent

$\frac{1}{10} = 10\%$ Transitive Property of Equality

5. $\frac{4}{10} = \frac{4}{10} \cdot \frac{10}{10}$ Identity Property of Multiplication

$\frac{4}{10} \cdot \frac{10}{10} = \frac{40}{100}$ Multiplication of fractions

$\frac{40}{100} = 40\%$ Definition of percent

$\frac{4}{10} = 40\%$ Transitive Property of Equality

For answers to Exercises 6–28, see Additional Answers.

TEACHING

1.4 LESSON

Solving Simple Equations Using Addition

> **SKILL A**   *Using addition to solve an equation in one step*

Focus: *Students solve simple equations in one variable by applying the Addition Property of Equality.*

EXAMPLES 1 and 2

PHASE 1: Presenting the Examples

Display the equation $9 = 9$ and ask students if it is true or false. [**true**] Tell them to write the result when 7 is added to each side of the equation. [**16 = 16**] Ask if the new equation is true or false. [**true**] Repeat this activity several times, each time adding a different number to each side of $9 = 9$. (Some samples are shown at right.) Lead students to see that the result will be true no matter what number they add to each side. Then discuss both the informal and formal statements of the Addition Property of Equality that appear at the top of the textbook page.

$$\begin{array}{r} 9 = 9 \\ +\ 7 \quad +\ 7 \\ \hline 16 = 16 \end{array}$$

$$\begin{array}{r} 9 = 9 \\ +\ 0.4 \quad +\ 0.4 \\ \hline 9.4 = 9.4 \end{array}$$

Now display the equation in Example 1, $x - 10 = 25$. Ask students if it is true or false. They should recall that the presence of the variable, x, makes it an open sentence, which is neither true nor false. Ask what value they could substitute for x to make a true statement. Most should see that the desired value is 35. So 35 is a solution of $x - 10 = 25$. In fact, it is the *only* solution.

$$\begin{array}{r} 9 = 9 \\ +\ 5\frac{1}{8} \quad +\ 5\frac{1}{8} \\ \hline 14\frac{1}{8} = 14\frac{1}{8} \end{array}$$

Refer again to $x - 10 = 25$. Ask students if it is permissible to add 10 to each side. [**yes**] Ask why it is permissible. [**The Addition Property of Equality**] The key question then becomes: Why would you choose to add 10 to each side? Lead students through the solution on the textbook page. The critical point is that in the resulting equation, $x = 35$, the variable stands alone, or is *isolated*, on one side of the equal sign. Note that the solution of this equation is quite clearly 35.

PHASE 2: Providing Guidance for **TRY THIS**

Stress the importance of checking a solution by substituting it for the variable *in the original equation.*

PHASE 3: ASSESSMENT AND CLOSURE for Lesson 1.4

- Give students the equation shown at right. Ask them to find two different ways to fill in the boxes so that the solution is 24.

- The process of finding all solutions of an equation is called *solving the equation*. Some equations can be solved easily by inspection, but many cannot. Several algebraic techniques are used in solving such equations, and the Addition Property of Equality is one of them. Students will learn additional techniques in the lessons that follow.

$$j - \square = \square$$

[**samples:** $j - 10 = 14$; $j - 20 = 4$ (Students should observe that they can use any two numbers whose sum is 24.)]

☞   *For a Lesson 1.4 Quiz, see* Assessment Resources *page 4.*

Solving Simple Equations Using Addition

SKILL A *Using addition to solve an equation in one step*

The *Addition Property of Equality* states the following:

If the same quantity is added to each side of a true equation, then the equation that results is also true. If a and b balance on the scale, then $a + c$ and $b + c$ will balance also.

balance ⟶ balance

Addition Property of Equality
If a, b, and c are numbers and $a = b$, then $a + c = b + c$.

Use the Addition Property of Equality to solve equations involving subtraction.

EXAMPLE 1 Solve $x - 10 = 25$. Check your solution.

▸ **Solution**

$x - 10 = 25$
$x - 10 + 10 = 25 + 10$ ⟵ *Apply the Addition Property of Equality.*
$x + 0 = 35$ ⟵ *Apply the Additive Identity Property.*
$x = 35$

Check: $35 - 10 = 25$ ✓

TRY THIS Solve $d - 5 = 4$. Check your solution.

The Symmetric Property of Equality allows you to switch the left and right sides of an equation.

EXAMPLE 2 Solve $12.5 = x - 4.6$. Check your solution.

▸ **Solution**

$12.5 = x - 4.6$
$x - 4.6 = 12.5$ ⟵ *Apply the Symmetric Property of Equality.*
$x - 4.6 + 4.6 = 12.5 + 4.6$ ⟵ *Apply the Addition Property of Equality.*
$x + 0 = 17.1$ ⟵ *Apply the Additive Identity Property.*
$x = 17.1$

Check: $17.1 - 4.6 = 12.5$ ✓

TRY THIS Solve $100 = t - 14.5$. Check your solution.

EXERCISES

KEY SKILLS

What number would you add to each side of the given equation to solve it? Do not solve.

1. $x - 2.3 = 12$ 2. $17 = t - 5$ 3. $22.5 = x - 3$ 4. $a - 4 = 19$

PRACTICE

Solve each equation. Check your solutions.

5. $x - 2 = 4$ 6. $x - 1 = 5$ 7. $6 = q - 3$ 8. $7 = q - 2$

9. $x - 2.5 = 5.0$ 10. $x - 3.5 = 12$ 11. $15 = d - 6.6$ 12. $20 = r - 6.9$

13. $x - 7.5 = 3.5$ 14. $x - 2.4 = 1.5$ 15. $1.75 = k - 0.75$ 16. $1.6 = k - 1.4$

17. $x - \frac{11}{3} = \frac{19}{3}$ 18. $x - \frac{10}{3} = \frac{7}{3}$ 19. $5\frac{1}{3} = z - 5\frac{1}{3}$ 20. $2\frac{3}{4} = z - 1\frac{3}{4}$

21. $x - 0 = 10\frac{1}{2}$ 22. $1\frac{3}{5} = k - 0$ 23. $x - 2.75 = 0$ 24. $0 = k - 7.2$

25. $x - 3 = 10 + 5$ 26. $x - 3.5 = 10.5 + 5.5$ 27. $3 + 5 = x - 5$

28. $13 + 13 = x - 7$ 29. $x - 2\frac{3}{4} = 2\frac{1}{4} + 3\frac{1}{4}$ 30. $x - 3\frac{3}{5} = 10\frac{1}{5} - 3\frac{1}{5}$

31. $9\frac{4}{5} - 4\frac{1}{5} = c - 2\frac{3}{5}$ 32. $12\frac{6}{7} - 3\frac{1}{7} = w - 8\frac{3}{5}$ 33. $12\frac{1}{4} - 3\frac{1}{4} = a - 6.6$

34. **Critical Thinking** Find b such that $x - b = 1$ has a solution of $4\frac{1}{2}$.

35. **Critical Thinking** Find c such that $x - 4\frac{1}{2} = c$ has a solution of $4\frac{1}{2}$.

MIXED REVIEW

Solve each problem. Round answers to two decimal places if necessary. (previous courses)

Sample: Find 45% of 20.   Solution: $0.45 \times 20 = 9$

36. Find 25% of 110.   37. Find 60% of 200.

38. What is 50% of 36?   39. What is 60% of 50?

40. 80% of 250   41. 120% of 18

Sample: What percent of 20 is 9?   Solution: $\frac{9}{20} = 0.45 = 45\%$

42. What percent of 100 is 55?   43. What percent of 24 is 16?

44. What percent of 75 is 90?   45. What percent of 15 is 20?

Answers for Lesson 1.4 Skill A

Try This Example 1
$d = 9$

Try This Example 2
$t = 114.5$

1. 2.3
2. 5
3. 3
4. 4
5. $x = 6$
6. $x = 6$
7. $q = 9$
8. $q = 9$
9. $x = 7.5$
10. $x = 15.5$
11. $d = 21.6$
12. $r = 26.9$
13. $x = 11$
14. $x = 3.9$
15. $k = 2.5$
16. $k = 3$
17. $x = 10$
18. $x = \frac{17}{3}$, or $5\frac{2}{3}$

19. $z = \frac{32}{3}$, or $10\frac{2}{3}$
20. $z = \frac{9}{2}$, or $4\frac{1}{2}$
21. $x = 10\frac{1}{2}$
22. $k = 1\frac{3}{5}$
23. $x = 2.75$
24. $k = 7.2$
25. $x = 18$
26. $x = 19.5$
27. $x = 13$
28. $x = 33$
29. $x = \frac{33}{4}$, or $8\frac{1}{4}$
30. $x = \frac{53}{5}$, or $10\frac{3}{5}$
31. $c = \frac{41}{5}$, or $8\frac{1}{5}$
32. $w = \frac{641}{35}$, or $18\frac{11}{35}$
33. $a = 15.6$

34. $b = \frac{7}{2}$, or $3\frac{1}{2}$
35. $c = 0$
36. 27.5
37. 120
38. 18
39. 30
40. 200
41. 21.6
42. 55%
43. 66.67%
44. 120%
45. 133.33%

★ **Advanced Exercises**

TEACHING 1.5 LESSON

Solving Simple Equations Using Subtraction

SKILL A ▸ *Using subtraction to solve an equation in one step*

Focus: *Students solve simple equations in one variable by applying the Subtraction Property of Equality.*

EXAMPLES 1 and 2

PHASE 1: Presenting the Examples

Display the six pairs of equations below. Ask students if they are *true* or *false*. [**All are true.**] For each pair, ask what action transforms $10 = 10$ into the second equation.

[**subtracting 7, 1, 10, 3.8, $8\frac{3}{4}$, and 9.5 from each side**]

$10 = 10$	$10 = 10$	$10 = 10$	$10 = 10$	$10 = 10$	$10 = 10$
$3 = 3$	$9 = 9$	$0 = 0$	$6.2 = 6.2$	$1\frac{1}{4} = 1\frac{1}{4}$	$0.5 = 0.5$

Have students write a possible statement for a Subtraction Property of Equality and compare to the statement on the textbook page. Then discuss Examples 1 and 2.

PHASE 2: Providing Guidance for TRY THIS

Some students focus on the + symbol in an equation and add the two numbers they see. Have them write their solutions in the form shown at right, emphasizing that each step is justified by a property.

$$
\begin{aligned}
a + 15.5 &= 40 \\
-15.5 &= -15.5 \quad &&\leftarrow \textit{Add. Prop. of} = \\
a + 0 &= 24.5 \quad &&\leftarrow \textit{Inverse Prop. of} + \\
a &= 24.5 \quad &&\leftarrow \textit{Identity Prop. of} +
\end{aligned}
$$

EXAMPLE 3

PHASE 1: Presenting the Example
PHASE 2: Providing Guidance for TRY THIS

Some students may feel disoriented in Example 3. Give them the two equations at right. Ask them to *verbalize* how the second is like the first and how it is different.

$$z + 14.95 = 20$$
$$14.95 + z = 20$$

[**Sample likenesses: The right side is 20. The left side is the sum of *z* and 14.95. Sample difference: The order of the left side is reversed.**]

PHASE 3: ASSESSMENT AND CLOSURE for Lesson 1.5

- Give students the equation at right. Ask them to find two different ways to fill in the boxes so that the solution is 16.

- Students have now learned to solve equations in one variable using the Addition and Subtraction Properties of Equality. In the next lesson, they will determine which method is appropriate to a given situation.

☞ *For a Lesson 1.5 Quiz, see* Assessment Resources *page 5.*

$$d + \square = \square$$

[**samples:** $d + 2 = 18$; $d + 10 = 26$ (**Students may use any two numbers for which the difference when the first is subtracted from the second is 16.)**]

1.5 **Solving Simple Equations Using Subtraction**

LESSON

SKILL A ► *Using subtraction to solve an equation in one step*

The *Subtraction Property of Equality* states the following:

If the same quantity is subtracted from each side of a true equation, then the equation that results is also true.

balance ⟶ balance

Subtraction Property of Equality
If a, b, and c are numbers and $a = b$, then $a - c = b - c$.

EXAMPLE 1 Solve $x + 12 = 80$.

► Solution

$x + 12 = 80$
$x + 12 - 12 = 80 - 12$ ⟵ *Apply the Subtraction Property of Equality.*
$x = 68$

TRY THIS Solve $a + 15.5 = 40$.

The Subtraction Property of Equality applies whether the variable is on the left side or the right side of the equation.

EXAMPLE 2 Solve $80 = t + 24$.

► Solution

$80 = t + 24$
$80 - 24 = t + 24 - 24$ ⟵ *Apply the Subtraction Property of Equality.*
$56 = t$

TRY THIS Solve $20 = c + 15.5$.

The Commutative Property of Addition allows you to rearrange equations.

EXAMPLE 3 Solve $14.95 + z = 20$.

► Solution

$14.95 + z = 20$
$z + 14.95 = 20$ ⟵ *Apply the Commutative Property of Addition.*
$z + 14.95 - 14.95 = 20 - 14.95$ ⟵ *Apply the Subtraction Property of Equality.*
$z = 5.05$

TRY THIS Solve $13.95 + c = 16$.

22 Chapter 1 Solving Simple Equations

 EXERCISES

KEY SKILLS

What number would you subtract from each side of the equation to solve it? Do not solve.

1. $p + 6.4 = 12$ 2. $7.2 + t = 12.3$ 3. $4 = 3.1 + y$

PRACTICE

Solve each equation. Check your solution.

4. $t + 4 = 19$ 5. $t + 6 = 16$ 6. $19 + a = 19$

7. $1.2 + a = 1.2$ 8. $1.7 = 1.4 + z$ 9. $100 = 99.95 + z$

10. $t + \frac{10}{3} = 6$ 11. $t + \frac{16}{7} = 7$ 12. $\frac{2}{3} + w = 2$

13. $3\frac{2}{3} + w = 4$ 14. $12\frac{5}{8} = 11\frac{5}{8} + x$ 15. $1\frac{2}{5} = t + \frac{3}{5}$

16. $11\frac{2}{5} = s + \frac{1}{5}$ 17. $t + 2.4 = 6 - 0.4$ 18. $z + 3.4 = 6 - 0.6$

19. $z + 1.5 = 6 - 3.5$ 20. $t + 2\frac{2}{5} = 6 - 0.4$ 21. $a + 2\frac{2}{5} = 6 - 2\frac{2}{5}$

22. Solve $z + \left(\frac{1}{8} + \frac{1}{4} + \frac{1}{2} + 1\right) = 1 + 2 + 4 + 8$.

23. Solve $z + \left(\frac{7}{4} + \frac{5}{4} + \frac{3}{4} + \frac{1}{4} + 1\right) = 7 + 5 + 3 + 1$.

24. **Critical Thinking** Find a such that $x + a = 3\frac{2}{3}$ has a solution of $\frac{1}{2}$.

MIXED REVIEW

Find each sum. (previous courses)

Sample: $\frac{3}{8} + \frac{5}{6}$ Solution: $\frac{3}{2 \times 4} + \frac{5}{2 \times 3} = \left(\frac{3}{3}\right)\left(\frac{3}{2 \times 4}\right) + \left(\frac{4}{4}\right)\left(\frac{5}{2 \times 3}\right) = \frac{9}{24} + \frac{20}{24} = \frac{29}{24}$

25. $\frac{1}{3} + \frac{1}{6}$ 26. $\frac{1}{4} + \frac{1}{12}$ 27. $\frac{2}{3} + \frac{5}{6}$ 28. $\frac{3}{4} + \frac{5}{12}$

29. $\frac{1}{6} + \frac{1}{9}$ 30. $\frac{5}{6} + \frac{2}{9}$ 31. $\frac{3}{8} + \frac{7}{12}$ 32. $\frac{5}{8} + \frac{11}{12}$

33. $\frac{1}{12} + \frac{1}{18}$ 34. $\frac{5}{12} + \frac{5}{18}$ 35. $\frac{1}{8} + \frac{1}{20}$ 36. $\frac{1}{12} + \frac{1}{20}$

Sample: $5\frac{2}{3} + 2\frac{2}{3}$ Solution: $5 + 2 = 7$ and $\frac{2}{3} + \frac{2}{3} = \frac{4}{3}$, or $1\frac{1}{3}$, so $7 + 1\frac{1}{3} = 8\frac{1}{3}$

37. $3\frac{3}{5} + 1\frac{1}{5}$ 38. $4\frac{3}{8} + 3\frac{3}{8}$ 39. $7\frac{1}{5} + 6\frac{4}{5}$ 40. $5\frac{7}{8} + 3\frac{3}{8}$

Solve each equation. Check your solution. (1.4 Skill A)

41. $x - 2.5 = 19$ 42. $0.1 = t - 15.9$ 43. $100 = w - 100$

Answers for Lesson 1.5 Skill A

Try This Example 1
$a = 24.5$

Try This Example 2
$c = 4.5$

Try This Example 3
$c = 2.05$

1. 6.4
2. 7.2
3. 3.1
4. $t = 15$
5. $t = 10$
6. $a = 0$
7. $a = 0$
8. $z = 0.3$
9. $z = 0.05$
10. $t = \frac{8}{3}$, or $2\frac{2}{3}$
11. $t = \frac{33}{7}$, or $4\frac{5}{7}$
12. $w = \frac{4}{3}$, or $1\frac{1}{3}$

13. $w = \frac{1}{3}$
14. $x = 1$
15. $t = \frac{4}{5}$
16. $s = 11\frac{1}{5}$
17. $t = 3.2$
18. $z = 2$
19. $z = 1$
20. $t = \frac{11}{6}$, or $1\frac{5}{6}$
21. $a = \frac{6}{5}$, or $1\frac{1}{5}$
22. $z = 13\frac{1}{8}$
23. $z = 11$
24. $a = \frac{19}{6}$, or $3\frac{1}{6}$
25. $\frac{1}{2}$
26. $\frac{1}{3}$

27. $\frac{3}{2}$, or $1\frac{1}{2}$
28. $\frac{7}{6}$, or $1\frac{1}{6}$
29. $\frac{5}{18}$
30. $\frac{19}{18}$, or $1\frac{1}{18}$
31. $\frac{23}{24}$
32. $\frac{37}{24}$, or $1\frac{13}{24}$
33. $\frac{5}{36}$
34. $\frac{25}{36}$
35. $\frac{7}{40}$
36. $\frac{2}{15}$
37. $\frac{24}{5}$, or $4\frac{4}{5}$
38. $\frac{31}{4}$, or $7\frac{3}{4}$

39. 14
40. $\frac{37}{4}$, or $9\frac{1}{4}$
41. $x = 21.5$
42. $t = 16$
43. $w = 200$

★ **Advanced Exercises**

1.6 Solving Simple Equations by Choosing Addition or Subtraction

SKILL A ▸ *Choosing addition or subtraction as the solution method*

Focus: *Students analyze a given equation and choose the Addition Property of Equality or the Subtraction Property of Equality to solve.*

EXAMPLE 1

PHASE 1: Presenting the Example

Display the equations from Example 1. Tell students to write a verbal expression of each equation. After all students have written their translations, discuss their responses. Summarize the correct responses by displaying the following:

$x - 12 = 18$: The difference of x minus 12 is 18.

$x + 12 = 18$: The sum when 12 is added to x is 18.

Now tell students to solve each equation. Give them a few minutes to do their work, and then check their answers. [**$x = 30$ and $x = 6$, respectively**] Ask for two volunteers to write step-by-step solutions on the chalkboard or on a transparency for all to see. [**See solutions at right.**]

$$x - 12 = 18$$
$$x - 12 + 12 = 18 + 12$$
$$x = 30$$

Discuss the solutions. Ask students how they decided which property of equality to use. Students should see that, if an equation involves subtraction, you solve by adding. Conversely, if an equation involves addition, you solve by subtracting. Because subtraction "undoes" addition, and vice versa, addition and subtraction are called *inverse operations.*

$$x + 12 = 18$$
$$x + 12 - 12 = 18 - 12$$
$$x = 6$$

PHASE 2: Providing Guidance for TRY THIS

The solution to the equation in part b, $x = 0$, may be disconcerting to students. When they arrive at a result like this, some feel that they have made a mistake, while others interpret it as "no solution." Remind students that 0 is a number and is therefore a valid solution to an equation. Ask them to write a similar equation that also has the solution $x = 0$. [**sample: $x + 7 = 7$**] Point out that the solution of any equation of the form $x + a = a$ will be $x = 0$.

EXAMPLE 2

PHASE 1: Presenting the Example

PHASE 2: Providing Guidance for TRY THIS

Students are encouraged to solve equations like these by first simplifying the expression on one side of the equal sign. However, alternative solution methods are acceptable. For instance, the Try This exercise is solved at right in two steps by first applying the Addition Property of Equality, and then applying the Subtraction Property of Equality.

$$4 = m + 2.3 - 1.6$$
$$4 + 1.6 = m + 2.3 - 1.6 + 1.6$$
$$5.6 = m + 2.3$$
$$5.6 - 2.3 = m + 2.3 - 2.3$$
$$3.3 = m$$

1.6 LESSON
Solving Simple Equations by Choosing Addition or Subtraction

Choosing addition or subtraction as the solution method

When you solve an equation, you need to choose a solution method that will leave the variable by itself on one side of the equation. This is sometimes called isolating the variable.

Example 1 part **a** shows an equation involving subtraction that is solved by addition. Example 1 part **b** shows an equation involving addition that is solved by subtraction. Addition and subtraction are *inverse operations*. That is, addition will undo subtraction and subtraction will undo addition.

EXAMPLE 1 Solve each equation. Check your solution.

  a. $x - 12 = 18$ **b.** $x + 12 = 18$

▶ **Solution**

  a.
$$x - 12 = 18$$
$$x - 12 + 12 = 18 + 12$$ ← Because 12 is subtracted from x, add 12 to each side.
$$x = 30$$

  Check: $30 - 12 = 18$ ✔

  b.
$$x + 12 = 18$$
$$x + 12 - 12 = 18 - 12$$ ← Because 12 is added to x, subtract 12 from each side.
$$x = 6$$

  Check: $6 + 12 = 18$ ✔

TRY THIS   Solve each equation. Check your solution.   **a.** $x - 9 = 9$ **b.** $x + 9 = 9$

EXAMPLE 2 Solve $10 = n + 3\frac{4}{5} - 2\frac{1}{5}$. Check your solution.

▶ **Solution**

$$10 = n + 3\frac{4}{5} - 2\frac{1}{5}$$
$$n + 3\frac{4}{5} - 2\frac{1}{5} = 10$$ ← Use the Symmetric Property of Equality to rewrite the equation with the variable on the left.
$$n + 1\frac{3}{5} = 10$$ ← Simplify: $3\frac{4}{5} - 2\frac{1}{5} = 1\frac{3}{5}$
$$n + 1\frac{3}{5} - 1\frac{3}{5} = 10 - 1\frac{3}{5}$$ ← Apply the Subtraction Property of Equality.
$$n = 8\frac{2}{5}$$

  Check: $8\frac{2}{5} + 3\frac{4}{5} - 2\frac{1}{5} = \left(8\frac{2}{5} + 3\frac{4}{5}\right) - 2\frac{1}{5} = 12\frac{1}{5} - 2\frac{1}{5} = 10$ ✔

TRY THIS   Solve $4 = m + 2.3 - 1.6$. Check your solution.

EXERCISES

Perform the given operation on each side of $x + 5 = 11$. Write the equation that results .

  1. Subtract 3.   **2.** Subtract 4.   **3.** Subtract 5.

Solve each equation. Check your solution.

  4. $x + 2.5 = 12$ **5.** $2.4 + c = 11$ **6.** $13 = z - 5.2$

  7. $y - 2.7 = 2.9$ **8.** $3.6 + y = 19$ **9.** $13 = 9 + w$

  10. $v - 6.2 = 19$ **11.** $22 = a - 2.2$ **12.** $b + 13.5 = 20$

  13. $d - 13.5 = 20$ **14.** $n - 3 = 4.5$ **15.** $m + \frac{10}{3} = \frac{14}{3}$

  16. $p + \frac{10}{3} = \frac{17}{3}$ **17.** $z - 5 = 3 + 2$ **18.** $12 - 3 = w + 5$

  19. $10 - 3 = 2 + h$ **20.** $\left(3\frac{1}{3} + 2\frac{1}{3}\right) = 3\frac{4}{5} + h$ **21.** $n - \left(7\frac{1}{3} + 5\frac{2}{3}\right) = 3\frac{4}{5}$

  22. $6 - 2\frac{1}{4} = 2\frac{1}{2} + z$ **23.** $7 = z + 4\frac{1}{2} - 3\frac{3}{4}$ **24.** $n - \left(3\frac{1}{3} + 2\frac{2}{3} - 1\frac{1}{3}\right) = 7\frac{2}{3}$

  25. Critical Thinking For what value of a will the solution to $x + a = b$ be the same as the solution to $x - a = b$?

  26. Critical Thinking Find the difference of the solution to $x + a = b$ less the solution to $x - a = b$.

Which fraction is greater? (previous courses)

Sample: $\frac{2}{5}$ or $\frac{3}{7}$ Solution: $\left(\frac{7}{7}\right)\frac{2}{5} = \frac{14}{35}$ $\left(\frac{5}{5}\right)\frac{3}{7} = \frac{15}{35}$ $\frac{15}{35} > \frac{14}{35} \rightarrow \frac{3}{7}$

  27. $\frac{1}{4}$ or $\frac{1}{3}$ **28.** $\frac{1}{4}$ or $\frac{1}{5}$ **29.** $\frac{3}{4}$ or $\frac{2}{3}$

  30. $\frac{3}{4}$ or $\frac{4}{5}$ **31.** $\frac{6}{7}$ or $\frac{7}{8}$ **32.** $\frac{6}{8}$ or $\frac{7}{9}$

Order each list of numbers from least to greatest. (previous courses)

  33. $\frac{3}{5}, \frac{1}{2}, \frac{5}{7}, \frac{2}{3}$ **34.** $0.145, 0.15, 0.155, 0.1$

  35. $\frac{1}{21}, 0.05, 0.45, 0.14, \frac{11}{25}, \frac{8}{50}$ **36.** $\frac{1}{12}, 0.12, \frac{1}{20}, 0.20, \frac{1}{6}, 0.6, 0.06, 0.16$

Answers for Lesson 1.6 Skill A

Try This Example 1
a. $x = 18$ **b.** $x = 0$

Try This Example 2
$m = 3.3$

  1. $x + 2 = 8$
  2. $x + 1 = 7$
  3. $x = 6$
  4. $x = 9.5$
  5. $c = 8.6$
  6. $z = 18.2$
  7. $y = 5.6$
  8. $y = 15.4$
  9. $w = 4$
  10. $v = 25.2$
  11. $a = 24.2$
  12. $b = 6.5$
  13. $d = 33.5$
  14. $n = 7.5$
  15. $m = \frac{4}{3}$, or $1\frac{1}{3}$

  16. $p = \frac{7}{3}$, or $2\frac{1}{3}$
  17. $z = 10$
  18. $w = 4$
  19. $h = 5$
  ★ **20.** $h = \frac{28}{15}$, or $1\frac{13}{15}$
  ★ **21.** $n = 16\frac{4}{5}$
  ★ **22.** $z = \frac{5}{4}$, or $1\frac{1}{4}$
  ★ **23.** $z = \frac{25}{4}$, or $6\frac{1}{4}$
  ★ **24.** $n = \frac{37}{3}$, or $12\frac{1}{3}$
  25. $a = 0$
  26. $-2a$
  27. $\frac{1}{3}$
  28. $\frac{1}{4}$
  29. $\frac{3}{4}$

  30. $\frac{4}{5}$
  31. $\frac{7}{8}$
  32. $\frac{7}{9}$
  33. $\frac{1}{2}, \frac{3}{5}, \frac{2}{3}, \frac{5}{7}$
  34. $0.1, 0.145, 0.15, 0.155$
  35. $\frac{1}{21}, 0.05, 0.14, \frac{8}{50}, \frac{11}{25}, 0.45$
  36. $\frac{1}{20}, 0.06, \frac{1}{12}, 0.12, 0.16, \frac{1}{6}, 0.20, 0.6$

★ **Advanced Exercises**

Focus: *Students use equations in one variable to represent real-world problems and choose the Addition or Subtraction Property of Equality to solve.*

EXAMPLES 1 and 2

PHASE 1: Presenting the Examples

Display the expression $t - 4$. Tell students that t represents an unknown temperature in degrees Fahrenheit. Have them write a verbal description for $t - 4$. [**sample: a decrease in temperature of 4°F**] Discuss the students' responses. Ask if they can identify the exact value of t. [**No; there is not enough information.**]

Now show the equation $t - 4 = 72$. Have students write a verbal description for it. [**sample: After a decrease of 4°F, the temperature was 72°F.**] Discuss the students' responses. Ask again if they can identify the value of t. This time they should recognize that, by adding 4 to each side of $t - 4 = 72$, they will arrive at $t = 76$. So the unknown temperature is 76°F.

Now discuss the problem in Example 1. Point out that here the process is reversed: Given a situation, they must write an equation for it.

Some students need a more structured approach to writing an equation. Use Example 2 as an opportunity to display the technique shown below. Emphasize the direct parallel between the mathematical language and the verbal language.

weight of box	plus	weight of marbles	equals	8.3 pounds
↓	↓	↓	↓	↓
1.2	+	m	=	8.3

PHASE 2: Providing Guidance for TRY THIS

When students are writing equations that represent real-world situations, emphasize the importance of stating what the variable represents. If it represents a measure, they must take care to specify the *unit* of measure. In the Example 2 Try This, for instance, encourage them to begin their work with a statement like the one at right.

Let w represent the weight of the contents in tons.

PHASE 3: ASSESSMENT AND CLOSURE for Lesson 1.6

- Give students the equations at right. Have them write a problem that each equation might represent and use the equation to solve the problem.

- Students have now seen how the operations of addition and subtraction are used in solving equations. In the three lessons that follow, they will learn about equations that are solved by using multiplication and division.

$m - 4 = 8$
$n + 6 = 16$

[**$m = 12$; $n = 10$; Problems will vary.**]

☞ *For a Lesson 1.6 Quiz, see* Assessment Resources *page 6.*

When you read a real-world problem, look for key words that suggest how to write the equation. Key words usually express an action indicating some kind of change. Key words may also describe a relationship between two quantities.

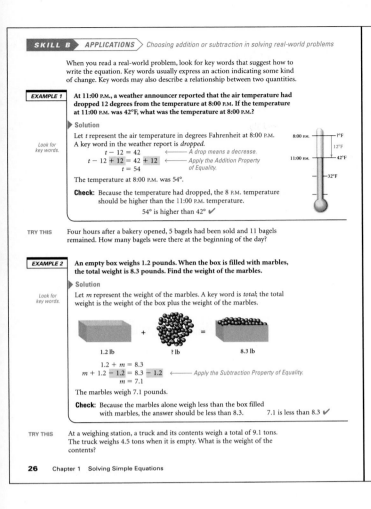

EXAMPLE 1

At 11:00 P.M., a weather announcer reported that the air temperature had dropped 12 degrees from the temperature at 8:00 P.M. If the temperature at 11:00 P.M. was 42°F, what was the temperature at 8:00 P.M.?

▶ **Solution**

Look for key words.

Let t represent the air temperature in degrees Fahrenheit at 8:00 P.M. A key word in the weather report is *dropped*.

$$t - 12 = 42 \longleftarrow \text{A drop means a decrease.}$$
$$t - 12 + 12 = 42 + 12 \longleftarrow \text{Apply the Addition Property of Equality.}$$
$$t = 54$$

The temperature at 8:00 P.M. was 54°.

Check: Because the temperature had dropped, the 8 P.M. temperature should be higher than the 11:00 P.M. temperature.

54° is higher than 42° ✔

TRY THIS Four hours after a bakery opened, 5 bagels had been sold and 11 bagels remained. How many bagels were there at the beginning of the day?

EXAMPLE 2

An empty box weighs 1.2 pounds. When the box is filled with marbles, the total weight is 8.3 pounds. Find the weight of the marbles.

▶ **Solution**

Look for key words.

Let m represent the weight of the marbles. A key word is *total*; the total weight is the weight of the box plus the weight of the marbles.

| 1.2 lb | + | ? lb | = | 8.3 lb |

$$1.2 + m = 8.3$$
$$m + 1.2 - 1.2 = 8.3 - 1.2 \longleftarrow \text{Apply the Subtraction Property of Equality.}$$
$$m = 7.1$$

The marbles weigh 7.1 pounds.

Check: Because the marbles alone weigh less than the box filled with marbles, the answer should be less than 8.3.   7.1 is less than 8.3 ✔

TRY THIS At a weighing station, a truck and its contents weigh a total of 9.1 tons. The truck weighs 4.5 tons when it is empty. What is the weight of the contents?

EXERCISES

KEY SKILLS

For each situation, write an equation and identify the key words. Do not solve.

1. Seven students left a room and 14 students remained. How many students were in the room before any students left?

2. After 7 students entered a room, there were 22 students in the room. How many students were in the room before any students entered?

PRACTICE

Write an equation to represent each situation. Solve the equation and then answer the question.

3. In 6 hours, the temperature has dropped 6 degrees Fahrenheit. If the current temperature is 72°F, what was the temperature 6 hours ago?

4. A customer gave a clerk 2 five-dollar bills and 2 one-dollar bills for a purchase of $11.52. How much change should the customer receive?

5. After 12 marbles were removed from a bag, 18 marbles remained. How many marbles were in the bag before any were removed?

6. A scale shows 1.2 pounds of flour. How much more flour should be added to make 1.5 pounds of flour on the scale?

7. A balance of $1245.35 remains in a bank account after a withdrawal of $320.39. How much was in the account before the withdrawal?

8. A motorist noticed that the odometer in her car read 12,345.2 miles. A few hours later, it read 12,511.1 miles. How far did she drive?

9. An empty container weighs 1.3 pounds. When filled with sand, the total weight is 10 pounds. How much does the sand weigh?

10. The following numbers represent the hourly changes in a stock market index during one trading day. If the index was 5420 at the beginning of the day, what was the final index?
$$+120, -80, +10, +35, +60, -5, -15, +10$$

MIXED REVIEW APPLICATIONS

Write each phrase as a mathematical expression. (1.1 Skill A)

11. the cost of a CD and 2 tapes at $19.95 and $7.95 each, respectively

12. total student attendance: three buses each containing 32 students and 7 cars each having 4 students

---

Answers for Lesson 1.6 Skill B

Try This Example 1
16 bagels

Try This Example 2
4.6 tons

1. A key word is *left*; let s represent the number of students originally in the room; $s - 7 = 14$

2. A key word is *entered*; let s represent the number of students originally in the room; $s + 7 = 22$

3. Let t represent the original temperature; $t - 6 = 72$; 78°F

4. Let c represent the amount of change; $11.52 + c = 12$; $0.48

5. Let m represent the number of marbles originally in the bag; $m - 12 = 18$; 30 marbles

6. Let f represent the amount of flour to be added; $1.2 + f = 1.5$; 0.3 pound

7. Let d represent the amount originally in the bank; $d - 320.39 = 1245.35$; $1565.74

8. Let d represent distance driven; $12,345.2 + d = 12,511.1$; 165.9 miles

9. Let s represent the amount of sand; $s + 1.3 = 10$; 8.7 pounds

10. 5555

11. $(1 \times 19.95) + (2 \times 7.95)$

12. $(3 \times 32) + (7 \times 4)$

Solving Simple Equations Using Multiplication

SKILL A ▷ *Using multiplication to solve an equation in one step*

Focus: *Students solve simple equations in one variable by applying the Multiplication Property of Equality.*

EXAMPLE 1

PHASE 1: Presenting the Example

Display the equation $100 = 100$. Tell students to choose any number, multiply both sides of the equation by that number, and write the result. Ask several students to share their answers with the class, recording each multiplication for all to see in the form shown at right.

$$\begin{array}{rcl} 100 & = & 100 \\ \times 4 & & \times 4 \\ \hline 400 & = & 400 \end{array}$$

Ask students to make a conjecture about the results. Elicit a statement of the Multiplication Property of Equality. Then discuss Example 1.

$$\begin{array}{rcl} 100 & = & 100 \\ \times \frac{1}{5} & = & \times \frac{1}{5} \\ \hline 20 & & 20 \end{array}$$

PHASE 2: Providing Guidance for TRY THIS

Some students may need a reminder that the fraction bar is a division symbol. Suggest that they rewrite the equation as $10.2 = t \div 3$.

$$\begin{array}{rcl} 100 & = & 100 \\ \times 8.3 & & \times 8.3 \\ \hline 830 & = & 830 \end{array}$$

EXAMPLE 2

PHASE 1: Presenting the Example

PHASE 2: Providing Guidance for TRY THIS

Some students may apply the Multiplication Property of Equality twice, as shown at right. Since the goal of this lesson is to arrive at the solution of an equation, allow students to use this method.

$$\frac{2n}{3} = 6.2$$
$$2n \div 3 = 6.2$$
$$2n \div 3 \times 3 = 6.2 \times 3$$
$$2n = 18.6$$
$$\frac{1}{2} \times 2n = \frac{1}{2} \times 18.6$$
$$n = 9.3$$

PHASE 3: ASSESSMENT AND CLOSURE for Lesson 1.7

- Give students the equations at right. Ask them how the process of solving the equations is similar x [**Both can be solved by multiplying each side by 6.**] Then have the students solve the equations. $\left[p = 32.4; q = 2\frac{2}{5} \right]$

$$p \div 6 = 5.4$$
$$\frac{2}{5} = \frac{q}{6}$$

- Students should now be able to summarize the Addition, Subtraction, and Multiplication Properties of Equality, and they should recognize that each is a tool used to solve equations. Point out that the next step will be to investigate whether there is a Division Property of Equality.

☞ *For a Lesson 1.7 Quiz, see* Assessment Resources *page 7.*

1.7 LESSON Solving Simple Equations Using Multiplication

SKILL A *Using multiplication to solve an equation in one step*

The Multiplication Property of Equality states the following:

> **Multiplication Property of Equality**
> If a, b, and c are numbers and $a = b$, then $ac = bc$.

Because multiplication and division are inverse operations, the Multiplication Property of Equality is useful when you solve an equation in which the variable is divided by a number.

EXAMPLE 1 Solve $\frac{x}{5} = 2.5$. Check your solution.

▸ Solution

$$\frac{x}{5} = 2.5$$
$$(5)\frac{x}{5} = (5)2.5 \quad \longleftarrow \textit{Apply the Multiplication Property of Equality.}$$
$$x = 12.5 \qquad \textbf{Check: } \frac{12.5}{5} = 2.5 \checkmark$$

TRY THIS Solve $10.2 = \frac{t}{3}$. Check your solution.

When a variable is isolated, as in $x = 5$, its coefficient is 1. You know from the Identity Property of Multiplication that you can obtain a coefficient of 1 by multiplying a fraction by its reciprocal. So, you can write $\frac{2n}{3}$ as $\frac{2}{3} \times n$ to identify its reciprocal, $\frac{3}{2}$, that can help you isolate the variable n.

EXAMPLE 2 Solve $\frac{2n}{3} = 6.2$. Check your solution.

▸ Solution

$$\frac{2}{3} \times n = 6.2 \quad \longleftarrow \textit{Write } \frac{2n}{3} \textit{ as } \frac{2}{3} \times n.$$
$$\left(\frac{3}{2}\right)\frac{2}{3} \times n = \left(\frac{3}{2}\right)6.2 \quad \longleftarrow \textit{Apply the Multiplication Property of Equality.}$$
$$n = 9.3 \qquad \textbf{Check: } \frac{2(9.3)}{3} = \frac{18.6}{3} = 6.2 \checkmark$$

TRY THIS Solve $\frac{5v}{3} = 20$. Check your solution.

28 Chapter 1 Solving Simple Equations

EXERCISES

KEY SKILLS

What number would you use to multiply each side of the equation by to isolate the variable? Do not solve.

1. $\frac{a}{3} = 12$ 2. $\frac{b}{4} = 1$ 3. $\frac{1}{2}x = 9$

4. $\frac{2x}{5} = 6$ 5. $\frac{4t}{9} = 8$ 6. $\frac{9y}{10} = 12$

PRACTICE

Solve each equation. Check your solution.

7. $\frac{t}{2} = 2$ 8. $\frac{1}{4}t = 4$ 9. $5 = \frac{a}{6}$

10. $6 = \frac{q}{8}$ 11. $\frac{v}{8} = 1.3$ 12. $\frac{w}{5} = 5.5$

13. $7.0 = \frac{x}{7}$ 14. $2.0 = \frac{y}{3}$ 15. $\frac{w}{6} = \frac{3}{4}$

16. $\frac{r}{10} = \frac{3}{5}$ 17. $\frac{3}{7} = \frac{z}{3}$ 18. $\frac{3}{16} = \frac{s}{8}$

19. $\frac{3r}{7} = \frac{3}{2}$ 20. $\frac{4r}{11} = \frac{11}{8}$ 21. $\frac{14}{8} = \frac{7s}{2}$

22. $\frac{14}{10} = \frac{7x}{10}$ 23. $\frac{3c}{10} = 1.5$ 24. $\frac{8c}{7} = 5.6$

25. $5.6 = \frac{7z}{10}$ 26. $12.1 = \frac{11m}{10}$ 27. $\frac{7c}{2} = 12.1 + 8.9$

28. $\frac{10z}{2} = 1.3 + 8.7$ 29. $1.3 + 7.1 = \frac{7n}{10}$ 30. $2.5 + 5.1 = \frac{2d}{6}$

31. $\left(\frac{7}{4}\right)\left(\frac{5}{2}\right)\frac{2x}{3} = 1$ 32. $\left(\frac{3}{2}\right)\left(\frac{5}{9}\right)\frac{9x}{3} = 14$ 33. $\left(\frac{7}{2}\right)\left(\frac{5}{9}\right)\frac{2t}{7} = 5$

34. $\frac{x}{4} + 5 = 6$ 35. $\frac{x}{3} - 2 = 4$ 36. $\frac{2y}{9} + 6 = 10$

37. **Critical Thinking** Suppose that a and b are nonzero numbers and c is any number. Solve $\frac{ax}{b} = c$ for x.

MIXED REVIEW

Solve each equation. Check your solution. (1.4 Skill A, 1.5 Skill A)

38. $c - 2 = 6$ 39. $g - 10 = 19$ 40. $18.1 = z - 10.1$

41. $0.3 = x - 0.8$ 42. $y + 0.8 = 1.2$ 43. $a + 11.3 = 11.4$

44. $x + \frac{2}{3} = \frac{11}{3}$ 45. $x + \frac{4}{5} = 10$ 46. $s - \frac{9}{2} = 5$

Lesson 1.7 Skill A **29**

Answers for Lesson 1.7 Skill A

Try This Example 1
$t = 30.6$

Try This Example 2
$v = 12$

1. 3
2. 4
3. 2
4. $\frac{5}{2}$
5. $\frac{9}{4}$
6. $\frac{10}{9}$
7. $t = 4$
8. $t = 16$
9. $a = 30$
10. $q = 48$
11. $v = 10.4$
12. $w = 27.5$
13. $x = 49.0$
14. $y = 6.0$

15. $w = \frac{18}{4}$, or $4\frac{1}{2}$
16. $r = 16$
17. $z = \frac{9}{7}$, or $1\frac{2}{7}$
18. $s = \frac{3}{2}$, or $1\frac{1}{2}$
19. $r = \frac{7}{2}$, or $3\frac{1}{2}$
20. $r = \frac{121}{32}$, or $3\frac{25}{32}$
21. $s = \frac{1}{2}$
22. $x = 2$
23. $c = 5$
24. $c = 4.9$
25. $z = 8$
26. $m = 11$
27. $c = 6$
28. $z = 2$
29. $n = 12$
30. $d = 22.8$
31. $x = \frac{12}{35}$

32. $x = \frac{28}{5}$, or $5\frac{3}{5}$
33. $t = 9$
34. $x = 4$
35. $x = 18$
36. $y = 18$
37. $x = \frac{bc}{a}$
38. $c = 8$
39. $g = 29$
40. $z = 28.2$
41. $x = 1.1$
42. $y = 0.4$
43. $a = 0.1$
44. $x = 3$
45. $x = \frac{46}{5}$, or $9\frac{1}{5}$
46. $s = \frac{19}{2}$, or $9\frac{1}{2}$

Lesson 1.7   Skill A   **29**

Solving Simple Equations Using Division

SKILL A ▶ *Using division to solve an equation in one step*

Focus: *Students solve simple equations in one variable by applying the Division Property of Equality.*

EXAMPLE 1

PHASE 1: Presenting the Example

Tell students to write one example that suggests the existence of a Division Property of Equality. Ask several students to share their examples, recording each division for all to see in the form shown at right.

$$\frac{35}{7} = \frac{35}{7}$$
$$5 = 5$$

Ask students if the following statement is true: *If both sides of a true equation are divided by the same real number, then the resulting equation is also true.* Lead them to see that it is true as long as zero is not the divisor, because division by zero is undefined. Then discuss Example 1.

$$\frac{4.8}{6} = \frac{4.8}{6}$$
$$0.8 = 0.8$$

PHASE 2: Providing Guidance for TRY THIS

After students have successfully completed the Try This exercise, ask a series of "what if" questions: What if the equation were $24 = c + 4$? [**The solution would be $c = 20$.**] What if it were $24 = c - 4$? [$c = 28$] What if it were $24 = \frac{c}{4}$? [$c = 96$]

$$\frac{\left(\frac{3}{4}\right)}{2} = \frac{\left(\frac{3}{4}\right)}{2}$$
$$\frac{3}{8} = \frac{3}{8}$$

EXAMPLE 2

PHASE 1: Presenting the Example

PHASE 2: Providing Guidance for TRY THIS

Some students may see that they can solve these problems by dividing the total length by the number of pieces. Therefore, they may view these as simple problems and question the use of an algebraic method. Stress that these familiar situations allow them to learn and practice a technique that they can use when a solution is not as obvious.

PHASE 3: ASSESSMENT AND CLOSURE for Lesson 1.8

- Give students the equation at right. Tell them that the circle and triangle are real numbers. Have them describe how to solve the equation. [**sample: Divide each side by "triangle," provided that it is not zero.**]

$$\bigcirc = \triangle w$$

- The Addition, Subtraction, Multiplication, and Division Properties of Equality are axioms. That is, these statements are *assumed* true for all real numbers. In coming chapters, students will continue to explore how such axioms form a foundation on which other statements can be *proved* true.

☞ *For a Lesson 1.8 Quiz, see* Assessment Resources *page 8.*

1.8 LESSON — Solving Simple Equations Using Division

SKILL A Using division to solve an equation in one step

Division Property of Equality

If a, b, and c are numbers, c is nonzero, and $a = b$, then $\dfrac{a}{c} = \dfrac{b}{c}$.

Use the Division Property of Equality to solve an equation in which the variable is multiplied by a number.

EXAMPLE 1 Solve $\dfrac{21}{5} = 7d$. Check your solution.

▸ **Solution**

$$\frac{21}{5} = 7d$$

$$\frac{21}{5} \div 7 = 7d \div 7 \quad \longleftarrow \text{Apply the Division Property of Equality.}$$

$$\frac{21}{5} \times \frac{1}{7} = d \quad \longleftarrow \begin{array}{l}\text{Recall that dividing by 7 is equivalent} \\ \text{to multiplying by } \frac{1}{7}.\end{array}$$

$$\frac{3}{5} = d \qquad \textbf{Check: } 7\left(\frac{3}{5}\right) = \frac{21}{5} \ ✔$$

TRY THIS Solve $24 = 4c$. Check your solution.

EXAMPLE 2 A carpenter wants to cut a 12 foot board into 4 shorter boards of equal length. How long will each board be?

▸ **Solution**

Look for key words.

Let v represent the length of each short board.

number of boards	times	length of each board	is	original board length
↓	↓	↓		↓
4	×	v	=	12

Write an equation.

Write and solve an equation.

$$4v = 12$$

$$\frac{4v}{4} = \frac{12}{4} \quad \longleftarrow \text{Apply the Division Property of Equality.}$$

$$v = 3$$

Each new board will be 3 feet long.

TRY THIS Rework Example 2 if the carpenter wants 5 boards of equal length cut from a board 12.5 ft long.

30   Chapter 1   Solving Simple Equations

EXERCISES

Which of the given numbers is the best divisor to use to isolate each variable?

1. $3x = 15$; 3 or 15
2. $2.5z = 13$; 2.5 or 13
3. $\dfrac{9}{2} = 5n$; $\dfrac{1}{2}$, $\dfrac{9}{2}$, or 5
4. $4t = 3.5$; 4, 3, or 3.5

Solve each equation. Check your solution.

5. $3x = 24$
6. $5x = 25$
7. $81 = 9b$
8. $121 = 11a$
9. $2s = 7$
10. $3s = 10$
11. $6.5 = 1.5d$
12. $28.7 = 0.7g$
13. $4r = \dfrac{9}{2}$
14. $7x = \dfrac{10}{3}$
15. $\dfrac{38}{5} = 8p$
16. $\dfrac{10}{7} = 2z$
17. $7x = 1\dfrac{3}{7} + 4\dfrac{4}{7}$
18. $3h = 10\dfrac{2}{5} - 4\dfrac{2}{5}$
19. $2\dfrac{9}{11} + 4\dfrac{2}{11} = 2c$
20. $12\dfrac{8}{13} + 4\dfrac{5}{13} = 5k$
21. $(3 + 1)x = 12$
22. $(15 - 12)x = 12$
23. $36 = (15 - 9)k$
24. $3 = (5 - 2)w$
25. $3n - 9 = 21$
26. $2x + 5 = 13$
27. $3 + 6x = 27$
28. $35 = 1.5x - 1$

Find the length of each board or the amount of each liquid.

29. a 12-foot board cut into 8 equal lengths
30. 18 gallons of water poured into 6 containers of equal capacity
31. a 12.6-foot board cut into 6 equal lengths
32. 20 gallons of water poured into 8 containers of equal capacity

33. **Critical Thinking** Is it possible to cut a 17-foot board into 4 pieces of equal length and have the length of each board be a whole number? Justify your response.

Evaluate each expression using the given value for each variable. (1.2 Skill C)

34. $3.5(x - 2) + 3$, $x = 4$
35. $3x + 4(x - 1)$, $x = 2.5$
36. $2.5(x + 3) - 10$, $x = 7$
37. $\dfrac{3}{4}(2c + 7) - 5$, $c = 0$
38. $\dfrac{3}{5}(2d + 7) - 4d$, $d = 1$
39. $0.6(3w + 1) - 0.6w$, $w = 3$

Lesson 1.8   Skill A   **31**

Answers for Lesson 1.8 Skill A

Try This Example 1
$c = 6$

Try This Example 2
2.5 feet

1. 3
2. 2.5
3. 5
4. 4
5. $x = 8$
6. $x = 5$
7. $b = 9$
8. $a = 11$
9. $s = 3.5$
10. $s = \dfrac{10}{3}$, or $3\dfrac{1}{3}$
11. $d = \dfrac{13}{3}$, or $4\dfrac{1}{3}$
12. $g = 41$
13. $r = \dfrac{9}{8}$, or $1\dfrac{1}{8}$
14. $x = \dfrac{10}{21}$
15. $p = \dfrac{19}{20}$
16. $z = \dfrac{5}{7}$
17. $x = \dfrac{6}{7}$
18. $h = 2$
19. $c = \dfrac{7}{2}$, or $3\dfrac{1}{2}$
20. $k = \dfrac{17}{5}$, or $3\dfrac{2}{5}$
21. $x = 3$
22. $x = 4$
23. $k = 6$
24. $w = 1$
25. $n = 10$
26. $x = 4$
27. $x = 4$
28. $x = 24$
29. $\dfrac{3}{2}$, or $1\dfrac{1}{2}$ feet
30. 3 gallons
31. 2.1 feet
32. 2.5 gallons

33. No; let x represent the length of each board. The solution to $4x = 17$ is $\dfrac{17}{4}$, or $4\dfrac{1}{4}$, which is not a whole number.
34. 10
35. 13.5
36. 15
37. $\dfrac{1}{4}$
38. $\dfrac{7}{5}$, or $1\dfrac{2}{5}$
39. 4.2

TEACHING
1.9
LESSON

Solving Simple Equations by Choosing Multiplication or Division

SKILL A *Choosing multiplication or division as the solution method*

Focus: *Students analyze a given equation and choose the Multiplication or Division Property of Equality to solve.*

EXAMPLES 1 and 2

PHASE 1: Presenting the Examples

Display the equations $3s = 24$ and $\frac{t}{3} = 24$. Tell students to solve each equation. Then have them guide you as you record the step-by-step solutions for all to see. The displayed results should appear as shown at right.

$$3s = 24$$
$$3s \div 3 = 24 \div 3$$
$$s = 8$$

$$\frac{t}{3} = 24$$
$$\frac{t}{3} \times 3 = 24 \times 3$$
$$t = 72$$

Now ask students to compare the two solution methods. Elicit the fact that the first solution is justified by the Division Property of Equality, while the second is justified by the Multiplication Property of Equality.

Ask students how they chose which property of equality to use. Lead them to see that, if an equation involves multiplication, you solve by dividing; if an equation involves division, you solve by multiplying. That is, division is used to "undo" a multiplication and vice versa. So, just as addition and subtraction are *inverse operations*, so too are multiplication and division.

Now discuss Examples 1 and 2. Note that each solution begins by translating the mathematical language of the equation into verbal language. Point out that this simple step aids in identifying the *given* operation, which in turn leads to choosing the correct *inverse* operation for the solution.

PHASE 2: Providing Guidance for **TRY THIS**

In the Try This exercise for Example 2, some students may note that $\frac{a}{4}$ can be rewritten as $\frac{1}{4}a$. As a result, they may choose the Division Property of Equality as the solution method, as shown at right. You may wish to use this as a lead-in to Example 3.

$$\frac{a}{4} = \frac{3}{2}$$
$$\frac{1}{4}a = \frac{3}{2}$$
$$\frac{1}{4}a \div \frac{1}{4} = \frac{3}{2} \div \frac{1}{4}$$
$$a = \frac{3}{2} \times \frac{4}{1} = \frac{12}{2} = 6$$

EXAMPLE 3

PHASE 1: Presenting the Example

Students may begin to notice that any equation of the form $ax = b$, $a \neq 0$, can be solved using either the Division Property of Equality or the Multiplication Property of Equality. You may want to present a simpler example, using whole-number values of a and b, before discussing Example 3. A sample using $7m = 42$ is shown at right.

Choice 1:	Choice 2:
Divide by 7.	Multiply by $\frac{1}{7}$.
$7m = 42$	$7m = 42$
$\dfrac{7}{m} = \dfrac{42}{7}$	$\dfrac{1}{7} \times 7m = \dfrac{1}{7} \times 42$
$m = 6$	$m = 6$

1.9
LESSON

Solving Simple Equations by Choosing Multiplication or Division

SKILL A › *Choosing multiplication or division as the solution method*

Multiplication and division are *inverse operations*. That is, multiplication will undo division and division will undo multiplication.

EXAMPLE 1 | Choose a method for solving $5h = 120$. Then solve and check.

› **Solution**
In words: 5 multiplied by h is 120. Choose division.
$$\frac{5h}{5} = \frac{120}{5} \quad \longleftarrow \text{Apply the Division Property of Equality.}$$
$$h = 24$$
Check: $5(24) = 120$ ✔

TRY THIS | Choose a method for solving $4z = 36$. Then solve and check.

EXAMPLE 2 | Choose a method for solving $\frac{n}{7} = 2.5$. Then solve and check.

› **Solution**
In words: n divided by 7 is 2.5. Choose multiplication.
$$(7)\frac{n}{7} = (7)2.5 \quad \longleftarrow \text{Apply the Multiplication Property of Equality.}$$
$$n = 17.5, \text{ or } 17\frac{1}{2}$$
Check: $\left(17\frac{1}{2}\right) \div 7 = \frac{35}{2} \times \frac{1}{7} = \frac{5}{2} = 2.5$ ✔

TRY THIS | Choose a method for solving $\frac{a}{4} = \frac{3}{2}$. Then solve and check.

Some equations can be solved by either multiplication or division.

EXAMPLE 3 | Choose a method for solving $2x = 4.6$. Then solve and check.

› **Solution**

Method 1: Choose division.
$$\frac{2x}{2} = \frac{4.6}{2}$$
$$x = 2.3$$

Method 2: Choose multiplication.
$$\left(\frac{1}{2}\right)2x = \left(\frac{1}{2}\right)4.6$$
$$x = 2.3$$

Check: $2(2.3) = 4.6$ ✔

TRY THIS | Choose a method for solving $12 = \frac{4x}{5}$. Then solve and check.

EXERCISES

KEY SKILLS

Write each equation in words.

Sample: $\frac{x}{2} = 10$ Solution: *x* divided by 2 is 10.

1. $3x = 22$
2. $\frac{c}{4} = 15$
3. $\frac{3c}{4} = 4$
4. $0.5d = 10$

PRACTICE

Write the operations you would choose to solve each equation. Then solve and check.

5. $3.5x = 7$
6. $\frac{1}{4}d = 1$
7. $4.3k = 8.6$
8. $\frac{1}{3}n = 13$
9. $\frac{4m}{3} = 10$
10. $\frac{3m}{4} = 10$
11. $\frac{n}{3.5} = 1$
12. $2.5p = 2.5$
13. $\frac{p}{2.5} = 2.5$
14. $(2.5 + 1.5)n = 8$
15. $\frac{n}{2.5 + 1.5} = 8$
16. $\frac{2n}{3} = \frac{2}{3}$
17. $\frac{3}{2}n = \frac{2}{3}$
18. $\frac{n}{\left(1\frac{1}{2} + 6\frac{1}{2}\right)} = 2$
19. $\left(1\frac{1}{2} + 6\frac{1}{2}\right)n = 32$

Solve each equation.

20. $5x + 6 = 11$
21. $3x - 4 = 8$
22. $9x + 1 = 10$
23. $\frac{x}{2} - 4 = 4$
24. $3 = \frac{x}{5} - 7$
25. $12 = \frac{3x}{4} + 8$

26. **Critical Thinking** Suppose that *a* and *b* are nonzero numbers. Show that $\frac{ax}{b} = 10$ and $\frac{x}{\left(\frac{b}{a}\right)} = 10$ have the same solution.

MIXED REVIEW

Solve each equation. Check your solution. (1.4 Skill A, 1.5 Skill A, 1.7 Skill A, 1.8 Skill A)

27. $4.5x = 27$
28. $z - 3.2 = 7.2$
29. $\frac{n}{7} = 3$
30. $z - \frac{5}{4} = \frac{31}{4}$
31. $c + 2.5 = 7.9$
32. $d + (4 - 3.8) = 10$
33. $2x = 7.8$
34. $\frac{3n}{2} = 3$
35. $t + 4.25 = 5$

Answers for Lesson 1.9 Skill A

Try This Example 1
$z = 9$

Try This Example 2
$a = 6$

Try This Example 3
$x = 15$

1. Three times *x* equals 22.
2. *c* divided by 4 equals 15.
3. Three times *c* divided by 4 equals 4.
4. One half of *d* equals 10.
5. Divide each side by 3.5; $x = 2$
6. Multiply each side by 4; $d = 4$
7. Divide each side by 4.3; $k = 2$
8. Multiply each side by 3; $n = 39$
9. Multiply each side by $\frac{3}{4}$ or divide each side by $\frac{4}{3}$; $m = 7.5$
10. Multiply each side by $\frac{4}{3}$ or divide each side by $\frac{3}{4}$; $m = 13\frac{1}{3}$
11. Multiply each side by 3.5; $n = 3.5$
12. Divide each side by 2.5; $p = 1$
13. Multiply each side by 2.5; $p = 6.25$
14. Divide each side by the sum 2.5 + 1.5, or 4; $n = 2$

★ **Advanced Exercises**

15. Multiply each side by the sum 2.5 + 1.5, or 4; $n = 32$
★16. Multiply each side by $\frac{3}{2}$; $n = 1$
★17. Divide each side by $\frac{3}{2}$; $n = \frac{4}{9}$
★18. Multiply each side by the sum $1\frac{1}{2} + 6\frac{1}{2}$, or 8; $n = 16$
★19. Divide each side by the sum $1\frac{1}{2} + 6\frac{1}{2}$, or 8; $n = 4$
20. $x = 1$
21. $x = 4$
22. $x = 1$
23. $x = 16$
24. $x = 50$
25. $x = \frac{16}{3}$, or $5\frac{1}{3}$

For answers to Exercises 26–35, see Additional Answers.

Focus: *Students use equations in one variable to represent real-world problems and choose the Multiplication or Division Property of Equality to solve.*

EXAMPLES 1 and 2

PHASE 1: Presenting the Examples

Display the four equations at right. Tell students that n represents an amount of money in dollars. Ask them to translate each mathematical equation into a verbal expression. [**samples, top to bottom: The result when n dollars is multiplied by 4 is \$6000. The result when n dollars is multiplied by 6000 is \$4. The result when n dollars is divided into 4 parts is \$6000. The result when n dollars is divided into 6000 parts is \$4.**]

$$4n = 6000$$
$$6000n = 4$$
$$\frac{n}{4} = 6000$$
$$\frac{n}{6000} = 4$$

Now display the problem in Example 1. Ask students which equation best represents it. [**$4n = 6000$**] Then discuss the solution on the textbook page.

Introduce Example 2 in the same way, this time working with $\frac{c}{3} = 8$, $\frac{c}{8} = 3$, $3c = 8$, and $8c = 3$, where c represents a number of crayons.

PHASE 2: Providing Guidance for TRY THIS

Some students look at problems like these and are confused by all the words. Have them copy the problem, cross out extraneous words, and highlight important facts.

~~Mrs. Gomez withdrew some money from her bank to~~ divide among ~~her~~ 6 children. Each ~~child~~ received \$35.50. How much money ~~did she withdraw?~~

PHASE 3: ASSESSMENT AND CLOSURE for Lesson 1.9

- Ask students to write an equation that can be solved in one step by applying the Division Property of Equality. Repeat the instruction for the Addition, Subtraction, and Multiplication Properties of Equality. Then have the students solve their equations. [**Answers will vary.**]

- After completing this chapter, students may feel as if they were exposed to a bewildering array of equations in a very short period of time. To dispel this notion, help them identify characteristics that were *common* to all the equations they solved. For instance: all the equations involved just one occurrence of just one variable; no equation had a variable term like x^2 or y^3, with an exponent other than 1; no equation had a negative solution.

 For a Lesson 1.9 Quiz, see Assessment Resources *page 9.*

 For a Chapter 1 Assessment, see Assessment Resources *pages 10–11.*

When you read a real-world problem, look for key words or phrases that suggest how to write an equation.

EXAMPLE 1

An investor normally deposits a fixed amount into a bank account each month. One month, the investor deposits four times the normal monthly amount. If the deposit was $6000 that month, what is the normal monthly deposit?

▶ Solution

Look for key words.

1. Let n represent the normal monthly deposit in dollars.
   The key phrase is *four times.*

Write an equation.

2. Write and solve an equation.
   current deposit = 4 times the normal monthly deposit
   $$6000 = 4n$$
   $$4n = 6000 \quad \longleftarrow \text{Apply the Symmetric Property of Equality.}$$
   $$\frac{4n}{4} = \frac{6000}{4} \quad \longleftarrow \text{Apply the Division Property of Equality.}$$
   $$n = 1500$$
   The normal monthly deposit is $1500.
   Check: $4(1500) = 6000$ ✓

TRY THIS

This week, Martha received $27, which is 3 times her normal allowance. How much is her normal allowance?

EXAMPLE 2

A box of crayons is evenly divided among 8 children. Each child receives 3 crayons. How many crayons were in the box?

▶ Solution

Look for key words.

1. Let c represent the total number of crayons in the box.
   The key word is *divided.*

Write an equation.

2. Write and solve an equation.
   $$\frac{\text{total number of crayons}}{\text{number of children}} = \text{number of crayons per child}$$
   $$\frac{c}{8} = 3$$
   $$(8)\frac{c}{8} = (8)3 \quad \longleftarrow \text{Apply the Multiplication Property of Equality.}$$
   $$c = 24$$

   There were 24 crayons in the box.
   Check: $\frac{24}{8} = 3$ ✓

TRY THIS

Mrs. Gomez withdrew some money from her bank to divide among her 6 children. Each child received $35.50. How much money did she withdraw?

EXERCISES

For each situation, identify the key word(s) and write an equation. Do not solve.

1. A father wants to divide a certain amount of money among 5 children and give each child $24. How much should he withdraw?

2. Jamal wants to divide 3 hours of study time so that he spends the same amount of time on each of 6 school subjects. How much time will he devote to each subject?

PRACTICE

For each situation, write and solve an equation.

3. A contractor wants to pay each of 6 workers an equal amount of money. How much should each worker get if the contractor distributes $1140?

4. A carpenter needs four boards of equal length cut from a board 16-foot long. How long will each board be?

5. A teenager wants to save $290 to buy a stereo. How many weeks will it take if he can save $25 every week? Round your answer to the nearest whole number.

6. How much time should Gina set aside so that she can spend 45 minutes studying each of 5 subjects? Give your answer in hours.

7. One bus can seat 32 students. How many buses will be needed to transport 380 students?

8. **Critical Thinking** Jack is twice the age of Alice. Sasha is twice the age of Jack. If Alice is 5, how old is Sasha?

9. **Critical Thinking** Deshon has finished 5 more homework exercises than Yuko. Yuko has finished 3 more than Andres. Andres has finished twice as many as Mary. If Mary has finished 7 exercises, how many has Deshon finished?

MIXED REVIEW

Solve each problem. (1.6 Skill A)

10. Rob's purchases total $11.53. How much change should he receive if he gives the cashier a ten-dollar bill and 2 one-dollar bills?

11. After a 12.5°F drop in temperature, a thermometer showed 63.4°F. What was the temperature before the drop?

12. A scale shows 1.13 pounds of shrimp on the tray. How much should be added so that the shopper will receive 1.25 pounds of shrimp?

Answers for Lesson 1.9 Skill B

Try This Example 1
$9

Try This Example 2
$213

1. divide; $\frac{d}{5} = 24$

2. divide; $\frac{3}{x} = 6$

3. Let x represent the amount of money each worker will receive; $6x = 1120$, $x = 190$. Each worker will receive $190.

4. Let b represent the length of each board; $4b = 16$, $b = 4$. Each board will be 4 feet long.

5. Let w represent the number of weeks needed; $25w = 290$; $w = 11.6$; 12 weeks

6. Let t represent the total study time in minutes; $\frac{t}{5} = 45$; $t = 225$. The student will need 225 minutes, or 3.75 hours.

7. Let n represent the number of buses needed; $\frac{380}{n} = 32$; $n = 11\frac{7}{8}$; 12 buses will be needed.

8. 20

9. 22 exercises

10. $0.47

11. 75.9°F

12. 0.12 pound

Write each phrase as a mathematical expression.

1. the difference of 4 minus 17 less the sum of 6 and 1

2. the cost of 4 bicycles at $165 each plus the cost of 3 tricycles at $54 each

Simplify each expression. Use mental math where possible. Identify the property or properties used.

3. $\frac{2}{3} \cdot \frac{3}{2} + \frac{5}{7} \cdot \frac{7}{5}$
4. $1 \cdot 6 + 1 \cdot 7$
5. $5(0 + 3)$

Simplify each expression by following the order of operations.

6. $2(10 - 5)^2 + \frac{12 - 9}{3}$
7. $\frac{30 + 6}{12 - 6} + \frac{(3 + 1)^2}{(3 - 1)^2}$

Evaluate each expression.

8. $\frac{n + 18}{n - 6} + \frac{2n + 1}{n - 7}; n = 12$
9. $\frac{3a + b}{a - 6} + \frac{2b + 6}{a - b}; a = 12, b = 6$

Find the values in the replacement set {0, 1, 2, 3, 4, 5, 6} that are solutions to each equation. If the equation has no solution, write none.

10. $3(x - 1) = 21$
11. $4(n - 1) + 3 = 23$
12. $7z - 5 = 9$

13. Give a counterexample to show that the statement below is false.
The sum of an even number and an odd number is always even.

14. Use Properties of Equality and properties of operations to show that the equation below is true.
$$3(a + 3) + 6 = 15 + 3a$$

Solve each equation. Check your solution.

15. $x - 2 = 6$
16. $4.6 = a - 6.2$
17. $t - 5 = 12.8$
18. $y + 19.95 = 20$
19. $18.65 = w + 12.3$
20. $x + 0.02 = 0.6$

Solve each equation. Check your solution.

21. $z + 75 = 138$
22. $19 = p + 7.5$
23. $12.5 = z - 7.4$
24. $2\frac{1}{2} = z - 5\frac{1}{3}$
25. $n - 4\frac{3}{5} = 4\frac{1}{5}$
26. $z + 13 = 7\frac{5}{8} + 9\frac{3}{8}$

27. Write an equation to represent the situation below. Solve the equation and answer the question.
After withdrawing $90 from a bank account, a customer had a balance of $1200.45. How much money did the customer have before the withdrawal?

Solve each equation. Check your solution.

28. $\frac{x}{4} = 13$
29. $2.4 = \frac{t}{12}$
30. $\frac{3a}{5} = 2.7$
31. $10 = \frac{10c}{3}$
32. $\frac{5}{2} = \frac{15n}{4}$
33. $\frac{3z}{7} = \frac{1}{2}$
34. $4k = 25$
35. $96 = 12h$
36. $4a = \frac{17}{2}$
37. $\frac{10}{3} = 9z$
38. $\frac{10}{3} - \frac{2}{3} = 8z$
39. $(6 + 5)b = 44$
40. $25 = \frac{2}{5}r$
41. $18b = 37$
42. $\frac{g}{2} = \frac{2}{5} + \frac{2}{5}$
43. $64 = (2.6 + 5.4)z$
44. $2s = 5\frac{2}{5} - 2\frac{1}{3}$
45. $\frac{x}{12.5 - 7} = 2$

Write an equation to represent each situation. Solve the equation and answer the question.

46. How much money should each of 7 workers receive if the boss distributes $1477 equally among them?

47. For how many weeks will a teenager need to save money if she can save $30 per week and she wants to save a total of $360?

Answers for Chapter 1 Assessment

1. $(4 - 17) - (6 + 1)$
2. $(4 \times 165) + (3 \times 54)$
3. 2; Inverse Property of Multiplication
4. 13; Identity Property of Multiplication
5. 15; Identity Property of Addition
6. 51
7. 10
8. 10
9. 10
10. none
11. 6
12. 2
13. Answers may vary. Sample answer: The sum of 2, an even number, and 3, an odd number, is 5, which is not an even number.

14.
$$3(a + 3) + 6 = 3a + 9 + 6$$
Distributive Property
$$= 3a + 15$$
$$= 15 + 3a$$
Commutative Property of Addition

15. $x = 8$
16. $a = 10.8$
17. $t = 17.8$
18. $y = 0.05$
19. $w = 6.35$
20. $x = 0.58$
21. $z = 63$
22. $p = 11.5$
23. $z = 19.9$
24. $z = \frac{47}{6}$, or $7\frac{5}{6}$
25. $n = \frac{44}{5}$, or $8\frac{4}{5}$
26. $z = 4$
27. $x - 90 = \$1200.45$; $\$1290.45$

28. $x = 52$
29. $t = 28.8$
30. $a = 4.5$
31. $c = 3$
32. $n = \frac{2}{3}$
33. $z = \frac{7}{6}$, or $1\frac{1}{6}$
34. $k = \frac{25}{4}$, or $6\frac{1}{4}$
35. $h = 8$
36. $a = \frac{17}{8}$, or $2\frac{1}{8}$
37. $z = \frac{10}{27}$
38. $z = \frac{1}{3}$
39. $b = 4$
40. $r = 62.5$
41. $b = \frac{37}{18}$, or $2\frac{1}{18}$
42. $g = \frac{8}{5}$, or $1\frac{3}{5}$

43. $z = 8$
44. $s = \frac{23}{15}$, or $1\frac{8}{15}$
45. $x = 11$
46. $7x = 1477$; $\$211$
47. $30x = 360$; 12 weeks

Chapter 2 Operations With Real Numbers	State or Local Standards	Corresponding Lessons in *Algebra 1*, Schultz et al.
2.1 Real Numbers		2.1
Skill A: Representing rational numbers as fractions and as decimals		
Skill B: Classifying numbers as members of subsets of the real numbers		
Skill C: Using opposites and absolute value		
2.2 Adding Real Numbers		2.2
Skill A: Using a number line to add real numbers		
Skill B: Adding real numbers		
2.3 Subtracting Real Numbers		2.3
Skill A: Subtracting real numbers		
Skill B: Adding and subtracting real numbers		
Skill C: APPLICATIONS Using addition and subtraction to solve real-world problems		
2.4 Multiplying and Dividing Real Numbers		2.4
Skill A: Multiplying real numbers		
Skill B: Dividing real numbers		
Skill C: Using multiplication and division with other operations		
2.5 Solving Equations Involving Real Numbers		3.1, 3.2
Skill A: Using properties of equality, addition rules, and subtraction rules to solve equations		
Skill B: Using properties of equality, multiplication rules, and division rules to solve equations		
2.6 The Distributive Property and Combining Like Terms		2.6, 2.7, 3.5
Skill A: Using the Distributive Property to combine like terms		
Skill B: Using the Distributive Property to multiply and divide		
2.7 Solving Two-Step Equations		3.3, 3.6
Skill A: Solving an equation in two steps		
Skill B: APPLICATIONS Solving a literal equation for a specified variable		
Skill C: APPLICATIONS Using a problem-solving plan		
2.8 Solving Multistep Equations		3.4, 3.6
Skill A: Solving an equation in multiple steps		
Skill B: Simplifying before solving multistep equations		
Skill C: APPLICATIONS Solving real-world problems with multistep equations		
2.9 Deductive and Indirect Reasoning in Algebra		11.6
Skill A: Judging the correctness of a solution		
Skill B: Writing a solution process as a logical argument		
Skill C: Showing that a statement is sometimes, always, or never true		

CHAPTER

2

Operations With Real Numbers

What You Already Know

In Chapter 1, you learned how to choose a single property of equality and apply it to solve a simple equation. When you applied the property, you performed addition, subtraction, multiplication, or division of positive numbers and found positive solutions.

What You Will Learn

In Chapter 2, you will have the opportunity to apply the same properties of equality to solve equations. Now, however, you will need to add, subtract, multiply, and divide both positive and negative numbers to arrive at the solutions. To help you, the chapter reviews the skills needed to perform operations on the integers.

Once you have extended your equation-solving skills to one-step equations involving both positive and negative numbers, you will then learn to solve equations that require two or more solution steps. You will also solve equations that involve the Distributive Property and the variable on both sides of the equation.

VOCABULARY	
absolute value	Multiplicative Property of −1
additive inverse	natural numbers
coefficient	negation
conclusion	Opposite of a Difference
conditional statement	Opposite of a Sum
deductive reasoning	opposites
equilateral triangle	origin
equivalent equations	rational numbers
hypothesis	real numbers
indirect reasoning	repeating decimal
infinite	scalene triangle
integers	subset
irrational numbers	terminating decimal
isosceles triangle	Transitive Property of Deductive
least common multiple (LCM)	Reasoning
like terms	whole numbers
monomial	

The diagram below shows how mathematical skills and mathematical reasoning are interrelated with the skills and concepts in Chapter 2. Notice that operations with real numbers are presented before you learn how to solve equations with real numbers.

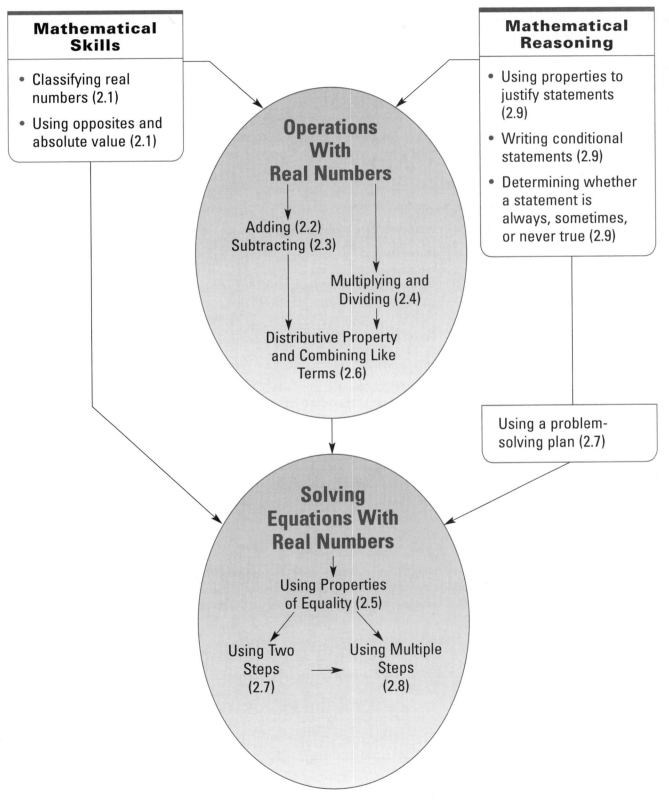

Mathematical Skills

- Classifying real numbers (2.1)
- Using opposites and absolute value (2.1)

Operations With Real Numbers

Adding (2.2)
Subtracting (2.3)

Multiplying and Dividing (2.4)

Distributive Property and Combining Like Terms (2.6)

Mathematical Reasoning

- Using properties to justify statements (2.9)
- Writing conditional statements (2.9)
- Determining whether a statement is always, sometimes, or never true (2.9)

Using a problem-solving plan (2.7)

Solving Equations With Real Numbers

Using Properties of Equality (2.5)

Using Two Steps (2.7)

Using Multiple Steps (2.8)

SKILL A ▶ *Representing rational numbers as fractions and as decimals*

Focus: *Students rewrite terminating decimals as fractions in simplest form, and they rewrite simple fractions as terminating or repeating decimals.*

EXAMPLE 1

PHASE 1: Presenting the Example

Discuss the definitions of natural number, whole number, integer, and rational number. Ask students to explain why all integers are rational numbers. [**Any integer n can be rewritten as $\dfrac{n}{1}$.**]

Now ask students to think of one rational number that is *not* an integer. They most likely will name fractions such as $\dfrac{2}{3}$ and $\dfrac{1}{2}$. Be sure they understand that a mixed number like $8\dfrac{1}{2}$ is also a rational number, because it can be rewritten as $\dfrac{17}{2}$.

Some students may have given decimal responses, providing a natural bridge to Example 1. If no student considered a decimal, ask whether they think decimals are rational numbers. Then introduce Example 1, which shows a method for rewriting a terminating decimal as a fraction.

PHASE 2: Providing Guidance for TRY THIS

You may need to remind students that a fraction is in simplest form when its numerator and denominator have no common factor other than 1. Therefore, although writing 0.625 as $\dfrac{625}{1000}$ demonstrates that it is a rational number, this is not the simplest form of the fraction.

EXAMPLE 2

PHASE 2: Providing Guidance for TRY THIS

Some students may wonder how to find the fraction equivalent for a repeating decimal. The examples below illustrate a technique used for blocks of one and two repeating digits. You may want to work through these examples with students who are interested and capable.

$$
\begin{aligned}
10N &= 2.2222\ldots \\
- \quad N &= -0.2222\ldots \\
\hline
9N &= 2 \\
N &= \frac{2}{9}
\end{aligned}
\qquad\qquad
\begin{aligned}
100N &= 18.1818\ldots \\
- \quad N &= -0.1818\ldots \\
\hline
99N &= 18 \\
N &= \frac{18}{99} = \frac{2}{11}
\end{aligned}
$$

So $0.\overline{2} = \dfrac{2}{9}$.   So $0.\overline{18} = \dfrac{2}{11}$.

2.1 LESSON — *Real Numbers*

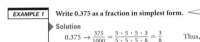 *Representing rational numbers as fractions and as decimals*

The numbers that you most commonly use are classified in these sets:

- **natural numbers:** the numbers that are used in counting
- **whole numbers:** the set of natural numbers plus 0
- **integers:** the set of whole numbers and their opposites
- **rational numbers:** the set of numbers that can be written as a fraction

  of the form $\frac{p}{q}$, where p and q are integers and q does not equal 0

Notice in the diagram that all integers are also rational numbers. This is because any integer can be written as a fraction. For example, 5 is also $\frac{5}{1}$.

Because a fraction can be written as a decimal, any rational number $\frac{p}{q}$ can be written in decimal form.

When you divide the denominator of a fraction into the numerator, the result is either a *terminating* or *repeating* decimal.

A **terminating decimal** has a *finite* number of nonzero digits to the right of the decimal point, such as 0.25 or 0.10. A **repeating decimal** has a string of one or more digits that repeat *infinitely*, such as 1.555555 . . ., or 0.345634563456 . . . These repeating decimals can be indicated by a bar over the repeating digits, $1.\overline{5}$ or $0.\overline{3456}$.

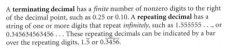

Number Sets

Rational Numbers $\frac{p}{q}$
Integers . . . , –3, –2, –1
Whole Numbers 0
Natural Numbers 1, 2, 3 . . .

EXAMPLE 1 | Write 0.375 as a fraction in simplest form.

> *A fraction in simplest form is the same as a fraction in lowest terms.*

▶ **Solution**

$0.375 \rightarrow \frac{375}{1000} = \frac{5 \cdot 5 \cdot 5 \cdot 3}{5 \cdot 5 \cdot 5 \cdot 8} = \frac{3}{8}$ Thus, $0.375 = \frac{3}{8}$.

TRY THIS | Write 0.625 as a fraction in simplest form.

EXAMPLE 2 | Write $\frac{1}{3}$ as a decimal.

▶ **Solution**

Divide 1 by 3, as shown at right. You can see that this division process will repeat infinitely. Thus, $\frac{1}{3} = 0.\overline{3}$.

$\begin{array}{r} 0.33 \\ 3\overline{)1.00} \\ \underline{9} \\ 10 \\ \underline{9} \\ 1 \end{array}$

TRY THIS | Write $\frac{5}{18}$ as a repeating decimal.

KEY SKILLS

Write each decimal as a fraction. Do not simplify.

Sample: 1.02   Solution: $\frac{102}{100}$

1. 0.35   2. 1.5   3. 13.755   4. 0.0054

PRACTICE

Write each decimal as a fraction in simplest form.

5. 0.2   6. 0.7   7. 1.4   8. 3.9
9. 0.25   10. 0.75   11. 1.25   12. 1.75
13. 0.155   14. 0.666   15. 0.111   16. 1.555

Write each fraction or mixed number as a decimal. Identify whether the decimal that results is terminating or repeating.

17. $\frac{1}{2}$ 18. $\frac{13}{100}$ 19. $4\frac{3}{5}$ 20. $5\frac{1}{4}$
21. $\frac{1}{9}$ 22. $\frac{2}{11}$ 23. $4\frac{2}{3}$ 24. $2\frac{1}{3}$
25. $\frac{1}{6}$ 26. $\frac{7}{18}$ 27. $10\frac{3}{11}$ 28. $10\frac{5}{6}$

29. **Critical Thinking** Show that the result of $\left(4\frac{1}{2}\right) \div \left(2\frac{3}{4}\right)$ is a rational number by
   a. writing it as a quotient of two integers.   **b.** writing it as a decimal.

MIXED REVIEW

Order each list of numbers from least to greatest. (previous courses)

30. $\frac{1}{30}, \frac{1}{33}, 0.30, \frac{1}{3}, 0.33, 0.03$ 31. $\frac{1}{2}, 0.2, \frac{1}{20}, 0.22, \frac{1}{200}$

Solve each problem. Round to two decimal places if necessary. (previous courses) (samples available on page 21)

32. Find 5% of 50.   33. Find 16% of 75.   34. Find 0.1% of 20.
35. Find 0.01% of 200.   36. Find 150% of 20.   37. Find 500% of 20.
38. What percent of 10 is 3?   39. What percent of 20 is 3?   40. What percent of 40 is 3?
41. What percent of 50 is 3?   42. What percent of 1 is 3?   43. What percent of $\frac{1}{2}$ is 3?

Solve each equation. Check your solution.
(1.4 Skill A, 1.5 Skill A, 1.7 Skill A, 1.8 Skill A)

44. $a - 3 = 8$ 45. $t + 3 = 18$ 46. $z + 1.2 = 10$ 47. $y - 18.2 = 105$
48. $18y = 54$ 49. $1.9y = 38$ 50. $\frac{3x}{5} = 2$ 51. $\frac{7x}{3} = 84$

Answers for Lesson 2.1 Skill A

Try This Example 1
$\frac{5}{8}$

Try This Example 2
$0.2\overline{7}$

1. $\frac{35}{100}$

2. $\frac{15}{10}$

3. $\frac{13,755}{1000}$

4. $\frac{54}{10,000}$

5. $\frac{1}{5}$

6. $\frac{7}{10}$

7. $\frac{7}{5}$, or $1\frac{2}{5}$

8. $\frac{39}{10}$, or $3\frac{9}{10}$

9. $\frac{1}{4}$

10. $\frac{3}{4}$

11. $\frac{5}{4}$, or $1\frac{1}{4}$

12. $\frac{7}{4}$, or $1\frac{3}{4}$

13. $\frac{31}{200}$

14. $\frac{333}{500}$

15. $\frac{111}{1000}$

16. $\frac{311}{200}$, or $1\frac{111}{200}$

17. 0.5; terminating
18. 0.13; terminating
19. 4.6; terminating
20. 5.25; terminating
21. $0.\overline{1}$; repeating
22. $0.\overline{18}$; repeating
23. $4.\overline{6}$; repeating
24. $2.\overline{3}$; repeating
25. $0.1\overline{6}$; repeating
26. $0.3\overline{8}$; repeating

27. $10.\overline{27}$; repeating
28. $10.8\overline{3}$; repeating
29. **a.** $\left(4\frac{1}{2}\right) \div \left(2\frac{3}{4}\right) =$

$\frac{9}{2} \div \frac{11}{4} =$

$\frac{9}{2} \times \frac{4}{11} = \frac{18}{11}$, a

quotient of two integers with nonzero denominator

b. $\left(4\frac{1}{2}\right) \div \left(2\frac{3}{4}\right) =$

$4.5 \div 2.75 =$
$1.\overline{63}$, a repeating decimal

30. $0.03, \frac{1}{33}, \frac{1}{30}, 0.30,$

$0.33, \frac{1}{3}$

31. $\frac{1}{200}, \frac{1}{20}, 0.2, 0.22, \frac{1}{2}$

32. 2.5
33. 12
34. 0.02
35. 0.02
36. 30
37. 100
38. 30%
39. 15%
40. 7.5%
41. 6%
42. 300%
43. 600%
44. $a = 11$
45. $t = 15$
46. $z = 8.8$
47. $y = 123.2$
48. $y = 3$
49. $y = 20$
50. $x = \frac{10}{3}$, or $3\frac{1}{3}$
51. $x = 36$

Classifying numbers as members of subsets of the real numbers

Focus: *Students identify all subsets of the real numbers to which a given number belongs. They also analyze relationships among the subsets.*

EXAMPLE 1

PHASE 1: Presenting the Example

Review the terms *terminating decimal* and *repeating decimal*. Challenge students to write a decimal that neither terminates nor repeats. Most students will probably write a long jumble of digits, such as 0.39140281395882479015… Point out that a response like this offers no guarantee against repetition. For instance, the digits of this particular number could begin repeating in the twenty-first decimal place.

Then discuss the decimal given in the text, 0.101001000100001. . . Students will probably notice that the digits in this number follow a pattern, but point out that this is not the same as having a repeating block of digits of a specific length. Therefore, this number is a valid example of a decimal that does not terminate or repeat. Now ask students to write a nonterminating, nonrepeating decimal using this one as a model. [**sample: 0.757557555755557. . .**] Tell them that all nonterminating, nonrepeating decimals form the set of *irrational numbers*. The rational numbers and irrational numbers together comprise the set of *real numbers*.

Point out to students that the set of real numbers is not a new concept—these are the numbers that they have been using in all of their math classes up to this point. All of the number rules and properties that they have learned in this course are rules and properties of real numbers.

EXAMPLE 2

PHASE 1: Presenting the Example

Remind students that, to prove a statement false, you need only show one counterexample. To prove a statement true, however, you must demonstrate in a logical argument that it is true for all numbers, and you must be able to justify each step in your argument.

PHASE 2: Providing Guidance for **TRY THIS**

In Example 1, students focused on a specific number and identified its place among the subsets of the real numbers. In contrast, Example 2 requires students to make generalizations about entire subsets of the real numbers. The thinking required here is a bit more abstract than in Example 1. Some students may need additional support in advancing from one type of thinking to the other. You can use Skill B Exercises 1–3 on the facing page for this purpose.

There are two more sets of numbers that are important to know.

The set of **irrational numbers** consists of all decimals that do not terminate and do not repeat. The number 0.101001000100001. . . is an example of an irrational number. The number pattern does not repeat exactly, and the *ellipsis* (. . .) indicates that the numbers do not terminate. Irrational numbers *cannot be written in form* $\frac{p}{q}$, *where p and q are integers and q ≠ 0.*

The set of **real numbers** consists of all rational and irrational numbers.

Notice that the sets of natural numbers, whole numbers, integers, rational numbers, and irrational numbers are all contained within the larger set of real numbers. When one set is contained in another set, we say that the smaller set is a **subset** of the larger set.

Number Sets

EXAMPLE 1

Classify each number in as many ways as possible.
  a. 4   b. −4   c. 0.$\overline{3}$

▸ Solution
  a. 4 is a natural number, a whole number, an integer, a rational number and a real number.
  b. −4 is an integer, a rational number, and a real number.
  c. 0.$\overline{3}$ is a rational number and a real number.

TRY THIS   Classify 10,000 in as many ways as possible.

EXAMPLE 2

Is each statement below true or false? Justify your answer.
  a. *All natural numbers are rational numbers.*
  b. *All integers are irrational numbers.*

▸ Solution
  a. The statement is true. If *n* is a natural number, then $n = \frac{n}{1}$, a rational number.
  b. The statement is false. For example, 2 is an integer and 2 can be written as $\frac{2}{1}$. Therefore, 2 is not irrational.

TRY THIS   Is the following statement true or false?
                 All integers are natural numbers.
             If the statement is false, give a counterexample.

EXERCISES

KEY SKILLS

1. Identify integers that are not whole numbers: 5, 2, −3, 7, 0, −12, −4, 17

2. Identify rational numbers that are not whole numbers: −4, 3.7, 0, 5, 12.67, −1.5

3. Identify rational numbers between 1 and 1.5: 1.55, 1.05, $\frac{4}{5}$, 1.$\overline{5}$, 1.$\overline{41}$, $\frac{7}{4}$

PRACTICE

Classify each number in as many ways as possible.

4. $\frac{1}{2}$ 5. $-\frac{1}{2}$ 6. 4.3

7. −6.125   8. −5   9. −105

10. 0.456456456. . .   11. 2.343434. . .   12. 0.151151115. . .

13. 3.334443334444. . .   14. 10,000,000   15. 1

Is each statement true or false? If the statement is true, give reasons for your response. If the statement is false, give a counterexample.

16. There is no number that is both rational and irrational.

17. The counting numbers include the integers.

18. Every mixed number is a rational number.

19. Whenever you divide a counting number by 3, the result is another counting number.

20. There are rational numbers between $\frac{1}{4}$ and $\frac{1}{3}$.

21. **Critical Thinking** A well-known irrational number is π, or *pi*. Pi represents the ratio $\frac{circumference}{diameter}$ of a circle. How can you explain that an irrational number can be represented by a ratio?

MIXED REVIEW

Evaluate each expression. (previous courses)

22. $8\frac{1}{11} + \frac{10}{11}$ 23. $1\frac{2}{9} - \frac{2}{9}$ 24. $4\frac{1}{3} + 3\frac{1}{2}$ 25. $8\frac{1}{3} - 5\frac{1}{2}$

Use the order of operations to simplify each expression. (1.2 Skill B)

26. $\frac{(2 \times 3 + 1)^2}{2 \times 7}$ 27. $\frac{(2 \times 3 - 1)^2}{4 \times 5 - 5}$ 28. $\frac{(3 \times 2)^2 - 4^2}{5^2 - 4}$ 29. $\frac{(3 \times 2)^2 - (2 + 3)^2}{(1 + 1)^2 - 3}$

Answers for Lesson 2.1 Skill B

Try This Example 1
natural number, whole number, integer, rational number, real number

Try This Example 2
False; sample counter-example: −2 is an integer but not a natural number.

1. −3, −12, −4
2. −4, 3.7, 12.67, −1.5
3. 1.05, 1.$\overline{41}$
4. rational, real
5. rational, real
6. rational, real
7. rational, real
8. integer, rational, real
9. integer, rational, real
10. rational, real
11. rational, real
12. irrational, real
13. irrational, real

14. natural number, whole number, integer, rational, real
15. natural number, whole number, integer, rational, real
★16. True; a number cannot be a repeating or terminating decimal and a nonrepeating, nonterminating decimal at the same time.
★17. False; sample counter-example: −7 is an integer but not a counting number.
★18. True; every mixed number can be written as an improper fraction.

★19. False; sample counter-example: 1 divided by 3 is $\frac{1}{3}$, which is not a counting number.
★20. True; sample answer: The average of $\frac{1}{4}$ and $\frac{1}{3}$, $\frac{7}{24}$, is a rational number and is between $\frac{1}{4}$ and $\frac{1}{3}$.
21. A ratio $\frac{p}{q}$ represents a rational number only when *p* and *q* are both integers. A ratio may be irrational if *p*, *q*, or both are not integers.
22. 9

23. 1
24. $\frac{47}{6}$, or $7\frac{5}{6}$
25. $\frac{17}{6}$, or $2\frac{5}{6}$
26. $\frac{7}{2}$, or $3\frac{1}{2}$
27. $\frac{5}{3}$, or $1\frac{2}{3}$
28. $\frac{20}{21}$
29. 11

★ **Advanced Exercises**

Focus: *Students represent real numbers as points on a number line. Then they use a number line to identify opposites and find absolute values.*

EXAMPLE 1

PHASE 1: Presenting the Example

Display the figure at right. Tell students that this is a *number line*. It serves as a visual means of representing real numbers and relationships among them. The number zero is its *origin*. Numbers to the right of zero are *positive numbers*; numbers to the left of zero are *negative numbers*.

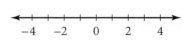

Have students copy the figure onto a sheet of paper. Point out that they must space the integers at equal intervals along the line. Each of the equal intervals is considered one *unit* of distance.

Discuss the correspondence between real numbers and a number line, which is stated in the box at the top of the textbook page. Have students draw a heavy dot at the point associated with 3. Tell them that the point is called the *graph* of 3, and 3 is called the *coordinate* of the point. Then have them graph the point whose coordinate is −3.

Now ask students to write a brief answer to these questions: How are the graphs of 3 and −3 different? How are they alike? Discuss their responses, and elicit the observations that the graphs of 3 and −3 are on different sides of zero, but they are the same distance from zero. Use this discussion to introduce the concepts of *opposite* and *absolute value*.

EXAMPLE 2

PHASE 2: Providing Guidance for **TRY THIS**

Many students read −n as "negative n," and so they believe that −n must represent a negative number. As you discuss the Try This exercises, insist that they use the correct verbal translation, "the opposite of n."

PHASE 3: ASSESSMENT AND CLOSURE for Lesson 2.1

- Ask students to list all the facts about −$\frac{2}{3}$ that they learned in this lesson.

  [It is a real number. It is a rational number. It can be written in decimal form as −0.$\overline{6}$. It can be graphed on a number line as a point two thirds of the way from 0 to −1. Its opposite is $\frac{2}{3}$. Its absolute value is $\frac{2}{3}$.]

- In Lesson 2.2, a number line will be used to visualize addition. In later lessons, students will use number lines to order real numbers and to identify solutions of an inequality.

☞ *For a Lesson 2.1 Quiz, see* Assessment Resources *page 12.*

Real Numbers and the Number Line

Every real number can be represented as a point on a number line. Every point on a number line represents a real number.

The number 0 is called the **origin** of the number line. Numbers to the left of 0 are negative, and numbers to the right of 0 are positive. Zero is neither positive nor negative.

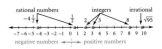

rational numbers   integers   irrational

$-4\frac{1}{2}$ $-\frac{1}{2}$ 2 √3   8 ↓√95

−7 −6 −5 −4 −3 −2 −1 0 1 2 3 4 5 6 7 8 9 10

negative numbers ← → positive numbers

Opposites

Two real numbers are opposites if they are on opposite sides of 0 and they are the same distance from 0 on a number line. Note: Zero is its own opposite.

EXAMPLE 1

Graph the number 3 and its opposite on a number line.

▶ Solution

The number which is 3 units from 0 in the opposite direction is −3. So, the opposite of 3 is −3.

−5 −4 −3 −2 −1 0 1 2 3 4 5

TRY THIS   Graph $\frac{9}{4}$ and its opposite on a number line.

The **absolute value** of a real number x, written $|x|$, is the distance from 0 to x on a number line. Because distance is always positive, the absolute value of a number is always positive.

Absolute Value

For any real number x,

$|x| = x$ if x is positive or zero, and $|x| = -x$ if x is negative.

EXAMPLE 2

Evaluate for $m = -4$.
a. $|m|$ b. $-m$ c. $-|m|$ d. $-(-m)$

▶ Solution

a. $|-4| = 4$ b. $-(-4) = 4$ c. $-|-4| = -|4| = 4$ d. $-[-(-4)] = -4$

TRY THIS   Evaluate for $n = -3$.   a. $|n|$ b. $-n$ c. $-|n|$ d. $-(-n)$

EXERCISES

KEY SKILLS

Graph all of the following numbers on the same number line.

1. 0   2. $3\frac{1}{2}$ 3. 4.5
4. −2   5. −3.5   6. −4.25

PRACTICE

Graph each number and its opposite on a number line.

7. 4   8. 1   9. 0
10. −5   11. $-2\frac{1}{2}$ 12. $-4\frac{1}{7}$
13. 1.5   14. −1.5   15. −3.25

Evaluate each expression for the given value of the variable.

16. $-r$; $r = 0$ 17. $-|v|$; $v = 5$ 18. $-|w|$; $w = -6$
19. $-(-n)$; $n = 5$ 20. $-|(-p)|$; $p = 5$ 21. $-d$; $d = 7$
22. $|-(-q)|$; $q = 0$ 23. $-(-h)$; $h = -1.3$ 24. $-|(-z)|$; $z = -3.7$
25. $-|-(-t)|$; $t = 3$ 26. $-|(-t)|$; $t = -0.8$ 27. $-[-(-y)]$; $y = -6$

Simplify each expression.

28. $|-2| + |2|$ 29. $|0| + |-3| + |5|$ 30. $(|-10| + |-9|) - (|-3| + |-1|)$
31. $(|-3| - |-1|) + (|-7| - |-1|)$ 32. $(|3| + |-5|)^2$ 33. $(|-8| - |-5|)^2$
34. $\left|\frac{3}{4} + \frac{1}{4}\right|^2$ 35. $\left|2\frac{3}{4} - \frac{1}{4}\right|^2$ 36. $\left|\frac{1}{2} + \frac{1}{4} + \frac{1}{8}\right|^2$

MIXED REVIEW

Evaluate each expression. (1.2 Skill C)

37. $3x^2 - 2x + 7$; $x = 2$ 38. $2.5(x - 5) + 2.5$; $x = 5.5$ 39. $3(x - 5)^2 + 3$; $x = 9$

Classify each number in as many ways as possible. (2.1 Skill B)

40. −3.9   41. 14.525252…   42. −120

Answers for Lesson 2.1 Skill C

Try This Example 1

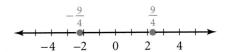

$-\frac{9}{4}$ $\frac{9}{4}$

−4 −2 0 2 4

Try This Example 2
a. 3   b. 3   c. 3   d. −3

1–6.

−4.25 −3.5 $3\frac{1}{2}$ 4.5

−4 −2 0 2 4

7.

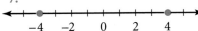

−4 −2 0 2 4

8.

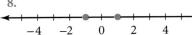

−4 −2 0 2 4

9.

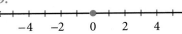

−4 −2 0 2 4

10.

−4 −2 0 2 4

11. $-2\frac{1}{2}$ $2\frac{1}{2}$

−4 −2 0 2 4

12. $-4\frac{1}{7}$ $4\frac{1}{7}$

−4 −2 0 2 4

13.   −1.5   1.5

−4 −2 0 2 4

14.   −1.5   1.5

−4 −2 0 2 4

15.   −3.25   3.25

−4 −2 0 2 4

16. 0
17. −5
18. −6

19. 5
20. −5
21. −7
22. 0
23. −1.3
24. −3.7
25. −3
26. −0.8
27. 6
28. 4
29. 8
30. 15
31. 8
32. 64
33. 9
34. 1
★**35.** $\frac{25}{4}$, or $6\frac{1}{4}$
★**36.** $\frac{49}{64}$

For answers to Exercises 37–42, see Additional Answers.

★ Advanced Exercises

Adding Real Numbers

SKILL A ▶ *Using a number line to add real numbers*

Focus: *Students use a series of moves on the number line to represent addition of real numbers.*

EXAMPLE 1

PHASE 1: Presenting the Example

Display the figure at right. Tell students that it illustrates movement on a number line. Ask them to write a brief description of the movement. Then ask several students to read their descriptions.

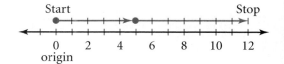

Although the students' exact words will vary, all correct responses should describe the same elements: a starting point, two moves to the right, and then a stop. Below the number line, summarize the movement with the four steps listed at right.

Start at the origin.
Move 5 units to the right.
Move 7 units to the right.
Stop at 12.

Tell students that this movement represents an addition of integers. Ask them to write the addition that they think it represents. Discuss their conjectures, finally displaying the correct response below the steps, as shown at right.

$5 + 7 = 12$

Now ask students how the movement would be different if the addition were $5 + (-7)$. [**The third step would be *Move 7 units to the left*. The fourth step would be *Stop at −2*.**] Use their responses as a lead-in to Example 1.

PHASE 2: Providing Guidance for **TRY THIS**

Tell students that when a negative number follows a plus sign, they should enclose that number in parentheses. For example, they should write $4 + (-5)$, not $4 + -5$.

EXAMPLE 2

PHASE 1: Presenting the Example

Display the addition $-2 + (-6)$. Ask students how it is different from the additions in Example 1. [**In Example 1, one addend was positive and one was negative. In Example 2, both addends are negative.**] Tell them to draw a number-line diagram that they think could represent this addition. Then compare their diagrams to the one given on the textbook page.

PHASE 2: Providing Guidance for **TRY THIS**

Watch for students who reverse the order, moving 4 units to the left and then moving 1 unit to the left. This diagram illustrates $-4 + (-1) = -5$ rather than $-1 + (-4) = -5$. Point out to students how the Commutative Property of Addition guarantees that they will get the same answer. However, not all operations are commutative, so it is important to develop the habit of performing operations in the correct order.

2.2 Adding Real Numbers

SKILL A Using a number line to add real numbers

You have already learned how to add positive numbers. How do you add two real numbers when one or both numbers are negative?

You can show addition of real numbers as a series of moves on the number line.

• Represent a positive number as a move to the right.
• Represent a negative number as a move to the left.

Note: If a number is not preceded by $+$ or $-$, it is positive. Negative numbers are always preceded by $-$.

EXAMPLE 1 Use a number line to find the sum $5 + (-7)$.

▶ **Solution**

Start at the origin. Move 5 units to the right.
Then move 7 units to the left.

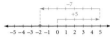

From the number line, $5 + (-7) = -2$.

TRY THIS Use a number line to find the sum of 4 and -5.

EXAMPLE 2 Use a number line to find the sum $-2 + (-6)$.

▶ **Solution**

Start at the origin. Move 2 units to the left.
Then move 6 units to the left.

From the number line, $-2 + (-6) = -8$.

TRY THIS Use a number line to find the sum $-1 + (-4)$.

46   Chapter 2   Operations With Real Numbers

EXERCISES

KEY SKILLS

Use a number line such as the one below to find the number that results when you start at the origin and follow the instructions.

1. Move 3 units to the right. Move 8 units to the left.
2. Move 4 units to the left. Move 8 units to the right.
3. Move 2 units to the left. Move 2 units to the right.
4. Move 5 units to the right. Move 5 units to the left.

$-5\ -4\ -3\ -2\ -1\ \ 0\ \ 1\ \ 2\ \ 3\ \ 4\ \ 5$

PRACTICE

Use a number line to find each sum.

5. $3 + (-4)$	6. $-4 + 5$	7. $-5 + 6$
8. $6 + (-9)$	9. $-3 + 3$	10. $-4 + 4$
11. $-3 + (-1)$	12. $-5 + (-1)$	13. $-6 + (-2)$
14. $-1 + (-1)$	15. $-3 + (-3)$	16. $-2 + (-2)$

Write the number that results when you follow the instructions.

17. Start at -3. Move 8 units to the right. Then move 2 units to the left.
18. Start at 5. Move 6 units to the right. Then move 2 units to the left.
19. Start at -2. Move 6 units to the left. Then move 2 units to the left.

Use a number line to find each sum.

20. $(-3 + 2) + 5$	21. $[-3 + (-2)] + 6$	22. $[4 + (-6)] + 3$
23. $[(-4) + 2] + 4$	24. $[(-4) + (-2)] + 3$	25. $[(-5) + (-1)] + (-4)$

26. **Critical Thinking**   Use a number line to show that when you add $\frac{1}{2} + \frac{1}{4} + \frac{1}{8} + \frac{1}{16} \cdots$, the sum is less than 1.

MIXED REVIEW

Simplify each expression. Use mental math where possible. Identify the properties used. (1.2 Skill A)

27. $1 \times \frac{3}{4} + 1 \times \frac{1}{4}$ 28. $\frac{4}{3} \times \frac{3}{4} + 1 \times 2$ 29. $\frac{5}{3} \times \frac{3}{5} + \frac{7}{4} \times \frac{4}{7}$ 30. $1 \times \frac{5}{3}(2 + 1)$

Use properties of equality and properties of numbers to show that each equation is true. (1.3 Skill C)

31. $\frac{9}{10} = 90\%$ 32. $6z = 4z + 2z$

Answers for Lesson 2.2 Skill A

Try This Example 1
-1

Try This Example 2
-5

1. -5
2. 4
3. 0
4. 0
5. -1
6. 1
7. 1
8. -3
9. 0
10. 0
11. -4
12. -6
13. -8
14. -2
15. -6
16. -4
17. 3
18. 9

19. -10
20. 4
21. 1
22. 1
23. 2
24. -3
25. -10
26.

Each addend increases the sum by one-half of the remaining distance to 1. Therefore, the sum approaches, but never reaches, 1.

27. 1; Identity Property of Multiplication
28. 3; Inverse and Identity Properties of Multiplication
29. 2; Inverse Property of Multiplication

30. 5; Inverse and Identity Properties of Multiplication

31. $\frac{9}{10} = \frac{9}{10} \cdot \frac{10}{10} = \frac{90}{100}$

$90\% = \frac{90}{100}$; Therefore, by the Transitive Property of Equality, $\frac{9}{10} = 90\%$.

32. $6z = z + z + z + z + z + z$;
$z + z + z + z + z + z = (z + z + z + z) + (z + z)$;
$(z + z + z + z) + (z + z) = 4z + 2z$;
Therefore, by the Transitive Property of Equality, $6z = 4z + 2z$.

Focus: *Students use algebraic rules to add real numbers.*

EXAMPLES 1 and 2

PHASE 1: Presenting the Examples

Display the statements at right. Have students copy them onto their papers, filling in each blank with the word *positive*, *negative*, or *zero* to make the statement true. Discuss their responses.

1. The sum of two positive numbers is _____?_____.

2. The sum of two negative numbers is _____?_____.

3. The sum of a positive number and a negative number is _____?_____.

Statements 1 and 2 should pose little difficulty. Students generally understand that the sum of two positive numbers is positive and the sum of two negative numbers is negative. However, students may have trouble with statement 3.

Lead students to see that, when two numbers have opposite signs, the sign of the sum depends on the absolute values of the numbers. Then replace statement 3 above with the expanded version at right. Again ask students to insert *positive*, *negative*, or *zero* to make true statements. [**a. positive; b. negative; c. zero**] Then discuss the rules for addition given in the box at the top of the textbook page.

3. **a.** If the positive number has the greater absolute value, then the sum of a positive number and a negative number is _____?_____.

b. If the negative number has the greater absolute value, then the sum of a positive number and a negative number is _____?_____.

c. If both numbers have the same absolute value, then the sum of a positive number and a negative number is _____?_____.

EXAMPLE 3

PHASE 2: Providing Guidance for **TRY THIS**

Given three or more numbers with mixed signs, some students find it easier to first group all the numbers with the same sign. For instance, given $-1 + 7 + (-2)$, they may find it helpful to rearrange the numbers as shown at right. Be sure they understand how the Commutative and Associative Properties of Addition justify this practice.

$$-1 + 7 + (-2)$$
$$= -1 + [7 + (-2)]$$
$$= -1 + [(-2) + 7]$$
$$= [-1 + (-2)] + 7$$
$$= -3 + 7$$
$$= 4$$

PHASE 3: ASSESSMENT AND CLOSURE for Lesson 2.2

- Give students the statement at right. Have them list ten different ways to fill in the boxes with integers to make a true statement.

- In this lesson, students used a number line to visualize addition of signed numbers, and then used their observations of this process to formulate algebraic rules for addition. In the lessons to come, students will develop rules for subtracting, multiplying, and dividing signed numbers.

$$\square + \square = -3$$

[**sample addends: 0, −3; −1, −2; −2, −1; −3, 0; −4, 1; −5, 2; −6, 3; −10, 7; 8, −11; 20, −23**]

☞ *For a Lesson 2.2 Quiz, see* Assessment Resources *page 13.*

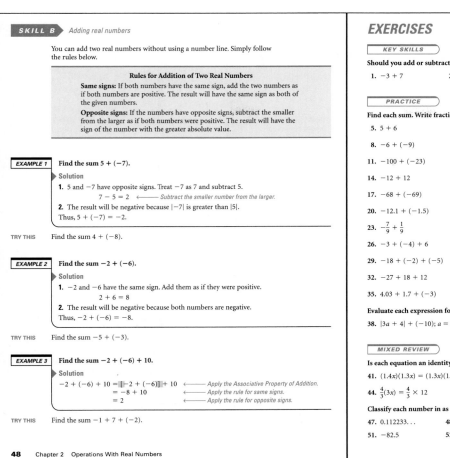

You can add two real numbers without using a number line. Simply follow the rules below.

Rules for Addition of Two Real Numbers

Same signs: If both numbers have the same sign, add the two numbers as if both numbers are positive. The result will have the same sign as both of the given numbers.

Opposite signs: If the numbers have opposite signs, subtract the smaller from the larger as if both numbers were positive. The result will have the sign of the number with the greater absolute value.

EXAMPLE 1 Find the sum $5 + (-7)$.

▶ **Solution**

1. 5 and -7 have opposite signs. Treat -7 as 7 and subtract 5.

$7 - 5 = 2$ ⟵ *Subtract the smaller number from the larger.*

2. The result will be negative because $|-7|$ is greater than $|5|$.
Thus, $5 + (-7) = -2$.

TRY THIS   Find the sum $4 + (-8)$.

EXAMPLE 2 Find the sum $-2 + (-6)$.

▶ **Solution**

1. -2 and -6 have the same sign. Add them as if they were positive.
$2 + 6 = 8$

2. The result will be negative because both numbers are negative.
Thus, $-2 + (-6) = -8$.

TRY THIS   Find the sum $-5 + (-3)$.

EXAMPLE 3 Find the sum $-2 + (-6) + 10$.

▶ **Solution**

$-2 + (-6) + 10 = [-2 + (-6)] + 10$ ⟵ *Apply the Associative Property of Addition.*
$= -8 + 10$ ⟵ *Apply the rule for same signs.*
$= 2$ ⟵ *Apply the rule for opposite signs.*

TRY THIS   Find the sum $-1 + 7 + (-2)$.

EXERCISES

KEY SKILLS

Should you add or subtract the absolute values of the given numbers?

1. $-3 + 7$ **2.** $-4 + (-5)$ **3.** $-1.4 + (-2.4)$ **4.** $4.2 + (-7.1)$

PRACTICE

Find each sum. Write fractions in simplest form.

5. $5 + 6$ **6.** $-8 + 5$ **7.** $-1 + 9$

8. $-6 + (-9)$ **9.** $-7 + (-5)$ **10.** $-3 + (-24)$

11. $-100 + (-23)$ **12.** $35 + (-42)$ **13.** $-55 + 18$

14. $-12 + 12$ **15.** $13 + (-13)$ **16.** $-37 + (-38)$

17. $-68 + (-69)$ **18.** $6.5 + (-5.5)$ **19.** $-4.2 + 5.7$

20. $-12.1 + (-1.5)$ **21.** $-1.23 + 5.15$ **22.** $\frac{4}{7} + \left(-\frac{2}{7}\right)$

23. $-\frac{7}{9} + \frac{1}{9}$ **24.** $-\frac{2}{5} + \left(-\frac{1}{5}\right)$ **25.** $-\frac{5}{12} + \left(-\frac{1}{12}\right)$

26. $-3 + (-4) + 6$ **27.** $-10 + (-3) + (-5)$ **28.** $30 + (-40) + (-20)$

29. $-18 + (-2) + (-5)$ **30.** $-10 + (-10) + 10$ **31.** $15 + (-15) + (-10)$

32. $-27 + 18 + 12$ **33.** $-32 + (-79) + (-19)$ **34.** $2.5 + (-3.5) + (-1)$

35. $4.03 + 1.7 + (-3)$ **36.** $1.1 + (-1.45) + 2.1$ **37.** $0.2 + (-0.25) + 0.2$

Evaluate each expression for the given value of the variable.

38. $|3a + 4| + (-10); a = 2$ **39.** $|4z - 3| + [-3 + (-5)]; z = 1$ **40.** $|9b - 8| + [4 + (-7)]; b = 2$

MIXED REVIEW

Is each equation an identity? If not, explain. (1.3 Skill A)

41. $(1.4x)(1.3x) = (1.3x)(1.4x)$ **42.** $3(x + 5) + 3 = 4$ **43.** $2(x + 3) + 5 = 5 + 2(x + 3)$

44. $\frac{4}{3}(3x) = \frac{4}{3} \times 12$ **45.** $d + 7 - 7 = 13 - 7$ **46.** $s + 7 - 7 = s + 7 - 7$

Classify each number in as many ways as possible. (2.1 Skill B)

47. $0.112233\ldots$ **48.** $3.313113111\ldots$ **49.** -13 **50.** 82.5

51. -82.5 **52.** $\frac{1}{3}$ **53.** $\frac{1}{101010}$ **54.** $\frac{1}{1010101}$

Answers for Lesson 2.2 Skill B

Try This Example 1
-4

Try This Example 2
-8

Try This Example 3
4

1. subtract
2. add
3. add
4. subtract
5. 11
6. -3
7. 8
8. -15
9. -12
10. -27
11. -123
12. -7
13. -37
14. 0
15. 0
16. -75

17. -137
18. 1
19. 1.5
20. -13.6
21. 3.92
22. $\frac{2}{7}$
23. $-\frac{2}{3}$
24. $-\frac{3}{5}$
25. $-\frac{1}{2}$
26. -1
27. -18
28. -30
29. -25
30. -10
31. -10
32. 3
33. -130
34. -2
★35. 2.73
★36. 1.75

★37. 0.15
38. 0
39. -7
40. 7
41. yes
42. No; for example, 0 is not a solution.
43. yes
44. No; for example, 1 is not a solution.
45. No; for example, 2 is not a solution.
46. yes
47. irrational, real
48. irrational, real
49. integer, rational, real
50. rational, real
51. rational, real
52. rational, real
53. rational, real
54. rational, real

★ **Advanced Exercises**

Subtracting Real Numbers

SKILL A   *Subtracting real numbers*

Focus: *Students learn to subtract a real number by adding its opposite.*

EXAMPLE 1

PHASE 1: Presenting the Example

Display the first five subtractions given in Column I at right. Tell students to copy them as shown and fill in the answer blanks. [**Answers are shown in red.**]

Ask students to describe any patterns that they observe. Elicit the response that, as the numbers being subtracted from 5 increase by 1, the differences decrease by 1.

Now write the next five subtractions in Column I directly below the first five. Point out that the numbers being subtracted continue to increase by 1. Tell students to assume that the pattern among the differences will also continue. Have them copy the subtractions and fill in each answer blank by continuing the pattern. [**Answers are shown in red.**]

Now display Column II. Have students copy it and fill in the sums. [**Answers are shown in red.**]

Point out the fact that the difference in each row is equal to the sum in the same row. Lead students to the conclusion that subtracting a positive number is equivalent to adding its opposite. Display the generalization $a - b = a + (-b)$. Then discuss Example 1.

Column I		Column II	
$5 - 1 =$	4	$5 + (-1) =$	4
$5 - 2 =$	3	$5 + (-2) =$	3
$5 - 3 =$	2	$5 + (-3) =$	2
$5 - 4 =$	1	$5 + (-4) =$	1
$5 - 5 =$	0	$5 + (-5) =$	0
$5 - 6 =$	-1	$5 + (-6) =$	-1
$5 - 7 =$	-2	$5 + (-7) =$	-2
$5 - 8 =$	-3	$5 + (-8) =$	-3
$5 - 9 =$	-4	$5 + (-9) =$	-4
$5 - 10 =$	-5	$5 + (-10) =$	-5

EXAMPLES 2 and 3

PHASE 1: Presenting the Examples

Repeat the activity for Example 1 using the six new subtractions and additions at right. If possible, position these directly *above* the subtractions and additions from Example 1, as shown, so that the *upward* pattern is obvious. That is, as you scan the subtractions in an upward direction, the numbers subtracted from 5 decrease by 1 as the differences increase by 1. As before, have students first find the new differences by continuing the pattern in Column I. Then have them calculate the new sums in Column II. [**Answers are shown in red.**]

Note that the results in each row again are equal. In fact, subtracting *any* real number is equivalent to adding its opposite, and the generalization $a - b = a + (-b)$ is true for all real numbers a and b. Then discuss Examples 2 and 3.

Column I		Column II	
$5 - (-5) =$	10	$5 + 5 =$	10
$5 - (-4) =$	9	$5 + 4 =$	9
$5 - (-3) =$	8	$5 + 3 =$	8
$5 - (-2) =$	7	$5 + 2 =$	7
$5 - (-1) =$	6	$5 + 1 =$	6
$5 - 0 =$	5	$5 + 0 =$	5
$5 - 1 =$	4	$5 + (-1) =$	4
$5 - 2 =$	3	$5 + (-2) =$	3
$5 - 3 =$	2	$5 + (-3) =$	2
$5 - 4 =$	1	$5 + (-4) =$	1
$5 - 5 =$	0	$5 + (-5) =$	0

2.3 LESSON *Subtracting Real Numbers*

SKILL A *Subtracting real numbers*

In Lesson 1.2, you learned various properties of numbers for addition and multiplication. In particular, you learned that every number a has an opposite, $-a$, which is also called an *additive inverse*.

For every real number a, there is a real number $-a$, called its opposite, such that $a + (-a) = 0$.

Because you can add any two real numbers, and every real number has an additive inverse, you can define subtraction by addition of the additive inverse.

Subtraction of Real Numbers
If a and b represent real numbers, then $a - b = a + (-b)$.

The following examples show how you can rewrite subtraction of a number as addition of its opposite.

EXAMPLE 1 Find the difference $13 - 25$.

> **Solution**
> $$13 - 25 = 13 + (-25) \quad \longleftarrow \text{ Rewrite subtraction as addition of the opposite.}$$
> $$= -12 \quad \longleftarrow \text{ Use the rule for addition of opposite signs.}$$

TRY THIS Find the difference $31 - 52$.

EXAMPLE 2 Find the difference $15 - (-21)$.

> **Solution**
> $$15 - (-21) = 15 + 21 \quad \longleftarrow \text{ Add 21, the opposite of } -21.$$
> $$= 36$$

TRY THIS Find the difference $150.4 - (-97.5)$.

EXAMPLE 3 Find the difference $-13 - (-27)$.

> **Solution**
> $$-13 - (-27) = -13 + 27 \quad \longleftarrow \text{ Add 27, the opposite of } -27.$$
> $$= 14 \quad \longleftarrow \text{ Use the rule for addition of opposite signs.}$$

TRY THIS Find the difference $-30 - (-19)$.

EXERCISES

KEY SKILLS

Rewrite each difference as a sum.

1. $12 - 15$
2. $10 - (-15)$
3. $1.2 - 0$
4. $-5 - (-7)$
5. $14 - 3$
6. $18 - 18$

Without performing the subtraction, tell whether the difference will be greater than 0, equal to 0, or less than 0.

7. $18 - 18$
8. $102 - 103$
9. $15 - 14$
10. $-18 - 4$
11. $-16 - (-12)$
12. $2.3 - (-2.3)$
13. $-33.5 - (-33.5)$
14. $7\frac{1}{3} - 7\frac{3}{5}$
15. $7\frac{3}{5} - 7\frac{1}{5}$

PRACTICE

Find each difference.

16. $10 - 4$
17. $15 - 10$
18. $-5 - 3$
19. $-9 - 6$
20. $4 - 10$
21. $12 - 18$
22. $-12 - 18$
23. $-15 - 10$
24. $6 - 12$
25. $15 - (-10)$
26. $-12 - (-18)$
27. $-15 - (-10)$
28. $-12 - (-7)$
29. $1.5 - 2.5$
30. $2.5 - 1.5$
31. $-2.45 - 1.5$
32. $-2.5 - (-1.15)$
33. $\frac{9}{2} - \frac{11}{2}$
34. $-\frac{9}{2} - \frac{10}{2}$
35. $-4\frac{1}{2} - \left(-5\frac{1}{2}\right)$
36. $4\frac{1}{2} - \left(-5\frac{1}{2}\right)$

Evaluate each expression for the given value of the variable.

37. $3x - |x - 4|$; $x = 2$
38. $2c - |4c - 5|$; $c = 3$
39. $|x - 3| - |x - 4|$; $x = 2$
40. $|t - 6| - |t - 9|$; $t = 0$
41. $3z - |6 + 5z|$; $z = 6$
42. $w - |6 - w|$; $w = -7$

MIXED REVIEW

Solve each equation. (1.4 Skill A, 1.5 Skill A)

43. $x - 1.5 = 7$
44. $x - 6 = 6$
45. $t + 7.3 = 8.9$

Solve each equation. (1.7 Skill A, 1.8 Skill A)

46. $\frac{n}{3} = 5$
47. $\frac{t}{4} = 6$
48. $3g = 27$

Answers for Lesson 2.3 Skill A

Try This Example 1
-21

Try This Example 2
247.9

Try This Example 3
-11

1. $12 + (-15)$
2. $10 + 15$
3. $1.2 + 0$
4. $-5 + 7$
5. $14 + (-3)$
6. $18 + (-18)$
7. equal to 0
8. less than 0
9. greater than 0
10. less than 0
11. less than 0
12. greater than 0
13. equal to 0
14. less than 0
15. greater than 0
16. 6

17. 5
18. -8
19. -15
20. -6
21. -6
22. -30
23. -25
24. -6
25. 25
26. 6
27. -5
28. -5
29. -1
30. 1
31. -3.95
32. -1.35
33. -1
34. $-\frac{19}{2}$, or $-9\frac{1}{2}$
35. 1
36. 10
37. 4
38. -1
39. -1

40. -3
41. -18
42. -20
43. $x = 8.5$
44. $x = 12$
45. $t = 1.6$
46. $n = 15$
47. $t = 24$
48. $g = 9$

Focus: *Students use the order of operations to simplify numerical and algebraic expressions involving addition, subtraction, and grouping symbols.*

EXAMPLE 1

PHASE 1: Presenting the Example

Display the expression $10 - (-5) + 3$. Have students work individually to simplify the expression. [**18**] Discuss any difficulties that students may have encountered.

Now display the expression $10 - (-5 + 3)$. Ask students how it is like the first expression. [**It involves a subtraction, an addition, and the integers 10, −5, and 3.**] Ask how it is different. [**In the second expression, parentheses group −5 + 3.**] Review the order of operations. Then discuss the simplification of $10 - (-5 + 3)$ that is presented on the textbook page.

EXAMPLES 2 and 3

PHASE 1: Presenting the Examples

Display each of the following statements, one at a time. As each statement appears, ask students to justify why it must be true.

1. $a + (-a) + b + (-b) = 0$ [**Inverse Property of Addition**]
2. $a + b + (-a) + (-b) = 0$ [**Commutative Property of Addition**]
3. $(a + b) + [-a + (-b)] = 0$ [**Associative Property of Addition**]
4. $(a + b) + (-a - b) = 0$ [**Definition of subtraction**]
5. $-a - b = -(a + b)$ [**Subtract $(a + b)$ from both sides.**]
6. $-(a + b) = -a - b$ [**Symmetric Property of Equality**]

Point out that the statements together prove that $-(a + b) = -a - b$, which is an important property of the real numbers. Ask students to give a verbal translation of this statement. [**The opposite of a sum of real numbers is equivalent to the sum of the opposites of the numbers.**]

Now ask students how this statement can be used to prove $-(a - b) = -a + b$. [**By the definition of subtraction, $-(a - b) = -[a + (-b)]$. By the statement just proved, $-[a + (-b)] = -a - (-b)$, or $-a + b$. So, by the Transitive Property of Equality, $-(a - b) = -a + b$.**] Then discuss how both of these statements are used to simplify the expressions in Examples 2 and 3.

PHASE 2: Providing Guidance for **TRY THIS**

The most common error that students make is to take the opposite of only the first term within parentheses. Encourage them to draw arrows like those at right to remind them to take the opposite of each term.

$$-[3 + (-7)]$$
$$= -3 + [-(-7)]$$
$$= -3 + \quad 7$$

When there is a negative sign in front of an expression in parentheses, such as $-(3 + 4)$, there are two methods that you can use to simplify the expression.

One method is to first simplify within the parentheses and then apply the rule for the addition of opposite signs, as shown in Example 1.

EXAMPLE 1 Simplify $10 - (-5 + 3)$.

▸ Solution

$$10 - (-5 + 3) = 10 - (-2) \longleftarrow \text{Work within parentheses. Add } -5 \text{ and 3 to get } -2.$$
$$= 10 + 2 \longleftarrow \text{Write subtraction as addition.}$$
$$= 12$$

TRY THIS   Simplify $-10 - [6 + (-43)]$.

A second method that can be used to simplify $-(3 + 4)$ is based upon the following property:

Multiplicative Property of -1
For all numbers a, $-1(a) = -a$, or the opposite of a.

This property can be extended to more than one term within the parentheses.

Let a and b represent real numbers.	
The Opposite of a Sum	$-(a + b) = -a - b$
The Opposite of a Difference	$-(a - b) = -a + b = b - a$

EXAMPLE 2 Simplify $-(-3 + 5)$.

▸ Solution

$$-(-3 + 5) = 3 - 5 \longleftarrow \text{The opposite of } -3 \text{ is 3 and the opposite of 5 is } -5.$$
$$= -2$$

TRY THIS   Simplify $-[3 + (-7)]$.

EXAMPLE 3 Simplify $-[-10 + (-12)] - [-7 - (-15)]$.

▸ Solution

$$-[-10 + (-12)] - [-7 - (-15)]$$
$$[10 - (-12)] + [7 + (-15)]$$
$$(10 + 12) + [7 + (-15)]$$
$$22 + (-8)$$
$$14$$

TRY THIS   Simplify $-[-15 + (-15)] - [(-17) + (-13)]$.

EXERCISES

| KEY SKILLS |

Use the Multiplicative Property of -1 as in the first step in Example 2.
Do not evaluate.

1. $-(2 + 1)$ 2. $-(3 - 5)$ 3. $-[-2 + (-3)]$
4. $-[3 + (-5)]$ 5. $-[-4 - (-7)]$ 6. $-[-1 - (-1)]$

| PRACTICE |

Simplify each expression.

7. $10 - (3 + 5)$ 8. $10 - (-3 + 6)$
9. $-12 - (4 + 9)$ 10. $-3 - (10 + 9)$
11. $-12 - (3 + 1)$ 12. $100 - (82 + 18)$
13. $-(3 + 4) - (5 + 6)$ 14. $-(3 - 7) - (5 - 6)$
15. $-(-3 - 7) - (-5 + 7)$ 16. $12 - (3 + 4) - (5 + 6)$
17. $-20 - (5 + 5) - (-5 + 6)$ 18. $-2 - (5 + 1) - (-5 - 8)$
19. $-(2x + 3)$ 20. $-(3 - 4a)$
21. $-(2d - 5)$ 22. $-(7z + 6)$
23. $-(x + 7)$ 24. $-(-5w + 11)$
25. $-(3 + 4) - (5 + 6) + (2 - 5)$ 26. $-(3 - 7) - (5 - 6) + (-2 + 4)$
27. $-(-3 - 7) - [(-5 + 7) - (1 - 9)]$ 28. $-20 - (5 + 5) - [(-5 + 6) - (-2 - 3)]$
29. $-[(-2 - 8) - (-5 + 11)] - [(1 - 10) - (-3 - 3)]$
30. $-[7 - (4 + 4)] - (-5 + 7) - [(-2 - 3) + (-8 + 8)]$

Evaluate each expression for the given values of the variables.

31. $|3a| + |2b| - |2a + 3b|$; $a = -1$ and $b = -3$ 32. $|-2r| + |5s| - |4r + 6s|$; $r = -2$ and $s = -5$
33. $(|x| - |y|)^2 - (|x| + |y|)^2$; $x = 1$ and $y = 1$ 34. $2w - 3|w| - |x - w|$; $x = -2$ and $w = -3$

| MIXED REVIEW |

Find each sum. (2.2 Skill B)

35. $5.5 + (-3)$ 36. $-5 + (-12)$ 37. $-12 + (-20)$
38. $-18 + 18$ 39. $-6 + (-9)$ 40. $-6.4 + 6.4$
41. $-11 + 18 + (-6)$ 42. $-7 + (-6) + (-2)$ 43. $-6.1 + 6.2 + (-3.2)$

Answers for Lesson 2.3 Skill B

Try This Example 1
27

Try This Example 2
4

Try This Example 3
60

1. $-2 - 1$
2. $-3 - (-5)$, or $-3 + 5$
3. $2 - (-3)$, or $2 + 3$
4. $-3 - (-5)$, or $-3 + 5$
5. $4 + (-7)$, or $4 - 7$
6. $1 + (-1)$, or $1 - 1$
7. 2
8. 7
9. -25
10. -22
11. -16
12. 0
13. -18
14. 5
15. 8
16. -6

17. -31
18. 5
19. $-2x - 3$
20. $4a - 3$
21. $5 - 2d$
22. $-7z - 6$
23. $-x - 7$
24. $5w - 11$
25. -21
26. 7
★27. 0
★28. -36
★29. 19
★30. 4
★31. -2
★32. -9
★33. -4
★34. -16
35. 2.5
36. -17
37. -32
38. 0
39. -15
40. 0

41. 1
42. -15
43. -3.1

★ **Advanced Exercises**

Focus: Students use real-number addition and subtraction to model problems.

EXAMPLE 1

PHASE 1: Presenting the Example

When reading the problem in Example 1, stress the phrases *lowered by 7°F* and *raised by 3°F*. Lead students to realize that, in the context of this problem, it does not matter which of these actions is performed first; instead we are concerned with the final outcome. For this reason, we may apply the Commutative Property, as shown in the Example, to simplify calculations.

PHASE 2: Providing Guidance for TRY THIS

When students simplify the problem, they may incorrectly apply the Associative Property and write $8 - (5 - 6)$. Stress to these students that they must be extremely careful when grouping in subtraction problems. Suggest that they rewrite the problem as $8 + (-5) + (-6)$ in order to see the correct grouping, $8 + [-5 + (-6)]$, or $8 + (-5 - 6)$.

EXAMPLE 2

PHASE 1: Presenting the Example

Be sure that students understand all of the terms used in the problem. They should know that the *balance* of an account is the amount of money in the account at a given time. Removing money from the account is called a *withdrawal*, which results in a subtraction from the balance. Adding money to the account is called a *deposit*. The difference of the deposits and the withdrawals is the *net change* in the balance. A positive net change means that the balance in the account has increased, while a negative net change means that the balance in the account has decreased.

Some students may believe that their answer represents the new balance in the account. Be sure they understand the difference between *a balance of* -4.71 and *a change of* -4.71.

PHASE 3: ASSESSMENT AND CLOSURE for Lesson 2.3

- Ask students to comment on the following: *If you know how to add real numbers, then you know how to subtract real numbers.* [**sample: This is true. Any subtraction can be rewritten as the addition of an opposite.**]

- Learning how to rewrite subtraction as addition allows you to apply the properties of the operations to a broader range of expressions. For instance, subtraction is not commutative, and so $7 - 5 \neq 5 - 7$. But rewriting $7 - 5$ as $7 + (-5)$ yields an expression that *is* commutative: $7 + (-5) = -5 + 7$.

☞ *For a Lesson 2.3 Quiz, see* Assessment Resources *page 14.*

EXAMPLE 1

The thermostat in Rachel's freezer was set at 5°F. She lowered the thermostat by 7°F. Later, she raised the thermostat by 3°F. At what temperature was the thermostat finally set?

▶ **Solution**

Look for key words.

Translate the phrases into a numerical expression. The key words are *lowered* and *raised*. *Lowered* indicates subtraction and *raised* indicates addition.

$$5 - 7 + 3$$
$$= 5 + 3 - 7 \longleftarrow \text{Commutative Property}$$
$$= (5 + 3) - 7 \longleftarrow \text{Associative Property}$$
$$= 8 - 7$$
$$= 1$$

> Notice that you can use the Commutative and Associative Properties to make your calculations easier. Adding all the positive numbers in an expression before subtracting can aid mental computation.

The thermostat was finally set at 1°F.

TRY THIS

One winter afternoon, the temperature was 8°F. By evening, the temperature had dropped 5°F, and by 11:00 P.M., the temperature had dropped another 6°F. What was the temperature at 11:00 P.M.?

EXAMPLE 2

A bank customer makes three withdrawals and two deposits. Find the resulting, or *net*, change in the customer's balance.

  deposits: $120.50 and $76.45
  withdrawals: $70.32, $18.59 and $112.75

▶ **Solution**

Find the sum of the deposits less the sum of the withdrawals.

deposits − withdrawals

$$= (120.50 + 76.45) - (70.32 + 18.59 + 112.75)$$
$$= 196.95 - 201.66$$
$$= -4.71$$

The net change is a decrease of $4.71 in the bank balance.

TRY THIS

A bank customer makes three withdrawals and two deposits. Find the net change in the customer's balance.

  deposits: $100.50 and $86.55
  withdrawals: $80.32, $20.55 and $112.45

You can also solve Example 2 by using the Multiplicative Property of −1.

$$(120.50 + 76.45) - (70.32 + 18.59 + 112.75)$$
$$= 120.50 + 76.45 - 70.32 - 18.59 - 112.75$$
$$= -4.71$$

EXERCISES

Translate each phrase into numerical expressions.

1. the total restaurant bill with the following items: the price of the meal, $6.75, a discount coupon for $1.50 off, tax of $0.42, and a tip of $1.00

2. the total bill at the clothing store after the following: a return of a shirt, $25.50, purchase of two shirts for $17.25 and $21.75, and $0.68 tax

3. the net change in altitude of an airplane after climbs of 450 ft and 275 ft and descents of 45 ft, 120 ft, and 230 ft

PRACTICE

Solve each problem.

4. The temperature was 16°F, and then it fell 19°F. What was the final temperature?

5. The temperature was −5°F at midnight. It rose 7°F by 3:00 A.M. Find the temperature at that time.

6. On Tuesday afternoon, the temperature was −3°F. That night it dropped 8°F, and the next morning it rose 15°F. What was the temperature on Wednesday morning?

7. The net change in stock market averages after daily changes of: +36, −107, −56, +76, and +45.

Find the net change in each bank balance.

8. deposits: $110.33, $202.56, and $53.70
   withdrawals: $33.55, $300.56, and $45.00

9. deposits: $300.55, $250.00, and $79.90
   withdrawals: $313.55, $256.56, and $65.00

10. **Critical Thinking** Write and evaluate an expression for the net value of the following transactions at the clothing store: the return of a shirt for $15, the purchase of two pairs of shoes at $30 a pair, a 30% discount on the shoes, and a 5% sales tax on the net purchase price.

MIXED REVIEW APPLICATIONS

Solve each problem. (1.9 Skill B)

11. Sam wants to deposit an equal amount of money into his new bank account each month. At the end of one year, he wants to have a total of $2700 in his account. How much money should he deposit each month?

12. **a.** How many people can be fed 6-ounce servings using 64 ounces of hamburger?

    b. How large is each portion of salad if there are 56 ounces of salad and 14 people to feed?

Answers for Lesson 2.3 Skill C

Try This Example 1
−3°F

Try This Example 2
decrease of $26.27

1. $6.75 - 1.50 + 0.42 + 1.00$
2. $-25.50 + 17.25 + 21.75 + 0.68$
3. $450 + 275 - 45 - 120 - 230$
4. −3°F
5. 2°F
6. 4°F
7. decrease of 6 points
8. decrease of $12.52
9. decrease of $4.66
10. Sample answer: $[-15 + 2(30) - 0.3(2)(30)]1.05 = \$28.35$
11. $225
12. **a.** 10 people
    b. 4 ounces

Multiplying and Dividing Real Numbers

SKILL A *Multiplying real numbers*

Focus: *Students use algebraic rules to multiply real numbers.*

EXAMPLES 1 and 2

PHASE 1: Presenting the Examples

Display the first five multiplications at right. Tell students to copy the multiplications as shown and to fill in the answer blanks. [**Answers are shown in red.**]

Ask students to describe any patterns that they observe. Elicit the response that, as the multipliers of 3 decrease by 1, the products decrease by 3.

Now display the next five multiplications directly below the first five. Point out that the multipliers of 3 continue to decrease by 1. Tell students to assume that the pattern among the products will also continue. Have them copy the multiplications and fill in each answer blank by continuing the pattern. [**Answers are shown in red.**]

Ask students to make a conjecture about the sign of the product of a negative number and a positive number. Lead them to the conclusion that the sign of the product is negative.

Repeat the activity, this time beginning with the next five multiplications at right $(4 \times (-3), 3 \times (-3), 2 \times (-3)$, and so on). Since students now know that the product of a negative number and a positive number is negative, they should easily find the products. [**Answers are shown in red.**] This time point out that, as the multipliers of -3 decrease by 1, the products *increase* by 3.

Display the final five multiplications directly below. Have students continue the pattern among the products. [**Answers are shown in red.**]

Ask students to make a conjecture about the sign of the product of two negative numbers. Lead them to the conclusion that the sign of the product is positive.

Summarize the activity by reviewing the rules for multiplying two real numbers as stated in the box at the top of the textbook page. Then discuss the examples.

$4 \times 3 =$	12
$3 \times 3 =$	9
$2 \times 3 =$	6
$1 \times 3 =$	3
$0 \times 3 =$	0
$-1 \times 3 =$	-3
$-2 \times 3 =$	-6
$-3 \times 3 =$	-9
$-4 \times 3 =$	-12
$-5 \times 3 =$	-15

$4 \times (-3) =$	-12
$3 \times (-3) =$	-9
$2 \times (-3) =$	-6
$1 \times (-3) =$	-3
$0 \times (-3) =$	0
$-1 \times (-3) =$	3
$-2 \times (-3) =$	6
$-3 \times (-3) =$	9
$-4 \times (-3) =$	12
$-5 \times (-3) =$	18

PHASE 2: Providing Guidance for **TRY THIS**

Sometimes students use the incorrect sign in the final product. Suggest that they begin any multiplication by first determining the sign of the product and writing *positive* or *negative* next to the multiplication. A sample is shown below.

$$(1.5) \times (-4) \rightarrow \underline{negative}$$
$$1.5 \times 4 = 6 \rightarrow \quad -6$$

56 Chapter 2 Operations With Real Numbers

2.4 LESSON
Multiplying and Dividing Real Numbers

SKILL A *Multiplying real numbers*

When you multiply two real numbers, use the following rules to find the sign of the product.

> **Rules for Multiplying Two Real Numbers**
> **Same signs:** If both numbers have the same sign, multiply as if the numbers were positive. The product will always be positive.
> **Opposite signs:** If the numbers have opposite signs, multiply as if the numbers were positive. The product will always be negative.

EXAMPLE 1 Find each product.   **a.** $(-26)(2)$ **b.** $(-26)(-2)$

▸ **Solution**
a. Multiply as if the numbers were positive. $26 \times 2 = 52$
Because -26 and 2 have opposite signs, use $-$. -52
b. Multiply as if the numbers were positive. $26 \times 2 = 52$
Because -26 and -2 have the same sign, no 52
change is necessary.

TRY THIS Find each product.   **a.** $(-120) \times (-4)$ **b.** $(1.5) \times (-4)$

EXAMPLE 2 Find $\left(-4\frac{1}{2}\right)^2$.

▸ **Solution**
Recall that $x^2 = x \cdot x$.
$$\left(-4\frac{1}{2}\right)^2 = \left(-4\frac{1}{2}\right)\left(-4\frac{1}{2}\right) = \left(-\frac{9}{2}\right)\left(-\frac{9}{2}\right) = \frac{81}{4}, \text{ or } 20\frac{1}{4}$$

TRY THIS Find $\left(-3\frac{1}{3}\right)^2$.

You can write a proof for the rules of multiplying two real numbers shown above. Let a and b represent positive real numbers. Then $-a$ is negative and $(-a)b$ is the product of a negative number and a positive number.
$(-a)b + ab = (-a + a)b$ ◀——— Distributive Property
$\qquad\qquad = 0 \cdot b$ ◀——— Inverse Property of Addition
$\qquad\qquad = 0$
This means ab is the additive inverse of $(-a)b$. So, $(-a)b = -(ab)$.

EXERCISES

KEY SKILLS

Write the sign of each product.

1. $(-4) \times (5)$ **2.** $(-4)(-5)$ **3.** $\left(4\frac{3}{5}\right)^2$ **4.** $\left(-\frac{3}{5}\right)\left(-2\frac{2}{3}\right)$

PRACTICE

Find each product. Write fractions in simplest form.

5. $6(-2)$ **6.** $9(-9)$ **7.** $(-5)(-5)$ **8.** $(-3)(-4)$

9. $4(-11)$ **10.** $(12)(-13)$ **11.** $(-36) \times (-3)$ **12.** $(-5) \times (-11)$

13. $2.5 \times (-2)$ **14.** $(-1.2) \times (-1.2)$ **15.** $\frac{-4}{7} \times \frac{7}{3}$ **16.** $\frac{-2}{7} \times \frac{3}{-5}$

17. $3 \times \left(-3\frac{1}{3}\right)$ **18.** $(-2) \times \left(-5\frac{1}{2}\right)$ **19.** $\left(1\frac{3}{5}\right)^2$ **20.** $\left(-\frac{3}{5}\right)\left(-\frac{2}{3}\right)$

21. $(-3)(-4)(-2)$ **22.** $(-10)(5)(-2)$ **23.** $(-10)(-6)(-3)$

24. $(-12)(-6)(-5)$ **25.** $[(-6) \times (-10)] \times 4$ **26.** $[(-7) \times (-2)] \times (-4)$

Determine whether each product is positive or negative.

27. $-4a$; a is positive   **28.** $-4a$; a is negative   **29.** $4b \times -\frac{1}{2}$; b is positive

30. $\frac{-5}{3} \times c$; c is negative   **31.** $\frac{3}{-4} \times n$; n is positive   **32.** $\frac{-4}{9} \times m$; m is negative

33. Critical Thinking   Use mental math to simplify the following expression and then describe your strategy. $\left(\frac{2}{3}\right)\left(-\frac{5}{4}\right)\left(-\frac{5}{6}\right)\left(\frac{3}{2}\right)\left(\frac{6}{5}\right)\left(-\frac{4}{5}\right)$

MIXED REVIEW

Find each sum. (2.2 Skill B)

34. $-14.1 + (-3.8)$ **35.** $2.5 + (-18)$ **36.** $5 + (-3)$ **37.** $-7 + (-4)$

38. $-5 + (-12.1)$ **39.** $-5.6 + (-9.3)$ **40.** $-13 + (-13)$ **41.** $-13 + 13$

Find each difference. (2.3 Skill A)

42. $2 - (-2)$ **43.** $-18 - 18$ **44.** $27.2 - 27.2$ **45.** $100 - 19.90$

46. $2.2 - (-2)$ **47.** $-1.9 - 19$ **48.** $27.2 - 7.2$ **49.** $10 - 9.8$

Answers for Lesson 2.4 Skill A

Try This Example 1
a. 480   **b.** –6

Try This Example 2
$\frac{100}{9}$, or $11\frac{1}{9}$

1. negative $(-)$
2. positive $(+)$
3. positive $(+)$
4. positive $(+)$
5. -12
6. -81
7. 25
8. 12
9. -44
10. -156
11. 108
12. 55
13. -5
14. 1.44
15. $-\frac{4}{3}$, or $-1\frac{1}{3}$
16. $\frac{6}{35}$

17. -10
18. 11
19. $\frac{64}{25}$, or $2\frac{14}{25}$
20. $\frac{2}{5}$
21. -24
22. 100
23. -180
24. -360
25. 240
26. -56
27. negative
28. positive
29. negative
30. negative
31. negative
32. positive
33. -1; strategies will vary. Sample strategy: Use the Commutative and Associative Properties of Multiplication to pair up multiplicative inverses.
$$\left(\frac{2}{3} \cdot \frac{3}{2}\right) \cdot \left(-\frac{5}{4} \cdot -\frac{4}{5}\right) \cdot \left(-\frac{5}{6} \cdot \frac{6}{5}\right) = 1 \cdot 1 \cdot (-1) = -1$$

34. -17.9
35. -15.5
36. 2
37. -11
38. -17.1
39. -14.9
40. -26
41. 0
42. 4
43. -36
44. 0
45. 80.1
46. 4.2
47. -20.9
48. 20
49. 0.2

Dividing real numbers

Focus: *Students use algebraic rules to divide real numbers.*

EXAMPLES 1 and 2

PHASE 1: Presenting the Examples

Display the five divisions given in Column I at right. Tell students to copy the divisions as shown and fill in the answer blanks. [**Answers are shown in red.**]

Now display the multiplications in Column II. Have students copy them and fill in the answer blanks. [**Answers are shown in red.**]

Point out to students that the quotient in each row is equal to the product in the same row. Lead students to the conclusion that dividing by a positive number is equivalent to multiplying by its reciprocal.

Column I	Column II
$16 \div 4 = \underline{\ 4\ }$	$16 \times \frac{1}{4} = \underline{\ 4\ }$
$12 \div 4 = \underline{\ 3\ }$	$12 \times \frac{1}{4} = \underline{\ 3\ }$
$8 \div 4 = \underline{\ 2\ }$	$8 \times \frac{1}{4} = \underline{\ 2\ }$
$4 \div 4 = \underline{\ 1\ }$	$4 \times \frac{1}{4} = \underline{\ 1\ }$
$0 \div 4 = \underline{\ 0\ }$	$0 \times \frac{1}{4} = \underline{\ 0\ }$

Display the generalization $a \div b = a \times \frac{1}{b}$. Tell students that, because of countless observations like those above, this statement is accepted as the algebraic definition of division for real numbers. Point out that there is one restriction on the definition: the divisor, b, cannot be zero.

Use a series of examples like those at right to illustrate why b cannot be zero. Tell students to suppose that a is a nonzero real number and that $a \div 0 = q$. Because multiplication and division are inverse operations, it would follow that $q \times 0 = a$. The product of zero and any real number is zero, so $a = 0$. But it was initially stated that a is nonzero. Because of this contradiction, mathematicians agree that division by zero is *undefined*.

$$a \div b = q \quad \Longleftrightarrow \quad q \times b = a$$
$$8 \div 8 = 1 \quad \Longleftrightarrow \quad 1 \times 8 = 8$$
$$8 \div 4 = 2 \quad \Longleftrightarrow \quad 2 \times 4 = 8$$
$$8 \div 2 = 4 \quad \Longleftrightarrow \quad 4 \times 2 = 8$$
$$8 \div 1 = 8 \quad \Longleftrightarrow \quad 1 \times 8 = 8$$
$$8 \div 0 = ? \quad \Longleftrightarrow \quad ? \times 0 = 8$$

Remind students that the product of a number and its reciprocal is 1. So, if a number is positive, its reciprocal is positive. If a number is negative, its reciprocal is negative. This means that the sign of b in the quotient $a \div b$ is the same as the sign of $\frac{1}{b}$ in the product $a \times \frac{1}{b}$. Therefore, the rules for division of two real numbers are the same as the rules for multiplication. Review the rules for division as stated in the box at the top of the textbook page. Then discuss the examples.

PHASE 2: Providing Guidance for TRY THIS

Given a fraction such as $-\frac{15}{8}$, some students mistakenly "distribute" the negative sign to the numerator and denominator and write $\frac{-15}{-8}$. Remind them that either the numerator or denominator can be considered negative, but not both. If students are confused, show a simpler example, such as the one below.

$$-\frac{20}{5} \quad \neq \quad \frac{-20}{-5}$$
$$\downarrow \qquad\qquad \downarrow$$
$$-(20 \div 5) \neq (-20) \div (-5)$$
$$\downarrow \qquad\qquad \downarrow$$
$$-4 \quad \neq \quad 4$$

Dividing real numbers

In Lesson 1.2, you learned that every nonzero real number a has a reciprocal, also called the multiplicative inverse.

Because you can multiply any two real numbers, and every nonzero real number has a multiplicative inverse, you can define division using multiplication and multiplicative inverses.

> **Division of Real Numbers**
>
> Let a and b represent real numbers, where b is nonzero.
>
> $$a \div b = a \times \frac{1}{b}$$
>
> **Same signs:** If a and b have the same sign, divide as if the numbers were positive. The quotient will always be positive.
>
> **Opposite signs:** If a and b have opposite signs, divide as if the numbers were positive. The quotient will always be negative.

EXAMPLE 1 Find $(-26) \div 2$.

▶ **Solution**
  1. Divide as if the numbers were positive. $26 \div 2 = 13$
  2. Make the quotient negative. -13

TRY THIS Find each quotient.   **a.** $(-120) \div (4)$ **b.** $(-120) \div (-4)$

EXAMPLE 2 Find $\frac{3}{4} \div \left(-\frac{15}{8}\right)$.

▶ **Solution**
$$\frac{3}{4} \div \left(-\frac{15}{8}\right) = \frac{3}{4} \times \left(-\frac{8}{15}\right) \quad \longleftarrow \text{Multiply by } -\frac{8}{15}, \text{ the reciprocal of } -\frac{15}{8}.$$
$$= -\left(\frac{\cancel{3}}{\cancel{4}} \times \frac{\cancel{8}^2}{\cancel{15}_5}\right) = -\frac{2}{5}$$

TRY THIS Find $\left(-3\frac{1}{3}\right) \div \left(1\frac{2}{3}\right)$.

The following is a proof of the rule for dividing a negative number by a positive number. Let a and b represent positive real numbers. Then $-a$ is negative.

$$\frac{-a}{b} = (-a) \times \frac{1}{b} \quad \longleftarrow \text{Definition of division}$$
$$= -\left(a \times \frac{1}{b}\right) \quad \longleftarrow \text{Multiplication with opposite signs}$$
$$= -\left(\frac{a}{b}\right) \quad \longleftarrow \text{Definition of division}$$

EXERCISES

KEY SKILLS

Write the sign of each quotient.

1. $(-45) \div (5)$ **2.** $(-15) \div (5)$ **3.** $\left(4\frac{3}{5}\right) \div 2$ **4.** $\left(-\frac{3}{5}\right) \div \left(-\frac{2}{3}\right)$

PRACTICE

Find each quotient. Write fractions in simplest form.

5. $(-60) \div 10$ **6.** $(-32) \div 8$ **7.** $(-72) \div 8$ **8.** $(-32) \div (-4)$

9. $(39) \div (-13)$ **10.** $(27) \div (-3)$ **11.** $\frac{-81}{9}$ **12.** $\frac{-54}{-27}$

13. $\frac{35}{-7}$ **14.** $(-55) \div (-11)$ **15.** $2.5 \div (-2)$ **16.** $(-1.2) \div (-1.2)$

17. $10 \div \left(-3\frac{1}{3}\right)$ **18.** $(-11) \div \left(-5\frac{1}{2}\right)$ **19.** $\frac{-4}{7} \div \frac{3}{7}$ **20.** $\frac{-2}{7} \div \frac{3}{-5}$

21. $\left(4\frac{3}{5}\right) \div \left(-4\frac{3}{5}\right)$ **22.** $\left(-\frac{3}{5}\right) \div \left(-\frac{3}{2}\right)$ **23.** $(-13) \div (-4)$ **24.** $(-17) \div (5)$

25. $(-25) \div (-5)$ **26.** $(22) \div (-5)$ **27.** $(-6) \div (-10)$ **28.** $(-12) \div (-15)$

Determine whether each quotient is positive or negative.

29. $\frac{a}{-2}$; a is positive   **30.** $\frac{-2}{a}$; a is negative   **31.** $\frac{4b}{-2}$; b is positive

32. $\frac{-5c}{-3}$; c is negative   **33.** $\frac{-3n-3}{-4}$; n is positive   **34.** $\frac{m+4}{-9}$; m is positive

MIXED REVIEW

Simplify. (previous courses)

35. 2^2 **36.** 3^2 **37.** 4^2 **38.** 5^2

39. 6^2 **40.** 7^2 **41.** 8^2 **42.** 9^2

43. $5^2 - 4^2$ **44.** $6^2 - 5^2$ **45.** $10^2 - 9^2$ **46.** $100^2 - 99^2$

Solve each equation. Check your solution. (1.6 Skill A)

47. $3.45 + g = 13.5$ **48.** $t - 2 = 18.5$ **49.** $100 = 13.8 + r$ **50.** $12.56 = s - 5$

Solve each equation. Check your solution. (1.9 Skill A)

51. $\frac{2w}{7} = 10$ **52.** $13 = \frac{13z}{6}$ **53.** $44 = \frac{4a}{11}$ **54.** $\frac{2n}{2} = 1$

Answers for Lesson 2.4 Skill B

Try This Example 1
a. -30 **b.** 30

Try This Example 2
-2

1. negative $(-)$
2. negative $(-)$
3. positive $(+)$
4. positive $(+)$
5. -6
6. -4
7. -9
8. 8
9. -3
10. -9
11. -9
12. 2
13. -5
14. 5
15. -1.25
16. 1
17. -3
18. 2

19. $-\frac{4}{3}$, or $-1\frac{1}{3}$
20. $\frac{10}{21}$
21. -1
22. $\frac{2}{5}$
23. $\frac{13}{4}$
24. $-\frac{17}{5}$, or $-3\frac{2}{5}$
25. 5
26. $-\frac{22}{5}$, or $-4\frac{2}{5}$
27. $\frac{3}{5}$
28. $\frac{4}{5}$
29. negative
30. positive
31. negative
32. negative
33. positive
34. negative

35. 4
36. 9
37. 16
38. 25
39. 36
40. 49
41. 64
42. 81
43. 9
44. 11
45. 19
46. 199
47. $g = 10.05$
48. $t = 20.5$
49. $r = 86.2$
50. $s = 17.56$
51. $w = 35$
52. $z = 6$
53. $a = 121$
54. $n = 1$

SKILL C Using multiplication and division with other operations

Focus: *Students use the order of operations to simplify numerical expressions involving signed numbers.*

EXAMPLE 1

PHASE 1: Presenting the Example

Remind students that, in Chapter 1, they learned a four-step order of operations. Tell them to write the four steps. [**The steps are described at right.**] Discuss the students' responses, having them make any necessary corrections on their papers.

1. Simplify expressions within grouping symbols, innermost groupings first.
2. Simplify expressions involving exponents.
3. Multiply/divide from left to right.
4. Add/subtract from left to right.

Now display the given expression, $(-8 + 5)^2 - \left(\frac{-24}{6}\right)$. Tell students to write the step numbers, 1 through 4, on their papers. Next to each number, have them describe the calculation(s) that correspond to each step. [**Calculations are shown at right.**]

1. Add $-8 + 5 = -3$; divide $-24 \div 6 = -4$.
2. Calculate $(-3)^2 = 9$.
3. none
4. Subtract $9 - (-4) = 9 + 4 = 13$.

PHASE 2: Providing Guidance for TRY THIS

Watch for students who simplify $(-5)^2$ as -25. There is a warning about this error on the textbook page. Suggest that they always expand such expressions as shown at right so that it becomes clear why the expression must represent a positive number.

$$(-5)^2$$
$$\downarrow$$
$$(-5) \times (-5)$$
$$\downarrow$$
$$25$$

EXAMPLE 2

PHASE 2: Providing Guidance for TRY THIS

Some students may correctly simplify the given expression as $\frac{25}{-10} \times \frac{-10}{25}$, and then incorrectly simplify the new expression and give the answer as 0. Remind them that they are dividing the numerators and denominators by common factors: $25 \div 25 = 1$, and $-10 \div (-10) = 1$. Suggest that they record these divisions as shown at right.

$$\frac{\overset{1}{\cancel{25}}}{\underset{1}{\cancel{-10}}} \times \frac{\overset{1}{\cancel{-10}}}{\underset{1}{\cancel{25}}}$$

PHASE 3: ASSESSMENT AND CLOSURE for Lesson 2.4

- Ask students why the rules for multiplication and division are so similar. [**sample: Any division of two numbers can be stated as a multiplication.**]
- Learning how to rewrite division as multiplication allows you to apply the properties of the operations to a broader range of expressions. For instance, division is not commutative, and so $8 \div 4 \neq 4 \div 8$. But rewriting $8 \div 4$ as $8 \times \frac{1}{4}$ yields an expression that *is* commutative: $8 \times \frac{1}{4} = \frac{1}{4} \times 8$.

☞ *For a Lesson 2.4 Quiz, see* Assessment Resources *page 15.*

You have learned the order of operations and the rules for adding, subtracting, multiplying, and dividing real numbers. You can now simplify many expressions that involve positive and negative numbers.

EXAMPLE 1   Simplify $(-8 + 5)^2 - \left(\frac{-24}{6}\right)$.

▶ **Solution**

$(-8 + 5)^2 - \left(\frac{-24}{6}\right)$

$= (-3)^2 - (-4)$ ⟵ *Perform the addition and division in parentheses.*

$= 9 - (-4)$ ⟵ *Calculate $(-3)^2 = (-3)(-3) = 9$.*

$= 13$ ⟵ *Subtract.*

TRY THIS   Simplify $\left(\frac{14 \times (-2)}{-7}\right) + (-5)^2 - 3$.

Notice in Example 1 that $(-3)^2$ equals 9, not -9. This can be generalized for any real number a: $(-a)^2 = a^2$, not $-a^2$.

When you simplify a product of fractions, look for common factors of the numerators and denominators. Divide both the numerator and a denominator by the common factors; this process is sometimes called *canceling*.

EXAMPLE 2   Simplify $\frac{(1-4)^2}{(-2)(5)} \times \frac{(-2)(-3)}{(6)(6)}$.

▶ **Solution**

$\frac{(1-4)^2}{(-2)(5)} \times \frac{(-2)(-6)}{(6)(6)}$

$= \frac{(-3)(-3)}{(-2)(5)} \times \frac{(-2)(-6)}{3 \times 2 \times 3 \times 2}$

$= \frac{(-3)(-3)}{(-2)(5)} \times \frac{(-2)(-6)}{3 \times 2 \times 3 \times 2}$

$= \frac{(-1)(-1)(-3)}{(5)(2)}$

$= -\frac{3}{10}$

TRY THIS   Simplify $\frac{(1-6)^2}{(-2)(5)} \times \frac{(2)(-5)}{(-5)(-5)}$.

EXERCISES

Simplify each power.

1. $(-5)^2$ 2. $-(-5)^2$ 3. $-(5)^2$ 4. 5^2

PRACTICE

Simplify each expression.

5. $3 - 5(1 - 8)$

6. $2 \times 8 \div [4 - (-4)]$

7. $2 \div [4 - (-4)] \times 8$

8. $\frac{2[-3 + (-2)]}{5}$

9. $\frac{-3[-3 - (-7)]}{-2}$

10. $\frac{(2-5)^2}{3}$

11. $\frac{[-4 - (-5)]^2}{2}$

12. $\frac{2 - (-12)}{12 - 7(-2)}$

13. $\frac{8 - (-12)}{12(-2) + 3(-2)}$

14. $\frac{(3-5)}{-2} \times \frac{(7-1)}{-4}$

15. $\frac{(7-10)}{-5} \times \frac{[7-(-3)]}{(12-13)}$

16. $\frac{(7-10)}{(3-7)} \times \frac{[8-(-4)]}{(14-17)}$

17. $(-3 - 3)^2(-2 + 4)^2$

18. $(-3 - 3)^2(-2 + 4)^2 \div (-4)^2$

19. $(-3 - 3)^2 \div (-2 + 6)^2 \div (-2)^2$

20. $\frac{-(10 - 1)^2}{(-2)(7)} \times \frac{(2)(-7)}{(-9) \times (9)}$

21. $\frac{(7-1)^2}{(-2)(11)} \times \frac{(3)(-11)}{(-3) \times (-6)}$

22. $\frac{(7-1)^2}{(-5)(10)} \times \frac{(3)(-10)}{(-3) \times (-6)}$

23. $\frac{(7-1)^2}{(-3)(10)} \times \frac{(3)(-1)}{(1 - 7)^2}$

24. $\frac{(2-5)^2}{[3 - (-1)]^2} \times \frac{(3)(-1)}{-(2 - 5)^2}$

25. **Critical Thinking**   Find a such that $\frac{(4 - a)^2}{100} \times \frac{[8 - (-2)]^2}{(5 - 10)^2} = 1$.

MID-CHAPTER REVIEW

For each number, find:
a. the opposite and   b. the absolute value. (2.1 Skill C)

26. -5.2 27. 0 28. 12.6 29. 2.3345

Find each sum or difference. (2.2 Skill B, 2.3 Skill A)

30. $-5.1 - (-12)$ 31. $18 + (-26)$ 32. $-11 + (-19)$ 33. $4 - 6.3$

Find each product or quotient. (2.4 Skills A and B)

34. $3.5 \times (-4)$ 35. $-2.5 \times (-6)$ 36. $28.4 \div (-4)$ 37. $30.6 \div (-6)$

Answers for Lesson 2.4   Skill C

Try This Example 1
26

Try This Example 2
1

1. 25
2. -25
3. -25
4. 25
5. 38
6. 2
7. 2
8. -2
9. 6
10. 3
11. $\frac{1}{2}$
12. $\frac{7}{13}$
13. $-\frac{2}{3}$
14. $-\frac{3}{2}$, or $-1\frac{1}{2}$

15. -6
16. -3
★17. 144
★18. 9
★19. $\frac{9}{16}$
★20. 1
★21. -3
★22. $\frac{6}{5}$, or $1\frac{1}{5}$
★23. $\frac{1}{10}$
★24. $\frac{3}{16}$
25. $a = -1$ or $a = 9$
26. **a.** 5.2   **b.** 5.2
27. **a.** 0   **b.** 0
28. **a.** -12.6 **b.** 12.6
29. **a.** -2.3345 **b.** 2.3345
30. 6.9
31. -8
32. -30
33. -2.3

34. -14
35. 15
36. -7.1
37. -5.1

★ **Advanced Exercises**

Solving Equations Involving Real Numbers

Using properties of equality, addition rules, and subtraction rules to solve equations

Focus: *Students apply their skills in adding and subtracting real numbers to solve simple equations.*

EXAMPLE 1

PHASE 1: Presenting the Example

Display the equation $t + 2 = 8$. Remind students that they solved equations like this in Chapter 1. Have them work individually to write a solution. [**See solution at right.**]

After all students have completed their work, discuss the solution process. Elicit the fact that the Subtraction Property of Equality justifies the method of subtracting 2 from each side of the equation. Remind them that it is important to always check a solution.

Now display the equation $t + 8 = 2$. Point out that this equation appears similar to the first, but the positions of 8 and 2 have been reversed. Once again have students work individually to solve. [**See solution at right.**]

Discuss the students' work. Ask them how the process of solving this equation differed from the process of solving the first. Lead them to see that, in this case, the number subtracted from each side was 8. Note that this resulted in a subtraction with a negative difference, that is, $2 - 8 = -6$. Again stress the importance of checking a solution.

Because subtraction of real numbers has now been defined in terms of addition, some students may note that it is possible to solve these equations by using the *Addition Property of Equality*. That is, the first equation can be solved by adding -2 to each side, while the second equation can be solved by adding -8 to each side. Be sure students understand that this is acceptable.

Now discuss the equation given in Example 1, $d + 5 = -4$.

$$t + 2 = 8$$
$$t + 2 - 2 = 8 - 2$$
$$t = 6$$

Check: $t + 2 = 8$
$$6 + 2 \stackrel{?}{=} 8$$
$$8 = 8 \checkmark$$

$$t + 8 = 2$$
$$t + 8 - 8 = 2 - 8$$
$$t = -6$$

Check: $t + 8 = 2$
$$-6 + 8 \stackrel{?}{=} 2$$
$$2 = 2 \checkmark$$

EXAMPLE 2

PHASE 2: Providing Guidance for **TRY THIS**

Have students do the Try This exercise in pairs. One student in each pair should solve the equation as it appears in the textbook. The other student should solve the equation using fractions, rewriting it as $v - 3\frac{1}{2} = -2$. They should arrive at the solutions $v = 1.5$ and $v = 1\frac{1}{2}$, respectively. Stress the fact that these solutions are equivalent, and so both are equally acceptable. Encourage the pairs of students to compare their solution methods and determine if one seems easier than the other.

2.5 LESSON
Solving Equations Involving Real Numbers

SKILL A *Using properties of equality, addition rules, and subtraction rules to solve equations*

You used the following properties of equality to solve simple equations in earlier lessons. These properties apply to both positive and negative numbers.

Addition and Subtraction Properties of Equality
Let a, b, and c be real numbers.
Addition   If $a = b$, then $a + c = b + c$.
Subtraction   If $a = b$, then $a - c = b - c$.

EXAMPLE 1   Solve $d + 5 = -4$. Check your solution.

▶ Solution
$$d + 5 = -4$$
$$d + 5 - 5 = -4 - 5 \quad \longleftarrow \text{ Apply the Subtraction Property of Equality.}$$
$$d + 0 = -4 - 5$$
$$d = -9 \quad \longleftarrow \text{ Use the rule for addition with same signs; } -4 + (-5) = -9$$

Check: Substitute -9 for d in the original equation.
$$-9 + 5 \overset{?}{=} -4$$
$$-4 \overset{?}{=} -4 \checkmark$$

TRY THIS   Solve $a + 2 = -4$.

EXAMPLE 2   Solve $z - 4.5 = -12$. Check your solution.

▶ Solution
$$z - 4.5 = -12$$
$$z - 4.5 + 4.5 = -12 + 4.5 \quad \longleftarrow \text{ Apply the Addition Property of Equality.}$$
$$z = -7.5$$

Check: Substitute -7.5 for z in the original equation.
$$-7.5 - 4.5 \overset{?}{=} -12$$
$$-12 \overset{?}{=} -12 \checkmark$$

TRY THIS   Solve $v - 3.5 = -2$.

EXERCISES

KEY SKILLS

a. Identify the property of equality needed to solve each equation.
b. Then state the number that you would add or subtract. Do not solve.

Sample: $x + (-4) = 7$ Solution: Change to $x - 4 = 7$.
a. Addition Property of Equality   b. 4

1. $d + 5 = -3$ 2. $r + (-3) = 6$ 3. $x + 4 = -11$ 4. $y - 5 = 12$

PRACTICE

Solve each equation. Check your solution.

5. $x - 23 = -7$ 6. $x - 35 = -18$ 7. $-12 = 18 - y$

8. $x - (-10) = 11$ 9. $a - (-2) = 3$ 10. $t + (-2) = -1$

11. $b + (-7) = -7$ 12. $-15 + a = 0$ 13. $-5 = p - 5$

14. $-7 = p + (-3)$ 15. $-2.2 + x = -5$ 16. $-11.2 = w + 5.6$

17. $7.5 = v - (-3.5)$ 18. $-4.3 - (-y) = -6.7$ 19. $4.8 - x = -0.5$

20. $g - \frac{2}{3} = 18$ 21. $d + \frac{1}{2} = 5$ 22. $6\frac{1}{2} - x = 4$

23. $4\frac{2}{3} + x = 5\frac{2}{3}$ 24. $-x - 4.5 = 10.7$ 25. $-y + 2.7 = 8.8$

MIXED REVIEW

Write each fraction as a decimal. (2.1 Skill A)

26. $\frac{7}{8}$ 27. $\frac{24}{25}$ 28. $\frac{2}{3}$ 29. $\frac{5}{12}$

Write each decimal as a fraction in simplest form. (2.1 Skill A)

30. 1.35   31. 0.04   32. 0.0404   33. 3.7575

Classify each number in as many ways as possible. (2.1 Skill B)

34. -0.4545 35. $\frac{30}{6}$ 36. $2.\overline{435}$ 37. $1.2334445555\ldots$

Evaluate each expression for $d = -6$. (2.1 Skill C)

38. d 39. $-d$ 40. $-(-d)$ 41. $|d|$

42. $|-d|$ 43. $-|-d|$ 44. $-|d|$ 45. $|-(-d)|$

Answers for Lesson 2.5 Skill A

Try This Example 1
$a = -6$

Try This Example 2
$v = 1.5$, or $1\frac{1}{2}$

1. a. Subtraction Property of Equality   b. 5
2. a. Addition Property of Equality b. 3
3. a. Subtraction Property of Equality   b. 4
4. a. Addition Property of Equality   b. 5
5. $x = 16$
6. $x = 17$
7. $y = 30$
8. $x = 1$
9. $a = 1$
10. $t = 1$

11. $b = 0$
12. $a = 15$
13. $p = 0$
14. $p = -4$
15. $x = -2.8$
16. $w = -16.8$
17. $v = 4$
18. $y = -2.4$
19. $x = 5.3$
20. $g = \frac{56}{3}$, or $18\frac{2}{3}$
21. $d = \frac{9}{2}$, or $4\frac{1}{2}$
22. $x = \frac{5}{2}$, or $2\frac{1}{2}$
23. $x = 1$
24. $x = -15.2$
25. $y = -6.1$
26. 0.875
27. 0.96
28. $0.\overline{6}$
29. $0.41\overline{6}$
30. $\frac{27}{20}$, or $1\frac{7}{20}$

31. $\frac{1}{25}$
32. $\frac{101}{2500}$
33. $\frac{1503}{400}$, or $3\frac{303}{400}$
34. rational, real
35. natural number, whole number, integer, rational, real
36. rational, real
37. irrational, real
38. -6
39. 6
40. -6
41. 6
42. 6
43. -6
44. -6
45. 6

SKILL B *Using properties of equality, multiplication rules, and division rules to solve equations*

Focus: *Students apply their skills in multiplying and dividing real numbers to solve simple equations.*

EXAMPLE 1

PHASE 1: Presenting the Example

Display the equation $6.5x = 26$. Have students work individually to solve it, and then discuss their solution methods. Most students will divide each side by 6.5 to arrive at $x = 4$.

Ask how the solution would be different if the equation were $-6.5x = 26$. Students who originally divided by 6.5 should recognize that they would instead divide by -6.5, giving $x = -4$. Discuss the solution method in the Example, in which the decimal is first converted to a fraction. Stress to students that both solution methods are correct, and allow them to use whichever method they feel most comfortable with.

PHASE 2: Providing Guidance for TRY THIS

Watch for students who solve by adding 5.5 to each side, arriving at $t = 27.5$. Suggest that they begin solving equations like these by making the multiplication explicit, as shown at right. Then it becomes clear that the operation needed for solving is the inverse of multiplication, or division.

$$\begin{array}{ccc} -5.5t & & = 22 \\ \downarrow & & \downarrow \;\; \downarrow \\ -5.5 \times t & & = 22 \end{array}$$

EXAMPLE 2

PHASE 1: Presenting the Example

Display the given equation, $\frac{z}{3 + (-5)} = 12$. Beneath it, write this incomplete sentence: *To solve this equation, the first thing I would do is* ___?___. Tell students to copy and complete the sentence on their papers. Discuss their responses. Then discuss the solution given on the textbook page.

PHASE 3: ASSESSMENT AND CLOSURE for Lesson 2.5

- Give students the equations at right. Have them fill in the boxes with real numbers so that the resulting equations each have a solution of -4.
  [samples: $a - 3 = -7$; $b - 2 = -6$; $5c = -20$; $\frac{d}{2} = -2$]
- Students solved simple equations in Chapter 1, but they worked with just positive numbers. In this lesson, they expanded their equation-solving skills to encompass both positive and negative numbers.

1. $a + \square = \square$

2. $b - \square = \square$

3. $\square c = \square$

4. $\frac{d}{\square} = \square$

☞ *For a Lesson 2.5 Quiz, see* Assessment Resources *page 16.*

You used the following properties of equality to solve simple equations in earlier lessons. These properties apply to both positive and negative numbers.

Multiplication and Division Properties of Equality

Let a, b, and c be real numbers.

Multiplication   If $a = b$, then $ac = bc$.

Division   If c is nonzero and $a = b$, then $\frac{a}{c} = \frac{b}{c}$.

When solving certain equations, changing a decimal to a fraction first may make the computations easier.

EXAMPLE 1   Solve $-6.5x = 26$. Check your solution.

▸ **Solution**

$$-6.5x = 26$$
$$-\frac{13}{2}x = 26 \qquad \longleftarrow \quad 6.5 = 6\frac{1}{2} = \frac{13}{2}$$
$$-\frac{2}{13}\left(-\frac{13}{2}\right)x = -\frac{2}{13} \times \overset{2}{26} \qquad \longleftarrow \text{Apply the Multiplication Property of Equality.}$$
$$x = -4$$

Check: $(-6.5)(-4) = (6.5)(4) = 26$ ✔

TRY THIS   Solve $-5.5t = 22$. Check your solution.

Before applying a property of equality to solve an equation, check to see if you can first simplify any expressions. In the following example, replacing $3 + (-5)$ with its sum, -2, gives a simpler equation to solve.

EXAMPLE 2   Solve $\frac{z}{3 + (-5)} = 12$. Check your solution.

▸ **Solution**

$$\frac{z}{3 + (-5)} = 12$$
$$\frac{z}{-2} = 12$$
$$(-2)\left(\frac{z}{-2}\right) = (-2)(12) \qquad \longleftarrow \text{Apply the Multiplication Property of Equality.}$$
$$z = -24$$

Check: $\frac{-24}{3 + (-5)} = \frac{-24}{-2} = 12$ ✔

TRY THIS   Solve $3a = -11 - 7$. Check your solution.

EXERCISES

Simplify and rewrite each equation. Do not solve.

1. $(2 + 4)x = 12$
2. $25 = \frac{4 - 3}{1 + 1}d$
3. $(1.5 + 1.5)a = 12 + 6$
4. $-5 = \frac{2 - 3}{5}d$
5. $\frac{2 + 11}{13 + 13}c = 1 - 2$
6. $\frac{2 + 11}{13 + 13}c = \frac{2.6 + 3.4}{2}$

Solve each equation. Check your solution.

7. $-6x = -42$
8. $-5a = -30$
9. $-\frac{1}{5}x = 7$
10. $-\frac{1}{7}g = -3$
11. $\frac{2}{3}y = -14$
12. $\frac{3}{4}x = 15$
13. $-2.5x = 5$
14. $\frac{d}{5 + 3} = -6$
15. $20 = -5z$
16. $-6 = \frac{d}{-6 + 3}$
17. $(-4.1 - 3.9)z = -64$
18. $\frac{d}{-6.2 - 3.8} = -2.6$
19. $(10 - 2.1 - 2.9)t = 100$
20. $-1 = \frac{(4 - 5)d}{-4.2 + 3.2}$
21. $2.9 = (100 - 28.1 - 61.9)t$
22. $\frac{d}{100 - 44.3 - 45.7} = 34.8$
23. $21 = \left(3 - \frac{1}{3} - \frac{1}{3}\right)r$
24. $\frac{d}{10 - \frac{22}{5} - \frac{33}{5}} = -2$

Is the solution positive or negative? Use mental math where possible. Do not solve.

25. $[4 + (-6)]x = -10$
26. $\frac{z}{10 - 5 + 2} = -2$
27. $(-2 - 2 - 2)r = 10$
28. $12 = [3 - 5 + (-7)]s$
29. $-2.1 = (-2 - 3 - 4)a$
30. $\frac{z}{1.2 + 1.7 + 2.2} = -2$

31. **Critical Thinking**   Use the equation $-\frac{a}{b} = \frac{-a}{b}$ to show that $-\frac{a}{b} = \frac{a}{-b}$.

    (*Hint:* Multiply each side of the equation by $\frac{-1}{-1}$, which equals 1.)

Give a counterexample to show that each statement is false. (1.3 Skill B)

32. The product of 1 and a number is 1 more than the number.

33. The quotient of a number divided by 1 equals the quotient of 1 divided by that number.

Answers for Lesson 2.5 Skill B

Try This Example 1

$t = -4$

Try This Example 2

$a = -6$

1. $6x = 12$
2. $25 = \frac{1}{2}d$
3. $3a = 18$
4. $-5 = -\frac{1}{5}d$
5. $\frac{13}{26}c = -1$, or $\frac{1}{2}c = -1$
6. $\frac{13}{26}c = 3$, or $\frac{1}{2}c = 3$
7. $x = 7$
8. $a = 6$
9. $x = -35$
10. $g = 21$
11. $y = -21$
12. $x = 20$
13. $x = -2$
14. $d = -48$
15. $z = -4$
16. $d = 18$
17. $z = 8$
18. $d = 26$
★ 19. $t = 20$
★ 20. $d = -1$
★ 21. $t = 0.29$
★ 22. $d = 348$
★ 23. $r = 9$
★ 24. $d = 2$
25. positive
26. negative
27. negative
28. negative
29. positive
30. negative
31. $-\frac{a}{b} = \frac{-a}{b} \cdot \frac{-1}{-1} = \frac{a}{-b}$
32. Sample counterexample: The product $1 \times 2 = 2$, but $2 + 1 = 3$.
33. Sample counterexample: $\frac{2}{1} = 2$, but $\frac{1}{2} \neq 2$.

★ Advanced Exercises

The Distributive Property and Combining Like Terms

> **SKILL A**   *Using the Distributive Property to combine like terms*

Focus: *Students combine like terms by connecting their skills in adding and subtracting real numbers to their knowledge of the Distributive Property.*

EXAMPLE 1

PHASE 1: Presenting the Example

Display the expression $4n + 2n$. Remind students that $4n$ means $4 \times n$ and $2n$ means $2 \times n$. Have them apply the definition of multiplication to $4n$ and $2n$. [$4n = n + n + n + n$; $2n = n + n$] Write the first two lines of the display at right. Ask students to give a simpler name for $n + n + n + n + n + n$. [$6n$] Expand the display to include $6n$ as shown.

$$\underbrace{4n}_{} \quad + \quad \underbrace{2n}_{}$$
$$\underbrace{n + n + n + n + n + n}_{6n}$$

Now display $6d + 3d$, $2r + 5r$, and $3y + y$. Have students rewrite each expression using the same method as used for $4n + 2n$. [$9d$; $7r$; $4y$]

Display the four statements shown at right. Point out that each is true based on the work students have just completed. Ask them to generalize the statements using the expression $ax + bx$. [$ax + bx = (a + b)x$] Lead students to see that this is an application of the Distributive Property.

$$4n + 2n = (4 + 2)n$$
$$6d + 3d = (6 + 3)d$$
$$2r + 5r = (2 + 5)r$$
$$3y + 1y = (3 + 1)y$$
$$\downarrow$$
$$ax + bx = (a + b)x$$

Introduce the terms *monomial, coefficient,* and *like terms,* as presented at the top of the textbook page. Then discuss Example 1.

EXAMPLE 2

PHASE 1: Presenting the Example

Display the expression $4y - 2y - 3y$. Point out that there are three terms in the expression. Tell students to work independently to combine the terms. They should arrive at $[4 + (-2) + (-3)]y$, or $-1y$. Point out that this is usually simplified further to $-y$, which is read "the opposite of y."

Now display the expression given in Example 2, which is $4y - 2 - 3y$. Ask students to explain how this expression is different from the one that they just simplified. [**There are only two like terms, $4y$ and $-3y$.**] Then discuss the process of simplification as shown on the textbook page.

EXAMPLE 3

PHASE 2: Providing Guidance for **TRY THIS**

Watch for students who give the response $5z - 12$. It is most likely that they forgot to take the opposite of each term within the parentheses. Suggest that they rewrite the subtractions as additions and organize their work as shown at right.

$$6z - \qquad (z - 5) \qquad -7$$
$$6z + (-1)[z + (-5)] - 7$$
$$6z + (-1)z + (-1)(-5) - 7$$
$$6z + \qquad \underbrace{(-1z) + 5}_{5z} \qquad \underbrace{-7}_{-2}$$

2.6 LESSON — The Distributive Property and Combining Like Terms

SKILL A *Using the Distributive Property to combine like terms*

A **monomial** is a real number or the product of a real number and a variable raised to a whole-number power. For example, 6, $-4c$, and $3x^2$ are monomials. When there is both a number and a variable in the product, the number is called the **coefficient**. In the monomials $-4c$ and $3x^2$, -4 and 3 are the coefficients.

Monomials are **like terms** if the variable parts are the same.

like terms: $3x$ and $-5x$ unlike terms: $3x$ and -5
like terms: 3 and -5 unlike terms: $3x^2$ and $-5x$

Recall that a monomial with no visible coefficient, such as x, actually has a coefficient of 1. Similarly, $-x$ has a coefficient of -1.

You can use the Distributive Property to combine like terms.

EXAMPLE 1   Simplify.   **a.** $3x + 5x$ **b.** $-4a - 6a$ **c.** $4n - n$

▸ **Solution**
a. $3x + 5x = (3 + 5)x$ **b.** $-4a - 6a = (-4 - 6)a$ **c.** $4n - n = (4 - 1)n$
$\quad = 8x$ $\quad = -10a$ $\quad = 3n$

TRY THIS   Simplify.   **a.** $-2x + 9x$ **b.** $-4.5r - 2.5r$ **c.** $11m - 10m$

EXAMPLE 2   Simplify $4y - 2 - 3y$.

▸ **Solution**
$4y - 2 - 3y = 4y + (-3y) + (-2)$ ⟵ *Apply the Commutative Property of Addition.*
$\quad = [4 + (-3)]y + (-2)$
$\quad = y - 2$

TRY THIS   Simplify $4x - 3 - 5x$.

When you have unlike terms in a sum or difference within parentheses, you may have to use the opposite of a sum or difference when simplifying.

EXAMPLE 3   Simplify $7c - (5c + 5) - 3$.

▸ **Solution**
$7c - (5c + 5) - 3 = 7c - 5c - 5 - 3$ ⟵ *Apply $-(a + b) = -a - b$.*
$\quad = (7 - 5)c - 5 - 3$
$\quad = 2c - 8$

TRY THIS   Simplify $6z - (z - 5) - 7$.

EXERCISES

KEY SKILLS

Is each set of monomials like terms?

1. $3x$ and $-4x$ **2.** $3x$ and -4 **3.** $2x^2$ and $3x^2$
4. x and $-2x$ **5.** $6, 2x$, and $3x^2$ **6.** 4 and 5

PRACTICE

Simplify each expression.

7. $-2x + 3x$ **8.** $4a + (-2a)$ **9.** $-8z + (-2z)$
10. $5x - 3x$ **11.** $3x - 5x$ **12.** $10a - 7a$
13. $7a - 10a$ **14.** $5.5d - 6.5d$ **15.** $7\frac{1}{2}t - 6\frac{1}{2}t$
16. $-3x - 1 + 5x$ **17.** $10a - 5 - 5a$ **18.** $12t + 3 - 5t$
19. $2y - 11 - 1 + 5y$ **20.** $-a - 5 - 5a - 3$ **21.** $12 + 5 - 8d + 7$
22. $(3y - 2y) - (4y + 5y)$ **23.** $(7w - 3w) - (4w + 6w)$
24. $-(3d + 7d) - (d + d)$ **25.** $10 - (3x + 5x) - (5x - 5x)$
26. $15 - (-3x + 7x) - (5x - 7x)$ **27.** $20 - (-3x - 8x) - (5x + 8x)$

Let a, b, c, and d represent real numbers. Let x be the variable.
Rewrite each expression so that it contains only one x-term.
Sample: $a(x + 3) + bx$ Solution: $(a + b)x + 3a$

28. $a(x + 3) - bx$ **29.** $a(2x + 1) - bx$
30. $a(3x - 1) + bx + 2$ **31.** $a(x - 1) + c(x + 3)$
32. $a(x - 2) + c(x + b)$ **33.** $a(x - 2) + c(x + b) + d(x + 1)$

MIXED REVIEW

Evaluate each expression. (2.3 Skill B)

34. $10 - (12 + 3)$ **35.** $-(3 + 9) - (3 + 10)$ **36.** $-(-2 - 5) + 3 - (-2)$
37. $12 - (-3 - 3) + (-9)$ **38.** $-10 - 12 - (25 - 13)$ **39.** $-(3 - 3) - (2 - 2) - (5 - 5)$

Simplify. (2.3 Skill B)

40. $6x - (x + 3x)$ **41.** $2x + 3x - (2x + 2x)$ **42.** $x + 2x + 3x - (6x - 3x)$

Answers for Lesson 2.6 Skill A

Try This Example 1
a. $7x$ **b.** $-7r$ **c.** m

Try This Example 2
$-x - 3$

Try This Example 3
$5z - 2$

1. yes
2. no
3. yes
4. yes
5. no
6. yes
7. x
8. $2a$
9. $-10z$
10. $2x$
11. $-2x$
12. $3a$

13. $-3a$
14. $-d$
15. t
16. $2x - 1$
17. $5a - 5$
18. $7t + 3$
19. $7y - 12$
20. $-6a - 8$
21. $-8d + 24$
22. $-8y$
23. $-6w$
24. $-12d$
25. $10 - 8x$
26. $15 - 2x$
27. $20 - 2x$
28. $(a - b)x + 3a$
29. $(2a - b)x + a$
★**30.** $(3a + b)x + (2 - a)$
★**31.** $(a + c)x + (3c - a)$
★**32.** $(a + c)x - (2a - bc)$
★**33.** $(a + c + d)x - (2a - bc - d)$

34. -5
35. -25
36. 12
37. 9
38. -34
39. 0
40. $2x$
41. x
42. $3x$

★ **Advanced Exercises**

SKILL B *Using the Distributive Property to multiply and divide*

Focus: *Students use the Distributive Property and real-number operations to simplify expressions of the form $a(x + b)$ and $a(x - b)$.*

EXAMPLE 1

PHASE 1: Presenting the Example

Display the expression $3(8 + 2)$. Tell students to apply the Distributive Property. [$\mathbf{3(8 + 2) = 3(8) + 3(2)}$] Point out that this result could be simplified to $24 + 6$, and then further simplified to 30, as shown at right.

$$\begin{array}{c} 3(8 + 2) \\ 3(8) + 3(2) \\ 24 \quad + \quad 6 \\ 30 \end{array}$$

Now display $3(x + 2)$. Ask students how it is different from $3(8 + 2)$. [**One of the terms inside the parentheses is a variable.**] Have them apply the Distributive Property to it. [$\mathbf{3(x) + 3(2)}$]. Note that this can be simplified to $3x + 6$. Ask why it cannot be simplified further. [**$3x$ and 6 are unlike terms.**]

$$\begin{array}{c} 3(x + 2) \\ 3(x) + 3(2) \\ 3x \quad + \quad 6 \end{array}$$

Have students simplify the second expression in Example 1, $-4(z - 7)$. Discuss their work, comparing it to the solution on the textbook page.

EXAMPLE 2

PHASE 2: Providing Guidance for TRY THIS

The most common error that students make is forgetting to distribute a *negative* multiplier to each term within the parentheses. Suggest that they rewrite all the subtractions as additions, and then organize their work as shown at right.

$$\begin{array}{ccccc}
-4(t - 7) & - & 2(3t + 9) & - & 3 \\
-4[t + (-7)] & + & (-2)(3t + 9) & & + (-3) \\
(-4)(t) + (-4)(-7) & + & (-2)(3t) + (-2)(9) & & + (-3) \\
(-4t) \quad + \quad 28 & + & (-6t) \quad + \quad (-18) & & + (-3) \\
[(-4t) + (-6t)] & + & [28 + (-18) & & + (-3)] \\
-10t & + & 7 & &
\end{array}$$

EXAMPLE 3

PHASE 1: Presenting the Example

Some students may be puzzled by the use of $-\frac{1}{10}$ in part b.

It may help to rewrite the division as shown at right.

$$\begin{aligned} \frac{10x - 25}{-10} &= (10x - 25) \div (-10) \\ &= (10x - 25) \times \left(-\frac{1}{10}\right) \end{aligned}$$

PHASE 3: ASSESSMENT AND CLOSURE for Lesson 2.6

- Ask students to compare the three expressions at right. [**sample: The first is in simplest form. The other two can be simplified using the Distributive Property; they are equivalent, respectively, to $5z + 10$ and $7z$.**]

$$\begin{array}{l} 5z + 2 \\ 5(z + 2) \\ 5z + 2z \end{array}$$

- In the lessons that follow, students will revisit equations and see how the Distributive Property can be used to simplify before solving.

☞ *For a Lesson 2.6 Quiz, see* Assessment Resources *page 17.*

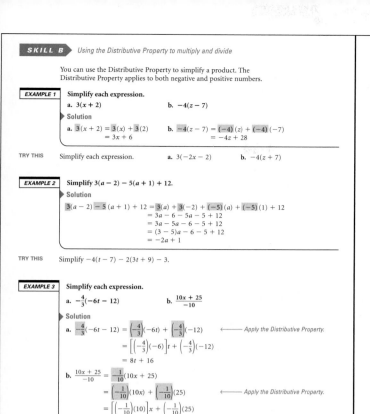

You can use the Distributive Property to simplify a product. The Distributive Property applies to both negative and positive numbers.

EXAMPLE 1 Simplify each expression.

　　a. $3(x + 2)$　　　　　b. $-4(z - 7)$

▶ Solution

a. $3(x + 2) = 3(x) + 3(2)$　　b. $-4(z - 7) = (-4)(z) + (-4)(-7)$
　　　　　　$= 3x + 6$　　　　　　　　　　　$= -4z + 28$

TRY THIS　Simplify each expression.　　a. $3(-2x - 2)$　　b. $-4(z + 7)$

EXAMPLE 2 Simplify $3(a - 2) - 5(a + 1) + 12$.

▶ Solution

$3(a - 2) - 5(a + 1) + 12 = 3(a) + 3(-2) + (-5)(a) + (-5)(1) + 12$
　　　　　　　　　　　　$= 3a - 6 - 5a - 5 + 12$
　　　　　　　　　　　　$= 3a - 5a - 6 - 5 + 12$
　　　　　　　　　　　　$= (3 - 5)a - 6 - 5 + 12$
　　　　　　　　　　　　$= -2a + 1$

TRY THIS　Simplify $-4(t - 7) - 2(3t + 9) - 3$.

EXAMPLE 3 Simplify each expression.

　　a. $-\frac{4}{3}(-6t - 12)$　　　　b. $\frac{10x + 25}{-10}$

▶ Solution

a. $-\frac{4}{3}(-6t - 12) = \left(-\frac{4}{3}\right)(-6t) + \left(-\frac{4}{3}\right)(-12)$ ⟵ Apply the Distributive Property.

　　　　$= \left[\left(-\frac{4}{3}\right)(-6)\right]t + \left(-\frac{4}{3}\right)(-12)$

　　　　$= 8t + 16$

b. $\frac{10x + 25}{-10} = -\frac{1}{10}(10x + 25)$

　　　　$= \left(-\frac{1}{10}\right)(10x) + \left(-\frac{1}{10}\right)(25)$ ⟵ Apply the Distributive Property.

　　　　$= \left[\left(-\frac{1}{10}\right)(10)\right]x + \left(-\frac{1}{10}\right)(25)$

　　　　$= -x - 2.5$

TRY THIS　Simplify each expression.　　a. $-\frac{1}{3}(6z + 9)$　　b. $\frac{10t - 5}{-5}$

EXERCISES

KEY SKILLS

In the following exercises, the first step(s) of simplifying the given expression are shown. Complete each simplification.

1. $-2(3g - 3) = (-2)(3g) + (-2)(-3)$　　　　2. $4(-3b - 5) = 4(-3b) + 4(-5)$

3. $\frac{3}{5}(10s - 20) = \frac{3}{5}(10s) + \frac{3}{5}(-20)$　　　4. $\frac{-4y + 16}{2} = \frac{1}{2}(-4y + 16) = \frac{1}{2}(-4y) + \left(\frac{1}{2}\right)(16)$

PRACTICE

Simplify each expression.

5. $-2(x + 5)$　　　　6. $-3(a + 4)$　　　　7. $7(t - 2)$

8. $-3(d - 10)$　　　9. $6(-2q - 3)$　　　10. $-2(-2d - 2)$

11. $2(a - 3)$　　　　12. $-3(4 - b)$　　　13. $2.5(4 - 2b)$

14. $2(h - 3) - 4(2h + 1) + 3$　　15. $2(h + 3) - 2(3h - 1) + 7$　　16. $-2(4h + 5) - (2h + 4) - 5$

17. $-(y - 3) - (2y + 1) + 1$　　18. $2(n + 1) + 5(3n - 3) + 1$　　19. $-9(-2m + 5) + (4m - 4) - 10$

20. $\frac{1}{3}(12r - 12)$　　　21. $-\frac{3}{4}(12s - 60)$　　　22. $-\frac{5}{6}(-9t + 18)$

23. $\frac{1}{2}(10g - 4)$　　　24. $\frac{3}{4}(12z - 4)$　　　25. $-\frac{1}{2}(2a + 4)$

26. $\frac{6x + 12}{3}$　　　27. $\frac{15z - 40}{-5}$　　　28. $\frac{-12x - 15}{-12}$

29. $3(2z - 4) + \frac{z}{2}$　　30. $2(z + 5) - \frac{z}{2}$　　31. $\frac{3}{2}a + 2(a - 5)$

32. $\frac{2}{5}b - 2(b + 5)$　　33. $3(c + 5) - 2(c - 3)$　　34. $-3(p + 5) + 2(p - 5)$

35. $-4(u + 1) + 2(u - 2) - \frac{u}{3}$　36. $\frac{2t}{3} - 2(t - 1) + 5(t - 2)$　37. $\frac{2k}{5} + \frac{3}{5}(10k - 15) + 5(k - 2)$

MIXED REVIEW

Evaluate each expression for the given value of the variable. (1.2 Skill C)

38. $6x + 5; x = \frac{2}{3}$　　　39. $14a + 7; a = \frac{3}{7}$　　　40. $45z + 20; z = \frac{3}{5}$

Solve each equation. (2.5 Skills A and B)

41. $x - 6.5 = -11$　　42. $y + 0.5 = -1$　　43. $a - 1\frac{1}{2} = 0$　　44. $b + 1\frac{1}{2} = -3\frac{2}{3}$

45. $-3x = 120$　　46. $\frac{-4b}{3} = 24$　　47. $-2.5x = -10$　　48. $\frac{4m}{-5} = -4$

Answers for Lesson 2.6 Skill B

Try This Example 1
a. $-6x - 6$　　b. $-4z - 28$

Try This Example 2
$-10t + 7$

Try This Example 3
a. $-2z - 3$　　b. $-2t + 1$

1. $-6g + 6$
2. $-12b - 20$
3. $6s - 12$
4. $-2y + 8$
5. $-2x - 10$
6. $-3a - 12$
7. $7t - 14$
8. $-3d + 30$
9. $-12q - 18$
10. $4d + 4$
11. $2a - 6$
12. $-12 + 3b$
13. $10 - 5b$
14. $-6h - 7$
15. $-4h + 15$
16. $-10h - 19$

17. $-3y + 3$
18. $17n - 12$
19. $22m - 59$
20. $4r - 4$
21. $-9s + 45$
22. $\frac{15}{2}t - 15$
23. $5g - 2$
24. $9z - 3$
25. $-a - 2$
26. $2x + 4$
27. $-3z + 8$
28. $x + \frac{5}{4}$
★ 29. $\frac{13}{2}z - 12$
★ 30. $\frac{3}{2}z + 10$
★ 31. $\frac{7}{2}a - 10$
★ 32. $-\frac{8}{5}b - 10$
★ 33. $c + 21$
★ 34. $-p - 25$

★ 35. $-\frac{7}{3}u - 8$

★ 36. $\frac{11}{3}t - 8$

★ 37. $\frac{57}{5}k - 19$

38. 9
39. 13
40. 47
41. $x = -4.5$
42. $y = -1.5$
43. $a = \frac{3}{2}$, or $1\frac{1}{2}$

44. $b = -\frac{31}{6}$, or $-5\frac{1}{6}$

45. $x = -40$
46. $b = -18$
47. $x = 4$
48. $m = 5$

★ Advanced Exercises

Solving Two-Step Equations

SKILL A ▶ *Solving an equation in two steps*

Focus: *Students solve equations in one variable by applying two properties of equality.*

EXAMPLE 1

PHASE 1: Presenting the Example

Display $\square + 5 = 1$. Tell students to assume that the box is holding the place of a variable. Have them write a solution to the equation. [**See the solution at right.**]

$$\square + 5 = 1$$
$$\square + 5 - 5 = 1 - 5$$
$$\square = -4$$

Ask students what the conclusion would be if the variable in the box were c. [$c = -4$] What if it were m? [$m = -4$] What if it were r? [$r = -4$]

$$\boxed{-3x} + 5 = 1$$
$$\boxed{-3x} + 5 - 5 = 1 - 5$$
$$\boxed{-3x} = -4$$

Tell students that they are now going to investigate a slightly different situation. Have them go back to their solutions and write "$-3x$" inside each box, as shown at right. Ask if they have identified a solution to the equation $-3x + 5 = 1$. Lead them to see that they have not, because another step is required. Specifically, they must now divide each side by -3. Tell them to complete the solution. They should arrive at $x = \frac{4}{3}$.

Note that some students may visualize the two key steps more easily if you use the vertical format shown at right.

Apply the Addition Property of Equality.

$$\begin{array}{rcr} -3x + 5 = & 1 \\ -5 = & -5 \\ \hline -3x = & -4 \end{array}$$

Apply the Division Property of Equality.

$$\frac{{}^{1}\cancel{-3}x}{\cancel{-3}_{1}} = \frac{-4}{-3}$$

PHASE 2: Providing Guidance for **TRY THIS**

Watch for students who first divide each side of the equation by 4. Be sure that they divide each term, not just $4x$, by 4. The resulting equation should be $\frac{4x}{4} - \frac{5}{4} = \frac{10}{4}$, or $x - \frac{5}{4} = -\frac{10}{4}$. This approach is valid, but many students will find it more difficult than first adding 5 to each side.

EXAMPLE 2

PHASE 1: Presenting the Example

Display the equation $5x = 7x + 6$. Tell half the class to subtract 6 from each side, and tell the other half to subtract $7x$ from each side. Discuss their results, displaying both for all to see. [**See the results at right.**]

$$5x = 7x + 6$$
$$5x - 6 = 7x + 6 - 6$$
$$5x - 6 = 7x$$

Point out that the goal is to solve the equation. Ask students which action seems to be a better first step and why. They should observe that, after subtracting $7x$, only one term remains on each side. Solving the resulting equation requires just one more step, which is dividing each side by -2. Then complete the solution as given on the textbook page.

$$5x = 7x + 6$$
$$5x - 7x = 7x + 6 - 7x$$
$$-2x = 6$$

2.7 LESSON
Solving Two-Step Equations

Solving an equation in two steps

Often you will need to use two properties of equality to solve a single equation.

EXAMPLE 1 Solve $-3x + 5 = 1$. Check your solution.

▶ **Solution**

$$-3x + 5 = 1$$
$$-3x + 5 - 5 = 1 - 5 \longleftarrow \text{Apply the Subtraction Property of Equality.}$$
$$-3x = -4$$
$$\frac{-3x}{-3} = \frac{-4}{-3} \longleftarrow \text{Apply the Division Property of Equality.}$$
$$x = \frac{4}{3}, \text{ or } 1\frac{1}{3}$$

Check: Use the fraction form rather than the mixed number.

$$-3\left(\frac{4}{3}\right) + 5 \stackrel{?}{=} 1$$
$$-4 + 5 \stackrel{?}{=} 1 ✔$$

TRY THIS   Solve $4x - 5 = -10$. Check your solution.

In Example 1, the equations $-3x + 5 = 1$, $-3x = -4$, and $x = \frac{4}{3}$ are *equivalent equations.* They are called **equivalent** because they all have the same solution.

EXAMPLE 2 Solve $5x = 7x + 6$. Check your solution.

▶ **Solution**

$$5x = 7x + 6$$
$$5x - 7x = 7x + 6 - 7x \longleftarrow \text{Apply the Subtraction Property of Equality.}$$
$$(5 - 7)x = 6 \longleftarrow \text{Apply the Distributive Property.}$$
$$-2x = 6$$
$$\frac{-2x}{-2} = \frac{6}{-2} \longleftarrow \text{Apply the Division Property of Equality.}$$
$$x = -3$$

Check: Substitute -3 for x in each place where x occurs.

$$5(-3) \stackrel{?}{=} 7(-3) + 6$$
$$-15 \stackrel{?}{=} -15 ✔$$

TRY THIS   Solve $-6x = -8x + 1$. Check your solution.

EXERCISES

KEY SKILLS

The equations in each exercise are equivalent. Which property of equality can be used to show the equivalence?

1. $9x = 3x + 1$ and $6x = 1$

2. $3x = 18$ and $x = 6$

3. $6x + 10 = 7x - 3$ and $6x + 13 = 7x$

4. $6x + 10 = 7x - 3$ and $10 = x - 3$

PRACTICE

Solve each equation. Check your solution.

5. $2x + 5 = 9$
6. $3m - 4 = 15$
7. $2 = 5x - 13$
8. $2 = 4n - 14$
9. $-2v + 5 = 9$
10. $-3x + 5 = 2$
11. $2 = -5f - 18$
12. $-8 = -2z + 4$
13. $2p + 5 = 3p - 1$
14. $7a = -7a - 4$
15. $-3b + 5 = 3b$
16. $2x = -7x - 5$
17. $-3t = -2t + 1$
18. $-2w = -6w + 1$
19. $2 = -3d - 6$
20. $10x = -9x - 38$
21. $\frac{2x}{3} - 3 = \frac{x}{3}$
22. $\frac{-2c}{5} + 4 = \frac{c}{5}$
23. $10.2x = -9.8x - 40$
24. $-2.7h = 7.3h - 13$
25. $-5.4j = -4.6j - 15$
26. $3 - \frac{1}{2}w = 12$
27. $7 - \frac{3}{4}q = 14$
28. $-11 = 7 - \frac{3}{5}y$
29. $-32 = -5 - \frac{7}{2}n$
30. $1 - \frac{3}{2}r = 1$
31. $-6 - \frac{3}{2}d = -6$

32. **Critical Thinking** Let a, b, and c represent real numbers, where $a \neq 0$. Solve $ax + b = c$ for x.

33. **Critical Thinking** Let a, b, and c represent real numbers, where $a \neq 0$. Solve $a(x + b) = c$ for x.

MIXED REVIEW

Simplify each expression. (2.6 Skill A)

34. $-2m + 6m$
35. $-n + 7n$
36. $-3p - 7p$
37. $-5g + 6g$
38. $-2.9h - 7.1h$
39. $-5.4k - 14.6k$

Simplify each expression. (2.6 Skill B)

40. $2(x - 3) - (x - 5)$
41. $-2(d + 5) - 3(d - 5)$
42. $-2(3n - 5) + 3(2n + 5)$
43. $-2(k + 1) - (k - 6)$
44. $-(p + 1.5) + 3(p - 1)$
45. $7(2u + 3) - 3(2u + 3)$

Answers for Lesson 2.7 Skill A

Try This Example 1

$x = -\frac{5}{4}$

Try This Example 2

$x = \frac{1}{2}$

1. Subtraction Property of Equality
2. Division Property of Equality
3. Addition Property of Equality
4. Subtraction Property of Equality
5. $x = 2$
6. $m = \frac{19}{3}$, or $6\frac{1}{3}$
7. $x = 3$
8. $n = 4$
9. $v = -2$
10. $x = 1$
11. $f = -4$

12. $z = 6$
13. $p = 6$
14. $a = -\frac{2}{7}$
15. $b = \frac{5}{6}$
16. $x = -\frac{5}{9}$
17. $t = -1$
18. $w = \frac{1}{4}$
19. $d = -\frac{8}{3}$, or $-2\frac{2}{3}$
20. $x = -2$
★21. $x = 9$
★22. $c = \frac{20}{3}$, or $6\frac{2}{3}$
23. $x = -2$
24. $h = 1.3$
25. $j = 18.75$
★26. $w = -18$
★27. $q = -\frac{28}{3}$, or $-9\frac{1}{3}$
★28. $y = 30$

★29. $n = \frac{54}{7}$, or $7\frac{5}{7}$
★30. $r = 0$
★31. $d = 0$
32. $x = \frac{c - b}{a}$
33. $x = \frac{c}{a} - b$
34. $4m$
35. $6n$
36. $-10p$
37. g
38. $-10h$
39. $-20k$
40. $x - 1$
41. $-5d + 5$
42. 25
43. $-3k + 4$
44. $2p - 4.5$
45. $8u + 12$

★ **Advanced Exercises**

SKILL B APPLICATIONS Solving a literal equation for a specified variable

Focus: *Given a literal equation, students apply the properties of equality to solve for one of the variables in terms of the other(s).*

EXAMPLE 1

PHASE 1: Presenting the Example

Display the equations $x + 5 = 9$, $x + 12 = 9$, and $x + 8.3 = 9$. Tell students to find a solution to each one. Discuss the results. [$\mathbf{x = 4; x = -3; x = 0.7}$]

Now display the first three statements shown at right. Point out to students that each is true based on the work they have just completed. Ask them to generalize the statements using the expression $x + a = 9$. [$\mathbf{x + a = 9}$ **is equivalent to** $\mathbf{x = 9 - a}$.]

$x + 5 = 9$	is equivalent to	$x = 9 - 5$.
$x + 12 = 9$	is equivalent to	$x = 9 - 12$.
$x + 8.3 = 9$	is equivalent to	$x = 9 - 8.3$.
	↓	
$x + a = 9$	is equivalent to	$x = 9 - a$.

Tell students that an equation involving two or more variables is called a *literal equation.* So $x + a = 9$ is an example of a literal equation. When you transform it to the equivalent equation $x = 9 - a$, you *solve for x in terms of a.* This means that the variable x is isolated on one side of the equal sign.

Point out that a literal equation can be solved for any one of its variables. Refer students again to $x + a = 9$. Tell them to solve for a in terms of x. [$\mathbf{a = 9 - x}$]

Tell students that solving for a specified variable sometimes requires taking two steps. Then discuss Example 1.

PHASE 2: Providing Guidance for TRY THIS

The most direct way to solve for C is to first subtract 32 from each side, and then multiply each side by $\frac{5}{9}$. The result is $C = \frac{5}{9}(F - 32)$. However, some students may choose to first multiply by $\frac{5}{9}$. Although this approach is valid, it is a bit more complicated. The steps of the solution are shown at right. Note that the resulting equation, $C = \frac{5}{9}F - \frac{160}{9}$, is equivalent to $C = \frac{5}{9}(F - 32)$.

$$F = \frac{9}{5}C + 32$$
$$\frac{5}{9}F = \frac{5}{9}\left(\frac{9}{5}C + 32\right)$$
$$\frac{5}{9}F = \frac{5}{9}\left(\frac{9}{5}C\right) + \frac{5}{9}(32)$$
$$\frac{5}{9}F = C + \frac{160}{9}$$
$$\frac{5}{9}F - \frac{160}{9} = C$$

EXAMPLE 2

PHASE 1: Presenting the Example

Display the equation $x + ax = 9$. Tell students to subtract ax from each side and write the result. [$\mathbf{x = 9 - ax}$] Ask if the equation has been solved for x. Lead students to see that it has not, because x is not isolated on one side.

Return to $x + ax = 9$. Remind students that they can apply the Distributive Property to the left side. That is, if $x + ax$ is considered as $1x + ax$, it can be rewritten as $(1 + a)x$. Then lead students through the steps at right, resulting in the equivalent equation $x = \frac{9}{1 + a}$. Note that x is now isolated on the left side, so the equation has been solved for x. Then discuss Example 2.

$$x + ax = 9$$
$$1x + ax = 9$$
$$(1 + a)x = 9$$
$$\frac{(1 + a)x}{(1 + a)} = \frac{9}{(1 + a)}$$
$$x = \frac{9}{(1 + a)}$$

Recall that a formula expresses a relationship between two or more variables in a mathematical or physical application.

When you solve a formula for one of the variables, the result will not be a number. Instead, you will have an algebraic expression that contains the other variable(s). This is called *solving for one variable in terms of another.*

EXAMPLE 1

The formula $C = \frac{5}{9}(F - 32)$ enables you to find a temperature, C, on the Celsius scale, given a temperature, F, on the Fahrenheit scale. Solve the formula for F in terms of C.

▶ **Solution**

$$C = \frac{5}{9}(F - 32)$$

$$\boxed{\frac{9}{5}} \cdot C = \boxed{\frac{9}{5}} \cdot \frac{5}{9}(F - 32) \longleftarrow \text{Apply the Multiplication Property of Equality.}$$

$$\frac{9}{5}C = F - 32$$

$$\frac{9}{5}C + 32 = F \longleftarrow \text{Apply the Addition Property of Equality.}$$

The formula is $F = \frac{9}{5}C + 32$.

TRY THIS Solve $F = \frac{9}{5}C + 32$ for C in terms of F.

If you deposit P dollars into an account paying simple interest, then the amount, A, that you will have after one year is given by $A = P + Pr$, where r is the annual interest rate as a decimal.

EXAMPLE 2

Solve $A = P + Pr$ for P.

▶ **Solution**

$$A = P + Pr$$

$$A = P(1 + r) \longleftarrow \text{Apply the Distributive Property.}$$

$$\frac{A}{1 + r} = \frac{P(1 + r)}{1 + r} \longleftarrow \text{Divide each side by } 1 + r.$$

$$\frac{A}{1 + r} = P$$

Thus, $P = \frac{A}{1 + r}$.

TRY THIS Solve $A = P + Pr$ for r.

EXERCISES

Identify the properties of equality that you would use to solve each equation for the variable in parentheses.

1. $(x)y = 4$

2. $\frac{2(a)}{5} = d$

3. $A = \frac{1}{2}b(h)$

4. $P = 2l + 2(w)$

PRACTICE

Solve each literal equation for the specified variable.

5. $x + a = b; a$

6. $x - r = c; x$

7. $rt = d; r\,(t \neq 0)$

8. $\frac{x}{r} = v; x$

9. $2x + a = 4; x$

10. $2a - 3x = 4; a$

11. $y = mx + b; x\,(m \neq 0)$

12. $y = mx + b; m\,(x \neq 0)$

13. $2x + 3y = 12; y$

14. $2x + 3y = 12; x$

15. $A = P + Prt; t\,(Pr \neq 0)$

16. $w = P + P(1 + c); P\,(c \neq -2)$

17. $2f = 3f + ag; g\,(a \neq 0)$

18. $4z = 2z + bd; z$

19. $2f = 3f + ag; f$

20. $4z = 2z + bd; d\,(b \neq 0)$

21. Twice the sum of x and y equals 12.
   a. Write an equation to represent this statement. **b.** Solve for x. **c.** Solve for y.

22. **Critical Thinking** Let a, b, and c be real numbers whose sum is nonzero. Solve $ax + bx + cx = 1$ for x.

MIXED REVIEW APPLICATIONS

Find the net change in each bank balance. (2.3 Skill C)

23. deposits: $210.30, $100.56, $17.95, and $34.22
   withdrawals: $43.50, $250.56, and $145.00

24. deposits: $1000.50, $100.50, $187.95, and $314.20
   withdrawals: $247.50, $274.56, and $305.00

25. The Lincoln High football team completes 2 series of downs with the gains and losses in yards shown below. Find the net gain (or loss) after these 2 series.

Series 1	Down	1	2	3	Series 2	Down	1	2	3
	Gain/Loss	8	-4	15		Gain/Loss	-3	-5	4

Answers for Lesson 2.7 Skill B

Try This Example 1

$C = \frac{5}{9}(F - 32)$

Try This Example 2

$r = \frac{A}{P} - 1$

1. Division Property of Equality

2. Multiplication and Division Properties of Equality

3. Division Property of Equality

4. Subtraction and Division Properties of Equality

5. $a = b - x$

6. $x = r + c$

7. $r = \frac{d}{t}$

8. $x = rv$

9. $x = \frac{4 - a}{2}$

10. $a = \frac{3x + 4}{2}$

11. $x = \frac{y - b}{m}$

12. $m = \frac{y - b}{x}$

13. $y = \frac{12 - 2x}{3}$

14. $x = \frac{12 - 3y}{2}$

15. $t = \frac{A - P}{Pr}$

★16. $P = \frac{w}{2 + c}$

17. $g = \frac{-f}{a}$

18. $z = \frac{bd}{2}$

19. $f = -ag$

20. $d = \frac{2z}{b}$

★21. **a.** $2(x + y) = 12$
   b. $x = 6 - y$
   c. $y = 6 - x$

22. $x = \frac{1}{a + b + c}$

23. decrease of $76.03

24. increase of $776.09

25. gain of 15 yards

★ **Advanced Exercises**

Focus: *Students develop a problem-solving plan and use that plan to write and solve an equation that models a real-world problem.*

EXAMPLE

PHASE 1: Presenting the Example

Ask students what measures can be used to indicate the size of a rectangle. Lead them to identify length (ℓ), width (w), perimeter (P), and area (A). Ask what formulas they know that relate these measures. [$P = 2\ell + 2w$; $A = \ell w$]

Have students read the problem in the Example. Point out that the garden is a rectangle. Ask them whether this is a problem about the perimeter or the area of the rectangle. [**perimeter**]

Now display the table at right and have students copy it. Be sure they understand that ℓ represents the length of a rectangle and w represents the width. Ask what information in the problem will help them find the value of w that corresponds to a given value of ℓ. [**A condition of the problem is that the length is twice the width.**] Then have them complete the table. [**Answers are shown in red at right.**]

ℓ	w	$2\ell + 2w$
10	5	30
20	10	60
30	15	90
40	20	120
50	25	150
60	30	180
70	35	210
80	40	240
90	45	270
100	50	300
110	55	330
120	60	360

Ask students to find the solution to the problem among the entries in the table. Lead them to the conclusion that the desired perimeter of 120 feet is obtained when the length is 40 feet and the width is 20 feet.

Now discuss the solution on the textbook page. After you have completed the presentation, ask students which solution method—table or equation—is more efficient. Point out that the equation method leads you directly to a solution, making equations a powerful problem-solving tool.

PHASE 2: Providing Guidance for TRY THIS

After students have completed the Try This exercise, have them write a literal equation for the perimeter, P, of any rectangle whose width is w feet and whose length is n times its width. [$P = 2w + 2nw$, or $P = 2w(1 + n)$]

PHASE 3: ASSESSMENT AND CLOSURE for Lesson 2.7

- Give students the equations at right. Ask them to compare the process of solving **1** to the process of solving **2**. [**Both begin with adding 7 to each side, but 2 requires the second step of dividing by 5.**] Now ask them to compare solving **2** for r to solving **3** for r. [**Both require two steps: adding, followed by dividing. In 2, there is a numerical solution. In 3, the solution is an algebraic expression.**]

- In this lesson, students advanced from solving one-step equations to solving two-step equations. In the lesson that follows, they will encounter equations for which the solution requires three or more steps.

Solve for r.

1. $r - 7 = 10$
2. $5r - 7 = 10$
3. $ar - b = 10$

For a *Lesson 2.7 Quiz*, see Assessment Resources *page 18*.

You can use a problem-solving plan to answer real-world questions.

A Problem-Solving Plan
1. **Understand the problem.** Read the problem carefully, perhaps more than once. Identify what information is given and what you are asked to find. Use a variable to represent any unknown quantity.
2. **Choose a strategy.** Make tables, diagrams, graphs, or other visual aids to help relate the known and the unknown quantities. Then express the relationships in an equation.
3. **Solve the problem.** Solve the equation using the properties of numbers and equality as you have learned. Express the answer using appropriate units.
4. **Check the results.** Check your calculations by substituting the solution in place of the variable in the equation. If the equation checks, then relate the solution to the statement of the original problem. If the solution is not reasonable, then repeat the four steps of the plan, looking for an error or faulty reasoning.

EXAMPLE

Jackie and Kevin want to design and build a rectangular garden. They have 120 feet of fencing for the border, and they want their garden's length to be twice its width. What will be the dimensions of their garden?

▶ **Solution**

1. Let w represent the width. Because the length is twice the width, let $2w$ represent the length.

Make a diagram.

2. Draw a diagram to represent the rectangular garden.

length, $2w$
border = 2 (width) + 2 (length)
= 2w + 2(2w)

Write an equation.

3. Write and solve an equation.
$$2w + 2(2w) = 120$$
$$2w + 4w = 120$$
$$6w = 120 \quad \longleftarrow \text{Combine like terms.}$$
$$w = 20 \quad \longleftarrow \text{Apply the Division Property of Equality.}$$

4. **Check:** $2(20) + 2[2(20)] = 40 + 80 = 120$ ✔

The solution is 20. So, the width of the garden will be 20 feet and the length will be twice the width, or 40 feet.

TRY THIS   Suppose that Jackie and Kevin want to use 90 feet of fencing and want the length to be 1.5 times the width. What will be the dimensions of their garden?

EXERCISES

KEY SKILLS

For each situation, choose a variable for the unknown quantity and write an equation. Do not solve.

1. Rochelle's age is twice Lin's age. If the sum of their ages is 27, how old is Lin?

2. A whole number plus one half itself less 2 equals 94. What is it?

PRACTICE

Solve each problem. Check your solution.

3. Ten ounces of water are removed from a container and two-thirds of the original amount remains. How much water did the container have originally?

4. Mickey has 160 yards of fencing to enclose a rectangular garden whose length is to be 4 times its width. What will be the dimensions of the garden?

5. A taxicab ride costs $1.10 plus $0.95 per mile. If the total cost of the trip is $12.50, how many miles did the passenger travel?

6. Karen has $120. If she saves $24 per week, how many weeks will it take her to have a total of $300?

7. In a *right triangle*, the sum of the two non-right angles is 90°. If the measure of one non-right angle is twice that of the other, what are the measures of the angles?

8. Maria's salary is three times Eileen's salary. They decide to share the cost of an $80 gift. Maria will contribute three times as much as Eileen. How much does each contribute?

9. **Critical Thinking** Jamie has a total of 40 coins that are all nickels and dimes. She has 12 more nickels than dimes. How many of each does she have?

10. **Critical Thinking** Show that two integers, one of which is three times the other, cannot add up to 13.

MIXED REVIEW APPLICATIONS

Solve the problem and answer the question. (2.3 Skill C)

11. The temperature was 15°F at 6:00 P.M. It dropped 4°F by midnight and then dropped another 6°F by 3:00 A.M. It then rose 12°F by 9:00 A.M. What was the temperature at 9:00 A.M.?

Answers for Lesson 2.7 Skill C

Try This Example
18 feet by 27 feet

1. Let ℓ represent Lin's age; $\ell + 2\ell = 27$, or $3\ell = 27$

2. Let n represent the whole number; $n + \frac{1}{2}n - 2 = 94$, or $\frac{3}{2}n - 2 = 94$

3. 30 ounces

4. 16 yards by 64 yards

5. 12 miles

6. 8 weeks

7. 30° and 60°

8. Maria contributes $60 and Eileen contributes $20.

9. 26 nickels and 14 dimes

10. Let x represent one integer. Then $3x$ will represent the other. The solution to $x + 3x = 13$ is $\frac{13}{4}$, or $3\frac{1}{4}$, which is not an integer.

11. 17°F

Solving Multistep Equations

SKILL A ▶ *Solving an equation in multiple steps*

Focus: *Students apply the properties of equality three times to solve equations in one variable.*

EXAMPLE 1

PHASE 1: Presenting the Example

Display $5x - 17 = 3x - 10$. Ask students how this equation is different from those in Lesson 2.7. Allow sufficient time for each student to jot down an answer. Then discuss their perceptions. Students should recognize that in this equation, there are variables and numbers on both sides of the equal sign.

For some students, it may be easier to visualize the three steps of the solution if you use the format shown at right.

Add 17 to each side.

$$\begin{array}{r} 5x - 17 = 3x - 10 \\ +\ 17 = \quad +\ 17 \\ \hline 5x \quad = 3x + 7 \end{array}$$

Subtract $3x$ from each side.

$$\begin{array}{r} 5x \quad = \quad 3x + 7 \\ -3x \quad = -3x \\ \hline 2x \quad = \quad 7 \end{array}$$

Divide each side by 2.

$$\frac{{}^{1}\cancel{2}x}{\cancel{2}_{1}} = \frac{7}{2}$$

PHASE 2: Providing Guidance for TRY THIS

Have students work in groups of four. Assign one group member to begin the solution by adding -8 to each side of the equation. Assign a second member to begin the solution by adding 10 to each side, the third by adding $2c$, and the fourth by subtracting $5c$. Each should complete the solution individually. Then the group can discuss whether one "first step" is more efficient than the others.

EXAMPLE 2

PHASE 1: Presenting the Example

Display $\frac{1}{4}d + 3 = \frac{1}{3}d + 1$. Ask students to multiply each side of the equation by 2 and simplify. Repeat the activity with 3, 4, 6, and 12 as multipliers. [**Answers are shown at right.**] Be sure that students distribute correctly.

Discuss the results. Point out that the last equation does not contain fractions, so it is probably the easiest to solve. Lead students to see that 12, the multiplier that "clears the fractions," is the LCM of the denominators 4 and 3.

$\times 2:\quad \frac{1}{2}d + 6 = \frac{2}{3}d + 2$

$\times 3:\quad \frac{3}{4}d + 9 = d + 3$

$\times 4:\quad d + 12 = \frac{4}{3}d + 4$

$\times 6:\quad \frac{3}{2}d + 18 = 2d + 6$

$\times 12:\ 3d + 36 = 4d + 12$

PHASE 2: Providing Guidance for TRY THIS

Watch for students who multiply only the r-terms by 6, obtaining $3r - 1 = 2r + 2$. Encourage them to enclose each side of the equation in parentheses and indicate the distribution with arrows, as shown at right.

$$6\left(\frac{1}{2}r - 1\right) = 6\left(\frac{1}{3}r + 2\right)$$

2.8 LESSON — Solving Multistep Equations

SKILL A — *Solving an equation in multiple steps*

In many situations, the solution process has more than two steps.

EXAMPLE 1 Solve $5x - 17 = 3x - 10$. Check your solution.

▸ **Solution**

$$5x - 17 = 3x - 10$$
$$5x - 17 \;\boxed{+\,17} = 3x - 10 \;\boxed{+\,17} \quad \longleftarrow \text{Apply the Addition Property of Equality.}$$
$$5x = 3x + 7$$
$$5x \;\boxed{-\,3x} = 3x + 7 \;\boxed{-\,3x} \quad \longleftarrow \text{Apply the Subtraction Property of Equality.}$$
$$2x = 7$$
$$x = \frac{7}{2}, \text{ or } 3.5 \quad \longleftarrow \text{Apply the Division Property of Equality.}$$

Check: $5(3.5) - 17 \overset{?}{=} 3(3.5) - 10 \;\rightarrow\; 0.5 = 0.5$ ✔

TRY THIS Solve $-2c + 8 = 5c - 10$. Check your solution.

Recall that the **least common multiple** (LCM) of two natural numbers is the smallest number divisible by the two numbers. You can eliminate fractions in an equation by multiplying both sides of the equation by the LCM of all the denominators (using the Multiplication Property of Equality). However, do *not* use this method to simplify an expression that is not part of an equation.

EXAMPLE 2 Solve $\frac{1}{4}d + 3 = \frac{1}{3}d + 1$. Check your solution.

▸ **Solution**

$$\frac{1}{4}d + 3 = \frac{1}{3}d + 1$$
$$\boxed{12}\left(\frac{1}{4}d + 3\right) = \boxed{12}\left(\frac{1}{3}d + 1\right) \quad \longleftarrow \begin{array}{l}\text{Apply the Multiplication Property of Equality.}\\ \text{The LCM of 3 and 4 is 12.}\end{array}$$
$$3d + 36 = 4d + 12$$
$$3d + 36 \;\boxed{-\,3d} = 4d + 12 \;\boxed{-\,3d} \quad \longleftarrow \text{Apply the Subtraction Property of Equality.}$$
$$36 = 12 + d$$
$$36 \;\boxed{-\,12} = 12 + d \;\boxed{-\,12} \quad \longleftarrow \text{Apply the Subtraction Property of Equality.}$$
$$24 = d$$
$$d = 24 \quad \longleftarrow \text{Apply the Symmetric Property of Equality.}$$

Check: $\frac{1}{4}(24) + 3 \overset{?}{=} \frac{1}{3}(24) + 1 \;\rightarrow\; 9 = 9$ ✔

TRY THIS Solve $\frac{1}{2}r - 1 = \frac{1}{3}r + 2$. Check your solution.

EXERCISES

KEY SKILLS

Write the equation that results from the specified step.

1. $-3x + 5 = 2x + 7$; add -5 to each side of the equation.

2. $5x - 6 = 7x + 7$; add $-5x$ to each side of the equation.

3. $\frac{5}{8}t - 5 = \frac{2}{3}t$; multiply each side of the equation by 24.

4. $0.3x - 1 = 0.5x$; multiply each side of the equation by 10.

PRACTICE

Solve each equation. Check your solution.

5. $x + 3 = 2x + 5$ 6. $3c - 5 = 2c + 5$ 7. $10t - 6 = -2t - 6$

8. $7x - 9 = 3x + 19$ 9. $6 + 10t = 8t + 12$ 10. $3x + 7 = 16 + 6x$

11. $18 + 3y = 5y - 4$ 12. $11a + 8 = -2 + 9a$ 13. $9x - 5 = 6x + 13$

14. $6x - 5 = 2x - 21$ 15. $8 - x = 5x - 4$ 16. $-17 - 2x = 6 - x$

17. $5y - 0.3 = 4y + 0.6$ 18. $10z + 1.3 = 5z - 1.6$ 19. $z - 2.7 = -1.5z + 1$

20. $-0.6d + 1.1 = d + 1$ 21. $\frac{3}{4}a + 1 = \frac{5}{4}a + 2$ 22. $3 + \frac{1}{3}d = 5 + \frac{7}{3}d$

23. $\frac{1}{2}z - 1 = \frac{5}{3}z$ 24. $\frac{3}{7}x + 3 = \frac{2}{7}x$ 25. $\frac{1}{3}x + \frac{2}{3} = \frac{5}{3}x$

26. $\frac{3}{7}x + \frac{4}{7} = \frac{2}{3}x$ 27. $\frac{10}{7}x = \frac{3}{4}x + \frac{1}{2}$ 28. $\frac{5}{16}x = \frac{1}{5}x + \frac{3}{40}$

29. **Critical Thinking** Solve for x in terms of a, b, and c: $\frac{1}{a}x + \frac{1}{b} = \frac{1}{c}x$

MIXED REVIEW

Use the order of operations to simplify each expression. (1.2 Skill B)

30. $\frac{3(5 - 2)}{9}$ 31. $\frac{1}{5}(2 - 1) + \frac{4}{5}$ 32. $\frac{4}{7}(12 - 5) - 4$

Evaluate each expression. (1.2 Skill C)

33. $-3(x + 5)$ given $x = -2$ 34. $4.5(t - 2)$ given $t = 0$ 35. $4.5\left(\frac{f + 1}{2}\right)$ given $f = 3$

Find the values in the replacement set $\{-3, -2, -1, 0, 1, 2, 3\}$ that are solutions to each equation. If the equation has no solution, write *none*. (1.2 Skill C)

36. $-2x + 3 = -1$ 37. $x - 1 = -8$ 38. $2(x + 3) = 12$

Answers for Lesson 2.8 Skill A

Try This Example 1
$c = \frac{18}{7}$, or $2\frac{4}{7}$

Try This Example 2
$r = 18$

1. $-3x = 2x + 2$
2. $-6 = 2x + 7$
3. $15t - 120 = 16t$
4. $3x - 10 = 5x$
5. $x = -2$
6. $c = 10$
7. $t = 0$
8. $x = 7$
9. $t = 3$
10. $x = -3$
11. $y = 11$
12. $a = -5$
13. $x = 6$
14. $x = -4$
15. $x = 2$

16. $x = -23$
17. $y = 0.9$
18. $z = -0.58$
19. $z = 1.48$
20. $d = -0.25$
21. $a = -2$
22. $d = -1$
23. $z = -\frac{3}{4}$
24. $x = -21$
25. $x = \frac{1}{2}$
★ 26. $x = \frac{12}{5}$, or $2\frac{2}{5}$
★ 27. $x = \frac{14}{19}$
★ 28. $x = \frac{2}{3}$
29. $x = \frac{-ac}{bc - ab}$, or $\frac{ac}{ab - bc}$

30. 1
31. 1
32. 0
33. 11
34. -2
35. 9
36. 2
37. none
38. 3

★ **Advanced Exercises**

Simplifying before solving multistep equations

Focus: *Students use the Distributive Property to simplify an equation before solving.*

EXAMPLE 1

PHASE 1: Presenting the Example

Display equations **1** and **2** given at right. Point out that $5n + 4 = 35$ is a familiar two-step equation. Have students solve it. [$n = 6.2$] Ask what must be done to $5(n + 4) = 35$ to make it a similar two-step equation. [**Simplify $5(n + 4)$ by applying the Distributive Property.**] Then discuss the textbook solution.

1. $5n + 4 = 35$

2. $5(n + 4) = 35$

PHASE 2: Providing Guidance for **TRY THIS**

Watch for students who simplify the left side as $-3a - 12$. Suggest that they first rewrite the given subtraction as addition, as shown at right.

$$-3(a - \quad 4) = 27$$
$$\downarrow \qquad \downarrow$$
$$-3[a + (-4)] = 27$$

EXAMPLE 2

PHASE 1: Presenting the Example

Begin by giving students the brief "quiz" at right. Use it to review the two ways of applying the Distributive Property: multiplying to remove parentheses, and combining like terms. Then display the given equation, $3(x - 2) - 5(2x + 1) = 3$. Lead students to discover the connections between this equation and the quiz that they just took.

Simplify.

1. $3(x - 2)$ [$3x - 6$]
2. $-5(2x + 1)$ [$-10x - 5$]
3. $3x + (-10x)$ [$-7x$]
4. $-6 + (-5)$ [-11]

PHASE 2: Providing Guidance for **TRY THIS**

Watch for students who become so absorbed in simplifying that they forget to solve. Suggest that, whenever they see the word *Solve*, they first write a reminder of their goal. A sample is shown at right.

1. goal: $x = \boxed{}$

EXAMPLE 3

PHASE 1: Presenting the Example

Display the given equation, $3m + 8 - 5m = 9 + 4m + 29$. Have students add $5m$ to each side and simplify. [$3m + 8 = 9m + 38$] Ask them why this is not an efficient way to begin solving. [**An m-term and a constant still remain on each side of the equal sign.**] Lead students to see that this situation can be avoided by combining like terms before solving. Then discuss the textbook solution.

PHASE 2: Providing Guidance for **TRY THIS**

Some students may add all the like terms from both sides and write $-3t = 20$. Suggest that they cover one side of the equation and simplify only the side that they can see. Then have them repeat the process, this time covering the other side. Once both sides have been simplified individually, they can then work with both sides of the equation to solve.

Simplifying before solving multistep equations

Sometimes the first step in solving an equation is simplifying the expression(s) on one or both sides.

EXAMPLE 1

Solve $5(n + 4) = 35$. Check your solution.

▶ **Solution**

$$5(n + 4) = 35$$
$$5n + 20 = 35 \longleftarrow \text{Simplify using the Distributive Property.}$$
$$5n = 15$$
$$n = 3$$

Check: $5(3 + 4) = 5(7) = 35$ ✔

TRY THIS Solve $-3(a - 4) = 27$. Check your solution.

EXAMPLE 2

Solve $3(x - 2) - 5(2x + 1) = 3$.

▶ **Solution**

1. Simplify the left side of the equation.
$$3(x - 2) - 5(2x + 1) = 3x - 6 - 10x - 5 \longleftarrow \text{Apply the Distributive Property twice.}$$
$$= -7x - 11$$

2. Solve $-7x - 11 = 3$.
$$-7x - 11 = 3$$
$$-7x = 14 \longleftarrow \text{Apply the Addition Property of Equality.}$$
$$x = -2 \longleftarrow \text{Apply the Division Property of Equality.}$$

Check: $3(-2 - 2) - 5[2(-2) + 1] = 3(-4) - 5(-4 + 1) = -12 + 15 = 3$ ✔

TRY THIS Solve $3(3x - 1) - 2(x + 2) = 28$.

EXAMPLE 3

Solve $3m + 8 - 5m = 9 + 4m + 29$. Check your solution.

▶ **Solution**

1. Simplify the left and right sides of the equation.
$$3m + 8 - 5m = \boxed{-2m + 8} \qquad 9 + 4m + 29 = \boxed{4m + 38}$$

2. Now solve an equivalent equation.
$$\boxed{-2m + 8} = \boxed{4m + 38}$$
$$-2m = 4m + 30$$
$$-6m = 30$$
$$m = -5$$

Check: $3(-5) + 8 - 5(-5) = 18$ ✔ $\quad 9 + 4(-5) + 29 = 18$ ✔

TRY THIS Solve $-2t + 7 - 5t = 3 + 4t + 10$. Check your solution.

EXERCISES

KEY SKILLS

Simplify each expression.

1. $2(x + 1) - 3$
2. $3 - 4(t + 1)$
3. $3(z + 1) + 4(z - 3)$
4. $-(a + 1) - 2(a + 1)$
5. $5t - 5 + t$
6. $-3y + 7 - 2y$

PRACTICE

Solve each equation. Check your solution.

7. $-3(h + 1) = 9$
8. $6(b - 5) = 10$
9. $2 = 4(z + 1)$
10. $-12 = 5(p - 2)$
11. $0 = 2(3 - q)$
12. $3 = 2(5 - r)$
13. $3(w + 1) - 2(w - 3) = 7$
14. $3(t - 1) - (t + 2) = 7$
15. $-2(3a + 1) - (a - 5) = -7$
16. $2(x - 1) - 3(2x + 2) = 10$
17. $14 = 3(d - 5) + 2(d - 3)$
18. $10 = -2(s + 2) + 3(s - 10)$
19. $4 + 2(w - 3) = 3w - 2(w + 5)$
20. $-3y + 2(y + 3) = -4 + 2(y - 3)$
21. $\frac{1}{2}y + \frac{1}{2}(y + 2) = 5 + 3(y - 3)$
22. $\frac{4}{5}a + \frac{1}{5}(a - 10) = 5 + 5(a - 5)$
23. $5n - 1 + n = 3 - 3n + 1$
24. $4k - 1 + k = 1 + 3k + 10$
25. $b + 1 + b = 10 - 5b - 12$
26. $3g - 5 - 3g = 7g + 5 - 6g$
27. $3r - 2 + 4r - 2 = 3r - 2r + 7r + 1$
28. $-2a + 3 + 3a - 2 = 5a + 2a - 7a + 9$

29. **Critical Thinking** Show that every real number is a solution to $2(s + 1) + 4(s + 1) + 6(s + 1) = 12(s + 1)$.

MIXED REVIEW

Find each amount. (previous courses)

30. 6% of 1200
31. (8% of 10,000) − 400
32. 10% of (10,000 − 500)

Simplify each expression. (2.6 Skill B)

33. $0.05x + 0.05(10,000 - x)$
34. $0.06x + 0.04(8000 - x)$
35. $0.08x + 0.08(8000 - x)$
36. $0.05(12,000 - x) + 0.06x$

Answers for Lesson 2.8 Skill B

Try This Example 1
$a = -5$

Try This Example 2
$x = 5$

Try This Example 3
$t = -\frac{6}{11}$

1. $2x - 1$
2. $-4t - 1$
3. $7z - 9$
4. $-3a - 3$
5. $6t - 5$
6. $-5y + 7$
7. $h = -4$
8. $b = \frac{20}{3}$, or $6\frac{2}{3}$
9. $z = -\frac{1}{2}$
10. $p = -\frac{2}{5}$
11. $q = 3$

12. $r = \frac{7}{2}$, or $3\frac{1}{2}$
13. $w = -2$
14. $t = 6$
15. $a = \frac{10}{7}$, or $1\frac{3}{7}$
16. $x = -\frac{9}{2}$, or $-4\frac{1}{2}$
17. $d = 7$
18. $s = 44$
19. $w = -8$
20. $y = \frac{16}{3}$, or $5\frac{1}{3}$
21. $y = \frac{5}{2}$, or $2\frac{1}{2}$
22. $a = \frac{9}{2}$, or $4\frac{1}{2}$
23. $n = \frac{5}{9}$
24. $k = 6$
25. $b = -\frac{3}{7}$
26. $g = -10$
27. $r = -5$

28. $a = 8$
29. For all real numbers s, combining like terms gives $2(s + 1) + 4(s + 1) + 6(s + 1) = 12(s + 1)$; thus $12(s + 1) = 12(s + 1)$. This is the Reflexive Property of Equality. So, the equation is always true, regardless of the value of s.
30. 72
31. 400
32. 950
33. 500
34. $0.02x + 320$
35. 640
36. $0.01x + 600$

Focus: *Students link the equation-solving process to real-life problems.*

EXAMPLE

PHASE 1: Presenting the Example

Remind students of the simple interest formula, shown at right. Have students work independently to find the amount of simple interest earned when $10,000 is invested for one year in an account that pays 6% annually and in an account that pays 8% annually. [**$600; $800**] Discuss their answers, checking that all students understand the concept of simple interest.

$I = prt$
I: interest
p: principal
r: rate of interest
t: time

Now present the given problem. If time permits, have students first work individually or in pairs to solve it by using the guess-and-check strategy. The time spent in this effort should lead students to appreciate the equation method that is taught in the textbook.

PHASE 2: Providing Guidance for TRY THIS

Ask students how the Try This problem differs from the problem in the Example. The only substantial difference is the desired amount of interest, so they should have little difficulty modeling their work on the textbook solution for the Example.

Some students may not understand how to work with just one variable when there are *two* groups of unknown size. Suggest that they consider the simpler problem of dividing six marbles into two groups. Lead them in making a table like the one at right and in looking for the pattern.

Dividing Six Marbles	
Group 1	**Group 2**
0	$6 = 6 - 0$
1	$5 = 6 - 1$
2	$4 = 6 - 2$
3	$3 = 6 - 3$
4	$2 = 6 - 4$
5	$1 = 6 - 5$
6	$0 = 6 - 6$
x	$6 - x$

PHASE 3: ASSESSMENT AND CLOSURE for Lesson 2.8

- Display the equations at right. Ask students to identify all equations that are of the type studied in this lesson, and have them justify their responses. [$2r + 10 = r - 15$, $2(r + 10) = -15, 2r + 10 + r = -15$; **Each is an equation for which the process of solving requires three or more steps.**] Then have students solve the three equations, showing all the steps of the solutions on their papers. $\left[r = -25; r = -17.5; r = -8\frac{1}{3} \right]$

$2r = -15$
$2r + 10 = -15$
$2r + 10 = r - 15$
$2r + 10 = s - 15$
$2(r + 10) = -15$
$2r + 10 + r = -15$

- In this lesson, students extended their skills to solve more complicated equations requiring three or more steps. Students should be aware that this concludes the sequence of instruction in solving equations in one variable. However, they will need to apply these skills throughout the course.

 For a Lesson 2.8 Quiz, see Assessment Resources *page 19.*

When you write an equation to solve a real-world problem, the equation you write may require a multistep solution process.

Recall from previous courses that simple interest on an investment is the product of the amount of money invested, P, the annual interest rate, r, and the time in years, t. The formula is $I = Prt$.

EXAMPLE

Mr. Shaw has $10,000 to invest. He wants to divide the money between two investments that earn interest. One investment pays 6% simple interest annually, and the other pays 8% simple interest annually. How much money should he invest in each account so that he will earn $720 of interest in one year?

▶ **Solution**

Write an equation.

1. Let x represent the amount Mr. Shaw will invest at 6%. He will invest a total of $10,000, so $(10,000 - x)$ represents the amount that he will invest at 8%.

2. Write an equation. The *total* interest earned equals the sum of the interest earned from each investment.

interest earned at 6%		interest earned at 8%		total interest
↓		↓		↓
6% of x	+	8% of $(10,000 - x)$	=	720
$0.06x$	+	$0.08(10,000 - x)$	=	720

3. Simplify the left side of the equation.
$0.06x + 0.08(10,000 - x) = 0.06x + 800 - 0.08x$ ◄── Apply the Distributive Property.
$= 0.06x - 0.08x + 800$
$= -0.02x + 800$

4. Solve the new equation.
$-0.02x + 800 = 720$
$-0.02x + 800 - 800 = 720 - 800$ ◄── Apply the Subtraction Property of Equality.
$-0.02x = -80$
$\dfrac{-0.02}{0.02}x = \dfrac{-80}{-0.02}$ ◄── Apply the Division Property of Equality.
$x = 4000$

Mr. Shaw should invest $4000 at 6% and $6000 at 8%.

TRY THIS

Ms. Moore has $10,000 to invest. She wants to divide the money between two investments that earn interest. One investment pays 6% simple interest annually, and the other pays 8% simple interest annually. How much money should she invest in each account so that she will earn $740 of interest in one year?

EXERCISES

KEY SKILLS

Write an equation to represent each situation. Do not solve.

1. You have $1500 to divide between two accounts, one paying 5% simple interest and the other paying 6% simple interest. You want to earn $85 of interest in one year. How much should you invest in each account?

2. Twenty-four children bought cookies. Some children bought plain cookies for $1 each. The others bought frosted cookies for $2 each. Together, they spent $30. How many of each kind of cookie did they buy?

PRACTICE

Solve each problem.

3. Suppose that you distribute $2000 between two interest-paying accounts. One account pays 5% simple interest, and the other pays 6% simple interest. How much should you invest in each account to earn $113 of interest in one year?

4. How many nickels and how many dimes does Shannon have if she has 40 coins whose total value is $4.00?

5. Mr. Suzuki drove 365 miles in 7 hours. During one part of the trip, he drove 50 miles per hour and during the other part, he drove 55 miles per hour. How far did he drive at each speed?

6. A chemist must divide 500 milliliters of a solution into two containers. One container will contain 50 milliliters more than twice the amount in the other container. How much will each container have?

7. A drama club sold 200 tickets and collected $640. If a child's ticket cost $2 and an adult ticket cost $5, how many of each ticket did the club sell?

8. **Critical Thinking** Suppose Mr. Shaw wants to divide $10,000 between two accounts that earn interest at 6% and 8%. Show that it is impossible for him to earn $850 of interest in one year.

MIXED REVIEW

Solve each equation. Identify the property used in each step. (2.5 Skills A and B)

9. $2(x - 5) = 11$
10. $2x - 5 = 12 + x$
11. $2x + x = -2$

Answers for Lesson 2.8 Skill C

Try This Example
Invest $3000 at 6% and $7000 at 8%.

1. Let x represent the amount invested at 5%;
$0.05x + 0.06(1500 - x) = 85$

2. Let n represent the number of children who bought plain cookies; $n + 2(24 - n) = 30$

3. $700 at 5% and $1300 at 6%

4. She has 40 dimes and no nickels.

5. 4 hours at 50 miles per hour and 3 hours at 55 miles per hour

6. 150 milliliters in one container and 350 milliliters in the other

7. 120 child tickets and 80 adult tickets

8. Let x represent the amount invested at 6%;
$0.06x + 0.08(10,000 - x) = 850$; The solution is -2500. Since investments must be positive amounts, there is no solution for x. The investment plan is not possible.

9. $2x - 10 = 11$ Distributive Property
$2x = 21$ Addition Property of Equality
$x = \dfrac{21}{2}$ Division Property of Equality

10. $x - 5 = 12$ Subtraction Property of Equality
$x = 17$ Addition Property of Equality

11. $3x = -2$ Distributive Property
$x = -\dfrac{2}{3}$ Division Property of Equality

TEACHING 2.9 LESSON

Deductive and Indirect Reasoning in Algebra

SKILL A *Judging the correctness of a solution*

Focus: *Students detect errors in solutions, and then correct the errors to create a correct solution.*

EXAMPLE 1

PHASE 1: Presenting the Example

Discuss with students how the process of simplifying an expression or solving an equation is like a chain. Each step in the solution forms a link; taken together, the statements form a chain that leads from the equation to its solution.

Tell students that, according to a common saying, a chain is only as strong as its weakest link. So, if one of the links is broken, the entire chain is broken. Ask students to interpret this saying in terms of the "solution chain." Lead them to the observation that each statement in the solution of an equation must be correct. If even one statement is not correct, the entire solution is "broken."

Students have solved many equations in this course so far, and they have undoubtedly made some errors. They should note that, whenever they arrived at an incorrect answer, they most likely broke a link in the chain by applying a property incorrectly. Therefore, it is important to learn to examine a chain of statements, locate the broken link, and fix it. That is what students will be doing in the examples on this page.

PHASE 2: Providing Guidance for TRY THIS

Some students may have difficulty seeing the individual "links" when a solution is presented horizontally, as in the Try This exercise. Suggest that they rewrite the simplification in a vertical format, as shown below. They can then write the name of the property being used next to each step. Point out that if they have trouble finding a property to justify a given step, that step may very well be the broken link.

$$\frac{4-10}{2} - 3 = \frac{10-4}{2} - 3 \quad \longleftarrow \text{ Commutative Property (applied incorrectly)}$$

$$= \frac{6}{2} - 3 \quad \longleftarrow \text{ Combine like terms; } 10 - 4 = 6$$

$$= 3 - 3 \quad \longleftarrow \text{ Simplify; } \frac{6}{2} = 3$$

$$= 0$$

EXAMPLE 2

PHASE 2: Proving Guidance for TRY THIS

Another method that students can use to find the broken link is to write their own solutions to $-2(x - 3) = 18$, without looking at the incorrect solution on the textbook page. They can then compare the two solutions to find any differences.

82 Chapter 2 Operations With Real Numbers

2.9 LESSON — Deductive and Indirect Reasoning in Algebra

SKILL A *Judging the correctness of a solution*

When you read or write a mathematical argument, you should check it carefully for mistakes in reasoning.

EXAMPLE 1

Find the error in the simplification below. Then give the correct value of
$\frac{14-3}{2} - 4$.

$$\frac{14-3}{2} - 4 = \frac{14}{2} - 3 - 4 = 7 - 3 - 4 = 4 - 4 = 0 \quad \text{Therefore, } \frac{14-3}{2} - 4 = 0. \; ✗$$

▶ **Solution**

In the first step, the Distributive Property was not applied correctly. The
value of $\frac{14-3}{2}$ should be $\frac{14}{2} - \frac{3}{2}$, or $5\frac{1}{2}$.

The correct value of $\frac{14-3}{2} - 4$ is $1\frac{1}{2}$.

TRY THIS Find the error in the simplification below. Then give the correct value of $\frac{4-10}{2} - 3$.

$$\frac{4-10}{2} - 3 = \frac{10-4}{2} - 3 = \frac{6}{2} - 3 = 3 - 3 = 0 \quad \text{Therefore, } \frac{4-10}{2} - 3 = 0. \; ✗$$

EXAMPLE 2

Find the error in the solution to $3(x - 1) = 12$ shown at right.
Then give the correct solution.

$$3(x-1) = 12$$
$$3x - 3 = 12$$
$$3x = 9$$
$$x = 3 \; ✗$$

▶ **Solution**

From the second step to the third step ($3x - 3 = 12$ to $3x = 9$), the
Addition Property of Equality was applied incorrectly; the number 3 was
added to the left side of the equation but subtracted from the right side
of the equation.

$$3x - 3 = 12$$
$$3x - 3 + 3 = 12 + 3$$
$$3x = 15$$
$$x = 5$$

The correct solution is 5.

TRY THIS Find the error in the solution to $-2(x - 3) = 18$ shown at right.
Then give the correct solution.

$$-2(x-3) = 18$$
$$-2x - 6 = 18$$
$$-2x = 24$$
$$x = -12 \; ✗$$

82   Chapter 2   Operations With Real Numbers

EXERCISES

KEY SKILLS

Substitute the value of the variable from the last line into each of the
equations above it. Which equations does it satisfy?

1. $3(x + 4) = -11$
    $3x + 4 = -11$
    $3x = -15$
    $x = -5$

2. $0 = 4(a - 2) + 12$
    $12 = 4(a - 2)$
    $12 = 4a - 8$
    $20 = 4a$
    $5 = a$

PRACTICE

Find the error(s) in each simplification. Then give the correct value.

3. $-(3 + 4) + 5 = -3 + 4 + 5$
    $= 1 + 5$
    $= 6$

4. $-2(3 + 4) + 5 = -6 + 4 + 5$
    $= -2 + 5$
    $= 3$

5. $\frac{3}{7} + \frac{2}{7} = \frac{3+2}{7+7} = \frac{5}{14}$

6. $\frac{48-5}{7} = \frac{48}{7} - 5 = 6\frac{6}{7} - 5 = 1\frac{6}{7}$

Find the error(s) in each solution. Then write the correct solution.

7. $3x + 5 - 4x = 19$
    $3x + 4x - 5 = 19$
    $7x + 5 = 19$
    $7x = 14$
    $x = 2$

8. $2(d - 3) + 3(d + 5) = 10$
    $2d - 6 + 3d + 5 = 10$
    $2d - 3d + 6 + 5 = 10$
    $-d + 11 = 10$
    $-d = -1$
    $d = -1$

Give a reason for each step in the proofs below.

9. $(a + b) + (-a) = b$
    $(a + b) + (-a) = a + [b + (-a)]$
    $= a + [(-a) + b]$
    $= [a + (-a)] + b$
    $= 0 + b$
    $= b$

10. $a(b + c + d) = ab + ac + ad$
    $a(b + c + d) = a[(b + c) + d]$
    $= a(b + c) + ad$
    $= ab + ac + ad$

MIXED REVIEW

Solve each equation for the specified variable. (2.7 Skill B)

11. $y = -2x + 5; x$

12. $y = -1.5x - 3; x$

13. $-3x + 5y = 12; x$

14. $7x - 5y = 15; y$

15. $y = -3(x - 2) + 3; x$

16. $y = 4(x + 5) + 3; x$

17. $-3x = -7y + 24; y$

18. $5x = -7y - 14; y$

19. $7y + 8x = 20; y$

Lesson 2.9   Skill A   83

Answers for Lesson 2.9 Skill A

Try This Example 1
$4 - 10 \neq 10 - 4; -6$

Try This Example 2
$-2(x - 3) = -2x + 6;$
$-2x + 6 = 18; -2x = 12; x = -6$

1. -5 satisfies $3x = -15$ and $3x + 4 = -11$, but not $3(x + 4) = -11$

2. 5 satisfies $20 = 4a$, $12 = 4a - 8$, and $12 = 4(a - 2)$, but not $0 = 4(a - 2) + 12$

3. $-(3 + 4) = -3 - 4$, not $-3 + 4$;
    $-(3 + 4) + 5 = -3 - 4 + 5 = -2$

4. $-2(3 + 4) = -6 - 8$, not $-6 + 4$;
    $-2(3 + 4) + 5 = -6 - 8 + 5 = -9$

5. $\frac{3}{7} + \frac{2}{7} = \frac{3+2}{7}$, not $\frac{3+2}{7+7}$; $\frac{3}{7} + \frac{2}{7} = \frac{3+2}{7} = \frac{5}{7}$

6. $\frac{48-5}{7} = \frac{48}{7} - \frac{5}{7}$, not $\frac{48}{7} - 5$; $\frac{48-5}{7} = \frac{43}{7} = 6\frac{1}{7}$

7. $3x + 5 - 4x = 3x - 4x + 5$; $3x - 4x + 5 = 19$;
    $-x + 5 = 19; -x = 14; x = -14$

8. $2(d - 3) + 3(d + 5) = 2d - 6 + 3d + 15$;
    $2d - 6 + 3d + 15 = 10; 5d + 9 = 10;$
    $5d = 1; d = \frac{1}{5}$

9. Associative Property of Addition; Commutative Property of Addition; Associative Property of Addition; Inverse Property of Addition; Identity Property of Addition

10. Associative Property of Addition; Distributive Property; Distributive Property

11. $x = \frac{y - 5}{-2}$, or $-\frac{1}{2}y + \frac{5}{2}$

12. $x = \frac{y + 3}{-1.5}$, or $-\frac{2}{3}y - 2$

13. $x = \frac{-5y + 12}{-3}$, or $\frac{5}{3}y - 4$

14. $y = \frac{7x - 15}{5}$, or $\frac{7}{5}x - 3$

15. $x = -\frac{y}{3} + 3$

16. $x = \frac{y - 3}{4} - 5$, or $\frac{y - 23}{4}$

17. $y = \frac{3x + 24}{7}$, or $\frac{3}{7}x + \frac{24}{7}$

18. $y = \frac{-5x - 14}{7}$, or $-\frac{5}{7}x - 2$

19. $y = \frac{-8x + 20}{7}$, or $-\frac{8}{7}x + \frac{20}{7}$

★ **Advanced Exercises**

Lesson 2.9   Skill A   83

Focus: *The step-by-step solution of an equation is examined as a logical argument that progresses from a given hypothesis to a justified conclusion.*

EXAMPLE

PHASE 1: Presenting the Example

For many students, the process of analyzing logical thinking and representing this analysis symbolically is a daunting task. You may find it helpful to introduce the topic by grounding it in everyday experiences.

Begin by discussing the definitions of *conditional statement*, *hypothesis*, and *conclusion* given at the top of the textbook page. Then display the following:

> If Ted mows his neighbor's lawn, then he can afford a concert ticket.

Point out that this is a conditional statement. Tell students to write the hypothesis and the conclusion. [**hypothesis: Ted mows his neighbor's lawn; conclusion: Ted can afford a concert ticket.**] Discuss their responses.

Ask students whether the conditional statement is true or false. Lead them to see that they cannot determine this without filling in some "gaps" in the reasoning. For example, how much will Ted earn? What is the cost of a concert ticket?

Now display the following set of statements:

I. If Ted mows his neighbor's lawn, then he will earn $25.
II. If Ted earns $25, then he can afford a concert ticket.

Ask whether this additional information helps decide whether the original statement is true or false. Students most likely will say it is true, perceiving statements I and II as a logical "chain" supporting it. Point out that it is still unknown whether statements I and II are true. Then display the following:

Conditional Statement	Justification
I. If Ted mows his neighbor's lawn, then he will earn $25.	Ted's neighbor has agreed to pay him $25 to mow the lawn.
II. If Ted earns $25, then he can afford a concert ticket.	The advertised cost of a concert ticket is $25.

Point out that each statement in the chain now has a basis in fact. That is, each statement has been *justified*.

Now introduce the Transitive Property of Deductive Reasoning. Tell students to identify p, q, and r in the argument about Ted and the concert ticket. [*p:* **Ted mows his neighbor's lawn;** *q:* **Ted will earn $25;** *r:* **Ted can afford a concert ticket.**] Since $p \Rightarrow q$ (statement I) and $q \Rightarrow r$ (statement II) have been justified, it follows that $p \Rightarrow r$ (If Ted mows his neighbor's lawn, then he can afford a concert ticket.) also is justified.

Now discuss the example. Throughout the discussion, stress the parallels between the logical argument that supports the solution of the equation and the argument that supports the statement about Ted and the concert ticket.

Writing a solution process as a logical argument

In everyday conversation, we often make *if-then* statements. Some examples are "If it is raining, then I carry an umbrella." and "If it is winter, then birds fly south."

In mathematics, an if-then statement is called a **conditional statement**. The part following the word *if* is the **hypothesis**, and the part following the word *then* is the **conclusion**. The Symmetric Property of Equality is an example of a conditional statement.

$$\underbrace{\text{If } a \text{ and } b \text{ are real numbers and } a = b,}_{\text{hypothesis}} \overset{\text{conditional statement}}{\text{then } \underbrace{b = a.}_{\text{conclusion}}}$$

The hypothesis of a statement is often represented by the letter p. The conclusion is often represented by q. A conditional statement is often abbreviated as $p \Rightarrow q$. You read $p \Rightarrow q$ as "If p then q" or "p implies q."

The process of beginning with a hypothesis and following a sequence of logical steps to reach a conclusion is called **deductive reasoning**. An important rule in deductive reasoning is the Transitive Property of Deductive Reasoning.

Transitive Property of Deductive Reasoning

Let p, q, and r represent statements. If $p \Rightarrow q$ and $q \Rightarrow r$, then $p \Rightarrow r$.

Solving an equation is actually a form of deductive reasoning.

EXAMPLE | Solve $2x + 5 = 11$. Write the steps in the solution as a logical argument. Write the solution as a conditional statement.

▶ **Solution**

Conditional Statement	Justification
If $2x + 5 = 11$, then $2x = 6$. $\underbrace{}_{p} \Rightarrow \underbrace{}_{q}$	Subtraction Property of Equality Subtract 5 from each side.
If $2x = 6$, then $x = 3$. $\underbrace{}_{q} \Rightarrow \underbrace{}_{r}$	Division Property of Equality Divide each side by 2.
If $2x + 5 = 11$, then $x = 3$. $\underbrace{}_{p} \Rightarrow \underbrace{}_{r}$	Transitive Property of Deductive Reasoning If $p \Rightarrow q$ and $q \Rightarrow r$, then $p \Rightarrow r$.

TRY THIS | Solve $\frac{x}{3} - 5 = 11$. Write the steps in the solution as a logical argument. Write the solution as a single conditional statement.

EXERCISES

Identify p, q, and r in each pair of conditional statements.

1. If $-2d + 7 = 11$, then $-2d = 4$.
   If $-2d = 4$, then $d = -2$.

2. If $4y - 6 = 11$, then $4y = 17$.
   If $4y = 17$, then $y = \frac{17}{4}$.

Justify each step in each logical argument.

3. If $2(3x - 1) = 7$, then $6x - 2 = 7$.
   If $6x - 2 = 7$, then $6x = 9$.
   If $6x = 9$, then $x = \frac{3}{2}$.
   If $2(3x - 1) = 7$, then $x = \frac{3}{2}$.

4. If $2v - 5 = 3v + 1$, then $-v - 5 = 1$.
   If $-v - 5 = 1$, then $v + 5 = -1$.
   If $v + 5 = -1$, then $v = -6$.
   If $2v - 5 = 3v + 1$, then $v = -6$.

5. If $3a - 4 = 7 + 5a$, then $-4 = 7 + 2a$.
   If $-4 = 7 + 2a$, then $-11 = 2a$.
   If $-11 = 2a$, then $-\frac{11}{2} = a$.
   If $-\frac{11}{2} = a$, then $a = -\frac{11}{2}$.
   If $3a - 4 = 7 + 5a$, then $a = -\frac{11}{2}$.

6. If $2w - 3w = 4w + 10$, then $-w = 4w + 10$.
   If $-w = 4w + 10$, then $-5w = 10$.
   If $-5w = 10$, then $w = -2$.
   If $2w - 3w = 4w + 10$, then $w = -2$.

Write the logical steps in the solution to each equation. Write the solution as a conditional statement.

7. $3(x - 5) = 5x$

8. $7h - (3h - 5) = 9$

9. $6(x - 3) + 4(x + 2) = 6$

10. **Critical Thinking** Find the missing step in the logical argument below.
    If $3(p + 1) - (p - 1) = 6$, then $3p + 3 - p + 1 = 6$.
    If $2p + 4 = 6$, then $2p = 2$. If $2p = 2$, then $p = 1$.

Write each set of numbers in order from least to greatest. (previous courses)

11. $-2, 9.2, 0, -6.1, 0.5$

12. $0, 0.1, 0.2, -0.1, -0.2, 5$

13. $3.7, 3.75, 3.754, 3.6, 3.65$

Replace each ___?___ with < (is less than), > (is greater than), or = to make the statement true. (previous courses)

14. $4\frac{1}{4}$ __?__ 4.25

15. 6.33 __?__ 6.43

16. $\frac{9}{5}$ __?__ 1.79

17. $-3\frac{1}{3}$ __?__ $3\frac{1}{3}$

Answers for Lesson 2.9 Skill B

Try This Example

$\frac{x}{3} - 5 = 11 \Rightarrow \frac{x}{3} = 16$ Addition Property of Equality

$\frac{x}{3} = 16 \Rightarrow x = 48$ Multiplication Property of Equality

$\frac{x}{3} - 5 = 11 \Rightarrow x = 48$ Transitive Property of Deductive Reasoning

1. p: $-2d + 7 = 11$; q: $-2d = 4$; r: $d = -2$

2. p: $4y - 6 = 11$; q: $4y = 17$; r: $y = \frac{17}{4}$

3. Distributive Property; Addition Property of Equality; Division Property of Equality; Transitive Property of Deductive Reasoning

4. Subtraction Property of Equality; Multiplication Property of Equality; Subtraction Property of Equality; Transitive Property of Deductive Reasoning

5. Subtraction Property of Equality; Subtraction Property of Equality; Division Property of Equality; Symmetric Property of Equality; Transitive Property of Deductive Reasoning

6. Combining like terms; Subtraction Property of Equality; Division Property of Equality; Transitive Property of Deductive Reasoning

★ 7. $3(x - 5) = 5x \Rightarrow 3x - 15 = 5x$ (Distributive Property)
   $3x - 15 = 5x \Rightarrow -15 = 2x$
   (Subtraction Property of Equality)
   $-15 = 2x \Rightarrow -\frac{7}{2} = x$
   (Division Property of Equality)
   $-\frac{7}{2} = x \Rightarrow x = -\frac{7}{2}$
   (Symmetric Property of Equality)
   $3(x - 5) = 5x \Rightarrow x = -\frac{15}{2}$
   (Transitive Property of Deductive Reasoning)

For answers to Exercises 8–17, see Additional Answers.

★ **Advanced Exercise**

SKILL C Showing that a statement is sometimes, always, or never true

Focus: *Students determine whether a given statement is true in some instances, in all instances, or in no instance. They also investigate indirect reasoning.*

EXAMPLE 1

PHASE 1: Presenting the Example

Display the statement $a^2 = 2a$. Ask students to name a counterexample for the statement. [**any real number except 0 or 2**] Point out that, because there is at least one counterexample, the statement is considered false.

Now ask: Is the statement $a^2 = 2a$ *ever* true? Elicit the two instances for which it is true: $0^2 = 2(0)$ and $2^2 = 2(2)$. So there are some instances when the statement is false, but other instances when it is true. That is, the statement is *sometimes true.*

PHASE 2: Providing Guidance for Try This

Point out a parallel to the statement in Example 1, as shown at right. Since $a^2 = 2a$ is true when $a = 0$ and when $a = 2$, $(a - 3)^2 = 2(a - 3)$ is true when $a - 3 = 0$ and when $a - 3 = 2$, or when $a = 3$ and $a = 5$.

$$\begin{array}{ccc} a^2 & = & 2a \\ \downarrow & & \downarrow \\ \overbrace{(a-3)}^{}{}^2 & = & 2\overbrace{(a-3)}^{} \end{array}$$

EXAMPLE 3

PHASE 1: Presenting the Example

First introduce the concept of $\sim p$, or "not p." Use non-mathematical examples, such as "Today is Monday," (p) and "Today is not Monday," ($\sim p$).

Indirect reasoning is challenging for most students. Again, begin with a non-mathematical example, such as the argument at right. Identify p and $\sim p$ for the students:

p: The defendant is innocent. $\sim p$: The defendant is not innocent.

Lead students to see how $\sim p$ leads to the false statement that the defendant was at the restaurant at 8:00 P.M.

"If the defendant is not innocent, then she must have been at the restaurant at 8:00 P.M. But three witnesses saw her at the mall at 8:00 P.M. Therefore, she is innocent."

PHASE 3: ASSESSMENT AND CLOSURE for Lesson 2.9

- Give students the five equations at right. Tell them to link the equations to form a logical argument and to justify each statement in the argument.

  [If $4(t + 1) = t - 2$, then $4t + 4 = t - 2$. If $4t + 4 = t - 2$, then $3t + 4 = -2$. If $t + 4 = -2$, then $3t = -6$. If $3t = -6$, then $t = -2$.]

- Students have now formalized logical principles. They will revisit these ideas throughout the course.

$3t = -6$
$4(t + 1) = t - 2$
$3t + 4 = -2$
$t = -2$
$4t + 4 = t - 2$

☞ *For a Lesson 2.9 Quiz, see* Assessment Resources *page 20.*

☞ *For a Chapter 2 Assessment, see* Assessment Resources *pages 21–22.*

There are various ways to determine whether a statement is sometimes, always, or never true.

EXAMPLE 1

Given that a is a real number, is the equation $a^2 = 2a$ sometimes, always, or never true? Justify your response.

▶ **Solution**

Guess and check.

Test values of a. $a = 0$: $0^2 = 2(0)$ ✔ True $a = 1$: $1^2 \neq 2(1)$ ✗ False

Because $a^2 = 2a$ is true for $a = 0$ but false for $a = 1$, it is sometimes true.

TRY THIS

Given that a is a real number, is the equation $(a - 3)^2 = 2(a - 3)$ sometimes, always, or never true? Justify your response.

EXAMPLE 2

Given that x is a real number, is the equation $2(2x + 3) = 5 + 4x + 1$ sometimes, always, or never true? Justify your response.

▶ **Solution**

$$2(2x + 3) = 4x + 6 \qquad \longleftarrow \text{Distributive Property}$$
$$4x + 6 = 5 + 4x + 1 \qquad \longleftarrow 6 = 5 + 1$$
$$2(2x + 3) = 5 + 4x + 1 \qquad \longleftarrow \text{Transitive Property of Equality}$$

The statement is always true.

TRY THIS

Given that b is a real number, is $2(b - 1) - 7(b + 1) = -3b - 9 - 2b$ sometimes, always, or never true? Justify your response.

If p represents a statement, then ~p ("not p") represents the **negation** of p. If p is true, then ~p is never true. So, another way to show that a statement, p, is true is to show that ~p is never true. When you do this, you are using a process called **indirect reasoning**.

EXAMPLE 3

Use indirect reasoning to show that $2(a + 1) \neq 2a + 1$.

▶ **Solution**

1. Identify p and ~p. p: $2(a + 1) \neq 2a + 1$ ~p: $2(a + 1) = 2a + 1$

2. Begin with ~p and use properties to arrive at a false statement.
$$2(a + 1) = 2a + 1$$
$$2a + 2 = 2a + 1 \qquad \longleftarrow \text{Distributive Property}$$
$$2 = 1 \text{ ✗} \qquad \longleftarrow \text{Subtraction Property of Inequality}$$

When you arrive at a false statement, the statement you began with must also be false. Since $2 = 1$ is false, ~p is never true.

3. Conclude that p is true: $2(a + 1) \neq 2a + 1$

TRY THIS

Use indirect reasoning to show that $3(t - 1) \neq 3t$.

EXERCISES

KEY SKILLS

1. Give one example and one counterexample to show that $n^2 = 3n$ is sometimes true.

2. Use properties to show that $n + (n + 1) = 2n + 1$ is always true.

3. Use indirect reasoning to show that $3(x + 5) + 2 \neq 3x + 20$.

PRACTICE

Is the given statement sometimes true or always true? Justify your response. (Assume that all variables represent real numbers.)

4. $3(n + 1) + (n + 1) = 4(n + 1)$

5. $m^2 = 4m$

6. $4(k - 3) = 4k - 12$

7. $4(k - 3) = 4(3 - k)$

8. $3x - 5 + 7x = -5 + 10x$

9. $x(y + 5) - 5 = xy$

10. $x(y + 5) - 5x = xy$

11. $zxy + zx + zy = z[x(y + 1) + y]$

12. Use indirect reasoning to show that $4(3 + y) \neq 4y - 4$.

13. Use indirect reasoning to show that $3x + 6 \neq 3(x + 4) + 2$.

14. Use indirect reasoning to show that $4y + 2 + 3(5 - y) \neq y + 1$.

15. **Critical Thinking** Refer to the geometry definitions below. Then decide whether the statement following them is sometimes, always, or never true.

> **scalene triangle:** no two sides have the same length.
> **isosceles triangle:** at least two sides have the same length.
> **equilateral triangle:** all three sides have the same length.

Statement: No scalene triangle is isosceles or equilateral and all equilateral triangles are isosceles.

MIXED REVIEW

Solve each equation. Check your solution. (2.8 Skill B)

16. $-2(x - 1) + 4x = 5x + 3$

17. $x - 1 - (4x + 3) = 5x$

18. $\frac{1}{2}(4x + 3) = \frac{3}{2}(5x - 1)$

19. $\frac{1}{2}(4x + 4) + \frac{1}{2}(4x - 8) = 0$

20. $c - 2c - 3c + 2 = 8$

21. $(r - 1) - (r - 2) + (r - 3) = 1$

22. $4(2n - 1) - 3(2n - 1) = 3(n - 1)$

23. $2m + 1 + 2(2m + 1) = -2(2m + 1)$

Answers for Lesson 2.9 Skill C

Try This Example 1
sometimes true; true when $a = 0$ or $a = 5$ but false otherwise

Try This Example 2
always true; $2(b - 1) - 7(b + 1) = -5b - 9$ and $-3b - 9 - 2b = -5b - 9$

Try This Example 3
$3(t - 1) = 3t$
$3t - 3 = 3t$
$-3 = 0$ ✗
Therefore, $3(t - 1) \neq 3t$.

1. true when $n = 0$ or $n = 3$ but false otherwise
2. always true; $n + (n + 1) = (n + n) + 1 = 2n + 1$
3. $3(x + 5) + 2 = 3x + 20$
$3x + 17 = 3x + 20$
$17 = 20$ ✗
Therefore, $3(x + 5) + 2 \neq 3x + 20$.
4. always true; $3(n + 1) + (n + 1) = 3n + 3 + n + 1 = (3n + n) + (3 + 1) = 4n + 4 = 4(n + 1)$
5. sometimes true; true when $m = 0$ or $m = 4$ but false otherwise
6. always true; $4(k - 3) = 4k - 12$
7. sometimes true; true when $k = 0$ but false otherwise

8. always true; $3x - 5 + 7x = 3x + 7x - 5 = 10x - 5$
9. sometimes true; true when y is any real number and $x = 1$
10. always true; $x(y + 5) - 5x = xy + 5x - 5x = xy$
★11. always true; $z(xy + x + y) = z(x[y + 1] + y) = zxy + zx + zy$
12. $4(3 + y) = 4y - 4$
$12 + 4y = 4y - 4$
$12 = -4$ ✗
Therefore, $4(3 + y) \neq 4y - 4$.
13. $3x + 6 = 3(x + 4) + 2$
$3x + 6 = 3x + 14$
$6 = 14$ ✗
Therefore, $3x + 6 \neq 3(x + 4) + 2$.
14. $4y + 2 + 3(5 - y) = y + 1$
$y + 17 = y + 1$
$17 = 1$ ✗
Therefore, $4y + 2 + 3(5 - y) \neq y + 1$.

For answers to Exercises 15–23, see Additional Answers.

★ **Advanced Exercise**

Write each fraction as a decimal and each decimal as a fraction.

1. $\frac{13}{25}$　　　　2. $\frac{5}{12}$　　　　3. 0.58　　　　4. 3.26

Classify each number in as many ways as possible.

5. $-\frac{7}{15}$　　　　6. 13　　　　7. -180　　　　8. 5.3030030003...

Evaluate each expression given $b = 12$.

9. $-|b + 20|$　　　　10. $|b + 3| + |b - 3|$

11. $-(-|b|)$　　　　12. $-(-|b - 3|)$

Use a number line to find each sum.

13. $-4 + 7$　　　　14. $-1 + (-5)$

Find each sum.

15. $-5 + 12$　　　　16. $-2\frac{1}{4} + 3\frac{3}{4}$

17. $-6.4 + (-7.3)$　　　　18. $1.2 + (-0.5) + (-3)$

Find each difference.

19. $-7 - 13$　　　　20. $12.4 - 18.1$

21. $-13 - (-19)$　　　　22. $-2\frac{1}{2} - 3\frac{1}{2}$

Simplify each expression.

23. $-(5 - 3) - (4 + 8)$　　　　24. $10 - (3 - 8) + (2 - 10)$

25. A bank customer deposited checks in the amounts of $45.20, $56.75, and $110.39. The customer also withdrew $36.75, $120.25, and $110.45. What was the net change in the account balance?

Find each product. Write fractions in simplest form.

26. $(-12)(12)$　　　　27. $\left(-2\frac{1}{2}\right)\left(-4\frac{1}{2}\right)$　　　　28. $\frac{-8}{15} \cdot \frac{3}{32}$

Find each quotient. Write fractions in simplest form.

29. $(-19) \div (19)$　　　　30. $\left(-3\frac{1}{2}\right) \div \left(-2\frac{2}{3}\right)$　　　　31. $\frac{-8}{15} \div \frac{4}{-35}$

Simplify each expression.

32. $\frac{(3 - 5)}{-2} \times \frac{-(9 - 3)}{-2}$　　　　33. $\frac{9 - (-15)}{12(-2) - 3(2)}$

Solve each equation. Check your solution.

34. $y + 5.6 = 2.5$　　　　35. $d - 3\frac{1}{3} = -1\frac{1}{3}$　　　　36. $-11.5 = d - \frac{1}{2}$

37. $\frac{x}{-5} = 2.5$　　　　38. $4g = -4.8$　　　　39. $\frac{z}{-2} - 5 = -3$

Simplify each expression.

40. $3a - 6 - 8a$　　　　41. $(4x - 5x) - (3x - x)$

42. $-3(3a - 5) - 8a + 12$　　　　43. $-2(6c - 8c) - 3(3c - 5c)$

Solve each equation. Check your solution.

44. $-2 - 5r = -7r$　　　　45. $-3a + 7 = -4a + 12$

46. Solve $3x - 5y = 12$ for y.

47. Tara has $200. If she can save $22 per week, how many weeks will it take her to save $420?

Solve each equation. Check your solution.

48. $z + 8.1 - 5z = -7z$　　　　49. $-3a - 4 = 5a + 1$

50. $12 = -3(x + 4) + 5(x - 1)$　　　　51. $4n - 6 - 7n = 2n - 5 + 7n$

52. A chemist must separate 770 milliliters of solution into two containers. One container will have 10 milliliters more than 3 times the amount of solution in the other container. How much will each container have?

53. Find the error(s) in the solution at right. Then write a correct solution.
$$-5(2 - 3) + 4 = -10 - 15 + 4$$
$$= -25 + 4$$
$$= 21$$

54. Write the steps in the solution of $4x = 2(x - 5)$ as a logical argument. Then write the solution as a single conditional statement.

55. Is the statement at right sometimes true or always true? Justify your response.　　$5(h + 3) - 2 = 5h + 13$

Answers for Chapter 2 Assessment

1. 0.52
2. $0.41\overline{6}$
3. $\frac{29}{50}$
4. $\frac{163}{50}$, or $3\frac{13}{50}$
5. rational, real
6. natural number, whole number, integer, rational, real
7. integer, rational, real
8. irrational, real
9. -14
10. 24
11. 12
12. 9
13. 3
14. -6
15. 7
16. $\frac{3}{2}$, or $1\frac{1}{2}$
17. -13.7
18. -2.3
19. -20

20. -5.7
21. 6
22. -6
23. -14
24. 7
25. decrease of $55.11
26. -144
27. $\frac{45}{4}$, or $11\frac{1}{4}$
28. $-\frac{1}{20}$
29. -1
30. $\frac{21}{16}$, or $1\frac{5}{16}$
31. $\frac{14}{3}$, or $4\frac{2}{3}$
32. 3
33. $-\frac{4}{5}$
34. $y = -3.1$
35. $d = 2$
36. $d = -11$
37. $x = -12.5$
38. $g = -1.2$

39. $z = 21$
40. $-5a - 6$
41. $-3x$
42. $-17a + 27$
43. $10c$
44. $r = 1$
45. $a = 5$
46. $y = \frac{12 - 3x}{-5}$, or $\frac{3x - 12}{5}$
47. 10 weeks
48. $z = -2.7$
49. $a = -\frac{5}{8}$
50. $x = \frac{29}{2}$, or $14\frac{1}{2}$
51. $n = -\frac{1}{12}$
52. 190 milliliters and 580 milliliters

For answers to Exercises 53–55, see Additional Answers.

Chapter 3 Linear Equations in Two Variables	State or Local Standards	Corresponding Lessons in *Algebra 1*, Schultz et al.
3.1 The Coordinate Plane		1.4, 1.5
Skill A: Graphing ordered pairs in the coordinate plane		
Skill B: Graphing solutions to equations in two variables		
3.2 Functions and Relations		1.5, 5.1, 14.1
Skill A: Differentiating between functions and relations		
Skill B: Finding the domain and range of a function or a relation		
Skill C: APPLICATIONS Using functions in real-world situations		
3.3 Ratio, Rate, and Direct Variation		4.1, 5.3
Skill A: APPLICATIONS Using direct-variation functions to describe real-world situations		
Skill B: APPLICATIONS Using proportions to solve problems that involve direct variation		
3.4 Slope and Rate of Change		5.2, 5.3
Skill A: Finding the slope of a line		
Skill B: Finding points on a line		
Skill C: APPLICATIONS Solving real-world problems that involve rate of change		
3.5 Graphing Linear Equations in Two Variables		1.5, 5.4, 5.5
Skill A: Using intercepts to graph a linear equation		
Skill B: Using slope and a point to graph a linear equation		
Skill C: APPLICATIONS Using linear equations to solve real-world problems		
3.6 The Point-Slope Form of an Equation		5.5
Skill A: Using slope and a given point to find the equation of a line		
Skill B: APPLICATIONS Using the point-slope form in solving real-world problems		
3.7 Finding the Equation of a Line Given Two Points		5.5
Skill A: Using two points to find the equation of a line		
Skill B: APPLICATIONS Modeling real-world data with linear equations		
3.8 Relating Two Lines in the Plane		5.2, 5.6
Skill A: Working with parallel lines in the coordinate plane		
Skill B: Working with perpendicular lines in the coordinate plane		
Skill C: APPLICATIONS Using facts about parallel lines and perpendicular lines to solve geometry problems		
3.9 Linear Patterns and Inductive Reasoning		1.1, 1.5, 11.6
Skill A: Extending a linear sequence given a table		
Skill B: Using logical reasoning to find algebraic patterns		
Skill C: Extending a linear sequence given a graph		

CHAPTER

3

Linear Equations in Two Variables

What You Already Know

In Chapters 1 and 2, you learned how to represent real numbers on a number line and how to solve linear equations in one variable. The connection between real numbers, solutions to linear equations in one variable, and points on a number line will help you take the next step in your study of equations.

What You Will Learn

In Chapter 3, you will first learn about the coordinate plane, linear equations in two variables, and graphs of ordered pairs. You will see that many real-world relationships involving ratio and rate can be represented by linear equations. You will also see that many linear relationships are examples of a more general concept, that of a function.

Your study of linear equations in two variables then takes a geometric point of view. You will learn how to sketch the graph of a linear equation in two variables and how to find an equation for a specified line.

Finally, you will use algebra to find a geometric relationship between two lines in the coordinate plane and you will model geometric patterns with linear equations.

VOCABULARY

collinear	mathematical model	right triangle
constant of variation	ordered pair	sequence
continuous graph	origin	slope
coordinate axes	parallel lines	slope-intercept form
coordinate plane	parallelogram	solution to an equation
deductive reasoning	parametric equations	standard form
dependent variable	perpendicular	term of a sequence
direct variation	point-slope form	trapezoid
discrete graph	proportion	vertical lines
domain	quadrilateral	vertical-line test
function	quadrants	x-axis
horizontal line	range	x-coordinate
independent variable	rate	x-intercept
inductive reasoning	rate of change	y-axis
linear equation	ratio	y-coordinate
linear function	relation	y-intercept
linear pattern		

The diagram below shows how mathematical skills and mathematical reasoning are interrelated with the skills and concepts in Chapter 3. Notice that this chapter has two major topics: 1) functions and relations, and 2) linear equations.

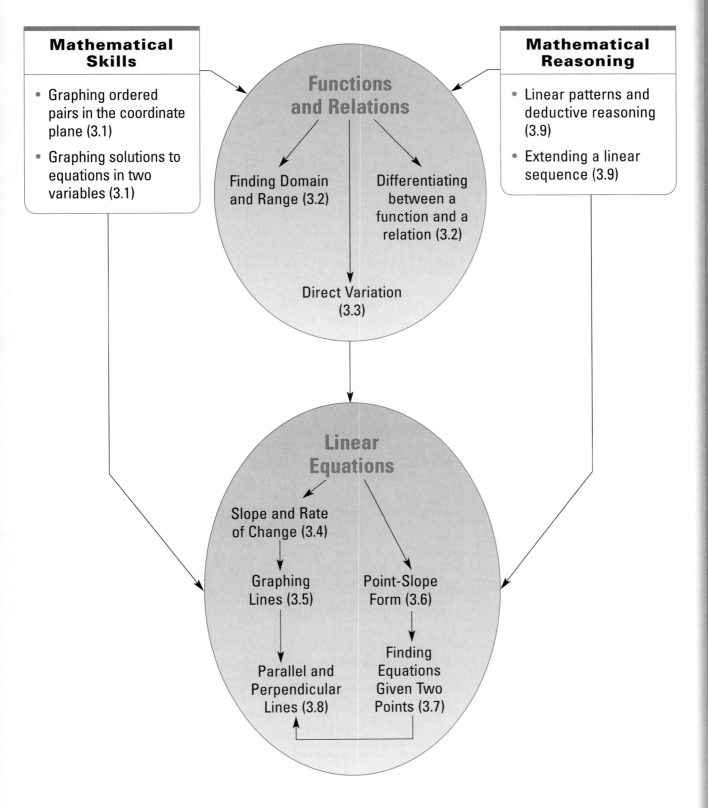

Mathematical Skills

- Graphing ordered pairs in the coordinate plane (3.1)
- Graphing solutions to equations in two variables (3.1)

Mathematical Reasoning

- Linear patterns and deductive reasoning (3.9)
- Extending a linear sequence (3.9)

Functions and Relations

Finding Domain and Range (3.2)

Differentiating between a function and a relation (3.2)

Direct Variation (3.3)

Linear Equations

Slope and Rate of Change (3.4)

Graphing Lines (3.5)

Point-Slope Form (3.6)

Parallel and Perpendicular Lines (3.8)

Finding Equations Given Two Points (3.7)

TEACHING LESSON 3.1

The Coordinate Plane

SKILL A *Graphing ordered pairs in the coordinate plane*

Focus: *Students locate a point in the plane by using an ordered pair to identify the point's distance from a set of coordinate axes.*

EXAMPLE

PHASE 1: Presenting the Example

Display the first figure at right. Tell students that the square represents a plane. They should recall from previous courses that a plane is a flat surface extending without end in all directions.

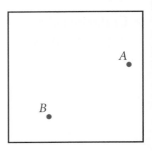

Tell students that the dots labeled *A* and *B* represent points in the plane. Ask them to write a sentence that describes the locations of points *A* and *B* in this plane. Discuss their responses. [**sample: Point *A* is above point *B* and to its right.**] Point out that the descriptions are fairly vague. That is, they do not give *exact* locations for points *A* and *B*.

Now display the second figure. Note that two lines, one vertical and one horizontal, have been superimposed on the plane. Ask students to describe the locations of points *A* and *B* relative to the two lines. [**sample: Point *A* is to the right of the vertical line; point *B* is to the left of it and a little nearer to it. Point *A* is above the horizontal line; point *B* is below it and a little farther away from it.**] Note that these descriptions come a little closer to locating points *A* and *B*, but they are still vague.

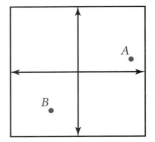

Finally, display the third figure. Here, in addition to the horizontal and vertical lines, a grid of equal units has been superimposed on the plane. Once again ask students to describe the locations of points *A* and *B*. [**sample: Point *A* is 4 grid units to the right of the vertical line and 1 grid unit above the horizontal line. Point *B* is 2 grid units to the left of the vertical line and 3 grid units below the horizontal line.**] Point out that this time it was possible to give *exact* locations for the points.

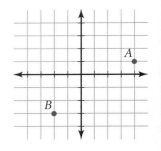

Now refer to the figure at the top of the textbook page. Note that scale numbers have been added to indicate distance on the lines, and the lines have been labeled *x* and *y*. Discuss the terms *coordinate plane, coordinate axes, x-axis, y-axis, origin, ordered pair, x-coordinate, y-coordinate,* and *quadrants*. Point out that an ordered pair is a "shorthand" way of describing the location of a point. Have students write ordered pairs for points *A* and *B* above. [*A*(4, 1), *B*(−2, −3)] Then discuss the Example.

PHASE 2: Providing Guidance for **TRY THIS**

Students may interpret the *x*-coordinate as a vertical move and the *y*-coordinate as a horizontal move. As a self-correcting activity, give them coordinates for a set of points that outline a figure when connected. For instance, give students the points at right, which outline a house. For further practice, have them name points that outline windows on the house.

(−5, 2)	(−5, −3)
(−1, −3)	(−1, 0)
(1, 0)	(1, −3)
(5, −3)	(5, 2)
(3, 4)	(3, 7)
(2, 7)	(2, 5)
(0, 7)	(−4, 3)

3.1 The Coordinate Plane

SKILL A *Graphing ordered pairs in the coordinate plane*

The **coordinate plane** is formed by placing two number lines called **coordinate axes** so that one is horizontal (**x-axis**) and one is vertical (**y-axis**). These axes intersect at a point called the **origin**, which is often labeled *O*.

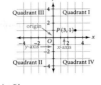

An **ordered pair** (x, y) is a pair of real numbers that correspond to a point in the coordinate plane. The first number in an ordered pair is the **x-coordinate** and the second number is the **y-coordinate**. The x-coordinate gives the distance right (positive) or left (negative) from the y-axis. The y-coordinate gives the distance up (positive) or down (negative) from the x-axis. In the figure shown above, point *P* has x-coordinate 3 and y-coordinate 1. The coordinates of the origin are (0, 0).

The coordinate axes divide the plane into four **quadrants**. Points on the coordinate axes are not in any quadrant. Point *P*, shown above, is in Quadrant I.

EXAMPLE

Graph each ordered pair. State the quadrant in which the point lies.

$$A(3, 4) \qquad B(-4, -4) \qquad C(0, -3)$$

▶ **Solution**

Begin at the origin, (0, 0). Count right or left (positive or negative) on the x-axis to locate the desired x-value. Then count up or down (positive or negative) to locate the desired y-value. Label this point with its x- and y-value.

$A(3, 4)$: Count 3 units right from *O* and then count 4 units up. Point *A* is in Quadrant I.

$B(-4, -4)$: Count 4 units left from *O* and then count 4 units down. Point *B* is in Quadrant III.

$C(0, -3)$: Because the x-coordinate is 0, do not move to the right or left. Count 3 units down from *O*. Point *C* is on the y-axis, so it is not in any quadrant.

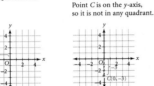

TRY THIS

Graph each ordered pair below. State the quadrant in which each point lies.
$D(-2, -5)$, $E(-1, -3)$, $F(0, -1)$, $G(1, 1)$, $H(2, 3)$, $J(3, 5)$

EXERCISES

KEY SKILLS

Examine the coordinates for each ordered pair. Is the corresponding point in a quadrant or on an axis?

1. $P(3, -2)$
2. $Q(0, 0)$
3. $R(3, 3)$
4. $B(0, -10)$

PRACTICE

Write the ordered pair that corresponds to each point.

5. point *F*
6. point *C*
7. point *X*
8. point *W*
9. point *Z*
10. point *A*
11. point *H*
12. point *J*
13. point *L*

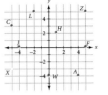

14. Write the coordinates of a point on the x-axis that lies to the right of point *H* and to the left of point *F*.

Graph and label each point on the same coordinate plane.

15. $A(-2, 0)$
16. $B(-2, -2)$
17. $C(0, -2)$
18. $D(4, -1)$
19. $W(-3, 5)$
20. $X(3, 5)$
21. $Y(-6, 2)$
22. $Z(-3, 0)$

Solve each problem.

23. If $N(a, 1)$ is in Quadrant II, what must be true about *a*?

24. If $C(a, b)$ is on the y-axis and above the x-axis, what must be true about *a* and *b*?

25. How are the locations of $P(a, b)$ and $Q(a, -b)$ related if $b \neq 0$?

26. How are the locations of $R(a, b)$ and $S(-a, b)$ related if $a \neq 0$?

27. **Critical Thinking** Let *r* and *s* be nonzero real numbers. What must be true about *r* and *s* if $X(|r|, |s|)$ and $Y(r, s)$ are the same point? In which quadrant is *X* found?

MIXED REVIEW

Find the values in the given replacement set that are solutions to the given equation. (1.3 Skill A)

28. $-2x + 5 = 12$; $\{-3, -2, -1, 0, 1, 2, 3\}$
29. $-2(t + 3) = -12$; $\{-3, -2, -1, 0, 1, 2, 3\}$
30. $2(a + 3) + 2a = 10$; $\{-3, -2, -1, 0, 1, 2, 3\}$
31. $-2(r + 3) = 0$; $\{-3, -2, -1, 0, 1, 2, 3\}$

Answers for Lesson 3.1 Skill A

Try This Example

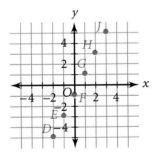

Points *D* and *E* are in Quadrant III; points *G*, *H*, and *J* are in Quadrant I; point *F* is on the y-axis and so is not in any quadrant.

1. *P* is in a quadrant.
2. *Q* is on both axes.
3. *R* is in a quadrant.
4. *B* is on the y-axis.
5. $F(5, 0)$
6. $C(-5, 3)$

7. $X(-5, -3)$
8. $W(0, -4)$
9. $Z(5, 5)$
10. $A(4, -4)$
11. $H(1, 2)$
12. $J(-4, 0)$
13. $L(-2, 5)$
14. The point chosen should be one of the following: $(2, 0)$, $(3, 0)$, or $(4, 0)$.

15–22.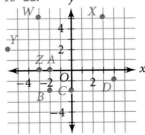

23. $a < 0$
24. $a = 0$ and $b > 0$

25. One point is directly above (below) the other.
26. One point is directly to the right (left) of the other.
27. Both *r* and *s* must be positive. *X* is in Quadrant I.
28. none
29. 3
30. 1
31. −3

Focus: *Students graph an equation in two variables, x and y, given a finite replacement set for x.*

EXAMPLES 1 and 2

PHASE 1: Presenting the Examples

Display the equation $x + 1 = 4$. Tell students to write the solution and graph it on a number line. [$x = 3$; **see the graph at right.**]

Next display $x + y = 4$. Ask students how this equation is different from the first. Elicit the fact that the first equation contains only one variable, x, while the second contains two variables, x and y. Ask students what they think is the solution to $x + y = 4$. Lead them to see that a solution is any pair of numbers whose sum is 4, and so there are infinitely many solutions. Record a few solutions for all to see, using the form $x = 1$ and $y = 3$; $x = 2$ and $y = 2$; $x = 0$ and $y = 4$; $x = 5$ and $y = -1$; and so on.

Now tell students that they may consider only $x = 0, 1, 2, 3, 4,$ and 5. Tell them to write all solutions of $x + y = 4$ given this restriction. [$x = 0$ and $y = 4$; $x = 1$ and $y = 3$; $x = 2$ and $y = 2$; $x = 3$ and $y = 1$; $x = 4$ and $y = 0$; $x = 5$ and $y = -1$] Ask if they know a "shorthand" way to write these solutions. [$(0, 4), (1, 3), (2, 2), (3, 1), (4, 0), (5, -1)$]

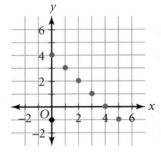

Ask students if they can graph these solutions on a number line. They should easily see that they cannot, because a number line only lets you graph one variable. Elicit from students the fact that, because the solutions can be given as ordered pairs, they can be graphed as points on a coordinate plane. Tell students to graph the solutions. [**See the graph at right.**] Check students' work. Then discuss the examples.

PHASE 2: Providing Guidance for Try This

All of the equations in this lesson are given in "$y =$" form, so y is the first variable that students see. As a result, some may mistakenly enter the y-value as the first coordinate of the ordered pair. To eliminate their confusion, you may want them to record their work in a table like the one started at right.

x	$-2(x) + 1 = y$	(x, y)
-3	$-2(-3) + 1 = 7$	$(-3, 7)$
-2	$-2(-2) + 1 = 5$	$(-2, 5)$
⋮	⋮	⋮

PHASE 3: ASSESSMENT AND CLOSURE for Lesson 3.1

- Ask students to give five solutions to $y = -x + 2$ and to graph their solutions. [**samples: (0, 2), (1, 1), (2, 0), (3, -1), (4, -2); check students' graphs.**]

- In the following lesson, a set of ordered pairs is viewed as a *relation*. Students will see how to use a graph to determine whether a relation is a *function*.

☞ *For a Lesson 3.1 Quiz, see* Assessment Resources *page 27.*

In Lesson 2.5, you solved equations in one variable. For example, given $x + 5 = -4$, you can find $x = -9$ as the solution. You can also graph this solution on a number line.

You will now learn how to solve equations in two variables, such as $y = 2x - 3$. The **solution to an equation in two variables** is all ordered pairs of real numbers (x, y) that satisfy the equation.

You can graph the solution to an equation in two variables on the coordinate plane, as shown below.

EXAMPLE 1

Let $y = 2x - 3$ and let $x = -3, -2, -1, 0, 1, 2, 3$. Graph the solutions.

▸ **Solution**

Substitute each value of x into $2x - 3$, evaluate the expression, and then make a table of ordered pairs.

Make a table.

x	-3	-2	-1	0	1	2	3
y	-9	-7	-5	-3	-1	1	3

Graph each ordered pair in the table. The points appear to lie on a line.

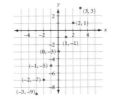

TRY THIS Let $y = -2x + 1$ and let $x = -2, -1, 0, 1, 2$. Graph the solutions.

EXAMPLE 2

Let $y = (x + 1)^2$ and let $x = -4, -3, -2, -1, 0, 1, 2$. Graph the solutions.

▸ **Solution**

Substitute each value of x into $(x + 1)^2$, evaluate the expression, and then make a table of ordered pairs.

Make a table.

x	-4	-3	-2	-1	0	1	2
y	9	4	1	0	1	4	9

Graph each ordered pair in the table. The points do not lie along a line.

TRY THIS Let $y = (x - 2)^2$ and let $x = -1, 0, 1, 2, 3, 4, 5$. Graph the solutions.

EXERCISES

KEY SKILLS

Make a table of ordered pairs for each equation.

1. $y = -3x + 5$; $x = -3, -2, -1, 0, 1, 2, 3$
2. $y = 3x - 1$; $x = -3, -2, -1, 0, 1, 2, 3$
3. $y = 2x^2 + 5$; $x = 0, 1, 2, 3, 4, 5, 6$
4. $y = 2x^2 - 3$; $x = -6, -4, -2, 0, 2, 4, 6$

PRACTICE

Graph the ordered pairs (x, y) in each table.

5.
x	-3	-2	-1	0	1	2	3
y	4	0	4	0	4	0	4

6.
x	-3	-2	-1	0	1	2	3
y	-5	-3	-1	3	5	7	9

7.
x	-3	-2	-1	0	1	2	3
y	8	6	4	2	0	-2	-4

8.
x	-3	-2	-1	0	1	2	3
y	4	4	4	4	4	4	4

Make a table of ordered pairs for each equation. Then graph the ordered pairs.

9. $y = x + 3$; $x = -3, -2, -1, 0, 1, 2, 3$
10. $y = x - 2$; $x = -3, -2, -1, 0, 1, 2, 3$
11. $y = 2x$; $x = 0, 1, 2, 3, 4, 5$
12. $y = -2x$; $x = -3, -2, -1, 0, 1, 2, 3$
13. $y = -x - 2$; $x = -3, -2, -1, 0, 1, 2, 3$
14. $y = -x - 1$; $x = -3, -2, -1, 0, 1, 2, 3$
15. $y = 2x^2 - 1$; $x = -2, -1, 0, 1, 2$
16. $y = x^2 - 4x$; $x = 0, 1, 2, 3, 4, 5$
17. $y = -2$; $x = 0, 1, 2, 3, 4, 5, 6$
18. $y = 3$; $x = -4, -2, 0, 2, 4, 6$

Let $x = -3, -2, -1, 0, 1, 2, 3$. For each value of x, find y and make a table of ordered pairs. Graph the ordered pairs.

19. $x + y = 3$
20. $x - y = 4$
21. $2x + y = 2$
22. $x - 2y = 6$

MIXED REVIEW

Write each phrase as a mathematical expression. Do not calculate. (1.1 Skill A)

23. 2 subtracted from the product of 5 and 9
24. the product of 3 and the difference of 12 minus 5
25. the difference of 5 minus the product of 2 and 5

Answers for Lesson 3.1 Skill B

Try This Example 1

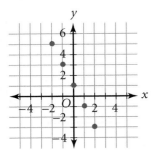

Try This Example 2

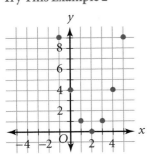

1.
x	-3	-2	-1	0	1	2	3
y	14	11	8	5	2	-1	-4

2.
x	-3	-2	-1	0	1	2	3
y	-10	-7	-4	-1	2	5	8

3.
x	0	1	2	3	4	5	6
y	5	7	13	23	37	55	77

4.
x	-6	-4	-2	0	2	4	6
y	69	29	5	-3	5	29	69

5.

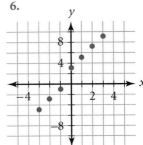

6.

For answers to Exercises 7–25, see Additional Answers.

Lesson 3.1 Skill B **95**

TEACHING 3.2 LESSON

Functions and Relations

SKILL A *Differentiating between functions and relations*

Focus: *Students determine whether a relation is a function by examining ordered pairs and by applying the vertical-line test to a graph.*

EXAMPLE 1

PHASE 1: Presenting the Example

Many students find the terms *relation* and *function* to be intimidating. Actually, the basic concepts are quite simple. To demonstrate this fact, display the following "shell:"

$$\{(\underline{\quad},\underline{\quad}), (\underline{\quad},\underline{\quad}), (\underline{\quad},\underline{\quad}), (\underline{\quad},\underline{\quad}), (\underline{\quad},\underline{\quad}), (\underline{\quad},\underline{\quad})\}$$

Tell students to copy the shell, filling in the blanks with any real numbers that they choose. After all students have completed the task, discuss the results.

Remind students that the brackets indicate a set. Any set of ordered pairs is called a *relation*. Therefore, each student has created a relation.

Now have each student examine the relation that he or she created. Do any of the ordered pairs have the same first coordinate? If not, then the ordered pairs form a special type of relation called a *function*.

It is most likely that the majority of students created functions. Ask if any created relations that are *not* functions. Have one or two of them write the relation on the chalkboard. Have the class consider what change(s) would transform these relations into functions. Then discuss Example 1.

PHASE 2: Providing Guidance for TRY THIS

Some students may become confused when several members of the domain or range are identical. You might want to have them use *mapping diagrams*. A mapping diagram for the relation in part b is shown at right. Using this visual image of the relation, it becomes clear whether each member of the domain is paired with exactly one member of the range.

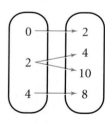

EXAMPLE 2

PHASE 2: Providing Guidance for TRY THIS

If students have difficulty visualizing the vertical-line test, have them hold a pencil vertically at the left of the graph. Then have them move the pencil across the graph from left to right, taking care to keep the pencil vertical. If at any time the pencil crosses the graph at more than one point, the graph does not represent a function.

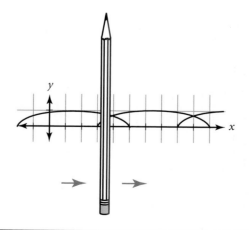

96 Chapter 3 Linear Equations in Two Variables

3.2 Functions and Relations
LESSON

SKILL A ▶ *Differentiating between functions and relations*

The equation $y = 2x - 3$ has an important characteristic. For each value of x, you will find *exactly one* value of y.

Definition of Relation and Function
A **relation** is a pairing between two sets. A **function** is a relation in which each member of the first set, the **domain**, is assigned exactly one member of the second set, the **range**.

A relation can be represented by a set of ordered pairs (x, y). The first number, x, is a member of the domain and the second number, y, is a member of the range.

EXAMPLE 1 Determine whether each relation is a function. Explain.
   a. $\{(-1, 7), (0, 3), (1, 5), (0, -3)\}$ **b.** $\{(-1, 7), (0, 3), (1, 5), (2, -3)\}$

▶ **Solution**
   a. Because 0 is paired with both 3 and -3, the relation is not a function.
   b. For each value of x, $\{-1, 0, 1, 2\}$, there is exactly one value of y, $\{7, 3, 5, -3\}$. This set of ordered pairs represents a function.

TRY THIS Determine whether each relation is a function. Explain.
   a. $\{(0, 2), (2, 4), (4, 8), (8, 10)\}$ **b.** $\{(0, 2), (2, 4), (4, 8), (2, 10)\}$

Use the following test to determine whether a graph represents a function.

Vertical-Line Test for a Function
If no vertical line in the coordinate plane intersects a graph in more than one point, then the graph represents a function.

EXAMPLE 2 Does the graph at right represent a function?

▶ **Solution**
Because it is impossible to draw a vertical line that will intersect the graph in more than one point, the graph represents a function.

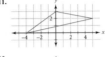

TRY THIS Does the graph at right represent a function?

96 Chapter 3 Linear Equations in Two Variables

EXERCISES

KEY SKILLS

Write the domain and range of each relation.
   1. $\{(3, 0), (-4, 1), (9, 2)\}$ **2.** $\{(2, 1), (3, 2), (7, 3)\}$

Does each situation describe a function or a relation that is not a function?
   3. Each car is assigned one license plate number.
   4. Each telephone owner is assigned two telephone numbers.

PRACTICE

Does each set of ordered pairs represent a function? Explain.
   5. $\{(1, 10), (2, 8), (3, 6), (4, 4), (5, 2)\}$ **6.** $\{(1, 10), (2, 8), (3, 6), (4, 8), (5, 2)\}$
   7. $\{(1, 1), (2, 1), (3, 1), (4, 1), (5, 1)\}$ **8.** $\{(1, 1), (1, 2), (1, 3), (1, 4), (1, 5)\}$
   9. $\{(1, 1), (4, 2), (9, 2), (16, 4), (16, -4)\}$ **10.** $\{(1, 1), (4, 2), (9, 3), (16, 4), (25, 5)\}$

Does each graph represent a function? Explain.

11. 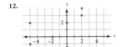   **12.**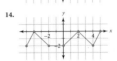

13.   **14.**

15. Critical Thinking For what value(s) of y will the list of ordered pairs $\{(2, y), (2, 4), (3, 5), (2, y + 1), (10, -3)\}$ be a function?

MIXED REVIEW

Graph all of the following points on the same coordinate plane. (3.1 Skill A)
   16. $A(-2, 3)$ **17.** $B(6, 0)$ **18.** $C(0, -3)$ **19.** $D(5, 2)$

Let $x = -3, -2, -1, 0, 1, 2, 3$. Graph the solutions to each equation that correspond to these values of x. (3.1 Skill B)
   20. $y = -x - 1.5$ **21.** $y = 2x + 1.5$

Answers for Lesson 3.2 Skill A

Try This Example 1
a. Yes; each member of the domain is assigned exactly one member of the range.
b. No; 2 is assigned both 4 and 10.

Try This Example 2
No; for example, the vertical line through $(3.5, 0)$ intersects the graph in two points.

1. domain: 3, -4, 9
   range: 0, 1, 2
2. domain: 2, 3, 7
   range: 1, 2, 3
3. function
4. not a function
5. Yes; each member of the domain is assigned exactly one member of the range.

6. Yes; each member of the domain is assigned exactly one member of the range.
7. Yes; each member of the domain is assigned exactly one member of the range.
8. No; the domain has one value, 1, that is assigned 1, 2, 3, 4, and 5.
9. No; 16 is assigned both 4 and -4.
10. Yes; each member of the domain is assigned exactly one member of the range.
11. No; for example, the y-axis intersects the graph in two points.

12. No; the vertical lines through $(-5, 0)$ and $(2, 0)$ intersect the graph in two points.
13. Yes; there is no vertical line that intersects the graph in more than one point.
14. Yes; there is no vertical line that intersects the graph in more than one point.
15. no value of y
16–19.

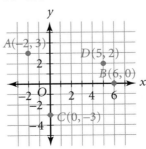

20.

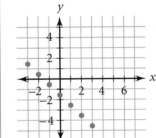

21.

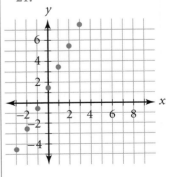

Focus: *Students find domains and ranges for relations that consist of finite sets of ordered pairs. Students examine the set of all real numbers as a domain or range.*

EXAMPLE 1

PHASE 1: Presenting the Example

Tell students that the domain of a relation is {−3, 0, 2}. Ask them to write a set of ordered pairs that could make up the relation. Then ask several students to write their answers on the chalkboard. The class should find that there are infinitely many possibilities, such as {(−3, 0), (0, 1), (2, 2)} and {(−3, −100), (0, 0), (2, 7.5)}. Be sure they are aware that it is not necessary for the relation to consist of just three ordered pairs. For example, {−3, 0, 2} is the domain of {(−3, 0), (−3, 1), (−3, 2), (−3, 3), (−3, 4), (0, 1), (2, 1)}.

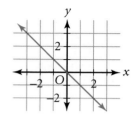

Now tell students that the domain of a relation is the set of all real numbers. Ask them if they think it is possible to write a set of ordered pairs that make up the relation. They should quickly arrive at the conclusion that there are infinitely many real numbers, and so the relation would consist of infinitely many ordered pairs.

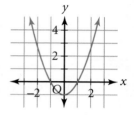

Point out that it is often possible to use a graph on a coordinate plane to show a relation whose domain is the set of all real numbers. Ask students what they think such a graph might look like, and brainstorm their ideas. Conclude the discussion by showing them several graphs like those at right, each picturing a relation whose domain is the set of all real numbers.

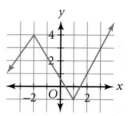

Now discuss Example 1. Note that here the process is reversed: Students are given the graph of a relation, and they are asked to identify its domain.

EXAMPLE 2

PHASE 1: Presenting the Example

Display the relation {(−3, −6), (−2, −4), (0, 0), (1, 2), (5, 10), (9, 18)}. Tell students to write its domain and range. [**domain: {−3, −2, 0, 1, 5, 9}; range: {−6, −4, 0, 2, 10, 18}**] Ask if they noticed a special relationship between the coordinates. Elicit the response that the second coordinate is always two times the first. Point out that, if (x, y) is the general form of the ordered pairs, the equation $y = 2x$ can be given as a *rule* for the relation.

PHASE 2: Providing Guidance for TRY THIS

Students may wonder how to determine when the domain is the set of all real numbers. Give them this guideline: When a rule is in "$y =$" form, the domain is all real numbers for which the expression on the right of the equal sign has meaning. For example, $y = \dfrac{1}{x}$ is restricted to a domain of nonzero real numbers because $\dfrac{1}{x}$ is undefined when x is zero.

Students will see more rules for relations with restricted domains later in the course.

Finding the domain and range of a function or a relation

EXAMPLE 1 Find the domain and range of each relation.

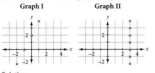

Graph I   Graph II   Graph III

▶ Solution

The domain consists of the *x*-values and the range consists of the *y*-values.

Graph I: domain: $\{-2, -1, 0, 1\}$ range: $\{-2, 0, 2, 4\}$
Graph II: domain: $\{3\}$ range: $\{-2, -1, 0, 1, 2, 3, 4\}$
Graph III: The arrows indicate that the graph continues indefinitely.
           domain: all real numbers   range: all real numbers

TRY THIS   Find the domain and range of the graph at right.

You can also find the domain and range of a relation or function represented by an equation. For the function $y = 3x - 5$, the variable *x* represents the members of the domain. The variable *y* represents the members of the range.

EXAMPLE 2 Find the domain and range of each function.   a. $y = x - 5$ b. $y = \frac{1}{x}$

▶ Solution

a. You can substitute any real number for *x*, and $x - 5$ will also be a real number (Closure Property). So, the domain and range are both all real numbers.

b. Because division by zero is undefined, *x* cannot be 0. So the domain is all real numbers except 0. The range is also all real numbers except 0 because there is no value of *x* that will make $y = 0$.

TRY THIS   Find the domain and range of each function.   a. $y = -2x + 3$ b. $y = x^2$

So far, you have seen functions represented in the following ways:
- list of ordered pairs
- equation, or assignment rule
- graph in the coordinate plane
- table of paired numbers

In Lesson 3.2 Skill C, you will see that functions can also be represented by word descriptions.

EXERCISES

KEY SKILLS

Refer to the diagram at right. Are the following statements true or false?

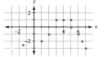

1. All values of the range are less than 2.
2. All values of the domain are integers.
3. The domain contains 0, but the range does not.

PRACTICE

Find the domain and range of each relation.

4.   5.   6.

7.   8.   9.

Find the domain and range of each function.

10. $y = \frac{5}{2}x$ 11. $y = -\frac{2}{3}x$ 12. $y = 3x + 4$

13. $y = -2x - 7$ 14. $y = x^2 - 1$ 15. $y = -2x^2$

16. **Critical Thinking**   Let *x* be any real number, let *a* be a fixed real number, and let $y = a|x|$. What is the range of the function?

MIXED REVIEW

Simplify each expression. (2.4 Skill C)

17. $\frac{8 + (-2)}{3 - 7} \times \frac{(-4)(-2)}{6(-5)}$ 18. $\frac{-2 + (-2)}{5 - 7} \times \frac{(-2)(-2)}{3(-8)}$ 19. $\frac{-2 + (-3)}{5(-5)} \times \frac{(-2)(-2)}{2(-2)}$

20. $\frac{(-2)(-3)}{5(-5)} \times \frac{(-1)(-1)}{5(-2)} \times \frac{-5}{3}$ 21. $\frac{(-5)(10)}{5(-10)} \times \frac{(-1)(-1)}{(-2)(-2)} \times \frac{-4}{7}$ 22. $\frac{(-5)(10)}{-1} \times \frac{(-1)(-1)}{(-6)(6)} \times \frac{-5}{5(10)}$

Answers for Lesson 3.2 Skill B

Try This Example 1
domain: $-4, -3, -2, -1,$ $0, 1, 2, 3$
range: 1

Try This Example 2
a. domain: all real numbers
   range: all real numbers
b. domain: all real numbers
   range: all nonnegative real numbers

1. true
2. false
3. false
4. domain: all real numbers
   range: all real numbers

5. domain: $-3, -2, -1,$ $0, 1, 2, 3$
   range: $-4, -3, -2,$ $-1, 0, 1, 2, 3$
6. domain: $-4, -3, -2,$ $-1, 0, 1, 2, 3, 4$
   range: 0, 1, 2, 3, 4
7. domain: all real numbers
   range: all real numbers greater than or equal to -2
8. domain: all real numbers
   range: 3
9. domain: all real numbers between -3 and 4, inclusive
   range: all real numbers between -2 and 3, inclusive

10. domain: all real numbers
    range: all real numbers
11. domain: all real numbers
    range: all real numbers
12. domain: all real numbers
    range: all real numbers
13. domain: all real numbers
    range: all real numbers
14. domain: all real numbers
    range: all real numbers greater than or equal to -1

15. domain: all real numbers
    range: all nonpositive real numbers
16. If $a > 0$, the range is all nonnegative real numbers. If $a < 0$, the range is all nonpositive real numbers.

17. $\frac{2}{5}$

18. $-\frac{1}{3}$

19. $-\frac{1}{5}$

20. $-\frac{1}{25}$

21. $-\frac{1}{7}$

22. $\frac{5}{36}$

Focus: *Students use function rules and graphs of functions to represent real-life situations in which one quantity depends on another.*

EXAMPLE 1

PHASE 1: Presenting the Example

Tell students that, in everyday life, one quantity often depends on another. The cost of mailing a package, for instance, generally depends on its weight. On an automobile trip, the amount of time that you spend driving depends on your speed. In cases like these, it is said that the second quantity *is a function of* the first. The second quantity is called *dependent*, while the first is *independent*. Work with students to make a list of several other real-life functions, identifying the dependent and independent quantities.

Remind students that many functions can be represented by an equation in two variables. The variable that represents the dependent quantity is called the *dependent variable*. The variable that represents the independent quantity is called the *independent variable*. When you write an ordered pair using these variables, the independent variable appears first.

Now discuss Example 1. Point out that the situation can be summarized as follows: *The total value of the nickels is a function of the number of nickels*, or *The total value of the nickels depends on the number of nickels.*

EXAMPLE 2

PHASE 1: Presenting the Example

Some students may need to write an equation that represents the function before they are able to identify its domain and range. If V represents volume in gallons and t represents time in minutes, an appropriate equation for the situation in Example 2 would be $V = 50t$.

PHASE 3: ASSESSMENT AND CLOSURE for Lesson 3.2

- Have students create two relations whose domain and range are the sets given at right. One relation should be a function, and one should not be a function. [**samples:** {(−1, −7), (0, 2), (5, −7)} (**function**); {(−1, −7), (0, 2), (5, −7), (5, 2)} (**not a function**)]

  domain: {−1, 0, 5}
  range: {−7, 2}

- Most students perceive mathematics as the study of numbers or quantities. Although this is true, it is far too limiting a description of the subject. It would be more accurate to portray mathematics as the study of numbers *and relationships among them*. Because a relation is a pairing of quantities, the study of relations— and functions—is fundamental to the study of mathematics. In this chapter, students will examine *linear functions*. In later chapters they will learn about *quadratic functions* and *polynomial functions*.

☞ *For a Lesson 3.2 Quiz, see* Assessment Resources *page 28.*

In a function such as $y = 2x$, each x-value is paired with exactly one y-value. Thus, the value of y *depends* on the value of x. In a function, the variable of the domain is called the **independent variable** and the variable of the range is called the **dependent variable**. On a graph, the independent variable is represented on the horizontal axis and the dependent variable is represented on the vertical axis.

EXAMPLE 1

The value v (in cents) of n nickels is given by $v = 5n$.
a. Identify the independent and dependent variables.
b. Represent this function in a table for $n = 0, 1, 2, 3, 4, 5, 6$.
c. Represent this function in a graph.

▶ **Solution**

a. The value, v, *depends* on the number of nickels, n. Therefore the independent variable is n and the dependent variable is v.

b.

n	0	1	2	3	4	5	6
v	0	5	10	15	20	25	30

c. Label the horizontal axis with the independent variable, n, and the vertical axis with the dependent variable, v. Graph the ordered pairs (n, v).
 (0, 0), (1, 5), (2, 10), (3, 15), (4, 20), (5, 25), (6, 30)

TRY THIS   The number of wheels, w, on n tricycles is given by $w = 3n$. Identify the independent and dependent variables. Represent the function in a table and in a graph for $n = 0, 1, 2, 3, 4, 5$.

EXAMPLE 2

Water is pumped through a pipe at the rate of 50 gallons per minute. The number of gallons pumped is a function of the number of minutes the pump operates. Find the domain and range of this function.

▶ **Solution**

Make a graph.

The domain, number of minutes, must be greater than or equal to zero. The range, number of gallons of water pumped, must also be greater than or equal to zero. Because time and water flow are continuous, the domain and range are all real numbers greater than or equal to zero.

TRY THIS   Find the domain and range of the function described in Example 1.

EXERCISES

A 1200-gallon tank is empty. A valve is opened and water flows into the tank at the constant rate of 25 gallons per minute.

1. How would you find the amount of water in the tank after 5 minutes, 6 minutes, and 7 minutes?

2. How would you find the time it takes for the tank to contain 500 gallons, 600 gallons, and 700 gallons of water?

Use both a table and a graph to represent each function for the given values of the independent variable.

3. The value v of 0, 1, 2, 3, 4, 5, and 6 dimes is a function of the number of dimes, n.

4. The distance d traveled by a jogger who jogs 0.1 mile per minute over 0, 1, 2, 3, 4, and 5 minutes is a function of time t.

5. One hat costs \$24. The total cost c of 1, 2, 3, 4, 5, and 6 hats is a function of the number n of hats purchased.

Find the domain and range of each function.

6. independent variable: number of dimes, d
 dependent variable: total value of the coins, v

7. One package of cookies contains 6 cookies.
 independent variable: number of packages of cookies, n
 dependent variable: number of cookies, c

8. A conveyor belt adds 300 pounds of sand to a pile each minute.
 independent variable: elapsed time, t
 dependent variable: total weight of the sand, w

Solve each problem. (2.8 Skill C)

9. How much of \$2000 should be put into an account paying 6% and how much should go into an account paying 12% simple interest to earn \$200 interest in one year?

10. A chemist divides 400 milliliters of a solution into two beakers. One beaker contains 70 milliliters more than the other beaker. How much will each beaker contain?

Answers for Lesson 3.2   Skill C

Try This Example 1
independent variable: n
dependent variable: w

n	0	1	2	3	4	5
w	0	3	6	9	12	15

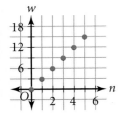

Try This Example 2
domain: all whole numbers
range: all nonnegative multiples of 5

1. Multiply 5, 6, and 7 by 25.

2. Let t represent the time it takes to achieve the specified volume. Solve $500 = 25t$, $600 = 25t$, and $700 = 25t$ for t.

3.

n	0	1	2	3	4	5	6
v	0	10	20	30	40	50	60

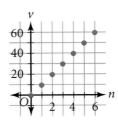

For answers to Exercises 4–10, see Additional Answers.

Ratio, Rate, and Direct Variation

SKILL A ▶ APPLICATIONS ⟩ *Using direct-variation functions to describe real-world situations*

Focus: *Students use function rules to solve direct-variation problems.*

EXAMPLE

PHASE 1: Presenting the Example

Display the table at right. Tell students that it shows data for an automobile trip. Have them find the ratio $\frac{distance}{time}$ for the numbers in each row of the table. In each case, they should arrive at the number 54. Ask what this number means. Elicit the response that the automobile is traveling at a constant speed of 54 miles per hour.

Time spent traveling (hours)	Distance traveled (miles)
0.5	27
1.0	54
1.5	81
2.0	108
2.5	135
3.0	162

Use this familiar situation to introduce the concept of direct variation. Point out that time and distance are changing, and so they are variable quantities. However, the quantities are not varying randomly: when the time increases by one hour, the distance always increases by 54 miles. In a situation like this, it is said that the distance *varies directly as* the time, and the relationship is called a *direct variation*. The constant rate of change, 54 miles per hour, is called the *constant of variation*.

Ask students if distance *depends on* time. They should quickly see that this is an accurate statement, and so it can further be said that distance *is a function of* time. Tell them to write a rule for the function. [$d = 54t$, where d is distance in miles and t is time in hours]

Now have students read the definition of direct variation given in the box on the textbook page, and answer any questions. Then discuss the Example.

PHASE 2: Providing Guidance for **TRY THIS**

The Try This exercise closely parallels the Example, and so most students should experience little difficulty. However, while the Example identifies the sales tax rate as "$0.05 per $1 of purchase," the Try This exercise expresses the rate as "6 cents per dollar." Therefore, watch for students who use 6 rather than 0.06.

Stress the fact that a constant of variation is a ratio that compares two quantities, and so it is important to consider the units in which the quantities are expressed. You can use the process of *dimensional analysis*, as shown below, to explain why students must use 0.06.

$$t = c \text{ dollars} + \frac{6 \text{ cents}}{1 \text{ dollar}} \times c \text{ dollars} \qquad t = c \text{ dollars} + \frac{0.06 \text{ dollars}}{1 \text{ dollar}} \times c \text{ dollars}$$

$$t = c \text{ dollars} + 6c \text{ cents} \qquad t = c \text{ dollars} + 0.06c \text{ dollars}$$

These expressions cannot be combined.

$$t = (1 + 0.06)c \text{ dollars}$$
$$t = 1.06c \text{ dollars}$$

3.3 LESSON — Ratio, Rate, and Direct Variation

SKILL A **APPLICATIONS** Using direct-variation functions to describe real-world situations

A **ratio** is a quotient that compares two quantities. A **rate** is a quotient that compares two different types of quantities.

ratio: 80 people out of 100 people   rate: 55 miles per hour

$$\frac{80}{100}, \text{ or } \frac{2}{5} \qquad\qquad \frac{55 \text{ miles}}{1 \text{ hour}}$$

Both ratios and rates are the basis for many simple but important functions. One type of function based on ratio or rate is *direct variation*.

> **Direct Variation**
>
> If there is a fixed nonzero number k such that $y = kx$, then you can say that y varies directly as x. The relationship between x and y is called a **direct-variation** relationship and k is called the **constant of variation**.

If you write $y = kx$ as $\frac{y}{x} = k$, you can see that k is a constant ratio or a constant rate. For example, if you drive for t hours at 55 miles per hour, then the distance d (in miles) that you travel is given by $d = 55t$. In other words, $\frac{d}{t} = 55$ miles per hour, which is a constant rate.

EXAMPLE In a certain state, the sales tax rate is $0.05 per $1 of purchase.
a. Use an equation to express the total cost of a purchase as a function of the purchase price.
b. Identify the constant of variation.
c. Find the domain and range.
d. Find the total cost of a purchase of $14.39.

▶ **Solution**
a. Let p represent the purchase price and let t represent the total cost.
  total cost = purchase price + 0.05 × purchase price
  $t = p + 0.05 \times p$
  $t = 1p + 0.05p = (1 + 0.05)p = 1.05p$ ⟵ Apply the Distributive Property.
b. The constant of variation is 1.05.
c. The domain is any nonnegative money amount in dollars and cents. The range is any nonnegative money amount in dollars and cents.
d. $t = 1.05p$
  $= 1.05(14.39)$ ⟵ Substitution
  ≈ 15.11 ⟵ Round to the nearest cent.

  The total cost is $15.11.

TRY THIS Rework the Example with a sales tax of 6 cents per dollar.

KEY SKILLS

Write the unit that would result from each product.

1. $\frac{5 \text{ dollars}}{1 \text{ package}} \times 6 \text{ packages}$

2. $\frac{55 \text{ miles}}{1 \text{ hour}} \times 8 \text{ hours}$

PRACTICE

For Exercises 3–5,
a. use an equation to express each direct variation as a function.
b. identify the constant of variation.
c. find the domain and range.
d. find the specified quantity.

3. The total price of a dinner, p, is the cost of the meal, m, plus a 15% tip. Find the total price of a dinner if the meal costs $19.50.

4. In a certain store, the retail price, p, of an item is the wholesale cost, c, of the item plus a 50% markup. Find the retail price of a coat whose wholesale cost is $98.00.

5. A pump fills an empty tank at the rate of 30 gallons per minute. Find the amount of liquid in the tank after 5.5 minutes.

Use a direct-variation equation to solve each problem.

6. A certain automobile can travel 22 miles on each gallon of gas.
a. How far will it travel on 36 gallons of gas?
b. How many gallons of gasoline are needed for an 800-mile trip?

7. On a certain map, a map distance of 1 inch corresponds to an actual distance of 50 miles.
a. What actual distance corresponds to a map distance of 2.5 inches?
b. What map distance would correspond to an actual distance of 300 miles?

8. A store has a sale and marks each item down to 80% of its original price.
a. Find the sale price of an appliance that cost $84 before the discount was taken.
b. If the sale price of an appliance is $64, what was its original price?

9. An employee's wage is $12.45 per hour.
a. How much money will the employee earn for 40 hours of work?
b. How many hours must the employee work to earn $240?

MIXED REVIEW APPLICATIONS

Solve each problem. (previous courses)

10. Find 60% of $18.

11. What percent of $140 is $210?

12. Find 20% of $20.

13. Find 150% of $200.

Answers for Lesson 3.3 Skill A

Try This Example
a. $t = 1.06c$
b. 1.06
c. The domain is any positive money amount in dollars and cents. The range is any positive money amount in dollars and cents.
d. $15.25

1. dollars
2. miles
3. a. $c = 1.15m$
  b. 1.15
  c. The domain is any positive money amount in dollars and cents. The range is any positive money amount in dollars and cents.
  d. $22.43

4. a. $c = 1.5p$
  b. 1.5
  c. The domain is any positive money amount in dollars and cents. The range is any positive money amount in dollars and cents.
  d. $147
5. a. Let v represent the amount of water in the tank after t minutes; $v = 30t$
  b. 30
  c. domain: all nonnegative numbers range: all nonnegative numbers
  d. 165 gallons
6. a. 792 miles
  b. about 36.4 gallons

7. a. 125 miles
  b. 6 inches
8. a. $67.20
  b. $80
9. a. $498
  b. about 19.3 hours
10. $10.80
11. 150%
12. $4
13. $300

Focus: *Students solve real-world problems involving direct-variation relationships by writing and solving proportions.*

EXAMPLE 1

PHASE 1: Presenting the Example

In Skill A, students used a given constant of variation to write a function rule, and then used the rule to solve a problem. Here they see that it is possible to solve direct-variation problems without ever calculating the constant of variation. That is, they learn to write and solve a proportion.

Be sure students understand how to read the notation x_1, x_2, y_1, and y_2. Tell them that the numbers are called *subscripts* and that x_1 is read as "x sub one," x_2 is read as "x sub two," and so on. Point out that this is easier to say than "the first x-value," "the second x-value," and so on.

From their work in previous courses, students may remember that they can use cross products to solve proportions as follows:

$$\frac{13}{3} = \frac{c}{6} \;\rightarrow\; \frac{13}{3} \times \frac{c}{6} \;\rightarrow\; 3c = 78 \;\rightarrow\; \frac{c}{5} = \frac{78}{3} \;\rightarrow\; c = 26$$

Allow students to use this method, but point out that using the Multiplication Property of Equality is more efficient:

$$\frac{13}{3} = \frac{c}{6} \rightarrow (6)\left(\frac{13}{3}\right) = (6)\left(\frac{c}{6}\right) \rightarrow 26 = c$$

EXAMPLE 2

PHASE 2: Providing Guidance for TRY THIS

Remind students that $\frac{a}{b} = \frac{c}{d}$ is read "*a* is to *b* as *c* is to *d*." If they have difficulty setting up their proportions, it may help to translate the problem into this language. For example, suggest that students write "125 gallons is to 2 minutes as how many gallons is to 9 minutes?" They should now easily see that the appropriate proportion is $\frac{125}{2} = \frac{v}{9}$, where *v* represents the unknown number of gallons.

PHASE 3: ASSESSMENT AND CLOSURE for Lesson 3.3

- Give students this situation: *When you mail-order an item from a certain company, a $5 handling charge is added to the item's price.* Ask them to explain why the total cost of an order does *not* vary directly as the item's price. [**sample: The total cost in dollars, *c*, is given by $c = p + 5$, where *p* is the item's price in dollars. This is not a relationship of the form $y = kx$.**]

- This begins a sequence of lessons that will relate direct variation, rate of change, slope of a line, and, ultimately, the equation of a line. In Chapter 9, students will learn about *inverse variation*.

☞   *For a Lesson 3.3 Quiz, see* Assessment Resources *page 29.*

A **proportion** is a statement that two ratios are equal. If y varies directly as x, then you can say that y is *proportional to* x.

Alternate Form of Direct Variation

If (x_1, y_1) and (x_2, y_2) satisfy the direct-variation relationship $y = kx$, then
$$\frac{y_1}{x_1} = \frac{y_2}{x_2}.$$

Proof: Suppose that (x_1, y_1) and (x_2, y_2) satisfy $y = kx$.

$y_1 = kx_1$ $y_2 = kx_2$ ⟵ *Definition of solution*

$\dfrac{y_1}{x_1} = k$ $\dfrac{y_2}{x_2} = k$ ⟵ *Multiplication Property of Equality*

$\dfrac{y_1}{x_1} = k = \dfrac{y_2}{x_2}$ ⟵ *Transitive Property of Equality*

EXAMPLE 1

The cost of lunch, c, varies directly as the number of people, n. If lunch for 3 people costs $13, find the cost of lunch for 6 people.

▶ **Solution**

Write and solve a proportion.

$\dfrac{13}{3} = \dfrac{c}{6}$ ⟵ *Set the ratios equal to each other.*

$6\left(\dfrac{13}{3}\right) = 6\left(\dfrac{c}{6}\right)$ ⟵ *Apply the Multiplication Property of Equality.*

$26 = c$

The cost of lunch for 6 people is $26.

TRY THIS   Refer to Example 1. Find the cost of lunch for 9 people.

EXAMPLE 2

An empty water tank is being filled by a pump. After 2 minutes, the tank contains 125 gallons. If the volume of water in the tank varies directly as the time the water is pumped, how much water is in the tank after 5 minutes?

▶ **Solution**

Let v represent the volume of water in the tank after 5 minutes.
Write and solve a proportion.

$\dfrac{volume}{time} \rightarrow \dfrac{125}{2} = \dfrac{v}{5}$

$5\left(\dfrac{125}{2}\right) = 5\left(\dfrac{v}{5}\right)$ ⟵ *Apply the Multiplication Property of Equality.*

$312.5 = v$

After 5 minutes, the tank contains 312.5 gallons of water.

TRY THIS   Refer to Example 2. How much water is in the tank after 9 minutes?

EXERCISES

Solve each proportion.

1. $\dfrac{12}{36} = \dfrac{n}{6}$ 2. $\dfrac{a}{42} = \dfrac{3}{14}$ 3. $\dfrac{10}{25} = \dfrac{n}{12}$ 4. $\dfrac{m}{21} = \dfrac{3}{10}$

Given that y varies directly as x, use a proportion to find y.

5. If $y = 33$ when $x = 3$, find y when $x = 22$.

6. If $y = 4.6$ when $x = 2.3$, find y when $x = 2$.

7. If $y = 8.4$ when $x = 3.5$, find y when $x = 12$.

8. If $y = 35.1$ when $x = 13.5$, find y when $x = 5.5$.

9. If $y = 109.56$ when $x = 8.3$, find y when $x = 2.2$.

Solve each direct-variation problem using:
   a. a direct-variation equation.   **b.** a proportion.

10. Simple interest in a period of time varies directly as the amount of the deposit. If a $2000 deposit earns $120, how much interest will $2400 earn in the same amount of time?

11. The number of gallons of paint needed is proportional to the area of the surface to be painted. If one gallon covers 250 square feet, how many gallons will be needed to cover 540 square feet? (Give one decimal answer and one whole number answer.)

12. The number of pages read is proportional to the time spent reading. If a student can read 3 pages in 10 minutes, how many pages will the student read in 30 minutes?

13. The number of students served in a cafeteria varies directly as time. If 10 students can be served in 6 minutes, how many students will be served in 48 minutes?

Solve each problem using an equation.   (1.9 Skill B)

14. Six workers together receive $111 for one hour of work. What is each worker's wage given that they all earn the same amount?

15. A stamp collector separated stamps into 5 equal groups of 31 stamps each. How many stamps did the collector have to sort?

Answers for Lesson 3.3   Skill B

Try This Example 1
$39

Try This Example 2
562.5 gallons

1. $n = 2$
2. $a = 9$
3. $n = 4.8$
4. $m = 6.3$
5. $y = 242$
6. $y = 4$
7. $y = 28.8$
8. $y = 14.3$
9. $y = 29.04$
10. $144
11. 2.16 gallons; 3 gallons
12. 9 pages
13. 80 students
14. $18.50
15. 155 stamps

Slope and Rate of Change

SKILL A   *Finding the slope of a line*

Focus: *Students use two points to find the slope of a line.*

EXAMPLE

PHASE 1: Presenting the Example

Provide each student with a sheet of square grid paper. Display line ℓ positioned on a grid exactly as shown at right. (There are no axes and there is no scale. These are deliberate omissions.) Tell students to draw a line on their grid paper that has exactly the same "slant" as line ℓ. Give them a few minutes to complete the task. Then discuss the results.

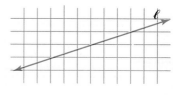

Ask students how they determined the slant. Consider all valid techniques, but ultimately direct the discussion toward the concept of slope. That is, the slant of the line can be described in terms of two components, the vertical change and the horizontal change. To draw a line with the same slant, you can start at the intersection of any two grid lines, count up one grid unit, count right three grid units, and then draw a second point. The line that contains these two points will have exactly the same slant as the given line.

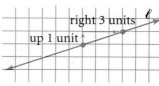

Now display line *m*, as shown at right. Tell students to draw another line on their grid paper, this one having exactly the same slant as line *m*. Give them a few minutes to complete the task, and then again discuss the results.

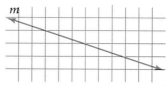

This time, to draw a line with the same slant as line *m*, you start at the intersection of any two grid lines, count up one grid unit, and then count *left* three grid units. So, although line *m* seems to have the same *amount* of slant as line ℓ, it has a different *direction* of slant. That is, line ℓ slants *up* to the right, while line *m* slants *down* to the right.

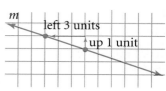

Next discuss with students the presentation of slope at the top of the textbook page. Stress that the slope of a line in the coordinate plane is a ratio of rise to run, $\frac{\text{rise}}{\text{run}}$.

Now discuss the Example. Here students will see how the coordinates of two points on a line are used to find the slope of the line. In part a, the line rises from left to right, and has a positive slope. The line in part b falls from left to right, signaling a negative slope.

PHASE 2: Providing Guidance for TRY THIS

Tell students that they may choose either ordered pair to represent (x_1, y_1), but that once the choice is made, they must subtract the *y*-coordinates and *x*-coordinates in the same order. If students confuse the coordinates, suggest that they write the coordinates on their papers using a different color for each ordered pair. They can then use the same colors to write the subtractions, making sure that the colors appear in the same order in each subtraction.

3.4
LESSON

Slope and Rate of Change

SKILL A *Finding the slope of a line*

Every nonvertical line in the coordinate plane has a steepness, or **slope**. The ratio $\frac{\text{rise}}{\text{run}}$, illustrated at right, is a measure of that steepness. If two points lie on the same line, then you can use the coordinates of those points to calculate the slope.

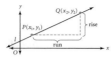

The Slope of a Line

If $P(x_1, y_1)$ and $Q(x_2, y_2)$ lie along a nonvertical line in the coordinate plane, the line has slope m, given by $m = \frac{\text{rise}}{\text{run}} = \frac{y_2 - y_1}{x_2 - x_1}$.

EXAMPLE

Find the slope of the line that contains each pair of points.
a. $P(-3, -1)$ and $Q(5, 6)$ **b.** $R(-2, 8)$ and $S(7, 3)$

▶ **Solution**

Choose (x_1, y_1) and (x_2, y_2) and apply the formula for slope.

a. $m = \frac{y_2 - y_1}{x_2 - x_1} \rightarrow \frac{6 - (-1)}{5 - (-3)} = \frac{7}{8}$ **b.** $m = \frac{y_2 - y_1}{x_2 - x_1} \rightarrow \frac{3 - 8}{7 - (-2)} = \frac{-5}{9} = -\frac{5}{9}$

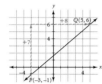 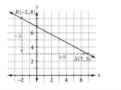

TRY THIS

Find the slope of the line that contains each pair of points.
a. $A(-2, 3)$ and $B(5, 4)$ **b.** $A(-2, 3)$ and $C(5, -4)$
Notice that the line graphed in part **a** *rises* from left to right and the line graphed in part **b** *falls* from left to right.

Slope and Orientation of a Line

In the coordinate plane, a line with positive slope *rises* from left to right. A line with negative slope *falls* from left to right. A line with 0 slope is horizontal. The slope of a vertical line is undefined.

EXERCISES

KEY SKILLS

Graph the line that contains each pair of points. Does the line rise from left to right, fall from left to right, or is it horizontal?

1. $A(2, 6)$ and $B(6, 0)$ **2.** $M(-3, 4)$ and $N(5, 4)$ **3.** $X(-1, -4)$ and $Y(4, 6)$

PRACTICE

Find the slope of the line that contains each pair of points.

4. $R(0, 0)$ and $S(6, 2)$ **5.** $G(0, 0)$ and $H(5, -1)$ **6.** $W(0, 0)$ and $Z(4, 0)$

7. $A(-2, 0)$ and $B(2, 0)$ **8.** $F(0, 4)$ and $G(-4, 0)$ **9.** $B(-1, -4)$ and $C(3, 9)$

10. $P(6, 2)$ and $Q(-6, 1)$ **11.** $M(7, -2)$ and $N(-5, 4)$ **12.** $X(4, -4)$ and $Y(-4, -4)$

13. $A(2.3, 0)$ and $B(6.6, 0)$ **14.** $D(3.5, 2)$ and $E(5.5, 7)$ **15.** $L(-1, -3)$ and $K(-4, -3)$

Use the figure at right to find the slope of each line.

16. the line that contains A and B
17. the line that contains X and Y
18. the line that contains R and S
19. the line that contains P and Q
20. the line that contains L and M

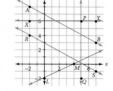

21. Points $P(5, 2)$ and $Q(-3, y)$ lie on a line with slope $\frac{3}{2}$. What is the y-coordinate of Q?

22. Critical Thinking You are given $A(a, b)$ and $R(r, s)$. What must be true of a, b, r, and s if the line that contains A and R rises from left to right?

MIXED REVIEW

Solve each equation. (2.9 Skill B)
  a. Write the steps in the solution as a logical argument.
  b. Write the solution as a conditional statement.

23. $-2(x + 5) = 16$ **24.** $3a - 2.6 = 4$ **25.** $-4 = 3x + 5x$

26. $-3r - r + 5 = 6$ **27.** $-2(x - 1) + 3 = 7$ **28.** $4b - 1 = 2b + 8$

Answers for Lesson 3.4 Skill A

Try This Example
a. $\frac{1}{7}$ **b.** -1

1. falls from left to right

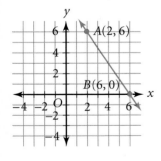

2. horizontal

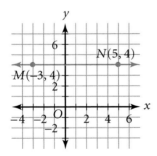

3. rise from left to right

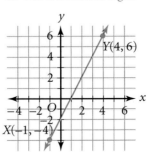

4. $\frac{1}{3}$

5. $-\frac{1}{5}$

6. 0

7. 0

8. 1

9. $\frac{13}{4}$

10. $\frac{1}{12}$

11. $-\frac{1}{2}$

12. 0

13. 0

14. $\frac{5}{2}$

15. 0

16. $-\frac{5}{9}$

17. 0

18. $-\frac{5}{9}$

19. undefined

20. $\frac{1}{2}$

★21. -10

22. Both $s - b$ and $r - a$ must have the same sign.

23. If $-2(x + 5) = 16$, then $x + 5 = -8$ by the Division Property of Equality.
If $x + 5 = -8$, then $x = -13$ by the Addition Property of Equality.
Therefore, by the Transitive Property of Equality, if $-2(x + 5) = 16$, then $x = -13$.

For answers to Exercises 24–28, see Additional Answers.

★ **Advanced Exercises**

Focus: *Students find points on a given line by using the fact that one point and a slope determine a unique line.*

EXAMPLE 1

PHASE 1: Presenting the Example

Display the figure at right. Ask students to name a line that contains the point $A(3, 2)$. They should quickly see that lines p, q, r, and s each satisfy that condition. Now ask them to name a line that contains $A(3, 2)$ *and has slope 3*. This time they should recognize that the description fits exactly one line, which is line r. Ask them to write similar descriptions to identify lines p, q, and s. [**line p: contains $A(3, 2)$ and has slope**

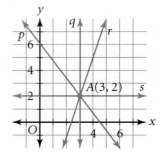

$-\dfrac{4}{3}$**; line q: contains $A(3, 2)$ and has no slope; line s: contains $A(3, 2)$ and has slope 0**]

Students should recognize that, to uniquely identify a line in the coordinate plane, it is sufficient to name one point on the line and the slope of the line.

Now discuss Example 1. Here students are given the slope of a line and one point on it, and they are asked to graph the line. They are also asked to find an additional point on the line.

PHASE 2: Providing Guidance for TRY THIS

Point out to students that they can check their answer to part b by using the point that they find along with $P(0, 5)$ to calculate slope. For example, if a student found a second point of $(4, 4)$, he or she could perform the following check:

$$\frac{5 - 4}{0 - 4} = \frac{1}{-4} = -\frac{1}{4}$$

If the result is not a slope of $-\dfrac{1}{4}$, the student knows that an error has been made.

After completing the Try This exercise, you can further reinforce the concept by having the students work in pairs. One partner writes a slope and a point; the other partner sketches the line and locates another point on it. The students then switch roles.

EXAMPLE 2

PHASE 1: Presenting the Example

PHASE 2: Providing Guidance for TRY THIS

Be sure that students are aware of the fundamental principle that underlies this type of problem. Remind them that, through any point P, there is exactly one line with slope m. Therefore, if the slope of the line through points P and Q is m, and if the slope of the line through points P and R is also m, then points P, Q, and R are all contained in the same unique line.

SKILL B Finding points on a line

EXAMPLE 1

A line with slope $\frac{3}{5}$ contains $P(1, -2)$.

a. Sketch the line.   b. Find the coordinates of a second point on the line.

▶ Solution

a. Graph $P(1, -2)$.
Because the slope is $\frac{3}{5}$, you can find a second point on
the line by counting up 3 units and then right 5 units.
Draw a line through $P(1, -2)$ and the new point. The
graph is shown at right.

b. **Method 1:** Use the graph from part **a.**
A second point on the line is $Q(1 + 5, -2 + 3)$, or
$Q(6, 1)$.

Method 2: Use the definition of slope.

Let $Q(x_2, y_2)$ be a second point on the line. Then $\frac{y_2 - (-2)}{x_2 - 1} = \frac{3}{5}$.

Choose a number other than 1 for x_2, because division by 0 is undefined.
For example, choose 0 for x_2. Then solve for y_2.

$$\frac{y_2 - (-2)}{0 - 1} = \frac{3}{5} \;\rightarrow\; \frac{y_2 + 2}{-1} = \frac{3}{5} \;\rightarrow\; y_2 = -\frac{13}{5}$$

Point $Q\left(0, -\frac{13}{5}\right)$ is another point on the line.

TRY THIS   A line with slope $-\frac{1}{4}$ contains $P(0, 5)$.

a. Sketch the line.

b. Find the coordinates of a second point on the line.

Three points are **collinear** if they lie on the same line. In the
figure at right, points P, Q, and R are collinear. Because the
points are collinear, the slopes of $\overleftrightarrow{PQ}$ and $\overleftrightarrow{QR}$ are equal.

EXAMPLE 2   Are $K(-3, -1)$, $L(-1, 3)$, and $M(1, 8)$ collinear? Explain.

▶ Solution

1. Calculate the slopes.

slope of $\overleftrightarrow{KL}$: $\frac{3 - (-1)}{-1 - (-3)} = \boxed{2}$ slope of $\overleftrightarrow{LM}$: $\frac{8 - 3}{1 - (-1)} = \boxed{\frac{5}{2}}$

> $\overleftrightarrow{KL}$ refers to the line that contains points K and L.

2. Compare the slopes.

The slopes, 2 and $\frac{5}{2}$, are not equal. Thus, the line that contains K and L
is not the same line that contains L and M. The three points are not
collinear.

TRY THIS   Are $A(-3, -1)$, $B(-1, 3)$, and $C(3, 7)$ collinear? Explain.

108   Chapter 3   Linear Equations in Two Variables

EXERCISES

KEY SKILLS

Graph the three given points. Are they collinear?

1. $P(0, 0)$, $B(3, 3)$, and $D(-2, -2)$
2. $R(2, 1)$, $E(6, 3)$, and $F(3, -2)$

PRACTICE

For Exercises 3–6,
a. sketch each line.
b. find a second point on the line.

3. containing $H(0, 0)$; slope: 2
4. containing $S(3, 3)$; slope: -3
5. containing $J(2, 5)$; slope: $-\frac{1}{2}$
6. containing $T(-3, 1)$; slope: $-\frac{2}{5}$

Are the given points collinear? Explain.

7. $A(0, 0)$, $B(2, 2)$, $C(4, 4)$
8. $D(3, 1)$, $E(6, 5)$, $F(9, 8)$
9. $P(6, 5)$, $Q(2, 2)$, $R(0, -2)$
10. $X(3, 7)$, $Y(4, 9)$, $Z(5, 11)$

Two distinct lines are *parallel* if they have the same slope. Determine if
the given lines are parallel.

11. Line s contains $(0, 5)$ and $(2, 7)$;
line t contains $(-1, 6)$ and $(2, 9)$.
12. Line q contains $(2, -3)$ and $(4, -7)$;
line r contains $(-1, 3)$ and $(0, 2)$.

13. **Critical Thinking**   A line has slope $\frac{m}{n}$ and contains $P(a, b)$.
Show that $Q(a + n, b + m)$ is also on the line.

MID-CHAPTER REVIEW

14. Graph the solutions to $y = -\frac{3}{2}x + 2$ that correspond to
$x = -4, -2, 0, 2, 4, 6$. **(3.1 Skill B)**

15. Does $\{(1, 5), (3, 2), (6, 2), (0, -2)\}$ represent a function? Explain.
(3.2 Skill A)

16. Find the domain and range of $y = \frac{2}{5}x + 3$. **(3.2 Skill B)**

17. The perimeter of a square is 4 times the length of one side. Find the
domain and range of the perimeter function. **(3.2 Skill C)**

18. Find the slope of the line that contains $A(-2, 1)$ and $B(5, 7)$.
(3.4 Skill A)

19. If y varies directly as x, and $y = 28$ when $x = 4$, find y when $x = 11$.
(3.3 Skill B)

Lesson 3.4   Skill B   **109**

Answers for Lesson 3.4 Skill B

Try This Example 1
a.

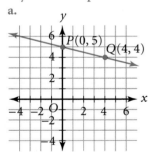

b. Method 1: $Q(4, 4)$
Method 2: Let $Q(x, y)$ be
the coordinates of a second
point on the line. Then
$\frac{y - 5}{x - 0} = -\frac{1}{4}$. Choose a
number besides 0 for x; for
example, let $x = 4$. Solve
for y:

$$\frac{y - 5}{4 - 0} = -\frac{1}{4}$$
$$y - 5 = -1$$
$$y = 4$$

So, a second point on the
line is $Q(4, 4)$.

Try This Example 2
slope of $\overleftrightarrow{AB}$:
$$\frac{3 - (-1)}{-1 - (-3)} = \frac{4}{2} = 2;$$

slope of $\overleftrightarrow{BC}$:
$$\frac{7 - 3}{3 - (-1)} = \frac{4}{4} = 1;$$
The points are not
collinear.

1. collinear

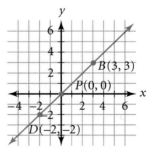

2. not collinear

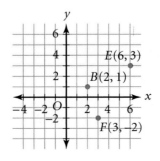

3. a.

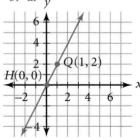

b. Answers may vary.
Sample answer:
$Q(1, 2)$

**For answers to Exercises
4–19 , see Additional
Answers.**

Lesson 3.4   Skill B   **109**

Focus: *Given two data points from real-life situations, students calculate the rate at which the data is changing.*

EXAMPLE

PHASE 1: Presenting the Example

PHASE 2: Providing Guidance for **TRY THIS**

Remind students that they have been solving problems in which quantities change since their earliest school days. For instance, here is a typical first-grade problem: *If I have two apples, and then I buy five more apples, how many apples do I have in all?* The situation concerns a change in the number of apples.

Now display the following problems:

One marker on a road showed 20 miles. The next marker showed 170 miles. How many miles have gone by?
The time was 2:00 P.M. Then it was 5:00 P.M. How many hours have passed?

Give students a minute to solve. Then point out that solving each problem requires just a simple subtraction. The answers, 150 miles and 3 hours, indicate a change in miles and a change in time, respectively.

Now display the answers as $\frac{150 \text{ miles}}{3 \text{ hours}}$. Point out that this expression compares the quantities and is called a *rate*. Since each quantity describes a change, the expression is a *rate of change*. You read the rate as *150 miles in 3 hours*, or *150 miles per 3 hours*.

In everyday life, most rates are given "per one unit" of a quantity. This is called a *unit rate*. So, it is more common to perform the division $150 \div 3 = 50$ to arrive at a unit rate of change: *50 miles per 1 hour*, or *50 miles per hour*.

Now discuss the Example and the Try This exercise. Be sure students understand the importance of the sign of the rate. For instance, the positive sign of the rate in the Example indicates that the bus is *gaining* mileage. In the Try This exercise, the negative sign of the rate indicates that the aircraft is *losing* altitude.

PHASE 3: ASSESSMENT AND CLOSURE for Lesson 3.4

- Ask students the following questions: How is a slope of $\frac{5}{2}$ different from a slope of $\frac{3}{5}$?

  [**slope $\frac{5}{2}$: constant rise of 5 for every run of 2; slope $\frac{2}{5}$: constant rise of 2 for every run of 5**] How is a slope of 4 different from a slope of -4? [**slope 4: a line that slopes up to the right; slope -4: a line that slopes down to the right**] How is a slope of 0 different from an undefined slope? [**slope 0: a horizontal line; undefined slope: a vertical line**]

- Students have now learned that, for a line in the coordinate plane, there is a constant rate of change between the y-coordinates and x-coordinates of the points. This rate is defined as the line's slope. In upcoming lessons, students will learn how slope is used to write a function rule.

☞ *For a Lesson 3.4 Quiz, see* Assessment Resources *page 30.*

The slope of a line is called *constant* because it is the same regardless of the points chosen to calculate it. This is illustrated at right by using the definition of slope as $\frac{\text{rise}}{\text{run}}$.

$$\frac{\text{rise}}{\text{run}} = \frac{2}{3} = \frac{4}{6} = \frac{-2}{-3}$$

Notice that rise represents the *change* in y and run represents the corresponding *change* in x. Because slope is the ratio $\frac{\text{rise}}{\text{run}}$, slope is sometimes called *rate of change*.

$$\text{slope} = m = \frac{\text{rise}}{\text{run}} = \frac{\text{change in } y}{\text{change in } x} = \text{rate of change}$$

Rate of change describes the relationship between two different quantities that are changing. An example of a rate of change is speed, or $\frac{\text{change in distance}}{\text{change in time}}$. Rates of change are usually expressed as *unit rates*. A unit rate has a denominator of one unit, such as 55 miles per hour, or $\frac{55 \text{ miles}}{1 \text{ hour}}$. So, if you graph the time and distance data for a person driving 55 miles per hour, the slope of the line is 55.

EXAMPLE

After stopping to buy gas, a bus driver drives at a constant rate as indicated in the table at right. Find the speed or rate of change for the bus and graph the situation.

Time t	0 hr	2 hr
Distance d	5 mi	120 mi

▶ **Solution**

Plot and connect the points in the table. The speed or rate of change is the slope of the line.

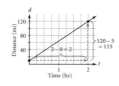

$$\text{slope} = \frac{\text{change in distance traveled}}{\text{corresponding time change}}$$
$$= \frac{d_2 - d_1}{t_2 - t_1} = \frac{120 - 5}{2 - 0} = \frac{115}{2} = 57.5$$

The bus traveled at a constant speed of 57.5 miles per hour.

TRY THIS

A small aircraft begins a descent to land as indicated in the table at right. If the plane descends at a constant rate, find its rate of descent and graph the situation.

Time t	1.5 min	3 min
Altitude a	5450 ft	4700 ft

EXERCISES

KEY SKILLS

Is the rate of change positive or negative?

1. $\frac{32 - 12}{10 - 5}$

2. $\frac{28 - 60}{100 - 80}$

3. $\frac{117.4 - 126.5}{7.0 - 6.9}$

4. $\frac{17.5 - 16.5}{3 - 2}$

PRACTICE

Calculate each rate of change.

5.
elapsed time t	3 hr	4.4 hr
distance d	140 mi	280 mi

6.
elapsed time t	3 min	4.6 min
altitude a	3450 ft	2350 ft

7.
elapsed time t	5 min	6.5 min
temperature T	35°F	32°F

8.
elapsed time t	1 min	5.4 min
temperature T	80°F	212°F

In Exercises 9–12, calculate each rate of change.

9. A hose flows 250 gallons of water in 2 minutes and 875 gallons in 7 minutes.

10. 32 people exit the stadium in 2.5 minutes and 384 people leave the stadium in 30 minutes.

11. After 3.5 minutes of descent, a small plane's altitude is 3425 feet. After 6.0 minutes, the plane's altitude is 2375 feet.

12. A racer drives 2 miles in 0.02 hours and 305 miles in 3.02 hours.

13. Does the data below indicate a constant speed? Explain.
   2 miles/4 minutes   8 miles/16 minutes   10 miles/24 minutes

14. **Critical Thinking**   Let m and b represent real numbers. Find the rate of change given $(x_1, mx_1 + b)$ and $(x_2, mx_2 + b)$.

MIXED REVIEW APPLICATIONS

Solve each direct-variation problem. (3.3 Skill B)

15. Pizza costs vary directly as quantity. Two pizzas costs $12.96 and 7 pizzas cost $45.36. How much will 10 pizzas cost?

16. The amount of an employee's paycheck, p, varies directly as the number of hours worked, h. If $p = 400$ when $h = 20$, find p when $h = 15$.

17. Interest earned varies directly as money invested. If $450 earns $15.75 and $2400 earns $84, how much interest will $6000 earn in the same amount of time?

Answers for Lesson 3.4 Skill C

Try This Example
500 feet per minute (downward)

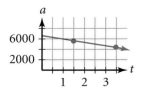

1. positive
2. negative
3. negative
4. positive
5. 100 miles per hour
6. 687.5 feet per minute (downward)
7. 2°F per minute (downward)
8. 30°F per minute (upward)
9. 125 gallons per minute
10. 12.8 people per minute (leaving)
11. 420 feet per minute (downward)
12. 101 miles per hour

★13. 2 miles/4 minutes
   8 miles/16 minutes
   rate of change: $\frac{8 - 2}{16 - 4} = \frac{1}{2}$
   8 miles/16 minutes
   10 miles/24 minutes
   rate of change: $\frac{10 - 8}{24 - 16} = \frac{1}{4}$
   The speed is not constant over the interval from 4 minutes to 24 minutes.

14. $\dfrac{(mx_2 + b) - (mx_1 + b)}{x_2 - x_1} = \dfrac{mx_2 - mx_1}{x_2 - x_1}$
   $= \dfrac{m(x_2 - x_1)}{x_2 - x_1}$
   $= m$

15. $64.80
16. $300
17. $210

★ **Advanced Exercises**

Graphing Linear Equations in Two Variables

SKILL A *Using intercepts to graph a linear equation*

Focus: *Students use the fact that two points determine a line to graph a linear equation in two variables.*

EXAMPLE

PHASE 1: Presenting the Example

Have students write the given equation, $2x + y = 4$, on a sheet of paper. Then guide them through the Example as follows:

Tell them to place one finger over the *x*-term. Ask what the visible equation tells them about the value of *y*.

Think: $x = 0$

$2x$ $+ y = 4$

So $y = 4$.

Then tell them to place one finger over the *y*-term. Ask what the visible equation tells them about the value of *x*.

Think: $y = 0$

$2x +$ y $= 4$

So $2x = 4$, and $x = 2$.

Point out that each case generates an ordered-pair solution:

$$\begin{matrix} x = 0 \\ y = 4 \end{matrix} \rightarrow (0, 4) \qquad \begin{matrix} y = 0 \\ x = 2 \end{matrix} \rightarrow (2, 0)$$

To demonstrate that the line through these two points is the graph of all the solutions of $2x + y = 4$, have students graph the equation on a sizable grid, perhaps 30 units by 30 units. [**The graph is shown at right.**] Have each student choose one point on the line other than $(0, 4)$ or $(2, 0)$ and show that its coordinates satisfy the equation. [**samples:** $(-3, 10), (-1, 6), (4, -4), (5, -6)$] Then have them choose any point *not* on the line and show that its coordinates do not satisfy the equation.

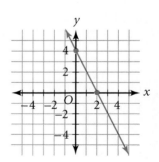

PHASE 2: Providing Guidance for **TRY THIS**

Check that students do not confuse the *x*- and *y*-intercepts. Many students try to find the *y*-intercept by covering up *y* and the *x*-intercept by covering up *x*. Stress the "opposite" relationship to these students—to find the *y*-intercept, you cover up *x*, and vice versa.

3.5 LESSON
Graphing Linear Equations in Two Variables

 SKILL A ▸ *Using intercepts to graph a linear equation*

A linear equation is an equation whose graph is a line. The following equations are examples of *linear equations*.

$$y = 3.5x - 5 \qquad 2x - 3y = 12 \qquad y = 5 \qquad x = -5.6 \qquad y + 2 = -3(x + 7)$$

Linear Equations in Two Variables

A **linear equation in *x* and *y*** is any equation of the form $Ax + By = C$, where *A*, *B*, and *C* are real numbers, and *A* and *B* are not both 0. The graph of $Ax + By = C$ is a straight line in the coordinate plane.

The **solution to a linear equation** in two variables is all ordered pairs of real numbers that satisfy the equation. One way to graph a linear equation is to use *intercepts* to find two points and draw a line through them.

The **y-intercept** is the *y*-coordinate of the point where a line crosses the *y*-axis. The **x-intercept** is the *x*-coordinate of the point where a line crosses the *x*-axis.

EXAMPLE   Let $2x + y = 4$. Use the *x*- and *y*-intercepts to graph the equation.

▸ **Solution**

To find the *y*-intercept, let $x = 0$ and solve for *y*.
$$x = 0 \rightarrow 2(0) + y = 4 \rightarrow y = 4$$
The *y*-intercept is 4, so $(0, 4)$ is on the graph.

To find the *x*-intercept, let $y = 0$ and solve for *x*.
$$y = 0 \rightarrow 2x + 0 = 4 \rightarrow x = 2$$
The *x*-intercept is 2, so $(2, 0)$ is on the graph.

Graph $(0, 4)$ and $(2, 0)$. Draw the line through $(0, 4)$ and $(2, 0)$.

TRY THIS   Let $-3x + y = 6$. Use the *x*- and *y*-intercepts to graph the equation.

Lines that rise or fall from left to right intersect both axes, as shown in the example above. Vertical lines only intersect the *x*-axis, and horizontal lines only intersect the *y*-axis.

Vertical and Horizontal Lines

The graph of $x = r$ is a vertical line that contains $(r, 0)$.
The graph of $y = s$ is a horizontal line that contains $(0, s)$.

---

112   Chapter 3   Linear Equations in Two Variables

KEY SKILLS

Find the *x*- and *y*-intercepts of the graph of each equation. If the graph has no *x*- or no *y*-intercept, write *none*.

1. $3x - 5y = 12$ **2.** $x = -2.5$ **3.** $y = 12\frac{3}{4}$ **4.** $3x + 5y = 25$

Tell whether the line is vertical, horizontal, or neither.

5. $6x + 2y = 8$ **6.** $y = -2$ **7.** $x = 5$ **8.** $4x - 3y = 11$

PRACTICE

Find the intercepts for the graph of each equation. Then graph the line.

9. $x + y = 3$ **10.** $x - y = 3$ **11.** $2x + y = 8$ **12.** $-2x + y = 8$

13. $3x + 4y = 12$ **14.** $2x + 3y = 12$ **15.** $y = -1$ **16.** $x = -1$

17. $y = \frac{5}{2}$ **18.** $x = \frac{5}{2}$ **19.** $x + y = 4$ **20.** $x - y = 4$

21. $2.5x + 2.5y = 5$ **22.** $1.5x + y = 9$ **23.** $3x + 5y = 15$ **24.** $3x - 6y = 15$

25. $y = 3x + 4$ **26.** $y = -2x - 3$ **27.** $y - 2 = 3(x + 1)$ **28.** $y + 1 = -2(x - 1)$

29. A line has *x*-intercept 3 and *y*-intercept 7. Is the line horizontal, vertical, or neither?

30. A line has *x*-intercept 2 and *y*-intercept 4. Write an equation for the line.

31. A line has *x*-intercept 3. Write an equation for the line given that
a. it rises from left to right.   **b.** it is vertical.

32. Critical Thinking   Let *a* and *b* be nonzero real numbers. Find the *x*- and *y*-intercepts of the graph of $\frac{x}{a} + \frac{y}{b} = 1$.

MIXED REVIEW

Write each decimal as a fraction. (previous courses)

33. 3.5 **34.** -1.5 **35.** 0.4 **36.** -0.6

Find the slope of the line that contains the given points. (3.4 Skill A)

37. $A(1, 2); B(3, 5)$ **38.** $K(1, -2); L(3, -6)$ **39.** $R(1, 2); S(5, 2)$ **40.** $X(5, -1); Y(-3, 0)$

Lesson 3.5   Skill A   **113**

Answers for Lesson 3.5 Skill A

Try This Example
x-intercept: -2
y-intercept: 6

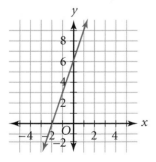

1. *x*-intercept: 4
y-intercept: $-\frac{12}{5}$

2. *x*-intercept: -2.5
y-intercept: none

3. *x*-intercept: none
y-intercept: $12\frac{3}{4}$

4. *x*-intercept: $\frac{25}{3}$
y-intercept: 5

5. neither
6. horizontal
7. vertical
8. neither
9. *x*-intercept: 3
y-intercept: 3

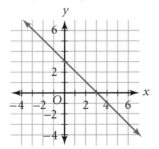

10. *x*-intercept: 3
y-intercept: -3

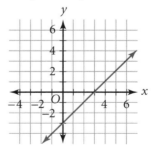

11. *x*-intercept: 4
y-intercept: 8

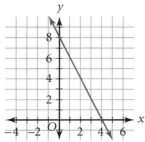

12. *x*-intercept: -4
y-intercept: 8

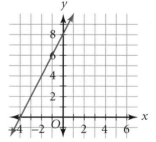

13. *x*-intercept: 4
y-intercept: 3

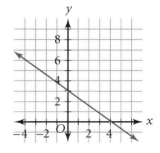

For answers to Exercises 14–40, see Additional Answers.

Lesson 3.5   Skill A   **113**

Focus: *Students graph a linear equation in two variables using the fact that a line is determined by its y-intercept and its slope.*

EXAMPLES 1 and 2

PHASE 1: Presenting the Examples

Display the two sets of equations at right. Have students graph the equations from Set I on one coordinate plane and the equations from Set II on another, using the two-intercept method of Skill A. [**The graph is given beneath each set of equations.**]

Discuss the results. Lead students to see that all of the graphs in Set I have a y-intercept of 3, which corresponds to the constant term on the right side of each equation. For Set II, note that the slope of each graph is 2, corresponding to the coefficient of the x-term of each equation.

Now display the equation from Example 1, $y = -\frac{3}{5} + 5$. Ask students to describe its

graph *without graphing*, that is, simply by examining the equation. Use their responses to introduce slope-intercept form.

When discussing Example 2, some students may notice that $-3x + 2y = 6$ could be graphed using the two-intercept method of Skill A. Although this is true, they should understand that an ability to read the slope and y-intercept of a line from an equation is an important skill.

PHASE 2: Providing Guidance for **TRY THIS**

When working with equations of the form $y = mx + b$, watch for students who read m as the y-intercept and b as the slope. In the Example 1 Try This exercise, for instance, they may graph the line with y-intercept 3 and slope −2. Stress that the coordinates of every point on the line graphed must satisfy $y = 3x - 2$. Tell them to identify the coordinates of any two points on the graph, such as (1, 1) and (2, −1), and check if they satisfy the equation. For the given choices, (1, 1) satisfies $y = 3x - 2$, while (2, −1) does not. This indicates that an error has been made.

Set I
$$y = -x + 3 \qquad y = x + 3$$
$$y = 2x + 3 \qquad y = 3$$

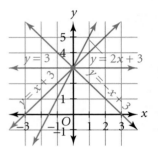

Set II
$$y = 2x - 2 \qquad y = 2x$$
$$y = 2x + 1 \qquad y = 2x + 3$$

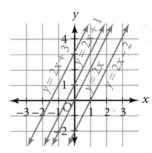

EXAMPLE 3

PHASE 2: Providing Guidance for **TRY THIS**

Watch for students who interpret the slope −2 as $\frac{-2}{-1}$.

Remind them that the fraction bar is a symbol for division, and the quotient of two negative numbers is *positive*. If the quotient is to be negative, either the numerator or the denominator may be negative, but not both. You can demonstrate this graphically as shown at right.

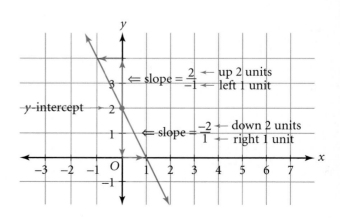

One useful form of a linear equation is $y = mx + b$, the *slope-intercept form*.

Slope-Intercept Form
If $y = mx + b$, the slope of the line is m and the y-intercept is b.

When the equation of a line is solved for y, you can quickly identify the slope and the y-intercept of the line.

EXAMPLE 1

Let $y = -\frac{3}{5}x + 5$. Use the slope and the y-intercept to graph the equation.

▶ **Solution**

1. Because the y-intercept is 5, plot $(0, 5)$.
2. Because the slope is $-\frac{3}{5}$, or $\frac{-3}{5}$, start at $(0, 5)$, count 3 units down, and then count 5 units right to arrive at $(5, 2)$.
3. Draw the line through $(0, 5)$ and $(5, 2)$.

TRY THIS Let $y = 3x - 2$. Use the slope and the y-intercept to graph the equation.

EXAMPLE 2

Let $-3x + 2y = 6$. Use the slope and the y-intercept to graph the equation.

▶ **Solution**

1. Rewrite the equation in the form $y = mx + b$.
$$-3x + 2y = 6$$
$$2y = 3x + 6 \quad \longleftarrow \text{ Add } 3x \text{ to both sides.}$$
$$y = \frac{3}{2}x + 3 \quad \longleftarrow \text{ Divide both sides by 2.}$$

2. Draw the line that contains $(0, 3)$ and that has slope $\frac{3}{2}$.

TRY THIS Let $3x - 2y = 6$. Use the slope and the y-intercept to graph the equation.

EXAMPLE 3

Sketch the line with slope 2.5 and y-intercept -1. Then write an equation for the line.

▶ **Solution**

Because the y-intercept is -1, plot $(0, -1)$.
Write 2.5 as $\frac{5}{2}$. Start at $(0, -1)$, count 5 units up, and then count 2 units right to arrive at $(2, 4)$. Draw the line through $(0, -1)$ and $(2, 4)$. An equation for the line is $y = 2.5x - 1$.

TRY THIS Rework Example 3 with slope -2 and y-intercept 2.

EXERCISES

KEY SKILLS

Start from $(0, -2)$ and find the coordinates of the point that results.

1. Count 2 units up and 3 units right.
2. Count 2 units up and 3 units left.
3. Count 2 units down and 3 units right.
4. Count 2 units down and 3 units left.

Describe each slope in terms of rise and run.

5. $\frac{3}{2}$ 6. $-\frac{3}{2}$ 7. $\frac{1}{3}$ 8. $-\frac{1}{3}$

PRACTICE

Use the slope and the y-intercept to graph each equation.

9. $y = -x + 2$ 10. $y = x + 3$ 11. $y = 3x + 1$ 12. $y = -2x + 4$
13. $y = -\frac{1}{3}x$ 14. $y = \frac{1}{3}x$ 15. $y = \frac{5}{3}x + 2$ 16. $y = -\frac{5}{3}x + 3$
17. $y = -1.5x$ 18. $y = 2.5x$ 19. $y = 0.5x + 1$ 20. $y = -0.6x - 1$

Write each equation in slope-intercept form. Then use the slope and the y-intercept to graph the equation.

21. $-\frac{1}{2}x + y = 4$ 22. $\frac{2}{3}x + y = 2$ 23. $2x - 3y = 6$ 24. $-3x + 5y = 15$

Graph the line with the given slope and y-intercept. Then write an equation for the line in slope-intercept form. (*Hint:* Write the slopes as fractions where necessary. Example: $6 = \frac{6}{1}$)

25. slope 3; y-intercept 0   26. slope -2; y-intercept 0
27. slope 1.5; y-intercept -2 28. slope -2.5; y-intercept 4
29. slope $\frac{7}{5}$; y-intercept 1   30. slope $-\frac{7}{4}$; y-intercept -2

MIXED REVIEW

Write each mixed number as a fraction and as a decimal. (previous courses)

31. $2\frac{1}{3}$ 32. $-3\frac{3}{5}$ 33. $10\frac{1}{2}$ 34. $-9\frac{1}{3}$

Solve each equation. (2.8 Skill B)

35. $2(x + 5) - 3 = 11$ 36. $2x + 5 = 11x + 32$ 37. $2(x + 3) + 5 = -2(x + 1) - 6$

Answers for Lesson 3.5 Skill B

Try This Example 1

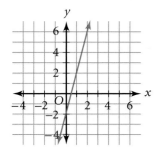

Try This Example 2

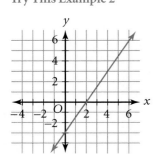

Try This Example 3

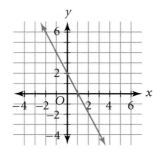

1. $(3, 0)$
2. $(-3, 0)$
3. $(3, -4)$
4. $(-3, -4)$
5. 3 units up and 2 units to the right, or 3 units down and 2 units to the left
6. 3 units down and 2 units to the right, or 3 units up and 2 units to the left

7. 1 unit up and 3 units to the right, or 1 unit down and 3 units to the left
8. 1 unit down and 3 units to the right, or 1 unit up and 3 units to the left

9.

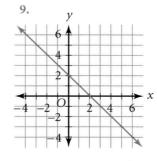

10.

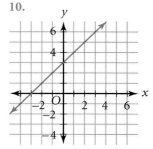

11.

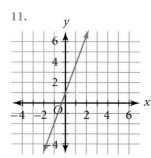

For answers to Exercises 12–37, see Additional Answers.

Focus: *Students use graphs of linear equations in two variables to model real-world situations.*

EXAMPLE 1

PHASE 1: Presenting the Example

Display the graph at right. Have each student write a sentence that describes the situation. [**A tank with 150 gallons of water empties at a constant rate of 25 gallons per minute.**] Discuss their responses. Then present Example 1, in which the process is reversed: Given the situation, students write and graph an equation to model it.

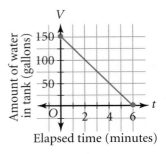

PHASE 2: Providing Guidance for **TRY THIS**

Have students use the same graph scales as in Example 1. Ask them to compare the steepness of the Try This graph to the Example 1 graph. [**It is steeper.**] Ask them to predict the steepness if the rate were 90 gallons per minute. [**It would fall between the graphs for Example 1 and Try This.**]

EXAMPLE 2

PHASE 1: Presenting the Example

Have each student write a possible number of nickels and dimes that Diego might have. Graph several of the students' responses for all to see. Use the graph to motivate the creation of the table on the textbook page.

PHASE 2: Providing Guidance for **TRY THIS**

Some students may graph just two points and connect them with a line. Point out that this line contains (0.5, 79). Ask why this is not reasonable. [**The first coordinate represents dimes. You cannot have half of a dime.**]

PHASE 3: ASSESSMENT AND CLOSURE for Lesson 3.5

• Give students each pair of equations below. Ask how the graphs in each pair are different.

$3x - 5y = 30, 5x - 3y = 30$
[**The first line passes through (0, −6) and (10, 0). The second passes through (0, −10) and (6, 0).**]

$y = 7x + 25, y = -7x + 25$
[**Both lines pass through (0, 25). The first slopes up to the right. The second slopes down to the right.**]

$x = 14, y = 14$
[**The first line is vertical and passes through (14, 0). The second is horizontal and passes through (0, 14).**]

• Students have now learned how to graph a line when given its equation. In the lessons that follow, they will learn how to find the equation for a given line.

 For a Lesson 3.5 Quiz, see Assessment Resources *page 31.*

EXAMPLE 1

A storage tank contains 300 gallons of water. A valve is opened and water is drained out at the rate of 60 gallons per minute. Represent the amount of water, V, in the tank after t minutes of drainage as a linear equation in V and as a graph.

▶ **Solution**

1. When the valve is opened, the volume is 300 gallons.
   discharge rate: $\dfrac{-60 \text{ gallons}}{1 \text{ minute}}$

2. Write a function for volume V in terms of elapsed time t.
   $V = -60t + 300$

3. When $t = 0$, $V = 300$. When $t = 5$, $V = 0$. Connect the points $(0, 300)$ and $(5, 0)$ with a line. Notice that the graph ends where it meets the axes.

TRY THIS   Rework Example 1 given a 600-gallon tank that is being drained at the rate of 120 gallons per minute.

The graph in Example 1 is a *continuous graph,* a graph that is unbroken.

EXAMPLE 2

Diego has $1.00 in nickels and dimes. Represent this with an equation and with a graph.

▶ **Solution**

1. Let n represent the number of nickels and let d represent the number of dimes. Make a table with amounts in cents.

Make a table.

Coins	n	d	$n + d$
Value	$5n$	$10d$	100

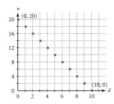

2. Write an equation for total value.
   $5n + 10d = 100$, or $n = 20 - 2d$
   d: whole number not more than 10
   n: positive even number 20 or less

3. Graph the intercepts, $(0, 20)$ and $(10, 0)$. Points (d, n) are solutions if d and n meet the conditions stated in Step **2**.

TRY THIS   Rework Example 2 given that Diego has $2.00.

The graph in Example 2 is a *discrete graph,* a graph of separate points. Because the variables in Example 2 represent coins and therefore must be whole numbers, we cannot connect the points with a line.

EXERCISES

KEY SKILLS

Write an algebraic expression for each situation.

1. The total value v of d dimes and n nickels

2. Due to a leak, a tank containing 1000 gallons loses 2 gallons per minute.

3. An airplane flying at an altitude of 1200 feet ascends 130 feet per minute.

PRACTICE

Represent each situation with an equation and with a graph. Is the graph discrete or continuous? Explain your response.

4. A box contains 15 marbles. Each minute, 3 marbles are removed from the box.

5. An airplane is flying at 1200 feet. Each minute, the plane descends 300 feet.

6. Jamie has $0.25 in pennies and nickels.

7. A tank contains 200 gallons of water when a discharge pipe is opened. Water is drained from the tank at the rate of 40 gallons per minute.

8. A clerk earns $12.50 per hour and works a maximum of 8 hours per day. Time worked is a whole number of hours.

9. The sale price is 75% of the original price.

10. **Critical Thinking**   An empty tub is filled with water at the rate of 3 gallons per minute for 5 minutes. Then the water is turned off and the water is allowed to sit for 5 minutes. At that time, the drain is opened and water is allowed to drain at the rate of 2 gallons per minute.

MIXED REVIEW APPLICATIONS

Represent volume as a function of time in a direct-variation relationship. Find the volume after 5.5 minutes. (3.3 Skill B)

11. A bin is initially empty. A conveyor belt pours 10 cubic feet of sand into the bin each minute.

12. A swimming pool is initially empty. A pump pours 125 gallons of water into the pool each minute.

Answers for Lesson 3.5 Skill C

Try This Example 1
$V = -120t + 600$

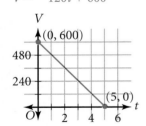

Try This Example 2
$5n + 10d = 200$, or $n = 40 - 2d$

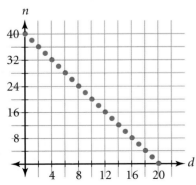

1. $10d + 5n$
2. $1000 - 2t$, where t represents elapsed time in minutes
3. $1200 + 130t$, where t represents elapsed time in minutes
4. Let m represent the number of marbles in the box after t minutes; $m = 15 - 3t$.

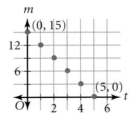

The graph is discrete.

5. Let a represent altitude after t minutes; $a = 1200 - 300t$.

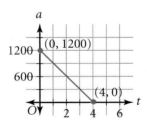

The graph is continuous.

6. Let p represent the number of pennies and n represent the number of nickels; $p + 5n = 25$.

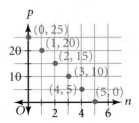

The graph is discrete.

For answers to Exercises 7–12, see Additional Answers.

3.6

The Point-Slope Form of an Equation

SKILL A ▶ *Using slope and a given point to find the equation of a line*

Focus: *Given two characteristics of a line, its slope and a point other than its y-intercept, students find the equation of the line in point-slope form.*

EXAMPLE 1

PHASE 1: Presenting the Example

Display the first graph at right. Instruct students to use the given information to find the equation for the line. Give them a minute or two to work on the question, and then discuss the results.

Remind students that they learned to write the slope-intercept form of an equation in Lesson 3.5. Most of them should have observed that, using the given slope of 2, you can start at $(2, 5)$ and count down 2 units for every 1 unit that you move to the left. By doing this, it is fairly easy to identify the y-intercept of the line as 1. Consequently, the equation of the line is $y = 2x + 1$.

Now display the second graph. Ask students if they think they can use the same method to find the y-intercept. A few students may see that it is possible to use the technique of counting down $\frac{7}{3}$ unit for every 1 unit that you move to the left, arriving at a y-intercept of $\frac{1}{3}$. However, the majority of students will most likely find it difficult to count fractions of units.

Use this problem to motivate the discussion of point-slope form that appears at the top of the textbook page. Demonstrate how point-slope form can be used to find the equation of the line in the second graph, as shown at right. Then discuss Example 1.

PHASE 2: Providing Guidance for **TRY THIS**

Watch for students who write $y - 3 = \frac{3}{5}(x - 2)$. Suggest that they begin each point-slope exercise by writing $y - \square = m(x - \square)$, with empty boxes holding the places of the coordinates. Then, as a second step, they can focus on inserting the coordinates into the boxes. This should help them arrive at $y - (-3) = \frac{3}{5}(x - 2)$, or $y + 3 = \frac{3}{5}(x - 2)$.

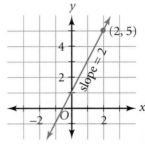

$$y = 2x + 1$$

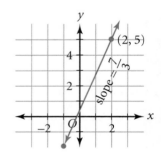

$$y - y_1 = m(x - x_1)$$
$$y - 5 = \frac{7}{3}(x - 2)$$
$$y - 5 = \frac{7}{3}x - \frac{7}{3}(2)$$
$$y - 5 = \frac{7}{3}x - \frac{14}{3}$$
$$y - 5 + 5 = \frac{7}{3}x - \frac{14}{3} + 5$$
$$y = \frac{7}{3}x + \frac{1}{3}$$

EXAMPLE 2

PHASE 1: Presenting the Example

Example 2 shows how to use point-slope form to find an equation of a horizontal line. Ask students why there is no similar method for a vertical line. [**It has an undefined slope, so there is no number to replace *m* in the equation.**]

3.6 The Point-Slope Form of an Equation
LESSON

SKILL A Using slope and a given point to find the equation of a line

If you know the coordinates of one point on a line and the slope of the
line, you can write an equation for the line.

Point-Slope Form

If $P(x_1, y_1)$ is a point on a nonvertical line with slope m, then the line can
be represented by $y - y_1 = m(x - x_1)$.
This is called the *point-slope form* of a linear equation.

Proof: Suppose $P_1(x_1, y_1)$ and $P(x, y)$ lie on a line with slope m.
Then $\frac{y - y_1}{x - x_1} = m$. Multiplying by $x - x_1$ results in $y - y_1 = m(x - x_1)$.

EXAMPLE 1 Point $Q(-3, -5)$ is on a line with slope $-\frac{5}{3}$. Write an equation for the
line in both point-slope form and slope-intercept form.

> **Solution**
> $$y - y_1 = m(x - x_1) \quad \longleftarrow \text{Use the point-slope form.}$$
> $$y - (-5) = -\frac{5}{3}[x - (-3)] \quad \longleftarrow \text{Substitute } -3 \text{ for } x_1, -5 \text{ for } y_1, \text{ and } -\frac{5}{3} \text{ for } m.$$
> $$y = -\frac{5}{3}x - 10 \quad \longleftarrow \text{Write in slope-intercept form.}$$

TRY THIS Point $Q(2, -3)$ is on a line with slope $\frac{3}{5}$. Write an equation for the line in
both point-slope form and slope-intercept form.

Recall from Lesson 3.5 Skill A that the graph of an equation of the form
$y = s$ is a horizontal line. Also recall that all horizontal lines have slope 0.

EXAMPLE 2 Point $Q(-3, -5)$ is on a line with slope 0. Write the equation for the line
in slope-intercept form.

> **Solution**
> $$y - y_1 = m(x - x_1) \quad \longleftarrow \text{Use the point-slope form.}$$
> $$y - (-5) = 0[x - (-3)] \quad \longleftarrow \text{Substitute } -3 \text{ for } x_1, -5 \text{ for } y_1, \text{ and 0 for } m.$$
> $$y + 5 = 0 \quad \longleftarrow \text{Simplify.}$$
> $$y = -5 \quad \longleftarrow \text{Write in slope-intercept form.}$$

TRY THIS Point $Q(7, 4)$ is on a line with slope 0. Write the equation for the line in
slope-intercept form.

EXERCISES

KEY SKILLS

1. Graph several different lines that all contain $Q(3, -3)$. Use the graph
   to show that one point alone is not enough to determine a unique line.

2. Graph several different lines that all have slope $-\frac{5}{3}$. Use the graph to
   show that slope alone is not enough to determine a unique line.

PRACTICE

Using the given information, write an equation in point-slope form and
in slope-intercept form.

3. $A(-4, 2)$ and slope 1   4. $B(5, -2)$ and slope -1 5. $V(5, 3)$ and slope 2

6. $K(-4, 0)$ and slope -3 7. $X(0, -3)$ and slope -0.5 8. $R(5, 1)$ and slope 0.5

9. $F(-4, -2)$ and slope $\frac{3}{7}$ 10. $B(-2, -2)$ and slope $-\frac{1}{7}$ 11. $A(5, -9)$ and slope 2.5

12. $C(7, 2)$ and slope 0   13. $P(6, -1)$ and slope 0   14. $N(0, 0)$ and slope 0

15. $S(0, 2)$ and slope 0   16. $Q(6, -1)$ and slope 1.6   17. $U(4, 10)$ and slope $\frac{11}{12}$

18. A bug begins traveling at $P(-5, 1)$ along a line with slope $-\frac{5}{7}$. The bug
    travels in the positive x-direction. Where will the bug cross the y-axis?

19. **Critical Thinking** Use the point-slope form to show that the line
    containing $P(a, b)$ with slope 0 has the form $y = b$.

20. **Critical Thinking** Use the point-slope form to show that the line
    containing $P(0, 0)$ with slope m has the form $y = mx$.

MIXED REVIEW

Find the slope of the line that contains the given points. (3.4 Skill B)

21. $T(0, 2.4)$ and $C(2.4, 0)$ 22. $M(5, -1)$ and $Y(6, -5)$ 23. $N(4.2, 8)$ and $X(4.4, 10)$

Use the x- and y-intercepts to graph each equation. (3.5 Skill A)

24. $-4.5x + 4.5y = 9$ 25. $5x - 8y = 40$ 26. $x - 8y = 2$

27. Is the graph of $x = r$ a horizontal or a vertical line? What point does
    the line contain? What can you say about the slope of the line?

Answers for Lesson 3.6 Skill A

Try This Example 1

$y + 3 = \frac{3}{5}(x - 2); y = \frac{3}{5}x - \frac{21}{5}$

Try This Example 2
$y = 4$

1.

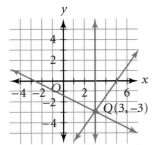

Answers may vary. Sample answer: Through a given point, you
can draw several lines. No unique
line is determined.

2.

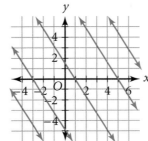

Answers may vary. Sample
answer: You can sketch several
lines that have a given slope. No
unique line is determined.

3. $y - 2 = x + 4; y = x + 6$
4. $y + 2 = -x + 5; y = -x + 3$
5. $y - 3 = 2(x - 5); y = 2x - 7$
6. $y - 0 = -3(x + 4);$
   $y = -3x - 12$
7. $y + 3 = -0.5(x - 0);$
   $y = -0.5x - 3$
8. $y - 1 = 0.5(x - 5);$
   $y = 0.5x - 1.5$

9. $y + 2 = \frac{3}{7}(x + 4); y = \frac{3}{7}x - \frac{2}{7}$

10. $y + 2 = -\frac{1}{7}(x + 2);$
    $y = -\frac{1}{7}x - \frac{16}{7}$

11. $y + 9 = 2.5(x - 5);$
    $y = 2.5x - 21.5$

12. $y = 2$
13. $y = -1$
14. $y = 0$
15. $y = 2$
16. $y + 1 = 1.6(x - 6);$
    $y = 1.6x - 10.6$

17. $y - 10 = \frac{11}{12}(x - 4);$
    $y = \frac{11}{12}x + \frac{19}{3}$

18. $\left(0, -2\frac{4}{7}\right)$

**For answers to Exercises 19–27, see
Additional Answers.**

★ **Advanced Exercises**

Focus: *Students use equations in point-slope form to represent real-world situations.*

EXAMPLE 1

PHASE 1: Presenting the Example

Have students read the problem. Ask them to write a list of the given information. [**Grace and Jamal drove at a constant speed of 55 miles per hour. They drove for 2.5 hours. They were 180 miles from their starting point.**] Ask them to write the question to be answered. [**How far will Grace and Jamal be from their starting point after 5 hours?**] Now work with students to convert their responses into a graphic display, as shown at right. Then discuss the solution on the textbook page.

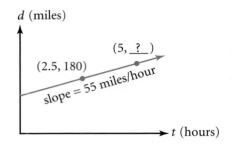

PHASE 2: Providing Guidance for TRY THIS

Students should arrive at $d = 52t + 42$ as the slope-intercept form of the equation. Point out that when $t = 0$, $d = 42$. Ask students to interpret these values of t and d. [**When Grace and Jamal begin driving at a constant speed of 52 miles per hour, they are 42 miles from their starting point.**]

EXAMPLE 2

PHASE 1: Presenting the Example

Have students read the problem. Ask them to compare it to the problem in Example 1. Ask how it is similar. [**sample: Starting from a nonzero amount, a quantity increases at a constant rate.**] To emphasize the similarities, again create a graphic display of the situation, as shown at right.

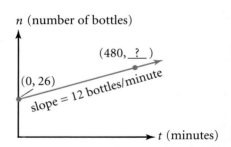

PHASE 3: ASSESSMENT AND CLOSURE for Lesson 3.6

- Give students the equation $y - 8 = -2(x + 10)$. Ask them what they can determine about its graph just by looking at the equation. [**The graph is a line with slope −2 that contains (−10, 8).**] If students have difficulty responding, repeat the question with other equations in point-slope form.

- Students have now learned how to find the equation of a line given its slope and *y*-intercept, and also given its slope and any point on it. In the next lesson, they will learn how to find the equation of a line given any two points on it.

☞ *For a Lesson 3.6 Quiz, see* Assessment Resources *page 32.*

The point-slope form of an equation can be used to write equations that represent real-world problems.

EXAMPLE 1

After driving for 2.5 hours, Grace and Jamal had traveled 180 miles from their starting point. If they drive 55 miles per hour from that point on, how far from their starting point will they be after 5 total hours of driving?

▶ **Solution**

1. Interpret the data. Let d represent the distance traveled and let t represent the elapsed time.
   (elapsed time, distance traveled) $\rightarrow$ (t_1, d_1) $\rightarrow$ $(2.5, 180)$
   constant speed of 55 mph $\rightarrow$ slope is 55 $\rightarrow$ $m = 55$

2. Write and simplify an equation. Because you know the coordinates of a point on the line and the slope, use the point-slope form.
   $$d - d_1 = m(t - t_1)$$
   $$d - 180 = 55(t - 2.5)$$
   $$d = 55t + 42.5$$

3. If $t = 5$, then $d = 55(5) + 42.5 = 317.5$.
   After 5 hours, they will be 317.5 miles from their starting point.

TRY THIS

Rework Example 1 given that they were 120 miles from their starting point after 1.5 hours and they drove 52 miles per hour.

EXAMPLE 2

A machine operator counted 26 full bottles in stock at the start of the business day. A filling machine is turned on and can fill 12 bottles every minute. Write an equation for the function. How many bottles will be filled after 8 hours?

▶ **Solution**

1. Interpret the data. Let b represent the number of bottles on hand and let t represent the elapsed time in minutes.
   (elapsed time, number of bottles) $\rightarrow$ (t_1, b_1) $\rightarrow$ $(0, 26)$
   12 bottles filled every minute $\rightarrow$ slope is 12 $\rightarrow$ $m = 12$

2. Write and simplify an equation.
   $$n - n_1 = m(t - t_1)$$
   $$n - 26 = 12(t - 0)$$
   $$n = 12t + 26$$

3. Because 8 hours = 480 minutes, let $t = 480$.
   $$n = 12t + 26 \rightarrow n = 12(480) + 26 = 5786$$
   There will be 5786 bottles filled after 8 hours (480 minutes).

TRY THIS

Suppose the filling rate is 15 bottles per minute and the day starts with 120 full bottles in stock. How many full bottles will be in stock after 4 hours?

EXERCISES

KEY SKILLS

Identify the point and slope found in each verbal description.

1. A small aircraft ascends at 350 feet per minute. After 4 minutes, the plane's altitude is 1900 feet.

2. Twenty students enter an auditorium each minute. After 5 minutes, there are 135 students in the room.

PRACTICE

Write an equation in point-slope form and solve each problem.

3. A bakery has 144 sugar cookies on hand. The baker makes 24 cookies every hour. How many cookies will be on hand after 6 hours?

4. After 2 years, a computer system is valued at $6200. Its value will diminish at the rate of $500 per year. What will its value be after 5 years?

5. A T-shirt company charges $50 for the first 5 T-shirts ordered, plus $6 for every additional shirt order after that. How much would ordering 20 shirts cost?

Find the domain and range of each function. Write an equation for the function. Then answer the question.

6. A wood-cutting machine has already cut 120 boards. If the machine can cut 12 boards per minute, how many cut boards will be on hand after 18 minutes?

7. After driving for 3 hours, Dave and Jenna have traveled 520 miles. If they drive at a constant speed of 50 miles per hour from that point on, how far will they have traveled after driving a total of 9 hours?

MIXED REVIEW APPLICATIONS

Find the domain and range of each function. (3.2 Skill C)

8. the cost of n dozen bagels if one dozen bagels costs $4.50

9. the quantity of r inches of rainfall in a storm when it rains 0.6 inch per minute

10. the cost of n nights in a hotel that costs $195 per night

11. the cost of p postage stamps if each stamp is valued at $0.33

Answers for Lesson 3.6 Skill B

Try This Example 1
302 miles

Try This Example 2
3720 bottles

1. point: (4, 1900); slope: 350
2. point: (5, 135); slope: 20
3. Let t represent hours spent baking and c represent the number of cookies on hand; $c - 144 = 24(t - 0)$; 288
4. Let v represent the value and t represent numbers of years; $v - 6200 = -500(t - 2)$; $4700
5. Let a represent the amount and let t represent the number of T-shirts; $a - 50 = 6(t - 5)$; $140
6. domain t: nonnegative real numbers; range b: whole numbers 120 or more; $b = 12t + 120$; 336 boards
7. domain t: all positive real numbers; range d: all positive real numbers; $d = 50t + 520$; 820 miles
8. domain: all whole numbers; range: all nonnegative multiples of 4.5

9. domain: all positive real numbers; range: all positive real numbers
10. domain: all whole numbers; range: all nonnegative multiples of 195
11. domain: all whole numbers; range: all nonnegative multiples of 0.33

TEACHING
3.7
LESSON

Finding the Equation of a Line Given Two Points

SKILL A ▶ *Using two points to find the equation of a line*

Focus: *Given the coordinates of two points on a line, students find the equation of the line by calculating slope and using the point-slope form.*

EXAMPLE 1

PHASE 1: Presenting the Example

Display the first figure at right. Do *not* mark a scale on either axis. Ask students to write all the facts they know about the line that is graphed. [**samples: It passes through *P*(−4, −2). The slope is positive. The *x*-intercept is negative. The *y*-intercept is positive.**]

Point out the fact that no one was able to give the equation of the line. Ask students what additional fact(s) about the line would provide enough information to write the equation. Because of their experiences in Lessons 3.5 and 3.6, they most likely will respond that they need to know either the *y*-intercept or the slope. Some students may perceive that knowing any other point on the line would give the information needed to calculate its slope.

Add point *Q* and its coordinates to the graph, as shown in the second figure at right. Ask students to write what additional facts about the line they now know.

[**It passes through *Q*(5, 3). Its slope is $\frac{5}{9}$.**] Using this new information, work with students to find the equation for the line in point-slope form and in slope-intercept form, as demonstrated in the Example on the textbook page.

PHASE 2: Providing Guidance for TRY THIS

Have students work in pairs. One student in each pair should find the equation using the slope and point *A*, while the other uses the slope and point *B*. The students should then compare their answers and discuss whether one solution method seemed easier than the other.

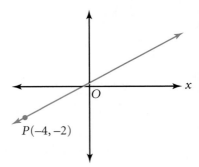

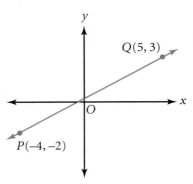

EXAMPLE 2

PHASE 1: Presenting the Example

PHASE 2: Providing Guidance for TRY THIS

Example 2 demonstrates how to write an equation for a vertical or horizontal line simply by inspecting the coordinates of two points. Students should be encouraged to use this "common sense" approach. However, be sure they know how to interpret the results if they first calculate slope, as shown at right.

A(4, 3) and *B*(4, −5):

slope $= \dfrac{-5 - 3}{4 - 4} = \dfrac{-8}{0} \rightarrow$ undefined

The line is vertical, with equation *x* = 4.

R(2.5, 3.5) and *S*(−1, 3.5):

slope $= \dfrac{3.5 - 3.5}{-1 - 2.5} = \dfrac{0}{-3.5} = 0$

The line is horizontal, with equation *y* = 3.5.

3.7 Finding the Equation of a Line Given Two Points
LESSON

SKILL A *Using two points to find the equation of a line*

If the coordinates of two points are given, you can write an equation for the line that passes through them.

EXAMPLE 1 Find an equation of the line that contains $P(-4, -2)$ and $Q(5, 3)$. Write the equation in slope-intercept form.

▶ **Solution**

1. Find the slope of $\overleftrightarrow{PQ}$. $m = \dfrac{3 - (-2)}{5 - (-4)} = \dfrac{5}{9}$

2. Use the point-slope form with either point P or point Q.

$$y - y_1 = m(x - x_1) \rightarrow y - \boxed{3} = \tfrac{5}{9}(x - \boxed{5})$$
$$y = \tfrac{5}{9}x + \tfrac{2}{9} \quad \longleftarrow \text{Slope-intercept form}$$

TRY THIS Find the equation of the line that contains $A(4, -2)$ and $B(-5, 1)$. Write the equation in slope-intercept form.

When two points are given, the line that passes through them might be vertical or horizontal. Recall from Lesson 3.5 that if a line is vertical, its equation has the form $x = a$. If it is horizontal, its equation has the form $y = b$.

EXAMPLE 2 Find an equation for the line that contains each pair of points.
   a. $A(4, 3)$ and $B(4, -5)$ **b.** $R(2.5, 3.5)$ and $S(-1, 3.5)$

▶ **Solution**

   a. Because the x-coordinates of A and B are equal, we know that the line is vertical. An equation for $\overleftrightarrow{AB}$ is $x = 4$. The slope is undefined.
   b. Because the y-coordinates of R and S are equal, we know that the line is horizontal. An equation for $\overleftrightarrow{RS}$ is $y = 3.5$. The slope is 0.

TRY THIS Find an equation for the line that contains each pair of points.
   a. $K(4, 3)$ and $L(-4, 3)$ **b.** $D(-1.4, 10)$ and $E(-1.4, -10)$

Summary of the Forms for the Equation of a Line		
Name	Form	Example
Standard	$Ax + By = C$	$2x + 3y = 7$
Slope-intercept	$y = mx + b$	$y = 5x + 1$
Point-slope	$y - y_1 = m(x - x_1)$	$y - 4 = -2(x - 3)$

KEY SKILLS

A line contains points $(2, 5)$ and (a, b), where a and b are real numbers. Tell whether the line is horizontal, vertical, or neither.

  1. $a = 2$ **2.** $b = 5$ **3.** $a \neq 2$ and $b \neq 5$

PRACTICE

Find an equation for the line that contains each pair of points. If the line is not vertical, write the equation in slope-intercept form.

  4. $A(0, 0)$ and $B(4, 5)$ **5.** $K(0, 0)$ and $L(-4, 5)$
  6. $X(4.5, 0)$ and $Y(0, 4.5)$ **7.** $P(-5.5, 0)$ and $Q(0, -5.5)$
  8. $G(4.5, 0)$ and $H(0, 4)$ **9.** $A(-2, 0)$ and $C(0, 7)$
10. $D(-2, 3)$ and $K(-2, 5)$ **11.** $W(-1, 2.4)$ and $S(5, 2.4)$
12. $X(-2, 3)$ and $Y(3, 6)$ **13.** $X(-1, 7)$ and $Y(3, -1)$
14. $Q(1, 3)$ and $B(10, 24)$ **15.** $T(-1, -9)$ and $R(5, 18)$

Tell whether the given point lies on, above, or below the line defined by A and C.

16. $P(0, 5)$; $A(-1, 3)$ and $C(2, 7)$ **17.** $Q(-2, 5)$; $A(-3, 3)$ and $C(-1, 3)$
18. $Z(-5, -3)$; $A(0, 3)$ and $C(2, 4)$ **19.** $D(7, 2)$; $A(-4, 7)$ and $C(5, -4)$

20. Critical Thinking Show that $y = \dfrac{y_2 - y_1}{x_2 - x_1}(x - x_1) + y_1$ represents the line that contains (x_1, y_1) and (x_2, y_2), where $x_1 \neq x_2$.

MIXED REVIEW

Use the x- and y-intercepts to graph each equation. (3.5 Skill A)

21. $-x + 2y = 4$ **22.** $-x = 3$ **23.** $4x + 6y = 24$

Use the slope and the y-intercept to graph each equation. (3.5 Skill B)

24. $y = 4x - 2$ **25.** $y = 2x - 6$ **26.** $2x + 3y = 9$

Find the equation in slope-intercept form for each line. (3.6 Skill A)

27. slope $-\dfrac{3}{2}$; contains $Q(-3, 7)$ **28.** slope $\dfrac{9}{20}$; contains $R(2, 12)$

Answers for Lesson 3.7 Skill A

Try This Example 1
$y = -\dfrac{1}{3}x - \dfrac{2}{3}$

Try This Example 2
a. $y = 3$ **b.** $x = -1.4$

  1. vertical
  2. horizontal
  3. neither
  4. $y = \dfrac{5}{4}x$
  5. $y = -\dfrac{5}{4}x$
  6. $y = -x + 4.5$
  7. $y = -x - 5.5$
  8. $y = -\dfrac{8}{9}x + 4$
  9. $y = \dfrac{7}{2}x + 7$
10. $x = -2$
11. $y = 2.4$
12. $y = \dfrac{3}{5}x + \dfrac{21}{5}$

13. $y = -2x + 5$
14. $y = \dfrac{7}{3}x + \dfrac{2}{3}$
15. $y = \dfrac{9}{2}x - \dfrac{9}{2}$
16. above
17. above
18. below
19. above
20. If (x_1, y_1) and (x_2, y_2), where $x_1 \neq x_2$, are on the line, then the slope is $\dfrac{y_2 - y_1}{x_2 - x_1}$. By the point-slope form, $y - y_1 = \dfrac{y_2 - y_1}{x_2 - x_1}(x - x_1)$. Thus, $y = \dfrac{y_2 - y_1}{x_2 - x_1}(x - x_1) + y_1$.

21. $(0, 2), (-4, 0)$

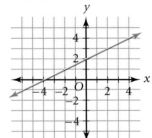

22. $x = -3$

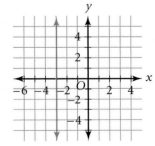

For answers to Exercises 23–28, see Additional Answers.

Focus: *Students use two data points to find linear equations that model real-world problems.*

EXAMPLE 1

PHASE 1: Presenting the Example

Students are well aware that mathematics can be used to solve real-world problems. Tell students that a *mathematical model* is something from mathematics that is used to represent an object or a situation. Ask students when they have used mathematical models.

Most students will probably respond with geometric models; for example, they have drawn rectangles to represent gardens or cylinders to represent cans. Point out to students that equations can also be used as mathematical models. Then discuss Example 1.

PHASE 2: Providing Guidance for **TRY THIS**

Students may write their ordered pairs in reverse order as (total cost, number of shirts). Ask these students which variable is independent and which is dependent: Do the number of shirts depend on the cost, or does the cost depend on the number of shirts? Then remind students that the independent variable is written first in an ordered pair.

EXAMPLE 2

PHASE 1: Presenting the Example

Display the graph at right. Tell students that it models the cost of having a party catered. Have each student write a list of facts that they can read directly from the graph. [**The cost is $570 for 24 people and $1035 for 55 people. There is a constant amount charged per person. There is a fixed fee regardless of the number of people.**] Ask students how they can find the cost per person. [**Find the slope of the line.**] Ask how they can determine the fixed fee. [**Find the y-intercept of the line.**] Then discuss Example 2.

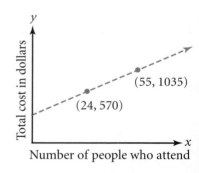

PHASE 3: ASSESSMENT AND CLOSURE for Lesson 3.7

- Have students summarize the methods for finding the equation of a line and give an example of each. [**The methods are: 1 using slope and y-intercept; 2 using slope and any point; 3 using any two points. Examples will vary.**]

- Students have now learned several techniques for writing a linear equation in two variables from information about its graph. In the lessons that follow, they will use equations like these to model sequences of numbers and to examine relationships between lines in the coordinate plane.

☞ *For a Lesson 3.7 Quiz, see* Assessment Resources *page 33.*

A **mathematical model** can describe the relationship between *data* taken from real-world situations. Some real-world situations can be modeled by linear equations, as shown below.

EXAMPLE 1

A T-shirt company charges a certain amount per T-shirt plus a handling fee for each order. Raul ordered 3 shirts and paid $32, and Maria ordered 5 shirts and paid $50.
a. How much does the company charge per shirt?
b. How much is the handling fee?
c. How much would 7 T-shirts cost?

▶ Solution

a. Use the given information to write data points.
   (number of shirts x, total cost y) → (3, 32) and (5, 50)
   Use the data points find the slope and write an equation.
   slope = $\frac{50-32}{5-3} = \frac{18}{2} = 9$ $y - y_1 = m(x - x_1)$ → $y - 50 = 9(x - 5)$
   $y = 9x + 5$
   The slope, 9, gives the cost per shirt, $9.
b. The y-intercept, 5, gives the handling fee for each order, $5.
c. If $x = 7$, then $y = 9(7) + 5$, or 68. It costs $68 for an order of 7 T-shirts.

TRY THIS   Refer to Example 1. Find the cost of 8 shirts if 3 shirts cost $25 and 5 shirts cost $39.

EXAMPLE 2

It costs $570 to cater an outdoor party for a group of 24 people and $1035 to cater the same party for a group of 55 people.
a. Find the slope and explain what it represents.
b. Find the cost to cater a party for 60 people.

▶ Solution

a. Use the given information to write two data points.
   (number of people x, party cost y) → (24, 570) and (55, 1035)
   Use the data points to write an equation for the line.
   $y - 570 = \frac{1035 - 570}{55 - 24}(x - 24)$, or $y = 15x + 210$
   The slope, 15, gives the cost per person, $15.
b. If $x = 60$, then $y = 15(60) + 210$, or 1110. It costs $1110 for a party of 60 people.

TRY THIS   After driving for 5 hours, Ben is 380 miles from home. After driving for 8 hours, Ben is 560 miles from home. Assume Ben's driving speed during this three-hour period is constant.
a. Find the slope and explain what it represents.
b. How far from home will Ben be after 10 hours?

EXERCISES

KEY SKILLS

Use the given information to create data points.

1. Four pounds of apples cost $3.60 and seven pounds of apples cost $6.30.

2. The cost of renting a garden tiller for 2 days is $55.00. The cost of renting a garden tiller for 3 days is $75.00.

PRACTICE

Solve each problem by finding a linear equation that models the data. Interpret the slope and the y-intercept of the line.

3. A home-health aide charges a fixed fee for each visit plus an hourly charge. A 2-hour visit costs $70 and a 4-hour visit costs $120. How much would a 5-hour visit cost?

4. The monthly charge for a cellular phone consists of a fixed fee plus a fee for every minute of calling time used. Vicki's monthly bill for 300 minutes of calling time was $35. Marty's monthly bill for 400 minutes of calling time was $40. How much would a monthly bill be for 500 minutes of calling time?

5. After a candle has been burning for 1 hour, its height is 10 cm. After three hours, the candle's height is 6 cm. What will the height of the candle be after it has been burning for 5 hours?

Solve each problem. Interpret the slope and the y-intercept of the line.

6. After driving for 2 hours, Lee is 220 from his starting point. After driving a total of 6 hours, Lee is 420 miles from his starting point. Assuming his speed is constant, how far is Lee from his starting point after a total of 8 hours?

7. It costs $77 to rent a car for 3 days and $115 to rent a car for 5 days. How much does it cost to rent a car for one week? Explain why the cost for 7.5 days might be the same as the cost for 8 days.

MIXED REVIEW APPLICATIONS

Solve each problem. (2.8 Skill C)

8. Jarrod needs 4 pounds of peanuts and cashews to put into a mix. Cashews cost $2.19 per pound and peanuts cost $1.98 per pound. How much of each will he buy if he spends $8.13?

9. Chen drives 375 miles in 7 hours. For the first part of the trip, he drives at 55 miles per hour. Then he drives at 50 miles per hour. How far did he drive at each speed?

Answers for Lesson 3.7 Skill B

Try This Example 1
$60

Try This Example 2
a. Let d represent distance from home after t hours of driving; $d = 60x + 80$; the slope gives the constant driving speed.
b. 680 miles

1. (4, 3.60) and (7, 6.30)
2. (2, 55) and (3, 75)
3. Let h represent the number of hours and let c represent the cost; $c = 25h + 20$; $145; the slope represents the cost per hour, $25; the y-intercept represents the fixed fee, $20.
4. Let m represent the number of minutes and let c represent the cost; $c = \frac{1}{20}m + 20$; $45; the slope represents the cost per minute, $\frac{1}{20}$ of a dollar (or 5 cents); the y-intercept represents the fixed fee, $20.

5. Let t represent time and let h represent height; $h = -2t + 12$; 2 cm; the slope represents how fast the candle is burning, 2 cm/hr; the y-intercept represents the candle's original height, 12 cm.
6. 520 miles; the slope gives the constant speed, 50 miles per hour; the y-intercept gives the distance Lee had traveled when he began driving at a constant speed, 120 miles.
7. $153; the slope gives the cost per day, $19; the y-intercept gives the fixed cost, $20; the cost for 7.5 days might be the same as the cost for 8 days if the rental company charges $19 for any part of a day.
8. 1 pound of cashews and 3 pounds of peanuts
9. 5 hours at 55 miles per hour and 2 hours at 50 miles per hour

Relating Two Lines in the Plane

Working with parallel lines in the coordinate plane

Focus: *Using the fact that parallel lines have equal slopes, students determine if two lines are parallel. They write an equation for a line parallel to a given line.*

EXAMPLE 1

PHASE 1: Presenting the Example

Display the first three equations shown at right. Tell students to graph them on the same coordinate plane. [**See the figure at right.**] Point out that the three graphs are *parallel lines*, lines in the same plane that never intersect.

Now display the next four equations. Ask students which equation has a graph that is parallel to the first three lines. Lead them to see that any line parallel to these has the same slope, 2. So the only parallel line is the graph of $y = 2x - 7$.

$y = 2x + 1$
$y = 2x$
$y = 2x - 3$

$y = -2x + 1$
$y = \frac{1}{2}x$
$y = \frac{1}{2}x - 3$
$y = 2x - 7$

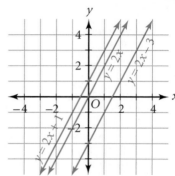

Discuss the box at the top of the textbook page, which summarizes the relationship between parallel lines and slope. Ask students why they think that parallelism of vertical lines is addressed separately. [**Vertical lines have undefined slope, so it makes no sense to talk about vertical lines having *equal* slope.**] Then discuss Example 1.

PHASE 2: Providing Guidance for TRY THIS

Watch for students who rewrite the equations as $3y = 5x + 15$ and $5y = 3x + 12$, and then decide that the lines are not parallel because the "slopes" 5 and 3 are not equal. Remind them that they must solve the equations for y, with a coefficient of 1.

EXAMPLE 2

PHASE 1: Presenting the Example

Display the problem at right. Ask students to write a brief description of the procedure to solve it. [**sample: I would write $y - 3 = 2(x - 4)$ as the point-slope form. Then I would solve for y to get $y = 2x - 5$, which is the slope-intercept form.**]

Now have students read Example 2. Ask them how this problem is similar to the one they just discussed. [**Any line parallel to $y = \frac{3}{2}x + 7$ must have slope $\frac{3}{2}$, so the problem can be rewritten as, *Write the equation in slope-intercept form for the line that has slope $\frac{3}{2}$ and contains P(4, 5).***] Then discuss the solution on the textbook page.

Write the equation in point-slope form and in slope-intercept form for the line that has slope 2 and contains P(4, 3).

3.8 LESSON — Relating Two Lines in the Plane

SKILL A — Working with parallel lines in the coordinate plane

Two lines in the plane are *parallel* if they never intersect. In the diagram at right, lines *m* and *n* are parallel and have the same slope. You can use slope to determine if two lines are parallel. Note: If two equations can be simplified to the same equation, then they will define the same line, not two distinct parallel lines.

Parallel Lines

All vertical lines are parallel.
If two distinct nonvertical lines are parallel, then they have equal slopes.
If two distinct lines have equal slopes, then they are parallel.

EXAMPLE 1

Are the lines with equations $2x - 3y = 10$ and $-3x + 2y = 2$ parallel? Justify your response.

▶ **Solution**

1. Write both equations in slope-intercept form.

$$2x - 3y = 10 \rightarrow y = \frac{2}{3}x - \frac{10}{3} \qquad -3x + 2y = 2 \rightarrow y = \frac{3}{2}x + 1$$

2. Compare the slopes. $y = \boxed{\frac{2}{3}}x - \frac{10}{3} \qquad y = \boxed{\frac{3}{2}}x + 1 \qquad \frac{2}{3} \neq \frac{3}{2}$

Because the slopes, $\frac{2}{3}$ and $\frac{3}{2}$, are not equal, the lines are not parallel.

TRY THIS Are the lines with equations $-5x + 3y = 15$ and $-3x + 5y = 12$ parallel?

EXAMPLE 2

Find the equation in slope-intercept form for the line parallel to $y = \frac{3}{2}x + 7$ that contains $P(4, 5)$.

▶ **Solution**

Any line parallel to $y = \frac{3}{2}x + 7$ must also have slope $\frac{3}{2}$. Use the slope and point $P(4, 5)$ to write an equation for the line.

$$y - y_1 = m(x - x_1) \rightarrow y - 5 = \frac{3}{2}(x - 4)$$

Simplify and write the equation in slope-intercept form: $y = \frac{3}{2}x - 1$.

TRY THIS Find the equation in slope-intercept form for the line parallel to $y = \frac{5}{4}x - 3$ that contains $P(-3, 1)$.

EXERCISES

KEY SKILLS

Compare the slopes and tell whether the lines are parallel.

1. $y = x + 2$ and $y = -x - 3$
2. $y = -\frac{2}{7}x - 1$ and $y = \frac{7}{2}x - 2$
3. $y = 2.5x + 2$ and $y = \frac{5}{2}x - 3$
4. $y = -\frac{1}{10}x + 5$ and $y = -0.1x + 5$

PRACTICE

Are the given lines parallel? Justify your response.

5. $x + 2y = 3$ and $x - 2y = 5$
6. $6x - 5y = 14$ and $6x - 5y = 13$
7. $y = 4$ and $x = 6$
8. $x = -2.6$ and $y = 7$

Find the equation in slope-intercept form for the line that contains the given point and that is parallel to the given line.

9. $P(0, 0)$; $7x + 2y = 14$
10. $Q(0, 0)$; $2x - 7y = 5$
11. $M(-5, 2)$; $-3x + 8y = 24$
12. $N(7, 3)$; $8x - 3y = 15$
13. $T(7, 12)$; $-7x - 8y = 15$
14. $Z(-3, 1)$; $11x - 3y = 1$
15. $V(0, 6)$; $2.5x - 1.5y = 1$
16. $J(3, 0)$; $10x + 15y = 50$

17. Find the equation in slope-intercept form for the line that passes through $(1, 6)$ and that is parallel to the line containing $(4, 3)$ and $(5, 1)$.

18. Find the equation in slope-intercept form for the line that passes through $(5, 2)$ and that is parallel to the line containing $(4, 2)$ and $(7, 1)$.

19. Line $\overleftrightarrow{PQ}$ contains $P(4, 3)$ and $Q(7, 5)$. Find the *x*-intercept of the line that is parallel to $\overleftrightarrow{PQ}$ and that contains $Z(5, 10)$.

20. **Critical Thinking** Are the lines with equations $35x + 14y = 21$ and $5x + 2y = 3$ parallel? Explain your response.

MIXED REVIEW

List in order from least to greatest. (previous courses)

21. 2.63, 2.62, 2.36, 2.26, 0, and 2.24
22. 10.5, 10, 11.1, 0.3, 0.03, and 1.05

Determine if each relation represents a function. (3.2 Skill A)

23. $\{(1, 2), (1, 3), (1, 4), (1, 5), (1, 6)\}$
24. $\{(3, 4), (4, 4), (5, 4), (6, 4), (7, 4)\}$
25. $\{(3, 5), (5, 8), (7, 11), (9, 14), (11, 17)\}$
26. $\{(1, 1), (2, 4), (3, 9), (2, 8), (3, 27)\}$

Answers for Lesson 3.8 Skill A

Try This Example 1
no

Try This Example 2
$y = \frac{5}{4}x + \frac{19}{4}$

1. not parallel
2. not parallel
3. parallel $(2.5 = \frac{5}{2})$
4. parallel $\left(-\frac{1}{10} = -0.1\right)$
5. $y = -\frac{1}{2}x + \frac{3}{2}$ and $y = \frac{1}{2}x - \frac{5}{2}$; not parallel
6. $y = \frac{6}{5}x - \frac{14}{5}$ and $y = \frac{6}{5}x - \frac{13}{5}$; parallel
7. not parallel; the slope of the first line is 0 and the slope of the second line is undefined.

8. not parallel; the slope of the first line is undefined and the slope of the second line is 0.
9. $y = -\frac{7}{2}x$
10. $y = \frac{2}{7}x$
11. $y = \frac{3}{8}x + \frac{31}{8}$
12. $y = \frac{8}{3}x - \frac{47}{3}$
13. $y = -\frac{7}{8}x + \frac{145}{8}$
14. $y = \frac{11}{3}x + 12$
15. $y = \frac{5}{3}x + 6$
16. $y = -\frac{2}{3}x + 2$
17. $y = -2x + 8$
18. $y = -\frac{1}{3}x + \frac{11}{3}$
19. -10

20. No; the equations have the same graph, so the lines are not distinct.
21. 0, 2.24, 2.26, 2.36, 2.62, 2.63
22. 0.03, 0.3, 1.05, 10, 10.5, 11.1
23. no
24. yes
25. yes
26. no

Focus: *Using the fact that slopes of perpendicular lines are negative reciprocals, students determine if two lines are perpendicular. They write an equation for a line perpendicular to a given line.*

EXAMPLE 1

PHASE 1: Presenting the Example

Distribute copies of the first graph shown at right. Tell students to write an equation of line ℓ in slope-intercept form. [$y = 2x$]

Remind students that *perpendicular lines* are two lines that intersect to form right angles. Tell them to choose any point on the y-axis and draw a line through it that appears perpendicular to line ℓ. Then have them write an equation of the new line in slope-intercept form.

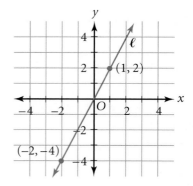

List several students' equations. Most students probably graphed a line with slope $-\frac{1}{2}$, such as $y = -\frac{1}{2}x + 3$, $y = -\frac{1}{2}x$, or $y = -\frac{1}{2}x - 2$. [**See the second graph at right.**] Circle all such equations in your list. Tell students that these lines not only *appear* perpendicular to ℓ, but in fact *are* perpendicular to it.

Note that the slope of each of these lines, $-\frac{1}{2}$, is the *negative reciprocal* of the slope of ℓ, which is 2.

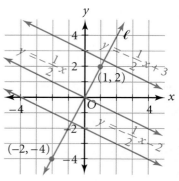

Discuss the box on the textbook page, which relates slope and perpendicular lines. Ask students why they think that vertical and horizontal lines are addressed separately. [**A vertical line has undefined slope, so it makes no sense to talk about multiplying its slope by another slope. A horizontal line has slope 0, so the product of its slope with another slope will never be −1.**] Then discuss Example 1.

PHASE 2: Providing Guidance for TRY THIS

Watch for students who rewrite $x - y = 5$ as $y = -x + 5$ and consequently decide that the lines are parallel. Suggest that they first rewrite $x - y$ as addition, as shown at right. Then it should become clear that the equivalent equation is $-y = -x + 5$, or $y = x - 5$.

$$x - y = 5$$
$$\downarrow$$
$$x + (-y) = 5$$

EXAMPLE 2

PHASE 1: Presenting the Example

Display the problem at right. Point out to students that they solved problems like this in Skill A. Ask them to write how they would solve it. [**sample: The slope of the given line is $\frac{5}{2}$, so write $y - 1 = \frac{5}{2}(x + 10)$ as the point-slope form of the parallel line. Then solve for y to get $y = \frac{5}{2}x + 26$, which is the slope-intercept form.**]

Write the equation in slope-intercept form for the line containing $Q(-10, 1)$ that is parallel to the line with equation $y = \frac{5}{2}x - 2$.

Now have students read Example 2. Ask them to compare this problem to the one they just discussed. They should quickly observe that the word *parallel* has been replaced by *perpendicular*. Then discuss the solution on the textbook page.

SKILL B *Working with perpendicular lines in the coordinate plane*

If two lines in the coordinate plane are not parallel, then they intersect at a point. If the lines intersect in such a way that they form a right angle, then the lines are *perpendicular*.

Perpendicular Lines
Every vertical line is perpendicular to every horizontal line.
Two lines are perpendicular if the product of their slopes is −1.

EXAMPLE 1

Are the lines with equations $3x - 5y = 5$ and $5x + 3y = -21$ perpendicular? Justify your response.

▶ **Solution**

1. Write both equations in slope-intercept form.

$3x - 5y = 5 \rightarrow y = \frac{3}{5}x - 1$ $5x + 3y = -21 \rightarrow y = -\frac{5}{3}x - 7$

2. Compare the slopes. $y = \boxed{\frac{3}{5}}x - 1$ $y = \boxed{-\frac{5}{3}}x - 7$ $\frac{3}{5} \times \left(-\frac{5}{3}\right) = -1$

Because the product of the slopes is −1, the lines must be perpendicular.

TRY THIS   Are the lines with equations $x - y = 5$ and $x + y = 3$ perpendicular? Justify your response.

EXAMPLE 2

Find the equation in slope-intercept form for the line perpendicular to $y = \frac{5}{2}x - 2$ that contains $Q(-10, 1)$.

▶ **Solution**

All lines perpendicular to $y = \frac{5}{2}x - 2$ must have slope $-\frac{2}{5}$, the negative reciprocal of $\frac{5}{2}$. Use the slope and point $Q(-10, 1)$ to write an equation.

$y - y_1 = m(x - x_1) \rightarrow y - 1 = -\frac{2}{5}[x - (-10)]$

Simplify and write the equation in slope-intercept form: $y = -\frac{2}{5}x - 3$.

TRY THIS   Find the equation in slope-intercept form for the line that contains $Z(3, -5)$ and that is perpendicular to $y = -\frac{7}{2}x + 6$.

128   Chapter 3   Linear Equations in Two Variables

EXERCISES

KEY SKILLS

Compare the slopes and tell whether the lines are perpendicular.

1. $y = \frac{1}{2}x - 5$ and $y = \frac{1}{2}x + 5$
2. $y = -\frac{4}{3}x - 2$ and $y = \frac{3}{4}x + 4$
3. $y = -\frac{1}{9}x + 8$ and $y = 9x - 2$
4. $y = 3x + 5$ and $y = -3x - 1$

PRACTICE

Are the given lines perpendicular? Justify your response.

5. $x = 4$ and $y = -3$
6. $x = -4$ and $y = 3.2$
7. $y = -\frac{1}{2}x + 3$ and $y = 2x + 3$
8. $y = 1.3x + 3$ and $y = -3.1x - 2$
9. $-7x + 2y = 7$ and $-4x - 7y = 10$
10. $3x + 4y = 13$ and $-4x + 3y = 10$

Find the equation in slope-intercept form for the line that is perpendicular to the given line and that contains the given point.

11. $y = -2$; $P(0, 0)$
12. $x = 5$; $Q(0, 0)$
13. $3x - y = 12$; $R(1, 0)$
14. $-2x - 5y = 10$; $Z(0, 1)$
15. $-6x - 5y = 3$; $C(1, 4)$
16. $9x + 2y = 9$; $D(4, 5)$

17. For what values of a and b will $ax + y = 10$ and $x + by = 10$ be perpendicular?
18. Write the equation of the line that contains $A(4, 2)$ and $B(8, 4)$. Then find the equation of the line that is perpendicular to $\overleftrightarrow{AB}$ and that contains $C(6, 3)$.
19. Write the equation of the line perpendicular to $3x - 11y = 12$ that contains $P(5, 2)$. Where does the line cross the x-axis?
20. **Critical Thinking** Without using pencil and paper, predict where the lines $y = \frac{7}{9}x + 4$ and $-4 = -\frac{9}{7}x - y$ will intersect.

MIXED REVIEW

Graph each set of numbers on a number line. (2.1 Skill C)

21. $-4, 3, 0, -2.5, 4$
22. $-4, 4, -2, 2, 0, 3$
23. $0, 1.5, 2.5, -3.5, -4.5, -5.5$

Let $x = -2, -1, 0, 1, 2$. Graph the solutions to each equation. (3.1 Skill B)

24. $y = -2.5x + 4$
25. $y = 2x - 2.5$
26. $y = -2.5x + 1$

Lesson 3.8   Skill B   **129**

Answers for Lesson 3.8 Skill B

Try This Example 1
Yes; $y = x - 5$ and $y = -x + 3$; the product of the slopes is −1. Therefore, the lines are perpendicular.

Try This Example 2
$y = \frac{2}{7}x - \frac{41}{7}$

1. not perpendicular
2. perpendicular
3. perpendicular
4. not perpendicular
5. Yes; one line is horizontal and the other is vertical.
6. Yes; one line is horizontal and the other is vertical.
7. Yes; the product of the slopes is −1.
8. No; the product of the slopes is not −1.

9. No; the product of $\frac{7}{2}$ and $-\frac{4}{7}$ is not −1.
10. Yes; the product of $-\frac{3}{4}$ and $\frac{4}{3}$ is −1.
11. $x = 0$
12. $y = 0$
13. $y = -\frac{1}{3}x + \frac{1}{3}$
14. $y = \frac{5}{2}x + 1$
15. $y = \frac{5}{6}x + \frac{19}{6}$
16. $y = \frac{2}{9}x + \frac{37}{9}$
17. $a = -b$
18. $y = \frac{1}{2}x$; $y = -2x + 15$
★ 19. $y = -\frac{11}{3}x + \frac{61}{3}$; $\left(\frac{61}{11}, 0\right)$
20. $(0, 4)$

21.

22.
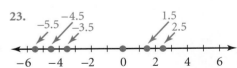

23.
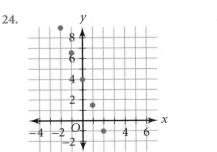

24.

For answers to Exercises 25–26, see **Additional Answers.**

★ **Advanced Exercises**

Lesson 3.8   Skill B   **129**

Focus: *Students solve problems in coordinate geometry by using slopes to determine relationships between lines.*

EXAMPLE 1

PHASE 1: Presenting the Example

Review the concept of right triangles, and give students the definition at the top of the textbook page. Then display the figure at right. Ask students to write an argument that explains why triangle *RST* must be a right triangle. [**$\overleftrightarrow{RT}$ is vertical because the *x*-coordinates of two points on it are equal. $\overleftrightarrow{ST}$ is horizontal because the *y*-coordinates of two points on it are equal. Any vertical line is perpendicular to any horizontal line. So angle *RTS* is a right angle, and triangle *RST* is a right triangle.**]

Direct students' attention to the figure in Example 1. Point out that triangle *ABC* appears to be a right triangle. However, you cannot assume that it is; you must justify the conclusion. Then discuss the solution on the textbook page.

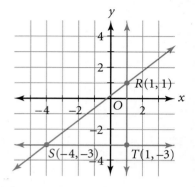

EXAMPLE 2

PHASE 2: Providing Guidance for TRY THIS

Be sure students understand the meaning of *exactly one* pair of parallel sides; if both pairs of opposite sides are parallel, the figure does not satisfy the definiton of trapezoid. This means that, unlike in Example 1, students must check the slopes of all four lines before they can determine whether the figure is a trapezoid; it is not sufficient to simply show that two sides are parallel.

PHASE 3: ASSESSMENT AND CLOSURE for Lesson 3.8

- Have students consider the line with equation $2x - 8y = 40$. Ask them to write equations of two lines parallel to it and two lines perpendicular to it.

  [**parallel: any equation of the form $y = \frac{1}{4}x + b$, where *b* is any real number; perpendicular: any equation of the form $y = -4x + b$, where *b* is any real number**]

- In Chapter 4, students will study linear inequalities in one variable. Then they resume their study of relationships among lines in the coordinate plane in Chapter 5, where they investigate systems of linear equations and inequalities in two variables.

  ☞ *For a Lesson 3.8 Quiz, see* Assessment Resources *page 34.*

3.9
LESSON

Linear Patterns and Inductive Reasoning

SKILL A ▶ Extending a linear sequence given a table

A **sequence** is an ordered set of real numbers. Each number in the sequence is called a **term of the sequence**. An example of a sequence is 2, 8, 14, 20, 26 . . . This can be represented in table form, as shown below.

term number x	1	2	3	4	5	... n
term y	2	8	14	20	26	...?

EXAMPLE 1 Find a pattern in the table above and use it to predict the 7th term.

▶ **Solution**

For each increase of 1 in the term number, the terms increase by 6.
6th term: 26 + 6, or 32
7th term: 32 + 6, or 38

	+1	+1	+1	+1
x	1	2	3	4	5
y	2	8	14	20	26
	+6	+6	+6	+6

TRY THIS Rework Example 1 using the table at right.

x	1	2	3	4	5
y	3	−1	−5	−9	−13

Linear Patterns and Linear Functions

Given a sequence of ordered pairs (x, y), if the differences between successive x-terms are the same and the differences between successive y-terms are the same, then the data can be modeled by a linear function.

EXAMPLE 2 Express the pattern in the table as a function with x as the independent variable. Predict the 80th term.

x	1	2	3	4	5
y	3	10	17	24	31

▶ **Solution**

The values of x increase by 1 and the values of y increase by 7. The pattern can be modeled by a linear function. Use data pairs (1, 3) and (2, 10) to write an equation.

	+1	+1	+1	+1
x	1	2	3	4	5
y	3	10	17	24	31
	+7	+7	+7	+7

$y - y_1 = m(x - x_1) \rightarrow y - 3 = \frac{7}{1}(x - 1)$, or $y = 7x - 4$

To predict the 80th term, let $x = 80$. $y = 7(80) - 4 = 556$
A prediction for the 80th term is 556.

TRY THIS Rework Example 2 using the table at right.

x	1	2	3	4	5
y	6	11	16	21	26

132   Chapter 3   Linear Equations in Two Variables

EXERCISES

KEY SKILLS

Tell whether the data in the table shows a linear pattern or a nonlinear pattern. Justify your response.

1.

x	1	2	3	4	5
y	−4	4	12	20	28

2.

x	1	2	3	4	5
y	−4	9	22	33	46

3. Verify that $y + 11 = 6(x - 1)$ models the data at right by verifying that each ordered pair is a solution to the equation.

x	1	2	3	4	5
y	−11	−5	1	7	13

PRACTICE

Find the specified terms in each linear sequence.

4. 7th and 8th terms

x	1	2	3	4	5
y	3	11	19	27	35

5. 7th, 8th, and 9th terms

x	1	2	3	4	5
y	19	9	−1	−11	−21

6. 7th and 9th terms

x	1	2	3	4	5
y	2.5	0.5	−1.5	−3.5	−5.5

7. 8th and 10th terms

x	1	2	3	4	5
y	0.3	2.8	5.3	7.8	10.3

Express each pattern as a function with x as the independent variable. Then predict the specified term of the sequence.

8. the table in Exercise 4; 90th term

9. the table in Exercise 5; 100th term

10. the table in Exercise 6; 29th term

11. the table in Exercise 7; 44th term

12. Critical Thinking The table at right contains a linear pattern. Find the missing terms.

x	1	2	3	4	5	6	7	8	9	10
y	4	?	?	?	16	?	?	25	?	?

MIXED REVIEW

Find the equation in slope-intercept form for the line that contains the given points. (3.7 Skill A)

13. $A(2.5, -4)$ and $B(5.5, 7)$

14. $X(0, -7)$ and $Y(-4, -7)$

15. $P(3.5, 0)$ and $Q(3.5, -8)$

16. $K(6.5, -4)$ and $L(-0.5, -4)$

17. $D(24, -2)$ and $E(27, 28)$

18. $R(98, -3)$ and $S(8, -13)$

Lesson 3.9   Skill A   133

Answers for Lesson 3.9 Skill A

Try This Example 1
For each increase of 1 in the term number, the terms decrease by 4; −21

Try This Example 2
$y = 5x + 1$; 401

1. The data shows a linear pattern; successive differences in x are 1 and successive differences in y are 8.

2. The data does not show a linear pattern; successive differences in x are 1 and successive differences in y are not constant.

3. $y + 11 = 6(x - 1)$
   $(1, -11)$: $-11 + 11 = 6(1 - 1)$; $0 = 0$ ✔
   $(2, -5)$: $-5 + 11 = 6(2 - 1)$; $6 = 6$ ✔
   $(3, 1)$: $1 + 11 = 6(3 - 1)$; $12 = 12$ ✔
   $(4, 7)$: $7 + 11 = 6(4 - 1)$; $18 = 18$ ✔
   $(5, 13)$: $13 + 11 = 6(5 - 1)$; $24 = 24$ ✔

4. 51, 59

5. −41, −51, −61

6. −9.5, −13.5

7. 17.8, 22.8

8. $y = 8x - 5$; 715

9. $y = -10x + 29$; −971

10. $y = -2x + 4.5$; −53.5

11. $y = 2.5x - 2.2$; 132.8

12. 7, 10, 13, 19, 22, 28, 31

13. $y = \frac{11}{3}x - \frac{79}{6}$

14. $y = -7$

15. $x = 3.5$

16. $y = -4$

17. $y = 10x - 242$

18. $y = \frac{1}{9}x - \frac{125}{9}$

Lesson 3.9   Skill A   133

Focus: *Students write linear equations in two variables to model geometric patterns and counting situations.*

EXAMPLE 1

PHASE 1: Presenting the Example

Display the figure below:

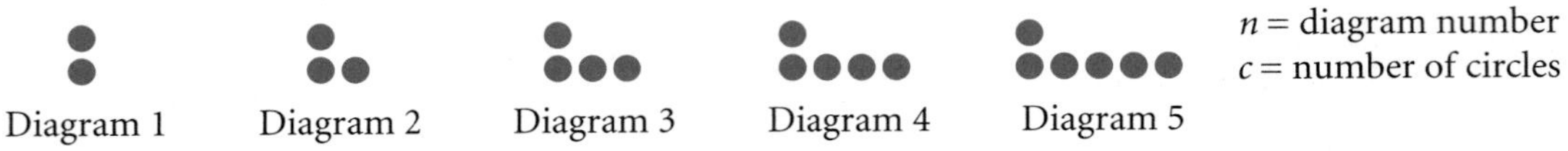

Diagram 1 Diagram 2 Diagram 3 Diagram 4 Diagram 5	n = diagram number c = number of circles

Tell students to inspect the diagrams for a pattern. [**The number of circles is one more than the diagram number.**] Have them write an equation that represents the pattern. [$c = n + 1$]

Repeat the activity with the following figure:

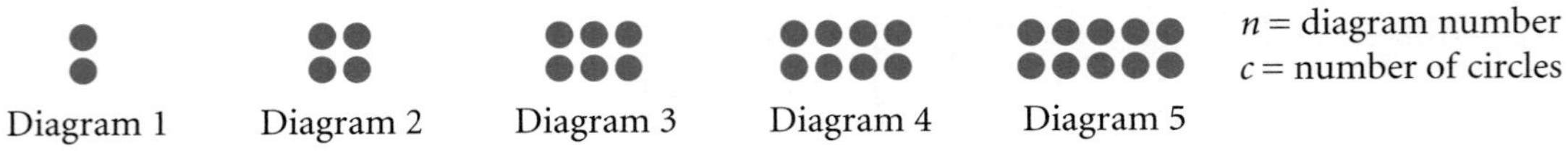

Diagram 1 Diagram 2 Diagram 3 Diagram 4 Diagram 5	n = diagram number c = number of circles

[**The number of circles is twice the diagram number; $c = 2n$**]

Now display the equation $c = 2n + 1$. Ask students to draw a set of diagrams that this equation might represent.

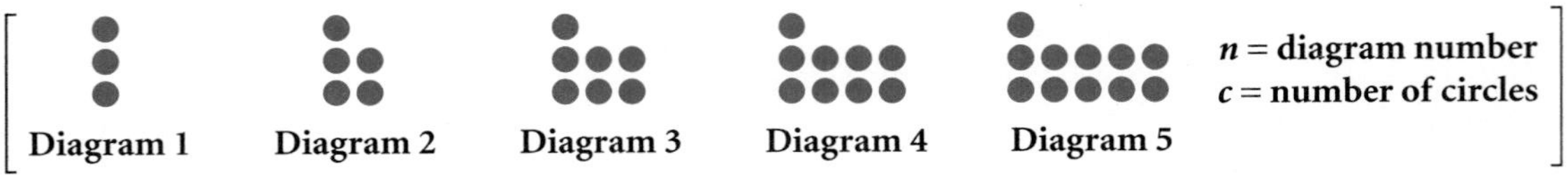

Diagram 1 **Diagram 2** **Diagram 3** **Diagram 4** **Diagram 5**	n = **diagram number** c = **number of circles**

Point out to students that they have now seen how a linear equation can model a geometric pattern. However, point out that there are many other arrangements in which the pattern is not as obvious. In these cases, it is often possible to use algebraic methods to find the numeric pattern. Then discuss Example 1, which shows such a pattern.

EXAMPLE 2

PHASE 2: Providing Guidance for **TRY THIS**

The solution method shown in Example 2 employs the point-slope form of a linear equation, which students learned earlier in this chapter. However, some students may intuitively perceive the alternative method that is shown at right.

1st term	$\rightarrow$	*Start with 12.*	$\rightarrow$ 12 $-$	(0)7 =	12
2nd term	$\rightarrow$	*Subtract 1 seven.*	$\rightarrow$ 12 $-$	(1)7 =	5
3rd term	$\rightarrow$	*Subtract 2 sevens.*	$\rightarrow$ 12 $-$	(2)7 =	-2
4th term	$\rightarrow$	*Subtract 3 sevens.*	$\rightarrow$ 12 $-$	(3)7 =	-9
5th term	$\rightarrow$	*Subtract 4 sevens.*	$\rightarrow$ 12 $-$	(4)7 =	-16
$\vdots$		$\vdots$		$\vdots$	$\vdots$
nth term	$\rightarrow$	*Subtract $(n-1)$ sevens.*	$\rightarrow$ 12 $- (n-1)7 =$		$-7n + 19$

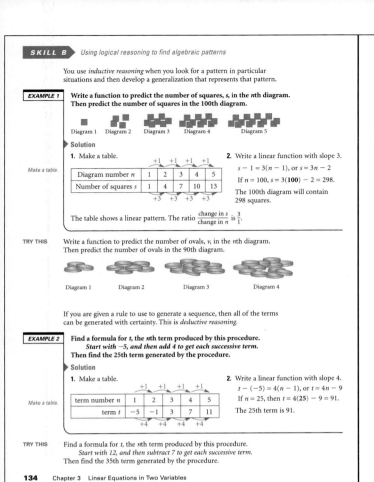

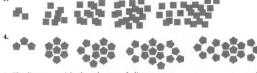

KEY SKILLS

Write the first five numbers generated by each procedure.

1. Start with -7, and then add 6 to get each successive term.

2. Start with 2.4, and then subtract 0.4 to get each successive term.

PRACTICE

Write a function to predict the number of objects, t, in the nth diagram. Then predict the number of objects in the 100th diagram.

3.

4.

5. The diagrams at right show the start of a linear pattern. After how many steps will all the stars disappear?

Find a formula for t, the nth term. Then find the 40th term.

6. Start with -13, and then add 8 to get each successive term.

7. Start with -12.4, and then subtract 3 to get each successive term.

8. Which type of reasoning, inductive or deductive, is needed to solve Exercises 3 and 4? Which type of reasoning is needed to solve Exercises 6 and 7? Explain your responses.

9. **Critical Thinking** Suppose you apply a procedure that determines a linear pattern, and the 4th number you get is 7 and the 8th number is 2. What is the starting number and how much do you add or subtract to get each successive term?

Step 1   Step 2

MIXED REVIEW

Find the slope of the line whose equation is given. (3.5 Skill B)

10. $-8x - y = 4$ 11. $7x - 8y = 56$ 12. $9x - 7y = 63$ 13. $x - y = 4.3$

Find the equation in slope-intercept form for the line that contains each pair of points. (3.7 Skill A)

14. $A(-9, -7)$ and $B(7, 9)$ 15. $J(-1, 11)$ and $K(1, -11)$ 16. $P(13, 2)$ and $Q(7, 10)$

Answers for Lesson 3.9 Skill B

Try This Example 1
$v = 3n$; 270

Try This Example 2
$t = -7n + 19$; -226

1. $-7, -1, 5, 11, 17$
2. $2.4, 2.0, 1.6, 1.2, 0.8$
3. $t = 5n - 2$; 498
4. $t = 4n - 1$; 399
5. The last star is removed at the ninth step.
6. $t = 8n - 21$; 299
7. $t = -3n - 9.4$; -129.4
8. Because Exercises 3 and 4 each require making a generalization about a pattern from specific terms in the pattern, these exercises use inductive reasoning. In Exercises 6 and 7, a given rule is used to find specific examples with certainty, so these exercises use deductive reasoning.
9. $\frac{43}{4}$; Subtract $\frac{5}{4}$.
10. -8
11. $\frac{7}{8}$

12. $\frac{9}{7}$
13. 1
14. $y = x + 2$
15. $y = -11x$
16. $y = -\frac{4}{3}x + \frac{58}{3}$

Focus: *Given a graph in the coordinate plane, students find horizontal and vertical rates of change and write a set of equations to model the changes.*

EXAMPLE

PHASE 1: Presenting the Example

Direct students' attention to the graph shown at the top of the textbook page. Then display the table shown at right. Have them copy and complete the table. [**Answers are shown in red at right.**]

position of point	1	2	3	4	5
x-coordinate	−4	−2	0	2	4
y-coordinate	7	4	1	−2	−5

Ask students if the entries in the table show any constant changes. [**Yes: first row — constant increase of 1; second row — constant increase of 2; third row — constant decrease of 3**] Ask why it is not possible to find a constant *rate* of change for all of the entries in the table. [**A rate compares just two quantities. There are three quantities displayed in the table.**]

Now discuss the Example, where this problem is resolved by creating two tables, one for *x*-coordinates and one for *y*-coordinates.

PHASE 2: Providing Guidance for **TRY THIS**

Some students may write ordered pairs for the points, and then get confused when separating the coordinates into two tables. Suggest that they instead read the coordinates in two stages: first read all *x*-coordinates from left to right (−5, −2, 1, 4, 7), and then all *y*-coordinates from bottom to top (−3, −1, 1, 3, 5). If they simultaneously use a finger to trace the movement across the plane, they will get visual reinforcement.

PHASE 3: ASSESSMENT AND CLOSURE for Lesson 3.9

- Have students work in pairs. Each student should write a table of ordered pairs that can be modeled by a linear function. Students should then exchange tables, and each should write an equation for the other's function.

- The process of *deductive reasoning* that students studied earlier is often described as "general to particular." That is, you use one or more general statements to arrive at a particular conclusion. If all the statements in a deductive argument are true, the conclusion is necessarily true.

  In contrast, the *inductive reasoning* process is often described as "particular to general." This means that you observe a series of particular cases and arrive at a general conclusion, often based on a pattern. It is important to note that the conclusion is *probably* true, but it is not necessarily true. For this reason, inductive reasoning is a valuable tool for making a conjecture, but deductive reasoning is most often used to prove the conjecture.

 For a Lesson 3.9 Quiz, see Assessment Resources *page 35.*

 For a Chapter 3 Assessment, see Assessment Resources *pages 36–37.*

You can also use inductive reasoning when you look for a pattern in graphical data. In the graph at right, you can see a set of points that suggest a pattern. In the Example, you will see how to represent the x- and y-coordinates using linear functions.

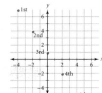

EXAMPLE Represent the pattern shown at right algebraically. Then use your equation to predict the coordinates of the 100th point.

▶ **Solution**

1. Let n represent the position of the point in the sequence of points. For instance, if $n = 1$, the point has coordinates $(-4, 7)$.
Make two tables.

Make two tables.

n	1	2	3	4	5
x	-4	-2	0	2	4

n	1	2	3	4	5
y	7	4	1	-2	-5

2. For each increase of 1 in n, there is an increase of 2 in x.

Thus, slope $= \frac{2}{1}$, or 2.

$x - (-4) = 2(n - 1)$
$x = 2n - 6$

For each increase of 1 in n, there is a decrease of 3 in y.

Thus, slope $= \frac{-3}{1}$, or -3.

$y - 7 = -3(n - 1)$
$y = -3n + 10$

Write two equations.

3. The 100th point corresponds to $n = 100$.

$x = 2(100) - 6$
$x = 194$

$y = -3(100) + 10$
$y = -290$

The coordinates of the 100th point are $(194, -290)$.

TRY THIS Represent the pattern shown at right algebraically. Then use your equation to predict the coordinates of the 100th point.

In the Example above, the equations for the x- and y-coordinates of the nth point are written in terms of the variable n.

$x = 2n - 6 \qquad y = -3n - 4$

Notice that, in the Example above, n is the independent variable. Both variables x and y are dependent variables.

Choose n. → This determines x.
→ This determines y.

Equations with this relationship are called *parametric equations*. The independent variable, in this case n, is called the parameter.

EXERCISES

Write a linear function to represent x in terms of n. Then write a second linear function to represent y in terms of n.

1.

n	1	2	3	4	5
x	0	3	6	9	12

n	1	2	3	4	5
y	0	-4	-8	-12	-16

2.

n	1	2	3	4	5
x	-5	-1	3	7	11

n	1	2	3	4	5
y	11	3	-5	-13	-21

PRACTICE

Write a linear function to represent x in terms of n. Write a second linear function to represent y in terms of n. Then find the coordinates of the 60th point in the pattern.

3.   4.   5.

6. The following table represents a graphical pattern in a set of points.

Point position	first	second	third	fourth	fifth
Coordinates	$(-15, 12)$	$(-11, 9)$	$(-7, 6)$	$(-3, 3)$	$(1, 0)$

a. Write a linear function to represent x in terms of n. Then write a second linear function to represent y in terms of n.
b. Find the 70th point. c. Is $(481, -347)$ part of the pattern?

7. Find the coordinates of the unknown points in this linear pattern.

Point position	first	second	third	fourth	fifth
Coordinates	$(3, 17)$	?	?	?	$(23, 37)$

MIXED REVIEW APPLICATIONS

Solve the problem by finding a linear equation that models the data. Interpret the slope and the y-intercept of the line. (3.7 Skill B)

8. It costs \$810 to make 30 vases and \$1242 to make 54 vases. What will it cost to make 120 vases? What will it cost to make 150 vases?

Answers for Lesson 3.9 Skill C

Try This Example
$x = 3n - 8; y = 2n - 5; (292, 195)$

1. $x = 3n - 3; y = -4n + 4$
2. $x = 4n - 9; y = -8n + 19$
3. $x = 3n - 8; y = -2n + 8; (172, -112)$
4. $x = n - 5; y = 2n - 6; (55, 114)$
5. $x = 2n - 6; y = 3; (114, 3)$
6. a. $x = 4n - 19; y = -3n + 15$
   b. $(261, -195)$
   c. No; if $x = 481$, then $n = 125$. If $n = 125$, then $y = -360 \neq -347$.
★ 7. $(8, 22), (13, 27),$ and $(18, 32)$
8. \$2430; \$2970; the slope gives the cost per vase; the y-intercept gives the fixed cost of producing vases.

★ **Advanced Exercises**

CHAPTER 3 ASSESSMENT

Write the ordered pair that corresponds to each point.

1. A   **2.** B   **3.** C   **4.** D

Graph each point on the same coordinate plane.

5. $W(3, -4)$ **6.** $X(-2, 0)$ **7.** $Y(4, 1)$ **8.** $Z(-4, -4)$

Let $x = -3, -2, -1, 0, 1, 2, 3$. For each value of x, find y and then make a table of ordered pairs. Then graph the ordered pairs.

9. $y = -2x + 3$ **10.** $y = x^2 - 2x + 1$

11. Does $\{(1, 6), (2, 1), (3, 1), (4, 6)\}$ represent a function? Explain.

12. Find the domain and range of the relation graphed at right.

13. Represent the situation below in a table and in a graph for the given values of the independent variable.
 One notebook costs $1.50. The total cost, c, of 1, 2, 3, 4, 5, and 6 notebooks is given by c = 1.5n, where n the number of notebooks.

14. Gasoline costs $1.39 per gallon. Use direct-variation equations to answer the following questions.
 a. What is the cost of 20 gallons?
 b. How many gallons can be purchased for $25?

15. If y varies directly as x and if $y = 40$ when $x = 2.5$, find y when $x = 9$.

16. A small plane begins its descent. If the plane has an altitude of 4300 feet after 2 minutes and an altitude of 3100 feet after 5 minutes, what is the plane's constant rate of descent?

Find the slope of the line that contains each pair of points.

17. $P(-5, 0)$ and $B(6, 3)$ **18.** $X(5, 9)$ and $Y(7, -4)$

19. A line has slope $-\frac{3}{5}$ and contains $K(-4, 7)$. Graph the line and write the coordinates of a second point on it.

Use the intercepts to graph each equation.

20. $-4x + 3y = 24$ **21.** $y = 3.6$

Use the slope and the y-intercept of the line to graph each equation.

22. $y = -\frac{3}{2}x + 3$ **23.** $-2x + 5y = 10$

24. Represent the following situation in an equation and in a graph.
 Michael has $0.40 in nickels and dimes.

Use the given information to write an equation in point-slope form and in slope-intercept form.

25. $K(-3, -5)$; slope $\frac{3}{7}$ **26.** $M(-5, -3)$; slope 0

27. Write a function to solve the problem below. Then solve.
 After 1.5 years, a copier's value is $4900, and after 2 years its value is $4700. What will its value be after 4 years?

Find the equation in slope-intercept form for $\overleftrightarrow{PQ}$.

28. $P(3, -2)$ and $Q(6, 7)$ **29.** $P(-2, 5)$ and $Q(7, 15)$

30. Write an equation to solve the problem below. Interpret the slope and the y-intercept.
 It costs $87 to rent a car for 3 days and $125 to rent the car for 5 days. What will the cost of renting the car for 7 days be?

31. Write the equation in slope-intercept form for the line that contains $Q(5, 2)$ and that is parallel to the graph of $3x - 5y = 15$.

32. Write the equation in slope-intercept form for the line that contains $S(3, 5)$ and that is perpendicular to the graph of $-3x - 2y = 4$.

33. Do $K(-3, 4)$, $L(0, 0)$, and $M(5, 2)$ determine a right triangle? Explain.

34. Express the pattern in the table at right as a function with x as the independent variable. Then predict the 50th term.

x	1	2	3	4	5
y	13	6	-1	-8	-15

35. Suppose you start with 10 and then subtract 2.5 to get each successive term. Find a formula for t, the nth term. Then predict the 55th term.

36. Refer to the pattern at right. Write two linear functions to represent x and y in terms of n, the point number. Then predict the coordinates of the 60th point.

138   Chapter 3   Linear Equations in Two Variables

Chapter 3   Assessment   **139**

Answers for Chapter 3 Assessment

1. $A(0, 5)$
2. $B(5, 3)$
3. $C(5, -3)$
4. $D(-1, -3)$

5–8.
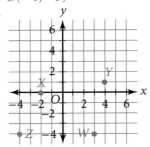

9.

x	-3	-2	-1	0	1	2	3
y	9	7	5	3	1	-1	-3

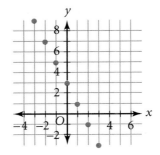

For answers to Exercises 10–36, see Additional Answers.

138   **Chapter 3**   **Linear Equations in Two Variables**

Chapter 4 *Solving Inequalities in One Variable*	State or Local Standards	Corresponding Lessons in *Algebra 1*, Schultz et al.
4.1 Introduction to Inequalities		6.1, 6.3
Skill A: Representing inequalities with mathematical symbols		
Skill B: Graphing simple inequalities		
Skill C: Graphing compound inequalities		
4.2 Solving Inequalities Using Addition or Subtraction		6.1
Skill A: Using addition to solve an inequality in one step		
Skill B: Using subtraction to solve an inequality in one step		
Skill C: APPLICATIONS Using addition or subtraction to solve real-world inequalities		
4.3 Solving Inequalities Using Multiplication or Division		6.2
Skill A: Using multiplication to solve an inequality in one step		
Skill B: Using division to solve an inequality in one step		
Skill C: APPLICATIONS Using multiplication or division to solve real-world inequalities		
4.4 Solving Simple One-Step Inequalities		6.1, 6.2
Skill A: Choosing an operation to solve an inequality in one step		
Skill B: APPLICATIONS Choosing an operation to solve a real-world inequality		
4.5 Solving Multistep Inequalities		6.2
Skill A: Solving an inequality in multiple steps		
Skill B: Simplifying expressions before solving inequalities		
Skill C: APPLICATIONS Using multistep inequalities to solve real-world problems		
4.6 Solving Compound Inequalities		6.3
Skill A: Solving compound inequalities joined by "and"		
Skill B: Solving compound inequalities joined by "or"		
Skill C: APPLICATIONS Using compound inequalities to solve real-world problems		
4.7 Solving Absolute-Value Equations and Inequalities		6.4, 6.5
Skill A: Solving absolute-value equations		
Skill B: Solving absolute-value inequalities		
Skill C: APPLICATIONS Using absolute-value inequalities to solve real-world problems		
4.8 Deductive Reasoning With Inequalities		11.6
Skill A: Showing that an inequality is always, sometimes, or never true		

CHAPTER

4

Solving Inequalities in One Variable

▶ What You Already Know

In earlier mathematics courses, you learned how to compare and order numbers. The skills you learned will be helpful in your study of inequalities involving variables. Your study of properties of equality from earlier chapters will be especially helpful.

▶ What You Will Learn

In Chapter 4, you will extend your equation-solving skills to solve linear inequalities in one variable. Just as you learned properties of equality to help solve equations, you will learn properties of inequality that help you solve inequalities.

In Chapter 4, you will gradually move from solving simple inequalities in one variable to more complicated ones. You will see that there are infinitely many solutions to a linear inequality in one variable and that its solution on a number line is a ray.

You will then have the opportunity to solve a pair of inequalities joined by *and* or *or,* as well as inequalities involving absolute value. Lastly, you will encounter mathematical statements involving inequalities that are sometimes, always, or never true.

VOCABULARY

absolute-value equation
absolute-value inequality
Addition Property of Inequality
compound inequality
conjunction
disjunction
Division Property of Inequality
inequality

maximum
minimum
Multiplication Property of Inequality
solution to an inequality
Subtraction Property of Inequality
tolerance
Transitive Property of Inequality

The diagram below shows how mathematical skills and mathematical reasoning are interrelated with the skills and concepts in Chapter 4. Notice that this chapter focuses on solving inequalities in one variable.

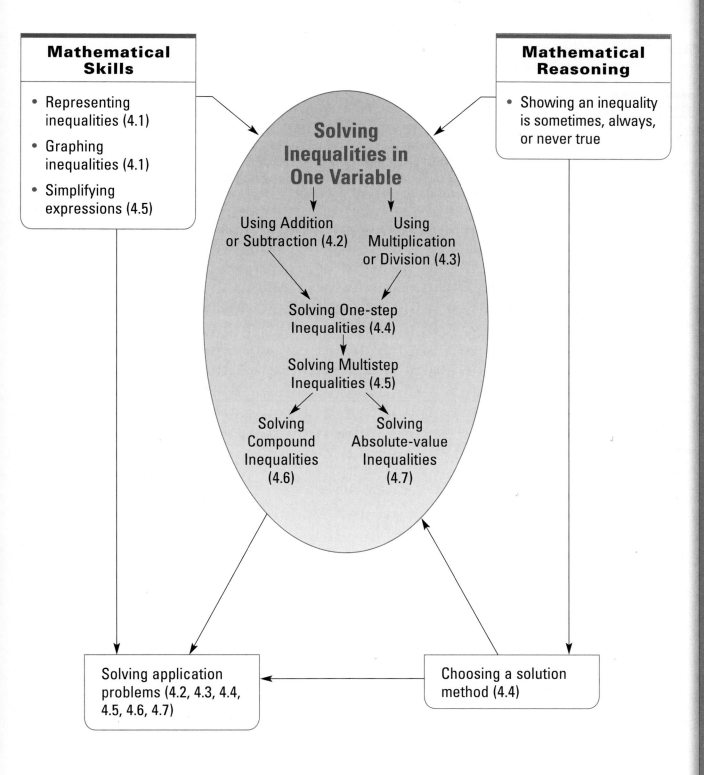

TEACHING

4.1

LESSON

Introduction to Inequalities

SKILL A ▷ *Representing inequalities with mathematical symbols*

Focus: *Students use inequalities to compare numbers on a number line and to represent real-world situations.*

EXAMPLE 1

PHASE 1: Presenting the Example

In Example 1, students must translate a visual image into the mathematical concept of inequality. Then they must translate the concept into symbols. Use a display such as the following to guide them through the translations.

visual image →	−6 is to the left of 2	2 is to the right of −6
concept →	−6 is less than 2	2 is greater than −6
symbols →	−6 < 2	2 > −6

PHASE 2: Providing Guidance for **TRY THIS**

Students often confuse the symbols < and >. If the confusion occurs when writing inequalities, point out that the larger side of the symbol, the "open" side, appears next to the greater number. So, when comparing −7 and −2, you can write either −7 > −2 or −2 < −7. If students have trouble reading inequalities (for example, they read < as "is greater than"), they can remember that "less than" begins with the letter *l* and the less than symbol, <, resembles the letter *l*.

EXAMPLE 2

PHASE 1: Presenting the Example

PHASE 2: Providing Guidance for **TRY THIS**

Students may have trouble distinguishing cases where the phrases *more than* and *less than* translate into expressions involving operation symbols (+, −) from cases where they translate into statements involving inequality symbols (>, <). Stress that an inequality is a mathematical *sentence*, and that a sentence involves a verb. Therefore, the phrases that translate into inequality symbols also include verbs: *is* more than and *is* less than. The display at right may help to make the distinction clear.

c more than 5
5 + *c*

c is more than 5
c > 5

Students sometimes have difficulty remembering whether the statement $a \le b$ is interpreted as $a < b$ or $a = b$ or as $a < b$ *and* $a = b$. Give them a statement such as $a \le 5$, and point out that it is not possible for a number to be equal to 5 and less than 5 at the same time. Therefore, the correct interpretation is $a < 5$ *or* $a = 5$.

4.1 LESSON — Introduction to Inequalities

SKILL A ▸ *Representing inequalities with mathematical symbols*

Recall that an equation is a statement that two expressions are equal. An **inequality**, such as $3 < 5$, is a statement that two expressions are *not* equal but are related to one another in one of these four ways:

$<$ is less than
$\leq$ is less than or equal to (is not more than, is at most)
$>$ is greater than
$\geq$ is greater than or equal to (is no less than, is at least)

Some examples of inequalities are:

$$-4 < -1 \qquad [3(5) - 10]^2 \leq 25 \qquad -2x + 5 \geq 10$$

Recall that on a number line the smaller of two numbers appears to the left. You can represent the relationship between numbers with an inequality.

EXAMPLE 1 Graph -6 and 2 on a number line and write two inequalities that describe their relationship.

▸ Solution

$$-7\ -6\ -5\ -4\ -3\ -2\ -1\ \ 0\ \ 1\ \ 2\ \ 3\ \ 4$$

Because -6 is to the left of 2, $-6 < 2$.
Because 2 is to the right of -6, $2 > -6$.

TRY THIS Graph -7 and -2 on a number line and write two inequalities that describe their relationship.

EXAMPLE 2 Write an inequality that represents the following sentence.
The cost of a purchase, c, is more than $5.

Look for key words.

▸ Solution

The cost of a purchase, c, **is more than** $5.
$c \qquad\qquad > \qquad 5$

The inequality is $c > 5$.

TRY THIS Write an inequality that represents the following sentence.
The cost, c, is not more than $12.95.

EXERCISES

KEY SKILLS

Write the number(s) from the set $\{-3, -2, -1, 0, 1, 2, 3\}$ that match each description.

1. the number(s) less than 2
2. the positive number(s)
3. the number(s) not more than -2
4. the number(s) not less than 0

PRACTICE

Graph each pair of numbers on a number line and write two inequalities that describe their relationship.

5. 0 and 3
6. 0 and -3
7. -3 and 3
8. 4 and -4
9. 4 and 3
10. -4 and -3
11. -4 and 3
12. 4 and -3
13. -2 and 3.5
14. -1.5 and 3
15. $-\frac{1}{2}$ and 4
16. $-\frac{1}{2}$ and $2\frac{1}{2}$

Write an inequality that represents each sentence.

17. A motorist drives at a speed, s, that is more than 55 miles per hour.
18. The amount of sugar, a, is no more than 2 ounces.
19. A child's height, h, must be at least 42 inches.
20. A voter's age, a, must be at least 18.
21. An athlete's weight, w, is more than 125 pounds.

In Exercises 22–25, determine whether each statement is true or false.

22. $|3.2| > 0$
23. $-|4.5| \geq 0$
24. $-|6| \leq -6$
25. $|9| \leq -(-8)$

26. Let x be the average of two numbers a and b. On a number line, where is x located with respect to a and b?

MIXED REVIEW

Solve each equation. Check your solution. (2.5 Skills A and B)

27. $x + 8 = 10$
28. $y - 5 = 4$
29. $z + 7 = -1$
30. $4 = y - 13$
31. $8x = 64$
32. $\frac{y}{4} = -5$

Answers for Lesson 4.1 Skill A

Try This Example 1
$-7 < -2$ and $-2 > -7$

$$-6 \quad -4 \quad -2 \quad 0 \quad 2$$

Try This Example 2
$c \leq 12.95$

1. $-3, -2, -1, 0, 1$
2. $1, 2, 3$
3. $-3, -2$
4. $0, 1, 2, 3$
5. $0 < 3$ and $3 > 0$

$$-4 \quad -2 \quad 0 \quad 2 \quad 4$$

6. $0 > -3$ and $-3 < 0$

$$-4 \quad -2 \quad 0 \quad 2 \quad 4$$

7. $-3 < 3$ and $3 > -3$

$$-4 \quad -2 \quad 0 \quad 2 \quad 4$$

8. $4 > -4$ and $-4 < 4$

$$-4 \quad -2 \quad 0 \quad 2 \quad 4$$

9. $4 > 3$ and $3 < 4$

$$-4 \quad -2 \quad 0 \quad 2 \quad 4$$

10. $-4 < -3$ and $-3 > -4$

$$-4 \quad -2 \quad 0 \quad 2 \quad 4$$

11. $-4 < 3$ and $3 > -4$

$$-4 \quad -2 \quad 0 \quad 2 \quad 4$$

12. $4 > -3$ and $-3 < 4$

$$-4 \quad -2 \quad 0 \quad 2 \quad 4$$

13. $-2 < 3.5$ and $3.5 > -2$

$$-4 \quad -2 \quad 0 \quad 2 \quad 4$$

14. $-1.5 < 3$ and $3 > -1.5$

$$-4 \quad -2 \quad 0 \quad 2 \quad 4$$

15. $-\frac{1}{2} < 4$ and $4 > -\frac{1}{2}$

$$-4 \quad -2 \quad 0 \quad 2 \quad 4$$

16. $-\frac{1}{2} < 2\frac{1}{2}$ and $2\frac{1}{2} > -\frac{1}{2}$

$$-4 \quad -2 \quad 0 \quad 2 \quad 4$$

17. $s > 55$
18. $a \leq 2$
19. $h \geq 42$
20. $a \geq 18$
21. $w > 125$
22. true
23. false
24. true
25. false
26. The average is halfway between a and b.
27. $x = 2$
28. $y = 9$
29. $z = -8$
30. $y = 17$
31. $x = 8$
32. $y = -20$

Focus: *Students graph simple inequalities on a number line.*

EXAMPLE 1

PHASE 1: Presenting the Example

Display the inequality $x \geq 3$. Tell students that a value of x that makes this inequality true is called a *solution of the inequality.* Now give students the replacement set $\{-1, 0, 1, 2, 3, 4, 5\}$. Ask which values in this replacement set are solutions to $x \geq 3$. [**3, 4, 5**]

Now tell students that the replacement set is all real numbers, and ask each student to come up with a different value for x that will make the inequality true. Display several solutions for all to see. Then ask how many solutions the inequality has. Lead students to realize that there are infinitely many solutions.

Because there are infinitely many solutions, we cannot continue to make a list of all of them. Ask students if they can think of a way to represent the solutions. If no student suggests a graph, remind them that a number line can be used to represent all the real numbers. So, we can use a number line to show all of the solutions to an inequality. Discuss Example 1 to show students how this is done.

PHASE 2: Providing Guidance for TRY THIS

Students are sometimes confused about the symbols $\geq$ and $\leq$; they may believe that $1 \geq 1$ is not a true statement, and so they will not include 1 in their graphs of $x \geq 1$. Emphasize to these students that the symbols $\geq$ and $\leq$ include the equal sign.

EXAMPLE 2

PHASE 1: Presenting the Example

Display number line **1**, shown at right. Tell students that the solid circle and arrow are the graph of a set of numbers. Ask them to write a verbal description of those numbers. [**all numbers greater than 1 or equal to 1**] Then display number line **2**. Ask students how this graph is different from the graph on number line **1**. [**It does not include the number 1.**]

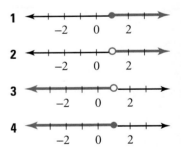

Now display number lines **3** and **4**. Ask students to write a verbal description of each graph. [**3** all numbers less than 1; **4** all numbers less than 1 or equal to 1] Then discuss Example 2.

PHASE 2: Providing Guidance for TRY THIS

Have students check their answers by substituting values that are on the given graph into their inequality.

The **solution to an inequality** in one variable is the set of all numbers that make the inequality true. For example, given the replacement set $\{1, 2, 3, 4\}$, the solutions to $5x + 2 \geq 12$ are 2, 3, and 4 as shown below.

$$x = 1 \rightarrow 5(1) + 2 \stackrel{?}{\geq} 12 \qquad x = 2 \rightarrow 5(2) + 2 \stackrel{?}{\geq} 12$$
$$7 \geq 12 \; ✗ \qquad\qquad 12 \geq 12 \; ✔$$
$$x = 3 \rightarrow 5(3) + 2 \stackrel{?}{\geq} 12 \qquad x = 4 \rightarrow 5(4) + 2 \stackrel{?}{\geq} 12$$
$$17 \geq 12 \; ✔ \qquad\qquad 22 \geq 12 \; ✔$$

If no replacement set is specified, you may assume that the replacement set is the set of real numbers.

You can graph the solution to an inequality in one variable on a number line.

EXAMPLE 1 Graph each inequality on a number line.
  a. $x \geq 2$ **b.** $x < -1$

▸ Solution

a. To graph the numbers greater than 2, draw a *ray* starting at 2 that points to the right. Notice that $x = 2$ satisfies $x \geq 2$. To show that 2 is included in the graph, draw a solid circle at 2.

b. To graph the numbers less than -1, draw a ray starting at -1 that points to the left. Notice that $x = -1$ does *not* satisfy $x < -1$. To show that -1 is *not* included in the graph, draw an open circle at -1.

TRY THIS Graph each inequality on a number line.
  a. $x < -2$ **b.** $x \geq 1$

EXAMPLE 2 Write an inequality that describes each graph.
  a.   **b.**

▸ Solution

a. The graph is a ray that includes all real numbers greater than -6. Because the endpoint at -6 is an open circle, -6 is not included in the solution. The inequality is $x > -6$.

b. The graph is a ray that includes all real numbers less than 3. Because the endpoint at 3 is a solid circle, 3 is included in the solution. The inequality is $x \leq 3$.

TRY THIS Write an inequality that describes each graph.
  a.   **b.**

EXERCISES

KEY SKILLS

Match each inequality on the left with its graph on the right.

  1. $x < 1$
  2. $x \geq 1$
  3. $x \leq 1$
  4. $x > 1$

  a.
  b.
  c.
  d.

PRACTICE

Graph each inequality on a number line.

  5. $x \geq 4$ 6. $t \leq 4$ 7. $s \geq -2$ 8. $s < 2$
  9. $d < 5$ 10. $k > 2$ 11. $a > 3$ 12. $x \leq 3$
  13. $k \geq -3$ 14. $b > 4$ 15. $w < -4$ 16. $y < 7$

Write an inequality that describes each graph.

  17.   18.
  19.   20.
  21.   22.
  23.   24.

MIXED REVIEW

Solve each equation. Check your solution. (2.5 Skill A)

  25. $x - 3.5 = -3$ 26. $d - 4.5 = -7$
  27. $v + 3\frac{1}{2} = -2\frac{1}{2}$ 28. $-4.25 + q = -0.5$
  29. $0 = -4\frac{1}{3} + p$ 30. $-5.5 = -4.25 + q$
  31. $-0.75 + x = -0.5$ 32. $-4.5 = t - 1.25$

Answers for Lesson 4.1 Skill B

Try This Example 1

a.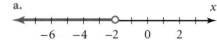

b.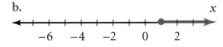

Try This Example 2
a. $x \leq 5$ b. $x > -3$

1. d
2. c
3. a
4. b

5.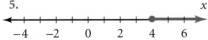

6.

7.

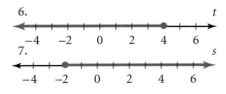

8.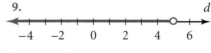

9.

10.

11.

12.

13.

14.

15.

16.

17. $x > -1$
18. $y \leq 2$
19. $w \geq 4$
20. $k > -4$
21. $y \leq 0$
22. $z < 1$
23. $k < 3$
24. $x \geq -2$
25. $x = 0.5$
26. $d = -2.5$
27. $v = -6$
28. $q = 3.75$
29. $p = 4\frac{1}{3}$

For answers to Exercises 30–32, see Additional Answers.

EXAMPLES 1 and 2

PHASE 1: Presenting the Examples

Compound inequalities will be taught more formally in Lesson 4.6. In this lesson, allow students to develop an intuitive understanding. In Example 1, for instance, lead them to describe the graph as *all numbers that are greater than or equal to −1 <u>and</u> that are less than 3*. Similarly, they can describe the graph in Example 2 as *all numbers that are less than −1 <u>or</u> that are greater than or equal to 2*. Emphasize the proper use of *and* and *or*, and point out to students the differences in the graphs of the two types of compound inequalities.

PHASE 2: Providing Guidance for **TRY THIS**

Many students can successfully graph each case of a compound inequality separately, but they may have difficulty graphing both cases on the same number line. Suggest that they use a different color to graph each case of a compound inequality. When *and* is used in the compound inequality, they should include in their solution the parts that are *common to both* of the colored graphs. When *or* is used in the compound inequality, they should include *all parts of each* colored graph.

Watch for students who include a boundary number in the graph when they should omit it, and vice versa. Suggest that they first copy the inequality and focus just on the inclusion or exclusion of the boundary numbers. A ✔ or ✘ placed below the numbers, as shown below, can serve as a visual reminder when students are checking their graphs.

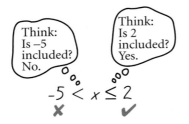

PHASE 3: ASSESSMENT AND CLOSURE for Lesson 4.1

- Ask students to compare and contrast inequalities and equations. [**samples: Both are mathematical sentences. Both can be open sentences in one variable, with all solutions graphed on a number line. An equation involves the symbol =, but an inequality involves <, ≤, >, or ≥.**]

- This lesson is the first in a sequence of lessons in which students learn about inequalities in one variable. As students encounter each new topic, have them continue to look for parallels to *equations* in one variable.

☞ *For a Lesson 4.1 Quiz, see* Assessment Resources *page 38.*

SKILL C *Graphing compound inequalities*

A **compound inequality** is a pair of inequalities joined by *and* or *or*. An example of each type of compound inequality follows.

$$x > 2 \text{ and } x \le 10 \qquad x < -5 \text{ or } x > 3$$

A compound inequality joined by *and* is true only if *both* inequalities are true as shown in Example 1.

EXAMPLE 1 Graph $x \ge -1$ *and* $x < 3$.

▶ **Solution**

First graph each inequality separately.

Because the solutions to the compound inequality must satisfy both inequalities, identify the region where the two graphs overlap.

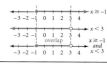

TRY THIS Graph $x > -5$ *and* $x \le 2$.

The Transitive Property of Inequality defined below allows us to write compound inequalities joined by *and* in a shortened form. We use this property to write $a < b$ and $b < c$ as $a < b < c$. Using this form, the compound inequality $x \ge -1$ *and* $x < 3$ in Example 1 can be shortened to $-1 \le x < 3$.

> **Transitive Property of Inequality**
> If a, b, and c are real numbers, $a < b$, and $b < c$, then $a < c$.
> This statement is also true if $<$ is replaced by $>$, $\le$, or $\ge$.

A compound inequality joined by *or* is true if *at least one* of the inequalities is true.

EXAMPLE 2 Graph $x < -1$ *or* $x \ge 2$.

▶ **Solution**

First graph each inequality separately.

Because the solutions to the compound inequality must satisfy at least one of the inequalities, identify the combination of the two graphs.

TRY THIS Graph $x \le -4$ *or* $x > 0$.

146 Chapter 4 Solving Inequalities In One Variable

EXERCISES

KEY SKILLS

Match each inequality on the left with its graph on the right.

1. $x > -2$ *and* $x < 1$
2. $x \le -2$ *or* $x > 1$
3. $x < -2$ *or* $x \ge 1$
4. $-2 \le x < 1$

a.
b.
c.
d.

PRACTICE

Graph each compound inequality.

5. $x \ge 1$ *or* $x < -1$
6. $g \ge 3$ *or* $g < -2$
7. $0 \le a \le 4$
8. $-4 \le b \le 4$
9. $c \ge 3$ *or* $c < 0$
10. $c \ge 0$ *or* $c < -2$
11. $q \ge 3$ *or* $q < 1$
12. $r \ge 4$ *or* $r \le 2$
13. $w \le 4$ *and* $w \ge 1$
14. $v \le 0$ *and* $v \ge -3$
15. $z > 2$ *and* $z \le 4$
16. $h < -2$ *or* $h \ge 2$
17. $u > 2$ *or* $u > 4$
18. $k \le 0$ *and* $k \le -3$
19. $d \le 0$ *and* $d \ge -4$

Write a compound inequality that describes each graph.

20.
21.

22. **Critical Thinking** What must be true of a and b if the compound inequality $x \ge a$ *and* $x < b$ has a solution?

23. **Critical Thinking** What must be true of r and s if the compound inequality $x > s$ *and* $x < r$ has no solution?

MIXED REVIEW

Solve each equation. Check your solution. (2.5 Skill B)

24. $-3z = -72$
25. $\frac{4}{3}b = -\frac{1}{2}$
26. $\frac{5d}{7} = -\frac{3}{7}$
27. $-\frac{5y}{7} = -4.5$
28. $4\frac{1}{2} = -\frac{2g}{7}$
29. $7.2 = -0.2w$
30. $2.5\,p = -10$
31. $0 = -4.875\,q$
32. $\frac{x}{-7} = 1$

Lesson 4.1 Skill C **147**

Answers for Lesson 4.1 Skill C

Try This Example 1 x
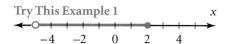

Try This Example 2 x
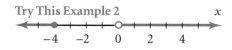

1. c
2. a
3. d
4. b

5. x
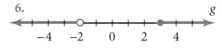

6. g

7. a

8. b

9. c
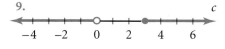

10. c
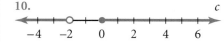

11. q
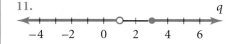

12. r

13. w
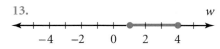

14. v

15. z
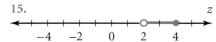

16. h
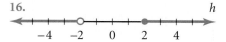

17. u
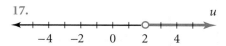

18. k
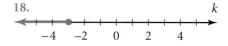

19. d

20. $-6 \le x \le 1$, or $x \ge 6$ *and* $x \le 1$
21. $x < -6$ *or* $x \ge -1$
22. $a < b$
23. $s \ge r$
24. $z = 24$
25. $b = -\dfrac{3}{8}$

For answers to Exercises 26–32, see Additional Answers.

Lesson 4.1 Skill C **147**

4.2 Solving Inequalities Using Addition or Subtraction

SKILL A | *Using addition to solve an inequality in one step*

Focus: *Students use the Addition Property of Inequality to solve inequalities in one variable.*

EXAMPLE 1

PHASE 1: Presenting the Example

Display the inequality $-2 < 5$. Ask students if this is true or false. [**true**] Now give students several different numbers, one at a time, each time telling them to add the number to each side of $-2 < 5$. (Some samples are shown at right.) Lead them to see that the resulting inequality will remain true, provided that the same number is added to each side. Repeat the activity with $6 > -4$.

$$\begin{array}{rcr} -2 & < & 5 \\ +4 & & +4 \\ \hline 2 & < & 9 \end{array}$$

$$\begin{array}{rcr} -2 & < & 5 \\ +(-5) & & +(-5) \\ \hline -7 & < & 0 \end{array}$$

Now display the following incomplete statement: *If the same quantity is added to both sides of a true inequality, then ___?___ .* Ask students to copy and complete it. Then discuss their responses. [**Most should arrive at a conclusion of this nature: . . .then the resulting inequality is also true.**]

$$\begin{array}{rcr} -2 & < & 5 \\ +1.7 & & +1.7 \\ \hline -0.3 & < & 6.7 \end{array}$$

Tell students that they have investigated a fundamental property of inequality. Ask them to make a conjecture about the name of the property. [**Addition Property of Inequality**] Then discuss the formal statement of the property that appears at the top of the textbook page.

Now display the inequality in Example 1, $x - 4.5 < -5$. Ask students how they think they could use the Addition Property of Inequality to solve it. Elicit the response that, if you add 4.5 to each side, the result will be an inequality in which x is isolated on one side of the inequality symbol. Have students perform the addition and write the result. [$x < -0.5$] Then have them graph the solution.

PHASE 2: Providing Guidance for **TRY THIS**

Some students will arrive at $x > -1.5$ and graph it correctly, yet they may not grasp exactly what this means. To check understanding, ask students how many solutions there are. [**infinitely many**] Ask if -4, -1.5, and -1 are solutions. [**no; no; yes**] Have them substitute these numbers into the *original inequality* to verify. Then ask them to name several other solutions.

EXAMPLE 2

PHASE 1: Presenting the Example

Some students may graph $-4 < w$ as all numbers less than -4. Encourage them to rewrite the inequality before graphing, as shown at right, with the variable on the left side. Since the variable and number have "flipped" (changed sides), the symbol itself must "flip" (change direction).

$$-4 < w$$
$$w > -4$$

4.2 LESSON — Solving Inequalities Using Addition or Subtraction

SKILL A *Using addition to solve an inequality in one step*

You solve an inequality by rewriting the inequality in a form whose solution set is easy to see. You can use the Addition Property of Inequality below to solve an inequality that involves subtraction, such as $x - 4.5 < -5$.

Addition Property of Inequality
If a, b, and c are real numbers and $a < b$, then $a + c < b + c$.
This statement is also true if $<$ is replaced by $>$, $\leq$, or $\geq$.

EXAMPLE 1 Solve $x - 4.5 < -5$. Graph the solution on a number line.

▶ Solution
$$x - 4.5 < -5$$
$$x - 4.5 + 4.5 < -5 + 4.5 \quad \longleftarrow \text{Apply the Addition Property of Inequality.}$$
$$x < -0.5$$

The solution is all real numbers less than -0.5.

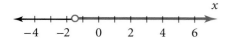

TRY THIS Solve $x - 2.5 > -4$. Graph the solution on a number line.

You can write any inequality in two ways. For example, $-3 \leq a$ is the same as $a \geq -3$. This is useful when you graph solutions.

EXAMPLE 2 Solve $-8 \leq a - 5$. Graph the solution on a number line.

▶ Solution
$$-8 \leq a - 5$$
$$-8 + 5 \leq a - 5 + 5 \quad \longleftarrow \text{Apply the Addition Property of Inequality.}$$
$$-3 \leq a$$
$$a \geq -3 \quad \longleftarrow \text{Rewrite the inequality so that } a \text{ is on the left.}$$

The solution is all real numbers greater than or equal to -3.

TRY THIS Solve $3 < w + 7$. Graph the solution on a number line.

EXERCISES

KEY SKILLS

Rewrite each inequality so that the variable is on the left.

1. $-4 \leq x$ 2. $3 < y$ 3. $-2 \geq d$ 4. $0 > w$

PRACTICE

Solve each inequality. Graph the solution on a number line.

5. $x - 6 < 5$ 6. $y - 2 \geq 10$ 7. $a - 1 \leq 9$

8. $x - 3 > -5$ 9. $h - 3 > -1$ 10. $5 \leq k - 3$

11. $-2 > m - 4$ 12. $-3.5 + x > -4$ 13. $-1.5 + b < -3.5$

14. $-0.5 < p - 3.5$ 15. $-1.25 \leq q - 3.0$ 16. $-\frac{1}{3} < z - \frac{7}{3}$

17. $-\frac{17}{4} \geq a - \frac{13}{4}$ 18. $y - 3.5 > -6.25$ 19. $s - 0.75 > -5.5$

Without solving, tell whether the solution contains only positive numbers, only negative numbers, or both positive and negative numbers.

20. $x - 4 > -10$ 21. $s - 10 \leq -12$ 22. $d - 5 > 2$ 23. $x - 3 \leq -2$

List all integers that meet each given restriction.

24. $d - 3.5 > 4$; solutions less than 10   25. $b - 2 \leq 6$; solutions greater than 0

26. $n - 3.5 < 4$; solutions greater than 4   27. $d - 1 \geq 6$; solutions less than 8

Find and graph all solutions.

28. $y - 3 > 5$ and $y - 3 < 6$ 29. $x - 3 > 5$ or $x - 3 < 3$

30. $h - 4 \geq 1$ or $h - 3 < 0$ 31. $n - 3 > -3$ and $n - 3 \leq 5$

32. **Critical Thinking** Solve $x + a > b$ for x. What must be true about a and b if the solution contains only positive numbers?

MIXED REVIEW

Find each difference. (2.3 Skill A)

33. $-4 - 7$ 34. $5 - 12$ 35. $-4 - 3.6$

36. $-4.25 - 0.75$ 37. $4.2 - 7.6$ 38. $-2\frac{4}{5} - 2\frac{3}{5}$

Answers for Lesson 4.2 Skill A

Try This Example 1
$x > -1.5$

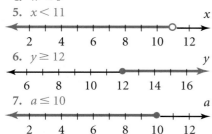

Try This Example 2
$w > -4$

1. $x \geq -4$
2. $y > 3$
3. $d \leq -2$
4. $w < 0$
5. $x < 11$

6. $y \geq 12$

7. $a \leq 10$

8. $x > -2$

9. $h > 2$

10. $k \geq 8$

11. $m < 2$

12. $x > -0.5$

13. $b < -2$

14. $p > 3$

15. $q \geq 1.75$

16. $z > 2$

17. $a \leq -1$

18. $y > -2.75$

19. $s > -4.75$

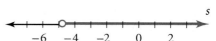

20. positive and negative numbers
21. negative numbers

For answers to Exercises 22–38, see Additional Answers.

SKILL B Using subtraction to solve an inequality in one step

Focus: *Students apply the Subtraction Property of Inequality to solve inequalities in one variable.*

EXAMPLE 1

PHASE 1: Presenting the Example

Display the inequality $9 > 4$. Ask students if this is true or false. [**true**] Tell them to choose any number, subtract it from each side of the inequality, and write the resulting inequality.

Ask several students, one by one, to state their new inequalities. Display each one directly beneath $9 > 4$, repeating $9 > 4$ as many times as necessary. (Several examples are shown at right.) Case by case, ask the class to identify the subtraction that transformed $9 > 4$ into the new inequality. [**for the examples at right: subtracting 1, 7, 4.7, $6\frac{1}{2}$, −1, and −3.5, respectively, from each side of $9 > 4$**] In each case, ask students if the new inequality is a true statement or a false statement. [**All are true.**]

$$\begin{aligned} 9 &> 4 \\ 8 &> 3 \end{aligned}$$

$$\begin{aligned} 9 &> 4 \\ 2 &> -3 \end{aligned}$$

$$\begin{aligned} 9 &> 4 \\ 4.3 &> -0.7 \end{aligned}$$

$$\begin{aligned} 9 &> 4 \\ 2\tfrac{1}{2} &> -2\tfrac{1}{2} \end{aligned}$$

$$\begin{aligned} 9 &> 4 \\ 10 &> 5 \end{aligned}$$

$$\begin{aligned} 9 &> 4 \\ 12.5 &> 7.5 \end{aligned}$$

Now display $3 < 10$. Ask students if it is true or false. [**true**] Ask what they think would happen if the same number were subtracted from each side. Most likely they will guess that the result will be another true inequality. Tell them to perform one experiment to test this conjecture.

After all students have written their new inequalities, ask if anyone found "evidence" to contradict the conjecture. They should conclude that any subtraction performed correctly will *support* the conjecture. Have students read the Subtraction Property of Inequality as presented at the top of the textbook page, and answer any questions. Then discuss Example 1.

EXAMPLE 2

PHASE 1: Presenting the Example

Ask students to write a statement that compares the numbers −1 and 8. Most students probably will write $-1 < 8$. Remind them that $8 > -1$ is also an acceptable way to write the comparison.

Now display $6 \le t + 4.2$. Point out that this statement compares the number 6 and the algebraic expression $t + 4.2$. Ask students to write the comparison in a different way. [**$t + 4.2 \ge 6$**] Then discuss the solution of the inequality as given on the textbook page.

PHASE 2: Providing Guidance for TRY THIS

Have students work in pairs. One student in each pair should solve $5 > a + 3.7$ *without* switching the order of the inequality. The other *should* first switch the order of the inequality, solving $a + 3.7 < 5$. The partners can then compare results and discuss whether one method seems preferable.

You can use the Subtraction Property of Inequality to solve an inequality that involves addition, such as $x + 6.5 > 5$.

Subtraction Property of Inequality

If a, b, and c are real numbers and $a < b$, then $a - c < b - c$. This statement is also true if $<$ is replaced by $>$, $\le$, or $\ge$.

EXAMPLE 1 Solve $x + 6.5 > 5$. Graph the solution on a number line.

▶ Solution

$$x + 6.5 > 5$$
$$x + 6.5 - 6.5 > 5 - 6.5 \quad \longleftarrow \text{ Apply the Subtraction Property of Inequality.}$$
$$x > -1.5$$

The solution is all real numbers greater than -1.5. Because -1.5 is not included in the solution, draw an open circle at -1.5.

$$\overset{-1.5}{\underset{-5\ -4\ -3\ -2\ -1\ \ 0\ \ 1\ \ 2\ \ 3\ \ 4\ \ 5}{\longleftrightarrow}} \ x$$

TRY THIS Solve $x + 3 \ge 5$. Graph the solution on a number line.

As you saw in Skill A, it is usually easier to graph an inequality when the variable is on the left. In Example 2, you can rewrite $6 \le t + 4.2$ as $t + 4.2 \ge 6$ and then solve the inequality.

EXAMPLE 2 Solve $6 \le t + 4.2$. Graph the solution on a number line.

▶ Solution

$$6 \le t + 4.2$$
$$t + 4.2 \ge 6 \quad \longleftarrow \text{ Rewrite the inequality so that } t \text{ is on the left.}$$
$$t + 4.2 - 4.2 \ge 6 - 4.2 \quad \longleftarrow \text{ Apply the Subtraction Property of Inequality.}$$
$$t \ge 1.8$$

The solution is all real numbers greater than or equal to 1.8. Because 1.8 is included in the solution, draw a closed circle at 1.8.

$$\overset{1.8}{\underset{-4\ -3\ -2\ -1\ \ 0\ \ 1\ \ 2\ \ 3}{\longleftrightarrow}} \ t$$

TRY THIS Solve $5 > a + 3.7$. Graph the solution on a number line.

EXERCISES

KEY SKILLS

Rewrite each inequality so that the variable is on the left. Do not solve.

1. $-14 < n + 3.2$
2. $-2 \ge 9 + n$
3. $4.7 > -2 + d$
4. $-10 \le -10 + w$

PRACTICE

Solve each inequality. Graph the solution on a number line.

5. $n + 5 \ge 5$ 6. $p + 4 \ge 8$ 7. $4 < -2 + m$

8. $-7 > -5 + k$ 9. $-5 < -5 + k$ 10. $3 < -3 + f$

11. $c + 5.1 \ge 10$ 12. $b + 3.1 \le 1$ 13. $0 < -2 + h$

14. $-7 < -3 + z$ 15. $z - 2.6 \ge 4.4$ 16. $z - (1.2 + 2.2) \le -2\frac{2}{5}$

17. $x - \left(4\frac{1}{3} - 2\frac{1}{3}\right) \ge -2\frac{2}{3}$ 18. $-6 > a - \left(2\frac{4}{5} + 2\frac{1}{5}\right)$ 19. $x - 1\frac{1}{3} \ge -4\frac{2}{3}$

Without solving, tell whether the solution contains only positive numbers, only negative numbers, or both positive and negative numbers.

20. $v + 5 \le -9$ 21. $a + 11 \le -12$

22. $q + 6 > 2$ 23. $n + 4 > -2$

List all integers that meet each given restriction.

24. $b - 6 \le -2$; solutions no less than -5 25. $n + 1 > 6$; solutions no more than 8

26. $c + 2.6 \ge -4$; solutions no more than 0   27. $m + 3 > -5$; solutions no more than 2

28. **Critical Thinking** Find and graph the integer solution(s) to the set of inequalities below.

$$x + 2 > -7 \text{ and } x - 3 \le 7 \text{ and } x + 2 > 5 \text{ and } x - 4 < 1$$

MIXED REVIEW

Find each product or quotient. (2.4 Skill A)

29. $(-3)(-11)$ 30. $(-4)(14)$ 31. $(7)(-1)$ 32. $4(4)$

33. $28 \div (-14)$ 34. $-108 \div (12)$ 35. $5 \div (-25)$ 36. $(-15) \div (15)$

37. $\left(-\frac{3}{5}\right)\left(-\frac{5}{21}\right)$ 38. $\left(-\frac{8}{3}\right)\left(\frac{15}{24}\right)$ 39. $\left(-\frac{14}{5}\right) \div \left(\frac{28}{25}\right)$ 40. $\left(\frac{35}{32}\right) \div \left(-\frac{7}{16}\right)$

Answers for Lesson 4.2 Skill B

Try This Example 1

$x \ge 2$

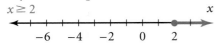

Try This Example 2

$a < 1.3$

1. $n + 3.2 > -14$
2. $9 + n \le -2$
3. $-2 + d < 4.7$
4. $-10 + w \ge -10$
5. $n \ge 0$

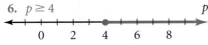

6. $p \ge 4$

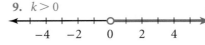

7. $m > 6$

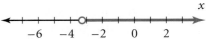

8. $k < -2$

9. $k > 0$

10. $f > 6$

11. $c \ge 4.9$

12. $b \le -2.1$

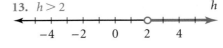

13. $h > 2$

14. $z > -4$

15. $z \ge 7$

16. $z \le 1$

17. $x > -\frac{2}{3}$

18. $a < -1$

19. $x > -3\frac{1}{3}$

For answers to Exercises 20–40, see Additional Answers.

Focus: *Students use inequalities in one variable to represent real-world problems. They then choose the Addition or Subtraction Property of Inequality to solve.*

EXAMPLE 1

PHASE 1: Presenting the Example

Display the inequality $m + 2 > 10$. Tell students that m represents the number of miles a person has walked. Have them write a verbal translation of the inequality. [**sample: When the number of miles is increased by 2, the result is more than 10 miles.**] Discuss the students' responses.

Now ask the students if they can identify the exact number of miles that this person has walked. They should recognize that, by subtracting 2 from each side of $m + 2 > 10$, they will arrive at the result $m > 8$. So they can tell that the person walked more than 8 miles. However, it is not possible to calculate an *exact* number of miles using just the given information.

Now discuss the problem in Example 1. Point out that here the process is reversed: Given a situation, the students must write an inequality to describe it.

EXAMPLE 2

PHASE 1: Presenting the Example

PHASE 2: Providing Guidance for **TRY THIS**

Some students may perceive that these problems can be solved with compound inequalities. In Example 2, for instance, the given conditions can be modeled by $0 \leq s - 25 \leq 20$. Similarly, the given conditions in the Try This exercise can be modeled by $0 \leq t - 40.2 < 42$, where t is the number of tons. Students will learn to solve compound inequalities like these in Lesson 4.6. However, you may wish to preview the technique with those students who are interested and capable. Solutions for the compound inequalities mentioned above are shown at right.

$$
\begin{array}{rcccc}
0 & \leq & s - 25 & \leq & 20 \\
+\,25 & & +\,25 & & +\,25 \\
\hline
25 & \leq & s & \leq & 45
\end{array}
$$

$$
\begin{array}{rcccc}
0 & \leq & t - 40.2 & \leq & 42 \\
+\,40.2 & & +\,40.2 & & +\,40.2 \\
\hline
40.2 & \leq & t & \leq & 82.2
\end{array}
$$

PHASE 3: ASSESSMENT AND CLOSURE for Lesson 4.2

- Have students write an inequality that can be solved in one step by applying the Addition Property of Inequality and a second that can be solved in one step by applying the Subtraction Property of Inequality. Then have them solve each of their inequalities. [**Answers will vary.**]

- In this lesson, students learned how to use addition and subtraction to solve inequalities. In the following lesson, they will learn how to solve inequalities by using multiplication and division.

☞ *For a Lesson 4.2 Quiz, see* Assessment Resources *page 39.*

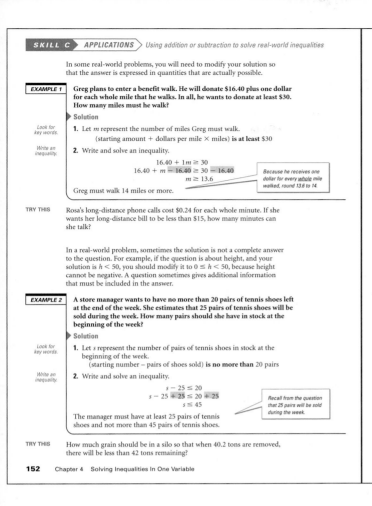

In some real-world problems, you will need to modify your solution so that the answer is expressed in quantities that are actually possible.

EXAMPLE 1

Greg plans to enter a benefit walk. He will donate $16.40 plus one dollar for each whole mile that he walks. In all, he wants to donate at least $30. How many miles must he walk?

▶ Solution

Look for key words.

1. Let m represent the number of miles Greg must walk.
   (starting amount + dollars per mile × miles) **is at least** $30

Write an inequality.

2. Write and solve an inequality.

$$16.40 + 1m \geq 30$$
$$16.40 + m - 16.40 \geq 30 - 16.40$$
$$m \geq 13.6$$

Because he receives one dollar for every _whole_ mile walked, round 13.6 to 14.

Greg must walk 14 miles or more.

TRY THIS

Rosa's long-distance phone calls cost $0.24 for each whole minute. If she wants her long-distance bill to be less than $15, how many minutes can she talk?

In a real-world problem, sometimes the solution is not a complete answer to the question. For example, if the question is about height, and your solution is $h < 50$, you should modify it to $0 \leq h < 50$, because height cannot be negative. A question sometimes gives additional information that must be included in the answer.

EXAMPLE 2

A store manager wants to have no more than 20 pairs of tennis shoes left at the end of the week. She estimates that 25 pairs of tennis shoes will be sold during the week. How many pairs should she have in stock at the beginning of the week?

▶ Solution

Look for key words.

1. Let s represent the number of pairs of tennis shoes in stock at the beginning of the week.
   (starting number − pairs of shoes sold) **is no more than** 20 pairs

Write an inequality.

2. Write and solve an inequality.

$$s - 25 \leq 20$$
$$s - 25 + 25 \leq 20 + 25$$
$$s \leq 45$$

Recall from the question that 25 pairs will be sold during the week.

The manager must have at least 25 pairs of tennis shoes and not more than 45 pairs of tennis shoes.

TRY THIS

How much grain should be in a silo so that when 40.2 tons are removed, there will be less than 42 tons remaining?

EXERCISES

1. Rework Example 1 given that Greg already has $15 to donate and wants to donate at least $35.

2. Rework Example 2 so that the manager wants to have no more than 5 pairs of tennis shoes at the end of the week.

Use an inequality to solve each problem. Be sure to include any restrictions.

3. Russell must have at least 510 points in his science class to receive an A. If he has 375 points, at least how many more points does he need to receive an A?

4. Casey bought a sandwich and a drink for $3.75. If she has $6.00 to spend, what is the most that she can spend on dessert?

5. Jenna has entered a benefit walk. Her parents gave her $15, and she has saved $24.25. If she receives one dollar for each whole mile that she walks, how far must she walk to donate at least $50?

6. A manager believes that his ski shop will sell 35 ski caps, 24 ski jackets, and 30 pairs of goggles. How many of each should he have at the beginning of the week so that he will have no more than 20 of each item at the end of the week?

7. A baker needs between 13 and 18 pounds of flour, and she already has 5.5 pounds of flour. How many 1-pound bags does she need to buy?

8. The Barlow family has less than $540 to spend for food and entertainment during their vacation. If they spend $365 for food, what is the most that they can spend for entertainment?

Use an equation to solve each problem. (2.7 Skill C)

9. Find the dimensions of a rectangular garden whose perimeter is 180 feet and whose length is 5 times its width.

10. Find two consecutive numbers whose sum is 47.

Answers for Lesson 4.2 Skill C

Try This Example 1
no more than 62 minutes

Try This Example 2
at least 40.2 tons and at most 82.2 tons

1. 20 miles or more
2. at least 25 and not more than 30 pairs of tennis shoes
3. at least 135 points
4. at most $2.25
5. at least 11 miles
6. at least 35 and no more than 55 ski caps, at least 24 and no more than 44 ski jackets, at least 30 and no more than 50 pairs of goggles
7. between 8 and 12 one-pound bags
8. less than $175
9. width $=$ 15 feet, length $=$ 75 feet
10. 23 and 24

TEACHING
4.3
LESSON

Solving Inequalities Using Multiplication or Division

SKILL A ▶ *Using multiplication to solve an inequality in one step*

Focus: *Students apply the Multiplication Property of Inequality to solve inequalities in one variable.*

EXAMPLES 1 and 2

PHASE 1: Presenting the Examples

Display the inequality $2 < 10$. Ask students if it is true or false. [**true**] Tell them to multiply each side of this inequality by 1, then by 2, by 3, and by 4. Check their answers. [$2 < 10, 4 < 20, 6 < 30, 8 < 40$] Ask if each of these new inequalities is true or false. [**All are true.**]

Now have students multiply each side of $2 < 10$ by -1, then by -2, by -3, and by -4. Check their answers. [$-2 < -10, -4 < -20, -6 < -30, -8 < -40$] Ask if each of these new inequalities is true or false. [**All are false.**] Ask students what could be done to make them true. Elicit the response that each is true when the order of the inequality symbol is reversed to $>$ rather than $<$.

Ask students to write a conjecture about the results. Then discuss their responses. Most should have perceived that they had to make one conjecture about multiplication by a positive number and a different conjecture about multiplication by a negative number. Lead the group to a consensus such as the following:

> *If both sides of a true inequality are multiplied by the same positive number, then the resulting inequality is true.*

> *If both sides of a true inequality are multiplied by the same negative number, then you must reverse the order of the inequality symbol to make a true inequality.*

Have students read the formal statement of the Multiplication Property of Inequality given near the top of the textbook page. Ask why multiplication by zero is not included. Lead students to see that multiplying each side of an inequality by zero would produce zero on each side. Then the only true conclusion would be $0 = 0$, which is not a very interesting or helpful result!

Now discuss Examples 1 and 2. Begin each by asking students how the Multiplication Property of Inequality can be applied.

PHASE 2: Providing Guidance for TRY THIS

The most common error made when solving inequalities of this type is failing to reverse the order of the inequality symbol when multiplying by a negative number. Alert students to this fact. Then lead a discussion in which they brainstorm ways that they can avoid the error. Some ideas are checking the order of the symbol at every step of the solution and substituting numbers from the set of solutions into the original inequality.

---

154   Chapter 4   Solving Inequalities In One Variable

4.3 LESSON
Solving Inequalities Using Multiplication or Division

SKILL A *Using multiplication to solve an inequality in one step*

You may need to multiply to solve an inequality.

> **Multiplication Property of Inequality**
> If a and b are real numbers, c is positive, and $a < b$, then $ac < bc$.
> If a and b are real numbers, c is negative, and $a < b$, then $ac > bc$.
> Similar statements can be written for $a > b$, $a \le b$, and $a \ge b$.

Notice that you must reverse the inequality symbol if you multiply both sides of an inequality by a negative number.

EXAMPLE 1 Solve each inequality. Graph the solution on a number line.

a. $\frac{x}{3} \le 12$ b. $2 > \frac{d}{-2}$

▶ **Solution**

a. $\frac{x}{3} \le 12$

$3 \cdot \frac{x}{3} \le 3 \cdot 12$

$x \le 36$

b. $2 > \frac{d}{-2}$

$(-2)2 < (-2)\left(\frac{d}{-2}\right)$ ← Reverse the inequality symbol.

$-4 < d$, or $d > -4$

TRY THIS Solve each inequality. Graph the solution on a number line.

a. $\frac{x}{4} > -\frac{1}{2}$ b. $-\frac{1}{3} > \frac{t}{-6}$

A multiplier can be a rational number other than an integer.

EXAMPLE 2 Solve $\frac{2}{5}y < -1$. Graph the solution on a number line.

▶ **Solution**

$\frac{2}{5}y < -1$

$\frac{5}{2} \cdot \frac{2}{5}y < \frac{5}{2}(-1)$ ← Multiply each side by $\frac{5}{2}$.

$y < -\frac{5}{2}$, or -2.5

TRY THIS Solve $-\frac{3}{2}y < -6$. Graph the solution on a number line.

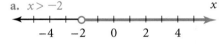
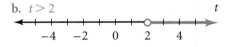

154   Chapter 4   Solving Inequalities In One Variable

EXERCISES

KEY SKILLS

What multiplier would you choose to solve each inequality? Will the inequality symbol be reversed?

1. $\frac{w}{4} \le -3$ 2. $5 > \frac{t}{-3}$ 3. $\frac{2}{-3}b \le -4$ 4. $\frac{12}{5} \ge \frac{-7}{5}c$

PRACTICE

Solve each inequality. Graph the solution on a number line.

5. $\frac{r}{2} \le -1$ 6. $\frac{t}{-3} \ge 4$ 7. $-3 \le \frac{x}{-2}$

8. $-1 \le \frac{f}{7}$ 9. $\frac{p}{6} > \frac{5}{-2}$ 10. $\frac{n}{-21} > \frac{2}{-7}$

11. $\frac{5}{6} > \frac{1}{-3}u$ 12. $-\frac{5}{2} \ge \frac{5}{2}c$ 13. $\frac{n}{-2.91} \ge 0$

14. $0 < \frac{k}{14.3}$ 15. $\frac{2x}{3} > 6$ 16. $18 \le \frac{9x}{5}$

17. $\frac{3}{5}p > -9$ 18. $-\frac{1}{8}x < 3$ 19. $-\frac{2}{3}w < -6$

20. $\frac{5k}{2.5} > -3$ 21. $-4 < \frac{-7k}{3.5}$ 22. $\left(-\frac{5}{2}\right)\left(-\frac{4}{15}\right) \ge -\frac{2}{5}z$

23. $\left(\frac{5}{2}\right)\left(-\frac{6}{25}\right)s \ge -\frac{2}{3}$ 24. $-\frac{2}{7} \le \left(\frac{1}{3}\right)\left(-\frac{9}{21}\right)a$ 25. $\left(\frac{5}{2}\right)\left(-\frac{6}{25}\right)s \ge \left(-\frac{2}{3}\right)\left(-\frac{3}{5}\right)$

Show that each pair of inequalities has the same solution.

26. $\frac{x}{-3} \le \frac{2}{5}$; $\frac{5}{2}x \ge -3$ 27. $\frac{7}{3} > \frac{x}{4}$; $\frac{3}{7}x < 4$

28. **Critical Thinking** Suppose that x is a variable, a and b are real numbers, and a is nonzero. Solve $\frac{x}{a} > b$ for x.

MIXED REVIEW

Find each quotient. (2.4 Skill B)

29. $\frac{-12}{12}$ 30. $\frac{-300}{-5}$ 31. $\frac{-12}{48}$

32. $\frac{7}{-3.5}$ 33. $\left(\frac{7}{24}\right) \div \left(-\frac{49}{12}\right)$ 34. $\left(-\frac{25}{24}\right) \div \left(-\frac{100}{96}\right)$

35. $\left(-\frac{3}{4}\right) \div \left(-\frac{4}{3}\right)$ 36. $\left(\frac{3}{4}\right) \div \left(-\frac{15}{20}\right)$ 37. $\left(-\frac{3}{4}\right) \div 4$

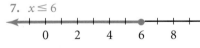
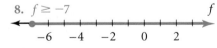
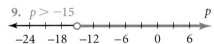
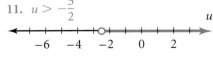

Answers for Lesson 4.3 Skill A

Try This Example 1

a. $x > -2$

b. $t > 2$

Try This Example 2

$y > 4$

1. 4; no
2. −3; yes
3. $-\frac{3}{2}$; yes
4. $-\frac{5}{7}$; yes
5. $r \le -2$

6. $t \le -12$
7. $x \le 6$
8. $f \ge -7$
9. $p > -15$
10. $n < 6$
11. $u > -\frac{5}{2}$
12. $c \le -1$

13. $n \le 0$
14. $k > 0$
15. $x > 9$
16. $x \ge 10$
17. $p > -15$
18. $x > -24$

For answers to Exercises 19–37, see Additional Answers.

Focus: *Students apply the Division Property of Inequality to solve inequalities in one variable.*

EXAMPLE

PHASE 1: Presenting the Example

Display the following statement:

> *If both sides of a true inequality are divided by the same real number, then the resulting inequality is also true.*

Tell students to write on their papers whether they believe that this statement is true or false. Then poll the class to find how many responded *true* and how many responded *false.*

Ask if a student from either group can give a logical argument to support her or his answer. It is most likely that one or more of the students who responded *false* will note that the resulting inequality is not true whenever the divisor is negative. For example, they might note that while $4 < 6$ is a true inequality, $4 \div (-2) < 6 \div (-2)$, or $-2 < -3$, is *not* true.

At some point in the discussion, most students should become aware that, in regard to inequalities, the rules for division parallel the rules for multiplication. The rules can be summarized as follows:

> *If both sides of a true inequality are divided by the same positive number, then the resulting inequality is true.*

> *If both sides of a true inequality are divided by the same negative number, then you must reverse the order of the inequality symbol to make a true inequality.*

Have students read the statement of the Division Property of Inequality near the top of the textbook page. Ask why zero has been excluded. Elicit the response that division by zero is undefined. Then discuss the Example.

PHASE 2: Providing Guidance for **TRY THIS**

Watch for students who give the solution to part a as $c \leq -2.5$. These students may have reversed the order of the inequality simply because they saw a negative number (-10) in the original inequality. Stress that the order is reversed only if they multiply or divide *by* a negative number.

When checking solutions to inequalities, remind students to use the *original inequality*. Encourage them to check one number that is in the set of proposed solutions and one that is not. For instance, the following checks support the conclusion that the solution of $4c \geq -10$ is $c \geq -2.5$.

Try $c = 0$. Since 0 is in the set of solutions, substituting 0 into the original inequality should give a true statement.	$4c \geq -10$ $4(0) \geq -10$ $0 \geq -10$ *True*	Try $c = -3$. Since -3 is *not* in the set of solutions, substituting -3 into the original inequality should give a false statement.	$4c \geq -10$ $4(-3) \geq -10$ $-12 \geq -10$ *False*

You may need to divide to solve an inequality.

Division Property of Inequality

If a and b are real numbers, c is positive, and $a < b$, then $\frac{a}{c} < \frac{b}{c}$.

If a and b are real numbers, c is negative, and $a < b$, then $\frac{a}{c} > \frac{b}{c}$.
Similar statements can be written for $a > b$, $a \leq b$, and $a \geq b$.

Notice that you must reverse the inequality symbol if you divide both sides of an inequality by a negative number.

EXAMPLE Solve each inequality. Graph the solution on a number line.
  a. $4v < 6$ 　　　　　　　　 **b.** $-4x < 6$

▸ **Solution**
  a. $4v < 6$
  $\frac{4v}{4} < \frac{6}{4}$ ⟵ Apply the Division Property of Inequality. Divide by 4.
  $v < \frac{3}{2}$, or 1.5

  $\begin{array}{c} 1.5 \\ \xleftarrow{\;\;\;\;\;\;\;\;} {\circ} \xrightarrow{\;\;\;\;\;\;\;\;} v \\ -5\;-4\;-3\;-2\;-1\;\;0\;\;1\;\;2\;\;3\;\;4\;\;5 \end{array}$

  b. $-4x < 6$
  $\frac{-4}{-4} \geq \frac{6}{-4}$ ⟵ Apply the Division Property of Inequality. Divide by –4. Change < to >.
  $x > -\frac{3}{2}$, or –1.5

  $\begin{array}{c} -1.5 \\ \xleftarrow{\;\;\;\;\;\;\;\;} {\circ} \xrightarrow{\;\;\;\;\;\;\;\;} x \\ -5\;-4\;-3\;-2\;-1\;\;0\;\;1\;\;2\;\;3\;\;4\;\;5 \end{array}$

TRY THIS Solve each inequality. Graph the solution on a number line.　**a.** $4c \geq -10$　**b.** $-4c \geq -10$

The multiplication inequality $ax < b$ and the division inequality $\frac{x}{a} < b$ can both be solved using multiplication. Suppose $a > 0$.

$ax < b$ 　　　　　　　　　　　　　 $\frac{x}{a} < b$

$\left(\frac{1}{a}\right)ax < \left(\frac{1}{a}\right)b$ ⟵ Multiply by $\frac{1}{a}$.　$a \cdot \frac{x}{a} < a \cdot b$ ⟵ Multiply by a.

$\left(\frac{1}{a} \cdot a\right) \cdot x < \left(\frac{1}{a}\right)b$ 　　　　　　$\left(a \cdot \frac{1}{a}\right) \cdot x < a \cdot b$ ⟵ Definition of division

$x < \frac{b}{a}$ 　　　　　　　　　　　 $x < ab$

EXERCISES

KEY SKILLS

By what number would you choose to divide each side of the inequality? Will the inequality symbol be reversed?

1. $4d \geq -12$ 　**2.** $-5r > 13$ 　**3.** $6t \leq 0$ 　**4.** $-7.5g < 3$

PRACTICE

Solve each inequality. Graph the solution on a number line.

5. $2r > 8$ 　　　**6.** $-3t \geq 12$ 　　**7.** $18 > -6x$ 　　**8.** $-25 \geq -5d$

9. $12x < -1$ 　　**10.** $-7 > 14t$ 　　**11.** $-1.5d > 3$ 　　**12.** $2.5h < -5$

13. $-1.5 \geq -3d$ 　**14.** $2.5 < -5h$ 　**15.** $\frac{1}{3}z \leq -2$ 　　**16.** $-6 \leq \frac{1}{5}n$

You can multiply by a fraction to solve each inequality below. Identify the fraction. Then use it to solve the inequality.

17. $\frac{2m}{-5} \leq 2$ 　　　**18.** $\frac{-3g}{7} > -9$ 　　　**19.** $\frac{-7g}{2} > -2$

Tell whether the solution contains only negative numbers, only positive numbers, or both positive numbers and negative numbers.

20. $-3k > 5$ 　　　**21.** $4h \geq 12$ 　　　**22.** $-3c \geq -12$

If $-1 < x < 3$, then we say "x is between -1 and 3." If $-1 \leq x \leq 3$, we say "x is between -1 and 3, **inclusive**," because -1 and 3 are *included* in the possible values for x.

Write and graph an inequality for each description.

23. x is between 0 and 3. 　　　　**24.** y is between -2 and 4.

25. x is between 0 and 3, inclusive. 　**26.** t is between 6 and 8, inclusive.

27. Critical Thinking Suppose that a is between 3 and -5, inclusive. Between what two numbers will the solution to $ax < 15$ be found?

MIXED REVIEW

Simplify by combining like terms. (2.6 Skill A)

28. $-3n + 6.3 - 4n$ 　**29.** $4j - 3j - 4j$ 　　**30.** $7.2c + (-5.8c) + 1$

Simplify each expression. (2.6 Skill B)

31. $-\frac{7}{3}(9x + 15)$ 　**32.** $\frac{y-4}{-2}$ 　　**33.** $\frac{-2x+18}{2}$

Answers for Lesson 4.3 Skill B

Try This Example
a. $c \geq -2.5$

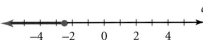

b. $c \leq -2.5$

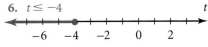

1. 4; no
2. -5; yes
3. 6; no
4. -7.5; yes
5. $r > 4$

$\xleftarrow{\;\;\;\;\;\;} \circ \xrightarrow{\;\;\;\;\;\;} r$
$-2\;\;0\;\;2\;\;4\;\;6$

6. $t \leq -4$

$\xleftarrow{\;\;\;\;\;\;} \bullet \xrightarrow{\;\;\;\;\;\;} t$
$-6\;-4\;-2\;\;0\;\;2$

7. $x > -3$

$\xleftarrow{\;\;\;\;\;\;} \circ \xrightarrow{\;\;\;\;\;\;} x$
$-6\;-4\;-2\;\;0\;\;2$

8. $d \geq 5$　　　　　　　　　d

$\xleftarrow{\;\;\;\;\;\;} \bullet \xrightarrow{\;\;\;\;\;\;}$
$-2\;\;0\;\;2\;\;4\;\;6$

9. $x < -\dfrac{1}{12}$　　　　　　x

$\xleftarrow{\;\;\;\;\;\;} \circ \xrightarrow{\;\;\;\;\;\;}$
$-6\;-4\;-2\;\;0\;\;2$

10. $t < -\dfrac{1}{2}$

$\xleftarrow{\;\;\;\;\;\;} \circ \xrightarrow{\;\;\;\;\;\;} t$
$-2\;\;0\;\;2\;\;4\;\;6$

11. $d < -2$　　　　　　　　d

$\xleftarrow{\;\;\;\;\;\;} \circ \xrightarrow{\;\;\;\;\;\;}$
$-6\;-4\;-2\;\;0\;\;2$

12. $h < -2$　　　　　　　　h

$\xleftarrow{\;\;\;\;\;\;} \circ \xrightarrow{\;\;\;\;\;\;}$
$-2\;\;0\;\;2\;\;4\;\;6$

13. $d \geq 0.5$　　　　　　　d

$\xleftarrow{\;\;\;\;\;\;} \bullet \xrightarrow{\;\;\;\;\;\;}$
$-2\;\;0\;\;2\;\;4\;\;6$

14. $h < -0.5$　　　　　　　h

$\xleftarrow{\;\;\;\;\;\;} \circ \xrightarrow{\;\;\;\;\;\;}$
$-6\;-4\;-2\;\;0\;\;2$

15. $z \leq -6$　　　　　　　　z

$\xleftarrow{\;\;\;\;\;\;} \bullet \xrightarrow{\;\;\;\;\;\;}$
$-8\;-6\;-4\;-2\;\;0$

16. $n \geq -30$　　　　　　　n

$\xleftarrow{\;\;\;\;\;\;} \bullet \xrightarrow{\;\;\;\;\;\;}$
$-32\;-24\;-16\;-8\;\;0\;\;8$

17. $-\dfrac{5}{2}$; $m \geq -5$

18. $-\dfrac{7}{3}$; $g < 21$

19. $-\dfrac{2}{7}$; $g < \dfrac{4}{7}$

20. negative numbers
21. positive numbers
22. negative and positive numbers

For answers to Exercises 23–33, see Additional Answers.

SKILL C APPLICATIONS *Using multiplication or division to solve real-world inequalities*

Focus: *Students use inequalities in one variable to represent real-world problems. They then choose the Multiplication or Division Property of Inequality to solve.*

EXAMPLE 1

PHASE 1: Presenting the Example

Display the inequalities shown at right. Tell students that d represents an amount of money in dollars. Ask them to translate each inequality into an English sentence. [**top to bottom: The result when *d* dollars is multiplied by 4 is at most \$300. The result when *d* dollars is multiplied by 4 is at least \$300. The result when *d* dollars is divided into 4 parts is at most \$300. The result when *d* dollars is divided into 4 parts is at least \$300.**]

$4d \leq 300$
$4d \geq 300$
$\frac{d}{4} \leq 300$
$\frac{d}{4} \geq 300$

Now display the problem in Example 1. Ask students which inequality best represents it. [$\mathbf{4d \geq 300}$] Then discuss the solution on the textbook page.

EXAMPLE 2

PHASE 1: Presenting the Example

Display the inequalities shown at right. Tell students that a represents an amount of money in dollars. Then display the problem in Example 2. Ask which inequality best represents it. Lead students to see that, in this case, two inequalities are needed: $\frac{a}{5} \geq 2$ and $\frac{a}{5} \leq 5$. Then discuss the solution on the textbook page.

$5a \leq 2$	$5a \leq 5$
$5a \geq 2$	$5a \leq 5$
$\frac{a}{5} \leq 2$	$\frac{a}{5} \leq 5$
$\frac{a}{5} \geq 2$	$\frac{a}{5} \geq 5$

PHASE 2: Providing Guidance for TRY THIS

Have students use dimensional analysis to decide whether an inequality they have written makes sense. For instance, labeling units as shown at right provides verification that $\frac{f}{6} \geq 5$ can be used in finding the number of flowers, f.

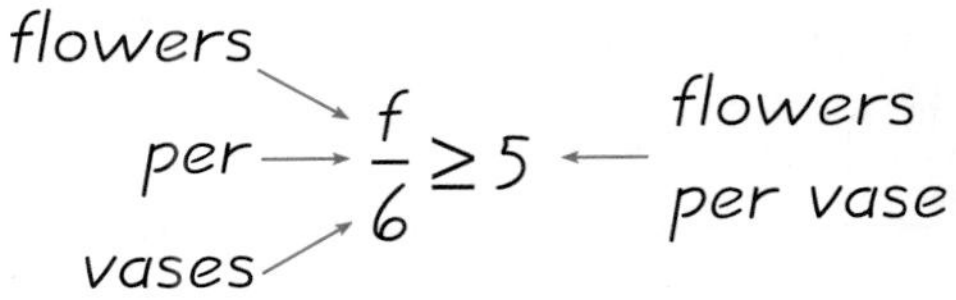
flowers
per
vases
f/6 ≥ 5
flowers
per vase

PHASE 3: ASSESSMENT AND CLOSURE for Lesson 4.3

- Give students the inequalities at right. Ask them to compare the process of solving one to the process of solving the other. Then have them solve. [**sample: Both can be solved by using the Division Property of Inequality, dividing by 3 and −3. In the first case, the inequality symbol is unchanged. In the second, it must be reversed. The solutions are $b > 5$ and $c < -5$.**]

  $3b > 15$
  $-3c > 15$

- Be sure students are aware that the Addition, Subtraction, Multiplication, and Division Properties of Inequality are axioms, meaning that they are *assumed* to be true for all real numbers.

For a Lesson 4.3 Quiz, see Assessment Resources *page 40.*

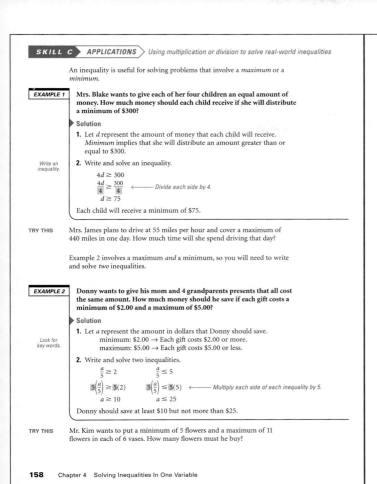

An inequality is useful for solving problems that involve a *maximum* or a *minimum*.

EXAMPLE 1

Mrs. Blake wants to give each of her four children an equal amount of money. How much money should each child receive if she will distribute a minimum of $300?

▶ **Solution**

Write an inequality.

1. Let d represent the amount of money that each child will receive. *Minimum* implies that she will distribute an amount greater than or equal to $300.

2. Write and solve an inequality.

$$4d \geq 300$$
$$\frac{4d}{4} \geq \frac{300}{4} \quad \longleftarrow \text{Divide each side by 4.}$$
$$d \geq 75$$

Each child will receive a minimum of $75.

TRY THIS   Mrs. James plans to drive at 55 miles per hour and cover a maximum of 440 miles in one day. How much time will she spend driving that day?

Example 2 involves a maximum *and* a minimum, so you will need to write and solve *two* inequalities.

EXAMPLE 2

Donny wants to give his mom and 4 grandparents presents that all cost the same amount. How much money should he save if each gift costs a minimum of $2.00 and a maximum of $5.00?

▶ **Solution**

Look for key words.

1. Let a represent the amount in dollars that Donny should save.
   minimum: $2.00 → Each gift costs $2.00 or more.
   maximum: $5.00 → Each gift costs $5.00 or less.

2. Write and solve two inequalities.

$$\frac{a}{5} \geq 2 \qquad\qquad \frac{a}{5} \leq 5$$
$$5\left(\frac{a}{5}\right) \geq 5(2) \qquad 5\left(\frac{a}{5}\right) \leq 5(5) \quad \longleftarrow \text{Multiply each side of each inequality by 5.}$$
$$a \geq 10 \qquad\qquad a \leq 25$$

Donny should save at least $10 but not more than $25.

TRY THIS   Mr. Kim wants to put a minimum of 5 flowers and a maximum of 11 flowers in each of 6 vases. How many flowers must he buy?

EXERCISES

Identify the minimum and maximum in each problem.

1. Each canning jar will contain between 15 ounces and 17 ounces, inclusive.

2. A certain truck driver may not drive more than 10 hours on a given day.

PRACTICE

Solve each problem.

3. How much time will a motorist spend driving if he drives between 500 miles and 600 miles at 50 miles per hour?

4. How much money must be budgeted if each of 6 school clubs receives an equal amount that is at least $2400 but no more than $3600?

5. What is the range of positive numbers, n, such that n divided by 5 is less than 32?

6. What is the range of negative numbers, n, such that the product of n and 7 is more than -63?

7. **a.** What is the range of positive real numbers, n, such that the product of n and 11 is less than 121?
   b. What are the positive integers, n, such that the product of n and 11 is less than 121?

8. **Critical Thinking**   When five objects with the same weight are put on a scale, the scale reads 1.5 kilograms. If the scale's reading is as much as 0.001 kilogram too little or too much, between what two numbers is the actual weight of one of the objects?

MIXED REVIEW APPLICATIONS

Find the domain and range of each function. (3.2 Skill C)

9. the weight, w, of n identical textbooks if each book weighs 2.3 pounds

10. the volume of water, v, in a tank that initially contains 800 gallons and drains at the rate of 20 gallons per minute for t minutes

11. the number of light bulbs, n, if there are c cartons that each contain 12 packages of 2 bulbs

Answers for Lesson 4.3 Skill C

Try This Example 1
no more than 8 hours

Try This Example 2
between 30 and 66 flowers, inclusive

1. minimum: 15 ounces
   maximum: 17 ounces
2. minimum: 0 hours
   maximum: 10 hours
3. between 10 hours and 12 hours
4. between $14,400 and $21,600, inclusive
5. between 0 and 160
6. between -9 and 0
7. **a.** between 0 and 11
   b. 1, 2, 3, 4, 5, 6, 7, 8, 9, 10
8. between 0.2998 and 0.3002 ounces, inclusive
9. domain: whole numbers; range: nonnegative multiples of 2.3
10. domain: all real numbers between 0 and 40, inclusive; range: all real numbers between 0 and 800, inclusive
11. domain: whole numbers; range: all nonnegative multiples of 24

TEACHING
4.4
LESSON

Solving Simple One-Step Inequalities

SKILL A ▶ *Choosing an operation to solve an inequality in one step*

Focus: *Students analyze a given inequality and choose the Addition, Subtraction, Multiplication, or Division Property of Inequality to solve.*

EXAMPLES 1 and 2

PHASE 1: Presenting the Examples

Display the inequalities shown at right. Point out that each can be solved in one step by applying one of the properties of inequality. Lead a discussion in which students brainstorm methods to decide which property to use. Have them assign an operation to each inequality and then solve all four. [**top to bottom: subtract 4 or add −4, $z > 8$;**

divide by −4 or multiply by $-\frac{1}{4}$, $z < -3$; add 4, $z > 16$; multiply by 4, $z > -48$]

Now discuss Examples 1 and 2.

$$z + 4 > 12$$
$$-4z > 12$$
$$z - 4 > 12$$
$$\frac{z}{4} > -12$$

EXAMPLE 3

PHASE 1: Presenting the Example

Have students draw a number line. Tell them to graph all real numbers whose *opposites* are greater than or equal to 5. If they have difficulty getting started, suggest that they begin by graphing just a few integers whose opposite is greater than or equal to 5, such as −5, −6, −7, and so on. They should eventually arrive at the following graph:

$$\longleftarrow \ \overset{\bullet}{\underset{-8 \quad -6 \quad -4 \quad -2 \quad 0 \quad 2 \quad 4 \quad 6 \quad 8}{\rule{0pt}{0pt}}} \longrightarrow$$

Now ask to write an open sentence that describes all the real numbers t graphed on their number lines. They will most likely write $t \leq -5$.

Remind students that you asked them to graph all real numbers whose opposites are greater than or equal to 5. Point out that an open sentence describing this condition is $-t \geq 5$. So $-t \geq 5$ must be equivalent to $t \leq -5$. Display these inequalities one above the other, as shown at right.

$$-t \geq 5$$
$$t \leq -5$$

Ask students how to transform the first inequality into the second inequality. Lead them to see that this goal is achieved by multiplying both sides of the first inequality by −1. Then discuss the solution that is outlined on the textbook page.

PHASE 2: Providing Guidance for **TRY THIS**

After students solve part a algebraically, note that $-j > 0$ describes all real numbers j such that the opposite of j is a positive number. Lead students to see that, since the opposite of j is any positive number, j itself must be any negative number. That is, $j < 0$. Point out that this "common sense" analysis verifies that $-j > 0$ and $j < 0$ are equivalent.

4.4
LESSON

Solving Simple One-Step Inequalities

SKILL A ▸ *Choosing an operation to solve an inequality in one step*

When you solve an inequality, it is important to think about which property of inequality you will use. Remember that you can rewrite inequalities so that the variable is on the left. For example, you can rewrite $5 < x$ as $x > 5$.

EXAMPLE 1 Choose a method for solving $8 \le t + 4.5$. Then solve.

▸ Solution

$$8 \le t + 4.5$$
$$t + 4.5 \ge 8 \qquad \longleftarrow \text{Rewrite the inequality so that } t \text{ is on the left.}$$
$$t + 4.5 - 4.5 \ge 8 - 4.5 \qquad \longleftarrow \text{Apply the Subtraction Property of Inequality.}$$
$$t \ge 3.5$$

TRY THIS Choose a method for solving $-6 < c + 1.5$. Then solve.

EXAMPLE 2 Choose a method for solving $25 > \frac{5}{2}d$. Then solve.

▸ Solution

$$25 > \frac{5}{2}d$$
$$\frac{5}{2}d < 25 \qquad \longleftarrow \text{Rewrite the inequality so that } d \text{ is on the left.}$$
$$\left(\frac{2}{5}\right)\frac{5}{2}d < \left(\frac{2}{5}\right)25 \qquad \longleftarrow \text{Apply the Multiplication Property of Inequality.}$$
$$d < 10$$

TRY THIS Choose a method for solving $\frac{1}{10} > \frac{2}{5}z$. Then solve.

EXAMPLE 3 Choose a solution method. Then solve.

a. $-t \ge 5$ b. $0 < 6 - s$

▸ Solution

a. $\quad -t \ge 5$
$\quad (-1)(-t) \le (-1)5 \qquad \longleftarrow \text{Multiply by } -1.$
$\quad t \le -5$

b. $\quad 0 < 6 - s$
$\quad 0 + s < 6 - s + s \qquad \longleftarrow \text{Add } s.$
$\quad s < 6$

TRY THIS Choose a solution method. Then solve. a. $-j > 0$ b. $0 \ge 4 - x$

EXERCISES

KEY SKILLS

What step would you take to solve each inequality? Do not solve.

1. $x - 6 < -3$ 2. $x + 3.5 > 8.2$

3. $5x \ge -40$ 4. $\frac{x}{7} > 2$

PRACTICE

For Exercises 5–22:
  a. What step would you take to solve each inequality?
  b. Will the inequality symbol be reversed?
  c. Solve the inequality.

5. $t - \frac{3}{4} \ge \frac{1}{4}$ 6. $j - 2.3 \le -6$ 7. $\frac{w}{-3} > 2.8$

8. $-3p < 3$ 9. $-y < 0$ 10. $-r \ge 9$

11. $j + 1.5 \le -2$ 12. $q + 2.6 < 6.6$ 13. $z - 6 \le -12.5$

14. $12.5 > 2.7 + w$ 15. $-12 < \frac{v}{-5}$ 16. $-\frac{3}{11} < \frac{a}{-22}$

17. $\frac{-3d}{11} > -3$ 18. $2 > \frac{-13f}{2}$ 19. $0 < 13 - h$

20. $-d > 4$ 21. $5 - z > 0$ 22. $-g \le -2$

Find the solution that is common to both inequalities. Graph the common solution on a number line.

23. $x - 5 > 1$ and $x + 6 < 13$ 24. $3w \ge -3$ and $3w \le 21$

25. Critical Thinking
  a. Find all real numbers that are solutions to all of the following inequalities: $x - 5 > -1$ and $x - 5 > -2$ and $x - 5 > -3$ and $x - 5 > -4$.
  b. Graph the common solution on a number line.

MIXED REVIEW

Solve each equation. (2.8 Skill B)

26. $5 - 3x = 23$ 27. $5a - 3a = 56$ 28. $-13 + 3x = 4x - 17$

29. $4(x + 3) = 3x$ 30. $-5(10 - z) = 4z$ 31. $11 - 4x = -3x - 1$

32. $3k - 2k - 6 = -7$ 33. $-3(n - 3n) = 54$ 34. $1 - 3x = -3x + x + 11$

Answers for Lesson 4.4 Skill A

Try This Example 1
Use the Subtraction Property of Inequality; $c > -7.5$

Try This Example 2
Use the Multiplication Property of Inequality; $z < \frac{1}{4}$

Try This Example 3
a. $j < 0$ b. $x \ge 4$

1. Add 6 to each side.
2. Subtract 3.5 from each side.
3. Divide each side by 5.
4. Multiply each side by 7.

5. a. Add $\frac{3}{4}$ to each side.
   b. no
   c. $t \ge 1$

6. a. Add 2.3 to each side.
   b. no
   c. $j \le -3.7$

7. a. Multiply each side by -3.
   b. yes
   c. $w < -8.4$

8. a. Divide each side by -3.
   b. yes
   c. $p > -1$

9. a. Multiply each side by -1.
   b. yes
   c. $y > 0$

10. a. Multiply each side by -1.
    b. yes
    c. $r \le -9$

11. a. Subtract 1.5 from each side.
    b. no
    c. $j \le -3.5$

12. a. Subtract 2.6 from each side.
    b. no
    c. $q < 4$

13. a. Add 6 to each side.
    b. no
    c. $z \le -6.5$

14. a. Subtract 2.7 from each side.
    b. no
    c. $w < 9.8$

15. a. Multiply each side by -5.
    b. yes
    c. $v < 60$

16. a. Multiply each side by -22.
    b. yes
    c. $a < 6$

17. a. Multiply each side by $-\frac{11}{3}$.
    b. yes
    c. $d < 11$

For answers to Exercises 18–34, see Additional Answers.

Focus: *Students use inequalities in one variable to represent real-world problems and choose the Addition, Subtraction, Multiplication, or Division Property of Inequality to solve.*

EXAMPLES 1 and 2

PHASE 1: Presenting the Examples

Remind students that there are many verbal phrases that translate into mathematical inequalities. Have students develop their own list of these phrases. Some examples are *less than, greater than, at least, at most, not less than, not greater than, fewer than,* and *more than.* Then work with students to match each phrase with the correct inequality symbol. A sample is shown at right.

When students are comfortable with these translations, discuss the examples. Remind students that when solving an inequality, it is not possible to get an exact answer. For instance, in Example 1, we cannot be sure of Jeff's exact score on last week's test; we only know that it was less than 85 points.

Phrase	Symbol
less than	
not greater than	$<$
fewer than	
greater than	
not less than	$>$
more than	
at least	$\geq$
at most	$\leq$

PHASE 2: Providing Guidance for TRY THIS

In problems such as these, where two quantities are described, some students may not understand which quantity the variable should represent. Emphasize to these students that the variable should always stand for the quantity that they are being asked to find. For instance, in the Example 2 Try This exercise, the variable should represent the cost of the radio. Then use the information in the problem to write the other quantity, the cost of the television, as an algebraic expression that contains the variable.

PHASE 3: ASSESSMENT AND CLOSURE for Lesson 4.4

- Ask students to write an inequality that can be solved in one step by applying the Addition Property of Inequality. Repeat the instruction for the Subtraction, Multiplication, and Division Properties of Inequality. Then have students solve their inequalities. [**Answers will vary.**]

- Ask students to explain the similarities and differences between solving equations and solving inequalities. [**Sample answer: The solution process is similar in that you apply Properties of Equality or Inequality to isolate the variable. However, when multiplying or dividing an inequality by a negative number, you must reverse the direction of the inequality. Also, the solution to an inequality is usually a set of numbers while the solution to an equation is usually just one number.**]

- Students have now seen how to solve simple inequalities in one variable by applying one of the basic properties of inequality to produce a one-step solution. In the lessons that follow, they will work with inequalities for which the solution process involves multiple steps, and they will work with compound inequalities.

☞ For a Lesson 4.4 Quiz, see Assessment Resources *page 41.*

SKILL B APPLICATIONS Choosing an operation to solve a real-world inequality

EXAMPLE 1

Jeff's score on this week's test is 10 points higher than his score on last week's test. His score this week is less than 95 points. What are Jeff's possible scores on last week's test?

▶ **Solution**

1. Let *t* represent Jeff's score on last week's test.
   Then *t* + 10 represents his score on this week's test.

Write an inequality.

2. Write and solve an inequality.
$$t + 10 < 95$$
$$t + 10 - 10 < 95 - 10 \quad \longleftarrow \text{Apply the Subtraction Property of Inequality.}$$
$$t < 85$$

3. **Check:** $85 + 10 \overset{?}{=} 95$ ✔
   A score of less than 85 plus 10 points will be less than 95.

Jeff's score on last week's test was less than 85 points.

TRY THIS   Marcy is driving on the highway. She notices that if she increases her speed by 15 miles per hour, her speed will still be under the speed limit of 65 miles per hour. How fast could Marcy be driving?

EXAMPLE 2

In the school election, the number of eighth graders who voted was one-half of the number of seventh graders who voted. If fewer than 220 eighth graders voted, how many seventh graders could have voted?

▶ **Solution**

1. Let *n* represent the number of seventh graders who voted.
   Then $\frac{1}{2}n$ is the number of eighth graders who voted.

Write an inequality.

2. Write and solve an inequality.
$$\frac{1}{2}n < 220$$
$$(2)\frac{1}{2}n < (2)220 \quad \longleftarrow \text{Apply the Multiplication Property of Inequality.}$$
$$n < 440$$

3. **Check:** $\frac{1}{2}(440) \overset{?}{=} 220$ ✔
   If fewer than 220 eighth graders voted, then the number of
   seventh graders must be less than 440.

Fewer than 440 seventh graders voted.

TRY THIS   The Hanes family bought a radio and a television set. The price of the television was four times the price of the radio. The television's price was less than $300. How much could the radio have cost?

EXERCISES

KEY SKILLS

Match each situation with its corresponding algebraic inequality.

1. The amount I have now and my next $5.00 allowance will give me at least $11.00.

2. I need a certain minimum amount to buy 5 CD's that cost at least $11.00 each.

3. I would have this amount if my four brothers and I split more than $11 equally.

a. $\frac{x}{5} \geq 11$

b. $5x > 11$

c. $x + 5 \geq 11$

PRACTICE

Solve each problem.

4. The difference of a number and 5 is less than -12. What are possible values for the number?

5. Three more than a number is greater than 23. What are possible values for the number?

6. Tina's softball glove cost $30 more than her softball bat. The price of the glove was less than $42. What are the possible prices of the bat?

7. The Martin family bought a new washer and dryer. The dryer cost $350, and the total cost was less than $600. What are the possible prices of the washer?

8. In a poll, the number of women surveyed was one-fourth of the number of men surveyed. If fewer than 100 women were surveyed, how many men could have been surveyed?

9. If the Smith family's monthly electric bill is less than $115, what is the range of their electric cost for one year?

MID-CHAPTER REVIEW

Graph each compound inequality on a number line. (4.1 Skill C)

10. $-2 \leq x < 3$

11. $0 < n$ or $n \leq -4$

Solve each inequality. Check your solution. (4.2 Skills A and B, 4.3 Skills A and B)

12. $-3.6 > x - 5$

13. $d + 4.3 \leq 1.1$

14. $-12.9 > h + 12.9$

15. $t - 3.5 \geq -3.6$

16. $4 > -6y$

17. $\frac{-9x}{17} > 1$

18. $-5g \geq 1$

19. $-2 \geq \frac{-x}{7}$

20. The amount of rainfall in March was 7.3 inches greater than the amount of rainfall in April. More than 9.9 inches of rain fell in March. How many inches of rain could have fallen in April? (4.4 Skill B)

Answers for Lesson 4.4 Skill B

Try This Example 1
less than 50 miles per hour

Try This Example 2
less than $75

1. c
2. a
3. b
4. less than -7
5. greater than 20
6. less than $12
7. less than $250
8. less than 400 men
9. less than $1380

10.

11.
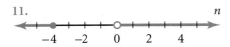

12. $x < 1.4$
13. $d \leq -3.2$

14. $h < -25.8$
15. $t \geq -0.1$
16. $y > -\frac{2}{3}$
17. $x < -\frac{17}{9}$
18. $g \leq -\frac{1}{5}$
19. $x \geq 14$
20. more than 2.6 inches

TEACHING LESSON 4.5

Solving Multistep Inequalities

SKILL A ▸ *Solving an inequality in multiple steps*

Focus: *Students apply two or more properties of inequality to solve inequalities in one variable.*

EXAMPLE 1

PHASE 1: Presenting the Example

Display inequality **1** at right. Have students solve and identify the property of inequality that they used. [$j \geq -4.5$; **Division or Multiplication Property of Inequality**] Then display inequality **2**. Lead students to see that, whereas inequality **1** is solved in one step, inequality **2** will require two steps. Then discuss the solution on the textbook page.

1. $4j \geq -18$

2. $4j + 2 \geq -18$

PHASE 2: Providing Guidance for TRY THIS

Watch for students who begin Try This exercise a by dividing each side by -3. The resulting inequality would be $\frac{-3t}{-3} + \frac{5}{-3} \geq \frac{-10}{-3}$, or $t + -\frac{5}{3} \geq \frac{10}{3}$. This approach is correct, but it provides many more opportunities for error. Most students will find it simpler to first subtract 5 from each side.

EXAMPLE 2

PHASE 2: Providing Guidance for TRY THIS

When given a two-step inequality (or equation), many students' first instinct is to work with a constant term. So, given $-2w \leq 6w + 16$, they may first subtract 16 from each side. While this is a valid action, point out that the resulting inequality, $-2w - 16 \leq 6w$, is no simpler than the given inequality. Because our final goal is to isolate w, a better first step is to collect all the terms containing w on one side of the inequality.

EXAMPLE 3

PHASE 1: Presenting the Example

To be sure that students understand the method used to check the solution, remind them that any point on a number line divides the real numbers into three distinct sets: the number that the point represents, all numbers that are greater than this number, and all numbers that are less than this number. By solving $2r + 5 = 3r - 1$, they are locating a point on a number line, which is 6. All numbers on one side of this *boundary number* will be solutions to the inequality $2r + 5 > 3r - 1$. All numbers on the other side will be solutions to the given inequality, $2r + 5 < 3r - 1$.

4.5 LESSON *Solving Multistep Inequalities*

SKILL A *Solving an inequality in multiple steps*

You may need to use more than one property to solve an inequality.

EXAMPLE 1 Solve $4j + 2 \geq -18$.

▶ **Solution**

$$4j + 2 \geq -18$$
$$4j \geq -20 \quad \longleftarrow \text{Subtract 2 from each side.}$$
$$j \geq -5 \quad \longleftarrow \text{Divide each side by 4.}$$

TRY THIS Solve each inequality. **a.** $-3t + 5 \leq -10$ **b.** $0 \geq 5 - 4v$

EXAMPLE 2 Solve $3a > 5a - 1$.

▶ **Solution**

$$3a > 5a - 1$$
$$-2a > -1 \quad \longleftarrow \text{Subtract 5a from each side.}$$
$$a < \frac{1}{2} \quad \longleftarrow \text{Divide each side by } -2. \text{ Change} > \text{to} <.$$

TRY THIS Solve each inequality. **a.** $-2w \leq 6w + 16$ **b.** $-3z + 2 < -7z$

EXAMPLE 3 Solve $2r + 5 < 3r - 1$. Check your solution.

▶ **Solution**

$$2r + 5 < 3r - 1$$
$$2r - 3r + 5 < 3r - 3r - 1 \quad \longleftarrow \text{Apply the Subtraction Property of Inequality.}$$
$$-r + 5 - 5 < -1 - 5 \quad \longleftarrow \text{Apply the Subtraction Property of Inequality.}$$
$$-r < -6 \quad \longleftarrow \text{Multiply by } -1 \text{ and change} < \text{to} >.$$
$$r > 6$$

Check: First substitute 6 into the related equation.

$$2(\textbf{6}) + 5 \overset{?}{=} 3(\textbf{6}) - 1$$
$$17 = 17 \checkmark$$

To check the direction of the inequality symbol, choose a number larger than 6 and substitute it into the original inequality.

$$r = 8 \quad \rightarrow \quad 2(\textbf{8}) + 5 < 3(\textbf{8}) - 1$$
$$21 < 23$$

The inequality is true, so the direction of the symbol in $r > 6$ is correct. ✔

TRY THIS Solve and graph $2 - 6z > 6 - 5z$. Check your solution.

EXERCISES

KEY SKILLS

Write the inequality that results when you perform each operation.
What is the next step you would take to solve the inequality?

1. $3n - 5 < -8$; add 5 to each side. **2.** $4 > 2t + 8$; subtract 8 from each side.

3. $\frac{2x}{3} \leq 6$; multiply each side by 3. **4.** $4x + 1 \geq 17$; subtract 1 from each side.

PRACTICE

Solve each inequality.

5. $-4r < -6r + 2$ **6.** $5z < -7z - 36$ **7.** $-a < -7a + 36$

8. $5z + 11 < -6z$ **9.** $-5y + 30 < 20$ **10.** $3f + 6 > 15$

11. $18 > 3g + 15$ **12.** $-2 > -2h + 2$ **13.** $0 > 4s - 8$

14. $4v + 20 \leq 0$ **15.** $-3u - 3 + 4u \leq 0$ **16.** $-4 > -4u + 4 + 5u$

17. $-8w - 8 + 9w > -4$ **18.** $-1 \leq -p + 1 + 2p$ **19.** $2q - 2 > 3q - 6$

20. $-10b - 1 > -11b + 3$ **21.** $-c - 1 \leq -2c + 2$ **22.** $\frac{1}{3}r - \frac{1}{3} < -\frac{2}{3}r$

Solve and graph each inequality. Check your solution.

23. $2x > 4$ **24.** $d - 4 > -1$ **25.** $3m - 2 \leq 7$ **26.** $-n + 2 \leq 3$

27. Critical Thinking For what value of n will the solution to $2(x - n) - 4 \geq 0$ be $x \geq 0$?

28. Critical Thinking Let a, b, and c be real numbers with $a \neq 0$. Solve $ax + b > c$ for x.

MIXED REVIEW

Graph each solution on a number line. (4.1 Skill B)

29. $x > -5$ **30.** $x \leq 5$

Graph each compound inequality. (4.1 Skill C)

31. $-2 < n < 4$ **32.** $n > 4 \text{ and } n \leq 6$

33. $k < 4 \text{ or } k \geq 6$ **34.** $m < 0 \text{ or } m > 3$

Answers for Lesson 4.5 Skill A

Try This Example 1
a. $t \geq 5$

b. $v \geq \frac{5}{4}$

Try This Example 2
a. $w \geq -2$

b. $z < -\frac{1}{2}$

Try This Example 3
$z < -4$

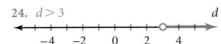

1. $3n < -3$; divide each side by 3.
2. $-4 > 2t$; divide each side by 2.
3. $2x \leq 18$; divide each side by 2.
4. $4x \geq 16$; divide each side by 4.
5. $r < 1$
6. $z < -3$
7. $a < 6$
8. $z < -1$
9. $y > 2$
10. $f > 3$

11. $g < 1$
12. $h > 2$
13. $s < 2$
14. $v \leq -5$
15. $u \leq 3$
16. $u < -8$
17. $w > 4$
18. $p \geq -2$
19. $q < 4$
20. $b > 4$
21. $c \leq 3$
22. $r < \frac{1}{3}$
23. $x > 2$

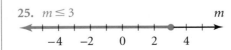

24. $d > 3$

25. $m \leq 3$

26. $n \geq -1$

27. $n = -2$

28. If $a > 0$, then $x > \frac{c - b}{a}$.

If $a < 0$, then $x < \frac{c - b}{a}$.

29.

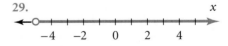

30.

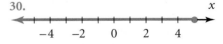

31.

For answers to Exercises 32–34, see Additional Answers.

SKILL B Simplifying expressions before solving inequalities

Focus: *Students use the Distributive Property to simplify one or both sides of an inequality before solving.*

EXAMPLE 1

PHASE 1: Presenting the Example

Display the three inequalities shown at right. Point out to students that inequality **1** is a two-step inequality like the ones that they studied in Skill A. Ask them to solve. [$\boldsymbol{x > 7}$]

1. $2x + 5 > 19$

2. $2(x + 5) > 19$

3. $2(x + 5) + 3 > 19$

Ask students how inequality **2** is different from inequality **1**. Lead them to see that the parentheses indicate a product, with 2 multiplying each term inside. So the left side is simplified as $2x + 10$, and the resulting inequality is $2x + 10 > 19$. Have students solve this simplified inequality. [$\boldsymbol{x > 4.5}$]

Now ask students to compare inequality **3** to inequality **2**. They should observe that solving this inequality will also involve rewriting $2(x + 5)$ as $2x + 10$. However, the left side will then be $2x + 10 + 3$. Point out that an additional step of combining the like terms, 10 and 3, is required. Then discuss the solution of the inequality as outlined on the textbook page.

EXAMPLE 2

PHASE 1: Presenting the Example

Display $a + 1 < a$ and ask students to solve. They should follow the steps at right, but they may be confused by the false statement $1 < 0$. Explain that this false result indicates that there is no solution to the original inequality.

$$a + 1 < a$$
$$a + 1 - a < a - a$$
$$1 < 0$$

Ask students to further verify that there is no solution by using their knowledge of the number line. [**sample: When you add 1 to a number, you move one unit to the right on a number line. So, no matter what the value of *a*, *a* + 1 will always be to its right. Therefore, *a* + 1 can never be less than *a*.**] Discuss their responses.

Now display $b + 1 > b$ and ask students to solve. This time they should arrive at the statement $1 > 0$, as shown at right. Explain that this statement, which is always true, indicates that all real numbers are solutions to the original inequality. Again ask students to verify this by using number-line reasoning similar to that above. [**When you add 1 to a number, you move one unit to the right on a number line. So, no matter what the value of *b*, *b* + 1 will always be to its right. Therefore, *b* + 1 will always be greater than *b*.**]

$$b + 1 > b$$
$$b + 1 - b > b - b$$
$$1 > 0$$

Have students read the text at the middle of the page. Answer any questions that they may have. Then discuss Example 2.

EXAMPLE 1 Solve and graph $2(x + 5) + 3 > 19$. Check your solution.

▶ **Solution**

$2(x + 5) + 3 > 19$
$2x + 10 + 3 > 19$ ⟵ Apply the Distributive Property.
$2x + 13 > 19$ ⟵ Combine like terms.
$2x > 6$ ⟵ Apply the Subtraction Property of Inequality.
$x > 3$ ⟵ Apply the Division Property of Inequality.

$$\xleftarrow{\hspace{1em}}\!\!\! \underset{-1\ 0\ 1\ 2\ 3\ 4\ 5\ 6}{\circ} \!\!\!\xrightarrow{\hspace{1em}} x$$

Check: First substitute 3 into the related equation. The true
statement confirms that the open circle is in the correct
position. To test the direction of the ray, test points on either
side of 3.

$2(3 + 5) + 3 \overset{?}{=} 19$
$16 + 3 \overset{?}{=} 19$
$19 = 19$ ✔

$x = 0 \rightarrow 2(0 + 5) + 3 = 13 \qquad 13 > 19$ ✘
$x = 4 \rightarrow 2(4 + 5) + 3 = 21 \qquad 21 > 19$ ✔

Because $13 > 19$ is not a true statement, 0 is not a solution. Because
$21 > 19$ is true, 4 is a solution and it should be included in the graph.

TRY THIS   Solve $3(p - 6) + 3 < -9$. Check your solution.

All of the inequalities that you have studied so far have had some real
numbers as their solution. But there are two other solution possibilities for
an inequality: all real numbers or no solution.

• If you arrive at an equivalent inequality that is always true, the original
inequality is satisfied by all real numbers.
• If you arrive at an equivalent inequality that is always false, the original
inequality is not satisfied by any real number; it has no solution.

EXAMPLE 2 Solve each inequality.
   a. $3(b + 2) - 7 \geq 2b - 5 + b$ **b.** $6h + 5 + 4h < 5(2h - 1)$

▶ **Solution**
   a. Simplify each side of the inequality.   **b.** Simplify each side of the inequality.

$3(b + 2) - 7 \geq 2b - 5 + b$ $6h + 5 + 4h < 5(2h - 1)$
$3b - 1 \geq 3b - 5$ $10h + 5 < 10h - 5$
$-1 \geq -5$ ✔ $5 < -5$ ✘

   This true statement indicates that   This false statement indicates that there
   the solution is all real numbers.   are no solutions to the original inequality.

TRY THIS   Solve each inequality.
   a. $-3(n - 1) - 5 > 2 - 3n$ **b.** $3 - (4t + 2) \leq 4 - 2t + 2(1 - t)$

EXERCISES

KEY SKILLS

Simplify each side of the inequalities below. Do not solve.

1. $5x - 4x > -7 + 3$ 2. $6x - 7 - x < 3 - 2$
3. $3(x + 2) \leq 5 + 1$ 4. $2(7 - x) + 1 \geq 4 - 6$

PRACTICE

Solve each inequality. Check your solution.

5. $-2(c - 5) - 7 > 10$ 6. $7 - 2(n - 5) \leq -5$
7. $0 \leq -5(q + 2) - 4$ 8. $4 \leq 2(w - 2) + 4$
9. $3 + 2(d - 1) > 3(d - 3) + 2$ 10. $5 - 4(d - 1) < -5(d - 3) + 2$
11. $5(n + 2) - 5 \leq 3(n - 1) + 4$ 12. $-(m + 2) - 2 \leq -2(m - 1) + 1$
13. $\frac{1}{2}(x + 1) + \frac{5}{2}(x - 1) > -3$ 14. $\frac{2}{3}(a + 2) < \frac{7}{3}(a + 2) + 5$
15. $\frac{1}{8}(t + 2) + \frac{7}{8}(t + 2) + 5t < 0$ 16. $\frac{1}{5}(5z + 2) > \frac{1}{5}(z - 2) + \frac{1}{5}(2z - 1)$

In Exercises 17–22, does the inequality have no solution or all real
numbers as its solution?

17. $5(c + 1) \leq 5(c - 1)$ 18. $-3(t - 2) \geq -3t$
19. $2(y + 9) \geq 2(y - 9)$ 20. $-3(v + 1) > -3(v - 1) + 1$
21. $-3(v + 1) \leq -3(v - 1) + 1$ 22. $-3(z + 9) \geq -2z - z$

23. Solve $\frac{6}{11}(y - 1) + \frac{4}{11}(y - 1) + \frac{1}{11}(y - 1) > 0$.

24. Solve $-\frac{1}{5}(z + 1) - \frac{2}{5}(z + 1) - \frac{2}{5}(z + 1) > 0$.

MIXED REVIEW

Show that each statement is either always true or never true. (2.9 Skill C)

25. $3(4x + 4) - 11 = 12x + 2$ 26. $3a + 4a - 11 + a = -9 + 8a - 2$
27. $3(t + t) - t = 5(t + 1)$ 28. $d(3 + 2) - 4d - 1 = d - 1$

Answers for Lesson 4.5 Skill B

Try This Example 1
$p < 2$

Try This Example 2
a. no solution
b. all real numbers

1. $x > -4$
2. $5x - 7 < 1$
3. $3x + 6 \leq 6$
4. $15 - 2x \geq -2$
5. $c < -\dfrac{7}{2}$
6. $n \geq 11$
7. $q \leq -\dfrac{14}{5}$
8. $w \geq 2$
9. $d < 8$
10. $d < 8$
11. $n \leq -2$
12. $m \leq 7$

13. $x > -\dfrac{1}{3}$
14. $a > -5$
15. $t < -\dfrac{1}{3}$
16. $z > -\dfrac{5}{2}$
17. no solution
18. all real numbers
19. all real numbers
20. no solution
21. all real numbers
22. no solution
★ 23. $y > 1$
★ 24. $z < -1$

25. If $3(4x + 4) - 11 = 12x + 2$, then
$12x + 12 - 11 = 12x + 2$. This implies that
$12x - 1 = 12x + 2$ and that $-1 = 2$. Since this is
impossible, the given equation is never true.
26. If $3a + 4a - 11 + a = -9 + 8a - 2$, then
$8a - 11 = 8a - 11$. Since this is true for all values of a,
the given equation is always true.
27. If $3(t + t) - t = 5(t + 1)$, then $5t = 5t + 5$ and $0 = 5$.
Since this is false, the given equation is never true.
28. If $d(3 + 2) - 4d - 1 = d - 1$, then $d - 1 = d - 1$. Since
this is true for all real values of d, the given equation is
always true.

★ **Advanced Exercises**

Focus: *Students use a multistep inequality to model a real-world problem.*

EXAMPLE 1

PHASE 1: Presenting the Example

Some students may find it easier to write the inequality if you use a verbal approach, like the one below, in place of the table on the textbook page.

amount of sugar in solution A	combined with	amount of sugar in solution B	must be no more than	20% of solution C
↓	↓	↓	↓	↓
10% of a mL	+	25% of $(500 - a)$ mL	≤	20% of 500 mL
↓	↓	↓	↓	↓
$0.10a$	+	$0.25(500 - a)$	≤	$0.20(500)$

PHASE 2: Providing Guidance for **TRY THIS**

Have students compare their answers to the Try This problem with the answer to Example 1 itself. Taken together, the two distinct answers should demonstrate why an inexact response such as *167 milliliters of one solution and no more than 333 milliliters of the other* is inadequate.

EXAMPLE 2

PHASE 1: Presenting the Example

Some students may be confused about the way the inequality is derived. It may help to first show a correlation to the formula $A = p(1 + rt)$, as follows:

$$
\begin{array}{ccccccc}
A & = & p & \cdot & (\ 1 \ + & rt &) \\
\downarrow & & \downarrow & & \downarrow & \downarrow & \\
2800 & = & (1200 + d) \cdot & & (\ 1 \ + & (0.06)(1) &)
\end{array}
$$

The equal sign is then changed to the appropriate inequality symbol.

PHASE 3: ASSESSMENT AND CLOSURE for Lesson 4.5

- Give students the inequalities at right, one at a time. Have them compare the solution process for each with the solution process for the one that immediately precedes it. Then have them solve all five inequalities. [**top to bottom:** $d > 3.5$; $d > 9$; $d > 4.5$; $d < -0.4$; $d < -0.8$]

  $2d > 7$
  $d - 2 > 7$
  $2d - 2 > 7$
  $2d - 2 > 7d$
  $2(d - 2) > 7d$

- Students have now learned all the basic techniques for solving inequalities in one variable. In Lesson 4.6, they will apply these techniques to the solution of compound inequalities in one variable.

 For a Lesson 4.5 Quiz, see Assessment Resources *page 42.*

Making a table is usually helpful in solving mixture problems.

EXAMPLE 1

A chemical technician has two solutions, A and B, that are to be mixed to make a third solution, C. Use the data shown at right. To the nearest milliliter, how much of each solution is needed to make 500 milliliters of solution C?

Solution	Sugar
A	10%
B	25%
C	no more than 20%

▶ Solution

Make a table.

1. Let a represent the amount of solution A. Then the amount of Solution B is $500 - a$.

2. Make a table.

	Solution A	+	Solution B	=	Solution C
Amount of Solution	a		$500 - a$		500
Amount of Sugar	$0.10a$		$0.25(500 - a)$		$\leq 0.20(500)$

3. Write and solve an inequality.
$$0.10a + 0.25(500 - a) \leq 0.20(500)$$
$$a \geq 166.67$$

The chemist needs at least 167 milliliters of solution A and no more than $500 - 167$, or 333 milliliters, of solution B.

TRY THIS Rework Example 1 if solution C is to be no more than 15% sugar.

Sometimes you can adapt a formula to help solve an inequality.

EXAMPLE 2

Mrs. James has \$1200 in a bank account that earns 6% simple interest annually. How much should she deposit now so that she will have at least \$2800 in one year?

▶ Solution

Use a formula.

1. The formula below gives amount, A, in terms of principal, p, the annual interest rate as a decimal, r, and time, t, in years.
$$A = p(1 + rt), \text{ or } A = p + prt$$
Let d represent the amount to add to the \$1200 already in the account.

2. Write and solve the inequality $p + prt \geq A$ using $p = 1200 + d$, $r = 0.06$, and $t = 1$.
$$(1200 + d) + (1200 + d)(0.06)(1) \geq 2800$$
$$d \geq 1441.51$$

Mrs. James should deposit at least \$1441.51.

TRY THIS Rework Example 2 if the goal is at least \$3000, the rate of interest is 5%, and there is \$2000 in the account.

EXERCISES

1. Refer to Example 1. Use the data below to set up an inequality. Do not solve.
   solution A: 20% sugar   solution B: 30% sugar
   solution C: 800 milliliters at no more than 25% sugar

2. Refer to Example 2. Use the data below to set up an inequality. Do not solve.
   current amount: \$2400   annual rate: 4%   savings goal: \$4000

3. Use the data below to rework Example 1.
   solution A: 10% sugar   solution B: 4% sugar
   solution C: 500 milliliters at no more than 6% sugar

4. Use an inequality to show that more of solution A than solution B can be used to make solution C, given the data below.
   solution A: 20% sugar   solution B: 40% sugar
   solution C: 450 milliliters at no more than 30% sugar

5. Use an inequality to show that it is impossible to make solution C given the data below.
   solution A: 20% sugar   solution B: 40% sugar
   solution C: 800 milliliters at no more than 15% sugar

6. Use an inequality to show that Mrs. James will need to double her current amount of \$1500 earning interest at 5% per year to reach a goal of \$3150 in one year.

7. Write a mixture problem similar to Example 1 that could be solved by this inequality: $0.15a + 0.30(600 - a) \leq 0.20(600)$.

8. Write a problem similar to Example 2 that could be solved by this inequality: $(1500 + d) + (1500 + d)(0.06) \geq 2000$.

9. **Critical Thinking** Use the table in Example 1 as a model to construct a table that illustrates Example 2.

Use an inequality to solve each problem. (4.2 Skill C)

10. Angela has saved \$112.45 and wants to have a total of at least \$180 by the end of the month. At least how much more money must she save?

11. What positive integers can you decrease by 9 and get an integer no more than -2?

Answers for Lesson 4.5 Skill C

Try This Example 1
at least 333 milliliters of solution A and no more than 167 milliliters of solution B

Try This Example 2
at least \$857.14

1. $0.20a + 0.30(800 - a) \leq 0.25(800)$
2. $(2400 + d) + (2400 + d)0.04 \geq 4000$
3. at least 333.33 milliliters of solution B and no more than 166.67 milliliters of solution A
4. Let a represent the amount of solution A that is needed. Then $0.20a + 0.40(450 - a) \leq 0.30(450)$. The solution is $a \geq 225$. Since 225 milliliters is one half the amount of the final solution, you can use more of solution A than solution B.
5. $0.20a + 0.40(800 - a) \leq 0.15(800)$
$$0.20a - 0.40a + 320 \leq 120$$
$$-0.20a \leq 200$$
$$a \geq 1000$$
This means that at least 1000 milliliters of solution A are needed, which is impossible because only 800 milliliters of solution C are to be mixed.

6. Let d represent the amount Mrs. James needs to deposit. The solution to $(1500 + d) + (1500 + d)(0.05) \geq 3150$ is $d \geq 1500$. This means she has to double the current amount of \$1500 and make it \$3000.

7. solution A: 15% sugar
   solution B: 30% sugar
   solution C: 600 milliliters at no more than 20% sugar

8. An investor has a savings goal of at least \$2000 and has \$1500 in the account that earns 6% annual interest.

9.
	Principal	+	Interest	=	Amount
now	1200		$.06(1200) = 72$		1272
after deposit	$1200 + d$		$.06(1200 + d)$		≥ 2800

10. at least \$67.55
11. any integer less than or equal to 7

TEACHING
4.6
LESSON

Solving Compound Inequalities

SKILL A ▶ *Solving compound inequalities joined by "and"*

Focus: *Students solve compound inequalities that are conjunctions and represent their solutions as the intersection of two graphs on a number line.*

EXAMPLE 1

PHASE 1: Presenting the Example

Display the two inequalities at right. Have half of the class solve and graph the first, while the other half of the class solves and graphs the second.

$$2x - 1 < 5$$
$$2x + 9 \geq 7$$

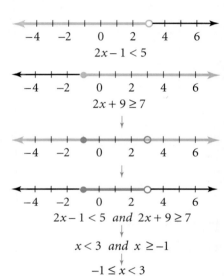

Prepare two transparencies, if possible, each displaying the graph of one of the inequalities, as shown at right. Draw the number lines and scale numbers in black, but use two different colors, perhaps red and blue, for the two graphs. Use these graphs, one at a time, to check the students' work.

Now slide the graphs together as shown at right. Ask students to describe where the graphs overlap. (If you graphed in red and blue, the overlap should appear purple.) Lead students to see that the overlap occurs where the numbers are greater than or equal to -1 *and* less than 3. Display a new number line with just these numbers graphed, as shown. Tell students that the overlap, or *intersection*, of the graphs is the graph of the compound inequality $2x - 1 < 5$ *and* $2x + 9 \geq 7$.

In an earlier lesson, students learned to describe this graph using the compound inequality $x < 3$ *and* $x \geq -1$, also written $-1 \leq x < 3$. Note that this simpler inequality is considered the solution of $2x - 1 < 5$ *and* $2x + 9 \geq 7$.

EXAMPLE 2

PHASE 1: Presenting the Example

Display the two inequalities shown at right. Point out to students that the left side of each is the same expression. Ask them to combine the inequalities into a single statement. **[$-5 \leq -2a + 5 \leq 10$]** Then discuss the solution of this compound inequality as given on the textbook page.

$$-2a + 5 \leq 10$$
$$-2a + 5 \geq -5$$

PHASE 2: Providing Guidance for TRY THIS

Working with compound inequalities in this concise form may overwhelm some students because they are, in effect, solving two inequalities at once. If you sense that this is happening, allow students to split the inequality into two parts, $2(d - 5) > -9$ *and* $2(d - 5) < 1$, and solve each individually.

4.6 LESSON — *Solving Compound Inequalities*

SKILL A *Solving compound inequalities joined by "and"*

Recall from Lesson 4.1 Skill C that a compound inequality may be a pair of inequalities joined by *and*. This type of compound inequality is called a **conjunction**. The solution to the conjunction $x > 1$ *and* $x < 7$ is graphed below.

$$x > 1 \quad \text{and} \quad x < 7$$

EXAMPLE 1

Solve $2x - 1 < 5$ *and* $2x + 9 \geq 7$. Graph the solution on a number line.

▶ **Solution**

$$
\begin{array}{ccc}
2x - 1 < 5 & \text{and} & 2x + 9 \geq 7 \\
2x - 1 + 1 < 5 + 1 & & 2x + 9 - 9 \geq 7 - 9 \\
2x < 6 & & 2x \geq -2 \\
x < 3 & \text{and} & x \geq -1
\end{array}
$$

$$-1 \leq x < 3$$

TRY THIS Solve $3b - 1 \leq 11$ *and* $5b > -25$. Graph the solution on a number line.

When a conjunction is written in shortened form, such as $a \leq x \leq b$, the pair can be solved simultaneously as shown below.

EXAMPLE 2

Solve $-5 \leq -2a + 5 \leq 10$. Graph the solution on a number line.

▶ **Solution**

Perform each operation on all three expressions simultaneously.

$$-5 \leq -2a + 5 \leq 10$$
$$-5 - 5 \leq -2a + 5 - 5 \leq 10 - 5 \qquad \longleftarrow \text{Subtract 5 from each of the three expressions.}$$
$$-10 \leq -2a \leq 5$$
$$\frac{-10}{-2} \geq \frac{-2a}{-2} \geq \frac{5}{-2} \qquad \longleftarrow \text{Divide each of the three expressions by } -2 \text{ and change } \leq \text{ to } \geq.$$
$$5 \geq a \geq -\frac{5}{2}$$
$$-\frac{5}{2} \leq a \leq 5 \qquad \boxed{\text{Compound inequalities are normally written with the least value on the left.}}$$

TRY THIS Solve $-9 < 2(d - 5) < 1$. Graph the solution on a number line.

170   Chapter 4   Solving Inequalities In One Variable

EXERCISES

KEY SKILLS

Write each pair of inequalities in shortened form as in Example 2.

1. $x > -7$ and $x \leq 8$
2. $-3n \geq 2$ and $-3n < 10$
3. $2(d + 1) \leq 3$ and $2(d + 1) > -2$
4. $-3a - 5 \geq 0$ and $-3a - 5 < 10$

PRACTICE

Solve each conjunction. Graph the solution on a number line.

5. $3x + 6 \geq 0$ and $x - 5 < 0$
6. $-2t - 2 \geq 0$ and $t + 5 > 3$
7. $g + 1 < 0$ and $-2(g + 3) \leq 0$
8. $p - 3 \geq 0$ and $-5(p - 5) > -15$
9. $2(n + 3) + 2 \geq 0$ and $-2(n + 3) - 2 \geq -2$
10. $-(w - 1) + 1 > 0$ and $-2(w + 1) - 3 \leq 1$
11. $13 > 3(b - 1) + 10$ and $0 \geq -(b - 4) - 5$
12. $-12 \geq -2(x + 4) + 1$ and $2(x + 2) - 8 \leq 3$

Solve each compound inequality. Graph the solution on a number line.

13. $-4 \leq 2d + 3 < 0$
14. $0 < 2d - 5 < 3$
15. $3 < -3(w - 1) \leq 12$
16. $-6 \leq -2(a + 3) \leq 8$
17. $-1 \leq -2(b - 1) + 3 \leq 7$
18. $-9 \leq -4(z + 4) - 1 \leq 7$
19. $0 < 3 + 3(k - 3) \leq 7$
20. $1 < 2(k - 1) - (k - 2) \leq 7$
21. $-2 < 3(q - 2) - 2(q - 3) \leq 2$

22. Show that $2(p - 3) + 3 < 3 - 2(p + 3)$ and $-4 + 2(p + 2) > 8 - (p + 1)$ has no solution.

23. Show that $-3 - 4(s + 3) > 0$ and $4 - 3(s - 3) < 10$ has no solution.

24. **Critical Thinking** Find a so that the pair of inequalities below has exactly one number as its solution.
$2(x + a) + 4 \geq 0$ and $-3(x - a) - 5 \geq -2$

MIXED REVIEW

Write a linear equation in two variables to represent the data in each table. Predict the 50th term. (3.8 Skill A)

25.
x	1	2	3	4	5
y	-30	-19	-8	3	14

26.
n	1	2	3	4	5
y	24	12	0	-12	-24

Represent each compound inequality on a number line. (4.1 Skill C)

27. $x < -2$ or $x \geq 3$
28. $y \leq 0$ or $y > 2$
29. $s \leq -2$ or $s \geq -2$
30. $z < -2$ or $z > -2$

Lesson 4.6   Skill A   **171**

Answers for Lesson 4.6 Skill A

Try This Example 1

$-5 < b \leq 4$

Try This Example 2

$\frac{1}{2} < d < \frac{11}{2}$

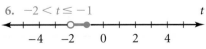

1. $-7 < x \leq 8$
2. $2 \leq -3n < 10$
3. $-2 < 2(d + 1) \leq 3$
4. $0 \leq -3a - 5 < 10$
5. $-2 \leq x < 5$

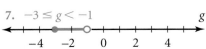

6. $-2 < t \leq -1$

7. $-3 \leq g < -1$

8. $3 \leq p < 8$

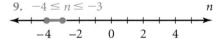

9. $-4 \leq n \leq -3$

10. $-3 \leq w < 2$

11. $-1 \leq b < 2$

12. $\frac{5}{2} \leq x \leq \frac{7}{2}$

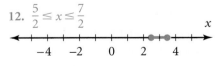

13. $-\frac{7}{2} \leq d \leq -\frac{3}{2}$

14. $\frac{5}{2} < d < 4$

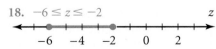

15. $-3 \leq w < 0$

16. $-7 \leq a \leq 0$

17. $-1 \leq b \leq 3$

18. $-6 \leq z \leq -2$

19. $2 < k \leq \frac{13}{3}$

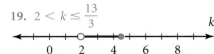

For answers to Exercises 20–30, see Additional Answers.

Lesson 4.6   Skill A   **171**

Focus: *Students solve compound inequalities that are disjunctions and represent their solutions as the union of two graphs on a number line.*

EXAMPLE 1

PHASE 1: Presenting the Example

Display the two inequalities at right. Again have half of the class solve and graph the first inequality while the other half solves and graphs the second.

$$2(x + 3) < 5$$
$$5x - 7 > 8$$

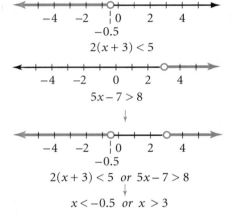

If possible, display the graph of each inequality, shown at right, on a separate transparency. Use these graphs to check the students' work, one graph at a time.

Now slide the graphs together, as shown at right. Ask students how this situation differs from the one that they encountered in Skill A. They should observe that, in this case, there is no overlap between the two graphs.

Lead students to see that the two graphs *combined* encompass all numbers that are either less than −1 *or* greater than 3. Tell them that the joining, or *union*, of the graphs is the graph of the compound inequality $2(x + 3) < 5$ or $5x - 7 > 8$.

In an earlier lesson, students learned to describe this graph by $x < -0.5$ or $x > 3$. Note that this simpler inequality is considered the solution of $2(x + 3) < 5$ or $5x - 7 > 8$.

EXAMPLE 2

PHASE 1: Presenting the Example

Display the statement at right. Tell students to draw a number line and graph *all* numbers that fit the description. Discuss their graphs. Students should have realized that every real number fits the description, and their graphs should cover the entire number line.

A number is either:

less than −3

or

greater than or equal to −4.

Now, if time permits, approach the inequality in Example 2 in the same manner as suggested for Example 1 above. Students should recognize that the solutions are all real numbers less than −3 or greater than or equal to −4. That is, all real numbers are solutions.

PHASE 2: Providing Guidance for **TRY THIS**

Watch for students who graph just an interval bounded by −3 and 0. Point out that they have graphed the conjunction $2a > -6$ *and* $0 \leq -5a$. Encourage them to focus on the connective before they begin to graph. That is, remind them that the word *and* indicates they will be graphing an intersection. If they see the word *or*, they will be graphing a union.

A pair of inequalities joined by *or* is called a **disjunction**. The solution to the disjunction $x \leq 0$ or $x > 4$ is shown on the number line below.

$$x \leq 0 \ \ or \ \ x > 4$$

←—●——————————○——→ x
−5 −4 −3 −2 −1 0 1 2 3 4 5

Notice that the graph is a pair of nonintersecting and opposite rays.

EXAMPLE 1   Solve $2(x + 3) < 5$ *or* $5x − 7 > 8$. Graph the solution on a number line.

▶ **Solution**

Solve each inequality separately.

$2(x + 3) < 5$	or	$5x − 7 > 8$
$2x + 6 < 5$	or	$5x > 15$
$x < −0.5$	or	$x > 3$

$$x < −0.5 \ or \ x > 3$$

←——————○——○——→ x
−5 −4 −3 −2 −1 0 1 2 3 4 5

TRY THIS   Solve $3(t + 1) \leq 3$ or $2t + 4 > 8$. Graph the solution on a number line.

Just as with simple inequalities, it is possible for a compound inequality to have no solution or a solution of all real numbers. The compound inequality $x < 0$ *and* $x > 5$ has no solution because x cannot be both less than zero and greater than 5.

Example 2 shows a compound inequality whose solution is all real numbers.

EXAMPLE 2   Solve $−6y + 5 > 23$ *or* $12 \geq −3y$. Graph the solution on a number line.

▶ **Solution**

Solve each inequality separately.

$−6y + 5 > 23$	or	$12 \geq −3y$
$−6y > 18$		$−4 \leq y$
$y < −3$	or	$y \geq −4$

$$y \geq −4$$
←————————————————→ y

$$y < −3$$
←——————○—————————→ y
−5 −4 −3 −2 −1 0 1 2 3 4 5

The solution to $−6y > 18$ or $12 \geq −3y$ is all real numbers.

←————————————————→ y
−5 −4 −3 −2 −1 0 1 2 3 4 5

TRY THIS   Solve $2a > −6$ or $0 \leq −5a$. Graph the solution on a number line.

EXERCISES

KEY SKILLS

Use mental math to decide whether the solution is a pair of non-intersecting opposite rays, the entire number line, or the entire number line with one point missing.

1. $x < 0$ or $x > 1$

2. $h < 2$ or $h > 1$

3. $a < −5.5$ or $a > 2.3$

4. $z < −3$ or $z > −3$

PRACTICE

Solve each disjunction. Graph the solution on a number line.

5. $2z − 5 > 1$ or $−3z \geq 9$

6. $2c + 1 < 1$ or $c − 3 > −2$

7. $2d − 1 \geq −4$ or $−2d − 5 > −2$

8. $−2v − 1 < 0$ or $3v + 1 < 7$

9. $2(f − 1) \geq 0$ or $−2(f + 3) > 3$

10. $−2(w − 1) < 0$ or $3(w + 3) < 6$

11. $2(g − 4) − 1 \geq 0$ or $−3(g + 1) + 5 > 3$

12. $−(y + 2) − 1 < 0$ or $2(y − 1) + 1 > 5$

13. $3 − 2(n − 1) > 0$ or $7 + 3(n + 1) > 0$

14. $11 − 3(r − 2) < 0$ or $2 − 3(r + 2) > 1$

Show that the solution to each compound inequality is a single ray.

Sample: $3x > −6$ or $−4x + 6 < 2$ Solution: $x > −2$ or $x > 1$

←——○————————————→ x
−4 −3 −2 −1 0 1 2 3 4

15. $2(x + 1) \geq 0$ or $−6(x − 1) \leq 3$

16. $−(p + 2) < 0$ or $2p − 1 > 5$

17. $1 − (n + 2) > 1$ or $2 − (n + 3) > 0$

18. $1 − (r − 3) < 1$ or $2 − (r + 1) < 5$

19. Show that, for all values of a, the solution to $−3(x + a) > 0$ or $−2(x − a) > 0$ is a single ray.

20. Critical Thinking   Show that, for all values of a, the solution to $−3(x + a) > 0$ or $2(x + a) > 0$ is the number line with one point missing.

MIXED REVIEW

Find the opposite and absolute value of each number. (2.1 Skill C)

21. $−12.9$

22. 111

23. $−(−2.7)$

24. $−(31.3)$

Solve each inequality. Graph the solutions on a number line. (4.1 Skill A)

25. $x − 2 > 2$

26. $d − 6 < −1$

27. $0 \geq a − 2.05$

28. $−1 > d − 1\frac{3}{4}$

Answers for Lesson 4.6 Skill B

Try This Example 1

$t \leq 0$ or $t > 2$

←———————●———○——————→ t
−4 −2 0 2 4

Try This Example 2

$a > −3$ or $a \leq 0$

←————————————————→ a
−4 −2 0 2 4

1. pair of non-intersecting opposite rays

2. entire number line

3. pair of non-intersecting opposite rays

4. entire number line with one point missing

5. $z \leq −3$ or $z > 3$

←————●—————————○———→ z
−4 −2 0 2 4

6. $c < 0$ or $c > 1$

←——————————○——○————→ c
−4 −2 0 2 4

7. $d \geq −\frac{3}{2}$ or $d < −\frac{3}{2}$

←————————————————→ d
−4 −2 0 2 4

8. $v > −\frac{1}{2}$ or $v < 2$

←————————————————→ v
−4 −2 0 2 4

9. $f \geq 1$ or $f < −\frac{9}{2}$

←——○——————————●————→ f
−4 −2 0 2 4

10. $w > 1$ or $w < −1$

←————————○———○———————→ w
−4 −2 0 2 4

11. $g \geq \frac{9}{2}$ or $g < −\frac{1}{3}$

←——————————○———●————→ g
−4 −2 0 2 4

12. $y > −3$ or $y > 3$

←————————○———————————→ y
−4 −2 0 2 4

13. $n < \frac{5}{2}$ or $n > −\frac{10}{3}$

←————○————————○——————→ n
−4 −2 0 2 4

14. $r > \frac{17}{3}$ or $r < −\frac{5}{3}$

←——○———————————○——————→ r
−2 0 2 4 6 8

15. $x \geq −1$ or $x \geq \frac{1}{2}$

←————————●——————————→ x
−4 −2 0 2 4

16. $p > −2$ or $p > 3$

←————————○——————————→ p
−4 −2 0 2 4

17. $n < −2$ or $n < −1$

←————————○——————————→ n
−4 −2 0 2 4

For answers to Exercises 18–28, see Additional Answers.

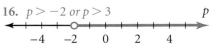

Focus: *Students use compound inequalities to model real-world problems.*

EXAMPLE 1

PHASE 1: Presenting the Example

If students have difficulty understanding how the inequality is formed, you might want to try the following approach:

least perimeter	cannot be greater than	actual perimeter	cannot be greater than	greatest perimeter
↓	↓	↓	↓	↓
4(10)	≤	$4s$	≤	200

PHASE 2: Providing Guidance for **TRY THIS**

You may need to remind students that an equilateral triangle has three sides of equal length. So, if the variable s represents the length of one side of an equilateral triangle, the expression $3s$ represents its perimeter.

EXAMPLE 2

PHASE 1: Presenting the Example

PHASE 2: Providing Guidance for **TRY THIS**

The given ratios relate two types of nuts, so some students may have difficulty translating them into expressions for the individual types of nuts. Suggest that they rewrite the ratios as shown at right. In this case, the ratio $1\frac{1}{2}$ to 1 represents $1\frac{1}{2}$ pounds of cashews for every 1 pound of peanuts. To write an equivalent ratio for p pounds of peanuts, both the numerator and the denominator of the given ratio must be multiplied by p. So the missing numerator of the second ratio must be $1\frac{1}{2} \times p$, or $1\frac{1}{2}p$.

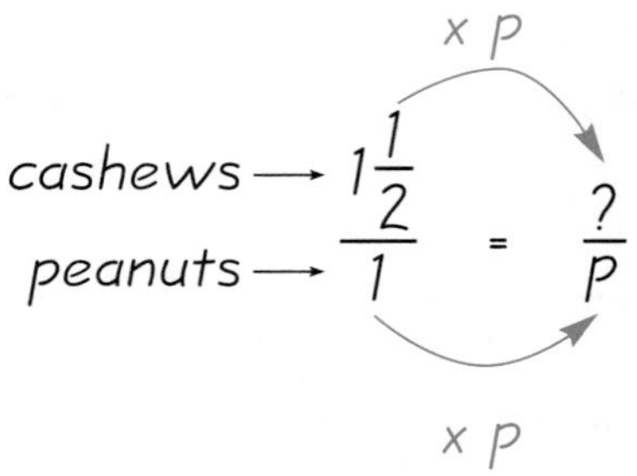

$$\times p$$
$$\text{cashews} \longrightarrow \frac{1\frac{1}{2}}{1} = \frac{?}{p} \longleftarrow \text{peanuts}$$
$$\times p$$

PHASE 3: ASSESSMENT AND CLOSURE for Lesson 4.6

- Display the inequalities shown at right. Ask students to identify similarities and differences. [**similarities: The boundary points for all of the graphs are −3 and 5. differences: The graph of the first is all numbers between −3 and 5; the graph of the second is all real numbers; the graph of the third is all numbers less than −3 combined with all numbers greater than 5. Also, the first inequality is the only one that can be rewritten, as −4 < t − 1 < 4.**]

  $t - 1 > -4$ *and* $t - 1 < 4$

  $t - 1 > -4$ *or* $t - 1 < 4$

  $t - 1 > -4$ *or* $t - 1 < 4$

- In the next lesson, students will learn to solve absolute-value equations and inequalities. In order to succeed at that task, students must have a firm grasp of compound inequalities as presented in this lesson, and they must master the skill of solving compound inequalities.

☞ *For a Lesson 4.6 Quiz, see* Assessment Resources *page 43.*

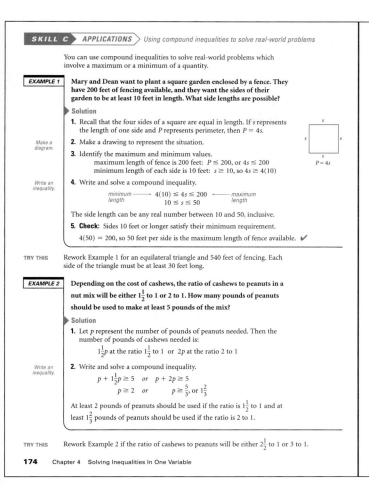

You can use compound inequalities to solve real-world problems which involve a maximum or a minimum of a quantity.

EXAMPLE 1

Mary and Dean want to plant a square garden enclosed by a fence. They have 200 feet of fencing available, and they want the sides of their garden to be at least 10 feet in length. What side lengths are possible?

▶ Solution

Make a diagram.

1. Recall that the four sides of a square are equal in length. If s represents the length of one side and P represents perimeter, then $P = 4s$.

2. Make a drawing to represent the situation.

3. Identify the maximum and minimum values.
   maximum length of fence is 200 feet: $P \leq 200$, or $4s \leq 200$
   minimum length of each side is 10 feet: $s \geq 10$, so $4s \geq 4(10)$

Write an inequality.

4. Write and solve a compound inequality.

   minimum length ⟶ $4(10) \leq 4s \leq 200$ ⟵ maximum length
   $10 \leq s \leq 50$

   The side length can be any real number between 10 and 50, inclusive.

5. **Check:** Sides 10 feet or longer satisfy their minimum requirement.
   $4(50) = 200$, so 50 feet per side is the maximum length of fence available. ✔

TRY THIS   Rework Example 1 for an equilateral triangle and 540 feet of fencing. Each side of the triangle must be at least 30 feet long.

EXAMPLE 2

Depending on the cost of cashews, the ratio of cashews to peanuts in a nut mix will be either $1\frac{1}{2}$ to 1 or 2 to 1. How many pounds of peanuts should be used to make at least 5 pounds of the mix?

▶ Solution

1. Let p represent the number of pounds of peanuts needed. Then the number of pounds of cashews needed is:
   $1\frac{1}{2}p$ at the ratio $1\frac{1}{2}$ to 1 or $2p$ at the ratio 2 to 1

Write an inequality.

2. Write and solve a compound inequality.
   $p + 1\frac{1}{2}p \geq 5$ or $p + 2p \geq 5$
   $p \geq 2$ or $p \geq \frac{5}{3}$, or $1\frac{2}{3}$

   At least 2 pounds of peanuts should be used if the ratio is $1\frac{1}{2}$ to 1 and at least $1\frac{2}{3}$ pounds of peanuts should be used if the ratio is 2 to 1.

TRY THIS   Rework Example 2 if the ratio of cashews to peanuts will be either $2\frac{1}{2}$ to 1 or 3 to 1.

EXERCISES

KEY SKILLS

Write an algebraic expression for each situation.

1. There are more than 3 times as many apples as oranges.

2. The ratio of science books to math books is at least 3 to 1.

PRACTICE

Solve each problem.

3. The sum of two consecutive integers is at least 51 but not more than 199. What might the integers be?

4. A triangular garden is shown at right. At least 490 feet of fencing but no more than 790 feet of fencing is to be used. What could be the length of the shortest side of the garden?

5. A motorist has driven 240 miles. If she is traveling at 50 miles per hour, how much more time will it take her to drive at least 480 miles but not more than 560 miles?

6. An engineer opens a valve to drain a tank that contains 1800 gallons of a mixture. If the mixture drains at the rate of 24 gallons per minute, how long will it take for the tank to be one-half to three-quarters empty, inclusive?

7. A party plannner wants to make a nut mix of peanuts and cashews. The ratio of cashews to peanuts will be either $1\frac{1}{2}$ to 1 or 2 to 1. How many pounds of peanuts should be used to make at least 9 pounds of the mix?

8. **Critical Thinking** Jaime has taken 3 tests. His score on the second test was 12 points higher than on the first test. The third score was 85. His average for the three tests was a B (80 to 89, inclusive). What is the possible range of his grade on the first test?

MIXED REVIEW APPLICATIONS

Use a direct-variation equation to solve each problem. (3.3 Skill B)

9. The distance that an object can be slid along the floor varies directly with the force applied. If 50 pounds of force move a box 4 feet, how far will the box move if 60 pounds of force are applied?

10. The distance that an object stretches a spring is directly proportional to the weight of the object. If a 25-pound object stretches a spring 0.01 inch, how heavy must an object be to stretch the spring 0.03 inch?

Answers for Lesson 4.6 Skill C

Try This Example 1
between 30 feet and 180 feet, inclusive

Try This Example 2
at least $1\frac{3}{7}$ pounds if the ratio is $2\frac{1}{2}$ to 1; at least $1\frac{1}{4}$ pounds if the ratio is 3 to 1

1. Let a represent the number of apples, and let g represent the number of oranges. Then $3g < a$.

2. Let s represent the number of science books, and let m represent the number of math books. Then $3m \leq s$.

3. The smaller number can be any integer from 25 to 99, inclusive. The larger number can be any integer from 26 to 100, inclusive.

4. The shortest side will be at least 80 feet and no more than 180 feet long.

5. between $4\frac{4}{5}$ hours and $6\frac{2}{5}$ hours, inclusive

6. between 37.5 minutes and 56.25 minutes

7. at least $3\frac{3}{5}$ pounds if the ratio is $1\frac{1}{2}$ to 1; at least 3 pounds if the ratio is 2 to 1.

8. between 71.5 points and 85 points, inclusive

9. 4.8 feet

10. 75 pounds

Solving Absolute-Value Equations and Inequalities

SKILL A *Solving absolute-value equations*

Focus: *Students solve absolute-value equations by interpreting $|x| = a$ as the compound sentence $x = a$ or $x = -a$.*

EXAMPLES 1 and 2

PHASE 1: Presenting the Examples

PHASE 2: Providing Guidance for TRY THIS

If students have trouble solving absolute-value equations, suggest that they use a number line to visualize the concept. The distance between two points on a number line is the absolute value of the difference of their coordinates. For instance, the first diagram at right shows that the distance between -5 and 4 is $|-5 - 4| = |4 - (-5)| = 9$.

Therefore, given the equation $|x - 5| = 3$ (part a of Example 1), students can use the second diagram. Looking at the number line, it is clear that $x = 2$ or $x = 8$.

When using this method, stress that the expression inside the absolute value symbols *must be a difference*. Thus, $|x + 4| = 3.5$ (Try This exercise a in Example 1) should be rewritten as $|x - (-4)| = 3.5$, as shown in the third diagram.

The distance between -5 and 4 is 9.

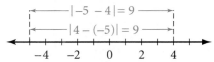

$|x - 5| = 3 \rightarrow$ The distance between x and 5 is 3.

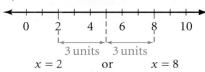

$|x + 4| = 3.5$
$|x - (-4)| = 3.5 \rightarrow$ The distance between x and -4 is 3.5.

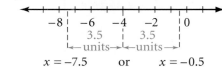

EXAMPLE 3

PHASE 1: Presenting the Example

Display equations **1** and **2** shown at right. Point out to students that equation **1** is similar to the equations in Examples 1 and 2. Ask them to solve it. [$m = 5$ or $m = -7$]

Now ask students how equation **2** is different. Elicit the response that "$+ 1$" appears outside the absolute-value symbol. Have students brainstorm possible approaches to solving the equation. Lead them to see that subtracting 1 from each side would result in the simple equation $|m| = 5$. Have them solve this equation. [$m = 5$ or $m = -5$] Then discuss Example 3.

1. $|m + 1| = 6$

2. $|m| + 1 = 6$

PHASE 2: Providing Guidance for TRY THIS

Watch for students who solve using the equations $|2z + 1| + 5 = 7$ and $|2z + 1| + 5 = -7$. Emphasize that their first step must be to isolate the absolute-value expression on one side of the equation.

4.7 Solving Absolute-Value Equations and Inequalities

LESSON

SKILL A *Solving absolute-value equations*

Recall from Lesson 2.1 Skill C that the absolute value of a number is the distance between the number and zero on a number line. Equations and inequalities involving absolute value are solved using compound sentences.

distance from 0 is 3 units

$-4\ -3\ -2\ -1\ \ 0\ \ 1\ \ 2\ \ 3\ \ 4$

$|x| = 3$ means $x = 3$ or $x = -3$

To solve an absolute-value equation such as $|x| = 3$, write it as two simpler equations joined by *or*. Then solve these two equations.

Solving Absolute-Value Equations
If $|x| = a$ and $a > 0$, then $x = a$ or $x = -a$.

EXAMPLE 1 Solve. **a.** $|x - 5| = 3$ **b.** $|-3t| = 18$

▶ **Solution**
a. If $|x - 5| = 3$, then $x - 5 = 3$ or $x - 5 = -3$. Thus, $x = 2$ or $x = 8$.
b. If $|-3t| = 18$, then $-3t = 18$ or $-3t = -18$. Thus, $t = -6$ or $t = 6$.

TRY THIS Solve. **a.** $|x + 4| = 3.5$ **b.** $|-2t| = 10$

EXAMPLE 2 Solve $|3r - 1| = 5$.

▶ **Solution**
If $|3r - 1| = 5$, then $3r - 1 = 5$ or $3r - 1 = -5$. Thus, $r = 2$ or $r = -\frac{4}{3}$.

TRY THIS Solve $|-3x - 5| = 7$.

Be sure to isolate the absolute-value expression on one side.

EXAMPLE 3 Solve $2|y - 1| - 2 = 4$.

▶ **Solution**
$2|y - 1| - 2 = 4$
$2|y - 1| = 6$ ⟵ Apply the Addition Property of Equality.
$|y - 1| = 3$ ⟵ Apply the Division Property of Equality.
$y - 1 = 3$ or $y - 1 = -3$
$y = 4$ or $y = -2$

TRY THIS Solve $2|z + 1| + 5 = 7$.

176 Chapter 4 Solving Inequalities In One Variable

EXERCISES

Use mental math to solve each equation.

1. $|w| = 3.6$ **2.** $|n| = 6$ **3.** $|z| = 0$

Write each equation in the form $|x| = a$. Do not solve.

4. $|x| - 5 = 13$ **5.** $2|x| = 16$ **6.** $3|x| + 5 = 8$

PRACTICE

Solve each equation. Check your solution.

7. $|x - 3| = 5$ **8.** $|w + 4| = 11$ **9.** $|2d| = 6$
10. $|3g| = 15$ **11.** $-|x + 2| = -2$ **12.** $-|k + 3| = -13$
13. $-5|h| = -25$ **14.** $-7|n| = -63$ **15.** $|3m - 5| = 11$
16. $|5x + 7| = 13$ **17.** $2|c + 1| - 7 = 10$ **18.** $3|z + 1| + 5 = 7$
19. $4|z - 4| - 3 = 7$ **20.** $3|y + 5| - 1 = 9$ **21.** $-7 = -2|k - 3| - 1$
22. $-11 = -4|h - 3| - 1$ **23.** $-4 = -|3z - 1| - 1$ **24.** $-11 = -2|2x - 3| - 1$

25. Solve $2|4(x - 2) - 3(x - 1)| = x$.
26. Show that $2|x + 3| = x$ has no solutions.
27. Find n such that $2|x| - 5 = n$ has exactly one solution.
28. Find m such that $3|x| - 4 = m$ has -2 and 2 as solutions.
29. For what values of n will $7|x| - 11 = n$ have no solutions?
30. Let a, b, and c be real numbers with $a \neq 0$ and $c \geq 0$.
Solve $|ax + b| = c$ for x.
31. **Critical Thinking** What does $|x - 5| = 3$ mean in terms of distance on a number line?

MIXED REVIEW

Solve each inequality and check your solution. Use mental math where possible. (4.4 Skill A)

32. $x + (-11) > -22$ **33.** $y - 5.9 \geq -11$ **34.** $7.5 \leq 9.1 + z$
35. $-3p > 18$ **36.** $4n > -1$ **37.** $-13 \geq -3y$

Solve each inequality. Check your solution. (4.5 Skill A)

38. $2x - 3x \geq 4x + 1$ **39.** $5 - 3y < 3 - 5y$ **40.** $-4(t - 3) \leq t$

Answers for Lesson 4.7 Skill A

Try This Example 1
a. $x = -7.5$ or $x = -0.5$
b. $t = -5$ or $t = 5$

Try This Example 2
$x = -4$ or $x = \frac{2}{3}$

Try This Example 3
$z = 0$ or $z = -2$

1. $w = -3.6$ or $w = 3.6$
2. $n = -6$ or $n = 6$
3. $z = 0$
4. $|x| = 18$
5. $|x| = 8$
6. $|x| = 1$
7. $x = -2$ or $x = 8$
8. $w = -15$ or $w = 7$
9. $d = -3$ or $d = 3$
10. $g = -5$ or $g = 5$
11. $x = 0$ or $x = -4$
12. $k = -16$ or $k = 10$
13. $g = -5$ or $g = 5$

14. $n = -9$ or $n = 9$
15. $m = -2$ or $m = \frac{16}{3}$
16. $x = -4$ or $x = \frac{6}{5}$
17. $c = -\frac{19}{2}$ or $c = \frac{15}{2}$
18. $z = -\frac{5}{3}$ or $z = -\frac{1}{3}$
19. $z = \frac{3}{2}$ or $z = \frac{13}{2}$
20. $y = -\frac{25}{3}$ or $y = -\frac{5}{3}$
21. $k = 0$ or $k = 6$
22. $h = \frac{1}{2}$ or $h = \frac{11}{2}$
23. $z = -\frac{2}{3}$ or $z = \frac{4}{3}$
24. $x = -1$ or $x = 4$
25. $x = 10$ or $x = \frac{10}{3}$

26. If $2|x + 3| = x$, then $|x + 3| = \frac{x}{2}$. This means that $x + 3 = \frac{x}{2}$ or $x + 3 = -\frac{x}{2}$. Then $x = -6$ or $x = -2$. So $|x + 3| = -3$ or $|x + 3| = -1$. Since absolute value must be nonnegative, neither solution for x will be acceptable. Therefore, the equation has no solution.
★**27.** $n = -5$
★**28.** $m = 2$
★**29.** $n < -11$
★**30.** $x = \frac{c - b}{a}$ or $x = -\frac{c + b}{a}$
31. The distance between x and 5 is 3.
32. $x > -11$
33. $y \geq -5.1$
34. $z \geq -1.6$
35. $p < -6$
36. $n > -\frac{1}{4}$
37. $y \geq \frac{13}{3}$
38. $x \leq -\frac{1}{5}$

For answers to Exercises 39–40, see Additional Answers.

★ **Advanced Exercises**

Focus: *Students solve absolute-value inequalities by interpreting $|x| < a$ as the compound inequality $-a < x < a$, and by interpreting $|x| > a$ as the compound inequality $x < -a$ or $x > a$.*

EXAMPLE 1

PHASE 1: Presenting the Example

Just as absolute-value equations can be solved using a number line, so too can absolute-value inequalities. The difference is that, rather than looking for numbers that are *exactly* a given distance from a number, you are looking for numbers that are *more than* or *less than* the given distance from a number. The diagrams below show how the technique can be used in Example 1.

a. $|z - 3| \leq 3 \rightarrow$ The distance between z and 3 is 3 or less than 3.

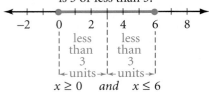

b. $|z - 3| \geq 3 \rightarrow$ The distance between z and 3 is 3 or more than 3.

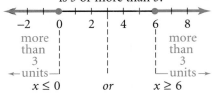

PHASE 2: Providing Guidance for **TRY THIS**

If students are using the distance method for solving the inequalities, remind them that the absolute-value expression must be rewritten as $|d - (-2)|$ in order to show a subtraction.

EXAMPLES 2 and 3

PHASE 1: Presenting the Examples

PHASE 2: Providing Guidance for **TRY THIS**

Students who are using the distance method may wonder if it will work when the variable has a coefficient other than 1. Examples 2 and 3 show how the method is used in such cases.

Example 2

$|2(x - 5)| \leq 2 \rightarrow |2x - 10| \leq 2 \rightarrow$ The distance between $2x$ and 10 is 2 or more than 2.

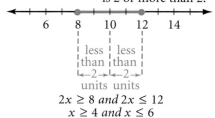

Example 3

$|12 - 3d| > 3 \rightarrow$ The distance between 12 and $3d$ is more than 3.

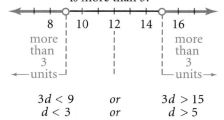

An absolute-value inequality can take one of two forms shown at right. (Similar inequalities can be written using $>$, $\le$, and $\ge$.) Use the following rules to write absolute-value inequalities as compound inequalities:

$|x| < a \qquad |x| > a$

Solving Absolute-Value Inequalities

If $|x| < a$ and $a > 0$, then $x > -a$ and $x < a$. That is, $-a < x < a$.
If $|x| > a$ and $a > 0$, then $x < -a$ or $x > a$.

If $|x| < 2$, then $-2 < x < 2$.   If $|x| > 2$, then $x < -2$ or $x > 2$.

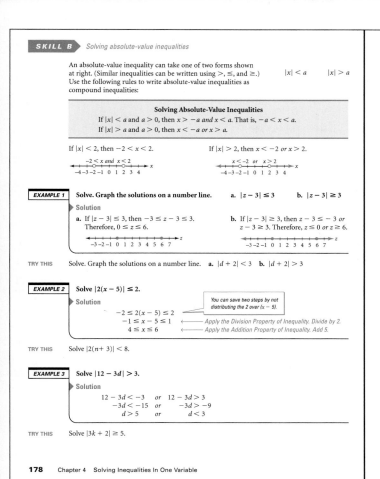

EXAMPLE 1   Solve. Graph the solutions on a number line.   a. $|z - 3| \le 3$ b. $|z - 3| \ge 3$

▸ Solution

a. If $|z - 3| \le 3$, then $-3 \le z - 3 \le 3$.
Therefore, $0 \le z \le 6$.

b. If $|z - 3| \ge 3$, then $z - 3 \le -3$ or $z - 3 \ge 3$. Therefore, $z \le 0$ or $z \ge 6$.

TRY THIS   Solve. Graph the solutions on a number line.   a. $|d + 2| < 3$ b. $|d + 2| > 3$

EXAMPLE 2   Solve $|2(x - 5)| \le 2$.

▸ Solution

You can save two steps by not distributing the 2 over $(x - 5)$.

$-2 \le 2(x - 5) \le 2$
$-1 \le x - 5 \le 1$ ⟵ *Apply the Division Property of Inequality. Divide by 2.*
$4 \le x \le 6$ ⟵ *Apply the Addition Property of Inequality. Add 5.*

TRY THIS   Solve $|2(n + 3)| < 8$.

EXAMPLE 3   Solve $|12 - 3d| > 3$.

▸ Solution

$12 - 3d < -3$ or $12 - 3d > 3$
$-3d < -15$ or $-3d > -9$
$d > 5$ or $d < 3$

TRY THIS   Solve $|3k + 2| \ge 5$.

EXERCISES

▸ KEY SKILLS

Write the next step in solving each absolute-value inequality. Do not solve.

1. $|x| \le 5$ 2. $|x| \ge 7$ 3. $|3y| < 9$
4. $|5z| > 12$ 5. $|3c - 5| > 11$ 6. $|2t + 1| \le 3$

▸ PRACTICE

Solve each inequality. Graph the solution on a number line.

7. $|d| \le 1$ 8. $|2x| \le 4$ 9. $|d| \ge 1$
10. $|2x| \ge 4$ 11. $|c - 2| < 3$ 12. $|z + 3| < 4$
13. $|2s - 1| \ge 5$ 14. $|2t + 3| > 5$ 15. $9 > |3(v + 1)|$
16. $6 \ge |4(y + 2)|$ 17. $2 \le |3(b - 2)|$ 18. $3 \le |-2(z + 2)|$
19. $|3(f + 1) - 2| > 0$ 20. $-|3(f + 1) - 2| > 0$ 21. $2|3(g + 2) - 2| \le 8$
22. $5|-2(n + 1) - 4| \le 10$ 23. $-|-3(m - 2) - 3| > -9$ 24. $-|1 - 2(t - 2)| \le -1$

25. Find r so that the solution to $|3x + 4| > r$ is the entire number line.

26. Show there is no nonzero number r such that the endpoints of the solution to $|x + 4| > r$ are opposites of one another.

27. Solve $|2x - 1| < |x|$.

28. Solve the conjunction at right. $|2x - 1| \le 11$ *and* $|3x| > 15$.

29. **Critical Thinking**   What does $|x| > 3$ mean in terms of distance on the number line?

30. **Critical Thinking**   What does $|x + 2| < 3$ mean in terms of distance on the number line?

31. **Critical Thinking**   Let a, b, and c be real numbers with a and $c > 0$. Prove that the solution to $|ax + b| \le c$ is $\dfrac{-b - c}{a} \le x \le \dfrac{c - b}{a}$.

▸ MIXED REVIEW

Is the given statement sometimes, always, or never true? Justify your response. (2.9 Skill C)

32. If x is any real number, then $-x + 4x - 1 = 3x - 1$.
33. If x is any real number, then $-(x - 4) = 3x - 1$.
34. If r is any real number, then $-(r - 4r) = 3r - 1$.

Answers for Lesson 4.7 Skill B

Try This Example 1

a. $-5 < d < 1$

b. $d < -5$ or $d > 1$

Try This Example 2

$-7 < n < 1$

Try This Example 3

$k \le -\dfrac{7}{3}$ or $k \ge 1$

1. $x \le 5$ *and* $x \ge -5$, or $-5 \le x \le 5$
2. $x \le -7$ *or* $x \ge 7$
3. $3y < 9$ *and* $3y > -9$, or $-9 < 3y < 9$
4. $5z < -12$ *or* $5z > 12$
5. $3c - 5 < -11$ *or* $3c - 5 > 11$
6. $-3 \le 2t + 1 \le 3$

7. $-1 \le d \le 1$

8. $-2 \le x \le 2$

9. $d \le -1$ or $d \ge 1$

10. $x \le -2$ or $x \ge 2$

11. $-1 < c < 5$

12. $-7 < z < 1$

13. $s \le -2$ or $s \ge 3$

14. $t < -4$ or $t > 1$

15. $-4 < v < 2$

16. $-\dfrac{7}{2} \le y \le -\dfrac{1}{2}$

17. $b \ge \dfrac{8}{3}$ or $b \le \dfrac{4}{3}$

18. $z \le -\dfrac{7}{2}$ or $z \ge -\dfrac{1}{2}$

For answers to Exercises 19–34, see Additional Answers.

Focus: Students use an absolute-value inequality to model real-world problems.

EXAMPLES 1 and 2

PHASE 1: Presenting the Examples

Although students may not be familiar with the phrase *manufacturing tolerance,* they are probably familiar with the idea from everyday experience. For instance, some may have bought two identical articles of clothing, in the same size, yet found that they fit differently. Others may have opened a food package that was supposed to contain a certain number of pieces, only to find one fewer or one more piece inside. Have students discuss their experiences with tolerance. Ask them to speculate about the *amount* of tolerance that the manufacturer allows in each case.

Students who have been using distance on a number line to interpret absolute values can still use the method in these real-world problems. Tell them to first graph the ideal value on the number line, and then graph the tolerance on the left and right sides of the ideal. The acceptable measurements will fall between the two tolerance values. One possible display of the acceptable weights for Example 1 is shown below.

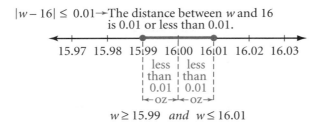

$|w - 16| \leq 0.01 \rightarrow$ The distance between w and 16 is 0.01 or less than 0.01.

$w \geq 15.99$ *and* $w \leq 16.01$

PHASE 2: Providing Guidance for **TRY THIS**

Stress that students should check their answers to real-world problems not only in the original inequality, but also in the conditions of the problem. For instance, in the Example 1 Try This, weight must be positive. Therefore, students should modify their answer to reflect this restriction.

PHASE 3: ASSESSMENT AND CLOSURE for Lesson 4.7

- Display $|r| < 10$. Then give students the four inequalities at right, one at a time. Have them compare each to $|r| < 10$, noting any similarities or differences. Then have them solve all five inequalities. [$|r| < 10$: $-10 < r < 10$; $|r| > 10$: $r < -10$ or $r > 10$; $|r| + 1 < 10$: $-9 < r < 9$; $|r + 1| < 10$: $-11 < r < 9$; $|2r| < 10$: $-5 < r < 5$]

- Students have now learned how to use basic properties of inequality to solve inequalities in one variable, and they have applied this skill to solving absolute-value inequalities. In the next lesson, they will use the properties of inequality to make generalizations about inequalities.

$|r| > 10$

$|r| + 1 < 10$

$|r + 1| < 10$

$|2r| < 10$

☞ *For a Lesson 4.7 Quiz, see* Assessment Resources *page 44.*

Tolerance describes the acceptable range that a measurement may differ from its ideal value. The examples show how to use absolute-value inequalities to solve problems involving tolerance.

EXAMPLE 1

The ideal weight of a can of vegetables is 16 ounces. The tolerance is 0.01 ounce. This means that a can is acceptable if it is within 0.01 ounce of 16 ounces and unacceptable otherwise. What are the acceptable weights and the unacceptable weights?

▶ **Solution**

Make a diagram.

1. Let w represent the actual weight of one can.
2. Make a drawing to represent the situation.
3. acceptable: $|w - 16| \leq 0.01$
   $16.00 - 0.01 \leq w \leq 16.00 + 0.01$
   $15.99 \leq w \leq 16.01$

   unacceptable: $|w - 16| > 0.01$
   $w < 16.00 - 0.01 \ or \ w > 16.00 + 0.01$
   $w < 15.99 \ or \ w > 16.01$

The acceptable weights are between 15.99 and 16.01 ounces, inclusive. The unacceptable weights are less than 15.99 ounces and greater than 16.01 ounces. Another restriction on the variable is that weight, w, has to be greater than or equal to zero.

TRY THIS   Rework Example 2 if the ideal weight is 12.5 ounces and the tolerance is 0.02 ounce.

EXAMPLE 2

At a watch factory, a consultant found that the weekly cost of producing w watches is $10,000 + 2.5w$ dollars. If the company's ideal weekly cost is $30,000 with a tolerance of $2000, how many watches can they produce?

▶ **Solution**

Write an inequality.

|weekly cost $- 30,000| \leq 2000$
$|(10,000 + 2.5w) - 30,000| \leq 2000$ ◀——— *Write an absolute-value inequality.*
$|2.5w - 20,000| \leq 2000$ ◀——— *Simplify within the absolute value.*
$-2000 \leq 2.5w - 20,000 \leq 2000$ ◀——— *Recognize the form $|x| \leq a$ and solve $-a \leq x \leq a$.*
$18,000 \leq 2.5w \leq 22,000$ ◀——— *Apply the Addition Property of Inequality.*
$7200 \leq w \leq 8800$ ◀——— *Apply the Division Property of Inequality.*

The factory can produce between 7200 and 8800 watches and stay within the cost limits.

TRY THIS   Rework Example 2 if the ideal weekly cost is $25,000 and the tolerance is $3000.

KEY SKILLS

Let w represent a weight. Write an absolute-value inequality to represent the acceptable weights for each ideal weight and tolerance.

1. ideal weight = 5 pounds; tolerance = 0.2 pound
2. ideal weight = 14 ounces; tolerance = 0.01 ounce

PRACTICE

Use absolute value to express the acceptable ranges and the unacceptable ranges.

3. plants whose height may vary at most 2 inches from the ideal height of 18 inches
4. a machine part whose length may vary at most 0.002 centimeter from the ideal length of 32 centimeters
5. glass whose thickness may vary at most 0.0025 centimeter from the ideal thickness of 1.2 centimeters

Solve each problem.

6. In Example 2, find the number of watches that can be produced if the weekly cost of producing w watches is $12,000 + 2.5w$, the ideal cost is $25,000, and the tolerance is $3000.
7. Margo's doctor tells her that her ideal weight is $3h - 56$, where h is Margo's height in inches. The tolerance is 9 pounds. The doctor then says that Margo's weight of 115 pounds is within this acceptable range. How tall can Margo be?

MIXED REVIEW APPLICATIONS

Solve each problem. (2.8 Skill C)

8. A chemist wants to separate 640 milliliters of solution into two containers in such a way that one container holds 200 milliliters less than twice the first container. How much should be put into each container?
9. How much money should be put into bank accounts paying 5% and 6% simple interest if $5000 is invested and the investor wants to earn $290 in one year?

Answers for Lesson 4.7 Skill C

Try This Example 1
acceptable: $12.48 \leq w \leq 12.52$
unacceptable: $0 < w < 12.48 \ or \ w > 12.52$

Try This Example 2
between 4800 watches and 7200 watches, inclusive

1. $|w - 5| \leq 0.2$
2. $|w - 14| \leq 0.01$
3. Let h represent height.
   acceptable: $16 \leq h \leq 20$
   unacceptable: $h < 16 \ or \ h > 20$
4. Let L represent length.
   acceptable: $31.998 \leq L \leq 32.002$
   unacceptable: $L < 31.998 \ or \ L > 32.002$
5. Let t represent thickness.
   acceptable: $1.1975 \leq t \leq 1.2025$
   unacceptable: $t < 1.1975 \ or \ t > 1.2025$
6. between 4000 watches and 6400 watches, inclusive
7. between 54 inches and 60 inches, inclusive
8. 280 milliliters and 360 milliliters
9. $1000 at 5% and $4000 at 6%

Deductive Reasoning with Inequalities

SKILL A *Showing that an inequality is always, sometimes, or never true*

Focus: *Using deductive reasoning, students determine whether a given inequality is true in some instances, in all instances, or in no instance.*

EXAMPLE 1

PHASE 1: Presenting the Example

Explain to students that determining whether an inequality is always, sometimes, or never true is not always a straightforward process. A good first step is to apply the properties of real numbers, as they have been doing throughout this chapter, and examine the results. Often this will lead to the correct answer. Then present the Example.

After completing Example 1, you may want to present an example in which the outcome is *never true*. One example is provided below:

Is $2x - 5 + 3x > 5x$ always, sometimes, or never true?

$$2x - 5 + 3x > 5x$$
$$5x - 5 > 5x$$
$$-5 > 0$$

$-5 > 0$ is never true, so the original inequality is never true.

EXAMPLE 2

PHASE 1: Presenting the Example

In Example 2, several students may assess $r > -r$ as always true. This is because students tend to think of every variable as positive. Point out that r could be negative, thus making $-r$ positive. After you discuss the solution, ask them how to adjust the given problem so that the statement *is* always true. [**Change the problem to "Given that r is a positive real number."**]

PHASE 3: ASSESSMENT AND CLOSURE for Lesson 4.8

- Give students the inequality shown at right. Tell them that c is a real number. Ask them to fill in the box to make a statement that is *sometimes true*. Repeat the question for *always true* and *never true*.

- In this lesson, students formalized the principles of logic that are used to solve inequalities. This concludes the sequence of instruction in solving inequalities in one variable. In Chapter 5, they will work with equations and inequalities in *two* variables.

$\square > c$

[samples:
sometimes — $c^2 > c$;
always — $c + 1 > c$;
never — $c - 1 > c$]

☞ *For a Lesson 4.8 Quiz, see* Assessment Resources *page 45.*

☞ *For a Chapter 4 Assessment, see* Assessment Resources *pages 46–47.*

4.8 Deductive Reasoning With Inequalities

LESSON

SKILL A ▸ Showing that an inequality is always, sometimes, or never true

To determine whether an inequality is true over the domain of all real numbers, simplify it. If the simplification is an inequality that is always true, then the original inequality is also always true.

EXAMPLE 1   Is $2(x + 5) > 2x + 9$ always, sometimes, or never true?

▸ Solution

$$2(x + 5) > 2x + 9$$
$$2x + 10 > 2x + 9 \longleftarrow \text{Apply the Distributive Property.}$$
$$2x + 1 > 2x \longleftarrow \text{Subtract 9 from each side.}$$
$$1 > 0 \longleftarrow \text{Subtract 2x from each side.}$$

Recall from Lesson 4.5 that if you arrive at an equivalent inequality that is always true, then the original inequality is true for all real numbers. $1 > 0$ is always true. Therefore, $2(x + 5) > 2x + 9$ is *always true*.

TRY THIS   Show whether $3(x - 4) > 3(x - 5)$ is always, sometimes, or never true.

If an inequality simplifies to a numerical inequality that is never true, then the original statement is never true. For example, $2x - 5 + 3x > 5x$ simplifies to $-5 > 0$, which is not true. Thus, $2x - 5 + 3x > 5x$ is *never true*.

EXAMPLE 2   Is $r > -r$ always, sometimes, or never true?

▸ Solution

$$r > -r$$
$$2r > 0 \longleftarrow \text{Add r to each side.}$$
$$r > 0 \longleftarrow \text{Divide each side by 2.}$$

The result shows that the inequality is true only when r is positive. When $r \leq 0$, the inequality is not true. Thus, the statement is only *sometimes true*.

TRY THIS   Show whether $r > 2r$ is always, sometimes, or never true.

In simplifying inequalities, *be careful of division by a variable*. Whenever you divide both sides of an inequality by a variable, you have to consider three cases: $r > 0$, $r < 0$ and $r = 0$.

In Example 2, if each side of $r > -r$ is divided by r, the result seems to be $1 > -1$ (true). However, if r is negative, the inequality sign must be reversed. Then the result is $1 < -1$ (false). And if $r = 0$, the division cannot be performed at all. So, $r > -r$ is true only when r is positive.

EXERCISES

KEY SKILLS

1. What can you conclude about an inequality that simplifies to $1 > -1$?

2. What can you conclude about an inequality that simplifies to $1 < -1$?

3. What can you conclude about an inequality that simplifies to $x > 0$?

PRACTICE

Show whether each statement is always, sometimes, or never true.

4. $3 - (x + 1) > 1 - x$ 5. $3x < 3(x + 1)$ 6. $-2(x + 1) < -2x$

7. $\frac{1}{2}r < r$ 8. $4x - 3 < 2(2x - 2)$ 9. $4(x - 3x) > 9 - 8x$

10. $-2(x + 1) > -2x$ 11. $3z - 5 + 4z \geq z + 6z - 5$ 12. $-2(3z - 2z) > z$

13. $9 - 2(z + 2) < -2z$ 14. $500x > 250x$ 15. $500x > 250x^2$

16. $-(x + 1) \geq x + 1$ 17. $-(x + 1) \leq x - 1$

Critical Thinking   Show whether each statement is always, sometimes or never true.

18. $x^2 > x$ 19. $x(-x) \leq 0$ 20. $x^2 + 1 \geq 1$

21. $x^2 \geq 2x$ 22. $\frac{x}{5} \geq \frac{5}{x}$ 23. $\frac{x}{3} \leq \frac{3}{x}$

24. **Critical Thinking**   Is the following statement always, sometimes, or never true? Justify your answer.
The reciprocal of a number is greater than the opposite of the number.

MIXED REVIEW

Use intercepts to graph each equation. (3.5 Skill A)

25. $-4x - 7y = 28$ 26. $x - y = 5.5$ 27. $3x - 7y = 21$

Use the slope and the y-intercept to graph each equation. (3.5 Skill A)

28. $y = \frac{-2}{5}x$ 29. $y = \frac{5}{2}x - 3.5$ 30. $y = -x - 4$

Answers for Lesson 4.8 Skill A

Try This Example 1
If $3(x - 4) > 3(x - 5)$, then $-12 > -15$. The inequality is always true.

Try This Example 2
If $r > 2r$, then $r < 0$. The statement is sometimes true.

1. The inequality is always true.
2. The inequality is never true.
3. The inequality is sometimes true.
4. If $3 - (x + 1) > 1 - x$, then $3 - x - 1 > 1 - x$. Thus, $2 - x > 1 - x$. This implies that $2 > 1$. The given statement is always true.
5. If $3x < 3(x + 1)$, then $3x < 3x + 3$. This implies that $0 < 1$. The given statement is always true.
6. If $-2(x + 1) < -2x$, then $-2x - 2 \leq -2x$. This implies that $-2 \leq 0$. The given statement is always true.
7. If $\frac{1}{2}r < r$, then $r < 2r$. Thus, $0 < r$. The given statement is sometimes true.
8. If $4x - 3 < 2(2x - 2)$, then $4x - 3 < 4x - 4$. This implies that $-3 \leq -4$. This is false. The given statement is never true.

9. If $4(x - 3x) > 9 - 8x$, then $-8x > 9 - 8x$. This implies that $0 > 9$. This is false. The given statement is never true.
10. If $-2(x + 1) > -2x$, then $-2x - 2 > -2x$. This implies that $-2 > 0$. This is false. The given statement is never true.
11. If $3z - 5 + 4z \geq z + 6z - 5$, then $-5 \geq -5$. The given statement is always true.
12. If $-2(3z - 2z) > z$, then $-2z > z$. So, $z < 0$. The statement is sometimes true.
13. If $9 - 2(z + 2) < -2z$, then $5 - 2z < -2z$. This means $5 < 0$. This is false. Therefore, the given statement is never true.
14. If $500x > 250x$, then $2x > x$. This implies that $x > 0$. The given statement is sometimes true.
15. If $500x > 250x^2$, then $2x > x^2$. The given statement is sometimes true.
16. If $-(x + 1) \geq x + 1$, then $-x - 1 \geq x + 1$. This implies that $x \leq -1$. The given statement is sometimes true.
17. If $-(x + 1) \leq x - 1$, then $-x - 1 \leq x - 1$. This implies that $x \geq 0$. The given statement is sometimes true.

For answers to Exercises 18–30, see Additional Answers.

1. Graph −3 and 5 on a number line and write two inequalities that describe their relationship.

2. Represent the following statement in symbols.
   The cost of a textbook, c, is more than $25.

3. Graph the solutions to $a \leq 1$ or $a > 2.5$ on a number line.

4. Graph the solutions to $t \geq -3$ and $t \leq -1.5$ on a number line.

5. Solve $a - 3.5 > 1$. Graph the solution on a number line.

6. Find and graph the solution to $x - 1 > 0$ and $x - 2 < 4$.

7. Solve $n + 2.5 \leq 6$. Graph the solution on a number line.

8. Find and graph the solution to $x + 1 > 3$ and $x + 2 < 8$.

9. Kathy wants to buy at least 2.5 pounds of hamburger. The scale shows 2.1 pounds. How much more should be put on the scale so that Kathy has at least 2.5 pounds of hamburger?

Solve. Graph each solution on a separate number line.

10. $\frac{k}{3} \geq -1$
11. $\frac{3}{5}d < 6$
12. $\frac{p}{-2} \geq -1.5$
13. $4g \geq 20$
14. $-4r \leq 20$
15. $-7w > -14$

16. The perimeter of a square playground is at least 225 feet. What is the length of each side?

Solve each inequality.

17. $7 \geq s + 2.5$
18. $-x > 5$
19. $2.5x \leq 2$

20. Ron wants to send roses to his wife, Sheila. Roses cost $1.50 each, and Ron has $25. What are the possible numbers of roses that Ron can send to Sheila?

Solve. Graph each solution on a separate number line.

21. $3b + 4 \leq 16 - b$
22. $4a + 4 > 3a + 1$
23. $7 > 5s - 3$
24. $5c - 3c > 10$
25. $7h - 6 < 6h$
26. $-7w > -14 - 5w$

Solve. Graph each solution on a separate number line.

27. $3(x - 1) - 2 \leq 19$
28. $4p + 4(p - 1) > 12$
29. $4r + 2(r + 1) < 3(r + 5)$
30. $-3(d - 1) + 5 > -2(d + 1)$
31. $-2(w + 1) - 3(w + 3) \geq 5 - (w + 6)$

32. Mr. Kent has $1500 in a bank account that pays 5% interest annually. How much should he deposit now so that he will have at least $1800 in one year?

Solve. Graph each solution on a separate number line.

33. $4(w + 1) < 3(w + 3) < 5 - (w + 6)$
34. $3a + 5 \leq 32$ and $4a - 3 \geq 21$
35. $-x - 2 \leq 6$ or $4x + 9 \leq 1$
36. $2(w - 1) \geq 3w + 1$ or $3(w - 2) \leq 5 - (w + 6)$
37. $3a + 5 \leq 23$ or $4a - 3 \leq -1$
38. $3x - 2 \geq 6$ and $4x + 7 \leq -3$

39. In a party mix, the ratio of peanuts to walnuts is 2 to 1. How many pounds of walnuts are needed to make at least 6.6 pounds of the mix but not more than 7.8 pounds of the mix?

Solve each absolute-value equation. Check your solution.

40. $|3x + 4| = 12$
41. $|-2d + 1| - 3 = 15$
42. $4 + |4x| = 16$

Solve. Graph each solution on a separate number line.

43. $|3(x + 1)| \geq 1$
44. $2|x + 1| - 2 < 4$
45. $5 - 2|3x| > -1$

46. The ideal weight of a can of oil is 32 ounces with a tolerance of 0.02 ounce. Find the acceptable weights and the unacceptable weights.

47. Is $2r > r$ always, sometimes, or never true?

Answers for Chapter 4 Assessment

1. $-3 < 5$ and $5 > -3$
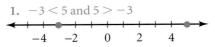

2. $c > 25$

3.
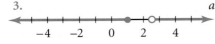

4.
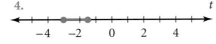

5. $a > 4.5$
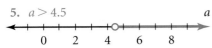

6. $1 < x < 6$

7. $n \leq 3.5$
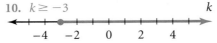

8. $2 < x < 6$
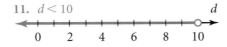

9. at least 0.4 pound of hamburger

10. $k \geq -3$

11. $d < 10$
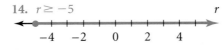

12. $p \leq 3$

13. $g \geq 5$

14. $r \geq -5$

15. $w < 2$
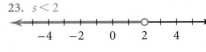

16. at least 56.25 feet

17. $s \leq 9.5$

18. $x < -5$

19. $x \leq 0.8$

20. less than 17 roses

21. $b \leq 3$
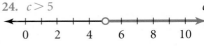

22. $a > -3$

23. $s < 2$

24. $c > 5$

25. $h < 6$

For answers to Exercises 26–47, see Additional Answers.

Chapter 5 *Solving Systems of Equations and Inequalities*	State or Local Standards	Corresponding Lessons in *Algebra 1*, Schultz et al.
5.1 Solving Systems of Equations by Graphing		7.1
Skill A: Using a graph to solve a system of linear equations		
5.2 Solving Systems of Equations by Substitution		7.2, 7.6
Skill A: Using the substitution method with one variable already isolated		
Skill B: Using the substitution method with no isolated variable		
Skill C: APPLICATIONS Using the substitution method in solving real-world problems		
5.3 Solving Systems of Equations by Elimination		7.3
Skill A: Using the elimination method with one multiplier		
Skill B: Using the elimination method with two multipliers		
Skill C: APPLICATIONS Using the elimination method in solving real-world problems		
5.4 Classifying Systems of Equations		7.4
Skill A: Classifying a system of equations as consistent or inconsistent		
Skill B: Classifying a system of equations as dependent or independent		
5.5 Graphing Linear Inequalities in Two Variables		7.5
Skill A: Graphing linear inequalities		
Skill B: APPLICATIONS Graphing linear inequalities with restrictions		
5.6 Graphing Systems of Linear Inequalities in Two Variables		7.5
Skill A: Graphing systems of linear inequalities		
Skill B: Graphing systems of inequalities given in standard form		
Skill C: APPLICATIONS Using systems of inequalities to solve geometry and real-world problems		

Solving Systems of Equations and Inequalities

▷ **What You Already Know**

In earlier chapters, you learned how to solve an equation for a specified variable, graph the equation, and find how pairs of lines represented by linear equations in two variables are related.

▷ **What You Will Learn**

In Chapter 5, you will be given a pair of linear equations in two variables, called a *system of equations*. The task will be to learn how to find any ordered pair of real numbers that satisfies both equations.

In particular, you will learn how to solve a system of linear equations in two variables by

• using graphs,

• using the substitution method, and

• using the elimination method.

Your study will also include ways to find out whether a system of equations has a solution at all or has infinitely many solutions.

Finally, you will have the opportunity to explore linear inequalities in two variables. You will see that the graph of such an inequality is a region in the coordinate plane rather than a line. How to find the solution to a pair of such inequalities graphically will also be explored.

VOCABULARY	
boundary line	linear inequality in two variables
consistent	solution of a system
dependent	solution region
elimination method	substitution
inconsistent	substitution method
independent	system of linear equations
intersection	system of linear inequalities in two
linear inequality in standard form	variables

The diagram below shows how mathematical skills and mathematical reasoning are interrelated with the skills and concepts in Chapter 5. Notice that in this chapter you will learn how to solve systems of equations and linear inequalities in two variables.

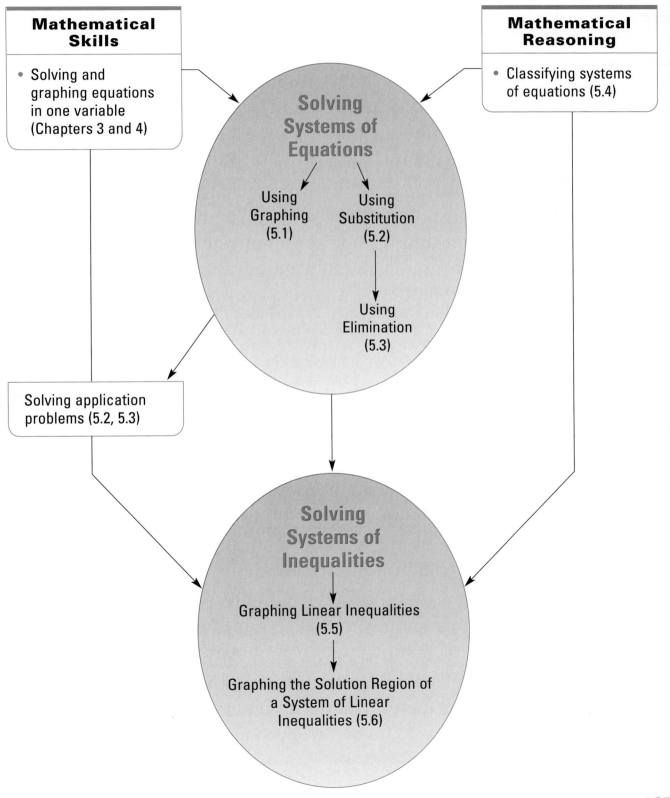

Solving Systems of Equations by Graphing

SKILL A ▶ *Using a graph to solve a system of linear equations*

Focus: *Students graph a system of linear equations in two variables and use the graph to locate the coordinates of the solution.*

EXAMPLE 1

PHASE 1: Presenting the Example

Display the figure at right. Ask students to write equations for lines ℓ and m.
[ℓ: $y = 2x - 3$; m: $y = -x + 3$] Note that lines ℓ and m intersect at $P(2, 1)$. Tell students to replace x and y with these coordinates in each equation. They should note that $(2, 1)$ is a solution of each. Point out that, in fact, it is the *only* ordered pair that is a solution of both equations.

Tell students that the two equations considered together form a *system* of equations in x and y, and $(2, 1)$ is the *solution of the system*. Show them how to write the system, as shown at right below the graph. Then discuss Example 1.

PHASE 2: Providing Guidance for TRY THIS

Emphasize how finding and plotting the x- and y- intercepts makes graphing the equations easier. For $x + y = 3$, both intercepts can be found and plotted. For $y = x - 7$, the y-intercept and slope can be efficiently used as well. Stress the importance of graphing carefully and of checking the solution found through graphical means, in case the graph is not precise.

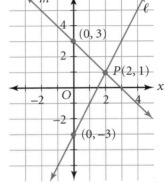

$$\begin{cases} y = 2x - 3 \\ y = -x + 3 \end{cases}$$

EXAMPLE 2

PHASE 1: Presenting the Example

Ask students if they think it is possible for a system of linear equations to have no solution. Lead them to a logical conclusion: The solution to a system occurs where the lines intersect. So, if the lines do not intersect, there is no solution. Now discuss Example 2.

PHASE 3: ASSESSMENT AND CLOSURE for Lesson 5.1

- Ask students to find the solution to the system of equations at right. [$(-3, 1)$]
- Drawing a graph to find the solution to a system of equations can be time consuming. It can also be difficult with noninteger solutions and equations with very large or very small numbers. In the lessons that follow, algebraic methods will be introduced that are much more practical for these situations.

$$\begin{cases} y = x - 4 \\ y = -x - 2 \end{cases}$$

 For a Lesson 5.1 Quiz, see Assessment Resources *page 48.*

5.1 LESSON

Solving Systems of Equations by Graphing

SKILL A *Using a graph to solve a system of linear equations*

A **system of linear equations** is a set of two or more equations with the same variables. The **solution of a system** in x and y is any ordered pair (x, y) that satisfies each of the equations in the system.

The solution of a system of equations is the *intersection* of the graphs of the equations.

EXAMPLE 1 Use a graph to solve the system of equations at right. Verify that your solution lies on both lines.

$$\begin{cases} x + y = 5 \\ y = 2x - 1 \end{cases}$$

▸ Solution

Graph both equations on the same coordinate plane.
• Graph $x + y = 5$ using the intercepts: $(5, 0)$ and $(0, 5)$.
• Graph $y = 2x - 1$ using the slope and y-intercept.
Locate the point where the lines intersect. From the graph, the solution appears to be $(2, 3)$.

Check: To be sure that $(2, 3)$ is the solution, substitute 2 for x and 3 for y into each equation.

$$x + y = 5 \qquad y = 2x - 1$$
$$2 + 3 \overset{?}{=} 5 ✔ \qquad 3 \overset{?}{=} 2(2) - 1 ✔$$

Because $(2, 3)$ satisfies both equations, it lies on both lines.

TRY THIS Use a graph to solve the system of equations at right. Verify that your solution lies on both lines.

$$\begin{cases} x + y = 3 \\ y = x - 7 \end{cases}$$

EXAMPLE 2 Use a graph to solve the system of equations at right. Check your solution.

$$\begin{cases} x + y = 4 \\ x + y = 2 \end{cases}$$

▸ Solution

Graph $x + y = 4$ using the intercepts: $(4, 0)$ and $(0, 4)$.
Graph $x + y = 2$ using the intercepts: $(2, 0)$ and $(0, 2)$.
The lines appear to be parallel. Because parallel lines never intersect, the system has no solution.

Check: To be sure that the lines are parallel, rewrite each equation in slope-intercept form and then compare the slopes.
$$x + y = 4 \;\rightarrow\; y = -x + 4 \qquad x + y = 2 \;\rightarrow\; y = -x + 2$$
The slopes, -1 and -1, are equal, so the lines are parallel. ✔

TRY THIS Use a graph to solve the system of equations at right. Check your solution.

$$\begin{cases} x + y = 4 \\ x + y = 3 \end{cases}$$

EXERCISES

KEY SKILLS

Compare slopes to determine whether each system of equations has a solution or not. Do not solve.

1. $\begin{cases} y = 3x - 2 \\ y = 3x - 1 \end{cases}$ 2. $\begin{cases} 2y = 10x - 4 \\ y = 5x - 1 \end{cases}$ 3. $\begin{cases} 4x + 2y = 8 \\ 4x + 3y = -2 \end{cases}$

PRACTICE

Use a graph to solve each system of equations. If the system has no solution, write *none*.

4. $\begin{cases} y = -x - 1 \\ y = 5 - x \end{cases}$ 5. $\begin{cases} y = 6x - 2 \\ y = 1 + 6x \end{cases}$ 6. $\begin{cases} y = -x - 2 \\ y = x + 6 \end{cases}$

7. $\begin{cases} y = -x - 5 \\ y = x + 1 \end{cases}$ 8. $\begin{cases} y = x + 11 \\ y + x = 3 \end{cases}$ 9. $\begin{cases} y = x + 5 \\ y + x = 3 \end{cases}$

10. $\begin{cases} x - y = 1 \\ x - y = 3 \end{cases}$ 11. $\begin{cases} -2x + y = 1 \\ -2x + y = -2 \end{cases}$ 12. $\begin{cases} y = \frac{2}{3}x - 1 \\ -2x + 3y = 6 \end{cases}$

13. $\begin{cases} y = -2x + 10 \\ y = -\frac{1}{3}x + 5 \end{cases}$ 14. $\begin{cases} y = \frac{1}{3}x - 3 \\ y = \frac{-5}{3}x - 9 \end{cases}$ 15. $\begin{cases} y = -\frac{1}{2}x - \frac{1}{2} \\ y = \frac{1}{2}x - \frac{1}{2} \end{cases}$

16. $\begin{cases} y = \frac{1}{4}x - 4 \\ y = -\frac{1}{2}x + 2 \end{cases}$ 17. $\begin{cases} 2x - y = -3 \\ 4x - 2y = -5 \end{cases}$ 18. $\begin{cases} 5y = x - 10 \\ x - y = 6 \end{cases}$

Use the slopes and the y-intercepts to show that each system has a solution. Then graph the system to verify your conclusion.

19. $\begin{cases} y = \frac{4}{3}x - 2 \\ y = 2 \end{cases}$ 20. $\begin{cases} y = -\frac{3}{5}x + 3 \\ y = 2x + 1 \end{cases}$ 21. $\begin{cases} -x + 5y = 5 \\ 2x + y = 4 \end{cases}$

MIXED REVIEW

Simplify each expression. (2.6 Skill B)

22. $\frac{2}{3}(3x - 6) - 2$ 23. $2 - \frac{4}{5}(x + 5)$ 24. $-2(5 + 4n) - 4$

25. $-3(3y - 1) + 1$ 26. $-0.5(3 - 5k) + 5$ 27. $-2.5(1 - 2x) + 5$

Answers for Lesson 5.1 Skill A

Try This Example 1
$(5, -2)$

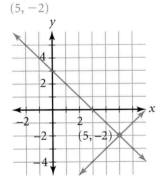

Try This Example 2
no solution

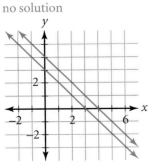

1. no solution
2. no solution
3. one solution
4. none

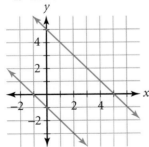

5. none

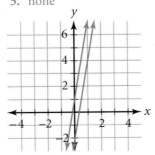

6. $(-4, 2)$

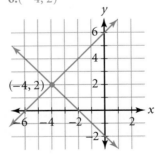

7. $(-3, -2)$

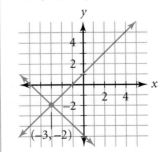

8. $(-4, 7)$

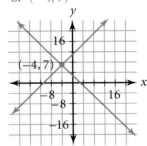

9. $(-1, 4)$

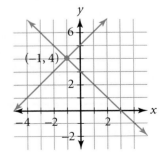

For answers to Exercises 10–27, see Additional Answers.

TEACHING

5.2

LESSON

Solving Systems of Equations by Substitution

 SKILL A   *Using the substitution method with one variable already isolated*

Focus: *Given a linear system in which one equation is either in the form y = ax + b or in the form x = cy + d, students solve the system algebraically by substituting the expression for y or x into the other equation.*

EXAMPLE

PHASE 1: Presenting the Example

Display the system shown at right. Tell students to graph it. [**The graph is given at right.**] Then ask whether any of the students can make a conjecture about the exact solution to the system. Several will probably guess that the solution is $(-2, 1)$. Have them verify that this is indeed the solution by substituting -2 for x and 1 for y in each equation.

Return to the system as given. Point out that $y = x + 3$ states that $x + 3$ is equivalent to y. Remind students that, by the Substitution Principle, any expression may be replaced by an equivalent expression. Tell them that you choose to replace y with $x + 3$ in the first equation, as shown below.

$$x \quad + \quad y \quad = \quad -1$$
$$\downarrow$$
$$x \quad + \quad (x+3) \quad = \quad -1$$

Point out that the new equation, $x + (x + 3) = -1$, is a one-variable linear equation that can easily be solved for x. Have students solve it. [$x = -2$] They should now observe that, given -2 as a real-number value for x, it is possible to solve for y by substituting -2 for x in either equation, as follows:

$$x \ + y = -1 \qquad\qquad\qquad y = \ x \ + 3$$
$$\downarrow \qquad\qquad\qquad\qquad\qquad\qquad \downarrow$$
$$-2 + y = -1 \qquad\qquad\qquad y = -2 + 3$$
$$-2 + y + 2 = -1 + 2 \qquad\qquad y = \quad 1$$
$$y = \quad 1$$

Lead students to see that the two values obtained algebraically, $x = -2$ and $y = 1$, yield the ordered pair obtained graphically, $(-2, 1)$. Tell students that this algebraic method of finding the solution is called the *substitution method*. Then discuss the Example.

PHASE 2: Providing Guidance for **TRY THIS**

At this point in the course, students have had considerable experience with substituting a given real-number value for a variable. However, they may have difficulty with the notion of substituting an *algebraic expression* for a variable, and they may need practice in this skill alone. You can use Skill A Exercises 1 through 3 on the facing page for this purpose.

$$\begin{cases} x + y = -1 \\ y = x + 3 \end{cases}$$

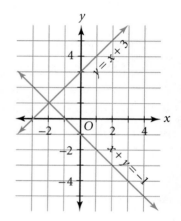

5.2 Solving Systems of Equations by Substitution

LESSON

SKILL A *Using the substitution method with one variable already isolated*

A second method for solving systems of equations is *substitution*. This method is quicker and more accurate than the graphing method. It is especially practical when one variable has a coefficient of 1.

The Substitution Method

Step 1. Solve one equation for x (or y).

Step 2. Substitute the expression from Step **1** into the other equation. The result is an equation in only one variable.

Step 3. Solve for y (or x).

Step 4. Take the value of y (or x) found in Step **3** and substitute it into one of the original equations. Then solve for the other variable.

Step 5. The ordered pair of values from Steps **3** and **4** is the solution. If the system has no solution, a contradictory statement will result in either Step **3** or **4**.

EXAMPLE   Solve $\begin{cases} y = 3x + 8 \\ x + 2y = 9 \end{cases}$ using the substitution method.

▶ **Solution**

1. The first equation is already solved for y, so Step **1** is already done.

2. In $x + 2y = 9$, substitute $3x + 8$ for y.
$$x + 2y = 9$$
$$x + 2(3x + 8) = 9$$

3. Solve $x + 2(3x + 8) = 9$ for x.
$$x + 2(3x + 8) = 9$$
$$x + 6x + 16 = 9$$
$$7x = -7$$
$$x = -1$$

4. Substitute -1 for x in $y = 3x + 8$ and then solve for y.
$$y = 3x + 8$$
$$y = 3(-1) + 8 = 5$$

5. The solution to the system is $(-1, 5)$.

Check: Check the values of x and y in the original equations.
$$y = 3x + 8 \qquad\qquad x + 2y = 9$$
$$5 \overset{?}{=} 3(-1) + 8 \qquad -1 + 2(5) \overset{?}{=} 9$$
$$5 \overset{?}{=} 5 ✔ \qquad\qquad 9 \overset{?}{=} 9 ✔$$

TRY THIS   Solve using the substitution method.   **a.** $\begin{cases} y = x - 9 \\ 4x + 3y = 22 \end{cases}$ **b.** $\begin{cases} x + y = -9 \\ 2y = 3x + 2 \end{cases}$

EXERCISES

KEY SKILLS

In each system below, one of the equations is solved for y. Replace that expression for y in the other equation and write the equation in x that results. Do not solve.

1. $\begin{cases} -2x + 3y = 1 \\ y = -3x + 1 \end{cases}$ 2. $\begin{cases} 2x + 3y = 6 \\ y = 2x - 5 \end{cases}$ 3. $\begin{cases} -\frac{1}{5}x + 3 = y \\ -3x + y = -5 \end{cases}$

PRACTICE

Use the substitution method to solve each system. Check your answers.

4. $\begin{cases} y = -5x - 1 \\ y = -1 \end{cases}$ 5. $\begin{cases} y = -5x + 1 \\ y = -4 \end{cases}$ 6. $\begin{cases} y = -2x + 4 \\ x + 3y = -3 \end{cases}$

7. $\begin{cases} y = -x + 4 \\ 2x + y = -5 \end{cases}$ 8. $\begin{cases} 2x + y = 5 \\ x = 2y - 4 \end{cases}$ 9. $\begin{cases} y = -\frac{3}{7}x + 2 \\ y = \frac{2}{7}x - 3 \end{cases}$

10. $\begin{cases} -2x + y = -11 \\ y = \frac{2}{5}x - 3 \end{cases}$ 11. $\begin{cases} \frac{1}{5}x + 2y = 8 \\ y = x - 7 \end{cases}$ 12. $\begin{cases} \frac{3}{4}x + 5y = 4 \\ y = 2x - \frac{7}{2} \end{cases}$

13. $\begin{cases} y = -3x + \frac{23}{7} \\ y = \frac{2}{7}x \end{cases}$ 14. $\begin{cases} y = -3x - 7 \\ 5x - 2y = 20 \end{cases}$ 15. $\begin{cases} 3 + 4y = x \\ -2x - 3y = 6 \end{cases}$

16. Let a and b be fixed real numbers. For what values of a and b will the solution to the system $\begin{cases} x + ay = 5 \\ x = by + 10 \end{cases}$ be $(2, -3)$?

17. Let a and b be fixed real numbers. Solve $\begin{cases} x = a + y \\ x + y = b \end{cases}$ for x and y.

18. **Critical Thinking**   Solve the system at right by substitution. $\begin{cases} x = z - 2 \\ y = z - 1 \\ z = x + y \end{cases}$

MIXED REVIEW

Solve each inequality. (4.5 Skill B)

19. $3d + 4(d + 1) \geq 4$ 20. $-4t - (t + 3) < 0$ 21. $3(g - 4) > -2(g - 1)$

22. $3z - 4(z + 2) \geq -3z$ 23. $\frac{1}{3}(a - 2) - \frac{1}{2}(a + 1) \geq 0$ 24. $4 - (w + 4) < -3(w - 1)$

25. $\frac{1}{3}(b - 2) > \frac{2}{5}(b + 2)$ 26. $\frac{c - 2}{3} > \frac{c + 2}{-5}$ 27. $\frac{-c}{3} \leq \frac{3(c + 2)}{-2}$

Answers for Lesson 5.2   Skill A

Try This Example
a. $(7, -2)$ **b.** $(-4, -5)$

1. $-2x + 3(-3x + 1) = 1$
2. $2x + 3(2x - 5) = 6$
3. $-3x + \left(-\frac{1}{5}x + 3\right) = -5$
4. $(0, -1)$
5. $(1, -4)$
6. $(3, -2)$
7. $(-9, 13)$
8. $\left(\frac{6}{5}, \frac{13}{5}\right)$
9. $(7, -1)$
10. $(5, -1)$
11. $(10, 3)$
12. $\left(2, \frac{1}{2}\right)$
13. $\left(1, \frac{2}{7}\right)$
14. $\left(\frac{6}{11}, -\frac{95}{11}\right)$

15. $\left(-\frac{15}{11}, -\frac{12}{11}\right)$
16. $a = -1$ and $b = \frac{8}{3}$
17. $\left(\frac{b + a}{2}, \frac{b - a}{2}\right)$
18. $(1, 2, 3)$
19. $d \geq 0$
20. $t > -\frac{3}{5}$
21. $g > \frac{14}{5}$
22. $z \geq 4$
23. $a \leq -7$
24. $w < \frac{3}{2}$
25. $b < -22$
26. $c > \frac{1}{2}$
27. $c \leq -\frac{18}{7}$

★ **Advanced Exercises**

Focus: *Given a linear system in two variables, students first rewrite one equation to isolate one of the variables. Then they use the substitution method to solve the system.*

EXAMPLE

PHASE 1: Presenting the Example

Display the following systems:

$$\mathbf{1}\ \begin{cases} x + y = 4 \\ 2x + y = 1 \end{cases} \qquad \mathbf{2}\ \begin{cases} x + y = 4 \\ y = -2x + 1 \end{cases}$$

Tell students that you want them to compare the two systems *without solving*. Ask them to write a brief statement explaining how the systems are different and how they are alike. Then discuss their responses.

Most students should observe that the only difference between **1** and **2** lies in the *form* of the second equation. That is, **1** and **2** are actually two forms of the same system. Ask them why one form might be preferred over the other. Lead them to see that, using form **2**, it is possible to substitute the expression $-2x + 1$ for y in the first equation. Tell students to make this substitution and find the solution to the system. $[(-3, 7)]$

Direct the students' attention back to form **1**. Ask if there is another way to rewrite the system in order to solve by the substitution method. Elicit the following two responses:

$$\mathbf{3}\ \begin{cases} y = -x + 4 \\ 2x + y = 1 \end{cases} \qquad \mathbf{4}\ \begin{cases} x = -y + 4 \\ 2x + y = 1 \end{cases}$$

Have half of the class solve the system using form **3**, while the other half solves using form **4**. Then have one student from each group write the solution for all to see. Students should note that each method leads to the same solution as before, $(-3, 7)$.

Now display the system given in the Example, which is $\begin{cases} 3u + 2v = 12 \\ u - 2v = 3 \end{cases}$. Point out that neither equation has the variable isolated on one side of the equal sign. Ask students which equation would be easier to rewrite in order to achieve this goal. $[u - 2v = 3]$ Tell them to rewrite the equation. $[u = 2v + 3]$ Then discuss the solution outlined on the textbook page.

Point out to students that the coordinates of the solution are not integers. Therefore, it would have been difficult to solve this system by graphing. The substitution method allows you to solve many more systems than you could solve by graphing.

PHASE 2: Providing Guidance for TRY THIS

In Lesson 5.1, systems were restricted to the variables x and y in order to locate the solution graphically in the conventional xy-plane. Therefore, some students may be confused by the appearance of other variables in these exercises. Assure them that the practice of using variables other than x and y not only is permitted, but actually may be more sensible when using a system to solve a real-world problem. They will see examples of this in Skill C.

A system of equations such as $\begin{cases} 3u + 2v = 12 \\ u - 2v = 3 \end{cases}$ may be given with neither variable already isolated.

In a case such as this, you must first isolate one of the variables before you can use substitution. It is easiest to isolate a variable that has a coefficient of 1, such as u in the second equation above.

EXAMPLE

Solve $\begin{cases} 3u + 2v = 12 \\ u - 2v = 3 \end{cases}$. Check your solution.

▶ **Solution**

1. Write the second equation so that u is isolated.

$$\begin{cases} 3u + 2v = 12 \\ u - 2v = 3 \end{cases} \rightarrow \begin{cases} 3u + 2v = 12 \\ u = 2v + 3 \end{cases}$$

2. Apply the substitution method.

$$\begin{cases} 3u + 2v = 12 \\ u = 2v + 3 \end{cases} \rightarrow 3(2v + 3) + 2v = 12$$

3. Solve $3(2v + 3) + 2v = 12$.

$$3(2v + 3) + 2v = 12$$
$$8v + 9 = 12 \rightarrow v = \frac{3}{8}$$

4. If $v = \frac{3}{8}$, then $u = 2\left(\frac{3}{8}\right) + 3$, so $u = \frac{15}{4}$.

5. The solution (u, v) to the given system is $\left(\frac{15}{4}, \frac{3}{8}\right)$.

Check: Check the values of x and y in both equations.

$$3u + 2v = 12 \qquad\qquad u - 2v = 3$$
$$3\left(\frac{15}{4}\right) + 2\left(\frac{3}{8}\right) \overset{?}{=} 12 \qquad \frac{15}{4} - 2\left(\frac{3}{8}\right) \overset{?}{=} 3$$
$$\frac{90 + 6}{8} \overset{?}{=} 12 \checkmark \qquad\qquad \frac{30 - 6}{8} \overset{?}{=} 3 \checkmark$$

TRY THIS

Solve each system. Check your solution.

a. $\begin{cases} a + 2b = -2 \\ 2a + 3b = 1 \end{cases}$ b. $\begin{cases} 2m - 5n = 4 \\ 7 = 3m + n \end{cases}$

Note that it would have been difficult to solve the system in the Example by graphing. The exact solution would be hard to determine from a graph because the coordinates are not integers. Solving a system algebraically is better than graphing when you need an accurate solution.

EXERCISES

Isolate one variable in one of the equations. Do not solve the system.

1. $\begin{cases} -2x + 3y = 5 \\ x + 5y = 2 \end{cases}$ 2. $\begin{cases} -2a - b = -2 \\ 2a - 3b = 7 \end{cases}$ 3. $\begin{cases} -2r - 3s = 2 \\ 7r - s = -3 \end{cases}$

Solve each system using the substitution method. Check your solution.

4. $\begin{cases} x - y = 2 \\ x + y = -4 \end{cases}$ 5. $\begin{cases} a + b = -2 \\ a - b = 4 \end{cases}$ 6. $\begin{cases} 2x + y = -1 \\ 3x - y = 11 \end{cases}$

7. $\begin{cases} -u + 3v = 1 \\ 2u - v = 5 \end{cases}$ 8. $\begin{cases} 4c + 3d = 12 \\ -2c + d = 1 \end{cases}$ 9. $\begin{cases} -2y + 5z = 16 \\ y + 3z = 3 \end{cases}$

10. $\begin{cases} 2x + 2y = 12 \\ x - y = 2 \end{cases}$ 11. $\begin{cases} x + 2y = 3 \\ 3x - \frac{2}{3}y = 1 \end{cases}$ 12. $\begin{cases} 6x + 8y = 9 \\ 3x + 2y = \frac{3}{2} \end{cases}$

13. $\begin{cases} \frac{1}{2}r + \frac{1}{2}s = -3 \\ \frac{1}{3}r + \frac{1}{2}s = 4 \end{cases}$ 14. $\begin{cases} \frac{3}{5}m + \frac{2}{3}n = -3 \\ \frac{3}{5}m - \frac{1}{3}n = 4 \end{cases}$ 15. $\begin{cases} a + \frac{2}{3}b = -3 \\ \frac{1}{3}a + \frac{2}{3}b = 1 \end{cases}$

16. $\begin{cases} 2m + 3n = 2 \\ 2m + 5n = 1 \end{cases}$ 17. $\begin{cases} 3c + 2d = 2 \\ -2c + 2d = 5 \end{cases}$ 18. $\begin{cases} 3a + 2b = 2 \\ 2a = 2b + 1 \end{cases}$

Simplify each expression. (2.6 Skill A)

19. $3k - 4k + 5 - k$ 20. $5 - n + 6 - 2n$ 21. $z - 2z - 3z - 1 - 2 - 3$

Simplify each expression. (2.6 Skill B)

22. $-2(d + 5) - 4(d - 1)$ 23. $-(n - 3) - (n + 1)$ 24. $4(a - 3) - 4(a + 3)$

25. $\frac{12w + 3}{3}$ 26. $\frac{-18b + 12}{-2}$ 27. $\frac{24b + 50}{-2}$

Answers for Lesson 5.2 Skill B

Try This Example

a. $(8, -5)$ b. $\left(\frac{39}{17}, \frac{2}{17}\right)$

1. $\begin{cases} -2x + 3y = 5 \\ x = -5y + 2 \end{cases}$

2. $\begin{cases} -2a + 2 = b \\ 2a - 3b = 7 \end{cases}$

3. $\begin{cases} -2r - 3s = 2 \\ 7r + 3 = s \end{cases}$

4. $(-1, -3)$

5. $(1, -3)$

6. $(2, -5)$

7. $\left(\frac{16}{5}, \frac{7}{5}\right)$

8. $\left(\frac{9}{10}, \frac{14}{5}\right)$

9. $(-3, 2)$

10. $(4, 2)$

11. $\left(\frac{3}{5}, \frac{6}{5}\right)$

12. $\left(-\frac{1}{2}, \frac{3}{2}\right)$

13. $(-42, 36)$

14. $\left(\frac{25}{9}, -7\right)$

15. $\left(-6, \frac{9}{2}\right)$

16. $\left(\frac{7}{4}, -\frac{1}{2}\right)$

17. $\left(-\frac{3}{5}, \frac{19}{10}\right)$

18. $\left(\frac{3}{5}, \frac{1}{10}\right)$

19. $-2k + 5$

20. $-3n + 11$

21. $-4z - 6$

22. $-6d - 6$

23. $-2n + 2$

24. -24

25. $4w + 1$

26. $9b - 6$

27. $-12b - 25$

★ **Advanced Exercises**

SKILL C APPLICATIONS *Using the substitution method in solving real-world problems*

Focus: *Students use a system of linear equations in two variables to model a problem, and then use the substitution method to solve the system.*

EXAMPLE 1

PHASE 1: Presenting the Example

PHASE 2: Providing Guidance for TRY THIS

Some students may have difficulty understanding how the equations were derived. A verbal explanation such as the following may be helpful:

drops of water	should be	75	times	drops of vitamins
↓	↓	↓	↓	↓
w	$=$	75	$\times$	v

$$w = 75v$$

drops of water	combined with	drops of vitamins	should be	1520 drops
↓	↓	↓	↓	↓
w	$+$	v	$=$	1520

$$w + v = 1520$$

EXAMPLE 2

PHASE 1: Presenting the Example

Draw a rectangle to represent the garden. Explain how the perimeter formula can be obtained by adding each of the sides and rewriting the expression as shown at right.

$$\begin{aligned} P &= l + w + l + w \\ &= 2l + 2w \\ &= 2(l + w) \end{aligned}$$

PHASE 2: Providing Guidance for TRY THIS

Students learned about inequalities quite recently, and so they may be inclined to translate *more than* as $>$. Remind them that this symbol is a translation of *is more than.* That is, a verb is required in order to use it. Suggest that students consider an alternative way to convey the same information, such as *The length of the garden is twice its width plus 10 feet.* Then it becomes more obvious that $+$ is the appropriate symbol to use.

PHASE 3: ASSESSMENT AND CLOSURE for Lesson 5.2

$$\begin{cases} x + y = -15 \\ x - y = 4 \end{cases}$$

- Give students the system at right. Have them describe four different ways that one equation can be rewritten to isolate a variable. [$\mathbf{x = -y - 15; y = -x - 15; x = y + 4; y = x - 4}$] Have them choose one of these equations, perform a substitution, and find the solution of the system. [**(−5.5, −9.5)**]
- Students probably have surmised that there are systems in which a variable is not easily isolated, and so performing a substitution is not convenient. In Lesson 5.3, they will learn an algebraic method to be used in such situations.

☞ *For a Lesson 5.2 Quiz, see* Assessment Resources *page 49.*

When there are two unknowns in a real-world problem, you can solve it using a system of two equations.

EXAMPLE 1

A veterinary assistant needs to prepare a mixture of a vitamin solution for parakeets. The ratio is 1 drop of vitamin solution to 75 drops of water. If the assistant needs to prepare 1520 drops of the combined solution, how much of each separate solution will she need?

> Solution

Write two equations.

1. Let v represent the number of drops of vitamin solution needed and let w represent the number of drops of water needed.
2. Write and solve a system of equations. Use the substitution method.

$$\begin{cases} w = 75v \\ w + v = 1520 \end{cases} \rightarrow 75v + v = 1520 \rightarrow v = 20$$

If $v = 20$, then $w = 75(20)$, or 1500.

3. **Check:**
$$w = 75v \qquad\qquad w + v = 1520$$
$$1500 \overset{?}{=} 75(20) \checkmark \qquad 1500 + 20 \overset{?}{=} 1520 \checkmark$$

The assistant needs to put 20 drops of vitamin solution into 1500 drops of water.

TRY THIS Rework Example 1 given that the assistant wants to make 5472 drops of combined solution.

Some geometry and perimeter problems can also be solved using systems.

EXAMPLE 2

A landscaper is creating a rectangular garden whose length is 10 feet more than twice its width. The available border material is 200 feet long. Find the dimensions of the garden.

> Solution

Use a formula.

1. Let P represent the perimeter of the garden, l the length, and w the width.
2. Recall that the perimeter P of a rectangle with length l and width w is given by $P = 2(l + w)$.
3. The length can be expressed as $10 + 2w$.

$$\begin{cases} 2(l + w) = 200 \\ l = 10 + 2w \end{cases} \rightarrow 2[(10 + 2w) + w] = 200 \rightarrow w = 30$$

If $w = 30$, then $l = 10 + 2(30)$, or 70.

4. **Check:**
$$2(l + w) = 200 \qquad\qquad l = 10 + 2w$$
$$2(70 + 30) \overset{?}{=} 200 \checkmark \qquad 70 \overset{?}{=} 10 + 2(30) \checkmark$$

The dimensions of the garden are 70 feet in length and 30 feet in width.

TRY THIS Rework Example 2 given that the length is 15 feet more than the width and there are 450 feet of border material available.

EXERCISES

KEY SKILLS

Refer to Example 1. Write a system of equations to describe each situation.

1. The ratio is 1 drop of vitamin solution to 75 drops of water. The assistant wants a total of 1824 drops of the combined solution.
2. The ratio is 1 drop of vitamin solution to 82 drops of water. The assistant wants a total of 1520 drops of the combined solution.

PRACTICE

Use a system of equations to solve each problem. Check your solution.

3. Rework Example 1 given that the assistant wants to make 5396 drops of a solution that is made by adding 1 drop of vitamin solution to 70 drops of water.
4. Rework Example 2 that the width is 20 feet shorter than three times length and there are 1200 feet of border material available.
5. April sold 75 tickets to a school play and collected a total of $495. If adult tickets cost $8 each and child tickets cost $5 each, how many adult tickets and how many child tickets did she sell?
6. In the last basketball game, George scored a total of 16 points on 11 baskets. He made only 2-point field goals and 1-point free throws. How many field goals and how many free throws did George make?
7. If two solutions are mixed in the ratio of 1.5 to 1 and together they make 600 milliliters of a new solution, how many milliliters of each solution were used?
8. Two bags of fertilizer and 3 bags of peat moss together weigh 140 pounds. If one bag of fertilizer weighs the same as 2 bags of peat moss, how much does each bag weigh?

MIXED REVIEW APPLICATIONS

Solve each problem. Check your solution. (4.5 Skill C)

9. A chemist has two solutions, A and B, that are to be mixed to make a third solution, C, that is no more than 18% salt. Solution A is 12% salt and solution B is 20% salt. How much of each solution is needed to make 250 milliliters of solution C?
10. Mr. Lyons wants to add money to his account, which currently contains $8000, so that two years from now he will have at least $10,000 in the account. If the bank pays 5% simple interest annually, how much should he add?

Answers for Lesson 5.2 Skill C

Try This Example 1
72 drops of vitamins and 5400 drops of water

Try This Example 2
105 feet by 120 feet

1. $\begin{cases} w = 75v \\ w + v = 152 \cdot 12 \end{cases}$

2. $\begin{cases} w = 82v \\ w + v = 152 \cdot 10 \end{cases}$

3. about 76 drops of vitamins and 5320 drops of water
4. 155 feet by 445 feet
5. 40 adult tickets and 35 child tickets
6. 5 field goals and 6 free throws
7. 240 milliliters and 360 milliliters
8. 20 pounds of peat moss and 40 pounds of fertilizer
9. at least 62.5 milliliters of the 12% solution and no more than 187.5 milliliters of the 20% solution
10. at least $1090.91

Solving Systems of Equations by Elimination

Focus: *Students solve a system of linear equations in two variables by adding the equations to eliminate one variable. They also learn to apply the method after first transforming one of the equations by multiplication.*

EXAMPLE 1

PHASE 1: Presenting the Example

Ask students to write a statement of the Addition Property of Equality. Discuss their responses, ultimately formalizing their statements as follows:

For all real numbers a, b, and c, if $a = b$, then $a + c = b + c$.

Now display the following statement:

For all real numbers a, b, c, and d, if $a = b$ <u>and $c = d$</u>, then $a + c = b + d$.

Ask students why this statement must also be true. Lead them to see that it is a direct consequence of the Addition Property of Equality and the Substitution Principle, as shown below.

$$a = b \qquad \longleftarrow \textit{hypothesis}$$
$$c = d \qquad \longleftarrow \textit{hypothesis}$$
$$a + c = b + c \qquad \longleftarrow \textit{Apply the Addition Property of Equality.}$$
$$a + c = b + d \qquad \longleftarrow \textit{Since } c = d, \textit{ substitute } d \textit{ for } c \textit{ on the right side.}$$

Display the system in Example 1, which is $\begin{cases} 2x + 3y = 4 \\ 3x - 3y = 11 \end{cases}$. Below it, write

$(2x + 3y) + (3x - 3y) = 4 + 11$. Ask students to justify this new equation. Elicit the response that the property just proved (*If $a = b$ and $c = d$, then $a + c = b + d$.*) was applied to the two equations in the system. Have them simplify the new equation to obtain $5x = 15$. Students should observe that they have arrived at a simple equation in one variable that is easily solved: $x = 3$. Point out that, from here, they can continue to solve the system as they did before, substituting 3 for x in either equation to obtain $y = -\dfrac{2}{3}$.

EXAMPLE 2

PHASE 2: Providing Guidance for **TRY THIS**

When using the elimination method, students generally have the greatest difficulty deciding which of the two equations to transform and what number to use as a multiplier. You may find it helpful to spend some time concentrating on this skill alone, temporarily freeing students from the additional task of solving the system. You can use Skill A Exercises 1 through 3 on the facing page for this purpose.

5.3 LESSON

Solving Systems of Equations by Elimination

SKILL A *Using the elimination method with one multiplier*

The system at right could be solved by substitution, but isolating one of the variables would involve several steps. Using the *elimination method* shortens the process.

$$\begin{cases} 2x + 3y = 4 \\ 3x - 3y = 11 \end{cases}$$

EXAMPLE 1   Solve $\begin{cases} 2x + 3y = 4 \\ 3x - 3y = 11 \end{cases}$ using the elimination method.

▸ **Solution**

Notice that the two coefficients of *y* are opposites.

1. Combine the two equations. The variable *y* is eliminated. Solve for *x*.

$$\begin{array}{r} 2x + 3y = 4 \\ 3x - 3y = 11 \\ \hline 5x + 0 \quad 15 \end{array} \to 5x = 15 \to x = 3$$

2. Replace *x* with 3 in either original equation. Solve for *y*.

$$3(3) - 3y = 11 \to y = -\tfrac{2}{3}$$

The solution to the system is $\left(3, -\tfrac{2}{3}\right)$.

TRY THIS   Solve using elimination.   **a.** $\begin{cases} -2x + 2y = 2 \\ 5x - 2y = 4 \end{cases}$ **b.** $\begin{cases} 3x + 5y = 20 \\ -3x + 7y = 16 \end{cases}$

If opposite terms do not already exist within a system, you can apply the Multiplication Property of Equality to the original equation(s) to create them.

EXAMPLE 2   Solve $\begin{cases} 2x + 5y = 14 \\ 6x + 7y = 10 \end{cases}$ using the elimination method.

▸ **Solution**

1. Multiply each side of the first equation by −3.

$$\begin{cases} (-3)(2x + 5y) = (-3)14 \\ 6x + 7y = 10 \end{cases} \to \begin{cases} -6x - 15y = -42 \\ 6x + 7y = 10 \end{cases}$$

2. Combine the two equations so that *x* is eliminated.

$$\begin{array}{r} -6x - 15y = -42 \\ 6x + 7y = 10 \\ \hline 0 - 8y = -32 \end{array}$$

3. Solve for *y*. $y = 4$

4. Replace *y* with 4 in either original equation. Solve for *x*.

$$2x + 5(4) = 14 \to x = -3$$

The solution to the system is $(-3, 4)$.

TRY THIS   Solve using elimination.   **a.** $\begin{cases} -8x + 3y = 34 \\ 4x + 7y = -34 \end{cases}$ **b.** $\begin{cases} 3x + 4y = 10 \\ 5x + 12y = 38 \end{cases}$

EXERCISES

KEY SKILLS

Explain how you would eliminate the specified variable. Do not solve the system.

1. $\begin{cases} -6x + 5y = 4 \\ 2x - 4y = 8 \end{cases}$; eliminate *x*   **2.** $\begin{cases} -3a + b = 4 \\ 2a + 3b = -3 \end{cases}$; eliminate *b*   **3.** $\begin{cases} 2r + 6s = 0 \\ 2r - 4s = -3 \end{cases}$; eliminate *r*

PRACTICE

Use elimination to solve each system of equations.

4. $\begin{cases} 2x - 5y = 1 \\ -2x - 3y = -9 \end{cases}$ **5.** $\begin{cases} 4m + 5n = 1 \\ -7m - 5n = 2 \end{cases}$ **6.** $\begin{cases} -4u + 3v = 7 \\ 4u + 3v = 23 \end{cases}$

7. $\begin{cases} 7p - 2q = 6 \\ 7p + 2q = 8 \end{cases}$ **8.** $\begin{cases} 2x + 6y = 0 \\ 2x - 4y = -3 \end{cases}$ **9.** $\begin{cases} 5c - 6d = 18 \\ 4c + 3d = 17 \end{cases}$

10. $\begin{cases} -3c - 8d = 4 \\ 6c - 2d = 1 \end{cases}$ **11.** $\begin{cases} -3c - 8d = 0 \\ 7c + 2d = 0 \end{cases}$ **12.** $\begin{cases} 5m - 8n = 0 \\ -10m + 7n = 0 \end{cases}$

13. $\begin{cases} \tfrac{1}{4}x + \tfrac{1}{3}y = 3 \\ \tfrac{1}{2}x - \tfrac{1}{6}y = 1 \end{cases}$ **14.** $\begin{cases} \tfrac{c}{3} + \tfrac{d}{3} = 5 \\ \tfrac{c}{2} - \tfrac{d}{9} = 2 \end{cases}$ **15.** $\begin{cases} \tfrac{c}{2} - \tfrac{d}{3} = 1 \\ \tfrac{c}{4} + \tfrac{d}{5} = 0 \end{cases}$

In Exercises 16–18, let *a*, *b*, and *c* represent real numbers. Solve each system for *x* and *y*.

16. $\begin{cases} 3ax - 3y = 1 \\ ax + 4y = 6 \end{cases}$, $a \neq 0$ **17.** $\begin{cases} 3x + 4by = 2 \\ 5x + by = 7 \end{cases}$, $b \neq 0$ **18.** $\begin{cases} 3x + 4y = c \\ 6x + 5y = c \end{cases}$

19. Critical Thinking Show that the system $\begin{cases} 3x + 2y = -c \\ 6x + 4y = 1 - 2c \end{cases}$ has no solution for any value of *c*.

MIXED REVIEW

Evaluate each expression. (1.2 Skill C)

20. $ad - bc$, given $a = 3$, $b = -3$, $c = 5$, and $d = 6$

21. $ad - bc$, given $a = -4$, $b = 14$, $c = -2$, and $d = 7$

Graph each equation. (3.5 Skill B)

22. $-2x + 3y = 9$ **23.** $3x + 5y = -10$ **24.** $2x - 4y = 8$

Answers for Lesson 5.3 Skill A

Try This Example 1

a. $(2, 3)$ **b.** $\left(\tfrac{5}{3}, 3\right)$

Try This Example 2

a. $(-5, -2)$ **b.** $(-2, 4)$

1. Multiply the second equation by 3.

2. Multiply the first equation by −3.

3. Multiply the first equation by −1 or multiply the second equation by −1.

4. $(3, 1)$

5. $(-1, 1)$

6. $(2, 5)$

7. $\left(1, \tfrac{1}{2}\right)$

8. $\left(-\tfrac{9}{10}, \tfrac{3}{10}\right)$

9. $\left(4, \tfrac{1}{3}\right)$

10. $\left(0, -\tfrac{1}{2}\right)$

11. $(0, 0)$

12. $(0, 0)$

13. $(4, 6)$

14. $(6, 9)$

15. $\left(\tfrac{12}{11}, -\tfrac{15}{11}\right)$

★**16.** $\left(\tfrac{22}{15a}, \tfrac{17}{15}\right)$

★**17.** $\left(\tfrac{26}{17}, -\tfrac{11}{17b}\right)$

★**18.** $\left(-\tfrac{c}{9}, \tfrac{c}{3}\right)$

19. Multiply the first equation by −2.

$$\begin{cases} 3x + 2y = -c \\ 6x + 4y = 1 - 2c \end{cases} \to$$

$$\begin{cases} -6x - 4y = 2c \\ 6x + 4y = 1 - 2c \end{cases}$$

Thus, $0 = 1$. Because this is false for all values of *c*, the given system has no solution.

20. 33

21. 0

22.

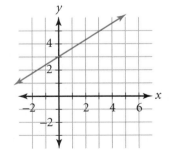

23.

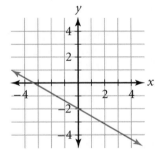

24.

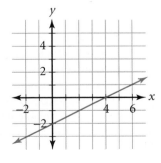

★ **Advanced Exercises**

Focus: *Students solve a system of linear equations in two variables by first multiplying to transform both equations, and then adding.*

EXAMPLE 1

PHASE 1: Presenting the Example

Display the system at right. Then display the following:

$$\begin{cases} 2a - 4b = -9 \\ -3a + 2b = 8 \end{cases}$$

> *To solve this system, I would multiply each side of the _____ equation by the integer _____. Then I could eliminate the variable _____, because _____ + _____ = 0.*

Tell students to copy the statement, filling in the blanks. Discuss their responses. [**The desired response is as follows: To solve this system, I would multiply each side of the** *second* (**or** *bottom*) **equation by the integer** 2. **Then I could eliminate the variable** *b*, **because** $-4b + 4b = 0$.]

As the students watch, change the 4 in the displayed system to 5, as shown at right. Again ask students to copy the statement and fill in the blanks.

$$\begin{cases} 2a - 5b = -9 \\ -3a + 2b = 8 \end{cases}$$

This time, there is no single integer multiplication that yields the desired result, and so students will likely be puzzled. Have them brainstorm ways to resolve the dilemma. Lead them to see that two multiplications are needed.

Now discuss the solution outlined in the textbook. Conclude the discussion by having students write a statement, similar to the one above, describing the method that was used. [**sample: To solve this system, we multiplied each side of the first equation by 3 and each side of the second equation by 2. Then we were able to eliminate the variable** *a*, **because** $6a + (-6a) = 0$.]

PHASE 2: Providing Guidance for TRY THIS

Have students work in pairs. One student in each pair should solve the system by eliminating the variable *r*, while the other solves by eliminating *s*. Then have the partners compare their results. Be sure they observe that, if the work is done correctly, both methods lead to exactly the same solution.

EXAMPLE 2

PHASE 1: Presenting the Example

PHASE 2: Providing Guidance for TRY THIS

You may wish to review with students the *standard form* and *slope-intercept form* of two-variable equations, which they studied in Chapter 3. Then note that, in Example 2, the variables were aligned by writing both equations in standard form. As students begin the Try This exercise, point out that the first equation is already close to slope-intercept form, and so a reasonable first step would be to rewrite each equation in slope-intercept form.

To solve a system of equations, you may need to choose a multiplier for each equation in the system.

EXAMPLE 1 Solve $\begin{cases} 2a - 5b = -9 \\ -3a + 2b = 8 \end{cases}$ using the elimination method.

▶ **Solution**

1. Multiply each side of the first equation by 3.
   Multiply each side of the second equation by 2.
   $$\begin{cases} 2a - 5b = -9 \\ -3a + 2b = 8 \end{cases} \rightarrow \begin{cases} 3(2a - 5b) = 3(-9) \\ 2(-3a + 2b) = 2(8) \end{cases} \rightarrow \begin{cases} 6a - 15b = -27 \\ -6a + 4b = 16 \end{cases}$$

2. Add the like terms of the two equations to eliminate a. Solve for b.
   $$-11b = -11 \rightarrow b = 1$$

3. Replace b with 1 in either original equation. Solve for a.
   If $b = 1$, then $2a - 5(1) = -9$. So, $a = -2$.

4. **Check:** Check using the other original equation. $-3(-2) + 2(1) = 8$ ✔

   The solution to the system is $(-2, 1)$.

TRY THIS Solve $\begin{cases} 6r + 8s = 4 \\ 9r + 10s = 7 \end{cases}$ using the elimination method.

Before adding the like terms of the equations, be sure that the corresponding variables are aligned.

EXAMPLE 2 Solve $\begin{cases} 2c + 3d = -10 \\ -2d + 3c = -2 \end{cases}$ using the elimination method.

▶ **Solution**

1. Align c terms and d terms.
   $$\begin{cases} 2c + 3d = -10 \\ -2d + 3c = -2 \end{cases} \rightarrow \begin{cases} 2c + 3d = -10 \\ 3c - 2d = -2 \end{cases}$$

2. Choose appropriate multipliers.
   $$\begin{cases} 2c + 3d = -10 \\ 3c - 2d = -2 \end{cases} \rightarrow \begin{cases} 3(2c + 3d) = 3(-10) \\ -2(3c - 2d) = -2(-2) \end{cases} \rightarrow \begin{cases} 6c + 9d = -30 \\ -6c + 4d = 4 \end{cases}$$

3. Combine the two equations so that c is eliminated. Solve for d.
   $$13d = -26 \rightarrow d = -2$$

   Replace d with -2 in either original equation. Solve for c.
   If $d = -2$, then $2c + 3(-2) = -10$. So, $c = -2$.

4. **Check:** $-2(-2) + 3(-2) \stackrel{?}{=} -2$ ✔

   The solution to the system is $(-2, -2)$.

TRY THIS Solve $\begin{cases} 4y = 11 - 4x \\ 3x - 5 = -5y \end{cases}$ using the elimination method.

EXERCISES

KEY SKILLS

Explain how you would eliminate the specified variable. Do not solve the system.

1. $\begin{cases} -3x + 2y = 4 \\ 2x + 4y = -3 \end{cases}$; eliminate x
2. $\begin{cases} -3a + 5b = -2 \\ 5a + 3b = -1 \end{cases}$; eliminate b
3. $\begin{cases} -7r - 6s = 10 \\ 2r - 5s = 7 \end{cases}$; eliminate r

PRACTICE

Use the elimination method to solve each system of equations.

4. $\begin{cases} 2x - 3y = -10 \\ 3x + 2y = -2 \end{cases}$
5. $\begin{cases} -3x + 7y = -14 \\ -4x - 2y = 4 \end{cases}$
6. $\begin{cases} 4c - 2d = 0 \\ 3c - 3d = 0 \end{cases}$

7. $\begin{cases} 5a - 2b = 0 \\ -4a - 5b = 0 \end{cases}$
8. $\begin{cases} 4f - 7g = 1 \\ -3f + 5g = 0 \end{cases}$
9. $\begin{cases} x - 3y = 17 \\ 3x + 2y = 18 \end{cases}$

10. $\begin{cases} 4a - 3b = 7 \\ 5b - 4a = -1 \end{cases}$
11. $\begin{cases} -2r + 2s = 4 \\ 5s - 7r = 0 \end{cases}$
12. $\begin{cases} -2h + 2n = 17 \\ 3n + 3h = 58 \end{cases}$

13. $\begin{cases} 4m - 7n = 37 \\ 2n + 3m = 6 \end{cases}$
14. $\begin{cases} -5a = 18 - 4z \\ -3z + 2a = -16 \end{cases}$
15. $\begin{cases} 5q - 8p = 7 \\ 4p = 7q - 17 \end{cases}$

16. $\begin{cases} \frac{x}{3} - \frac{y}{4} = 1 \\ \frac{x}{2} + \frac{y}{3} = 1 \end{cases}$
17. $\begin{cases} \frac{a}{5} - \frac{c}{3} = 1 \\ \frac{a}{4} - \frac{c}{5} = 1 \end{cases}$
18. $\begin{cases} \frac{m}{7} + \frac{n}{2} = 1 \\ \frac{m}{6} + \frac{n}{3} = 1 \end{cases}$

19. **Critical Thinking** Let m and n represent fixed real numbers.
    Solve $\begin{cases} 2x + my = 1 \\ 3x + ny = 1 \end{cases}$ for x and y. State restrictions on variables.

MIXED REVIEW

Are the given lines parallel? Justify your response. (3.8 Skill A)

20. $3x - 11y = 4$ and $-5x + 7y = 21$
21. $-4x - y = 14$ and $-8x - 2y = 0$

Are the given lines perpendicular? Justify your response. (3.8 Skill B)

22. $2.5x - y = 13$ and $-0.8x - 2y = 34$
23. $4x + 7y = -2$ and $7x + 4y = 3$

Answers for Lesson 5.3 Skill B

Try This Example 1
$\left(\dfrac{4}{3}, -\dfrac{1}{2}\right)$

Try This Example 2
$\left(\dfrac{35}{8}, -\dfrac{13}{8}\right)$

1. Multiply the first equation by 2 and the second equation by 3.
2. Multiply the first equation by -3 and the second equation by 5.
3. Multiply the first equation by 2 and the second equation by 7.
4. $(-2, 2)$
5. $(0, -2)$
6. $(0, 0)$
7. $(0, 0)$
8. $(-5, -3)$
9. $(8, -3)$
10. $(4, 3)$
11. $(5, 7)$
12. $\left(\dfrac{65}{12}, \dfrac{167}{12}\right)$

13. $(4, -3)$
14. $\left(\dfrac{10}{7}, \dfrac{44}{7}\right)$
15. $(1, 3)$
★16. $\left(\dfrac{42}{17}, -\dfrac{12}{17}\right)$
★17. $\left(\dfrac{40}{13}, -\dfrac{15}{13}\right)$
★18. $\left(\dfrac{14}{3}, \dfrac{2}{3}\right)$
19. $\left(\dfrac{m - n}{3m - 2n}, \dfrac{1}{3m - 2n}\right)$; $m \neq \dfrac{2}{3}n$
20. not parallel because the slopes are unequal
21. parallel because both lines have slope 4 and different y-intercepts
22. perpendicular because the product of the slopes is -1
23. not perpendicular because the product of the slopes is not -1

★ **Advanced Exercises**

Focus: *Students use a system of linear equations in two variables to model a problem, and then use the elimination method to solve the system.*

EXAMPLES 1 and 2

PHASE 1: Presenting the Examples

Students encountered "solution problems" like these when studying linear inequalities in one variable. Consequently, some may perceive that it is possible to solve these problems using a one-variable linear equation. A summary of a one-variable solution for the problem in Example 1 follows.

Let a represent the required number of milliliters of solution A.
Then $(500 - a)$ represents the required number of milliliters of solution B.
Write an equation in a to represent the conditions of the problem.

amount of sugar in solution A	combined with	amount of sugar in solution B	must be equal to	amount of sugar in solution C
$\downarrow$	$\downarrow$	$\downarrow$	$\downarrow$	$\downarrow$
$0.10a$	$+$	$0.25(500 - a)$	$=$	$0.20(500)$

The solution of $0.10a + 0.25(500 - a) = 0.20(500)$ is $a = 166\frac{2}{3}$. So the required quantity of solution A is $166\frac{2}{3}$ milliliters, and the required quantity of solution B is $\left(500 - 166\frac{2}{3}\right)$ milliliters, or $333\frac{1}{3}$ milliliters.

Discuss both solution methods with the students. Then ask them to analyze the advantages and disadvantages of each. If they indicate that they simply feel more comfortable with one method, ask if they can explain why.

PHASE 2: Providing Guidance for TRY THIS

After students have completed the Try This exercise for Example 2, point out that the algebraic solution provides mathematical verification for a common-sense observation: Since there is a 10% concentration of sugar in solution A and a 25% concentration in solution B, any combination of the two solutions has to result in a concentration somewhere between 10% and 25%. The required 2% concentration falls far below this range.

PHASE 3: ASSESSMENT AND CLOSURE for Lesson 5.3

- Ask students to write an original example of a system that they would solve by the elimination method rather than by substitution. Have them describe the steps they would take to solve it. [**Answers will vary.**]

- Students have now learned the two algebraic methods for solving systems that are presented in this course: substitution and elimination. In Lesson 5.4, they will investigate special types of systems.

☞ *For a Lesson 5.3 Quiz, see* Assessment Resources *page 50.*

Making a table can be helpful when you need to set up a system of equations for a problem.

Also, it is often helpful to change the fractions or decimals in an equation to integers. To do this, multiply both sides of the equation by a common denominator.

EXAMPLE 1

A chemist has two solutions, A and B, that are to be mixed to make a third solution, C. The data is shown at right. How many whole milliliters of solutions A and B are needed to make solution C?

solution A: 10% sugar
solution B: 25% sugar
solution C: 500 mL of 20% sugar

▶ Solution

1. Let a and b represent the numbers of milliliters of solution A and B, respectively, that are needed. Organize the data in a table.

Make a table.

	A	+	B	=	C
amount of solution	a		b		500
amount of sugar	$0.10a$		$0.25b$		$(0.20)500$

2. Write and solve a system of equations. Multiply the second equation by 20 so that it contains only integers. Multiply the first equation by -2.

$$\begin{cases} a + b = 500 \\ 0.10a + 0.25b = 0.20(500) \end{cases} \rightarrow \begin{cases} a + b = 500 \\ 2a + 5b = 2000 \end{cases} \rightarrow \begin{cases} -2a - 2b = -1000 \\ 2a + 5b = 2000 \end{cases}$$

Combine the equations to get $3b = 1000$. Thus, $b = 333\frac{1}{3}$, and then $a = 166\frac{2}{3}$.

The problem asks that the answers be rounded to whole milliliters. So, the chemist needs 167 mL of solution A and 333 mL of solution B.

TRY THIS

Rework Example 1 given that solution A is 12% sugar, solution B is 24% sugar, and 480 milliliters of solution C is 18% sugar.

EXAMPLE 2

Refer to Example 1. Can A and B be mixed to make 500 mL of a new solution, D, that is 5% sugar? Explain.

▶ Solution

The system is a modification of the system in Example 1.

$$\begin{cases} a + b = 500 \\ 0.10a + 0.25b = 0.05(500) \end{cases} \rightarrow \begin{cases} a + b = 500 \\ 2a + 5b = 500 \end{cases} \rightarrow 3b = -500 \rightarrow b = -\frac{500}{3}$$

Because a and b represent physical amounts, neither may be negative. So, the mixture cannot be made.

TRY THIS

Solution A is 10% sugar and solution B is 25% sugar. Can A and B be mixed to create 500 mL of a new solution, C, that is 2% sugar? Explain.

EXERCISES

KEY SKILLS

Refer to Example 1.

1. Represent the data below in a table.
   solution A: 18% sugar   solution B: 26% sugar
   solution C: 350 milliliters of 20% sugar

2. Write $0.18a + 0.26b = (0.20)(350)$ as an equation with only integer coefficients.

PRACTICE

Solve each problem using a system of equations.

3. Rework Example 1 with the data below.

   solution A: 5% sugar   solution B: 10% sugar

   solution C: 450 milliliters of 8% sugar

4. A shopkeeper wants to make 5 pounds of a candy mix from peppermint candy and lemon-lime candy. How many pounds of each candy should be put into the mix? Use the data below.
   peppermint: $0.75 per pound   lemon-lime: $0.50 per pound
   mix: sells for $0.65 per pound

5. A motorist drives for 8 hours and travels 390 miles. For some of that time, the speed is 55 miles per hour and for the rest of that time the speed is 45 miles per hour. How long does the motorist drive at each speed?

6. Is it possible to make a 600 milliliter solution that is 32% salt from a solution that is 30% salt and another solution that is 25% salt? Justify your response.

MID-CHAPTER REVIEW

7. Use a graph to solve the system of equations at right. (5.1 Skill A) $\begin{cases} y = x - 5 \\ y + x = 3 \end{cases}$

Solve each system using the substitution method. (5.2 Skills A and B)

8. $\begin{cases} y = 2x \\ -3x + 2y = 2 \end{cases}$

9. $\begin{cases} x + y = -8 \\ -3x + 2y = 9 \end{cases}$

Solve each system using the elimination method. (5.3 Skills A and B)

10. $\begin{cases} x - 2y = -3 \\ 3x + 2y = 7 \end{cases}$

11. $\begin{cases} 2x + 3y = 5 \\ 3x + 2y = 5 \end{cases}$

Answers for Lesson 5.3 Skill C

Try This Example 1
240 milliliters each of solution A and solution B

Try This Example 2
$a = 766\frac{2}{3}$, $b = -266\frac{2}{3}$; because b is negative and the amount of Solution A needed is more than the total amount, the solution cannot be made.

1.

A	B	C
a	b	350
$0.18a$	$0.26b$	$(0.20)350$

2. $18a + 26b = 7000$, or $9a + 13b = 3500$
3. 180 milliliters of solution A and 270 milliliters of solution B
4. 3 pounds of peppermint candy and 2 pounds of lemon-lime candy
5. 5 hours at 45 miles per hour and 3 hours at 55 miles per hour

6. No. Solving the system indicates that 840 milliliters of the 30% solution and -240 milliliters of the 25% solution are required. Neither amount is realistic since 840 milliliters is more than the total and -240 milliliters is less than none.
7. $(4, -1)$
8. $(2, 4)$
9. $(-5, -3)$
10. $(1, 2)$
11. $(1, 1)$

Classifying Systems of Equations

> **SKILL A**   *Classifying a system of equations as consistent or inconsistent*

Focus: *Students classify a system as consistent or inconsistent by its solution and by examining the relationship between the equations in the system.*

EXAMPLES 1 and 2

PHASE 1: Presenting the Examples

Introduce the lesson by reviewing with students the different ways that two lines in a plane can occur. Draw coordinate axes on the board and illustrate each possibility after students propose them. You many need to guide students by asking them for general ways two lines can occur *besides* intersecting in one point. [**The lines could be parallel or the same line.**] Ask students which of the graphs indicate solutions to a system. [**The intersecting lines (one solution) and the same line (infinite solutions).**]

Tell students that they can classify systems of equations based on their solutions. Introduce the term *consistent* for a system of equations with *one or more* solutions and the term *inconsistent* for a system of equations with *no* solution. You may wish to discuss the common usage of the terms *consistent* and *inconsistent* to help students remember their mathematical meanings. Next, display the graphs below, which illustrate the three possibilities and two classifications.

Consistent Systems   **Inconsistent Systems**

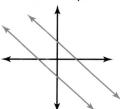

Tell students that they do not have to graph a system to classify it. In Example 1, the system is solved algebraically and classified as consistent by its unique solution. Example 2 is different because the system is not solved. It is simply rewritten in slope-intercept form and examined. Because the slopes are equal, the lines must be the same lines or parallel lines. Because the *y*-intercepts are different, the lines are parallel, forming an inconsistent system.

PHASE 2: Providing Guidance for TRY THIS

The Try This for Example 1 can most easily be solved using the elimination method; however, other methods can be used.

In the Try This for Example 2, encourage students to first focus on comparing the slopes of each equation. If the slopes are different, the system has one unique solution (consistent). If the slopes are the same, students must then compare the *y*-intercepts to determine whether the system has infinitely many solutions (consistent) or no solution (inconsistent).

A more detailed classification and examination of systems will be introduced in Skill B.

5.4 LESSON — Classifying Systems of Equations

SKILL A ▸ Classifying a system of equations as consistent or inconsistent

When a system of equations has at least one solution, the system is called **consistent**. When a system has no solution, it is called **inconsistent**.

EXAMPLE 1 Is $\begin{cases} -3x - y = -2 \\ 7x + 2y = -10 \end{cases}$ a consistent or inconsistent system? Explain.

▸ Solution
You can solve this system by graphing, substitution, or elimination. Here we solve using the elimination method.

$$\begin{cases} (2)(-3x - y) = (2)(-2) \\ 7x + 2y = -10 \end{cases} \rightarrow \begin{cases} -6x - 2y = -4 \\ 7x + 2y = -10 \end{cases} \rightarrow x = -14$$

If $x = -14$, then $y = 44$. The system has a unique solution, $(-14, 44)$. So the lines intersect at one point, $(-14, 44)$, and the system is consistent.

TRY THIS Is $\begin{cases} -16x + 4y = -8 \\ 12x - 3y = 6 \end{cases}$ a consistent or inconsistent system? Explain.

EXAMPLE 2 Is $\begin{cases} 5x + 10y = -2 \\ 2x + 4y = 10 \end{cases}$ a consistent or inconsistent system? Explain.

▸ Solution
Rewrite each equation in slope-intercept form.

$$\begin{cases} 5x + 10y = -2 \\ 2x + 4y = 10 \end{cases} \rightarrow \begin{cases} 10y = -2 - 5x \\ 4y = 10 - 2x \end{cases} \rightarrow \begin{cases} y = -\frac{1}{2}x - \frac{1}{5} \\ y = -\frac{1}{2}x + \frac{5}{2} \end{cases}$$

The slopes are equal but the y-intercepts, $-\frac{1}{5}$ and $\frac{5}{2}$, are different. So, the lines must be parallel. Because parallel lines do not intersect, the system has no solution and it is inconsistent.

TRY THIS Is $\begin{cases} -3x - y = -2 \\ 6x + 2y = 10 \end{cases}$ a consistent or inconsistent system? Explain.

EXERCISES

KEY SKILLS

Write the equations in each system in slope-intercept form.

1. $\begin{cases} x + y = 4 \\ x + 2y = 6 \end{cases}$
2. $\begin{cases} 3x - y = 2 \\ 6x - 2y = 8 \end{cases}$
3. $\begin{cases} -7x - 5y = -3 \\ y - 4x = -12 \end{cases}$

PRACTICE

Identify each system as consistent or inconsistent.

4. $\begin{cases} 3x - y = -5 \\ 9x - 3y = 1 \end{cases}$
5. $\begin{cases} 3u - v = 7 \\ -2u - 5v = 2 \end{cases}$
6. $\begin{cases} 6r - 2s = 6 \\ 2r + 5s = 0 \end{cases}$

7. $\begin{cases} 4u - 14v = -3 \\ -2u + 7v = 3 \end{cases}$
8. $\begin{cases} 4p - 4q = -5 \\ -4q + 7p = 3 \end{cases}$
9. $\begin{cases} -5g - 35h = 15 \\ -h + 7g = -12 \end{cases}$

Does each system have a unique solution? If it does, find it.

10. $\begin{cases} 2x + 3y = 6 \\ 2x - 3y = 6 \end{cases}$
11. $\begin{cases} 2x - y = -5 \\ 6x - 3y = 1 \end{cases}$
12. $\begin{cases} 2x - y = 3 \\ -2x + y = 2 \end{cases}$

13. $\begin{cases} -3m + 4n = 8 \\ -4m + 6n = 9 \end{cases}$
14. $\begin{cases} 5m - 5n = 8 \\ -4m + 4n = 9 \end{cases}$
15. $\begin{cases} 7p - 2q = -3 \\ -9p + 3q = 5 \end{cases}$

16. Consider $\begin{cases} ax - 2y = -2 \\ -5x + 3y = 7 \end{cases}$. Find a such that the system has no solution.

17. Consider $\begin{cases} 2x - 7y = 12 \\ -3x + dy = 11 \end{cases}$. Find d such that the system has exactly one solution.

18. Critical Thinking Let a, b, c, and d be real numbers. Solve $\begin{cases} ax + by = e \\ cx + dy = f \end{cases}$ for x. How does $ad - bc$ determine whether there is a solution for x?

MIXED REVIEW

Use intercepts to graph each equation. (3.5 Skill A)

19. $4x - 5y = 20$
20. $-x + 5y = 5$
21. $2x - 7y = 7$

Use the slope and y-intercept to graph each equation. (3.5 Skill B)

22. $y = -\frac{3}{4}x + 1$
23. $y = \frac{5}{4}x - 3$
24. $y = \frac{1}{5}x + 2.5$

Answers for Lesson 5.4 Skill A

Try This Example 1
consistent

Try This Example 2
$\begin{cases} -3x - y = -2 \\ 6x + 2y = 10 \end{cases} \rightarrow$

$\begin{cases} y = -3x + 2 \\ y = -3x + 5 \end{cases}$

The slopes, -3, are equal but the y-intercepts, 2 and 5, are different. So the lines must be parallel. The system is inconsistent.

1. $\begin{cases} y = -x + 4 \\ y = -\frac{1}{2}x + 3 \end{cases}$

2. $\begin{cases} y = 3x - 2 \\ y = 3x - 4 \end{cases}$

3. $\begin{cases} y = -\frac{7}{5}x + \frac{3}{5} \\ y = 4x - 12 \end{cases}$

4. inconsistent
5. consistent
6. consistent
7. inconsistent
8. consistent
9. consistent
10. $(3, 0)$
11. no unique solution
12. no unique solution
★ 13. $\left(-6, -\frac{5}{2}\right)$
★ 14. no unique solution
15. $\left(\frac{1}{3}, \frac{8}{3}\right)$
★ 16. $a = \frac{10}{3}$
★ 17. $d = \frac{21}{2}$
18. $x = \frac{ed - bf}{ad - bc}$
If $ad - bc \neq 0$, then x is a real number. If $ad - bc = 0$, the value for x is undefined.

19.

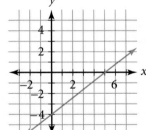

20.

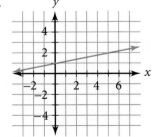

For answers to Exercises 21–24, see **Additional Answers.**

★ **Advanced Exercises**

Focus: *Students classify a system as independent or dependent by solving the system and examining the resulting solution or algebraic statement.*

EXAMPLE 1

PHASE 1: Presenting the Example

Solicit students' assistance in drawing on the board the three types of systems that were discussed in Skill A. [**intersecting lines, the same line, and parallel lines**] Then draw a concept map like the one shown at right to illustrate the two classifications introduced in Skill A, consistent and inconsistent. Ask students how many types of systems can be classified as consistent [**2**] and how many can be classified as inconsistent [**1**].

Introduce the term *independent* for a system that has *exactly* one solution and the term *dependent* for a system that has an *infinite* number of solutions. Add these terms to the concept map as shown at right. Point to each of the three types of systems that are now fully distinguished. You may want to add information about the solutions, slopes, *y*-intercepts, and graphs for each type of system as summarized in the table on the bottom of the textbook page.

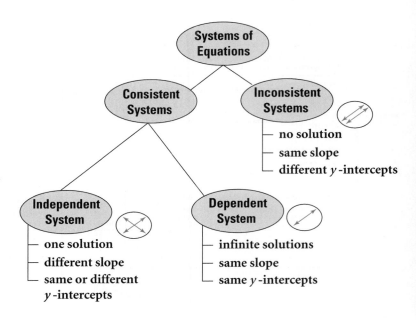

EXAMPLE 2

PHASE 2: Providing Guidance for TRY THIS

Students may have trouble interpreting algebraic solutions such as $0 = 0$ and $3 = 0$. Remind them that if the statement is always true, there are an infinite number of solutions to the system. If the statement is always false, there are no solutions.

PHASE 3: ASSESSMENT AND CLOSURE for Lesson 5.4

- Give students the partial system that is shown at right. Ask them to fill in the box with an equation that will make the system inconsistent, and then to fill it in with an equation that will make the system dependent.

- Students have now completed their study of systems of linear equations. In the lessons that follow, they will learn about linear inequalities, and then investigate *systems* of linear inequalities.

👉   *For a Lesson 5.4 Quiz, see* Assessment Resources *page 51.*

$\begin{cases} 2x - 5y = 9 \\ \boxed{} \end{cases}$

[samples:
inconsistent—
$4x - 10y = 10$;
dependent—
$4x - 10y = 18$]

Classifying a system of equations as dependent or independent

In Skill A, you learned that a system of equations with at least one solution is called consistent. Consistent systems can be further classified as **independent** and **dependent**. An independent system has exactly one solution. A dependent system has infinitely many solutions. All of the equations in a dependent system simplify to the same equation. Therefore, the lines of the equations coincide at every point and are considered to be the same line.

EXAMPLE 1 Classify the system $\begin{cases} y = 4 - x \\ 3x - y = 8 \end{cases}$ as specifically as possible.

▶ **Solution**

Solve the system using the substitution method by substituting $4 - x$ for y in the second equation.

$$\begin{cases} y = 4 - x \\ 3x - y = 8 \end{cases} \rightarrow 3x - (4 - x) = 8 \rightarrow x = 3$$

If $x = 3$, then $y = 4 - 3$, or 1. The solution is (3, 1). Because the solution is one point, (3, 1), the system is consistent and independent.

TRY THIS Classify the system $\begin{cases} -4x + 4y = 4 \\ 4x - y = -4 \end{cases}$ as specifically as possible.

EXAMPLE 2 Classify the system $\begin{cases} 6u + 10v = -2 \\ 3u + 5v = -1 \end{cases}$ as specifically as possible.

▶ **Solution**

Multiply the second equation by -2 and then combine the equations.

Use the elimination method.

$$\begin{cases} 6u + 10v = -2 \\ (-2)(3u + 5v) = (-2)(-1) \end{cases} \rightarrow \begin{cases} 6u + 10v = -2 \\ -6u + -10v = 2 \end{cases} \rightarrow 0 = 0 \ ✔ \text{ True}$$

The result is a statement that is true for any values of u or v, so all of the solutions of one equation are also solutions of the other equation. The equations describe the same line, so the system is consistent and dependent.

TRY THIS Classify the system $\begin{cases} 7r + 3s = 4 \\ 14r + 6s = 10 \end{cases}$ as specifically as possible.

If the result had been a contradictory statement such as $1 = 0$, then the system would have no solution and thus would be inconsistent.

Type of system	Solutions	Slopes	y-intercepts	Graphs
Consistent and independent	one	different	either same or different	
Consistent and dependent	infinitely many	same	same	
Inconsistent and independent	none	same	different	

204 Chapter 5 Solving Systems of Equations and Inequalities

EXERCISES

KEY SKILLS

Graph a pair of equations that satisfies the given condition.

1. There is no solution.
2. There are infinitely many solutions.

PRACTICE

Classify each system as specifically as possible.

3. $\begin{cases} x = 7 - 2y \\ 3x - 2y = -11 \end{cases}$

4. $\begin{cases} x = 14 - 5y \\ 2x - y = -5 \end{cases}$

5. $\begin{cases} 2x - y = 3 \\ 5x - 2y = 10 \end{cases}$

6. $\begin{cases} 3x + y = -3 \\ 2x - y = -7 \end{cases}$

7. $\begin{cases} x + y = 2 \\ 2x + 2y = 4 \end{cases}$

8. $\begin{cases} y - 2x = 7 \\ 2y - 4x = 14 \end{cases}$

9. $\begin{cases} 4x - y = 19 \\ 2x + 3y = -1 \end{cases}$

10. $\begin{cases} 5x + 2y = -7 \\ x + 3y = 9 \end{cases}$

11. $\begin{cases} -3.5m - 3n = -2 \\ 7m + 6n = 5 \end{cases}$

12. $\begin{cases} -28m - 31n = -20 \\ 56m + 62n = 40 \end{cases}$

13. $\begin{cases} 24x + 31y = 50 \\ 72x + 93y = 150 \end{cases}$

14. $\begin{cases} 75a - 33b = -120 \\ 25a - 11b = -50 \end{cases}$

15. $\begin{cases} -3x + 2y = 6 \\ -x + \frac{2}{3}y = 2 \end{cases}$

16. $\begin{cases} \frac{5}{2}u + 5v = 0 \\ \frac{2}{3}u + \frac{4}{3}v = -\frac{5}{3} \end{cases}$

17. $\begin{cases} \frac{2x}{4} - \frac{2y}{7} = -1 \\ \frac{3x}{4} - \frac{3y}{7} = 4 \end{cases}$

Critical Thinking Consider $\begin{cases} 5x + by = e \\ 3x + 2y = 10 \end{cases}$. Find b and e such that the system has:

18. no solution.
19. infinitely many solutions.

Critical Thinking Consider $\begin{cases} 5x + 4y = -2 \\ cx + 2y = f \end{cases}$. Find c and f such that:

20. the graphs coincide.
21. the graphs are distinct and parallel.

MIXED REVIEW

Solve using the substitution method. (5.2 Skill B)

22. $\begin{cases} 2x - y = -7 \\ 3x + y = -3 \end{cases}$

23. $\begin{cases} y = \frac{2}{3}x - 4 \\ 5y - x = 1 \end{cases}$

Solve using the elimination method. (5.3 Skill B)

24. $\begin{cases} 7x - 5y = 10 \\ 3x - 2y = 6 \end{cases}$

25. $\begin{cases} 7r - 2s = 9 \\ 2r + 3s = -1 \end{cases}$

Answers for Lesson 5.4 Skill B

Try This Example 1
consistent and independent

Try This Example 2
inconsistent

Answers for Exercises 1 and 2 may vary. Sample answers are given.

1.
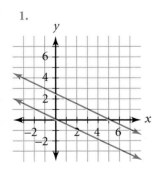

2.

3. consistent and independent
4. consistent and independent
5. consistent and independent
6. consistent and independent
7. consistent and dependent
8. consistent and dependent

9. consistent and independent
10. consistent and independent
11. inconsistent
12. consistent and dependent
13. consistent and dependent
14. inconsistent
15. consistent and dependent
16. inconsistent
17. inconsistent
18. $b = \frac{10}{3}$ and $e \neq \frac{50}{3}$
19. $b = \frac{10}{3}$ and $e = \frac{50}{3}$

20. $c = 2.5$ and $f = -1$
21. $c = 2.5$ and $f \neq -1$
22. $(-2, 3)$
23. $(9, 2)$
24. $(10, 12)$
25. $(1, -1)$

★ **Advanced Exercises**

Graphing Linear Inequalities in Two Variables

Focus: *Students graph a linear inequality in two variables as a region in the coordinate plane, with the graph of a linear equation as its boundary.*

EXAMPLE

PHASE 1: Presenting the Example

Tell students to graph $y = x$ in the coordinate plane. [**The graph is given at right.**] Discuss their results. Display the graph for all to see.

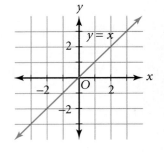

Ask students to write the coordinates of one point that is located to the left of the line $y = x$. Make a list of several of their answers. [**samples: $(2, 3), (−3, 1), (−2, −1)$**] Ask students what relationship between the x-coordinate and the y-coordinate is common to all the ordered pairs. Elicit the response that, in each ordered pair, the y-coordinate is greater than the x-coordinate. That is, each ordered pair is a solution to $y > x$.

Now display graph **1** below. Note that the shading covers *all* points in which the y-coordinate is greater than the x-coordinate. Therefore, the shaded region is the graph of $y > x$. Point out that the line $y = x$ must be drawn to show the *boundary* of the graph, but note that the line is dashed because it is not actually part of the solution. Ask students how they think the graph of $y \geq x$ would be different. Lead them to conclude that the graph would be similar, but the line would be solid. Display graph **2**, which illustrates this situation.

1 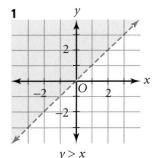 **2** 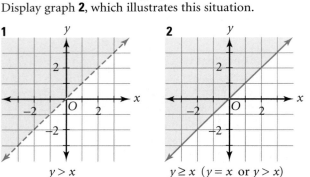 **3** 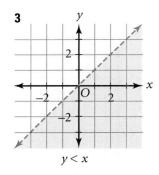 **4**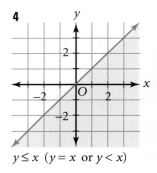

$y > x$ $y \geq x$ ($y = x$ or $y > x$) $y < x$ $y \leq x$ ($y = x$ or $y < x$)

Ask students to graph $y < x$, and then $y \leq x$. When they have finished, display graphs **3** and **4** so they can check their work. Then discuss the Example.

PHASE 2: Providing Guidance for **TRY THIS**

In Try This Exercise b, watch for students who shade the region to the left of the boundary simply because $4 < y + 2x$ contains the symbol $<$. Stress the importance of using the coordinates of points to determine which region contains the solutions. A sample test is shown below.

The point $(0, 0)$ is to the left of the boundary:	$4 < y + 2x$ $4 \overset{?}{<} 0 + 2(0)$ $4 < 0$ *False*	→	The point $(3, 3)$ is to the right of the boundary:	$4 < y + 2x$ $4 \overset{?}{<} 3 + 2(3)$ $4 < 9$ *True*	→ The solutions lie to the right of the boundary.

5.5 LESSON
Graphing Linear Inequalities in Two Variables

SKILL A ▶ *Graphing linear inequalities*

When the equal symbol in a linear equation is replaced with an inequality symbol, a *linear inequality* is formed. A **linear inequality in two variables**, x and y, is any inequality that can be written in one of the forms below.

$$y < mx + b \qquad y \leq mx + b$$
$$y > mx + b \qquad y \geq mx + b$$

A solution to a linear inequality in two variables is any ordered pair that makes the inequality true. For example, $(1, 3)$ is a solution to $y \geq x + 1$ because $3 \geq 1 + 1$. The set of all possible solutions of a linear inequality is called the **solution region**.

Graphing a linear inequality in x and y

Step 1. Isolate y on one side of the inequality.

Step 2. Substitute an equal symbol (=) for the inequality symbol in the given inequality. Graph the resulting linear equation to form the **boundary line**. For $\leq$ or $\geq$, use a solid line. For $<$ or $>$, use a dashed line. The dashed line shows that the points on the line are not included in the solution region.

Step 3. Shade the region that contains the solutions of the inequality. For $y >$, shade above the line. For $y <$, shade below the line. (Note: To graph one-variable inequalities on a coordinate plane, $x < a$ or $x > a$, the boundary is a vertical line and the shading is on the left or right, respectively.)

EXAMPLE Graph the solution region of $x - 2y < -6$.

▶ **Solution**

Recall that when you divide an inequality by a negative number you need to reverse the inequality sign.

1. $x - 2y < -6$ ←—— *Write y in terms of x.*
    $-2y < -x - 6$
    $y > \frac{1}{2}x + 3$ ←—— *Divide each side by −2.*

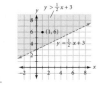

2. Graph $y = \frac{1}{2}x + 3$ with a dashed line.

3. Because $y > \frac{1}{2}x + 3$, the solution region should lie above the line. Shade this area.

4. Check: Choose any point from the shaded region to test. For example, choose $(1, 6)$:
$$1 - 2(6) < -6 \;\rightarrow\; -11 < -6 \;✔$$

TRY THIS   Graph the solution region.   **a.** $y \leq -2x + 4$ **b.** $4 < y + 2x$

KEY SKILLS

Write y in terms of x.

1. $5x + 2y \geq 10$ **2.** $4x - 3y < 9$

Is the given point above or below the graph of the given equation?

3. $A(0, 0)$; $3x - 2y = 3$ **4.** $B(1, -3)$; $-2x + 3y = 6$

PRACTICE

Graph the solution region of each inequality on a coordinate plane.

5. $x \geq 0$ **6.** $x > 4$ **7.** $y < 2$

8. $y \leq 0$ **9.** $x + y \geq -3$ **10.** $y - x \geq -5$

11. $3x + y < 0$ **12.** $-2x + y > 0$ **13.** $x + 3y > 2$

14. $x + 3y > -4$ **15.** $2x - y < 2$ **16.** $3x - y > 4$

17. $2x + 5y < 10$ **18.** $-5x + 4y \geq 12$ **19.** $2x + 4y \leq 0$

20. $3x - 4y \geq 0$ **21.** $-7x + 2y < 14$ **22.** $5x + 3y \geq 15$

Write a linear inequality in x and y that meets the given conditions.

23. The boundary contains $(2, 3)$ and $(5, 7)$ and the graph of the solution region contains $(3, 8)$.

24. The boundary contains $(-2, 3)$ and $(6, 9)$ and the graph of the solution region contains $(-2, -4)$ but does not contain the boundary.

Let $y > mx + b$, where m and b are real numbers and $m \neq 0$.
(*Hint:* If the boundary contains the origin, then $(0, 0)$ is a solution.)

25. Find m and b such that the boundary contains the origin and the solution region is the half-plane that contains $(1, 5)$.

26. Find m and b such that the boundary contains the origin and the solution region is the half-plane that contains $(-2, 7)$.

MIXED REVIEW

Solve each compound inequality. Graph the solution on a number line.
(4.6 Skill A)

27. $2x + 5 > 10 - 3x$ and $x - 3 < 9 - 3x$ **28.** $-2(a + 1) < -8$ and $-a + 4 \leq 16 - 3a$

Solve each compound inequality. Graph the solution on a number line.
(4.6 Skill B)

29. $y + 5 > 9$ or $3y + 4 < -8 - 3y$ **30.** $3(y + 5) \leq 9$ or $3(y - 4) \geq -3y$

Answers for Lesson 5.5   Skill A

Try This Example

a.

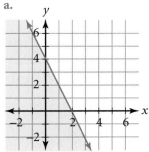

b.

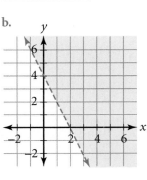

1. $y \geq -\dfrac{5}{2}x + 5$

2. $y > \dfrac{4}{3}x - 3$

3. above

4. below

5.

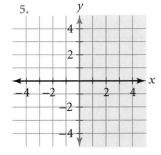

6.

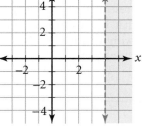

7.

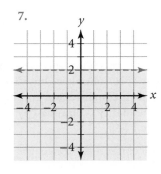

8.

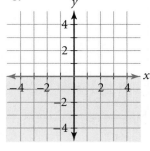

9.

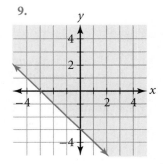

For answers to Exercises 10–30, see Additional Answers.

Focus: *Students graph a linear inequality in two variables when restrictions are imposed on the variables. They also investigate situations in which restrictions arise naturally from the context of a problem.*

EXAMPLE 1

PHASE 1: Presenting the Example

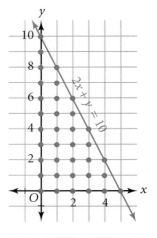

Review with students the idea of a *replacement set*. Then display $2t + 5 < 11$. Ask students to solve given that the replacement set for t is the set of all real numbers. [**t may be any real number less than 3.**] Ask how the answer would differ if the replacement set were restricted to just the set of whole numbers. [**The only possible values of t are 0, 1, and 2.**]

Point out that it is also possible to restrict the replacement set for each variable in a two-variable inequality. Then discuss Example 1.

PHASE 2: Providing Guidance for **TRY THIS**

Students can use the line with equation $2x + y = 10$ to check their work. As shown at right, all solutions must be points with whole-number coordinates that lie either on this line or to the left of it.

EXAMPLE 2

PHASE 1: Presenting the Example

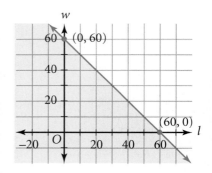

Introduce the problem that is given in Example 2. Ask students to graph $l + w \leq 60$ on a coordinate plane, graphing values of l along the horizontal axis and values of w along the vertical axis. [**The graph is given at right.**] Ask if their graphs accurately represent all possible lengths and widths for the rectangle. Lead them to see that, because the variables represent physical measures, the ordered pairs with negative coordinates have no meaning. Then discuss the graph on the textbook page.

PHASE 3: ASSESSMENT AND CLOSURE for Lesson 5.5

- Have students compare graphing a linear inequality in two variables to graphing a linear inequality in one variable. [**samples: One-variable inequalities are graphed on a number line; two-variable inequalities are graphed in the coordinate plane. The graph of a one-variable inequality is bounded by a point; the graph of a two-variable inequality is bounded by a line.**]

- In Lesson 5.6, students will connect the concept of a linear inequality in two variables to linear systems, which is the unifying theme of this chapter.

☞ *For a Lesson 5.5 Quiz, see* Assessment Resources *page 52.*

When restrictions are placed on one or both variables, the graph of a solution will not necessarily be a shaded region as in Lesson 5.5 Skill A.

EXAMPLE 1

Make an organized list.

Given that x and y are whole numbers, graph all solutions to $x + 2y \leq 8$.

▶ Solution

Isolate y in the given inequality: $y \leq -\frac{1}{2}x + 4$.

Examine possible x-values.

If $x = 0$, then $y \leq 4$. So, $y = 0, 1, 2, 3$, or 4.

If $x = 1$, then $y \leq 3\frac{1}{2}$. So, $y = 0, 1, 2$, or 3.

If $x = 2$, then $y \leq 3$. So, $y = 0, 1, 2$, or 3.

If $x = 3$, then $y \leq 2\frac{1}{2}$. So, $y = 0, 1$, or 2.

⋮

If $x = 7$, then $y \leq 0.5$. So, $y = 0$.

If $x = 8$, then $y \leq 0$. So, $y = 0$.

If $x = 9$, $y \leq -\frac{1}{2}$. ✘ There are no whole numbers less than or equal to $-\frac{1}{2}$, so x cannot equal 9.

The graph is shown at right.

TRY THIS   Given that x and y are whole numbers, graph all solutions to $2x + y \leq 10$.

In many problems, there are natural restrictions on variables. For example, if variables represent distances or measurements, their values must be nonnegative. The graphs of many real-world equations are often restricted to Quadrant I because it is the only quadrant that contains nonnegative values for both variables.

EXAMPLE 2

Use a formula.

Mickie and Michelle may use up to 120 feet of fencing to enclose a rectangular space. Draw a graph to represent the possible lengths and widths of the rectangular space.

▶ Solution

1. The perimeter of a rectangle, P, is given by the formula $P = 2(l + w)$, where l represents length and w represents width.

2. Write and solve an inequality in l and w.
   $2(l + w) \leq 120$, or $l + w \leq 60$

3. Graph $l + w = 60$ using the intercepts: $(0, 60)$ and $(60, 0)$. Because the inequality is *less than or equal to*, shade *below* the line. Since length and width must be nonnegative, shade only in Quadrant I.

TRY THIS   Rework Example 2 given that Mickie and Michelle may use up to 150 feet of fencing.

EXERCISES

KEY SKILLS

Given that x and y are whole numbers, graph all solutions to each equation below.

1. $x + y \leq 5$ 2. $x + y < 7$ 3. $y < x - 3$

4. Sketch the triangular region in the plane determined by $A(0, 0)$, $B(4, 0)$, and $C(0, 6)$ as the vertices of the triangle.

PRACTICE

Graph the solution of each inequality in two variables.

5. Graphically represent all pairs of whole numbers whose sum is less than 6.

6. One whole number is twice another. Represent all such pairs of numbers whose sum is less than or equal to 12.

7. Suppose that you have 5 blue chips and 5 red chips. Graphically represent the possibilities for a collection of blue chips and red chips if the collection contains less than 7 chips altogether.

8. Rework Example 2 given that 180 feet of fencing is available.

9. Graphically represent all lengths and widths of rectangles whose perimeter is less than 140 units. For which rectangles will the length be twice the width?

10. Graphically represent all pairs of whole numbers that are each greater than or equal to 4 and whose sum is less than 12.

11. Use a graph to show there are no pairs of whole numbers that are each greater than or equal to 5 and whose sum is less than 9.

MIXED REVIEW APPLICATIONS

How many milliliters of solution A and of solution B are needed to make solution C? (5.3 Skill C)

12. solution A: 20% salt
    solution B: 25% salt
    solution C: 200 milliliters of 24% salt

13. solution A: 5% salt
    solution B: 15% salt
    solution C: 300 milliliters of 8% salt

Answers for Lesson 5.5 Skill B

Try This Example 1

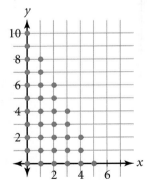

Try This Example 2

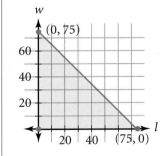

1.

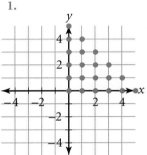

3.

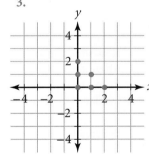

2.

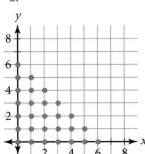

4.

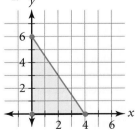

For answers to Exercises 5–13, see Additional Answers.

Graphing Systems of Linear Inequalities in Two Variables

SKILL A   *Graphing systems of linear inequalities*

Focus: *Students graph a system of linear inequalities in two variables when the boundaries of the inequalities are given in slope-intercept form.*

EXAMPLE

PHASE 1: Presenting the Example

Display the two inequalities at right. Have half of the class graph the first, while the other half graphs the second.

$$y \le -2x + 5$$
$$y < x - 1$$

Prepare two transparencies, if possible, each displaying the graph of one of the inequalities, as shown in **1** and **2** below. Draw the coordinate plane and lines in black, but use two different colors—perhaps yellow and blue—for the shading. Use these graphs to check students' work.

1 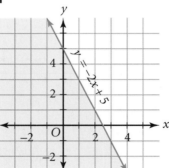   **2** 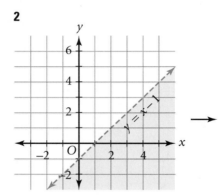   **3**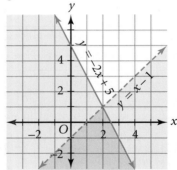

Now slide the graphs together so that the axes coincide, as shown in graph **3**. The region where the graphs overlap will appear to be a color that is a combination of the colors from graphs **1** and **2**. (For example, if you shaded graphs **1** and **2** in yellow and blue, the overlap will appear green.) Ask students what the overlap region represents. Lead them to

the conclusion that it is the graph of all solutions to the system $\begin{cases} y \le -2x + 5 \\ y < x - 1 \end{cases}$. Then

discuss the solution of the Example as outlined on the textbook page.

PHASE 2: Providing Guidance for **TRY THIS**

The Try This exercises illustrate the fact that two intersecting lines in the coordinate plane actually bound the solutions of several different systems. To check students' understanding, display the graph at right and ask them to

write the system whose solutions it represents. $\left[\begin{cases} y < -\frac{1}{2}x + 1 \\ y \ge -2x - 1 \end{cases} \right]$ Then ask

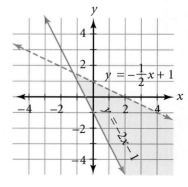

them to write and graph yet another system whose solutions are bounded by these same lines. [**Answers will vary.**]

5.6 LESSON
Graphing Systems of Linear Inequalities in Two Variables

SKILL A *Graphing systems of linear inequalities*

A **system of linear inequalities in two variables** is a set of two or more linear inequalities in those variables. A solution to a system of such inequalities is any ordered pair that makes all of the inequalities in the system true. The graph of the solution of a system is the intersection or common region of the combined graphs of the inequalities.

$x \geq 0$ $y \geq 0$ $\begin{cases} x \geq 0 \\ y \geq 0 \end{cases}$

EXAMPLE Graph the solution region of the system $\begin{cases} y \leq -2x + 5 \\ y > x - 1 \end{cases}$.

Solution

1. Graph each boundary line.
   $y = -2x + 5$ with a solid line $y = x - 1$ with a dashed line
2. Shade above or below the boundary as indicated by the inequality sign.
   Shade *below* the line for $y \leq -2x + 5$ because y is *less than* $-2x + 5$.
   Shade *above* the line for $y > x - 1$ because y is *greater than* $x - 1$.
3. The solution region is the area of overlap that is shown below at right.

Steps **1** and **2**   Step **3**

TRY THIS Graph the solution region of each system.
  a. $\begin{cases} y \geq -\frac{1}{2}x + 1 \\ y \geq -2x - 1 \end{cases}$ **b.** $\begin{cases} y < -\frac{1}{2}x + 1 \\ y > -2x - 1 \end{cases}$

EXERCISES

KEY SKILLS

Shade the described region.

1. Below the graph of $x + y = 8$ and above the graph of $x + y = 5$

2. Above the graph of $y = \frac{2}{3}x$ and below the graph of $y = 4$

3. What part of the coordinate plane is determined by $\begin{cases} x < 0 \\ y > 0 \end{cases}$?

PRACTICE

Graph the solution region of each system of inequalities.

4. $\begin{cases} y \leq 2x \\ x \geq 0 \end{cases}$ 5. $\begin{cases} y < -x \\ x \geq 0 \end{cases}$ 6. $\begin{cases} y < 2 \\ y > 4x - 1 \end{cases}$

7. $\begin{cases} y \leq -3x + 2 \\ y < -2 \end{cases}$ 8. $\begin{cases} y \leq 2x - 3 \\ x > 0 \end{cases}$ 9. $\begin{cases} y > -2x + 1 \\ x < 0 \end{cases}$

10. $\begin{cases} y < x + 2 \\ y \geq x - 2 \end{cases}$ 11. $\begin{cases} y < 2x + 1 \\ y \geq 2x - 3 \end{cases}$ 12. $\begin{cases} y > -2x + 1 \\ y \leq 3x - 1 \end{cases}$

13. $\begin{cases} y \leq 3x - 1 \\ y > -2x - 3 \end{cases}$ 14. $\begin{cases} y > -2x + 1 \\ y > 3x - 2 \end{cases}$ 15. $\begin{cases} y < -2x + 1 \\ y > -2x - 2 \end{cases}$

16. $\begin{cases} y \geq -3x \\ y < 2x + 3 \\ x \geq 0 \\ y \geq 0 \end{cases}$ 17. $\begin{cases} y > 2x - 1 \\ y < 3x + 2 \\ x \geq 0 \\ y \geq 0 \end{cases}$ 18. $\begin{cases} y > -x + 3 \\ y \leq x + 2 \\ x \geq 0 \\ y \geq 0 \end{cases}$

19. **Critical Thinking** Let $y > m_1 x + b_1$ and $y < m_2 x + b_2$, where m_1, b_1, m_2, and b_2 are real numbers. Under what conditions placed on m_1, b_1, m_2, and b_2 will the solution region be an infinite strip with parallel boundaries?

MIXED REVIEW

Classify each system as specifically as possible. (5.4 Skill B)

20. $\begin{cases} -9c + 3d = 10 \\ 12c - 4d = 1 \end{cases}$ 21. $\begin{cases} 7p + 7q = -3 \\ 6p + 6q = 5 \end{cases}$ 22. $\begin{cases} -3m - 2n = 2 \\ 12m + 8n = -8 \end{cases}$

Answers for Lesson 5.6 Skill A

Try This Example

a.
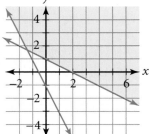

b.
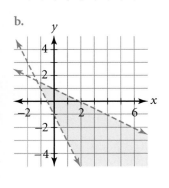

1.
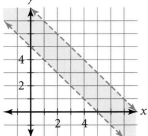

2.
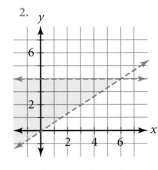

3. the second quadrant

4.
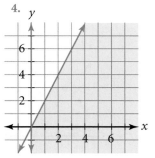

5.
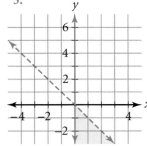

6.
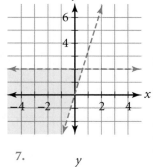

7.
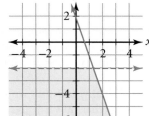

For answers to Exercises 8–22, see Additional Answers.

Focus: *Students graph a system of linear inequalities in two variables when the boundaries of the inequalities are given in standard form. They also graph systems whose solutions lie on a single line and systems with no solution.*

EXAMPLE 1

PHASE 2: Providing Guidance for TRY THIS

Students probably will notice that, in general, the solutions of $y > mx + b$ lie *above* the bounding line. Similarly, the solutions of $y < mx + b$ lie *below* the bounding line. However, remind them to always pick a point in the region that they graphed and to check that its coordinates satisfy the system. This serves as a safeguard against any errors made when rewriting the equations.

EXAMPLE 2

PHASE 1: Presenting the Example

Display graph **1** below. Tell students that it is the graph of all solutions to the system $\begin{cases} y \geq mx + a \\ y \leq mx + b \end{cases}$. Point out that the bounding lines of the inequalities have exactly the same slope. Ask students to describe the region that contains the solutions to the system. [**sample: It is an infinitely long strip bounded by two parallel lines.**] Have them suppose that the system were changed so that the line $y = mx + a$ moved closer to the line $y = mx + b$, but did not touch it or pass it. Ask what would happen to the region of solutions. [**It would become narrower.**] Display graph **2** to illustrate this idea.

1 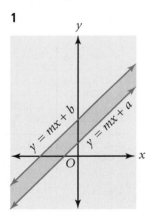 **2** 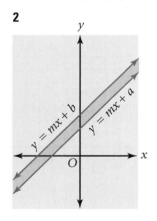 **3** 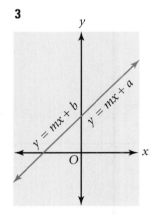 **4**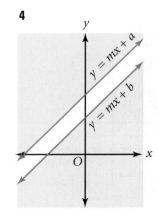

Now tell students to imagine that the line $y = mx + a$ were exactly the same line as $y = mx + b$. Ask how this would affect the region of solutions. [**It would be just a line.**] Display graph **3** to illustrate this situation.

Finally, ask what would happen to the solution region if the line $y = mx + a$ were moved further upward, above the line $y = mx + b$. Students should perceive that the solution region would be empty, and so the system would be inconsistent. Display graph **4** to illustrate. Then discuss Example 2.

Graphing systems of inequalities given in standard form

A **linear inequality in standard form** is any inequality of one of the forms below. In the inequalities below, A and B cannot both be 0.

$$Ax + By < C \qquad Ax + By \leq C \qquad Ax + B > C \qquad Ax + By \geq C$$

If you are given a system of inequalities in standard form, they will need to be rewritten in slope-intercept form before graphing.

EXAMPLE 1 Graph the solution region of $\begin{cases} 2x - 5y \leq 0 \\ x + 4y \leq -8 \end{cases}$.

▶ **Solution**

Rewrite each inequality so that y is isolated.

$$\begin{cases} 2x - 5y \leq 0 \\ x + 4y \leq -8 \end{cases} \rightarrow \begin{cases} y \geq \frac{2}{5}x \\ y \leq -\frac{1}{4}x - 2 \end{cases}$$

Graph each boundary line.
The solution is the region on or above the graph of $y = \frac{2}{5}x$ and on or below that of $y = -\frac{1}{4}x + 2$.

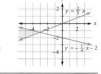

TRY THIS Graph the solution region of $\begin{cases} x + y \geq 2 \\ -2x + y \geq -3 \end{cases}$.

A system of inequalities may have no solution or may have a line as its solution. If there is no solution to the system, you will discover this when you graph the system.

EXAMPLE 2 Graph the solution region of $\begin{cases} y \leq \frac{2}{5}x + 2 \\ -2x + 5y \geq 20 \end{cases}$.

▶ **Solution**

The inequality $-2x + 5y \geq 20$ can be rewritten as $y \geq \frac{2}{5}x + 4$.
The graphs of the two inequalities are shown at right. Notice that there is no common region. This means that this system has no solution.

TRY THIS Graph the solution region. **a.** $\begin{cases} y \leq \frac{1}{3}x + 2 \\ 3y - 6 \geq x \end{cases}$ **b.** $\begin{cases} y < \frac{1}{3}x + 2 \\ 3y - 9 \geq x \end{cases}$

EXERCISES

KEY SKILLS

Write each system so that y is expressed in terms of x in both inequalities.

1. $\begin{cases} y \geq \frac{2}{5}x \\ 3x - 5y \leq 2 \end{cases}$ 2. $\begin{cases} 2x + 7y < 0 \\ -2x - y < 3 \end{cases}$ 3. $\begin{cases} -3x + 4y \geq 1 \\ -2x + 3y < 2 \end{cases}$

PRACTICE

Graph the solution region of the system of inequalities.

4. $\begin{cases} y < 2x \\ 2x - y < 3 \end{cases}$ 5. $\begin{cases} y \leq 3x \\ 2x + y < 5 \end{cases}$ 6. $\begin{cases} x + y \leq 5 \\ x + y > 3 \end{cases}$

7. $\begin{cases} x + y \leq 6 \\ x - y \geq 2 \end{cases}$ 8. $\begin{cases} 2x + y \leq -2 \\ x - 2y < 4 \end{cases}$ 9. $\begin{cases} x - 3y \geq -3 \\ 2x + y < 4 \end{cases}$

10. $\begin{cases} 3x - 5y < 15 \\ 2x + 3y \leq 6 \end{cases}$ 11. $\begin{cases} -5x - 2y \leq 10 \\ 2x + 7y \leq 14 \end{cases}$ 12. $\begin{cases} 4x - 5y < 20 \\ 2x - 7y > 14 \end{cases}$

13. Use algebraic reasoning to show that the graph of the solution of $\begin{cases} 2x + 3y \leq -1 \\ 2x + 3y \geq -1 \end{cases}$ is a line.

14. Use algebraic reasoning to show that the solution to the system at right is a single point. $\begin{cases} y \geq x + 1 \\ y \leq x + 1 \\ y \geq -x + 1 \\ y \leq -x + 1 \end{cases}$

15. **Critical Thinking** How must a and b be related if $\begin{cases} y \geq -2x + a \\ y \leq -2x + b \end{cases}$ has a solution?

16. **Critical Thinking** Use algebraic reasoning to show that the graph of the solution to the system at right is a strip with parallel boundaries for all real numbers a. $\begin{cases} y \geq -2x + a \\ y \leq -2x + (a + 1) \end{cases}$

MIXED REVIEW

Simplify. (previous courses)

17. 2^3 18. 2^5 19. $(-3)^2$ 20. $(-5)^2$

21. $2^1 \times 2^3$ 22. $3^2 \times 3^1$ 23. $1^2 \times 1^2$ 24. $10^2 \times 10^2$

25. $\frac{5^3}{5}$ 26. $\frac{5^4}{5^2}$ 27. $\frac{2^5}{2^2}$ 28. $\frac{3^7}{3^7}$

Answers for Lesson 5.6 Skill B

Try This Example 1

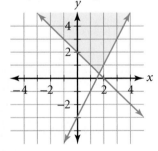

Try This Example 2
a. The solutions form a line.

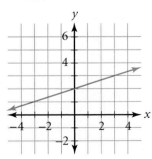

b. The solution areas do not intersect.

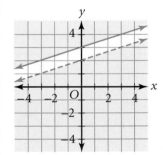

1. $\begin{cases} y \geq \frac{2}{5}x \\ y \geq \frac{3}{5}x - \frac{2}{5} \end{cases}$

2. $\begin{cases} y < -\frac{2}{7}x \\ y > -2x - 3 \end{cases}$

3. $\begin{cases} y \geq \frac{3}{4}x + \frac{1}{4} \\ y < \frac{2}{3}x + \frac{2}{3} \end{cases}$

4.

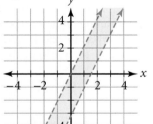

5.

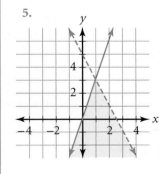

6.

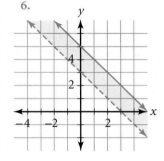

7.

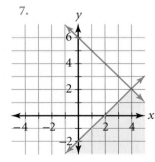

For answers to Exercises 8–28, see Additional Answers.

Focus: *Students use a system of three or more linear inequalities to define geometric figures and to identify feasible choices in a real-life situation.*

EXAMPLE 1

PHASE 1: Presenting the Example

This example introduces students to the notion that a system may contain three or more inequalities. When discussing the solution on the textbook page, point out that the answer was given *without* using system notation. Tell students that this practice is acceptable as long as care is taken to use the connector *and*, because a system is a conjunction of sentences.

PHASE 2: Providing Guidance for TRY THIS

Before they can write a system of inequalities, some students may need to brush up on the skill of writing equations for the bounding lines. You can use Skill C Exercises 1 and 2 on the facing page for this purpose.

EXAMPLE 2

PHASE 1: Presenting the Example

The four inequalities that restrict the farmer's choices, $c + s \leq 30$, $c \geq s$, $c \geq 0$, and $s \geq 0$, are called the *constraints* of the situation. To find the number of acres of each crop to plant, you consider the constraints as a system of linear inequalities. The region of the graph that contains all the solutions to the system is called the *feasible region* because it represents all the suitable, or *feasible*, choices that the farmer can make.

After you have discussed the example, have each student pick one point from within the feasible region and interpret its meaning in the context of the problem. [**sample: (10, 15); The farmer can plant 10 acres of soybeans and 15 acres of corn.**] Make a list of several points that the students picked in order to illustrate the broad range of the farmer's choices.

PHASE 3: ASSESSMENT AND CLOSURE for Lesson 5.6

- Give students the partial system shown at right. Ask them to fill in the box with an inequality that makes a system with infinitely many solutions, and then with an inequality that makes a system with no solutions.

$$\begin{cases} y \leq x + 3 \\ \boxed{} \end{cases}$$

- This lesson concludes the study of systems in this course, which is restricted to linear systems. Students who take further courses in algebra will learn about *nonlinear* systems. They will also see systems and feasible regions used in an optimization method called *linear programming*.

[**samples: infinitely many solutions—** $y \geq x - 2$; **no solutions—** $y \geq x + 4$]

☞ *For a Lesson 5.6 Quiz, see* Assessment Resources *page 53.*

☞ *For a Chapter 5 Assessment, see* Assessment Resources *pages 54–55.*

EXAMPLE 1

Represent the pentagon-shaped region graphed at right as a system of inequalities.

▶ Solution

The shaded region is partly determined by $x \geq 2$, $x \leq 8$, and $y \geq 0$. In addition, the shaded region is bounded by l_1 and l_2.

Using slope and y-intercept, the equations for l_1 and l_2 are:

$$l_1: y = -\frac{1}{3}x + 6 \qquad l_2: y = \frac{1}{2}x + 2$$

The system is $x \geq 2$, $x \leq 8$, $y \geq 0$, $y \leq -\frac{1}{3}x + 6$, and $y \leq \frac{1}{2}x + 2$.

TRY THIS   Represent this pentagon-shaped region as a system of inequalities.

EXAMPLE 2

A farmer has a total of 30 acres available for planting corn and soybeans. He wants to plant at least as much corn as soybeans. Represent the possible quantities of corn and soybeans in a graph.

▶ Solution

Look for key words.

1. Information about the total number of acres available indicates an inequality. The phrase "at least as much corn as soybeans" indicates a second inequality. The natural restriction is that the number of acres planted must be nonnegative.

   Let c represent the number of acres of corn and let s represent the number of acres of soybeans to be planted.

Write inequalities.

2. Write and graph a system of inequalities.

$$\begin{cases} c + s \leq 30 \\ c \geq s \end{cases} \text{ and } \begin{cases} c \geq 0 \\ s \geq 0 \end{cases}$$

The graph is shown at right.

TRY THIS   A farmer has a total of 40 acres available for planting corn and soybeans. He wants to plant no more corn than soybeans. Write and graph a system of inequalities to represent this situation.

EXERCISES

KEY SKILLS

Refer to the diagram at right.

1. Write equations for the lines that contain the vertical sides and the horizontal side of the polygon.

2. Write equations for the slanted sides of the polygon.

PRACTICE

Write a system of linear inequalities to represent each polygon and its interior.

3.

4.

5.

6.

A farmer wants to plant up to 25 acres of corn and soybeans. Write a system of linear inequalities and graph the possible choices.

7. The acreage for soybeans is to be at least twice that of corn.

8. At least 5 acres of corn are to be planted. The acreage for soybeans is to be at least twice that of corn.

MIXED REVIEW APPLICATIONS

Solve each problem. (4.6 Skill C)

9. Jesse and Maria have 600 feet of fencing to enclose a square at least 20 feet on a side. What are the possible side lengths?

10. How many pounds of peanuts are needed to make between 5 pounds and 6 pounds of a mix if the ratio of cashews to peanuts is 2.5 to 1? Round quantities to the nearest tenth.

Answers for Lesson 5.6 Skill C

Try This Example 1

$$\begin{cases} x \geq 3 \\ x \leq 7 \\ y \geq 0 \\ y \leq \frac{1}{3}x + 3 \\ y \geq -\frac{3}{2}x + 5 \end{cases}$$

Try This Example 2

$$\begin{cases} c + s \leq 40 \\ c \leq s \\ c \geq 0 \\ s \geq 0 \end{cases}$$

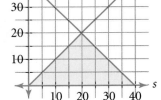

1. $x = 1, x = 8,$ and $y = 0$

2. $m_1: y = \frac{1}{6}x + \frac{8}{3};$
   $m_2: y = \frac{1}{2}x + 2$

3. $\begin{cases} x \geq 0 \\ x \leq 5 \\ y \geq 0 \\ y \leq 6 \end{cases}$

4. $\begin{cases} x \geq -2 \\ x \leq 8 \\ y \geq 1 \\ y \leq 5 \end{cases}$

5. $\begin{cases} y \leq \frac{1}{2}x + 2 \\ y \geq -\frac{3}{4}x + 3 \\ x \leq 4 \end{cases}$

6. $\begin{cases} y \leq \frac{2}{3}x + 3 \\ y \geq \frac{2}{3}x - 1 \\ y \geq -\frac{2}{3}x + 3 \\ y \leq -\frac{2}{3}x + 7 \end{cases}$

7. $\begin{cases} c + s \leq 25 \\ s \geq 2c \\ c \geq 0 \\ s \geq 0 \end{cases}$

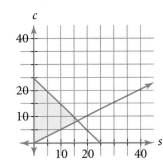

8. $\begin{cases} c + s \leq 25 \\ s \geq 2c \\ c \geq 5 \\ c \geq 0 \\ s \geq 0 \end{cases}$

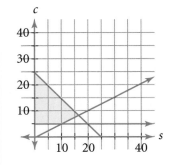

9. between 20 feet and 150 feet, inclusive

10. between 1.4 pounds and 1.7 pounds of peanuts

CHAPTER **5** ASSESSMENT

Use a graph to solve each system of equations.

1. $\begin{cases} 2x + 3y = 0 \\ x - y = 5 \end{cases}$

2. $\begin{cases} 3x + y = -1 \\ -2x - y = 5 \end{cases}$

3. $\begin{cases} -5x + 2y = -15 \\ y = \frac{2}{5}x + 3 \end{cases}$

4. $\begin{cases} x = -2y + 1 \\ y = -\frac{1}{2}x + \frac{3}{2} \end{cases}$

Solve each system using the substitution method.

5. $\begin{cases} 3m + 7 = n \\ m - 2n = 11 \end{cases}$

6. $\begin{cases} x = -3y + 2 \\ 3x + 2y = 6 \end{cases}$

7. $\begin{cases} y = -2z + 5 \\ 5 - 3z = y \end{cases}$

8. $\begin{cases} 2.5c + 1 = d \\ d = -\frac{3}{7}c + 1 \end{cases}$

9. $\begin{cases} j - 7k = 0 \\ -3j + 5k = -16 \end{cases}$

10. $\begin{cases} 2x - 3y = -2 \\ -2x + y = -6 \end{cases}$

11. $\begin{cases} \frac{a}{2} - \frac{b}{2} = 3 \\ \frac{a}{2} + \frac{b}{2} = 6 \end{cases}$

12. $\begin{cases} -1 = \frac{u}{5} - \frac{v}{2} \\ 0 = \frac{u}{5} + \frac{v}{3} \end{cases}$

Use a system of equations to solve each problem.

13. The ratio of milk to water for a certain recipe is 1 cup of milk to 4 cups of water. If the milk and water together make a 10 cup solution, how many cups of milk and how many cups of water are needed?

14. Find two whole numbers such that:
   a. one is twice the other, and
   b. five times the smaller minus two times the larger equals 7.

Solve each system using the elimination method.

15. $\begin{cases} 2g + 7h = 10 \\ g - 7h = 5 \end{cases}$

16. $\begin{cases} -3r - 3s = 8 \\ 3r + 5s = -6 \end{cases}$

17. $\begin{cases} 11y - 6z = 10 \\ 5y - 3z = 7 \end{cases}$

18. $\begin{cases} 2.5c + 5d = -1.5 \\ 5c - 6d = 5 \end{cases}$

Solve each system using the elimination method.

19. $\begin{cases} 5x + 7y = 1 \\ 3x + 4y = 1 \end{cases}$

20. $\begin{cases} 3x + 5y = -4 \\ 4x + 3y = 2 \end{cases}$

21. $\begin{cases} 5a + 4b = 0 \\ 3a - 6b = 4 \end{cases}$

22. $\begin{cases} -11j + 8k = 5 \\ 3j - 3k = -4 \end{cases}$

23. Is it possible to make a 720 milliliter salt solution that is 28% salt from one solution that is 20% salt and another solution that is 30% salt? Justify your response.

Solve, if possible.

24. $\begin{cases} 3x - 2y = 12 \\ x - 2y = 6 \end{cases}$

25. $\begin{cases} 3c + d = 3.5 \\ 9c + 3d = 7.5 \end{cases}$

26. $\begin{cases} 2c - 2d = 0.5 \\ 2c + 2d = -1.5 \end{cases}$

Classify each system as specifically as possible.

27. $\begin{cases} 3x + 2y = 6 \\ -6x - 4y = -0.5 \end{cases}$

28. $\begin{cases} 14m + 4n = -3 \\ 7m + 2n = -1.5 \end{cases}$

29. $\begin{cases} 9u - 3v = -1 \\ 27u - 9v = -3 \end{cases}$

30. $\begin{cases} b - 5b = 1 \\ 3b - 15b = -3 \end{cases}$

Graph the solution region of each linear inequality.

31. $y \geq \frac{3}{4}x - 4$

32. $y < -4$

33. $3x + 5y < 20$

34. Given that x and y are whole numbers, graph the solutions to $x + 3y \leq 9$.

Graph the solution region of each system of linear inequalities.

35. $\begin{cases} y < 5x - 3 \\ y > -3x - 3 \end{cases}$

36. $\begin{cases} y \geq 2 \\ y \leq -2x + 2 \end{cases}$

37. $\begin{cases} x < 3 \\ y > 2x - 3 \end{cases}$

38. $\begin{cases} 3x - y \geq 1 \\ 2x + 4y \leq 8 \end{cases}$

39. $\begin{cases} x - y < 0 \\ x + y > 0 \end{cases}$

40. $\begin{cases} 4x - y < 0 \\ 5x + 2y < 0 \end{cases}$

41. The perimeter of a rectangle may not exceed 20 inches and its length must be at least twice its width. Graph the possible lengths and widths of the rectangle.

Answers for Chapter 5 Assessment

1. $(3, -2)$
2. $(4, -13)$
3. $(5, 5)$
4. no solution
5. $(-5, -8)$
6. $(2, 0)$
7. $(5, 0)$
8. $(0, 1)$
9. $(7, 1)$
10. $(5, 4)$
11. $(9, 3)$
12. $\left(-2, \frac{6}{5}\right)$
13. 2 cups of milk and 8 cups of water
14. 7 and 14
15. $(5, 0)$
16. $\left(-\frac{11}{3}, 1\right)$
17. $(-4, -9)$
18. $(0.4, -0.5)$

19. $(3, -2)$
20. $(2, -2)$
21. $\left(\frac{8}{21}, -\frac{10}{21}\right)$
22. $\left(\frac{17}{9}, \frac{29}{9}\right)$
23. Yes; use 144 milliliters of the 20% solution and 576 milliliters of the 30% solution.
24. $\left(3, -\frac{3}{2}\right)$
25. no unique solution
26. $\left(-\frac{1}{4}, -\frac{1}{2}\right)$
27. inconsistent
28. consistent and dependent
29. consistent and dependent

30. inconsistent
31.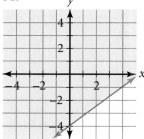
32.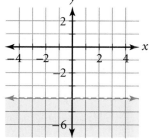

33.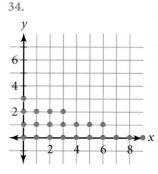

34.

For answers to Exercises 35–41, see Additional Answers.

Chapter 6 *Operations With Polynomials*	State or Local Standards	Corresponding Lessons in *Algebra 1,* Schultz et al.
6.1 Integer Exponents		8.1, 8.2, 8.3, 8.4, 8.5
Skill A: Using properties of exponents for powers and products		
Skill B: Using properties of exponents for quotients, zero exponents, and negative exponents		
Skill C: Using scientific notation		
6.2 The Power Functions $y = kx$, $y = kx^2$ and $y = kx^3$		5.3, 9.4, 10.2
Skill A: Using power functions to solve direct-variation problems		
Skill B: APPLICATIONS Using $y = kx^2$ to solve real-world problems		
6.3 Polynomials		9.1
Skill A: Classifying polynomials by degree and number of terms		
Skill B: APPLICATIONS Using polynomials to solve geometry problems		
6.4 Adding and Subtracting Polynomials		9.1, 9.4
Skill A: Adding polynomials		
Skill B: Subtracting polynomials		
Skill C: APPLICATIONS Adding and subtracting polynomials to solve geometry and real-world problems		
6.5 Multiplying and Dividing Monomials		2.7, 8.1, 8.3, 8.4
Skill A: Multiplying monomials		
Skill B: Dividing monomials		
Skill C: APPLICATIONS Multiplying and dividing monomials to solve geometry problems		
6.6 Multiplying Polynomials		9.2, 9.3
Skill A: Multiplying a polynomial by a monomial		
Skill B: Multiplying polynomials		
Skill C: Using special products		
6.7 Dividing Polynomials		8.3
Skill A: Dividing a polynomial by a monomial		
Skill B: Dividing a polynomial by a binomial		

CHAPTER

6

Operations With Polynomials

What You Already Know

Adding, subtracting, multiplying, and dividing whole numbers are skills that you have learned and practiced for many years now. Your experience with whole numbers and operations on whole numbers will come into play when you study polynomials.

What You Will Learn

To work successfully with polynomials, you first need to know about laws of exponents. The chapter begins with a study of them and their use in simplifying expressions involving exponents.

After studying laws of exponents but before getting into a study of polynomials, you will have the opportunity to extend your knowledge of functions. You can study simple polynomial functions called power functions and explore applications of them.

You then proceed to study vocabulary that is needed to differentiate one type of polynomial from another. However, a great deal of your work with polynomials will involve addition, subtraction, multiplication, and division of them. It is not unusual to learn how to perform operations on newly created mathematical objects.

VOCABULARY

base	leading coefficient	Exponents
binomial	leading term	Product Property of
congruent polygons	linear	Exponents
constant	monomial	quadratic
cubic	negative exponent	Quotient Property of
degree	perfect-square trinomial	Exponents
descending order	polynomial	scientific notation
difference of two	power function	term
squares	Power-of-a-Fraction	trinomial
evaluate	Property of	varies directly as the
exponent	Exponents	cube
exponential expression	Power-of-a-Power	varies directly as the
FOIL method	Property of	square
horizontal-addition	Exponents	vertical-addition format
format	Power-of-a-Product	zero exponent
	Property of	

The diagram below shows how mathematical skills and mathematical reasoning are interrelated with the skills and concepts in Chapter 6. Notice that this chapter involves operations with polynomials such as addition and multiplication.

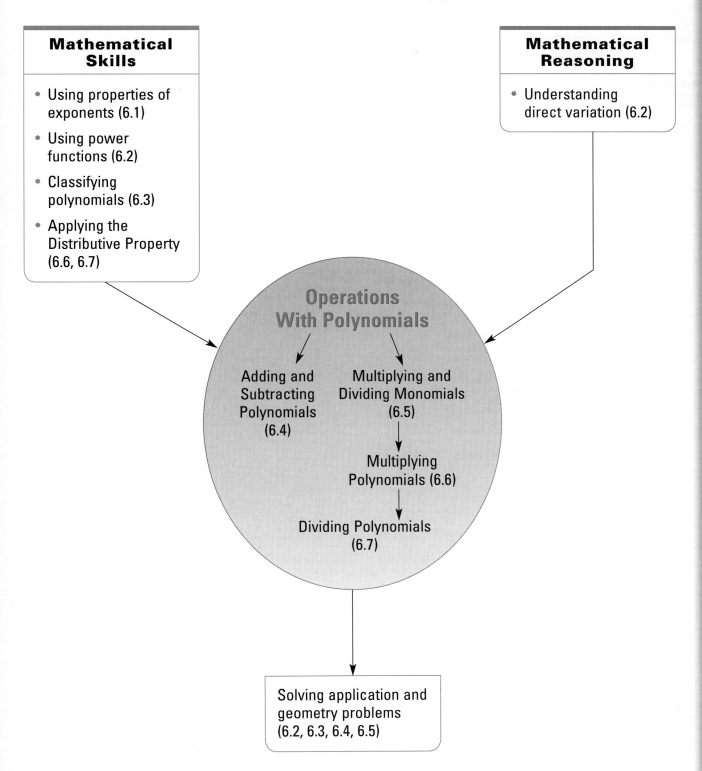

Mathematical Skills

- Using properties of exponents (6.1)
- Using power functions (6.2)
- Classifying polynomials (6.3)
- Applying the Distributive Property (6.6, 6.7)

Mathematical Reasoning

- Understanding direct variation (6.2)

Operations With Polynomials

Adding and Subtracting Polynomials (6.4)

Multiplying and Dividing Monomials (6.5)

Multiplying Polynomials (6.6)

Dividing Polynomials (6.7)

Solving application and geometry problems (6.2, 6.3, 6.4, 6.5)

SKILL A ▶ *Using properties of exponents for powers and products*

Focus: *Students use the Product Property, the Quotient Property, the Power-of-a-Power Property, and the Power-of-a-Product Property to simplify numerical expressions.*

EXAMPLE

PHASE 1: Presenting the Example

Present the definition of exponent in the following form:

$$a^m = \underbrace{a \cdot a \cdot a \cdot \cdots \cdot a}_{m \text{ factors}}$$

Using this definition, work with students to develop the three properties of exponents that are related to products, as shown below:

Product Property

$$a^m a^n = \overbrace{(a \cdot a \cdot a \cdot \cdots \cdot a)}^{m \text{ factors}} \underbrace{\overbrace{(a \cdot a \cdot a \cdot \cdots \cdot a)}^{n \text{ factors}}}_{m + n \text{ factors}} = a^{m+n}$$

Power-of-a-Power Property

$$(a^m)^n = \overbrace{(a^m \cdot a^m \cdot a^m \cdot \cdots \cdot a^m)}^{n \text{ factors}} = a^{\overbrace{m + m + m + \cdots + m}^{n \text{ terms}}} = a^{m \times n} = a^{mn}$$

Power-of-a-Product Property

$$(ab)^m = \overbrace{(ab) \cdot (ab) \cdot (ab) \cdot \cdots \cdot (ab)}^{m \text{ factors}} = \overbrace{(a \cdot a \cdot a \cdot \cdots \cdot a)}^{m \text{ factors}} \overbrace{(b \cdot b \cdot b \cdot \cdots \cdot b)}^{m \text{ factors}} = a^m b^m$$

PHASE 2: Providing Guidance for **TRY THIS**

A common error that students make when applying the Product Property of Exponents is to multiply rather than add exponents. For example, watch for students who write 10^{12} in response to Try This part a. Show them how to simplify by successively applying the definition of exponent, as shown below. Then it becomes clear why the answer must be 10^7.

$$\overbrace{10^3}^{} \qquad \times \qquad \overbrace{10^4}^{}$$
$$\overbrace{10 \times 10 \times 10}^{} \quad \times \quad \overbrace{10 \times 10 \times 10 \times 10}^{}$$
$$10^7$$

6.1 Integer Exponents

SKILL A *Using properties of exponents for powers and products*

You can use *exponents* to express a repeated multiplication, such as $2 \cdot 2 \cdot 2 \cdot 2 \cdot 2$, in a shorter form.

Exponents

For any real number a and any positive integer m, $a^m = \overbrace{a \cdot a \cdot a \cdot \cdots \cdot a}^{m \text{ factors}}$.

a^m is called an **exponential expression**; a is the **base** of the expression, and m is the **exponent**.

Using this definition, you can write $2 \cdot 2 \cdot 2 \cdot 2 \cdot 2$ as 2^5, which is read *two to the fifth power* or *two raised to the fifth power*.

To multiply exponential expressions with the same base, add the exponents.

$$2^4 \cdot 2^3 = \overbrace{(2 \cdot 2 \cdot 2 \cdot 2)}^{4} \cdot \overbrace{(2 \cdot 2 \cdot 2)}^{3} = \overbrace{2 \cdot 2 \cdot 2 \cdot 2 \cdot 2 \cdot 2 \cdot 2}^{4+3\,=\,7} = 2^7$$

To raise an exponential expression to a power, multiply the exponents.

$$(3^2)^4 = (3^2)(3^2)(3^2)(3^2) = \overbrace{3 \cdot 3 \cdot 3 \cdot 3 \cdot 3 \cdot 3 \cdot 3 \cdot 3}^{2 \times 4 = 8} = 3^8$$

To raise a product to a power, distribute the exponent.

$$(4 \cdot 3)^2 = (4 \cdot 3)(4 \cdot 3) = (4 \cdot 4)(3 \cdot 3) = 4^2 \cdot 3^2$$

Let a and b be nonzero real numbers and let m and n be integers.

Product Property of Exponents $\quad a^m a^n = a^{m+n}$

Power-of-a-Power Property of Exponents $\quad (a^m)^n = a^{m \cdot n}$

Power-of-a-Product Property of Exponents $\quad (ab)^m = a^m b^m$

EXAMPLE Write in the form a^b. **a.** $3^4 \cdot 3^2$ **b.** $(5^2)^6$ **c.** $2^3 \cdot 4^3$

▸ **Solution**

a. $3^4 \cdot 3^2 = 3^{4+2} = 3^6$ ⟵ *Apply the Product Property.*

b. $(5^2)^6 = 5^{2 \cdot 6} = 5^{12}$ ⟵ *Apply the Power-of-a-Power Property.*

c. $2^3 \cdot 4^3 = (2 \cdot 4)^3 = 8^3$ ⟵ *Apply the Power-of-a-Product Property.*

TRY THIS Write in the form a^b. **a.** $10^3 \cdot 10^4$ **b.** $(3^2)^4$ **c.** $2^4 \cdot 3^4$

Be careful with powers of negative numbers. $(-3)^2 = (-3)(-3) = 9$, but $-3^2 = -(3^2) = -(3)(3) = -9$. The parentheses in $(-3)^2$ indicate that the exponent applies to the negative sign as well as to 3. However, in -3^2, the exponent applies only to 3.

EXERCISES

KEY SKILLS

Rewrite each expression as repeated multiplication.

1. 2^3 **2.** 3^4 **3.** 1^5 **4.** 10^3

Which property of exponents would you use to simplify each expression?

5. $5^3 \cdot 4^3$ **6.** $(4 \cdot 6)^3$ **7.** $5^2 \cdot 5^5$ **8.** $(7^2)^2$

PRACTICE

Write each expression in the form a^b.

9. $5^5 \cdot 5^4$ **10.** $2^2 \cdot 2^4$ **11.** $(3^2)^3$ **12.** $(2^3)^2$

13. $(10^7)^5$ **14.** $(10^6)^7$ **15.** $3^2 \cdot 3^2$ **16.** $1^3 \cdot 1^3$

17. $3^6 \cdot 9^6$ **18.** $3^9 \cdot 6^9$ **19.** $6^5 \cdot 6^4$ **20.** $5^7 \cdot 5^3$

21. $(3^2)^2$ **22.** $(7^3)^5$ **23.** $4^3 \cdot 4^2$ **24.** $5^2 \cdot 5^7$

25. $2^3 \cdot 2^5$ **26.** $3^2 \cdot 3^5$ **27.** $7^2 \cdot 2^2$ **28.** $3^4 \cdot 2^4$

29. $(-2)^3 \cdot (-3)^3$ **30.** $(-3)^3 \cdot (-3)^3$ **31.** $(-2)^4 \cdot 3^4$ **32.** $6^2 \cdot (-3)^2$

33. $(-3^4)^2$ **34.** $(-2^2)^5$ **35.** $(-2)^1(-2)^3(-2)^2$ **36.** $(-7)^2(-7)^1(-7)^5$

37. Explain why $(2x)^3$ is not the same as $2x^3$.

38. Can $3^5 \cdot 4^3$ be simplified using the properties of exponents? Explain your response.

39. Critical Thinking You have seen that $(-a)^2 \neq -a^2$. Is it also true that $(-a)^3 \neq -a^3$? Try this for several different powers. For what values of n does $(-a)^n = -a^n$?

MIXED REVIEW

Use the substitution method to solve each system of equations. (5.2 Skill A)

40. $\begin{cases} y = x + 3 \\ y = 2x - 4 \end{cases}$ **41.** $\begin{cases} x = y + 4 \\ 2x + 3y = 43 \end{cases}$ **42.** $\begin{cases} 4x + 2y = 20 \\ y = x - 2 \end{cases}$ **43.** $\begin{cases} x - y = 3 \\ 2x + 2y = 2 \end{cases}$

Use the elimination method to solve each system of equations. (5.3 Skill A)

44. $\begin{cases} 3x - y = 0 \\ 3x + 2y = -7 \end{cases}$ **45.** $\begin{cases} 2x + 2y = 3 \\ 5x - 2y = 0 \end{cases}$ **46.** $\begin{cases} x - 3y = 0 \\ 5x + 3y = 6 \end{cases}$ **47.** $\begin{cases} 2x - 2y = 10 \\ 4x + 3y = 12 \end{cases}$

Answers for Lesson 6.1 Skill A

Try This Example
a. 10^7 **b.** 3^8 **c.** 6^4

1. $2 \cdot 2 \cdot 2$
2. $3 \cdot 3 \cdot 3 \cdot 3$
3. $1 \cdot 1 \cdot 1 \cdot 1 \cdot 1$
4. $10 \cdot 10 \cdot 10$
5. Power-of-a-Product Property
6. Power-of-a-Product Property
7. Product Property
8. Power-of-a-Power Property
9. 5^9
10. 2^6
11. 3^6
12. 2^6
13. 10^{35}
14. 10^{42}
15. 3^4
16. 1^3
17. 27^6
18. 18^9
19. 6^9
20. 5^{10}
21. 3^4

22. 7^{15}
23. 4^5
24. 5^9
25. 2^8
26. 3^7
27. 14^2
28. 6^4
29. 6^3
30. 9^3
31. $(-6)^4$
32. $(-18)^2$
33. 3^8
34. -2^{10}
35. $(-2)^6$
36. $(-7)^8$
37. $(2x)^3 = 2x \cdot 2x \cdot 2x = 2^3 x^3 = 8x^3$
$2x^3 = x^3 + x^3$
38. No; neither the bases nor the exponents are the same.
39. n is an odd integer or $n = 0$.
40. $(7, 10)$
41. $(11, 7)$
42. $(4, 2)$
43. $(2, -1)$

44. $\left(-\dfrac{7}{9}, -\dfrac{7}{3} \right)$

45. $\left(\dfrac{3}{7}, \dfrac{15}{14} \right)$

46. $\left(1, \dfrac{1}{3} \right)$

47. $\left(\dfrac{27}{7}, -\dfrac{8}{7} \right)$

SKILL B Using properties of exponents for quotients, zero exponents, and negative exponents

Focus: *Students use the Quotient Property of Exponents, the definition of zero exponents, and the definition of negative exponents to simplify numerical expressions.*

EXAMPLE 1

PHASE 1: Presenting the Example

Ask students to use the definition of exponent to write an expression for $\left(\frac{a}{b}\right)^m$.

$\left[\underbrace{\frac{a}{b} \cdot \frac{a}{b} \cdot \frac{a}{b} \cdot \cdots \cdot \frac{a}{b}}_{m \text{ factors}}\right]$ Ask students if they can simplify this expression. Lead them through the process shown below.

$$\underbrace{\frac{a}{b} \cdot \frac{a}{b} \cdot \frac{a}{b} \cdot \cdots \cdot \frac{a}{b}}_{m \text{ factors}} = \frac{\overbrace{a \cdot a \cdot a \cdot \cdots \cdot a}^{m \text{ factors}}}{\underbrace{b \cdot b \cdot b \cdot \cdots \cdot b}_{m \text{ factors}}} = \frac{a^m}{b^m}$$

Use the properties of exponents learned in Skill A to develop the Quotient Property of Exponents as follows:

$$\frac{a^m}{a^n} = \frac{a^{m+0}}{a^n} = \frac{a^{m+(-n+n)}}{a^n} = \frac{a^{(m-n)+n}}{a^n} = \frac{a^{m-n}a^n}{a^n} = a^{m-n}\left(\frac{a^n}{a^n}\right) = a^{m-n}(1) = a^{m-n}$$

PHASE 2: Providing Guidance for TRY THIS

Students may have trouble with part c. Lead them to see that, before they can apply the Quotient Property of Exponents, they must first apply the Product Property to the numerator, as shown at right.

$$\frac{5^6 \cdot 5^4}{5^6} = \frac{5^{6+4}}{5^6} = \frac{5^{10}}{5^6} = 5^{10-6} = 5^4$$

Some students may also notice that you can divide both the numerator and denominator by 5^6 to find the answer 5^4.

EXAMPLE 2

PHASE 1: Presenting the Example

You can use properties of exponents in the following manner to explain zero and negative exponents:

$$\begin{aligned} a^0 &= a^{n-n} \\ &= \frac{a^n}{a^n} \quad \longleftarrow \textit{Quotient Property of Exponents} \\ &= 1 \end{aligned}$$

$$\begin{aligned} a^{-m} &= a^{0-m} \\ &= \frac{a^0}{a^m} \quad \longleftarrow \textit{Quotient Property of Exponents} \\ &= \frac{1}{a^m} \quad \longleftarrow \textit{Definition of zero exponent} \end{aligned}$$

PHASE 2: Providing Guidance for TRY THIS

Some of the expressions in this lesson may be difficult to simplify completely if the students are not using calculators. If this is the case, you may wish to allow your students to leave their answers in the form a^b or $\frac{1}{a^b}$. For example, you might accept $\frac{1}{4^3}$ for Try This part c.

You can use the definition of exponent to raise a fraction to a power. For example, $\left(\frac{1}{2}\right)^3 = \left(\frac{1}{2}\right)\left(\frac{1}{2}\right)\left(\frac{1}{2}\right) = \frac{1 \cdot 1 \cdot 1}{2 \cdot 2 \cdot 2} = \frac{1^3}{2^3} = \frac{1}{8}$. So, $\left(\frac{1}{2}\right)^3 = \frac{1^3}{2^3}$.

To divide exponential expressions, subtract the exponents. For example, $\frac{2^5}{2^3} = \frac{2 \cdot 2 \cdot 2 \cdot 2 \cdot 2}{2 \cdot 2 \cdot 2} = \frac{2 \cdot 2}{1} = 2^2 = 4$. So $\frac{2^5}{2^3} = 2^{5-3} = 2^2$.

> Let a and b be nonzero real numbers and let m and n be integers.
>
> **Power-of-a-Fraction Property of Exponents** $\left(\frac{a}{b}\right)^m = \frac{a^m}{b^m}$
>
> **Quotient Property of Exponents** $\frac{a^m}{a^n} = a^{m-n}$

EXAMPLE 1 Simplify. **a.** $\left(\frac{3}{4}\right)^2$ **b.** $\frac{4^8}{4^6}$ **c.** $\frac{3^5 \cdot 3^2}{3^4}$

▸ Solution

a. $\left(\frac{3}{4}\right)^2 = \frac{3^2}{4^2} = \frac{9}{16}$ **b.** $\frac{4^8}{4^6} = 4^{8-6}$ **c.** $\frac{3^5 \cdot 3^2}{3^4} = \frac{3^{5+2}}{3^4} = \frac{3^7}{3^4}$
$= 4^2 = 16$ $= 3^{7-4} = 3^3 = 27$

TRY THIS Simplify. **a.** $\left(\frac{2}{3}\right)^3$ **b.** $\frac{3^8}{3^4}$ **c.** $\frac{5^6 \cdot 5^4}{5^6}$

Exponents can also be zero or negative. Examine the pattern below.

$6^2 = 6 \cdot 6 = 36$ $6^{-1} = \frac{1}{6}$

$6^1 = 6$ $6^{-2} = \frac{1}{6 \cdot 6} = \frac{1}{6^2} = \frac{1}{36}$

$6^0 = \frac{6}{6} = 1$

> Let a be a nonzero real number and let n be an integer.
>
> **Zero Exponent** $a^0 = 1$ (0^0 does not exist.)
>
> **Negative Exponents** $a^{-n} = \frac{1}{a^n}$

EXAMPLE 2 Simplify. **a.** 54^0 **b.** 5^{-2} **c.** $\frac{3^4 \cdot 3^{-2}}{3^3}$

▸ Solution

a. $54^0 = 1$ **b.** $5^{-2} = \frac{1}{5^2} = \frac{1}{25}$ **c.** $\frac{3^4 \cdot 3^{-2}}{3^3} = \frac{3^{4+(-2)}}{3^3} = \frac{3^2}{3^3} = 3^{2-3} = 3^{-1} = \frac{1}{3}$

TRY THIS Simplify. **a.** 16^0 **b.** 10^{-2} **c.** $\frac{4^5 \cdot 4^{-2}}{4^6}$

EXERCISES

KEY SKILLS

Which property or definition of exponents would you use to simplify each expression?

1. $\frac{3^5}{3^2}$ **2.** 8^{-12} **3.** 7^0 **4.** $\left(\frac{9}{2}\right)^3$

5. $(6^2)^3$ **6.** $\frac{4^3}{4^2}$ **7.** $3^2 \cdot 3^5$ **8.** $(3 \cdot 5)^2$

PRACTICE

Simplify each expression.

9. $\left(\frac{1}{3}\right)^2$ **10.** $\left(\frac{1}{8}\right)^2$ **11.** $\left(\frac{3}{5}\right)^3$ **12.** $\left(\frac{4}{3}\right)^3$

13. $\frac{3^5}{3^2}$ **14.** $\frac{5^6}{5^3}$ **15.** $\frac{2^3}{2}$ **16.** 10^{-1}

17. 2^{-3} **18.** 8^{-2} **19.** $\frac{3^6}{3^9}$ **20.** $\frac{2^5}{2^8}$

21. $\frac{3^2}{3^0}$ **22.** $\frac{2^0}{2^2}$ **23.** $\frac{3^2 \cdot 3^3}{3^5}$ **24.** $\frac{2^3 \cdot 2^3}{2^5}$

25. $\frac{4^6}{4^2 \cdot 4^3}$ **26.** $\frac{5^8}{5^3 \cdot 5^4}$ **27.** $\frac{7^3 \cdot 7^3}{7^8}$ **28.** $\frac{(-2)^2 \cdot (-2)^3}{(-2)^7}$

29. $\frac{(-5)^9}{(-5)^7 \cdot (-5)^2}$ **30.** $\frac{(-10)^3}{(-10) \cdot (-10)^2}$ **31.** $[(-1)^3 (-2)^3]^3$ **32.** $(3^{-1} \cdot 2^{-1})^2$

33. $[(-3)^{-2} \cdot (-3)^{-1}]^2$ **34.** $[(-4)^{-1} \cdot (-2)^{-1}]^3$ **35.** $\left(\frac{10^3 \cdot 10^2}{10^4 \cdot 10^2}\right)^2$ **36.** $\left(\frac{4^2 \cdot 4^2}{4^4 \cdot 4^3}\right)^2$

37. $\left(\frac{3^{-2} \cdot 3^{-2}}{2^4 \cdot 2^3}\right)^{-1}$ **38.** $\left(\frac{2^{-1} \cdot 3^{-1}}{5^4 \cdot 5^3}\right)^{-2}$ **39.** $\left(\frac{1^{-1} \cdot 1^{-1}}{2^{-1} \cdot 3^3}\right)^{-1}$ **40.** $\left(\frac{5^{-1} \cdot 5^{-1}}{2^4 \cdot 3^3}\right)\left(\frac{2^{-2} \cdot 3^{-1}}{5^2 \cdot 5^2}\right)^{-1}$

41. **Critical Thinking** How are the expressions a^n and a^{-n} related?

MIXED REVIEW

Find each product or quotient. Write your answer as a decimal. Use mental math where possible. (previous courses)

42. 10×235 **43.** $\frac{118}{100}$ **44.** 100×1389 **45.** $\frac{0.78}{10}$

46. 100×0.56 **47.** $\frac{2.57}{100}$ **48.** 1000×0.232 **49.** $\frac{18.3}{1000}$

Answers for Lesson 6.1 Skill B

Try This Example 1

a. $\frac{2^3}{3^3}$, or $\frac{8}{27}$ **b.** 3^4, or 81

c. 5^4, or 625

Try This Example 2

a. 1 **b.** $\frac{1}{100}$ **c.** $\frac{1}{4^3}$, or $\frac{1}{64}$

1. Quotient Property
2. Definition of negative exponent
3. Definition of zero exponent
4. Power-of-a-Fraction Property
5. Power-of-a-Power Property
6. Quotient Property
7. Product Property
8. Power-of-a-Product Property
9. $\frac{1}{9}$

10. $\frac{1}{64}$
11. $\frac{3^3}{5^3}$, or $\frac{27}{125}$
12. $\frac{4^3}{3^3}$, or $\frac{64}{27}$
13. 27
14. 125
15. 4
16. $\frac{1}{10}$
17. $\frac{1}{8}$
18. $\frac{1}{64}$
19. $\frac{1}{27}$
20. $\frac{1}{8}$
21. 9
22. $\frac{1}{4}$
23. 1
24. 2

25. 4
26. 5
27. $\frac{1}{49}$
28. $\frac{1}{4}$
29. 1
30. 1
31. 2^9, or 512
32. $\frac{1}{36}$
33. $\frac{1}{(-3)^6}$, or $\frac{1}{729}$
34. $\frac{1}{2^9}$, or $\frac{1}{512}$
35. $\frac{1}{100}$
36. $\frac{1}{4^6}$, or $\frac{1}{4096}$
★ 37. $3^4 \cdot 2^7$, or 10,368
★ 38. $2^2 \cdot 3^2 \cdot 5^{14}$
★ 39. $\frac{27}{2}$

★ 40. $\frac{25}{36}$
41. Each is the multiplicative inverse of the other.
42. 2350
43. 1.18
44. 138,900
45. 0.078
46. 56
47. 0.0257
48. 232
49. 0.0183

★ **Advanced Exercises**

Focus: *Students learn to write numbers in scientific notation. They then find products and quotients of numbers written in scientific notation.*

EXAMPLE 1

PHASE 1: Presenting the Example

PHASE 2: Providing Guidance for TRY THIS

Scientific notation was devised as a more practical means of expressing the very large and very small numbers involved in scientific measurements. To pique the students' interest, begin the lesson by displaying the following two statements:

The radius of an electron is about 0.0000000000000028 meter.
The mass of Earth is about 5,974,000,000,000,000,000,000,000 kilograms.

Keep the statements displayed as you work through Example 1 and the Try This exercises. Then, when you feel students are comfortable with the technique, have them rewrite these measurements in scientific notation. [2.8×10^{-15} **meters;** 5.974×10^{24} **kilograms**]

EXAMPLE 2

PHASE 1: Presenting the Example

Be sure students understand that the reordering and regrouping of factors is justified by the Commutative and Associative Properties of Multiplication.

PHASE 2: Providing Guidance for TRY THIS

Watch for students who correctly perform all the operations, and then forget to rewrite the result in scientific notation. For instance, in part a, they may give the answer as 73.8×10^4, and in part b, they may write 0.125×10^4. Remind them that the result is not in scientific notation until there is exactly one nonzero digit to the left of the decimal point.

PHASE 3: ASSESSMENT AND CLOSURE for Lesson 6.1

- Give students the three expressions shown at right. Have them fill in the boxes with nonzero integers so that the value of the resulting expression is 64. [**samples: 1.** 2, 4

  2. 2, 3 **3.** 8, 2] Repeat the activity, this time asking for the value of the expression to be $\frac{1}{4}$.
  [**samples: 1.** 2, -4 **2.** -1, 2 **3.** 2, 4]

- In this lesson, students used basic properties of exponents to simplify numerical expressions. In the upcoming lessons, they will apply the properties to variable expressions while learning about operations with polynomials.

  ☞ *For a Lesson 6.1 Quiz, see* Assessment Resources *page 60.*

1. $2^{\square} \cdot 2^{\square}$

2. $\left(2^{\square}\right)^{\square}$

3. $\dfrac{2^{\square}}{2^{\square}}$

A number is written in **scientific notation** when it has the form $a \times 10^n$, where n is an integer and $1 \leq a < 10$. Scientific notation is a shortened way of writing very large or very small numbers. For example:

$$1{,}210{,}000{,}000{,}000{,}000{,}000{,}000{,}000{,}000{,}000{,}000 = 1.21 \times 10^{33}$$
$$0.0000000000000000000000000000368 = 3.68 \times 10^{-29}$$

To write a number in scientific notation, use the following procedure.

Scientific Notation
1. Move the decimal point so that the number is between 1 and 10 (excluding 10).
2. Multiply by a power of 10. To find the exponent of 10, count the number of places the decimal point must move to return to its original position. The exponent is negative if the decimal point must move to the left, and positive if it must move to the right.

EXAMPLE 1 Write each number in scientific notation. **a.** 1364 **b.** 0.0258

▷ **Solution**

First move the decimal point so that the number is between 1 and 10. Then multiply by the correct power of 10.

a. $1364 = 1.364 \times 10^3$

> To return to its original position, the decimal point must move 3 places to the right.

b. $0.0258 = 2.58 \times 10^{-2}$

> To return to its original position, the decimal point must move 2 places to the left.

TRY THIS Write each number in scientific notation. **a.** 26.5 **b.** 0.0014

EXAMPLE 2 Find each product or quotient. Write the answer in scientific notation.

a. $(1.2 \times 10^3)(9.0 \times 10^4)$ **b.** $\frac{1.8 \times 10^3}{3.0 \times 10^5}$

▷ **Solution**

a. $(1.2 \times 10^3)(9.0 \times 10^4)$
$= (1.2 \times 9.0)(10^3 \times 10^4)$
$= (1.2 \times 9.0)(10^{3+4})$
$= 10.8 \times 10^7$
$= (1.08 \times 10^1) \times 10^7$
$= 1.08 \times 10^8$

b. $\frac{1.8 \times 10^3}{3.0 \times 10^5} = \left(\frac{1.8}{3.0}\right)\left(\frac{10^3}{10^5}\right)$
$= \left(\frac{1.8}{3.0}\right) \times 10^{3-5}$
$= 0.6 \times 10^{-2}$
$= (6 \times 10^{-1}) \times 10^{-2}$
$= 6 \times 10^{-3}$

TRY THIS Find each product or quotient. Write the answer in scientific notation.
a. $(8.2 \times 10^2)(9.0 \times 10^2)$ **b.** $\frac{1.20 \times 10^6}{9.60 \times 10^2}$

EXERCISES

KEY SKILLS

Is each number in scientific notation? If not, write it in scientific notation.

1. 3.45×10^3 **2.** 34.5×10^3 **3.** 345 **4.** 0.345×10^{-2}

PRACTICE

Write each number in scientific notation.

5. 12.4 **6.** 15,334 **7.** 0.012 **8.** 0.00025

9. 10,050 **10.** 1,200,000 **11.** 0.0165 **12.** 0.00335

Find each product or quotient. Write the answer in scientific notation.

13. $(1.5 \times 10^2)(2.0 \times 10^2)$ **14.** $(1.4 \times 10^3)(5.0 \times 10^3)$

15. $(6.5 \times 10^1)(3.0 \times 10^3)$ **16.** $(7.6 \times 10^4)(3.0 \times 10^1)$

17. $(5.4 \times 10^{-2})(6.0 \times 10^2)$ **18.** $(6.5 \times 10^{-3})(4.0 \times 10^{-2})$

19. $\frac{6.0 \times 10^4}{3.0 \times 10^2}$ **20.** $\frac{7.5 \times 10^5}{5.0 \times 10^4}$

21. $\frac{7.5 \times 10^3}{3.0 \times 10^5}$ **22.** $\frac{8.4 \times 10^4}{4.0 \times 10^7}$

23. $\frac{2.4 \times 10^2}{4.8 \times 10^3}$ **24.** $\frac{2.1 \times 10^2}{8.4 \times 10^5}$

25. Critical Thinking Assume $1 \leq a < 10$. When $(a \times 10^3)(5.0 \times 10^2)$ is written in scientific notation, the power of 10 is 5. What are the possible values of a?

26. Critical Thinking Assume $1 \leq a < 10$. When $(a \times 10^4) \div (5.0 \times 10^2)$ is written in scientific notation, the power of 10 is 2. What are the possible values of a?

MIXED REVIEW

Assume that y varies directly as x. (3.3 Skill B)

27. $y = 12$ when $x = 2.5$. Find y when $x = 6$.

28. $y = 3$ when $x = 4$. Find y when $x = 12.5$.

29. $y = 13$ when $x = 5$. Find y when $x = 0.5$.

30. $y = 2$ when $x = 7$. Find y when $x = 1$.

Answers for Lesson 6.1 Skill C

Try This Example 1
a. 2.65×10^1
b. 1.4×10^{-3}

Try This Example 2
a. 7.38×10^5
b. 1.25×10^3

1. yes
2. no; 3.45×10^4
3. no; 3.45×10^2
4. no; 3.45×10^{-3}
5. 1.24×10^1
6. 1.5334×10^4
7. 1.2×10^{-2}
8. 2.5×10^{-4}
9. 1.005×10^4
10. 1.2×10^6
11. 1.65×10^{-2}
12. 3.35×10^{-3}
13. 3.0×10^4
14. 7.0×10^6
15. 1.95×10^5

16. 2.28×10^6
17. 3.24×10^1
18. 2.6×10^{-4}
19. 2.0×10^2
20. 1.5×10^1
21. 2.5×10^{-2}
22. 2.1×10^{-3}
23. 5.0×10^{-2}
24. 2.5×10^{-4}
25. $1 \leq a < 2$
26. $5 \leq a < 10$
27. $y = 28.8$
28. $y = 9.375$
29. $y = 1.3$
30. $y = \frac{2}{7}$

The Power Functions $y = kx$, $y = kx^2$, and $y = kx^3$

SKILL A   *Using power functions to solve direct-variation problems*

Focus: *Students use power functions to solve direct-variation problems.*

EXAMPLES 1 and 2

PHASE 1: Presenting the Examples

Display the first table at right. Have students calculate the ratio $\frac{C}{r}$ for each row.

[**6.28, 6.28, 6.28, 6.28, 6.28**] Use their results to review the concept of *direct variation*. That is, the circumference, C, of a circle is said to *vary directly as* its radius, r. The number 6.28 is called the *constant of variation*, and an equation that represents this relationship is $\frac{C}{r} = 6.28$, or $C = 6.28r$. (Students who have previously studied geometry may recall the exact formula, $C = 2\pi r$.)

Display the second table. Ask students to calculate the ratio $\frac{A}{r}$ for each row. [**3.14, 6.28, 9.42, 12.56, 15.70**] In this case, the ratios are not constant, so the area, A, of a circle does *not* vary directly as its radius, r.

Refer students back to the second table and ask them to calculate the ratio $\frac{A}{r^2}$ for each row. [**3.14, 3.14, 3.14, 3.14, 3.14**] Now there is a constant value, 3.14, which means that the area, A, of a circle *varies directly as the square* of its radius, r. The number 3.14 is the constant of variation, and an equation that represents the relationship is $\frac{A}{r^2} = 3.14$, or $A = 3.14r^2$. (Again, some students may recall the exact formula, $A = \pi r^2$.)

Display the third table. This time, lead students to observe the relationship $\frac{V}{r^3} = 4.19$, or $V = 4.19r^3$. That is, the volume of a sphere *varies directly as the cube* of its radius. (The exact formula from geometry is $V = \frac{4}{3}\pi r^3$.)

Summarize by reviewing with students the definition of *power function* given at the top of the textbook page. Then discuss the examples.

PHASE 2: Providing Guidance for **TRY THIS**

Although students should recognize the relationships on this page as power functions, some will have greater success in solving the problems if they set up and solve proportions by cross-multiplying, as shown below.

Radius of a circle (r)	Circumference of the circle (C)
1	6.28
2	12.56
3	18.84
4	25.12
5	31.40

Radius of a circle (r)	Area of the circle (A)
1	3.14
2	12.56
3	28.26
4	50.24
5	78.50

Radius of a sphere (r)	Volume of the sphere (V)
1	4.19
2	33.52
3	113.13
4	268.16
5	523.75

Example 1
Try This:
$$\begin{array}{c} y \\ x^2 \end{array} \to \frac{36}{12^2} = \frac{y}{10^2} \begin{array}{c} \leftarrow y \\ \leftarrow x^2 \end{array}$$

$$144y = 3600$$
$$y = 25$$

Example 2
Try This:
$$\begin{array}{c} y \\ x^3 \end{array} \to \frac{1}{2^3} = \frac{y}{12^3} \begin{array}{c} \leftarrow y \\ \leftarrow x^3 \end{array}$$

$$8y = 1728$$
$$y = 216$$

6.2 LESSON
The Power Functions $y = kx$, $y = kx^2$, and $y = kx^3$

SKILL A — *Using power functions to solve direct-variation problems*

A **power function** is any function of the form $y = kx^n$, where k is nonzero and n is a positive integer.

Recall that if $y = kx$ and $k \neq 0$, then y varies directly as x and the constant of variation is k. This is a power function with $n = 1$.

If $y = kx^2$ and $k \neq 0$, then y **varies directly as the square** of x. The constant of variation is k. This is a power function with $n = 2$.

EXAMPLE 1 If y varies directly as the square of x, and $y = 36$ when $x = 3$, find y when $x = 8$.

▶ Solution

Write an equation. Because y varies directly as the square of x, you can write $y = kx^2$.
$$y = kx^2$$
$$36 = k \cdot 3^2 \quad \longleftarrow \quad y = 36 \text{ when } x = 3$$
$$4 = k$$
$$k = 4 \quad \rightarrow \quad y = 4x^2$$
If $x = 8$, then $y = 4 \cdot 8^2 = 256$.

TRY THIS If y varies directly as the square of x, and $y = 36$ when $x = 12$, find y when $x = 10$.

If $y = kx^3$ and $k \neq 0$, then y **varies directly as the cube** of x. The constant of variation is k. This is a power function with $n = 3$.

EXAMPLE 2 If y varies directly as the cube of x, and $y = 36$ when $x = 2$, find y when $x = 5$.

▶ Solution

Write an equation. Because y varies directly as the cube of x, you can write $y = kx^3$.
$$y = kx^3$$
$$36 = k \cdot 2^3 \quad \longleftarrow \quad y = 36 \text{ when } x = 2$$
$$4.5 = k$$
$$k = 4.5 \quad \rightarrow \quad y = 4.5x^3$$
If $x = 5$, then $y = 4.5 \cdot 5^3 = 562.5$.

TRY THIS If y varies directly as the cube of x, and $y = 1$ when $x = 2$, find y when $x = 12$.

EXERCISES

KEY SKILLS

Let $x = 0, 1, 2, 3, 4$, and 5. Make a table of ordered pairs for each function. Find the difference in successive y values. Is this difference constant?

1. $y = 3x$
2. $y = 3x^2$
3. $y = 3x^3$

PRACTICE

In Exercises 4–9, find k and write an equation for y in terms of x. Then find the specified value of y.

4. y varies directly as the square of x. If $y = 18$ when $x = 3$, find y when $x = 5$.

5. y varies directly as the square of x. If $y = 40$ when $x = 4$, find y when $x = 8$.

6. y varies directly as the square of x. If $y = 36$ when $x = 3$, find y when $x = 4$.

7. y varies directly as the cube of x. If $y = 40$ when $x = 2$, find y when $x = 6$.

8. y varies directly as the cube of x. If $y = 94.5$ when $x = 3$, find y when $x = 10$.

9. y varies directly as the cube of x. If $y = 48$ when $x = 2$, find y when $x = 3$.

10. The length of one side of a square is x units.
    a. Write a direct-variation equation in the form $y = kx$ for the perimeter, P, of the square.
    b. Write a direct-variation equation in the form $y = kx^2$ for the area, A, of the square.
    c. Write a direct-variation equation in the form $y = kx^3$ for the volume, V, of a cube whose side length is x units.

11. **Critical Thinking** Suppose that y varies directly as the square of x and that (x_1, y_1) and (x_2, y_2) satisfy this direct-variation relationship. Find a relationship among x_1, x_2, y_1, and y_2 that does not include the constant of variation.

MIXED REVIEW

Simplify. (2.6 Skill A)

12. $\frac{3}{4}a + 5 - \frac{1}{4}a$
13. $3s + 4s - 3s + 4$
14. $4 - (y + 2) - 3y + 1$
15. $3(u - 3) + 4(u - 3)$
16. $\frac{2}{3}t - 3 - \frac{1}{3}t + 2$
17. $0.6z + 4(0.2z - 2)$

Answers for Lesson 6.2 Skill A

Try This Example 1
$y = 25$

Try This Example 2
$y = 216$

1.

x	0	1	2	3	4	5
y	0	3	6	9	12	15

Yes, the differences are constant, 3.

2.

x	0	1	2	3	4	5
y	0	3	12	27	48	75

No, the differences are not constant.

3.

x	0	1	2	3	4	5
y	0	3	24	81	192	375

No, the differences are not constant.

4. $k = 2$; $y = 2x^2$; $y = 50$
5. $k = 2.5$; $y = 2.5x^2$; $y = 160$
6. $k = 4$; $y = 4x^2$; $y = 64$
7. $k = 5$; $y = 5x^3$; $y = 1080$
8. $k = 3.5$; $y = 3.5x^3$; $y = 3500$
9. $k = 6$; $y = 6x^3$; $y = 162$
10. a. $y = 4x$
    b. $y = x^2$
    c. $y = x^3$
11. $\dfrac{y_2}{y_1} = \left(\dfrac{x_2}{x_1}\right)^2$
12. $\dfrac{1}{2}a + 5$
13. $4s + 4$
14. $-4y + 3$
15. $7u - 21$
16. $\dfrac{1}{3}t - 1$
17. $1.4z - 8$

Focus: *Students solve real-world problems involving direct variation as the square by using given information to write a variation equation.*

EXAMPLE

PHASE 1: Presenting the Example

Stress the fact that the given speed and stopping distance apply to an unspecified set of conditions. As noted in the text, there are many different variation equations relating speed of travel and stopping distance. On an icy road, for instance, the stopping distance for a given car will be greater than it is for the same car on a dry road, and different equations will apply.

Some students may find it easier to understand the variation equation if you use a verbal approach like the one below.

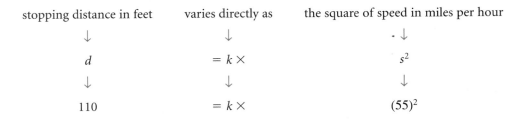

stopping distance in feet	varies directly as	the square of speed in miles per hour
↓	↓	· ↓
d	$= k \times$	s^2
↓	↓	↓
110	$= k \times$	$(55)^2$

PHASE 2: Providing Guidance for TRY THIS

Have students use the variation equation to make a table like the one below. Then ask several questions such as the following: How is stopping distance affected if the speed doubles from 10 miles per hour to 20 miles per hour? [**Stopping distance is multiplied by 4.**] What is the effect if the speed triples from 30 miles per hour to 90 miles per hour? [**Stopping distance is multiplied by 9.**] What is the effect if the speed quadruples from 10 miles per hour to 40 miles per hour? [**Stopping distance is multiplied by 16.**]

Speed (miles per hour)	10	20	30	40	50	60	70	80	90	100
Stopping distance (feet)	4.3	17.2	38.7	68.7	107.4	154.7	210.6	275.0	348.1	429.8

PHASE 3: ASSESSMENT AND CLOSURE for Lesson 6.2

- Each table at right shows a variation relationship between y and x. Have students describe and compare the relationships. [**Each is a direct variation with constant of variation 3. In 1, y varies directly as x, with equation $y = 3x$. In 2, y varies directly as the square of x, with equation $y = 3x^2$. In 3, y varies directly as the cube of x, with equation $y = 3x^3$.**]

- The notion of a direct variation function, first introduced in Chapter 3 with the rule $y = kx$, has now been extended to encompass all functions of the form $y = kx^n$, where $k \neq 0$ and n is a positive integer. In Chapter 9, students will learn about *inverse variation*.

☞ *For a Lesson 6.2 Quiz, see Assessment Resources page 61.*

1

x	1	2	3	4
y	3	6	9	12

2

x	1	2	3	4
y	3	12	27	48

3

x	1	2	3	4
y	3	24	81	194

The distance required for a moving car to stop varies directly as the square of the car's speed. Therefore, we can use an equation of the form $y = kx^2$ to represent this situation. The value of k will change for different types of cars, road conditions, etc.

EXAMPLE

A car traveling at 55 miles per hour had a stopping distance of 110 feet.
a. Write an equation that describes the car's stopping distance in terms of its speed.
b. If the speed is reduced by half, will the stopping distance also be reduced by half? Justify your response.

▶ **Solution**

a. We know that stopping distance, d, varies directly as the square of the speed, s, so let $d = ks^2$ for some constant k.

Write an equation.

$$110 = k(55)^2$$
$$k = \frac{110}{(55)^2} = \frac{2 \cdot 55}{55 \cdot 55} = \frac{2}{55}$$
$$k = \frac{2}{55} \rightarrow d = \frac{2}{55}s^2$$

b. One-half of the original speed, 55 mph, is 27.5 mph.
If $s = 27.5$, then $d = \frac{2}{55} \times (27.5)^2 = 27.5$.

When the speed is reduced by half to 27.5 mph, the stopping distance is reduced to 27.5 ft, which one-fourth of the original stopping distance of 110 ft. So, the stopping distance is reduced by more than one-half.

TRY THIS

A car traveling at 55 miles per hour had a stopping distance of 130 feet.
a. Find the stopping distance for a speed of 60 miles per hour.
b. Will the stopping distance be reduced by half if the speed is reduced by half? Justify your response.

You can show that when speed is reduced by half, stopping distance is always reduced by more than half. Let d_s represent the stopping distance at a speed of s miles per hour, and let $d_{\frac{1}{2}s}$ represent the stopping distance at a speed of $\frac{1}{2}s$ miles per hour.

$$\frac{d_{\frac{1}{2}s}}{d_s} = \frac{k\left(\frac{1}{2}\right)^2}{ks^2} = \frac{\left(\frac{1}{2}\right)^2}{s^2} = \frac{\left(\frac{1}{2}\right)^2 s^2}{s^2} = \frac{1}{4}$$

When the speed is reduced by half, the stopping distance is reduced to $\frac{1}{4}$ of the original stopping distance. Speed and stopping distance are clearly *not* linearly related.

EXERCISES

KEY SKILLS

1. Copy and complete the following table for the car in the Example. Round answers to the nearest tenth.

s	0	5	10	15	20	25	30
d	?	?	?	?	?	?	?

2. Are the differences in successive values of d constant?

PRACTICE

A car traveling at 50 miles per hour had a stopping distance of 135 feet. Find the stopping distance for each speed. Round to the nearest tenth.

3. 45 miles per hour
4. 65 miles per hour
5. 55 miles per hour
6. 100 miles per hour

7. **Critical Thinking** Let $d = ks^2$, where k is a nonzero constant. If d_s is the stopping distance for a speed of s, find and simplify an expression for $\frac{d_{rs}}{d_s}$, where $0 < r < 1$.

The vertical distance that a falling object travels varies directly as the square of time. Let d represent distance in feet, and let t represent time in seconds. The constant of variation is 16 feet per second squared. Find d for each value of t.

8. 2 seconds
9. 3 seconds
10. 5 seconds
11. 10 seconds
12. 15 seconds
13. 30 seconds

MIXED REVIEW APPLICATIONS

Use an inequality to solve each problem. (4.4 Skill B)

14. Lin spent $50 on a skirt and a sweater. The sweater cost less than $22.50. How much could the skirt have cost?

15. Video games cost $15 each. If Percy spent less than $120, how many video games could he have bought?

16. The Gibbons family spent more than $220 for a barbecue grill and patio furniture. The cost of the grill was $70 more than twice the cost of the patio furniture. How much could the patio furniture have cost?

Answers for Lesson 6.2 Skill B

Try This Example
a. about 154.7 feet

b. No; $\dfrac{d_{0.5s}}{d_s} = \dfrac{\frac{130}{55^2}(0.5s)^2}{\frac{130}{55^2}s^2} = \dfrac{(0.5s)^2}{s^2} = \dfrac{1}{4}$

The stopping distance is reduced to $\frac{1}{4}$ of the original stopping distance when the speed is reduced by $\frac{1}{2}$.

1.

s	0	5	10	15	20	25	30
d	0	0.9	3.6	8.2	14.5	22.7	32.7

2. no
3. 109.4 feet
4. 228.2 feet
5. 163.4 feet
6. 540 feet
7. $\dfrac{d_{rs}}{d_s} = \dfrac{k(rs)^2}{ks^2} = \dfrac{kr^2s^2}{ks^2} = \dfrac{r^2}{1} = r^2$
8. 64 feet
9. 144 feet
10. 400 feet
11. 1600 feet
12. 3600 feet
13. 14,400 feet
14. more than $27.50
15. less than 8 games
16. more than $50

TEACHING
6.3
LESSON

Polynomials

SKILL A ▷ *Classifying polynomials by degree and number of terms*

Focus: *Students learn to recognize polynomials and are introduced to the vocabulary of polynomials.*

EXAMPLE 1

PHASE 1: Presenting the Example

Display the following:

These are monomials in *x*.	These are not monomials in *x*.
$-4 \quad x \quad 9x^2 \quad -x^5 \quad \frac{3}{5}x^7$	$x + 6 \quad x^2 - 1 \quad \frac{1}{x} \quad -\frac{4}{x^3} \quad 2x - 7$

If possible, have students work in pairs. Tell them to study the two groups of expressions, and then write what they think is the definition of a *monomial in x.* Ask several pairs to share their definitions with the class. As a class, reach a consensus on the definition. Then compare the class results to the definition at the top of the textbook page.

Now display $-3x + 5x^2 - 7x^3 + 5$ and use it to illustrate the definition of a *polynomial in x.* Then refer to the vocabulary on the textbook page. Have a student read each definition aloud. After each is read, ask the class to apply it to $-3x + 5x^2 - 7x^3 + 5$. They should arrive at the conclusions at right.

Its *terms* are $-3x, 5x^2, -7x^3$, and 5.
Its *degree* is 3.
Its *leading term* is $-7x^3$.
Its *leading coefficient* is -7.
Its *constant term* is 5.

Finally, have students read the definition of *descending order*. Then discuss Example 1 and its solution.

PHASE 2: Providing Guidance for **TRY THIS**

Some students may be confused because there is no visible exponent in $15y$, nor is *y* visible at all in the constant term 5. As a result, they may misplace $15y$ and 5 in their responses, or omit 5 altogether. Suggest that they begin by rewriting these terms with the "phantom exponents" in place, as shown at right.

$$13y^2 + 15y - 7y^3 + 5$$
$$\downarrow \qquad\qquad \downarrow$$
$$13y^2 + 15y^1 - 7y^3 + 5y^0$$

EXAMPLE 2

PHASE 1: Presenting the Example

Have students review the tables that precede Example 2. Then display expression **1**, shown at right. Ask them to classify it by degree and by number of terms. [**cubic binomial**] Now display expression **2**. Ask if they think this is also a cubic binomial. Lead them to see that the two terms could actually be combined by applying the Distributive Property: $-7x^3 + 5x^3 = (-7 + 5)x^3 = -2x^3$. Point out that, despite its appearance, expression **2** is actually a cubic *monomial*. Then discuss Example 2.

1. $-7x^3 + 5x^2$
2. $-7x^3 + 5x^3$

6.3 LESSON *Polynomials*

SKILL A ▶ Classifying polynomials by degree and number of terms

Recall that a monomial in x is the product of a number and a whole-number power of x. For example, $-3x^2, \frac{2}{3}x^3, 2x, 10$ are each monomials. (Note that $2x = 2x^1$ and $10 = 10x^0$.)

A **polynomial in x** is a sum of one or more monomials in x. For example, $5x^2 - 4x + 7$ is a polynomial with three *terms*, $5x^2$, $-4x$, and 7. A **term** is any monomial in a polynomial. The **degree** of a polynomial is the greatest exponent in any of its terms, which in $5x^2 - 4x + 7$ is 2. The term with the greatest exponent, $5x^2$, is called the **leading term**. The **leading coefficient** is the numerical part of the leading term, in this case 5.

A polynomial is written in **descending order** if the exponents decrease as you read the polynomial from left to right. For example, $3x^4 + 5x^2 + 2x + 5$ is written in descending order.

EXAMPLE 1   Write $-3x + 5x^2 - 7x^3 + 5$ in descending order.

▶ Solution

$-3x + 5x^2 - 7x^3 + 5 = -7x^3 + 5x^2 - 3x + 5$ ◀——— $-3x = -3x^1$ and $5 = 5x^0$

TRY THIS   Write $13y^2 + 15y - 7y^3 + 5$ in descending order.

Polynomials have special names based on degree and number of terms.

Degree	Name	Example
0	constant	6
1	linear	$3x - 5$
2	quadratic	$-4x^2 + x + 1$
3	cubic	$9x^3 + x^2 - 5$

Terms	Name	Examples
1	monomial	$-5x, 2x^3, x^8, 4$
2	binomial	$3x - 1, x^2 + 6x$
3	trinomial	$x^2 - 5x + 1$

EXAMPLE 2   Classify $-2a^3 + 2a^3 + 4a^2 - 3a - 5$ by degree and number of terms.

▶ Solution

First simplify the polynomial.
$-2a^3 + 2a^3 + 4a^2 - 3a - 5$
$= (-2 + 2)a^3 + 4a^2 - 3a - 5$ ◀——— Combine like terms.
$= 4a^2 - 3a - 5$ ◀——— The greatest exponent is 2.
The degree is 2 and there are 3 terms. It is a quadratic trinomial.

TRY THIS   Classify $5b^3 - 3.5b^3 - 3b + 3b^2 - b^2$ by degree and number of terms.

KEY SKILLS

Write the degree, number of terms, leading term, and leading coefficient for each polynomial.

1. $r^3 + r^2 - 5$

2. $a^3 + a^2 - 2a + 1$

3. $3k^4 + k^3 - 2k^2 + k$

PRACTICE

Write each polynomial in descending order.

4. $2x + 3x^2 - 1$

5. $5g - 7 + g^2$

6. $-2n + 1 - n^2$

7. $-m + 7 - 3m^2$

8. $3x^2 + 5x - 4 + 5x^3$

9. $3c^2 + 5c^2 - 4 + 5c^4$

10. $-2p + 6p - p^2 - p^3$

11. $y^2 + y + y + 5y^3 - y^2$

12. $-a^3 + 5a^2 - 4 + 5a^2 - a$

Classify each polynomial by degree and number of terms.

13. $2x^2 + 6x + x^3$

14. $-3s^2 + 6 - s^3 - 3s^4$

15. $q^2 + 6 - q^3 + 3q^4$

16. $c^2 + 7 - 2c^3 + 2c^2$

17. $-1 - 2z^3 + z^2 + 2z^3$

18. $-y^2 - 4y^3 + 2y^2 + 4y^3$

19. $5k^2 + 7k^3 - 2k^3 + 2k^2$

20. $5k + k^2 - 5k^2 + 2k$

21. $-3v + 6v^2 - 5v^2 + 3v$

22. Classify $2(3n - 2) - 3(2n^2 + 2n) + 6(n^2 - 2)$ by degree and number of terms.

23. Every linear function can be written in the form $y = mx + b$, where m and b are real numbers. Suppose that $m \neq 0$ and $b \neq 0$. Classify $mx + b$ by degree and number of terms.

24. Let k be a nonzero real number. Classify kx^2 and kx^3 by degree and number of terms.

25. **Critical Thinking**   Find real numbers a and b such that the polynomial $ax + 6x^5 - bx^3 + 3x + x - 5$ is a linear binomial.

26. **Critical Thinking**   Find the missing polynomial in the equation below.
$(\underline{\quad ? \quad}) - (2k^4 - 2k^2 - 3) = 4k^4 + k^3 + 2k^2 + 7$

MIXED REVIEW

Evaluate each formula for the given value(s) of the variables. (1.2 Skill C)

27. $S = 2r^2 + 20r$ given $r = 3$

28. $S = 2r^2 + 20r$ given $r = 4$

29. $W = 4.5x + 5y$ given $x = 6$ and $y = 6$

30. $z = \frac{2}{3}x + \frac{3}{7}y$ given $x = 3$ and $y = 7$

31. $t = x^2 + y^2$ given $x = \frac{1}{2}$ and $y = \frac{1}{3}$

32. $y = (x - 2)^2 + 2$ given $x = 4$

Answers for Lesson 6.3 Skill A

Try This Example 1
$-7y^3 + 13y^2 + 15y + 5$

Try This Example 2
cubic trinomial

1. degree: 3
   number of terms: 3
   leading term: r^3
   leading coefficient: 1

2. degree: 3
   number of terms: 4
   leading term: a^3
   leading coefficient: 1

3. degree: 4
   number of terms: 4
   leading term: $3k^4$
   leading coefficient: 3

4. $3x^2 + 2x - 1$

5. $g^2 + 5g - 7$

6. $-n^2 - 2n + 1$

7. $-3m^2 - m + 7$

8. $5x^3 + 3x^2 + 5x - 4$

9. $5c^4 + 8c^2 - 4$

10. $-p^3 - p^2 + 4p$

11. $5y^3 + 2y$

12. $-a^3 + 10a^2 - a - 4$

13. cubic trinomial

14. quartic polynomial

15. quartic polynomial

16. cubic trinomial

17. quadratic binomial

18. quadratic monomial

19. cubic binomial

20. quadratic binomial

21. quadratic monomial

22. constant monomial

23. linear binomial

24. quadratic monomial and cubic monomial

25. $a \neq -4$ and $b = 6$

26. $6k^4 + k^3 + 4$

27. 78

28. 112

29. 57

30. 5

31. $\frac{13}{36}$

32. 6

SKILL B APPLICATIONS Using polynomials to solve geometry problems

Focus: *Students use polynomials to model geometry problems.*

EXAMPLE 1

PHASE 1: Presenting the Example

In Skill A, students learned that a polynomial in one variable is a specific type of algebraic expression. Here they see how such expressions can be used to determine the value of a second variable.

Note that the general formula for the volume, V, of a cylinder is $V = \pi r^2 h$, where r is its radius and h is its height. In other words, the volume of a cylinder usually depends on *two* variables, radius and height. In the situation described here, however, the height is held constant at 15 centimeters. Thus the formula used is $V = \pi r^2(15)$, or $V = 15\pi r^2$ and the volume depends only on the radius.

As you work through Example 1 with the students, you may wish to display the results as a set of ordered pairs, as shown at right.

(r, V)
(4, 753.6)
(5, 1177.5)
(6, 1695.6)

EXAMPLE 2

PHASE 2: Providing Guidance for TRY THIS

Be sure students understand the difference between volume and surface area. The volume of a cylinder is the amount of space that the cylinder encloses, and it is measured in cubic units. The surface area of a cylinder is the area of the boundary surface enclosing the cylinder, and its measured in square units. You may want to present the concepts to students in a more intuitive way. For example, knowing the volume of a cylinder would be useful if you wanted to fill it; knowing the surface area would be useful if you wanted to paint the outside of it.

PHASE 3: ASSESSMENT AND CLOSURE for Lesson 6.3

- Ask students to state as many facts about the expression $-9x^2 + 7 - 2x^3$ as they can. **[samples: It is a cubic trinomial. In descending order of exponents of x, it is rewritten as $-2x^3 - 9x^2 + 7$. The leading term is $-2x^3$, and the leading coefficient is -2. The constant term is 7.]**
- In this lesson, students investigated fundamental polynomial concepts. This sets the stage for the sequence of lessons that follow, where they will learn how to add, subtract, multiply, and divide polynomials.

☞ *For a Lesson 6.3 Quiz, see* Assessment Resources *page 62.*

Polynomials are used in many geometry formulas. In the examples below, we will *evaluate* polynomials to answer questions about the surface area and volume of containers. To evaluate a polynomial, substitute the given value(s) for the variable into the expression and then simplify.

Note: You may find it helpful to use a calculator for the exercises in this section.

The **volume** of an object is the amount of space the object occupies. Volume is measured in *cubic units*.

EXAMPLE 1

The total volume of a cylinder whose height is 15 centimeters can be represented by the polynomial $15\pi r^2$, where r represents the radius. Find the volume of the container at right if the radius is 4 cm, 5 cm, and 6 cm. Use 3.14 to approximate π.

▶ **Solution**

Evaluate $15\pi r^2$ for $r = 4$, 5, and 6.

$r = 4$: $(15)(3.14)(4)^2$ ⟵ Replace r with 4.
         $= 753.6 \text{ cm}^3$
$r = 5$: $(15)(3.14)(5)^2$ ⟵ Replace r with 5.
         $= 1177.5 \text{ cm}^3$
$r = 6$: $(15)(3.14)(6)^2$ ⟵ Replace r with 6.
         $= 1695.6 \text{ cm}^3$

TRY THIS   Rework Example 1 if the radius is 1 cm, 2 cm, and 3 cm.

The **surface area** of an object is the total area of all the outer surfaces of the object. Surface area is measured in square units.

EXAMPLE 2

The total surface area of a cylinder whose height is 10 feet can be represented by the polynomial $2\pi r^2 + 20\pi r$. Find the surface area of the cylindrical tank at right if the radius is 5 ft, 6 ft, and 7 ft. Use 3.14 to approximate π.

▶ **Solution**

Evaluate $2\pi r^2 + 20\pi r$ for $r = 5$, 6, and 7.

$r = 5$: $(2)(3.14)(5)^2 + (20)(3.14)(5) = 471 \text{ ft}^2$
$r = 6$: $(2)(3.14)(6)^2 + (20)(3.14)(6) = 602.88 \text{ ft}^2$
$r = 7$: $(2)(3.14)(7)^2 + (20)(3.14)(7) = 747.32 \text{ ft}^2$

TRY THIS   Rework Example 2 if the radius is 8 ft, 9 ft, and 10 ft.

EXERCISES

Evaluate each polynomial for the given value of the variable.

1. $x^2 + 3x - 2$; $x = 3$
2. $2s^2 - 5s + 2$; $s = 4$
3. $v^3 + v^2 + v + 1$; $v = 3$
4. $2n^3 - 5n^2 - 5$; $n = -2$

Refer to Example 1. Find the volume of the container for each radius. Use 3.14 to approximate π.

5. 7 cm
6. 8 cm
7. 8.5 cm
8. 9.5 cm

Refer to Example 2. Find the surface area of the tank for each radius. Use 3.14 to approximate π.

9. 1 foot
10. 2 feet
11. 3 feet
12. 3.5 feet

Sports balls come in many different sizes. Use the given polynomials to complete the table below. Use 3.14 to approximate π. Round answers to the nearest hundredth.

	Type of ball	Radius, r	Volume $\frac{4\pi r^3}{3}$	Surface Area $4\pi r^2$
13.	Golf ball	0.8 inches	?	?
14.	Tennis ball	1.3 inches	?	?
15.	Softball	1.75 inches	?	?
16.	Soccer ball	4.5 inches	?	?
17.	Basketball	4.8 inches	?	?

Simplify each exponential expression. (6.1 Skill B)

18. $\frac{7^1 \cdot 7^2}{7^3}$
19. $[(-2)^2]^{-2}$
20. $\frac{(-3)^6 \cdot (-3)^3}{(-3)^{10}}$
21. $5^3 \times (5^2)^{-2}$

22. If y varies directly as the square of x, and $y = 11$ when $x = 2.5$, find y when $x = 15$. (6.2 Skill A)

Classify by degree and number of terms. (6.3 Skill A)

23. $-5x^3 + 2x + 5x^2 + 5x^3$
24. $3n^2 + n - 3n^2 + 1 + 5n^3$

Answers for Lesson 6.3 Skill B

Try This Example 1
47.1 cm³; 188.4 cm³;
423.9 cm³

Try This Example 2
904.32 ft²; 1073.88 ft²;
1256 ft²

1. 16
2. 14
3. 40
4. −41
5. 2307.9 cm³
6. 3014.4 cm³
7. 3402.975 cm³
8. 4250.775 cm³
9. 69.08 ft²
10. 150.72 ft²
11. 244.92 ft²
12. 296.73 ft²
13. 2.14 in³; 8.04 in²
14. 9.20 in³; 21.23 in²
15. 22.44 in³; 38.47 in²

16. 381.51 in³; 254.34 in²
17. 463.01 in³; 289.38 in²
18. 1
19. $\frac{1}{16}$
20. $-\frac{1}{3}$
21. $\frac{1}{5}$
22. $y = 396$
23. quadratic binomial
24. cubic trinomial

Adding and Subtracting Polynomials

SKILL A *Adding polynomials*

Focus: *Students simplify a sum of two polynomials by combining like terms, employing both vertical-addition and horizontal-addition techniques.*

EXAMPLE 1

PHASE 1: Presenting the Example

Display the four polynomials at right. Ask students to classify each by degree and number of terms. Give them a minute to jot down their responses, and then discuss the results.

1	$-2x^3 + 5$
2	$-2x^3 + 5x$
3	$-2x^3 + 5x^2$
4	$-2x^3 + 5x^3$

Students should arrive at the conclusion that polynomials **1**, **2**, and **3** are cubic binomials, whereas polynomial **4** is a cubic *monomial*. Point out that polynomial **4** looks very similar to the other three polynomials. Ask students to explain why it is different. Elicit the response that polynomial **4** consists of two *like terms* that can be combined, whereas the other three polynomials each consist of two *unlike terms*. Emphasize the fact that like terms involve the same variable(s) raised to exactly the same power(s).

Now display $(-2x^3 + 2.5x^2 - 5x + 3) + (5x^2 + 4x - 5)$, which is the addition given in Example 1. Ask students to identify all pairs of like terms that they see. [**$2.5x^2$ and $5x^2$; $-5x$ and $4x$; 3 and -5**] Show them how to rewrite the addition vertically, aligning the like terms that they identified. Then discuss the solution given on the textbook page.

PHASE 2: Providing Guidance for TRY THIS

Each addend consists of three terms, and so some students may simply line up the terms from left to right and attempt to add. Suggest that they rewrite each polynomial as shown at right, inserting a zero placeholder for the "missing" term in each.

$$2.6a^3 - 1.5a^2 - 5a + 0$$
$$0.4a^3 + 0a^2 - 5a + 3$$

EXAMPLE 2

PHASE 1: Presenting the Example

PHASE 2: Providing Guidance for TRY THIS

Some students may observe that it is often possible to apply the Distributive Property to terms that are not like terms. For example, the sum $-2x^3 + 5x^2$ could be rewritten as follows:

$$-2x^3 + 5x^2 = -2x \cdot x^2 + 5x^2 = (-2x + 5)x^2$$

Point out to students that the goal when adding polynomials is to write the sum in the simplest form possible. Thus, although rewriting $-2x^3 + 5x^2$ as $(-2x + 5)x^2$ is correct, this result would be considered more complex rather than simplified.

6.4 LESSON Adding and Subtracting Polynomials

SKILL A *Adding polynomials*

Just as you can perform operations on integers, you can perform operations on polynomials.

The following statements say that the set of polynomials in the same variable(s) is *closed* under addition and under subtraction.

> **Closure of Polynomials Under Addition and Subtraction**
> The sum of two polynomials is a polynomial.
> The difference of two polynomials is a polynomial.

You can add polynomials by using the **vertical-addition format.** When you perform addition in this format, place like terms vertically above one another.

EXAMPLE 1 Use the vertical-addition format to find the sum
$(-2x^3 + 2.5x^2 - 5x + 3) + (5x^2 + 4x - 5)$.

▶ Solution

Align corresponding terms.
Add corresponding coefficients.

$$\begin{array}{r} -2x^3 + 2.5x^2 - 5x + 3 \\ + \quad\quad\; 5x^2 + 4x - 5 \\ \hline -2x^3 \;\; 7.5x^2 (-1)x - 2 \end{array}$$

The sum is $-2x^3 + 7.5x^2 - x - 2$.

TRY THIS Use the vertical-addition format to find $(2.6a^3 - 1.5a^2 - 5a) + (0.4a^3 - 5a + 3)$.

Example 2 shows how to perform addition using the Distributive Property and what is called the **horizontal-addition format.** When you use this format, you add by grouping and combining like terms.

EXAMPLE 2 Use the horizontal-addition format to find the sum in Example 1.

▶ Solution

$(-2x^3 + 2.5x^2 - 5x + 3) + (5x^2 + 4x - 5)$
$= (-2 + 0)x^3 + (2.5 + 5)x^2 + (-5 + 4)x + (3 - 5)$
$= -2x^3 + (7.5)x^2 + (-1)x + (-2)$
$= -2x^3 + 7.5x^2 - x - 2$

TRY THIS Use the horizontal-addition format to find $(2.6a^3 - 1.5a^2 - 5a) + (0.4a^3 - 5a + 3)$.

Add the polynomials.

1. $\begin{array}{r} -3x^2 - 5x + 1 \\ + \quad 2x^2 \quad\quad - 5 \end{array}$

2. $\begin{array}{r} a^3 - 3a^2 - a + 1 \\ + -4a^3 - 5a^2 \quad\quad - 7 \end{array}$

3. $\begin{array}{r} -3y^3 \quad\quad - 5y + 1 \\ + \quad 4y^3 - y^2 \quad\quad + 1 \end{array}$

Find each sum.

4. $(2x^2 - 2x + 1) + (5x^2 - x - 2)$

5. $(-b^2 - 2b) + (5b^2 - b - 1)$

6. $(c^3 - 2c^2 - 5c + 3) + (c^2 + 2c - 1)$

7. $(z^3 + 3z^2) + (-3z^2 + z)$

8. $(9n^4 - n^2 - 7n) + (5n^3 - n^2 + 9)$

9. $(3k^4 - k^2 - 1) + (k^3 - k + 1)$

10. $(h^4 - h^2 + 1) + (-h^2 + 3h - 5)$

11. $(-b^5 - b^3 + b) + (b^5 + b^3 - 1)$

12. $(z^5 + z^3 - z^2 + 1) + (z^3 + z^2 + 3z - 1)$

13. $(5d^5 - d^3 + d^2 + 2d) + (d^5 + d^3 - 2d)$

14. $3x^3 - 4x^2 + 2x - 1, x^2 + 2x + 1,$ and $2x^4 - 3x^2 - 4x + 5$

15. $-5x^3 - 3x + 3, 2x^4 + 3x^2 + 3x + 2,$ and $-2x^4 - 4x^2 + 3$

16. Find a, b, and c such that the sum of $ax^2 + bx + c$ and $-2x^2 + 3x + 7$ equals 0.

17. Let $a, b, c, k, m,$ and n represent real numbers. When is the sum $(ax^2 + bx + c) + (kx^2 + mx + n)$ only a constant term?

18. **Critical Thinking** Let $a, b, r,$ and s represent real numbers, where $a \neq -r$ and $b \neq -s$. Show that the sum of $ax + b$ and $rx + s$ is a linear binomial.

Find each product or quotient. Write the answer in scientific notation. (6.1 Skill C)

19. $(2.1 \times 10^{-1})(3.0 \times 10^2)$

20. $(1.5 \times 10^{-1})(4.0 \times 10^{-1})$

21. $(6.0 \times 10^2)(6.0 \times 10^3)$

22. $(2.5 \times 10^3)(2.5 \times 10^{-1})$

23. $\dfrac{9.0 \times 10^5}{2.0 \times 10^2}$

24. $\dfrac{7.5 \times 10^2}{3.0 \times 10^{-1}}$

25. $\dfrac{3.5 \times 10^3}{7.0 \times 10^4}$

Answers for Lesson 6.4 Skill A

Try This Example 1

$$\begin{array}{r} 2.6a^3 - 1.5a^2 - 5a \\ 0.4a^3 \quad\quad - 5a + 3 \\ \hline 3.0a^3 - 1.5a^2 - 10a + 3 \end{array}$$

Try This Example 2

$(2.6 + 0.4)a^3 + (-1.5)a^2 +$
$(-5-5)a + 3 = 3a^3 - 1.5a^2 - 10a + 3$

1. $-x^2 - 5x - 4$
2. $-3a^3 - 8a^2 - a - 6$
3. $y^3 - y^2 - 5y + 2$
4. $7x^2 - 3x - 1$
5. $4b^2 - 3b - 1$
6. $c^3 - c^2 - 3c + 2$
7. $z^3 + z$
8. $9n^4 + 5n^3 - 2n^2 - 7n + 9$
9. $3k^4 + k^3 - k^2 - k$
10. $h^4 - 2h^2 + 3h - 4$
11. $b - 1$
12. $z^5 + 2z^3 + 3z$
13. $6d^5 + d^2$
14. $2x^4 - 3x^3 - 6x^2 + 5$
15. $-5x^3 - x^2 + 8$

16. $a = 2, b = -3,$ and $c = -7$
17. when $a = -k$ and $b = -m$
18. $(ax + b) + (rx + s) = (a + r)x + (b + s)$
   Since $a \neq -r$, the coefficient of x is nonzero. Since $b \neq -s$, the constant term is nonzero. Thus, the sum is a linear binomial.
19. 6.3×10^1
20. 6.0×10^{-2}
21. 3.6×10^6
22. 6.25×10^2
23. 4.5×10^3
24. 2.5×10^3
25. 5.0×10^{-2}

Focus: *Students simplify a difference of two polynomials by finding the opposite of the polynomial being subtracted, and then adding.*

EXAMPLE 1

PHASE 1: Presenting the Example

Display the three polynomials at right. Ask students to evaluate each given $r = 3$. [**1.** -33 **2.** -3 **3.** -33] Repeat the activity, this time given $r = -5$. [**1.** -25 **2.** -75 **3.** -25]

1. $-(2r^2 + 5r)$
2. $-2r^2 + 5r$
3. $-2r^2 - 5r$

Have students examine the results. Based on their observations, ask if they can make a conjecture about any relationships among the given polynomials. Elicit the response that polynomials **1** and **3** are most likely equivalent.

Now display the statement $-(2r^2 + 5r) = (-1)(2r^2 + 5r)$. Ask students how they can proceed from here to *prove* that polynomials **1** and **3** are equivalent. Lead them to the procedure at right. Then discuss Example 1.

$$\begin{aligned} -(2r^2 + 5r) &= (-1)(2r^2 + 5r) \\ &= (-1)(2r^2) + (-1)(5r) \\ &= -2r^2 + (-5r) \\ -(2r^2 + 5r) &= -2r^2 - 5r \end{aligned}$$

PHASE 2: Providing Guidance for TRY THIS

Some students become confused about the signs of terms when subtracting. Suggest that they begin their work by rewriting all subtractions as additions, as shown below.

$$-3a^3 - 3.5a^2 - 0.5a = -3a^3 + (-3.5a^2) + (-0.5a)$$

EXAMPLE 2

PHASE 1: Presenting the Example

Display polynomial **1**, shown at right, and have students simplify. [$-4m^2 + 7m$] Now display polynomial **2**. Ask students to compare it to polynomial **1**. Lead them to see that, in this case, $(4m^2 - 7m)$ is being subtracted from $8m$. Point out that simplifying this polynomial will also involve taking the opposite of $(4m^2 - 7m)$, but that the result must then be combined with $8m$. Ask students to simplify. [$-4m^2 + 15m$] Then discuss Example 2, which shows a subtraction involving many more terms.

1. $-(4m^2 - 7m)$
2. $8m - (4m^2 - 7m)$

PHASE 2: Providing Guidance for TRY THIS

When subtracting polynomials, students frequently forget to find the opposite of each term of the polynomial being subtracted. Alert students to this potential source of error, and lead a discussion in which they brainstorm ways that they might avoid it. For instance, they could rewrite the subtraction to show a multiplication by -1 and then draw arrows as a reminder to apply the Distributive Property.

$$(3n^3 - n - 1) - (3n^2 - 5n - 2)$$

$$= (3n^3 - n - 1) + (-1)(3n^2 - 5n - 2)$$

Recall from Lesson 2.3 the Opposite of a Sum Property and the Opposite of a Difference Property.

$$-(a + b) = -a - b \qquad -(a - b) = -a + b$$

Notice that to find the opposite of a sum or the opposite of a difference, you take the opposite of *each term* in the expression. To find the opposite of a polynomial, you must find the opposite of each of its terms.

EXAMPLE 1 Find the opposite of $-5x^3 + 3x^2 - 5x - 3$.

▶ **Solution**

$-(-5x^3 + 3x^2 - 5x - 3)$
$= -(-5x^3) - (3x^2) - (-5x) - (-3)$ ⟵ Find the opposite of each term.
$= 5x^3 - 3x^2 + 5x + 3$

TRY THIS Find the opposite of $-3a^3 - 3.5a^2 - 0.5a$.

Recall the definition of subtraction for real numbers: $a - b = a + (-b)$. In other words, subtracting is the same as adding the opposite. This is also true for polynomials. To subtract one polynomial from another, add the opposite.

Because subtraction is really addition of the opposite, you can use the vertical-addition format when you subtract. Subtraction in vertical format for $(-5x^3 + 2x^2 - x + 5) - (-5x^3 + 3x^2 - 5x - 3)$ is shown below.

$$\begin{array}{r} -5x^3 + 2x^2 - x + 5 \\ -\ -5x^3 + 3x^2 - 5x - 3 \\ \hline \end{array} \longrightarrow \begin{array}{r} -5x^3 + 2x^2 - x + 5 \\ +\ 5x^3 - 3x^2 + 5x + 3 \\ \hline -x^2 + 4x + 8 \end{array}$$

The following example shows how to use the Distributive Property to subtract polynomials.

EXAMPLE 2 Find $(-5x^3 + 2x^2 - x + 5) - (-5x^3 + 3x^2 - 5x - 3)$.

▶ **Solution**

$(-5x^3 + 2x^2 - x + 5) - (-5x^3 + 3x^2 - 5x - 3)$
$= (-5x^3 + 2x^2 - x + 5) + (-1)(-5x^3 + 3x^2 - 5x - 3)$ ⟵ Rewrite subtraction as addition of the opposite.
$= (-5x^3 + 2x^2 - x + 5) + (5x^3 - 3x^2 + 5x + 3)$ ⟵ Distribute -1 to each term.
$= (-5 + 5)x^3 + (2 - 3)x^2 + (-1 + 5)x + (5 + 3)$ ⟵ Combine like terms.
$= -x^2 + 4x + 8$

TRY THIS Find $(3n^3 - n - 1) - (3n^2 - 5n - 2)$.

KEY SKILLS

Write the opposite of each expression.

1. $(a + b)$
2. $(a - b)$
3. $(a - b + c)$

Rewrite each subtraction as addition of the opposite.

4. $(2x^2 - x + 4) - (x^2 + x + 2)$
5. $(3z^3 + z^2 - z + 5) - (4z^3 + z^2 - z + 5)$

PRACTICE

Find the opposite of each polynomial.

6. $-3x^3 + 2x^2 - x + 7$
7. $6x^3 - 5x^2 - 3x + 1$
8. $9x^4 + 3x^3 + x^2 - 2$
9. $-7x^5 + 8x^4 - 3x + 5$

Find each difference.

10. $(4n^3 + n^2 - 2n + 1) - (n^2 - 3n + 1)$
11. $(4m^3 - 2m + 1) - (5m^3 + 4m + 6)$
12. $(6p^4 - 3p^3 + 5p + 2) - (5p^2 + p - 1)$
13. $(6b^4 - 3b^3 - 5b + 1) - (6b^4 + b^3 - 5b + 1)$
14. $(s^4 - 5s^3 + s^2) - (-s^4 + s^3 - s^2 + 3)$
15. $(a^4 - a^3 + a^2 - a) - (a^4 + a^3 - a^2 - a)$
16. $(2c^3 - 3c^4 + 5c + 1) - (5c + c^2 - 3)$
17. $(d^3 + 3d^4 + 5d + 2) - (2d^3 - 5d + d^2 - 1)$
18. Subtract $2a^2 + 3ab - b^2$ from $6a^2 - 5b^2$.
19. Subtract $3x^2 - 3xy - y^2$ from $9x^2 + 7y^2$.
20. Subtract $3x^4 + x^3 + x^2 - 5x - 2$ from $2x^4 - x^3 + 2x^2 - 3x + 2$.
21. Subtract $x^4 - x^3 - x^2 - x - 1$ from $x^4 - x^3 + x^2 - x + 1$.
22. Subtract $4a^3 + 2a^2 - a + 1$ from $-2a^3 + a^2 + 2a + 1$.

MIXED REVIEW

Find an equation in slope-intercept form for the line that contains each point and that is parallel to the given line. (3.9 Skill A)

23. $A(-3, -3); y = \frac{1}{2}x - 2$
24. $P(0, 5); 2x - 3y = 7$
25. $Q(2, 7); y = 4$

Find an equation in slope-intercept form for the line that contains each point and that is perpendicular to the given line. (3.9 Skill B)

26. $Z(3, 6); y = -\frac{3}{4}x - \frac{3}{2}$
27. $N(2, 0); 4x + 7y = 1$
28. $Q(-1, 4); x = -3$

Answers for Lesson 6.4 Skill B

Try This Example 1
$3a^3 + 3.5a^2 + 0.5a$

Try This Example 2
$3n^3 - 3n^2 + 4n + 1$

1. $-(a + b) = -a - b$
2. $-(a - b) = -a + b$
3. $-(a - b + c) = -a + b - c$
4. $(2x^2 - x + 4) + (-x^2 - x - 2)$
5. $(3z^3 + z^2 - z + 5) +$
   $(-4z^3 - z^2 + z - 5)$
6. $3x^3 - 2x^2 + x - 7$
7. $-6x^3 + 5x^2 + 3x - 1$
8. $-9x^4 - 3x^3 - x^2 + 2$
9. $7x^5 - 8x^4 + 3x - 5$
10. $4n^3 + n$
11. $-m^3 - 6m - 5$
12. $6p^4 - 3p^3 - 5p^2 + 4p + 3$
13. $-4b^3$
14. $2s^4 - 6s^3 + 2s^2 - 3$
15. $-2a^3 + 2a^2$
16. $-3c^4 + 2c^3 - c^2 + 4$

17. $3d^4 - d^3 - d^2 + 10d + 3$
18. $4a^2 - 3ab - 4b^2$
19. $6x^2 - 3xy + 8y^2$
20. $-x^4 - 2x^3 + x^2 + 2x + 4$
21. $2x^2 + 2$
22. $-6a^3 - a^2 + 3a$
23. $y = \frac{1}{2}x - \frac{3}{2}$
24. $y = \frac{2}{3}x + 5$
25. $y = 7$
26. $y = \frac{4}{3}x + 2$
27. $y = \frac{7}{4}x - \frac{7}{2}$
28. $y = 4$

Focus: *Students use sums or differences of polynomials to model geometry and real-world problems.*

EXAMPLE 1

PHASE 1: Presenting the Example

The solution of Example 1 uses the formula $P = 2\ell + 2w$, where P is the perimeter of a rectangle, ℓ is its length, and w is its width. Although students are most likely familiar with this formula from their earlier classes in mathematics, some may find its use a little overwhelming in this context. As an alternative, show them how to find the perimeter of each rectangle simply by adding the lengths of the four sides, as shown below.

$$P = x + 2x + x + 2x$$
$$P = 1x + 2x + 1x + 2x$$
$$P = 6x$$

$$P = \frac{x}{2} + 3x + \frac{x}{2} + 3x$$
$$P = \frac{x + x}{2} + 6x$$
$$P = \frac{2x}{2} + 6x = x + 6x = 7x$$

EXAMPLE 2

PHASE 2: Providing Guidance for TRY THIS

If students have difficulty understanding how to develop the expression for profit, suggest a more structured approach such as the following:

income per pound (dollars)	times	number of pounds	minus	expenses (dollars)
↓	↓	↓	↓	↓
5.39	×	p	−	$(-0.02p^2 + 4p + 610)$

Note that students are expected to generate the expression for income, $5.39p$, because income is a simple linear function. However, the expression for the expense function is given. Expense functions in general are more complex, and in real life such a model would be generated by a cost analyst.

PHASE 3: ASSESSMENT AND CLOSURE for Lesson 6.4

- Ask students several questions like the following: Write an addition of two trinomials for which the sum is $-6s^3 + 9s^2 + 1$. [**sample:** $(-5s^3 + 3s^2 + 2) + (-s^3 + 6s^2 - 1)$] Write a subtraction of two trinomials for which the difference is $5k^3 - 3$. [**sample:** $(6k^3 + 2k^2 + 2) - (k^3 + 2k^2 + 5)$]

- Students have now used the Distributive Property and the process of combining like terms to add and subtract polynomials. In the lessons that follow, they will see how other familiar properties of real numbers are used to multiply and divide polynomials.

 For a Lesson 6.4 Quiz, see Assessment Resources *page 63.*

EXAMPLE 1

Represent the perimeter of the geometric figure below as a polynomial. In the figure, the four gray rectangles are congruent.

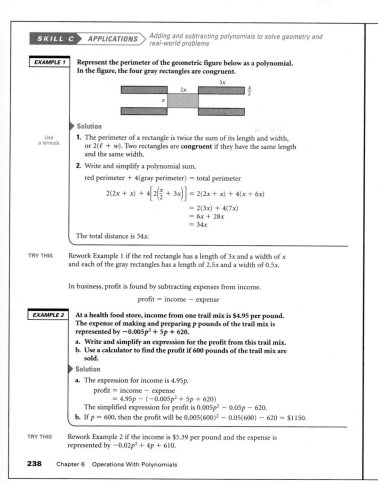

Use a formula.

▶ **Solution**

1. The perimeter of a rectangle is twice the sum of its length and width, or $2(\ell + w)$. Two rectangles are **congruent** if they have the same length and the same width.

2. Write and simplify a polynomial sum.

red perimeter + 4(gray perimeter) = total perimeter

$$2(2x + x) + 4\left[2\left(\frac{x}{2} + 3x\right)\right] = 2(2x + x) + 4(x + 6x)$$
$$= 2(3x) + 4(7x)$$
$$= 6x + 28x$$
$$= 34x$$

The total distance is $34x$.

TRY THIS   Rework Example 1 if the red rectangle has a length of $3x$ and a width of x and each of the gray rectangles has a length of $2.5x$ and a width of $0.5x$.

In business, profit is found by subtracting expenses from income.

profit = income − expense

EXAMPLE 2

At a health food store, income from one trail mix is $4.95 per pound. The expense of making and preparing p pounds of the trail mix is represented by $-0.005p^2 + 5p + 620$.

a. Write and simplify an expression for the profit from this trail mix.
b. Use a calculator to find the profit if 600 pounds of the trail mix are sold.

▶ **Solution**

a. The expression for income is $4.95p$.
   profit = income − expense
   $= 4.95p - (-0.005p^2 + 5p + 620)$
   The simplified expression for profit is $0.005p^2 - 0.05p - 620$.

b. If $p = 600$, then the profit will be $0.005(600)^2 - 0.05(600) - 620 = \$1150$.

TRY THIS   Rework Example 2 if the income is $5.39 per pound and the expense is represented by $-0.02p^2 + 4p + 610$.

EXERCISES

Refer to the geometric figure at right.
Rectangles of the same color are congruent.

1. Write an expression for the perimeter of
   a. one red region.   b. all the red regions.

2. Write an expression for the perimeter of
   a. one gray region.   b. all the gray regions.

3. Write and simplify an expression for the entire perimeter.

PRACTICE

Write and simplify an expression for the total perimeter of each figure. Rectangles of the same color are congruent.

4.

5.

6.

7.

At a health food store, one trail mix sells for $5.50 per pound. The polynomial $-0.005m^2 + 4m + 500$ represents the expense in dollars of making p pounds of the mix. Use a calculator to find the profit or loss on each sale.

8. 300 pounds   9. 400 pounds   10. 100 pounds   11. 150 pounds

MIXED REVIEW APPLICATIONS

Solve each problem. (6.2 Skills A and B)

12. If y varies directly as the square of x, and $y = 12$ when $x = 2$, find y when $x = 3$.

13. The bending of a beam varies directly as the mass of the load it supports. A beam is bent 20 mm by a mass of 40 kg. How much will the beam bend when supporting a mass of 100 kg?

Answers for Lesson 6.4 Skill C

Try This Example 1
$32x$

Try This Example 2
a. $0.02p^2 + 1.39p - 610$
b. $\$7424$

1. a. $7.2a$ b. $14.4a$
2. a. $8a$ b. $48a$
3. $14.4a + 48a = 62.4a$
4. $21x$
5. $25x$
6. $25x$
7. $22b$
8. profit of $400
9. profit of $900
10. loss of $300
11. loss of $162.50
12. 27
13. 50 mm

Multiplying and Dividing Monomials

SKILL A ▶ *Multiplying monomials*

Focus: *Students simplify products of monomials by applying the Commutative and Associative Properties of Multiplication, the Product Property of Exponents, and the Power-of-a-Power Property of Exponents.*

EXAMPLES 1, 2, and 3

PHASE 1: Presenting the Examples

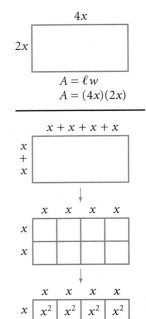

Display the first figure at right. Tell students that it is a rectangle. Remind them that the area, A, of a rectangle with length ℓ and width w is given by $A = \ell w$. So the area of this rectangle is $(4x)(2x)$, as shown.

Point out the fact that $4x = x + x + x + x$ and $2x = x + x$. Then display the second figure, in which these equivalences are used to rename the sides of the rectangle.

Note that the rectangle can be divided into squares with sides of length x. Display the third figure, in which this has been done. Ask students to write an expression for the area of each of the small squares. [x^2]

Now display the fourth figure. Ask students to write an expression for the area of the entire rectangle. [$8x^2$]

Summarize the results. That is, by directly applying the area formula for a rectangle, the area of the given rectangle must be $(4x)(2x)$. Using the areas of the small squares, the area must be $8x^2$. So the logical conclusion is that $(4x)(2x) = 8x^2$. This geometric model suggests the following algebraic method for simplifying the product $(4x)(2x)$.

$$(4x)(2x) = (4 \cdot x)(2 \cdot x) = (4 \cdot 2)(x \cdot x) = 8x^2$$

That is, to simplify a product of monomials:

- use the Commutative and Associative Properties of Multiplication to group the numerical coefficients and to group like variables;
- calculate the product of the numerical coefficients; and
- use the properties of exponents to simplify the variable product.

Now discuss Examples 1–3, which demonstrate how the algebraic method of multiplying is applied to several products of monomials.

PHASE 2: Providing Guidance for TRY THIS

Students often forget to apply the Power-of-a-Power Property to a numerical coefficient. For instance, in part a of the Example 2 Try This, some students may simplify $\left(\frac{1}{4}t^3\right)^2$ as $\frac{1}{4}t^6$. Encourage them to rewrite the given expression as shown at right. In this way, it becomes clear that $\frac{1}{4}$ must be considered as a factor two times.

$$\left(\frac{1}{4}t^3\right)^2$$
$$\downarrow$$
$$\left(\frac{1}{4}t^3\right)\left(\frac{1}{4}t^3\right)$$

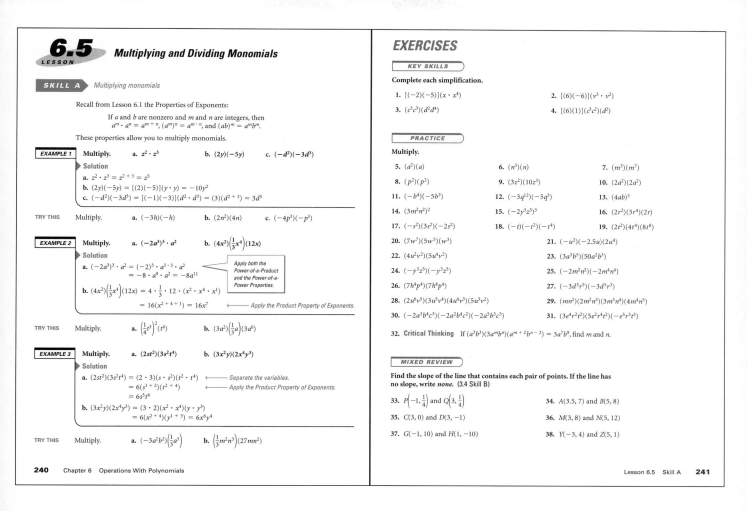

6.5 LESSON
Multiplying and Dividing Monomials

SKILL A *Multiplying monomials*

Recall from Lesson 6.1 the Properties of Exponents:

If a and b are nonzero and m and n are integers, then
$$a^m \cdot a^n = a^{m+n}, (a^m)^n = a^{m \cdot n}, \text{ and } (ab)^m = a^m b^m.$$

These properties allow you to multiply monomials.

EXAMPLE 1 Multiply. **a.** $z^2 \cdot z^3$ **b.** $(2y)(-5y)$ **c.** $(-d^2)(-3d^3)$

▶ Solution
a. $z^2 \cdot z^3 = z^{2+3} = z^5$
b. $(2y)(-5y) = [(2)(-5)](y \cdot y) = -10y^2$
c. $(-d^2)(-3d^3) = [(-1)(-3)](d^2 \cdot d^3) = (3)(d^{2+3}) = 3d^5$

TRY THIS Multiply. **a.** $(-3h)(-h)$ **b.** $(2n^2)(4n)$ **c.** $(-4p^3)(-p^3)$

EXAMPLE 2 Multiply. **a.** $(-2a^3)^3 \cdot a^2$ **b.** $(4x^2)\left(\frac{1}{3}x^4\right)(12x)$

▶ Solution
a. $(-2a^3)^3 \cdot a^2 = (-2)^3 \cdot a^{3 \cdot 3} \cdot a^2$ *Apply both the Power-of-a-Product and the Power-of-a-Power Properties.*
$= -8 \cdot a^9 \cdot a^2 = -8a^{11}$
b. $(4x^2)\left(\frac{1}{3}x^4\right)(12x) = 4 \cdot \frac{1}{3} \cdot 12 \cdot (x^2 \cdot x^4 \cdot x^1)$
$= 16(x^{2+4+1}) = 16x^7$ ◀ *Apply the Product Property of Exponents.*

TRY THIS Multiply. **a.** $\left(\frac{1}{4}t^3\right)^2(t^4)$ **b.** $(3a^2)\left(\frac{1}{3}a\right)(3a^0)$

EXAMPLE 3 Multiply. **a.** $(2st^2)(3s^2t^4)$ **b.** $(3x^2y)(2x^4y^3)$

▶ Solution
a. $(2st^2)(3s^2t^4) = (2 \cdot 3)(s \cdot s^2)(t^2 \cdot t^4)$ ◀ *Separate the variables.*
$= 6(s^{1+2})(t^{2+4})$ ◀ *Apply the Product Property of Exponents.*
$= 6s^3t^6$
b. $(3x^2y)(2x^4y^3) = (3 \cdot 2)(x^2 \cdot x^4)(y \cdot y^3)$
$= 6(x^{2+4})(y^{1+3}) = 6x^6y^4$

TRY THIS Multiply. **a.** $(-3a^2b^2)\left(\frac{1}{3}a^3\right)$ **b.** $\left(\frac{1}{3}m^2n^3\right)(27mn^2)$

EXERCISES

Complete each simplification.

1. $[(-2)(-5)](x \cdot x^4)$ **2.** $[(6)(-6)](v^3 \cdot v^2)$

3. $(c^3c^3)(d^2d^4)$ **4.** $[(6)(1)](c^3c^2)(d^2)$

PRACTICE

Multiply.

5. $(a^2)(a)$ **6.** $(n^3)(n)$ **7.** $(m^3)(m^7)$

8. $(p^2)(p^2)$ **9.** $(3z^2)(10z^3)$ **10.** $(2a^2)(2a^2)$

11. $(-b^4)(-5b^3)$ **12.** $(-3q^{12})(-5q^3)$ **13.** $(4ab)^3$

14. $(3m^2n^2)^2$ **15.** $(-2y^3z^5)^3$ **16.** $(2r^2)(3r^4)(2r)$

17. $(-s^2)(3s^2)(-2s^2)$ **18.** $(-t)(-t^2)(-t^4)$ **19.** $(2t^2)(4t^4)(8t^8)$

20. $(7w^7)(5w^5)(w^3)$ **21.** $(-u^2)(-2.5u)(2u^4)$

22. $(4u^2v^2)(5u^4v^2)$ **23.** $(3a^3b^5)(50a^2b^3)$

24. $(-y^3z^5)(-y^3z^5)$ **25.** $(-2m^2n^2)(-2m^4n^4)$

26. $(7h^4p^4)(7h^4p^4)$ **27.** $(-3d^3r^5)(-3d^5r^3)$

28. $(2u^6v^5)(3u^5v^4)(4u^4v^3)(5u^3v^2)$ **29.** $(mn^2)(2m^2n^3)(3m^3n^4)(4m^4n^5)$

30. $(-2a^3b^4c^3)(-2a^2b^4c^2)(-2a^5b^3c^5)$ **31.** $(3e^4r^2t^2)(5e^2r^4t^2)(-e^5r^3t^3)$

32. Critical Thinking If $(a^3b^3)(3a^mb^n)(a^{m+2}b^{n-2}) = 3a^7b^9$, find m and n.

MIXED REVIEW

Find the slope of the line that contains each pair of points. If the line has no slope, write *none*. (3.4 Skill B)

33. $P\left(-1, \frac{1}{4}\right)$ and $Q\left(3, \frac{1}{4}\right)$ **34.** $A(3.5, 7)$ and $B(5, 8)$

35. $C(3, 0)$ and $D(3, -1)$ **36.** $M(3, 8)$ and $N(5, 12)$

37. $G(-1, 10)$ and $H(1, -10)$ **38.** $Y(-3, 4)$ and $Z(5, 1)$

Answers for Lesson 6.5 Skill A

Try This Example 1
a. $3h^2$ **b.** $8n^3$ **c.** $4p^6$

Try This Example 2
a. $\frac{1}{16}t^{10}$ **b.** $3a^3$

Try This Example 3
a. $-a^5b^2$ **b.** $9m^3n^5$

1. $10x^5$

2. $-36v^5$

3. c^6d^6

4. $6c^5d^2$

5. a^3

6. n^4

7. m^{10}

8. p^4

9. $30z^5$

10. $4a^4$

11. $5b^7$

12. $15q^{15}$

13. $64a^3b^3$

14. $9m^4n^4$

15. $-8y^9z^{15}$

16. $12r^7$

17. $6s^6$

18. $-t^7$

19. $64t^{14}$

20. $35w^{15}$

21. $5u^7$

22. $20u^6v^4$

23. $150a^5b^8$

24. y^6z^{10}

25. $4m^6n^6$

26. $49h^8p^8$

27. $9d^8r^8$

28. $120u^{18}v^{14}$

29. $24m^{10}n^{14}$

30. $-8a^{10}b^{11}c^{10}$

31. $-15e^{11}r^9t^7$

32. $m = 1$ and $n = 4$

33. 0

34. $\frac{2}{3}$

35. none

36. 2

37. -10

38. $-\frac{3}{8}$

Focus: *Students simplify quotients of monomials by applying the Quotient Property of Exponents, the Power-of-a-Power Property of Exponents, and the Power-of-a-Product Property of Exponents.*

EXAMPLE 1

PHASE 1: Presenting the Example

Simplifying the first two quotients in this example involves a straightforward application of the Quotient Property of Exponents. However, students are learning a variety of monomial operations, and they may occasionally become confused. Show them how to check that they have applied the Quotient Property of Exponents properly by using the following extended technique.

$$\frac{n^5}{n^3} = \frac{n \cdot n \cdot n \cdot n \cdot n}{n \cdot n \cdot n} = \frac{\overset{1}{\cancel{n}} \cdot \overset{1}{\cancel{n}} \cdot \overset{1}{\cancel{n}} \cdot n \cdot n}{\underset{1}{\cancel{n}} \cdot \underset{1}{\cancel{n}} \cdot \underset{1}{\cancel{n}}} = \frac{1 \cdot 1 \cdot 1 \cdot n \cdot n}{1 \cdot 1 \cdot 1} = \frac{n^2}{1} = n^2$$

$$\frac{m^2}{m^5} = \frac{m \cdot m}{m \cdot m \cdot m \cdot m \cdot m} = \frac{\overset{1}{\cancel{m}} \cdot \overset{1}{\cancel{m}}}{\underset{1}{\cancel{m}} \cdot \underset{1}{\cancel{m}} \cdot m \cdot m \cdot m} = \frac{1 \cdot 1}{1 \cdot 1 \cdot m \cdot m \cdot m} = \frac{1}{m^3}$$

PHASE 2: Providing Guidance for **TRY THIS**

Some students may write their answers as b^1, $\frac{1}{d^1}$, and $-\frac{8}{5y^1}$. Although these forms are correct, be sure students understand that the exponent 1 is generally omitted in a final answer.

EXAMPLE 2

PHASE 1: Presenting the Example

Ask students to write statements of the Power-of-a-Power Property of Exponents and of the Power-of-a-Product Property of Exponents. [$(a^m)^n = a^{mn}$; $(ab)^n = a^n b^n$] Discuss their responses. Tell them that they are now going to see how to use these in combination with the Quotient Property of Exponents.

PHASE 2: Providing Guidance for **TRY THIS**

Watch for the response $100u^2v^4$. Most likely students correctly rewrote $\frac{(5u^2v^2)^3}{(5u^2v)^2}$ as $\frac{125u^6v^6}{25u^4v^2}$, and then mistakenly applied the Quotient Property of Exponents to the numerical coefficients. It may help if they leave the numerical coefficients in exponential form until the final step, as shown below.

$$\frac{(5u^2v^2)^3}{(5u^2v)^2} = \frac{5^3 u^6 v^6}{5^2 u^4 v^2}$$
$$= 5^{3-2} u^{6-4} v^{6-2}$$
$$= 5^1 u^2 v^4$$
$$= 5u^2 v^4$$

You can use the Quotient Property of Exponents to simplify quotients containing monomials. In this section, you may assume that no denominator has a value of 0.

EXAMPLE 1 Divide. Write the answer with positive exponents.

a. $\frac{n^5}{n^3}$ b. $\frac{m^2}{m^5}$ c. $\frac{10r^2}{2r^6}$

▸ Solution

a. $\frac{n^5}{n^3} = n^{5-3} = n^2$ b. $\frac{m^2}{m^5} = m^{2-5} = m^{-3} = \frac{1}{m^3}$ c. $\frac{10r^2}{2r^6} = 5r^{2-6} = \frac{5}{r^4}$

TRY THIS Divide. Write the answer with positive exponents. a. $\frac{b^5}{b^4}$ b. $\frac{d^6}{d^7}$ c. $\frac{-16y}{10y^2}$

You can use the Power-of-a-Product Property along with the Quotient Property to simplify expressions that contain both products and quotients.

EXAMPLE 2 Divide. Write the answer with positive exponents.

a. $\frac{(2a^2b)^2}{(4ab^2)^3}$ b. $\frac{(a^2b^3)^2}{(5a^6b^5)^2}$

▸ Solution

a. $\frac{(2a^2b)^2}{(4ab^2)^3} = \frac{4a^4b^2}{64a^3b^6}$ ⟵ Apply the Power-of-a-Product Property of Exponents.

$= \frac{4}{64} \cdot \frac{a^4}{a^3} \cdot \frac{b^2}{b^6}$ ⟵ Separate the variables.

$= \frac{4}{64} \cdot a^{4-3} \cdot b^{2-6}$ ⟵ Apply the Quotient Property of Exponents.

$= \frac{1}{16} \cdot a^1 \cdot b^{-4}$

$= \frac{a}{16b^4}$ ⟵ Write with positive exponents.

b. $\frac{(a^2b^3)^2}{(5a^6b^5)^2} = \frac{a^4b^6}{25a^{12}b^{10}}$ ⟵ Apply the Power-of-a-Product Property.

$= \frac{1}{25} \cdot \frac{a^4}{a^{12}} \cdot \frac{b^6}{b^{10}}$ ⟵ Separate the variables.

$= \frac{1}{25} \cdot a^{4-12} \cdot b^{6-10}$ ⟵ Apply the Quotient Property of Exponents.

$= \frac{1}{25} \cdot a^{-8} \cdot b^{-4}$

$= \frac{1}{25a^8b^4}$ ⟵ Write with positive exponents.

TRY THIS Divide $\frac{(5u^2v^2)^3}{(5u^2v)^2}$. Write the answer with positive exponents.

EXERCISES

KEY SKILLS

Rewrite each expression so that all exponents are positive.

1. a^{7-3} 2. b^{3-7} 3. m^{2-3} 4. v^{3-2}

PRACTICE

Divide. Write each answer with positive exponents.

5. $\frac{d^7}{d^2}$ 6. $\frac{c^5}{c^6}$ 7. $\frac{w^5}{w^7}$ 8. $\frac{z^6}{z}$

9. $\frac{12a^5}{-3a^4}$ 10. $\frac{-2f^4}{6f^3}$ 11. $\frac{15g^4}{6g^5}$ 12. $\frac{96q^7}{-12q^3}$

13. $\frac{(2s^2)^3}{s^2}$ 14. $\frac{(-3t^2)^3}{(2t^2)^2}$ 15. $\frac{(4a^3)^2}{(2a^2)^4}$ 16. $\frac{(5n^3)^3}{(2n^2)^5}$

17. $\frac{24p^3q^2}{12p^9q^3}$ 18. $\frac{12a^4d^4}{24a^3d^2}$ 19. $\frac{y^3z^3}{5yz}$ 20. $\frac{100y^3z}{25y^4z^3}$

21. $\frac{(2mn^3)^3}{(mn)^2}$ 22. $\frac{(3a^3z)^4}{(9az)^3}$ 23. $\frac{(2a^3b^2)^2}{(2a^4b^3)^2}$ 24. $\frac{(2kp^2)^3}{(2k^4p)^2}$

Simplify the individual quotients. Then multiply.

25. $\left(\frac{a^3}{a^2}\right)\left(\frac{c^2}{c^3}\right)$ 26. $\left(\frac{m^4}{m^3}\right)\left(\frac{n^5}{n^3}\right)$ 27. $\left(\frac{-q^5}{-q^4}\right)\left(\frac{k^4}{k^3}\right)$ 28. $\left(\frac{-d^3}{d^5}\right)\left(\frac{s^4}{s^3}\right)$

29. $\left(\frac{12x^3}{7x^3}\right)\left(\frac{21y^5}{15y^4}\right)$ 30. $\left(\frac{-2a^7}{3a^4}\right)\left(\frac{27z^5}{8z^5}\right)$ 31. $\left(\frac{-7b^7}{15b^5}\right)\left(\frac{5c^5}{-14c^3}\right)$ 32. $\left(\frac{-49u^7}{11u^5}\right)\left(\frac{22v^7}{-7v^8}\right)$

33. $\left(\frac{16p^4}{9p^2}\right)\left(\frac{3a^3}{4a^2}\right)^2$ 34. $\left(\frac{h^6r^3}{2h^2r^2}\right)^2\left(\frac{4h^3r^4}{r^3}\right)^2$ 35. $\left(\frac{f^4b^3}{4f^3b^2}\right)^2\left(\frac{4f^3b^3}{b^2}\right)^2$ 36. $\left(\frac{-9a^5b}{3a^2b}\right)^2\left(\frac{2a^2b^3}{ab^2}\right)^3$

MIXED REVIEW

Find the equation in point-slope form for the line with the given slope that contains the given point. (3.6 Skill A)

37. slope -2.5; $P(-3, -5)$ 38. slope 6; $N(2.5, -3)$ 39. slope $\frac{3}{7}$; $Z(0, 0)$

40. slope $-\frac{7}{2}$; $Q(2, 0)$ 41. slope 0; $W(3, 5)$ 42. slope $\frac{1}{5}$; $A(0, 3)$

Answers for Lesson 6.5 Skill B

Try This Example 1

a. b b. $\frac{1}{d}$ c. $-\frac{8}{5y}$

Try This Example 2

$5u^2v^4$

1. a^4
2. $\frac{1}{b^4}$
3. $\frac{1}{m}$
4. v
5. d^5
6. $\frac{1}{c}$
7. $\frac{1}{w^2}$
8. z^5
9. $-4a$
10. $-\frac{f}{3}$
11. $\frac{5}{2g}$

12. $-8q^4$
13. $8s^4$
14. $\frac{-27t^2}{4}$
15. $\frac{1}{a^2}$
16. $\frac{125}{32n}$
17. $\frac{2}{q}$
18. $\frac{d^2}{2a}$
19. $\frac{y^2z^2}{5}$
20. $\frac{4}{yz^2}$
21. $8mn^7$
22. $\frac{a^9z}{9}$
23. $\frac{1}{a^2b^2}$

24. $\frac{2p^4}{k^5}$
25. $\frac{a}{c}$
26. mn^2
27. qk
28. $-s$
29. $\frac{12y}{5}$
30. $-\frac{9a^4}{4}$
31. $\frac{b^2c^2}{6}$
32. $14u^2v$
33. $\frac{16p^4a^2}{9}$
34. $4h^{14}r^4$
35. f^8b^4
36. $72a^9b^3$
37. $y + 5 = -2.5(x + 3)$
38. $y + 3 = 6(x - 2.5)$
39. $y = \frac{3}{7}x$

40. $y = -\frac{7}{2}(x - 2)$
41. $y - 5 = 0$
42. $y - 3 = \frac{1}{5}x$

Focus: *Students use products or quotients of monomials to model geometry problems.*

EXAMPLE 1

PHASE 1: Presenting the Example

Display a sketch of the box that is shown on the textbook page. Give students the table at right and have them complete each row for the given value of w. [**Answers are given in red at right.**]

width (w)	length ($2w$)	height ($3w$)	volume of box
1	2	3	6
2	4	6	48
3	6	9	162
4	8	12	384
5	10	15	750

Ask students if it is possible to calculate the volume for a given width without first calculating the length and height. Lead them to the conclusion that, since the length is always twice the width, and the height is always three times the width, the volume can always be represented by the product $(w)(2w)(3w)$, or $6w^3$. Have students verify that evaluating this expression for $w = 1$, $w = 2$, $w = 3$, $w = 4$, and $w = 5$ yields the same volume as the table value. [$6(1)^3 = 6(1) = 6$; $6(2)^3 = 6(8) = 48$; $6(3)^3 = 6(27) = 162$; $6(4)^3 = 6(64) = 384$; $6(5)^3 = 6(125) = 750$] Then discuss Example 1.

EXAMPLE 2

PHASE 1: Presenting the Example

PHASE 2: Providing Guidance for TRY THIS

Lead a discussion in which students consider why a manufacturer, when packaging a product, might be interested in the ratio of volume to surface area. [**sample: A greater ratio of product volume to package surface area is more cost-effective than a lower ratio of volume to surface area.**]

PHASE 3: ASSESSMENT AND CLOSURE for Lesson 6.5

- Give students the two statements at right. Tell them to fill in the boxes with positive integers to make each statement true. [**samples: 1.** $(2jk^4)(9j^2k^3)$ **2.** $\dfrac{8r^5s^4}{10r^3s^7}$]

  1. $(\Box j^{\Box}k^{\Box})(\Box j^{\Box}k^{\Box}) = 18j^3k^7$

  2. $\dfrac{\Box r^{\Box}s^{\Box}}{\Box r^{\Box}s^{\Box}} = \dfrac{4r^2}{5s^3}$

- Students should be aware that knowing how to multiply and divide monomials is critical to the process of multiplying and dividing *polynomials*, which will be covered in the next two lessons.

☞ *For a Lesson 6.5 Quiz, see* Assessment Resources *page 64.*

Recall that the volume of a rectangular solid is given by the following equation.

$$\text{Volume} = \text{length} \times \text{width} \times \text{height}$$

EXAMPLE 1

A box's length is twice its width, and its height is three times its width. Write an expression for the total volume of 24 boxes.

▶ Solution

Use a formula.

Let w represent the width. Then length is $2w$ and height is $3w$.

Write and simplify an expression for volume.

$$24(w)(2w)(3w) = (24 \cdot 1 \cdot 2 \cdot 3)(w \cdot w \cdot w) = 144w^3$$

The total volume of 24 boxes is $144w^3$ cubic units.

TRY THIS Find the total volume of 48 boxes whose dimensions are x, $5x$, and $10x$.

The volume and surface area of a cube with edge length l are given by the following equations.

$$\text{Volume} = l^3 \qquad \text{Surface area} = 6l^2$$

Recall from Lesson 3.3 that a ratio is the comparison of two quantities by division. In the next example, we will find the ratio of volume to surface area.

EXAMPLE 2

A cube has an edge length of $3d$ units. Write a ratio to compare the cube's volume to its surface area.

▶ Solution

Use a formula.

1. Write expressions for volume and surface area.

   Volume: $(3d)^3$ Surface area: $6(3d)^2$

2. Write and simplify a ratio.

$$\frac{\text{Volume}}{\text{Surface area}} \rightarrow \frac{(3d)^3}{6(3d)^2} = \frac{27d^3}{54d^2}$$

$$= \frac{27}{54} \cdot d^{3-2} = \frac{1}{2} \cdot d = \frac{d}{2}$$

Therefore, the ratio of volume to surface area for this cube is $\frac{d}{2}$.

TRY THIS A cube has an edge length of $4d$ units. Write a ratio to compare the cube's volume to its surface area.

Monomials can be found in several other geometry formulas, including some that you will study in later chapters as well as in future courses.

EXERCISES

Refer to the diagram at right.

1. Identify the length, width, and height.

2. Write an expression for the volume.

Solve each problem.

3. The dimensions of a rectangular box are x units, $3x$ units, and $5x$ units. Write an expression for the total volume of 36 boxes.

4. A cube has an edge length of $2x$ units. Write a ratio to compare the cube's volume to its surface area.

5. The sides of a square are x units long. The sides of a second square are kx units long, where k is a positive number. Write a ratio to compare the area of the second square to the area of the first square.

6. The area, A, of a circle whose radius is r units is $A = \pi r^2$. Let k be a positive number. Write a ratio to compare the area of a circle whose radius is kr with the area of a circle whose radius is r.

7. **Critical Thinking** The surface area, S, of a rectangular box with length l, width w, and height h is $S = 2(lw + lh + hw)$. Write a ratio to compare volume to surface area when $w = 2l$ and $h = 5l$.

8. Write a ratio to compare the total area of the red squares with the total perimeter of the figure below. (All red squares are congruent.)

An automobile traveling at 55 miles per hour had a stopping distance of 125 feet. (6.2 Skill B)

9. Find the stopping distance if the speed is 60 miles per hour.

10. Write a ratio to compare the stopping distance for 60 miles per hour to the stopping distance for 30 miles per hour.

Answers for Lesson 6.5 Skill C

Try This Example 1

$2400x^3$

Try This Example 2

$\dfrac{2d}{3}$

1. length $= 2a$
   width $= a$
   height $= 2a$
2. $V = (2a)(2a)(a) = 4a^3$
3. $540x^3$
4. $\dfrac{x}{3}$
5. k^2
6. k^2
7. $\dfrac{5l}{17}$
8. $\dfrac{5a}{11}$
9. 148.8 feet
10. 4 to 1

TEACHING 6.6 LESSON

Multiplying Polynomials

SKILL A   *Multiplying a polynomial by a monomial*

Focus: *Students use the Distributive Property and the properties of exponents to find the product when a polynomial is multiplied by a monomial.*

EXAMPLE 1

PHASE 1: Presenting the Example

Display the first figure at right. Tell students that it is a rectangle. Ask them to write the multiplication that must be performed in order to find its area. Be sure they understand that they do not need to perform the multiplication at this time. [$2m(m + 3)$]

Now display the second figure. Point out that the side labeled $2m$ has been renamed $m + m$, and the original rectangle has been divided into four smaller rectangles. Ask students to write expressions for the areas of the small rectangles. [**clockwise from top left:** $m^2, 3m, 3m, m^2$]

Now display the third figure, in which the areas of the small rectangles are labeled. Ask students to write a simplified expression for the area of the entire original rectangle. [$m^2 + 3m + 3m + m^2 = 2m^2 + 6m$]

Note that this geometric model suggests that $2m(m + 3)$ must be equal to $2m^2 + 6m$. Elicit students' ideas about how they might obtain this result algebraically. Lead them to see that they can apply the Distributive Property to $2m(m + 3)$. Then discuss the solution on the textbook page.

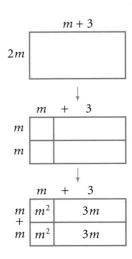

EXAMPLE 2

PHASE 1: Presenting the Example

Some students may wonder why this is considered an extended form of the Distributive Property. Remind them that the basic statement of the Distributive Property involves only two terms inside the parentheses: $a(b + c) = ab + ac$. However, applying the basic assumption twice, as shown at right, easily leads to the extended form used in Example 2.

$$a(b + c + d)$$
$$= a[(b + c) + d]$$
$$= a(b + c) + ad$$
$$= ab + ac + ad$$

EXAMPLES 1, 2, and 3

PHASE 2: Providing Guidance for **TRY THIS**

Students commonly forget to multiply *each* factor inside parentheses by the factor outside. As they do the Try This exercise for each example, you may want to suggest that they multiply using the vertical format shown below.

Example 1:
$$\begin{array}{r} n - 4 \\ \times 2n^2 \\ \hline 2n^3 - 8n^2 \end{array}$$

Example 2:
$$\begin{array}{r} 5x^2 - x - 5 \\ \times 2x^3 \\ \hline 10x^5 - 2x^4 - 10x^3 \end{array}$$

Example 3:
$$\begin{array}{r} 3r^2s - r^3s \\ \times r^2s \\ \hline 3r^4s^2 - r^5s^2 \end{array}$$

6.6
LESSON

Multiplying Polynomials

SKILL A ▶ Multiplying a polynomial by a monomial

To multiply a polynomial by a monomial, apply the Distributive Property.
Then use the Product Property of Exponents to simplify the expression.

EXAMPLE 1

Simplify $2m(m + 3)$.

▶ Solution

$2m(m + 3) = 2m(m) + 2m(3)$ ⟵ Apply the Distributive Property.
$\quad\quad\quad\quad = 2m^2 + 6m$

TRY THIS   Simplify $2n^2(n - 4)$.

The Distributive Property can be extended to sums with more than two terms. In the examples below, 2 is distributed to three terms, and $3x$ is distributed to four terms.

$2(x^2 + x + 3)$ $3x(x^3 + 2x^2 + 4x + 1)$
$2(x^2) + 2(x) + 2(3)$ $3x(x^3) + 3x(2x^2) + 3x(4x) + 3x(1)$
$2x^2 + 2x + 6$ $3x^4 + 6x^3 + 12x^2 + 3x$

Use this extended form of the Distributive Property to multiply a monomial by a polynomial with many terms.

EXAMPLE 2

Simplify $2x^2(3x^3 + 5x^2 - 4)$.

▶ Solution

$2x^2(3x^3 + 5x^2 - 4)$
$= 2x^2(3x^3) + 2x^2(5x^2) + 2x^2(-4)$ ⟵ Apply the Distributive Property.
$= 6x^5 + 10x^4 - 8x^2$

TRY THIS   Simplify $2x^3(5x^2 - x - 5)$.

EXAMPLE 3

Simplify $2a^2b(3ab^2 - 4ab)$.

▶ Solution

$2a^2b(3ab^2 - 4ab) = 2a^2b(3ab^2) + 2a^2b(-4ab)$
$\quad\quad\quad\quad\quad = 6(a^2 \cdot a)(b \cdot b^2) - 8(a^2 \cdot a)(b \cdot b)$ ⟵ Separate the variables.
$\quad\quad\quad\quad\quad = 6a^3b^3 - 8a^3b^2$

TRY THIS   Simplify $r^2s(3r^2s - r^3s)$.

EXERCISES

KEY SKILLS

Write the sum that results when you apply the Distributive Property.
Do not simplify further.

1. $3z^2(2z + 4)$ 2. $3p^2(p^2 + 3p - 1)$ 3. $uv(2uv^2 + 3u^2v)$

Complete each simplification.

4. $(5x^3)(-2x^3) + (5x^3)(2x^2)$ 5. $(y^3)(-2y^4) + (y^3)(4y)$ 6. $(5s^2)(-2s^4) + (5s^2)(4s)$

PRACTICE

Find each product.

7. $3x(4x + 5)$ 8. $4y(5y - 2)$
9. $-2a^2(3a - 1)$ 10. $3b^3(3b - 2)$
11. $c^3(10c^2 + 5c)$ 12. $2n^3(-3n^2 - n)$
13. $q^3(3q^2 - q + 3)$ 14. $r^3(2r^2 - 5r - 1)$
15. $w(2w^4 - 5w^3 - w^2)$ 16. $-2d^2(d^4 - 3d^3 + 2d^2)$
17. $(2k^5 + 4k^3 - 2k^2)(4k^3)$ 18. $(12h^4 + 4h^2 - 3h)(-h^2)$
19. $rs(rs^2 - rs)$ 20. $m^2n(m^2n^2 + mn^2)$
21. $-2x^2y^2(3x^2y - 5x^2y^2)$ 22. $-2a^4b^2(7a^3b + 4a^4b^3)$
23. $2u^2v^2(7u^3 + 4u^2v^2 - 4v^3)$ 24. $u^3v^2(-3u^2 - 5u^3v^3 + v^2)$

25. Simplify $2m^2(3m^2 - m) + 3m(2m^3 + 2m^2)$.
26. Simplify $3z^2(2z^2 - z) - 3z(5z^3 + z^2)$.
27. Find a such that $2x^a(5x^{2a-3} + 2x^{2a+2}) = 10x^3 + 4x^8$.
28. **Critical Thinking** Find m and n such that
    $x^my^n(x^ny^m + x^{n-2}y) = x^5y^5 + x^3y^3$.
29. **Critical Thinking** Write $2x(a + b) + 3y(a + b)$ as a product of two binomials.

MIXED REVIEW

Graph each inequality in the coordinate plane. (5.5 Skill A)

30. $y \geq 4$ 31. $x < -3$ 32. $y > -\frac{2}{3}x + 1$
33. $y < \frac{5}{4}x - 1$ 34. $y \leq \frac{1}{6}x$ 35. $y \geq -3.5x$

Answers for Lesson 6.6 Skill A

Try This Example 1
$2n^3 - 8n^2$

Try This Example 2
$10x^5 - 2x^4 - 10x^3$

Try This Example 3
$3r^4s^2 - r^5s^2$

1. $3z^2(2z) + 3z^2(4)$
2. $3p^2(p^2) + 3p^2(3p) + 3p^2(-1)$
3. $uv(2uv^2) + uv(3u^2v)$
4. $-10x^6 + 10x^5$
5. $4y^4 - 2y^7$
6. $-10s^6 + 20s^3$
7. $12x^2 + 15x$
8. $20y^2 - 8y$
9. $-6a^3 + 2a^2$
10. $9b^4 - 6b^3$
11. $10c^5 + 5c^4$
12. $-6n^5 - 2n^4$

13. $3q^5 - q^4 + 3q^3$
14. $2r^5 - 5r^4 - r^3$
15. $2w^5 - 5w^4 - w^3$
16. $-2d^6 + 6d^5 - 4d^4$
17. $8k^8 + 16k^6 - 8k^5$
18. $-12h^6 - 4h^4 + 3h^3$
19. $r^2s^3 - r^2s^2$
20. $m^4n^3 + m^3n^3$
21. $-6x^4y^3 + 10x^4y^4$
22. $-14a^7b^3 - 8a^8b^5$
23. $14u^5v^2 + 8u^4v^4 - 8u^2v^5$
24. $-3u^5v^2 - 5u^6v^5 + u^3v^4$
25. $12m^4 + 4m^3$
26. $-9z^4 - 6z^3$
★27. $a = 2$
28. $m = 3$ and $n = 2$
29. $(2x + 3y)(a + b)$

30.
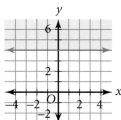

31.
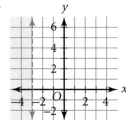

32.
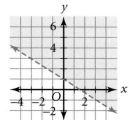

33.
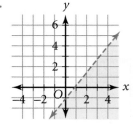

For answers to Exercises 34–35, see Additional Answers.

★ **Advanced Exercises**

SKILL B *Multiplying polynomials*

Focus: *Students multiply polynomials by applying the Distributive Property two or more times in succession.*

EXAMPLES 1 AND 2

PHASE 1: Presenting the Examples

Display ($\boxed{}$)$(3r + 2)$. Tell students to "distribute" the box to each term inside the parentheses. Display the result as shown below.

$$(\boxed{})(3r + 2) = (\boxed{})(3r) + (\boxed{})(2)$$

Ask students what the simplified expression would be if the box contained the number 5. What if it contained the variable r? What if it contained the expression $4r$? One by one, display the results as shown below.

$$(\boxed{5})(3r + 2) = (\boxed{5})(3r) + (\boxed{5})(2) = 15r + 10$$

$$(\boxed{r})(3r + 2) = (\boxed{r})(3r) + (\boxed{r})(2) = 3r^2 + 2r$$

$$(\boxed{4r})(3r + 2) = (\boxed{4r})(3r) + (\boxed{4r})(2) = 12r^2 + 8r$$

Now ask students to write the result if the box contained the expression $2r - 3$. Display it as shown below.

$$(\boxed{2r - 3})(3r + 2) = (\boxed{2r - 3})(3r) + (\boxed{2r - 3})(2)$$

Ask students how this situation is different from those above. Elicit the response that, in order to simplify, they must now apply the Distributive Property two additional times. Then lead them through the process.

$$\begin{aligned}(\boxed{2r - 3})(3r + 2) &= (\boxed{2r - 3})(3r) + (\boxed{2r - 3})(2) = \\ &= (2r)(3r) - 3(3r) + (2r)(2) - 3(2) \\ &= 6r^2 - 9r \quad + \quad 4r - 6 \\ &= 6r^2 - 5r - 6\end{aligned}$$

Now discuss the examples. In Example 1, students will see how to perform the same multiplications using a vertical format. In Example 2, they will see another horizontal multiplication method, the FOIL method.

EXAMPLE 3

PHASE 2: Providing Guidance for **TRY THIS**

Finding the product of a trinomial and a binomial requires six different monomial multiplications, and some students may start to feel overwhelmed. As a result, they may overlook one or more of the multiplications. Suggest that students use a rectangular grid like the one shown at right as a means of keeping track of all of the products.

	$-5n^2$	n	3
$3n$	$(3n)(-5n^2)$	$(3n)(n)$	$(3n)(3)$
2	$(2)(-5n^2)$	$(2)(n)$	$(2)(3)$

You can use a vertical format to multiply polynomials.

EXAMPLE 1 Multiply $(2r - 3)(3r + 2)$.

▸ Solution

Multiply $2r - 3$ by 2.   Multiply $2r - 3$ by $3r$.   Add.

$$\begin{array}{r} 2r - 3 \\ \times\ 3r + 2 \\ \hline 4r - 6 \end{array} \qquad \begin{array}{r} 2r - 3 \\ \times\ 3r + 2 \\ \hline 4r - 6 \\ 6r^2 - 9r \end{array} \qquad \begin{array}{r} 2r - 3 \\ \times\ 3r + 2 \\ \hline 4r - 6 \\ +\ 6r^2 - 9r \\ \hline 6r^2 - 5r - 6 \end{array}$$

Therefore, $(2r - 3)(3r + 2) = 6r^2 - 5r - 6$.

TRY THIS   Multiply $(3z + 2)(-z + 5)$.

Example 2 shows a horizontal format for multiplying polynomials.

EXAMPLE 2 Multiply $(2x - 1)(3x + 2)$.

▸ Solution

Method 1

$(2x - 1)(3x + 2)$
$= (2x - 1)(3x) + (2x - 1)(2)$
$= 2x(3x) - 1(3x) + 2x(2) - 1(2)$
$= 6x^2 - 3x + 4x - 2$
$= 6x^2 + x - 2$

Method 2

$(2x - 1)(3x + 2)$
$= 2x(3x) + 2x(2) - 1(3x) - 1(2)$
$= 6x^2 + 4x - 3x - 2$
$= 6x^2 + x - 2$

TRY THIS   Multiply $(3n + 2)(n + 3)$.

Method 2 above is called the **FOIL method**. In this
method, multiply the First terms, the Outer terms, the
Inner terms, and the Last terms.

$(2x - 1)(3x + 2)$ *First terms*
$(2x - 1)(3x + 2)$ *Outer terms*
$(2x - 1)(3x + 2)$ *Inner terms*
$(2x - 1)(3x + 2)$ *Last terms*

The FOIL method can only be used to multiply two
binomials. If either polynomial has more than two terms, use the Distributive Property.

EXAMPLE 3 Multiply $(2x - 1)(3x^2 - 3x + 2)$.

▸ Solution

$(2x - 1)(3x^2 - 3x + 2) = (2x - 1)(3x^2) + (2x - 1)(-3x) + (2x - 1)(2)$
$= 2x(3x^2) - 1(3x^2) + 2x(-3x) - 1(-3x) + 2x(2) - 1(2)$
$= 6x^3 - 3x^2 - 6x^2 + 3x + 4x - 2 = 6x^3 - 9x^2 + 7x - 2$

TRY THIS   Multiply $(3n + 2)(-5n^2 + n + 3)$.

KEY SKILLS

Complete each multiplication.

1. $(x + 3)(x + 10) = (x + 3)(x) + (x + 3)(10)$
2. $(a - 2)(a - 7) = (a - 2)(a) + (a - 2)(-7)$
3. $(w + 3)(w^2 - 5w) = (w + 3)(w^2) - (w + 3)(5w)$
4. $(n - 1)(n^2 + 7n) = (n - 1)(n^2) + (n - 1)(7n)$

PRACTICE

Multiply.

5. $(x + 2)(x + 3)$
6. $(a + 2)(a + 3)$
7. $(b + 2)(b - 3)$
8. $(r + 1)(-r + 5)$
9. $(2z + 2)(z - 3)$
10. $(n + 2)(3n + 3)$
11. $(2m + 2)(4m + 3)$
12. $(3s + 1)(3s + 5)$
13. $(-v - 4)(v + 4)$
14. $(-k + 1)(-k + 7)$
15. $(3p - 3)(-2p + 5)$
16. $(-2s - 3)(-2s + 4)$
17. $(x + 1)(x^2 + x + 1)$
18. $(w + 2)(w^2 + 3w + 2)$
19. $(2r - 1)(r^2 - r + 1)$
20. $(a - 4)(2a^2 - a - 3)$
21. $(-c + 3)(-c^2 + c + 3)$
22. $(-d + 1)(-d^2 - 2d + 5)$
23. $(-5n + 4)(3n^2 + 2n - 7)$
24. $(3q - 5)(-4q^2 + 5q + 7)$
25. $(-5h + 5)(5h^2 + 5h + 5)$
26. $x(3x - 5)(x - 2)$
27. $(2z - 1)(z + 2)(z - 3)$
28. $(3x + 1)(2x - 3)(x - 1)$
29. $3ab(a + b)(a - b)$
30. $2mn(2n + m)(2m + n)$
31. $2c^2d(2d^2 - c)(2c^2 - d)$

32. Find a so that $(a + 2)(2a + 2) = (a - 2)(2a + 2)$.
33. Find r so that $(r - 3)(2r + 1) = (r - 2)(2r - 3)$.
34. **Critical Thinking** Show that $(a + n)[a + (n + 1)] = a^2 + n^2 + 2an + a + n$.

MIXED REVIEW

Find each sum. (6.4 Skill A)

35. $(4c^2 + 2c - 2) + (-4c^2 - 2c + 3)$
36. $(n^3 + 5n - 1) + (-4n^2 + 2n)$
37. $(-n^2 + 7n) + (n - 3)$
38. $(t^2 + 7t + 2) + (t^2 - 7t - 2)$

Find each difference. (6.4 Skill B)

39. $(2d^3 - 7d^2 + 2) - (d^3 - 7d^2 - 1)$
40. $(-2m^3 - m^2 + m) - (m^3 - 7m^2 + 4m + 1)$

Answers for Lesson 6.6 Skill B

Try This Example 1
$-3z^2 + 13z + 10$

Try This Example 2
$3n^2 + 11n + 6$

Try This Example 3
$-15n^3 - 7n^2 + 11n + 6$

1. $x^2 + 13x + 30$
2. $a^2 - 9a + 14$
3. $w^3 - 2w^2 - 15w$
4. $n^3 + 6n^2 - 7n$
5. $x^2 + 5x - 6$
6. $a^2 + 5a + 6$
7. $b^2 - b - 6$
8. $-r^2 + 4r + 5$
9. $2z^2 - 4z - 6$
10. $3n^2 + 9n + 6$
11. $8m^2 + 14m + 6$
12. $9s^2 + 18s + 5$
13. $-v^2 - 8v - 16$

14. $k^2 - 8k + 7$
15. $-6p^2 + 21p - 15$
16. $4s^2 - 2s - 12$
17. $x^3 + 2x^2 + 2x + 1$
18. $w^3 + 5w^2 + 8w + 4$
19. $2r^3 - 3r^2 + 3r - 1$
20. $2a^3 - 9a^2 + a + 12$
21. $c^3 - 4c^2 + 9$
22. $d^3 + d^2 - 7d + 5$
23. $15n^3 + 2n^2 + 43n - 28$
24. $-12q^3 + 35q^2 - 4q - 35$
25. $-25h^3 + 25$
26. $3x^3 - 11x^2 + 10x$
27. $2z^3 - 3z^2 - 11z + 6$
28. $6x^3 - 13x^2 + 4x + 3$
29. $3a^3b - 3ab^3$
30. $4m^3n + 10m^2n^2 + 4mn^3$

31. $-4c^5d + 8c^4d^3 + 2c^3d^2 - 4c^2d^4$
32. $a = -1$
33. $r = \dfrac{9}{2}$
34. $(a + n)[a + (n + 1)]$
$= (a + n)a + (a + n)(n + 1)$
$= a^2 + an + a(n + 1) + n(n + 1)$
$= a^2 + an + an + a + n^2 + n$
$= a^2 + n^2 + n + an + an + a$
$= a^2 + n^2 + n + 2an + a$
$= a^2 + n^2 + 2an + a + n$
35. 1
36. $n^3 - 4n^2 + 7n - 1$
37. $-n^2 + 8n - 3$
38. $2t^2$

39. $d^3 + 3$
40. $-3m^3 + 6m^2 - 3m - 1$

Using special products

Focus: *Students perform multiplications of two binomials that result in perfect square trinomials and differences of two squares.*

EXAMPLES 1 and 2

PHASE 1: Presenting the Examples

Give students the following sets of binomial multiplications, one set at a time, and have them find the products.

Set 1		Set 2		Set 3	
$(h+3)(h+3)$	$[h^2+6h+9]$	$(r-2)(r-2)$	$[r^2-4r+4]$	$(c+4)(c-4)$	$[c^2-16]$
$(m+7)(m+7)$	$[m^2+14h+49]$	$(z-5)(z-5)$	$[z^2-10z+25]$	$(y+6)(y-6)$	$[y^2-36]$
$(a+1)(a+1)$	$[a^2+2a+1]$	$(n-1)(n-1)$	$[n^2-2n+1]$	$(q+1)(q-1)$	$[q^2-1]$

Ask students to describe any patterns they see. [**Answers will vary.**] After discussing Sets 1 and 2, introduce the term *perfect-square trinomial* and the generalization $(a+b)^2 = a^2 + 2ab + b^2$. Be sure to point out that the middle term of the trinomial can be either positive or negative (depending on the signs of a and b), but the last term must be positive; a polynomial such as $4x^2 - 12x - 9$ does not fit the pattern.

After discussing Set 3, introduce the term *difference of two squares* and the generalization $(a+b)(a-b) = a^2 - b^2$. Remind students that a variable raised to any even power is also a perfect square. For example, $c^8 = (c^4)(c^4)$; so $(x - c^4)(x + c^4)$ can be simplified as $x^2 - c^8$. Then discuss Examples 1 and 2.

PHASE 2: Providing Guidance for TRY THIS

Students frequently simplify $(a+b)^2$ as $a^2 + b^2$. Until they recognize the correct pattern for this product, suggest that they write the expanded form $(a+b)(a+b)$ and multiply using one of the methods in Skill B.

Students may also try to simplify the sum of two squares, $a^2 + b^2$. Stress that a polynomial in this form cannot be simplified any further.

PHASE 3: ASSESSMENT AND CLOSURE for Lesson 6.6

- Give students the statements at right. Tell them that each is false. Have them correct the right side of each to make it true. [**1.** $3k^2 + 12k$ **2.** $-4g^3 + 4g$ **3.** $6t^2 + 4t - 16$ **4.** $25p^2 + 30pq + 9q^2$]

1. $3k(k+4) = 3k^2 + 4$
2. $-4g(g^2-1) = -4g^3 - 4g$
3. $(2t+4)(3t-4) = 6t^2 - 16$
4. $(5p+3q)^2 = 5p^2 + 3q^2$

- In Chapter 8, students will "reverse" the process of multiplying polynomials. That is, given a simplified polynomial product, they will work backward to find its monomial and polynomial *factors*.

☞ *For a Lesson 6.6 Quiz, see* Assessment Resources *page 65.*

A **perfect-square trinomial** is the result of squaring a binomial. The **difference of two squares** is an expression of the form $a^2 - b^2$.

Special Products
Perfect-square trinomial: $a^2 + 2ab + b^2 = (a + b)^2$
Difference of two squares: $a^2 - b^2 = (a + b)(a - b)$

EXAMPLE 1 Write $(3x - 4)^2$ as a perfect-square trinomial.

▶ Solution
Rewrite $(3x - 4)^2$ as $[3x + (-4)]^2$.
Use the form for a perfect-square trinomial with $a = 3x$ and $b = -4$.
$(3x - 4)^2 = (3x)^2 + 2(3x)(-4) + (-4)^2$
$= 9x^2 - 24x + 16$

TRY THIS   Write as a perfect-square trinomial.   **a.** $(-2x + 5)^2$ **b.** $(5d - 2)^2$

EXAMPLE 2 Multiply $(5st - 4)(5st + 4)$.

▶ Solution
Use the form for a difference of two squares with $a = 5st$ and $b = 4$.
$(5st - 4)(5st + 4) = (5st)^2 - (4)^2$
$= 25s^2t^2 - 16$

TRY THIS   Multiply $(5c + 6d)(5c - 6d)$.

You can use a vertical format to find the special products given above.

$$\begin{array}{r} a + b \\ \times\ a + b \\ \hline ab + b^2 \\ +\ a^2 +\ ab \\ \hline a^2 + 2ab + b^2 \end{array} \qquad \begin{array}{r} a + b \\ \times\ a - b \\ \hline -ab - b^2 \\ +\ a^2 +\ ab \\ \hline a^2 \qquad - b^2 \end{array}$$

You can also use an area model to find special products.

Area: $(a + b)(a + b)$
Area: $a^2 + ab + ab + b^2$
$= a^2 + 2ab + b^2$

Area: $a^2 - b^2$
Area: $(a - b)^2 + b(a - b) + b(a - b) = (a - b)(a - b) +$
$2b(a - b) = (a - b)(a - b + 2b) = (a - b)(a + b)$

KEY SKILLS

Identify a and b in each special product.

1. $(2d + 4)^2$ 2. $(5z - 3)^2$ 3. $(3k + 5)(3k - 5)$ 4. $(-m + 3)(-m - 3)$

PRACTICE

Find each special product.

5. $(k + 1)^2$ 6. $(v + 2)^2$ 7. $(x - 2)^2$
8. $(g - 5)^2$ 9. $(2h + 3)^2$ 10. $(2h - 1)^2$
11. $(-2t - 3)^2$ 12. $(-3x + 5)^2$ 13. $(t + 1)(t - 1)$
14. $(z + 5)(z - 5)$ 15. $(2z - 1)(2z + 1)$ 16. $(3z - 2)(3z + 2)$
17. $(3 - d)(3 + d)$ 18. $(5 + d)(5 - d)$ 19. $(2d + 3)(2d - 3)$
20. $(7v - 5)(7v + 5)$ 21. $(2u + 5v)^2$ 22. $(3m - 2n)^2$
23. $(5r + 4s)(5r - 4s)$ 24. $(2c - 3d)(2c + 3d)$ 25. $(2rs + 3)^2$
26. $(3xy - 5)^2$ 27. $(4pq + 1)(4pq - 1)$ 28. $(3yz - 2)(3yz + 2)$

29. Use multiplication to show that $(a - b)^2 = a^2 - 2ab + b^2$.
30. Find m such that $(m - 3)^2 = m^2 - 3^2$.
31. Find n such that $(n + 3)^2 = n^2 + 3^2$.
32. How are m and n related if $[(m + n) + 2]^2 = (m + n)^2 + 2^2$?
33. Simplify $[(a + b) + c][(a + b) - c]$.
34. Use $(a + b)^3 = (a + b)(a + b)^2$ to write $(a + b)^3$ as a sum.
35. **Critical Thinking** Use mental math to find the product of 38 and 42. (*Hint:* Use the difference of two squares.)

MIXED REVIEW

Find each quotient. Write the answers with positive exponents. Use mental math where possible. (6.5 Skill B)

36. $\frac{4f^2}{2f}$ 37. $\frac{21y^6}{3y^3}$ 38. $\frac{5h^2}{25h^4}$ 39. $-\frac{ab}{2a^2b^2}$
40. $\frac{(2w^2)^3}{2w^7}$ 41. $\frac{(5mn)^2}{25m^3n}$ 42. $\frac{6(uv)^4}{36(uv)^3}$ 43. $\frac{(-2)^4a}{32ab^3}$

Answers for Lesson 6.6 Skill C

Try This Example 1
a. $4x^2 - 20x + 25$
b. $25d^2 - 20d + 4$

Try This Example 2
$25c^2 - 36d^2$

1. $a = 2d$ and $b = 4$
2. $a = 5z$ and $b = -3$
3. $a = 3k$ and $b = 5$
4. $a = -m$ and $b = 3$
5. $k^2 + 2k + 1$
6. $v^2 + 4v + 4$
7. $x^2 - 4x + 4$
8. $g^2 - 10g + 25$
9. $4h^2 + 12h + 9$
10. $4h^2 - 4h + 1$
11. $4t^2 + 12t + 9$
12. $9x^2 - 30x + 25$
13. $t^2 - 1$
14. $z^2 - 25$
15. $4z^2 - 1$

16. $9z^2 - 4$
17. $9 - d^2$
18. $25 - d^2$
19. $4d^2 - 9$
20. $49v^2 - 25$
21. $4u^2 + 20uv + 25v^2$
22. $9m^2 - 12mn + 4n^2$
23. $25r^2 - 16s^2$
24. $4c^2 - 9d^2$
25. $4r^2s^2 + 12rs + 9$
26. $9x^2y^2 - 30xy + 25$
27. $16p^2q^2 - 1$
28. $9y^2z^2 - 4$
29. $(a - b)(a - b) =$
   $a(a - b) - b(a - b)$
   $= a^2 - ab - ba + b^2$
   $= a^2 - 2ba + b^2$
   $= a^2 - 2ab + b^2$
30. $m = 3$
31. $n = 0$

32. $m = -n$
33. $a^2 + 2ab + b^2 - c^2$
34. $a^3 + 3a^2b + 3ab^2 + b^3$
35. $38 \cdot 42 = (40 - 2)$
   $(40 + 2) = 40^2 - 2^2$
   $= 1600 - 4 = 1596$
36. $2f$
37. $7y^3$
38. $\dfrac{1}{5h^2}$
39. $-\dfrac{1}{2ab}$
40. $\dfrac{4}{w}$
41. $\dfrac{n}{m}$
42. $\dfrac{uv}{6}$
43. $\dfrac{1}{2b^3}$

6.7 Dividing Polynomials

Focus: *Students use the Distributive Property and the properties of exponents to find the quotient when a polynomial is divided by a monomial.*

EXAMPLE 1

PHASE 1: Presenting the Example

Discuss with students the presentation of the Distributive Property that appears at the top of the textbook page. Answer any questions that they may have.

Display the monomial divisions $\frac{45x^3}{5x}$ and $\frac{-25x^2}{5x}$. Have students find each quotient.

$[9x^2; -5x]$ Record their results as shown at right.

$$\frac{45x^3}{5x} \qquad \frac{-25x^2}{5x}$$
$$\downarrow \qquad\qquad \downarrow$$
$$9x^2 \qquad\quad -5x$$

Now display $\frac{45x^3 - 25x^2}{5x}$. Ask students how the quotients they just found relate to

simplifying this expression. Elicit the response that, by applying the Distributive Property,

this expression is equivalent to the sum of the two monomial divisions, $\frac{45x^3}{5x} + \frac{-25x^2}{5x}$.

Consequently, the expression can be simplified as the sum of the quotients, $9x^2 + (-5x)$, or $9x^2 - 5x$.

$$\frac{45x^3 - 25x^2}{5x}$$
$$\downarrow$$
$$9x^2 - 5x$$

PHASE 2: Providing Guidance for TRY THIS

Watch for students who give the answer to part b of the Try This exercise as *uv*. Remind them that division and multiplication are inverse operations. Consequently, if they multiply their quotients by the monomial divisor, the result must be the polynomial dividend. Encourage them to check their quotients as shown at right.

Check the quotient *uv*:

$$(4u^2v^2)(uv) = 4u^3v^3 \; ✗$$

Check the quotient *uv* − 1:

$$(4u^2v^2)(uv - 1) = (4u^2v^2)(uv) - (4u^2v^2)(1)$$
$$= 4u^3v^3 - 4u^2v^2 \; ✔$$

EXAMPLE 2

PHASE 1: Presenting the Example

Ask students whether the set of whole numbers is closed under division. In other words, when you divide two whole numbers, is the result always another whole number? Students should perceive that the answer is no. For example, 2 ÷ 8 is not a whole number. Point out that the set of polynomials is not closed under division either. When you divide two polynomials, it is possible to get a result that is not a polynomial. Then discuss the Example.

6.7 LESSON — Dividing Polynomials

SKILL A Dividing a polynomial by a monomial

To divide an expression by a monomial, use the following forms of the Distributive Property.

Division and the Distributive Property
If a, b, and c are real numbers and $c \neq 0$, then
$$\frac{a+b}{c} = \frac{a}{c} + \frac{b}{c} \quad \text{and} \quad \frac{a-b}{c} = \frac{a}{c} - \frac{b}{c}.$$

You can prove the first part of the statement above as shown below.

$$\frac{a+b}{c} = \frac{1}{c}(a+b) \qquad \longleftarrow \text{Definition of division}$$
$$= \frac{1}{c} \cdot a + \frac{1}{c} \cdot b \qquad \longleftarrow \text{Distributive Property}$$
$$= \frac{a}{c} + \frac{b}{c} \qquad \longleftarrow \text{Definition of division}$$

EXAMPLE 1
a. Divide $45x^3 - 25x^2$ by $5x$.   **b.** Divide $4a^3b^2 + 8a^2b$ by $2ab$.

▸ **Solution**

a. $\dfrac{45x^3 - 25x^2}{5x}$

$= \dfrac{45x^3}{5x} - \dfrac{25x^2}{5x}$

$= \dfrac{45}{5} \cdot \dfrac{x^3}{x} - \dfrac{25}{5} \cdot \dfrac{x^2}{x}$

$= 9x^2 - 5x$

b. $\dfrac{4a^3b^2 + 8a^2b}{2ab}$

$= \dfrac{4a^3b^2}{2ab} + \dfrac{8a^2b}{2ab}$

$= \dfrac{4}{2} \cdot \dfrac{a^3}{a} \cdot \dfrac{b^2}{b} + \dfrac{8}{2} \cdot \dfrac{a^2}{a} \cdot \dfrac{b}{b}$

$= 2a^2b + 4a$

TRY THIS   **a.** Divide $32n^4 + 8n^2$ by $16n^2$.   **b.** Divide $4u^3v^3 - 4u^2v^2$ by $4u^2v^2$.

When you divide one polynomial by another, you may get an expression that is not a polynomial. The set of polynomials is not *closed* under division. For example, $\frac{x}{x^2} = x^{1-2} = x^{-1} = \frac{1}{x}$, which is *not* a polynomial.

EXAMPLE 2   Divide $20x^2 - 15$ by $5x$.

▸ **Solution**

$\dfrac{20x^2 - 15}{5x} = \dfrac{20x^2}{5x} - \dfrac{15}{5x}$ $\qquad \longleftarrow$ Recall that $\frac{a-b}{c} = \frac{a}{c} - \frac{b}{c}$.

$= \dfrac{20}{5} \cdot \dfrac{x^2}{x} - \dfrac{15}{5} \cdot \dfrac{1}{x} = 4x - \dfrac{3}{x}$ $\qquad \longleftarrow$ This is not a polynomial.

TRY THIS   Divide $30m^4 - 15m$ by $5m^2$.

EXERCISES

KEY SKILLS

Complete each division.

1. $\dfrac{15n^2 + 6n}{3n} = \dfrac{15n^2}{3n} + \dfrac{6n}{3n}$

2. $\dfrac{5k^3 - 12k^2}{5k} = \dfrac{5k^3}{5k} - \dfrac{12k^2}{5k}$

3. $\dfrac{y^3z^3 - yz^2}{yz} = \dfrac{y^3z^3}{yz} - \dfrac{yz^2}{yz}$

4. $\dfrac{3p^3q^2 + 6p^2q^2}{2pq} = \dfrac{3p^3q^2}{2pq} + \dfrac{6p^2q^2}{2pq}$

PRACTICE

Find each quotient below. Is the quotient a polynomial?

5. $\dfrac{12a + 15}{3}$

6. $\dfrac{14y + 7}{7}$

7. $\dfrac{20x - 8}{-4}$

8. $\dfrac{-100d + 50}{25}$

9. $\dfrac{8c^2 - 8c}{2c}$

10. $\dfrac{-10r^2 + 2r}{-2r}$

11. $\dfrac{10s^3 - 20s^2}{4s}$

12. $\dfrac{6v^3 - 24v^2}{-6v}$

13. $\dfrac{a^3b^3 + a^2b^2}{ab}$

14. $\dfrac{c^3d^3 - 3cd}{cd}$

15. $\dfrac{7k^3p - 49k^2p^3}{7kp}$

16. $\dfrac{6r^2d + 27r^2d^2}{3rd}$

17. $\dfrac{s^2 - 1}{s}$

18. $\dfrac{2r^2 + 1}{r}$

19. $\dfrac{-12d^2 + 6}{3d}$

20. $\dfrac{-10q^2 + 2}{-q}$

21. $\dfrac{14z^3 - 28}{7z}$

22. $\dfrac{15c^3 - 27}{3c}$

23. $\dfrac{x^3 + x}{x^2}$

24. $\dfrac{y^4 - y}{y^2}$

25. $\dfrac{2z^3 - 6z}{z}$

26. When $24y + 36$ is divided by a, the quotient has only positive integer coefficients. What are the possible values of a?

27. **Critical Thinking** Use $\frac{a-b}{c} = \frac{a+(-b)}{c}$ to show that $\frac{a-b}{c} = \frac{a}{c} - \frac{b}{c}$.

MIXED REVIEW

Multiply. **(6.5 Skill A)**

28. $(-2a^2)(5a)(-2a^2)$

29. $(-3u)(-3u)(-3u^3)$

30. $(a^2b^2)(a^2b)(ab^2)$

Multiply. **(6.6 Skill A)**

31. $(2d^2)(3d^2 + 2d + 1)$

32. $(-v^2)(3v^3 - 2v^2 + 2v)$

33. $(a^2)(a^3 + a^2 + a + 1)$

Answers for Lesson 6.7 Skill A

Try This Example 1

a. $2n^2 + \dfrac{1}{2}$

b. $uv - 1$

Try This Example 2

$6m^2 - \dfrac{3}{m}$

1. $5n + 2$

2. $k^2 - \dfrac{12k}{5}$

3. $y^2z^2 - z$

4. $\dfrac{3}{2}p^2q + 3pq$

5. $4a + 5$; yes

6. $2y + 1$; yes

7. $-5x + 2$; yes

8. $-4d + 2$; yes

9. $4c - 4$; yes

10. $5r - 1$; yes

11. $\dfrac{5}{2}s^2 - 5s$; yes

12. $-v^2 + 4v$; yes

13. $a^2b^2 + ab$; yes

14. $c^2d^2 - 3$; yes

15. $k^2 - 7kp^2$; yes

16. $2r + 9rd$; yes

17. $s - \dfrac{1}{s}$; no

18. $2r + \dfrac{1}{r}$; no

19. $-4d + \dfrac{2}{d}$; no

20. $10q - \dfrac{2}{q}$; no

21. $2z^2 - \dfrac{4}{z}$; no

22. $5c^2 - \dfrac{9}{c}$; no

23. $x + \dfrac{1}{x}$; no

24. $y^2 - \dfrac{1}{y}$; no

25. $2z^2 - 6$; yes

26. $1, 2, 3, 4, 6, 12$

27. $\dfrac{a-b}{c} = \dfrac{a+(-b)}{c}$

$= \dfrac{1}{c}[a + (-b)]$

$= \dfrac{1}{c} \cdot a + \dfrac{1}{c} \cdot (-b)$

$= \dfrac{a}{c} + \dfrac{-b}{c}$

$= \dfrac{a}{c} - \dfrac{b}{c}$

28. $20a^5$

29. $-27u^5$

30. a^5b^5

31. $6d^4 + 4d^3 + 2d^2$

32. $-3v^5 + 2v^4 - 2v^3$

33. $a^5 + a^4 + a^3 + a^2$

Focus: *Students use a division algorithm to find the quotient when a polynomial is divided by a binomial.*

EXAMPLES 1 and 2

PHASE 1: Presenting the Examples

The division algorithm for polynomials is similar to the familiar long division algorithm for whole numbers. Ask students to find the quotient $1472 \div 64$. [**23**] Then review the division with the class, as shown below.

1. Divide 14 by 6.

$$64)\overline{1472} \quad \frac{2}{}$$

2. Multiply 64 by 2.

$$\frac{2}{64)\overline{1472}}$$
$$128$$

3. Subtract.

$$\frac{2}{64)\overline{1472}}$$
$$\frac{128}{19}$$

4. Bring down 2.

$$\frac{23}{64)\overline{1472}}$$
$$\frac{128\downarrow}{192}$$

5. Divide 19 by 6.

$$\frac{23}{64)\overline{1472}}$$
$$\frac{128}{192}$$

6. Multiply 64 by 3.

$$\frac{23}{64)\overline{1472}}$$
$$\frac{128}{192}$$
$$192$$

7. Subtract.

$$\frac{23}{64)\overline{1472}}$$
$$\frac{128}{192}$$
$$\frac{192}{0}$$

8. Check.

$$23$$
$$\times 64$$
$$\overline{92}$$
$$\underline{138}$$
$$1472 \checkmark$$

Keeping this division prominently displayed, discuss Examples 1 and 2 with the class. Have them compare each step of those divisions to the corresponding steps of the whole-number division.

PHASE 2: Providing Guidance for **TRY THIS**

When the expression being subtracted involves a negative number, students often mistakenly add. In the Try This for Example 1, for instance, they may add $-y + (-3y)$, as shown at right. Suggest that they write the four recurring steps of the algorithm—*divide, multiply, <u>subtract</u>, bring down*—in a prominent position on their papers. They should then check off each step as they perform it.

$$\frac{3y}{y-1)\overline{3y^2 - y - 2}}$$
$$\frac{3y^2 - 3y}{-4y} \; \textsf{X}$$

PHASE 3: ASSESSMENT AND CLOSURE for Lesson 6.7

- Give students the division at right. Ask them to fill in the box with a monomial divisor that will yield a polynomial quotient, and find the quotient. [**sample: $2s$; $6s^2 + 2s - 3$**] Repeat the question, this time asking for a nonpolynomial quotient.

$$\frac{12s^3 + 4s^2 - 6s}{\boxed{}}$$

[**sample: $2s^2$; $6s + 2 - \dfrac{3}{s}$**]

- This lesson completes the study of operations with polynomials. In Chapter 8, students will see how these operations relate to the process of factoring.

☞ *For a Lesson 6.7 Quiz, see* Assessment Resources *page 66.*

☞ *For a Chapter 6 Assessment, see* Assessment Resources *pages 67–68.*

To divide a polynomial by a binomial, you can use long division, as shown in Example 1.

EXAMPLE 1 Divide $2x^2 + 5x - 3$ by $x + 3$.

▶ Solution

1. Divide x into $2x^2$.

$$x + 3\overline{)2x^2 + 5x - 3}$$ with $2x$ above

2. Multiply $x + 3$ by $2x$.

$$x + 3\overline{)2x^2 + 5x - 3}$$
$$2x^2 + 6x$$

3. Subtract.

$$x + 3\overline{)2x^2 + 5x - 3}$$
$$-(2x^2 + 6x)$$
$$-x$$

4. Bring down -3.

$$x + 3\overline{)2x^2 + 5x - 3}$$
$$2x^2 + 6x \quad\downarrow$$
$$-x - 3$$

5. Divide x into $-x$.

$$x + 3\overline{)2x^2 + 5x - 3}$$ with $2x - 1$ above
$$2x^2 + 6x$$
$$-x - 3$$

6. Multiply $x + 3$ by -1. Then subtract.

$$x + 3\overline{)2x^2 + 5x - 3}$$ with $2x - 1$ above
$$2x^2 + 6x$$
$$-x - 3$$
$$-(-x - 3)$$
$$0$$

The quotient is $2x - 1$.

TRY THIS Divide $3y^2 - y - 2$ by $y - 1$.

The dividend must contain every possible power of the variable. If one is missing, include a coefficient of 0 for that term. For example, to divide $x^2 - 4$ by $x - 2$, rewrite $x^2 - 4$ as $x^2 + 0x - 4$, and then proceed as in Example 1. The quotient is $x + 2$.

$$x - 2\overline{)x^2 - 4} \rightarrow x - 2\overline{)x^2 + 0x - 4}$$
$$-(x^2 - 2x)\quad\downarrow$$
$$2x - 4$$
$$-(2x - 4)$$
$$0$$

EXAMPLE 2 The product of two polynomials is $a^2 + 6a - 16$. If one polynomial is $a - 2$, what is the other?

▶ Solution

$$(a - 2)(\text{polynomial}) = a^2 + 6a - 16$$

Divide $a^2 + 6a - 16$ by $a - 2$ to find that the unknown polynomial is $a + 8$.

$$a - 2\overline{)a^2 + 6a - 16}$$ with $a + 8$ above
$$-(a^2 - 2a)$$
$$8a - 16$$
$$-(8a - 16)$$
$$0$$

TRY THIS The product of two polynomials is $m^2 - 49$. If one polynomial is $m - 7$, what is the other?

254 Chapter 6 Operations With Polynomials

EXERCISES

In Exercises 1–3, check the division by multiplying the divisor and the quotient. If the division is correct, your answer will be the dividend.

1. $x + 2\overline{)x^2 + 5x + 6}$ with $x + 3$ above

2. $x - 3\overline{)x^2 + 2x - 15}$ with $x + 5$ above

3. $x - 2\overline{)x^2 - 9x + 14}$ with $x - 7$ above

Divide.

4. $\dfrac{z^2 + 3z - 10}{z - 2}$

5. $\dfrac{b^2 + 2b - 8}{b - 2}$

6. $\dfrac{c^2 + 8c + 7}{c + 7}$

7. $\dfrac{u^2 - 7u - 30}{u - 10}$

8. $\dfrac{n^2 - 5n - 36}{n - 9}$

9. $\dfrac{w^2 - w - 42}{w + 6}$

10. $\dfrac{3a^2 - 13a - 10}{a - 5}$

11. $\dfrac{3h^2 - 4h - 15}{h - 3}$

12. $\dfrac{2v^2 + 15v + 28}{v + 4}$

13. $4k^2 - 19k - 5$ by $k - 5$

14. $3c^2 + 17c - 56$ by $c + 8$

15. $5p^2 + 18p - 8$ by $p + 4$

16. $n^2 - 25$ by $n + 5$

17. $x^2 - 64$ by $x + 8$

18. $25p^2 - 9$ by $5p + 3$

Solve each problem.

19. The product of two polynomials is $100z^2 - 49$. If one polynomial is $10z - 7$, what is the other?

20. The product of two polynomials is $81n^2 - 121$. If one polynomial is $9n + 11$, what is the other?

21. The product of two polynomials is $6a^2 + ab - b^2$. If one polynomial is $2a + b$, what is the other?

22. Show that if $6x^2 + 11x - 35$ is divided by $3x - 5$, the quotient is $2x + 7$.

Solve each absolute-value inequality. Graph the solutions on a number line. (4.7 Skill B)

23. $3|a - 2| \geq 6$

24. $|6(t + 1) - 5t| \geq 0$

25. $|4t - (t + 1)| < 0$

26. $|-2u + 5| < 5$

27. $3 + 3|x| \leq 12$

28. $4|r + 5| > 8$

Answers for Lesson 6.7 Skill B

Try This Example 1
$3y + 2$

Try This Example 2
$m + 7$

1. $(x + 2)(x + 3) = x(x + 3)$
$+ 2(x + 3)$
$= x^2 + 3x + 2x + 6$
$= x^2 + 5x + 6$

2. $(x - 3)(x + 5) = x(x + 5)$
$- 3(x + 5)$
$= x^2 + 5x - 3x + 15$
$= x^2 + 2x + 15$

3. $(x - 2)(x - 7) = x(x - 7)$
$- 2(x - 7)$
$= x^2 - 7x - 2x + 14$
$= x^2 - 9x + 14$

4. $z + 5$

5. $b + 4$

6. $c + 1$

7. $u + 3$

8. $n + 4$

9. $w - 7$

10. $3a + 2$

11. $3h + 5$

12. $2v + 7$

13. $4k + 1$

14. $3c - 7$

15. $5p - 2$

16. $n - 5$

17. $x - 8$

18. $5p - 3$

19. $10z + 7$

20. $9n - 11$

21. $3a - b$

22. $(3x - 5)(2x + 7) = 3x(2x + 7)$
$- 5(2x + 7)$
$= 3x(2x) + (3x)(7) - 5(2x)$
$- (5)(7)$
$= 6x^2 + 21x - 10x - 35$
$= 6x^2 + 11x - 35$

23. $a \leq 0 \text{ or } a \geq 4$

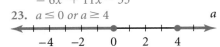

24. all real numbers

25. no solution

26. $0 < u < 5$

27. $-3 \leq x \leq 3$

28. $r < -7 \text{ or } r > -3$

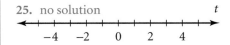

Write each expression in the form a^b.

1. $3^2 \cdot 3^5$ 2. $\dfrac{2.5^4}{2.5^4}$ 3. $6^{-2} \cdot 6 \cdot 6^0$ 4. $\dfrac{(-7)^3}{(-7)^5}$

Simplify.

5. $\dfrac{(-2)^3(-5)^5}{(-2)^5(-5)^3}$ 6. $[(3^{-2} \cdot 3^2)^2 \cdot 3^2]^2$ 7. $\dfrac{3^3 \cdot 3^{-5}}{(-2)^{-5}(-2)^3}$

Write each number in scientific notation.

8. 0.000312   9. 18,560

Write each product or quotient in scientific notation.

10. $(2.1 \times 10^2)(5.0 \times 10^3)$ 11. $\dfrac{6.9 \times 10^3}{3.0 \times 10^5}$

12. If y varies directly as the square of x, and $y = 20$ when $x = 4$, find y when $x = 10$.

13. If y varies directly as the cube of x, and $y = 250$ when $x = 5$, find y when $x = 12$.

The distance, d, that a falling object travels downward varies directly as the square of elapsed time, t. The constant of variation is 16 feet per second squared. Find d for each value of t.

14. 4.5 seconds   15. 12 seconds

Classify each polynomial by degree and number of terms.

16. $3a^2 - 2a^3 + a^2 - a^2$

17. $t - 3 - 2t^3 + 2t^2 + 4$

The surface area, s, of a cylindrical container with a height of 6 feet can be represented by $s = 2\pi r^2 + 12\pi r$. Find the surface area of each cylindrical container with the given radius. Use 3.14 to approximate π.

18. 6.5 feet   19. 10 feet

Simplify.

20. $3g^5 + 2g^3 - g^2 + g^3 + g^2 + 3g - 1$

21. $x^3 + 5x^2 + 2x - 2 + 3x^2 + 5x + 2 + 2x^4 - 3x^2 - 7x$

Simplify.

22. $(2b^4 - 2b^3 + 3b^2) - (-2b^4 + b^3 - 2b^2 + 5)$

23. $(2v^4 - 5v^3 - v^2 + 5v + 3) - (3v^4 + v^3 + v^2 - 5v - 2)$

24. Write an expression for the perimeter of the geometric figure below. Rectangles of the same color are congruent.

Simplify. Write each answer with positive exponents only.

25. $\left(\dfrac{3y^2}{6}\right)\left(\dfrac{4y^3}{5y^3}\right)(15y^2)$ 26. $(3a^3b^2)(4a^5b)$

27. $\dfrac{(3x^2)^3}{(6x^2)^2}$ 28. $\dfrac{(9m^2n^2)^2}{(3m^2n)^3}$

29. A rectangular box has dimensions $1.5c$, $3c$, and $9c$. Write an expression for the total volume of 36 of these boxes.

Find each product.

30. $3g^3(g^2 - 3g + 7)$ 31. $2n^3p(np^2 + 6n^2p^2)$

32. $(3s + 5)(-2s + 11)$ 33. $(-2t + 7)(2t^2 + 11t - 1)$

34. $(-5z + 7)(5z + 7)$ 35. $(11a + 1)^2$

Find each quotient.

36. $\dfrac{5n^2 + 30n}{15n}$ 37. $\dfrac{16y^5z^3 - 10y^2z^4}{2y^2z^3}$

38. $\dfrac{2b^3 + 2b^2 + 3}{2b}$ 39. $\dfrac{2d^4 - 2d^2h + 5d^2h^2}{2d^2h}$

40. $(a^2 - 15a + 56) \div (a - 8)$ 41. $(4q^2 - 44q + 121) \div (2q - 11)$

42. The product of two polynomials is $12n^2 + 5n - 72$. If one polynomial is $3n + 8$, what is the other?

Answers for Chapter 6 Assessment

1. 3^7
2. 2.5^0
3. 6^{-1}
4. $(-7)^{-2}$
5. $\dfrac{25}{4}$
6. 81
7. $\dfrac{4}{9}$
8. 3.12×10^{-4}
9. 1.856×10^4
10. 1.05×10^6
11. 2.3×10^{-2}
12. $y = 125$
13. $y = 3456$
14. 324 feet
15. 2304 feet
16. cubic binomial
17. cubic polynomial
18. 510.25 square feet
19. 1004.8 square feet

20. $3g^5 + 3g^3 + 3g - 1$
21. $2x^4 + x^3 + 5x^2$
22. $4b^4 - 3b^3 + 5b^2 - 5$
23. $-v^4 - 6v^3 - 2v^2 + 10v + 5$
24. $35x$
25. $6y^2$
26. $12a^8b^3$
27. $\dfrac{3x^2}{4}$
28. $\dfrac{3n}{m^2}$
29. $1458c^3$
30. $3g^5 - 9g^4 + 21g^3$
31. $2n^4p^3 + 12n^5p^3$
32. $-6s^2 + 23s + 55$
33. $-4t^3 - 8t^2 + 79t - 7$
34. $49 - 25z^2$
35. $121a^2 + 22a + 1$

36. $\dfrac{n + 6}{3}$, or $\dfrac{n}{3} + 2$
37. $8y^3 - 5z$
38. $b^2 + b + \dfrac{3}{2b}$
39. $\dfrac{d^2}{h} - 1 + \dfrac{5h}{2}$
40. $a - 7$
41. $2q - 11$
42. $4n - 9$

Chapter 7 *Factoring Polynomials*	State or Local Standards	Corresponding Lessons in *Algebra 1*, Schultz et al.
7.1 Using the Greatest Common Factor to Factor an Expression		9.5
Skill A: Factoring integers		
Skill B: Finding the greatest common factor		
Skill C: Factoring the greatest common monomial factor from a polynomial		
7.2 Factoring Special Polynomials		9.6
Skill A: Factoring polynomials with special forms		
Skill B: Factoring by grouping		
7.3 Factoring $x^2 + bx + c$		9.7
Skill A: Factoring $x^2 + bx + c$ when c is positive		
Skill B: Factoring $x^2 + bx + c$ when c is negative		
7.4 Factoring $ax^2 + bx + c$		9.7
Skill A: Factoring $ax^2 + bx + c$ when c is positive		
Skill B: Factoring $ax^2 + bx + c$ when c is negative		
Skill C: Simplifying before factoring a polynomial		
7.5 Solving Polynomial Equations by Factoring		9.8, 10.4
Skill A: Solving quadratic equations by factoring		
Skill B: Solving quadratic equations in multiple steps by factoring		
Skill C: Solving cubic equations by factoring		
7.6 Quadratic Functions and Their Graphs		10.1
Skill A: Graphing a quadratic function		
Skill B: Finding and counting x-intercepts		
7.7 Analyzing the Graph of a Quadratic Function		10.1
Skill A: Finding the range of a quadratic function		
Skill B: Writing an equation for a quadratic function given sufficient information		
Skill C: APPLICATIONS Using the characteristics of a quadratic function to solve real-world problems		

7

Factoring Polynomials

▶ What You Already Know

In Chapter 6, you learned how to multiply two binomial expressions and write the product of them in simplest form. In earlier mathematics courses, you learned what factoring a positive integer means and you learned how to do it.

▶ What You Will Learn

In Chapter 7, you are going to learn how to factor polynomials.

Studying how to factor polynomials involves many procedures.

- Recognize factoring patterns such as the difference of two squares and perfect-square trinomials.

- Factoring expressions of the form $ax^2 + bx + c$, where $a = 1$.

- Factoring expressions of the form $ax^2 + bx + c$, where $a \neq 1$.

- Factoring expressions by finding the greatest common factor.

- Factoring polynomial expressions by grouping.

Finally, you will learn how to apply your factoring skills to

- find the number of x-intercepts of the graph of a quadratic function, and

- find the number of real solutions to a quadratic equation.

VOCABULARY

axis of symmetry	prime factorization
common binomial factor	prime number
composite number	quadratic equation
cubic equation	quadratic function
difference of two squares	relatively prime
double root	root
factor	splitting the middle term
factored completely	standard form of a cubic equation
greatest common factor (GCF)	standard form of a quadratic equation
parabola	vertex
perfect-square trinomials	Zero-Product Property

The diagram below shows how mathematical skills and mathematical reasoning are interrelated with the skills and concepts in Chapter 7. Notice that the focus of this chapter is on factoring polynomials.

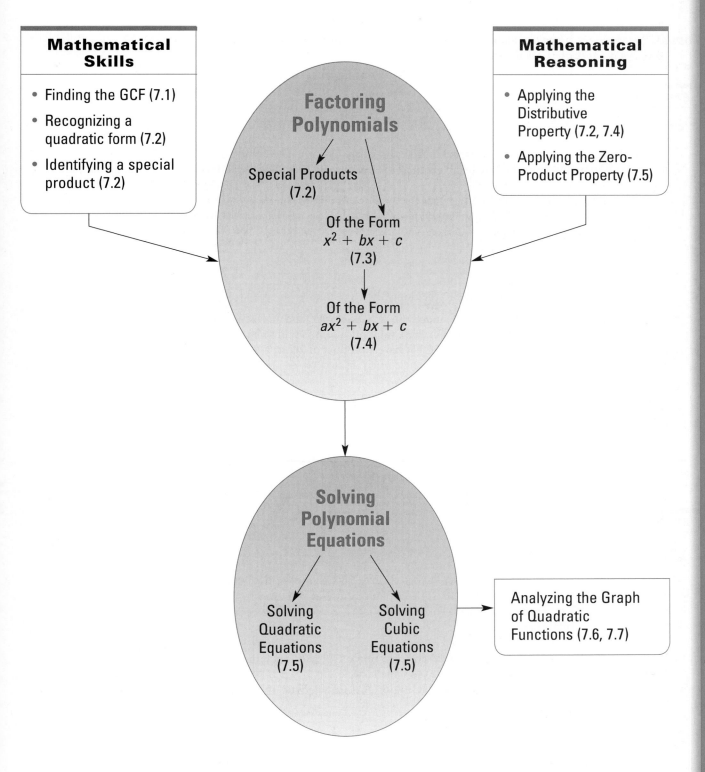

Using the Greatest Common Factor to Factor an Expression

SKILL A   *Factoring integers*

Focus: *Students write a positive integer as a product of two natural-number factors and as the product of its prime factors.*

EXAMPLE 1

PHASE 1: Presenting the Example

Display the figure at right. Tell students that it shows 24 square tiles arranged to form a rectangle. Ask them to write the length and width of the rectangle. [**4 tiles wide, 6 tiles long**]

Now ask students to sketch all other ways that the 24 tiles could be arranged to form a rectangle. Tell them to write the length and width of each rectangle. They should arrive at the following three arrangements:

3 tiles wide, 8 tiles long   2 tiles wide, 12 tiles long   1 tile wide, 24 tiles long

Point out that each rectangle has area 24, and 4×6, 3×8, 2×12, and 1×24 are all equal to 24. Each of the numbers involved in these multiplications is said to be a *factor* of 24. There are no other natural-number pairs that have a product of 24, so 1, 2, 3, 4, 6, 8, 12, and 24 are the only factors of 24.

Now discuss Example 1. Students should note that the problem can be solved by using the same model as above. That is, given 16 square tiles, they could look for different ways to make a rectangle. Note that there are exactly three ways: 1 tile by 16 tiles, 2 tiles by 8 tiles, and 4 tiles by 4 tiles.

EXAMPLE 2

PHASE 1: Presenting the Example

Ask students to write one natural-number factorization of 300. [**samples: 1 × 300, 2 × 150, 3 × 100, 5 × 60, 6 × 50, 10 × 30, 15 × 20**] Point out that you can also write factorizations with more than one factor. For example, $1 \times 2 \times 150$ and $2 \times 3 \times 5 \times 10$ would also be acceptable answers.

Now ask students to write one *prime-number* factorization of 300. Although the order may vary, every correct response will involve the same factors: $2 \times 2 \times 3 \times 5 \times 5$. Tell students that a *prime factorization* like this is an important characteristic of a number because it is unique.

PHASE 2: Providing Guidance for **TRY THIS**

Some students may have difficulty using a factor tree like the one used in the text. These students may prefer to use the "ladder" shown at right.

$$\begin{array}{r|r} 2 & 180 \\ \hline 2 & 90 \\ \hline 3 & 45 \\ \hline 3 & 15 \\ \hline & 5 \end{array}$$

7.1 LESSON
Using the Greatest Common Factor to Factor an Expression

SKILL A *Factoring integers*

From your study of multiplication with real numbers, recall the following facts:

If $3 \times 4 = 12$, then

- 12 is the product of 4 and 3.
- 4 and 3 are *factors*, or *divisors*, of 12.
- The quotient of 12 divided by 4 is 3.
- The quotient of 12 divided by 3 is 4.

To **factor** a number is to write it as the product of two or more numbers, usually natural numbers. Factoring and division are closely related.

EXAMPLE 1 Write three different factorizations of 16. Use natural numbers.

▶ **Solution**

Make a table.

Divide 16 by several natural numbers to find divisors.

16 divided by	1	2	3	4	5	6	7	8
natural number?	16 ✔	8 ✔	no	4 ✔	no	no	no	2 ✔

Three factorizations of 16 are 1×16, 2×8, and 4×4.

TRY THIS Write four different factorizations of 36. Use natural numbers as factors.

A **prime number** is any natural number greater than 1 whose only factors are 1 and itself. For example, 7 is prime because it has no factors besides 1 and 7. A number that has additional factors is called a **composite number**. An example of a composite number is 24; its factors are 1, 2, 3, 4, 6, 8, 12, and 24. Note that the number 1 is neither prime nor composite.

The **prime factorization** of a natural number is the factorization that contains only prime numbers or powers of prime numbers.

EXAMPLE 2 Write the prime factorization of 300.

▶ **Solution**

Make a diagram.

$300 = 3 \times 100$ ⟵ Factor 300 as 3×100.

10×10 ⟵ 100 is not prime, so factor 100 as 10×10.

$5 \times 2 \quad 5 \times 2$ ⟵ 10 is not prime, so factor 10 as 5×2.

The prime factorization of 300 is $2 \times 2 \times 3 \times 5 \times 5$, or $2^2 \times 3 \times 5^2$.

TRY THIS Write the prime factorization of 180.

EXERCISES

KEY SKILLS

Identify each number as prime, composite, or neither.

1. 1   2. 100   3. 31   4. 75

PRACTICE

Write three different factorizations of each number. Use natural numbers as factors.

5. 24   6. 30   7. 40   8. 48

Write the prime factorization of each number, or write *prime*.

9. 24   10. 32   11. 19   12. 17
13. 25   14. 36   15. 72   16. 27
17. 100   18. 81   19. 51   20. 35
21. 125   22. 8   23. 64   24. 441

Use prime factorization to show that each number is a power of a single prime number.

25. 243   26. 343   27. 625   28. 1024

Two numbers are *relatively prime* if their only common factor is 1. Write the prime factorization of both numbers in each pair below. Then use the prime factorizations to show that the two numbers are relatively prime.

29. 12 and 13   30. 10 and 21   31. 18 and 25   32. 36 and 49

33. How many factors, other than 1, does 1024 have?

34. **Critical Thinking** In the table in Example 1, why would it be necessary to use only the natural numbers up through 7 in the search for factor pairs?

MIXED REVIEW

Solve each equation. Check your solution. (2.5 Skill A)

35. $x - 7 = -10$ 36. $z + 2.5 = -0.5$ 37. $x - 6.4 = -10.5$
38. $x + 6.4 = -4.3$ 39. $2 = a + 6.9$ 40. $-22.6 = t - 100$

Solve each equation. Check your solution. (2.5 Skill B)

41. $(2 - 6)z = -18$ 42. $20 = \dfrac{c}{-2 - 3}$ 43. $\left(4 - \dfrac{1}{2}\right)z = -14$

Answers for Lesson 7.1 Skill A

Try This Example 1
Answers should include four of the following:
$1 \times 36, 2 \times 18, 3 \times 12,$
$4 \times 9, 6 \times 6$

Try This Example 2
$2^2 \times 3^2 \times 5^1$

1. neither
2. composite
3. prime
4. composite
5. Answers should include three of the following: $1 \times 24,$
$2 \times 12, 3 \times 8, 4 \times 6$
6. Answers should include three of the following: $1 \times 30,$
$2 \times 15, 3 \times 10, 5 \times 6$

7. Answers should include three of the following: $1 \times 40,$
$2 \times 20, 4 \times 10, 5 \times 8$
8. Answers should include three of the following: $1 \times 48,$
$2 \times 24, 3 \times 16,$
$4 \times 12, 6 \times 8$
9. $2^3 \times 3^1$
10. 2^5
11. prime
12. prime
13. 5^2
14. $3^2 \times 2^2$
15. $2^3 \times 3^2$
16. 3^3
17. $2^2 \times 5^2$
18. 3^4
19. $3^1 \times 17^1$
20. $5^1 \times 7^1$
21. 5^3
22. 2^3

23. 2^6
24. $3^2 \times 7^2$
25. $243 = 3^5$
26. $343 = 7^3$
27. $625 = 5^4$
28. $1024 = 2^{10}$
★29. $12 = 2^2 \times 3^1$;
$13 = 13^1$; the only common factor is 1.
★30. $10 = 2^1 \times 5^1$;
$21 = 3^1 \times 7^1$; the only common factor is 1.
★31. $18 = 2^1 \times 3^2$; $25 = 5^2$;
the only common factor is 1.
★32. $36 = 2^2 \times 3^2$; $49 = 7^2$;
the only common factor is 1.
33. 10
34. After 7, the factor pairs begin to repeat.
35. $x = -3$
36. $z = -3$

37. $x = -4.1$
38. $x = -10.7$
39. $a = -4.9$
40. $t = -77.4$
41. $z = 4.5$
42. $c = -100$
43. $z = -4$

★ **Advanced Exercises**

Focus: *Students use prime factorizations to find the greatest common factor (GCF) of a set of integers or monomials.*

EXAMPLE 1

PHASE 1: Presenting the Example

Ask students to make a list of all the factors of 36, written in order from least to greatest. Repeat the instruction for the number 54. Display the results as shown at right.

Tell students to examine the two lists for factors that appear in both. [**1, 2, 3, 6, 9, 18**] Highlight these numbers in your display, as shown at right. Tell students that these numbers are called *common factors* of 36 and 54.

Point out that the greatest highlighted number is 18. So 18 is called the *greatest common factor*, or *GCF*, of 36 and 54. Then discuss Example 1, where students see how to find the GCF more directly.

36: 1, 2, 3, 4, 6, 9, 12, 18, 36
54: 1, 2, 3, 6, 9, 18, 27, 54

36: 1 , 2 , 3 ,4, 6 , 9 , 12, 18 , 36
54: 1 , 2 , 3 , 6 , 9 , 18 , 27, 54

common factors: 1, 2, 3, 6, 9, 18
GCF: 18

EXAMPLE 2

PHASE 1: Presenting the Example

Ask students, "What if the numbers in Example 1 were the expressions $36c^3$ and $54c^2$?" Demonstrate how the method for finding a GCF is easily extended to variable expressions, as shown at right.

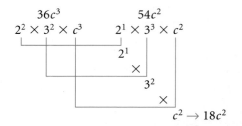

EXAMPLE 3

PHASE 1: Presenting the Example

Display the expressions a^3b and $5a^2b^2$ and ask: How do you find the GCF of variable expressions if one of them has no visible constant? Elicit the fact that the unseen constant is 1, and so a^3b should be considered as $1a^3b$.

EXAMPLE 4

PHASE 1: Presenting the Example

PHASE 2: Providing Guidance for **TRY THIS**

This example extends the concept of GCF to three expressions, and some students may begin to feel overwhelmed. It may help to organize the task in columns, as shown at right.

$3y^3z$ →	3^1	5^0	y^3	z^1
$15y^2z^2$ →	3^1	5^1	y^2	z^2
$-45y^2z$ →	-3^2	5^1	y^2	z^1
	3^1	5^0	y^2	z^1

GCF → $3y^2z$

The **greatest common factor**, or **GCF**, of two or more numbers is the greatest integer that is a common factor to all of those numbers.

EXAMPLE 1   Find the GCF of 36 and 54.

▸ **Solution**

Write the prime factorization of each number. Choose the smallest power of each prime factor that appears in both numbers. The GCF will be the product of those prime factors.

$$36 = 2^2 \times 3^2 \qquad 54 = 2^1 \times 3^3$$

The GCF of 36 and 54 is $2^1 \times 3^2$, or 18.

TRY THIS   Find the GCF of 60 and 140.

The GCF of two or more monomials is the product of all the integer and variable factors that are common to those monomials. To find the GCF, first find the GCF of the coefficients. Then find the product of the smallest power of each variable factor that appears in all the monomials. The GCF is the product of these two results.

EXAMPLE 2   Find the GCF of $36c^3$ and $54c^2$.

▸ **Solution**

From Example 1, the GCF of 36 and 54 is 18.
The factor c appears in both monomials; the smallest power of c that occurs is c^2.
Therefore, the GCF of $36c^3$ and $54c^2$ is $18c^2$.

TRY THIS   Find the GCF.   **a.** $18d^5$ and $108d$ **b.** $18d$ and 5

EXAMPLE 3   Find the GCF of a^3b and $5a^2b^2$.

▸ **Solution**

GCF of 1 and 5 is 1.   GCF of a^3 and a^2 is a^2.   GCF of b and b^2 is b.
Therefore, the GCF of a^3b and $5a^2b^2$ is $1a^2b$, or a^2b.

TRY THIS   Find the GCF of $3m^3n^3$ and $9m^2n^2$.

EXAMPLE 4   Find the GCF of $3y^3z$, $15y^2z^2$, and $-45y^2z$.

▸ **Solution**

GCF of 3, 15, and -45 is 3.   GCF of y^3 and y^2 is y^2.   GCF of z and z^2 is z.
Therefore, the GCF of $3y^3z$, $15y^2z^2$, and $-45y^2z$ is $3y^2z$.

TRY THIS   Find the GCF of $4mn^3$, $4m^2n^3$, and $16m^2n^2$.

EXERCISES

KEY SKILLS

List the variables common to both monomials in each pair.

1. $3d^3z$ and $12z$ **2.** $2a^2n$ and $2mn$ **3.** a^3n and $2a^2n^2$ **4.** $3a^3$ and $2n^2$

PRACTICE

Find the GCF of each pair of numbers.

5. 6 and 9   **6.** 27 and 54   **7.** 14 and 8   **8.** 14 and 16
9. 16 and 32   **10.** 15 and 45   **11.** 14 and 51   **12.** 16 and 27
13. 100 and 2   **14.** 11 and 77   **15.** 63 and 81   **16.** 19 and 57

Find the GCF of each pair of monomials.

17. $3x^3$ and $6x^2$ **18.** $-4a^3$ and $8a^2$ **19.** $6c^3$ and $9c$
20. $6c^3$ and $15c$ **21.** $4t^3$ and 6   **22.** $-5z^3$ and 30
23. $14x^3$ and 21   **24.** $10y^3$ and 25   **25.** $2x^3y$ and $4x^2$
26. $-4b^3t$ and $4b^2$ **27.** $5d^2z$ and $12dz$ **28.** $3c^2r$ and $-15cr$
29. $-4t^2y^2$ and 6   **30.** $-10z^2y^3$ and 20   **31.** $21a^3b^3$ and $3a^3b^3$
32. $10a^2v^3$ and $10a^2v^2$

Find the GCF of each set of monomials.

33. $3a^2b$, $15a^2b^2$, and $15ab$ **34.** $4c^2d$, $6c^2d^2$, and $8cd$ **35.** $6r^2s$, $6rs^2$, and 11
36. $7u^3v$, $7uv^2$, and 1   **37.** $3x^3y$, $6xy^2$, and $8x^2y^2$ **38.** $5m^3n$, $5m^3n^2$, and $7m^2n$
39. p^3q, $-p^3q^2$, and $2p^2q$ **40.** $3b^3t$, $-3b^3t^2$, and 15   **41.** $-4b^9t$, $-8b^8t^7$, and 32
42. 1, x, x^2, x^3, x^4, and x^5 **43.** 1, $2x$, 2^2x, 2^3x, 2^4x, and 2^5x

MIXED REVIEW

Find an equation in point-slope form for the line that contains the given point and has the given slope. (3.6 Skill A)

44. $A(3, -4)$; slope: $\frac{3}{4}$ **45.** $B(-2, 4)$; slope: $-\frac{1}{3}$ **46.** $T(3, -1)$; slope: 0

Find the equation in slope-intercept form for the line that contains each pair of points. (3.7 Skill A)

47. $A(3, 4)$ and $B(5, -2)$ **48.** $K(-2, 4)$ and $L(5, 4)$ **49.** $P(-3, -1)$ and $Q(6, 8)$

Answers for Lesson 7.1 Skill B

Try This Example 1
20

Try This Example 2
a. $18d$ **b.** 1

Try This Example 3
$3m^2n^2$

Try This Example 4
$4mn^2$

1. z
2. n
3. a and n
4. none
5. 3
6. 27
7. 2
8. 2
9. 16
10. 15
11. 1
12. 1
13. 2

14. 11
15. 9
16. 19
17. $3x^2$
18. $4a^2$
19. $3c$
20. $3c$
21. 2
22. 5
23. 7
24. 5
25. $2x^2$
26. $4b^2$
27. dz
28. $3cr$
29. 2
30. 10
31. $3a^3b^3$
32. $10a^2v^2$
33. $3ab$
34. $2cd$
35. 1
36. 1

37. xy
38. m^2n
39. p^2q
40. 3
41. 4
42. 1
43. 1
44. $y + 4 = \frac{3}{4}(x - 3)$
45. $y - 4 = -\frac{1}{3}(x + 2)$
46. $y + 1 = 0$
47. $y = -3x + 13$
48. $y = 4$
49. $y = x + 2$

Focus: *Students factor a polynomial by identifying the GCF of all of its terms.*

EXAMPLE 1

PHASE 1: Presenting the Example

Display $6x^2 - 4x$. Have students identify it as a binomial. Emphasize that it is a *difference* of two monomials. Tell students that their goal is to rewrite it as a *product*. Brainstorm their ideas about how this might be done. Use this discussion as a lead-in to the method of common monomial factoring.

PHASE 2: Providing Guidance for **TRY THIS**

Watch for answers such as $3(-v^2 + 5v)$ and $v(-3v + 15)$. Point out that the expressions inside parentheses can be factored further. Stress that the GCF of all terms inside the parentheses must be 1.

EXAMPLE 2

PHASE 2: Providing Guidance for **TRY THIS**

Some students may write $13t^2(t^2 + 2t)$, forgetting that 1 is the unseen factor that must remain as the third term inside the parentheses. Point out that the polynomial factor must have the same number of terms as the original expression. Encourage them to check their answers by multiplying. Lead them through the two multiplications at right to demonstrate that $13t^2(t^2 + 2t)$ does not yield the original expression when multiplied.

$13t^2(t^2 + 2t)$
$= 13t^4 + 26t^3$ ✗

$13t^2(t^2 + 2t + 1)$
$= 13t^4 + 26t^3 + 13t^2$ ✓

EXAMPLE 3

PHASE 1: Presenting the Example

Display $2ab^2 + 6a^2b^2 + 9a^2b$. Ask students how this expression differs from the expressions in Examples 1 and 2. Lead them to observe that two variables are involved and the GCF of the coefficients is 1. Then discuss the Example.

PHASE 3: ASSESSMENT AND CLOSURE for Lesson 7.1

- Display the four expressions at right. Ask students how the concept of GCF relates all the expressions to each other. [**The GCF of $-6m$ and $10m^2$ is $2m$. You can use this fact to factor the expression $10m^2 - 6m$ as $2m(5m + 3)$.**]

$-6m$
$10m^2$
$2m$
$5m + 3$

- Common monomial factoring is the most fundamental method for factoring a polynomial. In the lessons that follow, students will learn additional factoring methods and will see how to combine them with common monomial factoring.

☞ *For a Lesson 7.1 Quiz, see* Assessment Resources *page 69.*

To factor a polynomial means to write it as a product of two or more polynomials. The first step is to find the GCF of all terms.

EXAMPLE 1

Use the GCF to factor $6x^2 - 4x$.

▶ Solution

1. Find the GCF of all the terms. The GCF of $6x^2$ and $-4x$ is $2x$.
2. Write each term in $6x^2 - 4x$ as a product involving the GCF.
$$6x^2 = (2x)(3x) \qquad -4x = (2x)(-2)$$
3. Apply the Distributive Property.
$$6x^2 - 4x = (2x)(3x) + (2x)(-2)$$
$$= 2x(3x - 2)$$

TRY THIS   Use the GCF to factor $-3v^2 + 15v$.

EXAMPLE 2

Use the GCF to factor $3p^3 + 12p^2 - 3p$.

▶ Solution

1. The GCF of $3p^3$, $12p^2$, and $-3p$ is $3p$.
2. Write each term in $3p^3 + 12p^2 - 3p$ as a product involving the GCF.
$$3p^3 = (3p)(p^2) \qquad 12p^2 = (3p)(4p) \qquad -3p = (3p)(-1)$$
3. Apply the Distributive Property.
$$3p^3 + 12p^2 - 3p = (3p)(p^2) + (3p)(4p) + (3p)(-1)$$
$$= 3p(p^2 + 4p - 1)$$

TRY THIS   Use the GCF to factor $13t^4 + 26t^3 + 13t^2$.

EXAMPLE 3

Use the GCF to factor $2ab^2 + 6a^2b^2 + 9a^2b$.

▶ Solution

1. The GCF of $2ab^2$, $6a^2b^2$, and $9a^2b$ is ab.
2. Write each term in $2ab^2 + 6a^2b^2 + 9a^2b$ as a product involving the GCF.
$$2ab^2 = (ab)(2b) \qquad 6a^2b^2 = (ab)(6ab) \qquad 9a^2b = (ab)(9a)$$
3. Apply the Distributive Property.
$$2ab^2 + 6a^2b^2 + 9a^2b = (ab)(2b) + (ab)(6ab) + (ab)(9a)$$
$$= ab(2b + 6ab + 9a)$$

TRY THIS   Use the GCF to factor $-3c^3d^2 + 6c^2d^2 + 9c^2d$.

EXERCISES

KEY SKILLS

Use multiplication to verify that the factorization shown is correct.

1. $2x^2 + 6x = 2x(x + 3)$
2. $2x^3 - 10x = 2x(x^2 - 5)$
3. $3x^3 + 3x^2 = 3x^2(x + 1)$
4. $x^3 - 4x = x(x^2 - 4)$

PRACTICE

Use the GCF to factor each expression.

5. $7x + 14$
6. $5a - 10$
7. $10b^2 + 12b$
8. $-6y^2 + 24y$
9. $6z^2 - 9z$
10. $r^2 + r$
11. $12n^3 + 4n^2 - 4n$
12. $5m^3 + 5m^2 + 15m$
13. $5k^3 - 25k^2 - 75k$
14. $2n^3 + 17n^2 - 41n$
15. $2w^4 + 18w^3 - 18w^2$
16. $x^4 - 11x^3 - 7x^2$
17. $-12z^5 + 15z^4 - 18z^3$
18. $7v^5 + 49v^4 - 98v^2$
19. $11v^5 - 11v^4 + 22v^3$
20. $2ab^2 - 8a^2b^2 - 8ab$
21. $r^2s^2 - 8rs^2 + 8r^2s$
22. $3d^2n^2 - 8dn^2 + 8d^2n$
23. $3mn^2 + 10m^2n^2 - 7mn - 2n$
24. $5h^3p^3 + 10hp^2 - 15hp - 15h$
25. $7u^2v + 28uv^2 - 7uv^4 + 7uv$
26. $h^5p^3 + 10hp^2 - 15hp^4 - 15h^3$
27. $ab^2 + a^2b^2 - a^3b^3 + a^2b^5$
28. $-4sc^3 + 10s^2c^2 - 16sc^4 - 10sc^2$
29. $3(a + b)^3 + 3a(a + b)^2$
30. $3(x + y)^3 - 5x(x + y)^2$
31. $3(m + n)^3 - 5m(m + n)^2 + 2n(m + n)^2$
32. $(p + q)^3 - 5p(p + q)^2 + 2q(p + q)^2$
33. $(x^2y^3)^2 + (x^3y^2)^3$

MIXED REVIEW

Find each product. (6.6 Skill B)

34. $(2x + 4)(x - 5)$
35. $(2x - 5)(2x + 1)$
36. $(3r + 1)(3r + 2)$
37. $(x + 7)(2 + x)$
38. $(2 - 5x)(1 + 6x)$
39. $(3 - 2y)(3y - 2)$

Find each product. (6.6 Skill C)

40. $(2a + 4)^2$
41. $(2x - 3)(2x + 3)$
42. $(3y + 1)^2$

Answers for Lesson 7.1 Skill C

Try This Example 1
$3v(-v + 5)$

Try This Example 2
$13t^2(t^2 + 2t + 1)$

Try This Example 3
$3c^2d(-cd + 2d + 3)$

1. $2x(x + 3) = (2x)(x) + (2x)(3)$
$= 2x^2 + 6x$
2. $2x(x^2 - 5) = (2x)(x^2) + (2x)(-5)$
$= 2x^3 - 10x$
3. $3x^2(x + 1) = (3x^2)(x) + (3x^2)(1)$
$= 3x^3 + 3x^2$
4. $x(x^2 - 4) = (x)(x^2) + (x)(-4)$
$= x^3 - 4x$
5. $7(x + 2)$
6. $5(a - 2)$
7. $2b(5b + 6)$
8. $6y(-y + 4)$
9. $3z(2z - 3)$
10. $r(r + 1)$
11. $4n(3n^2 + n - 1)$

12. $5m(m^2 + m + 3)$
13. $5k(k^2 - 5k - 15)$
14. $n(-n^2 + 17n - 41)$
15. $2w^2(w^2 + 9w - 9)$
16. $x^2(x^2 - 11x - 7)$
17. $3z^3(-4z^2 + 5z - 6)$
18. $7v^2(v^3 + 7v^2 - 14)$
19. $11v^3(v^2 - v + 2)$
20. $2ab(b + 4ab - 4)$
21. $rs(rs - 8s + 8r)$
22. $dn(3dn - 8n + 8d)$
23. $n(3mn + 10m^2n - 7m - 2)$
24. $5h(h^2p^3 + 2p^2 - 3p - 3)$
25. $7uv(u + 4v - v^3 + 1)$
26. $h(h^4p^3 + 10p^2 - 15p^4 - 15h^2)$
27. $ab^2(1 + a - a^2b + ab^3)$
28. $2sc^2(-2c + 5s - 8c^2 - 5)$
★ 29. $3(a + b)^2(2a + b)$
★ 30. $(x + y)^2(-2x + 3y)$
★ 31. $(m + n)^2(-2m + 5n)$
★ 32. $(p + q)^2(-4p + 3q)$
★ 33. $x^4y^6(1 + x^5)$
34. $2x^2 - 6x - 20$

35. $4x^2 - 8x - 5$
36. $9r^2 + 9r + 2$
37. $x^2 + 9x + 14$
38. $-30x^2 + 7x + 2$
39. $-6y^2 + 13y - 6$
40. $4a^2 + 16a + 16$
41. $4x^2 - 9$
42. $9y^2 + 6y + 1$

★ **Advanced Exercises**

7.2

LESSON

Factoring Special Polynomials

Factoring polynomials with special forms

Focus: *Students use pattern recognition to find the binomial factors of perfect-square trinomials and of differences of two squares.*

EXAMPLE 1

PHASE 1: Presenting the Example

Display the expressions shown at right. Have students simplify each one. [**1.** $p^2 + 10p + 25$
2. $a^2 - 24a + 144$ **3.** $4v^2 + 12v + 9$ **4.** $49j^2 - 56jk + 16k^2$] Remind them that the simplified expressions are *perfect-square trinomials* because they are the result when a binomial is multiplied by itself, or *squared*.

1. $(p + 5)(p + 5)$
2. $(a - 12)(a - 12)$
3. $(2v + 3)(2v + 3)$
4. $(7j - 4k)(7j - 4k)$

Ask students to examine the perfect-square trinomials and describe what characteristics they share. Lead them to observe the following pattern:

$p^2 + 10p + 25$	$a^2 - 24a + 144$	$4v^2 + 12a + 9$	$49j^2 - 56jk + 16k^2$
$p^2 = (p)^2$	$a^2 = (a)^2$	$4v^2 = (2v)^2$	$49j^2 = (7j)^2$
$25 = (5)^2$	$144 = (12)^2$	$9 = (3)^2$	$16k^2 = (4k)^2$
$10p = 2(p)(5)$	$24a = 2(a)(12)$	$12v = 2(2v)(3)$	$56jk = 2(7j)(4k)$

Now display the three trinomials shown at right. Ask students which, if any, is a perfect square. They should conclude that $t^2 + 6t + 9$ is the only perfect square because it is the only one that follows the perfect square pattern. That is, $t^2 = (t)^2$, $9 = (3)^2$, and $6t = 2(t)(3)$.

$t^2 + 6t + 10$
$t^2 + 5t + 9$
$t^2 + 6t + 9$

Ask students what binomial factors must have been multiplied to produce $t^2 + 6t + 9$. From the pattern, they should infer that the factors must be $(t + 3)(t + 3)$, or $(t + 3)^2$. Use this conclusion as a lead-in to Example 1.

EXAMPLE 2

PHASE 1: Presenting the Example

Display the expressions shown at right. Have students simplify each one. [**1.** $p^2 - 25$
2. $a^2 - 144$ **3.** $4v^2 - 9$ **4.** $49j^2 - 16k^2$] Compare this *difference of two squares* pattern to the perfect-square trinomial pattern that they observed in Example 1. Then discuss Example 2.

1. $(p + 5)(p - 5)$
2. $(a - 12)(a + 12)$
3. $(2v + 3)(2v - 3)$
4. $(7j - 4k)(7j + 4k)$

PHASE 2: Providing Guidance for **TRY THIS**

Some students will factor $h^2 - 100$ as $(h - 10)(h - 10)$ and $16x^2 - 49$ as $(4x - 7)(4x - 7)$. Encourage them to check by multiplying their two binomial factors. A quick application of the FOIL method, as shown at right, will show that they have actually factored the perfect-square trinomials $h^2 - 20h + 100$ and $16x^2 - 56x + 49$.

$$h^2 \qquad 100$$
$$(h - 10)(h - 10)$$
$$-10h$$
$$-10h$$

$$16x^2 \qquad 49$$
$$(4x - 7)(4x - 7)$$
$$-28x$$
$$-28x$$

7.2 Factoring Special Polynomials

SKILL A ▸ *Factoring polynomials with special forms*

Some expressions can be quickly factored if you can recognize special patterns. For example, perfect-square trinomials and differences of two squares are easily factorable.

Factoring Special Products
Perfect-square trinomial: $a^2 + 2ab + b^2 = (a + b)^2$
Difference of two squares: $a^2 - b^2 = (a + b)(a - b)$

The polynomial $49x^2 + 14x + 1$ is a perfect-square trinomial.
$$49x^2 + 14x + 1 = 7^2x^2 + (2)(7)x + 1^2$$
$$= (7x)^2 + 2(7x)(1) + 1^2 \quad \longleftarrow \text{Power-of-a-Product Property}$$
$$= (7x + 1)^2$$

EXAMPLE 1   **Factor.**   **a.** $m^2 + 12m + 36$ **b.** $4n^2 - 20n + 25$

▸ **Solution**

a. $m^2 + 12m + 36 = m^2 + 2(m)(6) + 6^2$
$$= (m + 6)^2 \quad \longleftarrow a = m \text{ and } b = 6$$

b. $4n^2 - 20n + 25 = (2n)^2 + 2(2n)(-5) + (-5)^2 \quad \longleftarrow 4n^2 = (2n)^2 \text{ and } 25 = (-5)^2$
$$= [2n + (-5)]^2 \quad \longleftarrow a = 2n \text{ and } b = -5$$
$$= (2n - 5)^2$$

TRY THIS   **Factor.**   **a.** $d^2 - 18d + 81$ **b.** $9z^2 + 6z + 1$

The polynomial $25x^2 - 81$ is a difference of two squares.
$$25x^2 - 81 = 5^2x^2 - 9^2$$
$$= (5x)^2 - 9^2 \quad \longleftarrow \text{Power-of-a-Product Property}$$
$$= (5x + 9)(5x - 9)$$

EXAMPLE 2   **Factor.**   **a.** $v^2 - 121$ **b.** $9x^2 - 25$

▸ **Solution**

a. $v^2 - 121 = v^2 - 11^2 \quad \longleftarrow 121 = 11^2$
$$= (v + 11)(v - 11) \quad \longleftarrow a = v \text{ and } b = 11$$

b. $9x^2 - 25 = (3x)^2 - 5^2 \quad \longleftarrow 9x^2 = (3x)^2 \text{ and } 25 = 5^2$
$$= (3x + 5)(3x - 5) \quad \longleftarrow a = 3x \text{ and } b = 5$$

TRY THIS   **Factor.**   **a.** $h^2 - 100$ **b.** $16x^2 - 49$

EXERCISES

KEY SKILLS

Each polynomial below is in the form of a perfect-square trinomial, $a^2 + 2ab + b^2$, or a difference of two squares, $a^2 - b^2$. Write the values of a and b. Do not factor.

1. $x^2 + 6x + 9$ **2.** $64x^2 - 16x + 1$

3. $k^2 - 9$ **4.** $25n^2 - 36$

PRACTICE

Factor.

5. $d^2 + 2d + 1$ **6.** $c^2 + 10c + 25$ **7.** $u^2 - 8u + 16$

8. $y^2 - 10y + 25$ **9.** $h^2 - 144$ **10.** $g^2 - 4$

11. $k^2 - 225$ **12.** $q^2 - 400$ **13.** $4y^2 + 20y + 25$

14. $16y^2 - 40y + 25$ **15.** $49m^2 + 14m + 1$ **16.** $16y^2 - 72y + 81$

17. $16x^2 - 121$ **18.** $64c^2 - 1$ **19.** $u^2v^2 + 2uv + 1$

20. $y^2z^2 - 8yz + 16$ **21.** $4m^2n^2 - 16mn + 16$ **22.** $3m^2n^2 + 6mn + 3$

23. $64c^2d^2 - 25$ **24.** $100k^2q^2 - 1$ **25.** $121s^2t^2 - 81$

26. $4m^2n^2 + 24mnxy + 36x^2y^2$ **27.** $25m^2n^2 - 49x^2y^2$

28. Suppose that a is an integer between 1 and 100 inclusive. Find all values of a such that $ak^2 - 49$ is the difference of two squares.

29. Critical Thinking   Suppose that $25y^2 - 50y + 25 = (my + n)^2$. Find m and n.

30. Critical Thinking   Suppose that $36x^2y^2 + 12xy + 1 = (mxy + n)^2$. Find m and n.

MIXED REVIEW

Find each product. (6.6 Skill B)

31. $(3a - 10)(2a + 3)$ **32.** $(11t - 1)(2t - 1)$ **33.** $(-2n + 3)(2n + 5)$

Find each product. (6.6 Skill C)

34. $(3nt + 4)(3nt - 4)$ **35.** $(3t - 4c)^2$ **36.** $(6xy + 1)(6xy - 1)$

Write the prime factorization of each number. (7.1 Skill A)

37. 154   **38.** 104   **39.** 231   **40.** 112

Answers for Lesson 7.2 Skill A

Try This Example 1
a. $(d - 9)^2$ **b.** $(3z + 1)^2$

Try This Example 2
a. $(h - 10)(h + 10)$
b. $(4x - 7)(4x + 7)$

1. $a = x, b = 3$
2. $a = 8x, b = -1$
3. $a = k, b = 3$
4. $a = 5n, b = 6$
5. $(d + 1)^2$
6. $(c + 5)^2$
7. $(u - 4)^2$
8. $(y - 5)^2$
9. $(h - 12)(h + 12)$
10. $(g - 2)(g + 2)$
11. $(k - 15)(k + 15)$
12. $(q - 20)(q + 20)$
13. $(2y + 5)^2$
14. $(4y - 5)^2$
15. $(7m + 1)^2$
16. $(4y - 9)^2$

17. $(4x - 11)(4x + 11)$
18. $(8c - 1)(8c + 1)$
19. $(uv + 1)^2$
20. $(yz - 4)^2$
21. $(4mn - 2)^2$
22. $3(mn + 1)^2$
23. $(8cd - 5)(8cd + 5)$
24. $(10kq - 1)(10kq + 1)$
25. $(11st - 9)(11st + 9)$
★ **26.** $4(mn + 3xy)^2$
★ **27.** $(5mn - 7xy)(5mn + 7xy)$
★ **28.** 1, 4, 9, 16, 25, 36, 49, 64, 81, 100
29. $m = 5$ and $n = -5$
30. $m = 6$ and $n = 1$
31. $6a^2 - 11a - 30$
32. $22t^2 - 13t + 1$
33. $-4n^2 - 4n + 15$
34. $9n^2t^2 - 16$
35. $9t^2 - 24tc + 16c^2$
36. $36x^2y^2 - 1$
37. $2 \times 7 \times 11$

38. $2^3 \times 13$
39. $3 \times 7 \times 11$
40. $2^4 \times 7$

★ **Advanced Exercises**

Focus: *Students use grouping to find the binomial factors of expressions that are not of the form* $ax^2 + bx + c$.

EXAMPLE 1

PHASE 1: Presenting the Example

Display $3(\boxed{}) + g(\boxed{})$. Tell students to think of the box as a variable and to factor the expression. Display the result as shown at right.

$$3(\boxed{}) + g(\boxed{}) = (3 + g)(\boxed{})$$

$$3(\boxed{x}) + g(\boxed{x}) = (3 + g)(\boxed{x})$$

Now ask students what the factorization would be if the box contained the variable x. Display the result beneath the first statement, as shown.

$$3(\boxed{a + b}) + g(\boxed{a + b}) = (3 + g)(\boxed{a + b})$$

Next, ask what the factorization would be if the box contained $a + b$. Display the result as shown.

EXAMPLES 2 and 3

PHASE 1: Presenting the Examples

Point out to students that different groupings are possible, but that all correct groupings will lead to the same factorization. For instance, a different grouping of the expression in Example 2 is shown at right.

$$ax + 2a + bx + 2b$$
$$= ax + bx + 2a + 2b$$
$$= (ax + bx) + (2a + 2b)$$
$$= x(a + b) + 2(a + b)$$
$$= (x + 2)(a + b)$$

PHASE 2: Providing Guidance for TRY THIS

In the Example 2 Try This, watch for students who correctly group and factor the expression as $3(3m + 5n) + y(3m + 5n)$ and then write $3y(3m + 5n)$. Encourage them to check their factorizations by multiplying.

In the Example 3 Try This, some students may instinctively factor d from each of the first three terms and write $d(d - 8 + b) - 8b$ as a first step. Point out that the expression inside parentheses is a trinomial, and their goal is to find *binomial* factors.

PHASE 3: ASSESSMENT AND CLOSURE for Lesson 7.2

- Give students the expressions at right. Have them fill in the boxes to make a perfect square trinomial or a difference of two squares. [**samples: 1.** $r^2 + 10r + 25$ **2.** $a^2 - 8ab + 16b^2$ **3.** $4x^2y^2 + 12xy + 9$ **4.** $36d^2 - 1$]

- Perfect square trinomials and differences of two squares are special cases of polynomials of the form $ax^2 + bx + c$. In Lessons 7.3 and 7.4, students will learn to factor the general polynomial of this form.

1. $r^2 + \Box r + \Box$

2. $a^2 - \Box ab + \Box$

3. $\Box x^2y^2 + \Box + 9$

4. $\Box d^2 - 1$

☞ *For a Lesson 7.2 Quiz, see* Assessment Resources *page 70.*

A polynomial may have a common factor that is a binomial instead of a monomial.

EXAMPLE 1

Factor $3(a + b) + g(a + b)$.

▶ **Solution**

Notice that $3(a + b)$ and $g(a + b)$ each contain the **common binomial factor** $(a + b)$. Use the Distributive Property to write

$$3(a + b) + g(a + b) = (3 + g)(a + b).$$

common factor

$3\underbrace{(a + b)} + g\underbrace{(a + b)}$

TRY THIS   Factor $3(2a - b) - y(2a - b)$.

You may need to group terms before you can find a common binomial factor.

EXAMPLE 2

Factor $ax + 2a + bx + 2b$ by grouping.

▶ **Solution**

Group the terms that have a common number or variable as a factor. Treat $ax + 2a$ as one expression, and treat $bx + 2b$ as another expression.

$$ax + 2a + bx + 2b = (ax + 2a) + (bx + 2b) \quad \longleftarrow \text{Group terms.}$$
$$= a(x + 2) + b(x + 2) \quad \longleftarrow \text{Factor each group.}$$
$$= (a + b)(x + 2) \quad \longleftarrow \text{Apply the Distributive Property.}$$

TRY THIS   Factor $9m + 15n + 3my + 5ny$ by grouping.

When factoring by grouping, be careful when the terms involve subtraction. Notice that

$$x^3 + 3x^2 - ax - 3a = (x^3 + 3x^2) - (ax + 3a) \quad ✔$$

and *not* $x^3 + 3x^2 - ax - 3a = (x^3 + 3x^2) - (ax - 3a)$. ✗

You may find it helpful to rewrite each subtraction as addition of the opposite.

EXAMPLE 3

Factor $x^3 + 3x^2 - ax - 3a$ by grouping.

▶ **Solution**

$$x^3 + 3x^2 - ax - 3a = (x^3 + 3x^2) - (ax + 3a) \quad \longleftarrow \text{Group terms.}$$
$$\qquad\qquad\text{Apply the Multiplicative Property of } -1.$$
$$= x^2(x + 3) - a(x + 3) \quad \longleftarrow \text{Factor each group.}$$
$$= (x^2 - a)(x + 3) \quad \longleftarrow \text{Apply the Distributive Property.}$$

TRY THIS   Factor $d^2 - 8d + bd - 8b$ by grouping.

EXERCISES

KEY SKILLS

What is the common binomial factor in each polynomial?

1. $2x(3x - 7) - 5(3x - 7)$
2. $x^2(x + 2) + 4(x + 2)$
3. $2(x - y) - g(x - y)$
4. $5(p + q) + 3g(p + q)$

PRACTICE

Factor.

5. $2(2p - 5) + h(2p - 5)$
6. $9(2n + 1) - s(2n + 1)$
7. $-2(m + 7) + t(m + 7)$
8. $-(a - 9) + 3z(a - 9)$
9. $-(v - 3t) - w(v - 3t)$
10. $-(-u + s) - w(-u + s)$
11. $cx + cy + 3x + 3y$
12. $4a + 4b + ta + tb$
13. $x^2 - 10x + xy - 10y$
14. $x^2 + x + xw + w$
15. $y^2 - 3y + yd - 3d$
16. $t^2 - 9t + 3t - 27$
17. $9cn + 12cm + 15dn + 20dm$
18. $4mna + 4mnb + 5na + 5nb$
19. $15a - 10b + 21at - 14tb$
20. $6x - 2y + 12xz - 4yz$
21. $2x^2 - 4x + xy - 2y$
22. $7y^2 - 14y + by - 2b$
23. $4ab - 2c - 8a + bc$
24. $12xy - z - 4x + 3yz$
25. $9x + 6 - 6ax - 4a$
26. $20a + 12 - 25ax - 15x$

27. Find a such that $3(2y - a) + n(2y + 5)$ can be factored. Then factor.
28. Find a such that $3(2y - a) + n[2y + 2(a + 1)]$ can be factored. Then factor.
29. If $4z(3x - 7) + 3y(ax + b) = (4z + 3y)(3x - 7)$, what are a and b?

MIXED REVIEW

Graph the solution to each inequality. (5.5 Skill A)

30. $y \geq -2.5x + 1$
31. $y < -3x + 5$
32. $y \leq 5x - 1.5$

Graph the solution to each system of inequalities. (5.6 Skill A)

33. $\begin{cases} y < 2x \\ y \geq -2x - 3 \end{cases}$
34. $\begin{cases} y > 2x - 1 \\ y \leq -2.5x + 3 \end{cases}$
35. $\begin{cases} y < 1.5x + 1 \\ y < -1.5x + 1 \end{cases}$

Answers for Lesson 7.2 Skill B

Try This Example 1
$(3 - y)(2a - b)$

Try This Example 2
$(3 + y)(3m + 5n)$

Try This Example 3
$(d + b)(d - 8)$

1. $3x - 7$
2. $x + 2$
3. $x - y$
4. $p + q$
5. $(2 + h)(2p - 5)$
6. $(9 - s)(2n + 1)$
7. $(-2 + t)(m + 7)$
8. $(-1 + 3z)(a - 9)$
9. $(-1 - w)(v - 3t)$
10. $(-1 - w)(-u + s)$
11. $(c + 3)(x + y)$
12. $(4 + t)(a + b)$
13. $(x + y)(x - 10)$
14. $(x + w)(x + 1)$
15. $(y + d)(y - 3)$
16. $(t + 3)(t - 9)$

17. $(3c + 5d)(3n + 4m)$
18. $(4mn + 5n)(a + b)$
19. $(5 + 7t)(3a - 2b)$
20. $(2 + 4z)(3x - y)$
21. $(2x + y)(x - 2)$
22. $(7y + b)(y - 2)$
23. $(4a + c)(b - 2)$
24. $(4x + z)(3y - 1)$
25. $(3 - 2a)(3x + 2)$
26. $(4 - 5x)(5a + 3)$
27. $a = -5; (3 + n)(2y + 5)$
★28. $a = -\dfrac{2}{3}; (3 + n)\left(2y + \dfrac{2}{3}\right)$
★29. $a = 3$ and $b = -7$

30.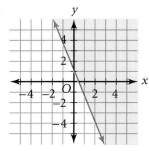

31.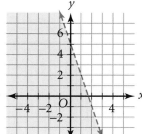

32.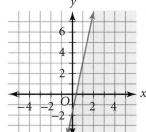

For answers to Exercises 33–35, see Additional Answers.

★ **Advanced Exercises**

Factoring $x^2 + bx + c$

 Factoring $x^2 + bx + c$ when c is positive

Focus: *Students find the binomial factors of a quadratic trinomial whose leading coefficient is 1 and whose constant term is positive.*

EXAMPLE 1

PHASE 1: Presenting the Example

Display the multiplications shown in the first column at right. Have students find each product. Then display the products in a second column, as shown. Ask students to describe any patterns they see. Elicit observations such as the following: The form of each product is $x^2 + \square x + 30$. The product of the constants in each binomial pair is 30. The sum of the constants in each binomial pair is the coefficient of the x-term of the product.

1 $(x + 5)(x + 6)$ $[x^2 + 11x + 30]$

2 $(x + 3)(x + 10)$ $[x^2 + 13x + 30]$

3 $(x + 2)(x + 15)$ $[x^2 + 17x + 30]$

4 $(x + 1)(x + 30)$ $[x^2 + 31x + 30]$

Remove the display and replace it with just the product $x^2 + 17x + 30$. Remind students of the patterns that they observed. Tell them to write the binomial multiplication that gives this product. $[(x + 2)(x + 15)]$ Check their answers, and discuss any questions they may have. Then discuss Example 1, which shows a direct method for finding factors.

PHASE 2: Providing Guidance for **TRY THIS**

Some students may have trouble finding the required factor pair. These students may have greater success if they use a geometric model, as shown at right.

	a	20
a	a^2	$20a$
1	$1a$	20

$\times$

$1a + 20a \neq 9a$

	a	10
a	a^2	$10a$
2	$2a$	20

$\times$

$2a + 10a \neq 9a$

	a	5
a	a^2	$5a$
4	$4a$	20

✔

$4a + 5a = 9a$

EXAMPLE 2

PHASE 1: Presenting the Example

Display $y^2 + 10y + 24$ and ask students to factor. $[(y + 4)(y + 6)]$ Now, as they watch, alter the expression so that it becomes $y^2 - 10y + 24$. Ask how they must change their factors to give this new product. Lead them to see that, if they rewrite their factors as $(y - 4)$ and $(y - 6)$, the product of the constants remains 24, but the sum of the constants is now -10 instead of 10. So the product becomes $y^2 - 10y + 24$ instead of $y^2 + 10y + 24$.

PHASE 2: Providing Guidance for **TRY THIS**

Some students may identify 4 and 9 as the required factors of 36, and then factor $n^2 - 13n + 36$ as $(n + 4)(n + 9)$. Encourage them to begin by identifying the signs within the binomial factors, as shown at right.

$n^2 - 13n + 36$

$(\quad - \quad)(\quad - \quad)$

7.3 Factoring $x^2 + bx + c$

SKILL A ▸ Factoring $x^2 + bx + c$ when c is positive

Suppose that you want to factor $x^2 + 5x + 6$.

- Because $x^2 + 5x + 6$ has no common monomial factor other than 1, you cannot factor out a monomial.
- Because $x^2 + 5x + 6$ is not a perfect-square trinomial or a difference of squares, you cannot write a special product.

The following fact will help you factor quadratic trinomials like $x^2 + 5x + 6$.

> **Factoring $x^2 + bx + c$**
> For $x^2 + bx + c$, if there are two real numbers r and s such that $c = rs$ and $b = r + s$, then $x^2 + bx + c = (x + r)(x + s)$.

In other words, to factor $x^2 + bx + c$, look for factor pairs, two numbers whose product is c. Choose the pair whose sum is b.

EXAMPLE 1   Factor $x^2 + 5x + 6$.

Make an organized list.

▸ **Solution**

1. $c = 6$: write factor pairs of 6. Remember to include negative factors.
$$1 \times 6 \qquad 2 \times 3 \qquad -1 \times -6 \qquad -2 \times -3$$
2. $b = 5$: choose the pair whose sum is 5.
$$\cancel{1 + 6 = 7} \qquad 2 + 3 = 5 \checkmark \qquad \cancel{-1 + -6 = -7} \qquad \cancel{-2 + -3 = -5}$$
3. Write the product using 2 and 3. Thus, $x^2 + 5x + 6 = (x + 2)(x + 3)$.

TRY THIS   Factor $a^2 + 9a + 20$.

EXAMPLE 2   Factor $y^2 - 10y + 24$.

Make an organized list.

▸ **Solution**

1. $c = 24$: write factor pairs of 24.
$$\begin{array}{llll} 1 \times 24 & -1 \times -24 & 4 \times 6 & -4 \times -6 \\ 2 \times 12 & -2 \times -12 & 3 \times 8 & -3 \times -8 \end{array}$$
2. $b = -10$: choose the pair whose sum is -10.
$$\cancel{1 + 24 = 25} \qquad \cancel{-1 + (-24) = -25} \qquad \cancel{4 + 6 = 10} \qquad -4 + (-6) = -10 \checkmark$$
3. Write the product using -4 and -6. $\quad [y + (-4)][y + (-6)] = (y - 4)(y - 6)$
Thus, $y^2 - 10y + 24 = (y - 4)(y - 6)$.

TRY THIS   Factor $n^2 - 13n + 36$.

EXERCISES

KEY SKILLS

Given each quadratic trinomial and list of factor pairs, which pair can you use to write the factorization? Write the factorization.

1. $m^2 + 12m + 20$
$1 \times 20, -1 \times -20,$
$2 \times 10, -2 \times -10,$
$4 \times 5, -4 \times -5$

2. $n^2 - 9n + 20$
$1 \times 20, -1 \times -20,$
$2 \times 10, -2 \times -10,$
$4 \times 5, -4 \times -5$

3. $k^2 - 21k + 20$
$1 \times 20, -1 \times -20,$
$2 \times 10, -2 \times -10,$
$4 \times 5, -4 \times -5$

4. $z^2 + 11z + 18$
$1 \times 18, -1 \times -18,$
$2 \times 9, -2 \times -9,$
$3 \times 6, -3 \times -6$

5. $d^2 - 12d + 35$
$1 \times 35, 5 \times 7,$
$-1 \times -35, -5 \times -7$

6. $h^2 - 15h + 36$
$2 \times 18, -2 \times -18,$
$3 \times 12, -3 \times -12,$
$4 \times 9, -4 \times -9$

Verify each factorization by multiplying.

7. $x^2 + 12x + 35 = (x + 5)(x + 7)$

8. $z^2 - 12z + 35 = (z - 5)(z - 7)$

PRACTICE

Factor.

9. $m^2 + 5m + 4$
10. $x^2 + 7x + 12$
11. $z^2 + 8z + 15$

12. $g^2 + 6g + 5$
13. $k^2 - 11k + 30$
14. $m^2 - 4m + 3$

15. $m^2 - 8m + 7$
16. $m^2 - 12m + 11$
17. $x^2 + 13x + 40$

18. $x^2 + 19x + 48$
19. $y^2 + 26y + 48$
20. $w^2 + 15w + 56$

21. $t^2 + 23t + 42$
22. $s^2 - 16s + 63$
23. $t^2 - 23t + 42$

24. $m^2 - 20m + 64$
25. $p^2 - 50p + 400$
26. $m^2 - 30m + 144$

27. Factor $a^2b^2 + 10ab + 21$.

28. **Critical Thinking**   In $x^2 + bx + c$, if c is positive, what do you know about the factors of c?

29. **Critical Thinking**   In $x^2 + bx + c$, if c is positive and bx is positive, what do you know about the factors of c? What do you know if bx is negative?

MIXED REVIEW

Find the GCF of each pair of monomials. (7.1 Skill B)

30. $3a^3$ and a
31. $8v^3$ and $8v^2$
32. $3d^2$ and $9d$
33. $5n^2$ and $15n^3$

34. $3m^3$ and m^2
35. $4k^2$ and $8k^3$
36. z^5 and $9z$
37. $6h^4$ and $15h$

Use the GCF to factor each expression. (7.1 Skill C)

38. $n^2 - 2n$
39. $3p^3 - 15p^2 + 3p$
40. $2z^3 - 6z^2$

Answers for Lesson 7.3   Skill A

Try This Example 1
$(a + 5)(a + 4)$

Try This Example 2
$(n - 9)(n - 4)$

1. 2 and 10;
$(m + 2)(m + 10)$
2. -4 and -5;
$(n - 4)(n - 5)$
3. -1 and -20;
$(k - 1)(k - 20)$
4. 2 and 9; $(z + 2)(z + 9)$
5. -5 and -7;
$(d - 5)(d - 7)$
6. -3 and -12;
$(h - 3)(h - 12)$
7. $(x + 5)(x + 7) =$
$x(x + 7) + 5(x + 7) =$
$x^2 + 7x + 5x + 35 =$
$x^2 + 12x + 35$

8. $(z - 5)(z - 7) =$
$z(z - 7) - 5(z - 7) =$
$z^2 - 7z - 5z + 35 =$
$z^2 - 12z + 35$
9. $(m + 1)(m + 4)$
10. $(x + 3)(x + 4)$
11. $(z + 3)(z + 5)$
12. $(g + 1)(g + 5)$
13. $(k - 5)(k - 6)$
14. $(m - 1)(m - 3)$
15. $(m - 1)(m - 7)$
16. $(m - 1)(m - 11)$
17. $(x + 5)(x + 8)$
18. $(x + 3)(x + 16)$
19. $(y + 2)(y + 24)$
20. $(w + 7)(w + 8)$
21. $(t + 2)(t + 21)$
22. $(s - 7)(s - 9)$
23. $(t - 2)(t - 21)$
24. $(m - 4)(m - 16)$
25. $(p - 10)(p - 40)$
26. $(m - 6)(m - 24)$
27. $(ab + 3)(ab + 7)$

28. The factors of c are either both positive or both negative.
29. Both factors are positive; both factors are negative.
30. a
31. $8v^2$
32. $3d$
33. $5n^2$
34. m^2
35. $4k^2$
36. z
37. $3h$
38. $n(n - 2)$
39. $3p(p^2 - 5p + 1)$
40. $2z^2(z - 3)$

Focus: *Students find the binomial factors of a quadratic trinomial whose leading coefficient is 1 and whose constant term is negative.*

EXAMPLES 1 and 2

PHASE 1: Presenting the Examples

Display multiplications **1** through **4** at right. Have students find each product, and display the products in a second column. Lead students to see that the constant terms of these products are negative rather than positive, as was the case in Skill A.

Now display multiplications **5** through **8**. Once again have students find each product, and then display the products. Ask students to compare these trinomials to the those in multiplications **1** through **4**. They should observe that the constant term remains -30. However, the coefficients of the x-terms are now $-1, -7, -13,$ and -29 rather than $1, 7, 13,$ and 29.

Highlight the constants in all the binomial factors: -5 and 6; -3 and 10; -2 and 15; -1 and 30; 5 and -6; 3 and -10; 2 and -15; and 1 and -30. Lead students to see that the product of each pair is -30, and the sum of each is the coefficient of the x-term of the product. Then discuss the examples.

1 $(x - 5)(x + 6)$ $[x^2 + x - 30]$

2 $(x - 3)(x + 10)$ $[x^2 + 7x - 30]$

3 $(x - 2)(x + 15)$ $[x^2 + 13x - 30]$

4 $(x - 1)(x + 30)$ $[x^2 + 29x - 30]$

5 $(x + 5)(x - 6)$ $[x^2 - x - 30]$

6 $(x + 3)(x - 10)$ $[x^2 - 7x - 30]$

7 $(x + 2)(x - 15)$ $[x^2 - 13x - 30]$

8 $(x + 1)(x - 30)$ $[x^2 - 29x - 30]$

EXAMPLE 3

PHASE 2: Providing Guidance for **TRY THIS**

Students sometimes overlook one or more of the factor pairs that they need to check, and thus may incorrectly decide that a given polynomial is not factorable. For these students to be sure of their answer, suggest that they organize their work in a table, as shown.

Factors of -8	Sum of the factors	Sum equal to 5?
1, -8	1 + (-8) = -7	X
-1, 8	-1 + 8 = 7	X
2, -4	2 + (-4) = -2	X
-2, 4	-2 + 4 = -2	X

PHASE 3: ASSESSMENT AND CLOSURE for Lesson 7.3

- Give students the expression at right. Tell them to find two different positive integers to place in the box to make a trinomial that is factorable. [**samples: 8, 20**] Repeat the activity, this time asking them to find two different negative integers. [**samples: $-10, -22$**]

  $s^2 - 9s + \square$

- In this lesson, students factored trinomials with leading coefficient 1. They will next learn to factor when the leading coefficient is not 1.

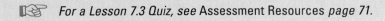 *For a Lesson 7.3 Quiz, see* Assessment Resources *page 71.*

SKILL B Factoring $x^2 + bx + c$ when c is negative

In Skill A, you learned how to factor expressions of the form $x^2 + bx + c$ when c is positive. You may be also able to factor such expressions when c is negative.

EXAMPLE 1 Factor $x^2 + x - 20$.

▶ Solution

Make an organized list.

1. $c = -20$: write factor pairs of -20.

$$\begin{array}{ccc} -1 \times 20 & -2 \times 10 & -4 \times 5 \\ 1 \times -20 & 2 \times -10 & 4 \times -5 \end{array}$$

2. $b = 1$: find the pair whose sum is 1.
$$-4 + 5 = 1$$

3. Write the product using -4 and 5.
$$[x + (-4)](x + 5)$$

Thus, $x^2 + x - 20 = (x - 4)(x + 5)$.

TRY THIS Factor $n^2 + 3n - 40$.

EXAMPLE 2 Factor $z^2 - 4z - 12$.

▶ Solution

Make an organized list.

1. $c = -12$: write factor pairs of -12.

$$\begin{array}{ccc} -1 \times 12 & -2 \times 6 & -3 \times 4 \\ 1 \times -12 & 2 \times -6 & 3 \times -4 \end{array}$$

2. $b = -4$: find the pair whose sum is -4.
$$2 + (-6) = -4$$

3. Write the product using 2 and -6.
$$(z + 2)[z + (-6)]$$

Thus, $z^2 - 4z - 12 = (z + 2)(z - 6)$.

TRY THIS Factor $n^2 - 3n - 40$.

Not every polynomial of the form $x^2 + bx + c$ is factorable.

EXAMPLE 3 Show that $v^2 + 3v - 1$ cannot be factored.

▶ Solution

$c = -1$: the only factors of -1 are 1 and -1.
$b = 3$: because $-1 + 1 \neq 3$, $v^2 + 3v - 1$ cannot be factored.

TRY THIS Show that $q^2 + 5q - 8$ cannot be factored.

EXERCISES

KEY SKILLS

Each expression is in the form $ax^2 + bx + c$. Identify b and c, including their signs, in the following quadratic trinomials.

1. $x^2 - 7x - 18$ 2. $x^2 + 5x - 24$
3. $x^2 - 2x + 35$ 4. $t^2 - 3t - 1$

PRACTICE

Factor.

5. $x^2 - 6x - 7$ 6. $y^2 + 10y - 11$ 7. $v^2 + 4v - 5$
8. $n^2 - 4n - 5$ 9. $k^2 - k - 2$ 10. $a^2 + 6a - 7$
11. $w^2 - w - 42$ 12. $p^2 - 2p - 63$ 13. $d^2 + 2d - 63$
14. $b^2 + 31b - 32$ 15. $h^2 - 17h - 38$ 16. $x^2 - 10x - 96$
17. $z^2 + 10z - 96$ 18. $s^2 - 10s - 75$ 19. $t^2 + 4t - 45$
20. $d^2 + 9d - 36$ 21. $m^2 - 42m - 43$ 22. $s^2 + 11s - 80$

Show that the given polynomial cannot be factored.

23. $x^2 - 9x + 7$ 24. $a^2 + 8a + 5$ 25. $b^2 + 3b + 1$
26. $q^2 - 6q - 3$ 27. $k^2 - 5k - 3$ 28. $z^2 + 2z - 10$

29. Given $x^2 + bx - c$, what do you know about the signs of the factors of $-c$?

30. **Critical Thinking** Let r and s represent real numbers. Show that if $x^2 + bx + c = (x + r)(x + s)$, then $c = rs$ and $b = r + s$.

MIXED REVIEW

Factor. (7.2 Skill A)

31. $n^2 + 22n + 121$ 32. $k^2 - 26k + 169$ 33. $4a^2 + 20a + 25$
34. $36z^2 - 36$ 35. $9y^2 - b^2$ 36. $u^2v^2 - 9$

Factor. (7.2 Skill B)

37. $2(z + 1) + 3z(z + 1)$ 38. $3(a + 2) - 2a(a + 2)$ 39. $-(n - 1) + 3n(n - 1)$
40. $x^3y + x^2 - 3y - 3$ 41. $x^2 - 20x - yx + 20y$ 42. $4wx^2 + 5w - 4x^2 - 5$

Answers for Lesson 7.3 Skill B

Try This Example 1
$(n + 8)(n - 5)$

Try This Example 2
$(n - 8)(n + 5)$

Try This Example 3
List all factor pairs for -8: -1×8, 1×-8, -2×4, 2×-4. None of these pairs has a sum of 5. Thus, $q^2 + 5q - 8$ cannot be factored.

1. $b = -7, c = -18$
2. $b = 5, c = -24$
3. $b = -2, c = 35$
4. $b = -3, c = -1$
5. $(x - 7)(x + 1)$
6. $(y + 11)(y - 1)$
7. $(v + 5)(v - 1)$
8. $(n - 5)(n + 1)$
9. $(k - 2)(k + 1)$
10. $(a + 7)(a - 1)$
11. $(w - 7)(w + 6)$
12. $(p - 9)(p + 7)$

13. $(d + 9)(d - 7)$
14. $(b + 32)(b - 1)$
15. $(h - 19)(h + 2)$
16. $(x - 16)(x + 6)$
17. $(z + 16)(z - 6)$
18. $(s - 15)(s + 5)$
19. $(t + 9)(t - 5)$
20. $(d + 12)(d - 3)$
21. $(m - 43)(m + 1)$
22. $(s + 16)(s - 5)$
23. List all factor pairs for 7: 1×7, -1×-7. None of these pairs has a sum of -9. Thus, $x^2 - 9x + 7$ cannot be factored.
24. List all factor pairs for 5: 1×5, -1×-5. None of these pairs has a sum of 8. Thus, $a^2 + 8a + 5$ cannot be factored.
25. List all factor pairs for 1: 1×1, -1×-1. None of these pairs has a sum of 3. Thus, $b^2 + 3b + 1$ cannot be factored.

26. List all factor pairs for -3: 1×-3, -1×3. None of these pairs has a sum of -6. Thus, $q^2 - 6q - 3$ cannot be factored.
27. List all factor pairs for -3: 1×-3, -1×3. None of these pairs has a sum of -5. Thus, $k^2 - 5k - 3$ cannot be factored.
28. List all factor pairs for -10: 1×-10, 2×-5, -1×10, -2×5. None of these pairs has a sum of 2. Thus, $z^2 + 2z - 10$ cannot be factored.
29. The factor with the larger absolute value is positive and the factor with the smaller absolute value is negative.

For answers to Exercises 30–42, see Additional Answers.

Factoring $ax^2 + bx + c$

Factoring $ax^2 + bx + c$ when c is positive

Focus: *Students find the binomial factors of a quadratic trinomial whose constant term is positive and whose leading coefficient is not 1.*

EXAMPLE 1

PHASE 1: Presenting the Example

Display multiplications **1** and **2** at right. Have students find each product. [**1** $x^2 + 8x + 7$ **2** $x^2 + 8x + 7$] Ask them to explain why the products are the same. Elicit the response that the factors of each are identical, with only their order being reversed. So the Commutative Property of Multiplication guarantees that the products will be the same.

1 $(x + 1)(x + 7)$

2 $(x + 7)(x + 1)$

Display multiplications **3** and **4**. Again have students find each product. [**3** $2x^2 + 15x + 5$ **4** $2x^2 + 9x + 5$] Point out that, in this case, the two products have the same leading coefficient and the same constant term, but they have different x-terms. Ask students what accounts for this difference. Lead them to see that, although the binomial factors *appear* similar, they are in fact not identical. In **3**, the linear term $15x$ is the sum of $2x \cdot 7$ and $x \cdot 1$. In **4**, the linear term $9x$ is the sum of $2x \cdot 1$ and $x \cdot 7$.

3 $(2x + 1)(x + 7)$

4 $(2x + 7)(x + 1)$

Now display $2x^2 + 7x + 3$ and discuss the process of factoring that is outlined in Example 1.

Point out that in Step 3, the required factors of 6 were both positive. This will always be the case when both b and c are positive.

Be sure students understand Step 6. Each group must be factored, but the only common factor of x and 3 is 1. Therefore, 1 is factored out of the second group.

EXAMPLE 2

PHASE 1: Presenting the Example

PHASE 2: Providing Guidance for **TRY THIS**

Point out to students that in Step 3, the required factors of 12 were both negative. This will always be the case when b is negative and c is positive.

Be sure students understand that the process of splitting the middle term contains the process of factoring by grouping. So, just as with factoring by grouping, they must be careful to use the correct signs when terms are subtracted. If students have trouble, suggest that they begin each problem by rewriting each subtraction as addition of the opposite. Also stress that students should check their answers by multiplying the factors to get back the original polynomial.

7.4 LESSON Factoring $ax^2 + bx + c$

SKILL A Factoring $ax^2 + bx + c$ when c is positive

When the x^2-term in a quadratic trinomial has a coefficient other than 1, the task of factoring is a little more involved. However, the process uses what you have learned about factoring by grouping.

EXAMPLE 1 Factor $2x^2 + 7x + 3$.

▶ **Solution**

1. Identify a, b, and c. $a = 2$, $b = 7$, $c = 3$
2. Find the product ac. $(2)(3) = 6$
3. Find factors of ac that add to b. $6 \times 1 = 6$ $6 + 1 = 7$
4. Rewrite the original equation, rewriting bx
   as the sum of the factors you found in Step 2. $2x^2 + 6x + 1x + 3$
   (The order of the factors does not matter.)
5. Now apply factoring by grouping. Using
   parentheses, group the four terms into $(2x^2 + 6x) + (x + 3)$
   two groups of two terms.
6. Factor each group. (Factor out 1 if it is the GCF.)
   The expressions remaining in parentheses will $2x(x + 3) + 1(x + 3)$
   be equal.
7. Apply the Distributive Property. $(2x + 1)(x + 3)$

TRY THIS   Factor $2n^2 + 11n + 5$.

Because you rewrite bx as two terms in Step 4, this method is sometimes called *splitting the middle term*.

Remember to group carefully when terms involve subtraction.

EXAMPLE 2 Factor $12y^2 - 7y + 1$.

▶ **Solution**

1. $a = 12$, $b = -7$, $c = 1$
2. $ac = (12)(1) = 12$
3. Factors of 12 that add to -7 are -3 and -4.
4. Rewrite the middle term: $12y^2 - 3y - 4y + 1$
5. Make two groups: $(12y^2 - 3y) - (4y - 1)$ *Remember that* $-4y + 1 = -(4y - 1)$.
6. Factor each group: $3y(4y - 1) - 1(4y - 1)$
7. Apply the Distributive Property: $(3y - 1)(4y - 1)$

TRY THIS   Factor $8a^2 - 14a + 3$.

EXERCISES

KEY SKILLS

For each quadratic trinomial, identify a, b, and c. Then list the possible factor pairs of ac and find the pair whose sum is b.

1. $4x^2 + 9x + 2$ 2. $8x^2 + 27x + 9$
3. $2x^2 + 8x + 8$ 4. $3x^2 - 10x + 8$

PRACTICE

In Exercises 5–12, $ac = 120$. List the factor pairs of 120 and their sums. Then, for each expression, rewrite the middle term as two terms whose coefficients multiply to 120.

Sample: $3x^2 + 23x + 40$ Solution: $8x + 15x$

5. $8x^2 + 26x + 15$ 6. $5x^2 + 34x + 24$
7. $40x^2 + 43x + 3$ 8. $15y^2 + 22y + 8$
9. $2y^2 - 23y + 60$ 10. $120x^2 - 121x + 1$
11. $3v^2 - 29v + 40$ 12. $4v^2 - 62v + 30$

Factor.

13. $2a^2 + 7a + 3$ 14. $3n^2 + 7n + 2$ 15. $3k^2 + 5k + 2$
16. $5a^2 + 7a + 2$ 17. $3h^2 + 8h + 4$ 18. $7h^2 + 8h + 1$
19. $5z^2 + 17z + 6$ 20. $3v^2 + 11v + 10$ 21. $4v^2 + 7v + 3$
22. $8p^2 - 14p + 5$ 23. $7x^2 - 15x + 2$ 24. $8c^2 - 6c + 1$
25. $8c^2 - 11c + 3$ 26. $12a^2 - 8a + 1$ 27. $-9m^2 - 3m + 2$
28. $-6x^2 - 11x + 2$ 29. $-2x^2 - 4x + 6$ 30. $-8b^2 - 7b + 1$
31. $4x^2y^2 + 16xy + 15$ 32. $4a^2b^2 + 16abmn + 15m^2n^2$

MIXED REVIEW

Solve each system of equations. (5.3 Skills A and B)

33. $\begin{cases} 5a - 2b = 0 \\ 2a + 2b = 3 \end{cases}$ 34. $\begin{cases} 7m - 2t = 1 \\ 6m - 2t = 0 \end{cases}$ 35. $\begin{cases} -2s - 2t = 1 \\ 5s - 6t = 1 \end{cases}$

36. $\begin{cases} \frac{x}{2} + \frac{y}{3} = 0 \\ \frac{x}{3} - \frac{y}{5} = 0 \end{cases}$ 37. $\begin{cases} \frac{a}{2} - \frac{b}{3} = 1 \\ \frac{a}{3} + \frac{b}{2} = -1 \end{cases}$ 38. $\begin{cases} \frac{m}{2} - \frac{n}{5} = 2 \\ \frac{m}{2} - \frac{n}{2} = 1 \end{cases}$

Answers for Lesson 7.4 Skill A

Try This Example 1
$(n + 5)(2n + 1)$

Try This Example 2
$(4a - 1)(2a - 3)$

1. $a = 4$, $b = 9$, $c = 2$, $ac = 8$
   List all factor pairs for 8: 1×8, 2×4, -1×-8, -2×-4. The factor pair whose sum is 9 is 1 and 8.
2. $a = 8$, $b = 27$, $c = 9$, $ac = 72$
   List all factor pairs for 72: 1×72, 2×36, 3×24, 4×18, 6×12, 8×9, -1×-72, -2×-36, -3×-24, -4×-18, -6×-12, -8×-9.
   The factor pair whose sum is 27 is 3 and 24.
3. $a = 2$, $b = 8$, $c = 8$, $ac = 16$
   List all factor pairs for 16: 1×16, 2×8, 4×4, -1×-16, -2×-8, -4×-4. The factor pair whose sum is 8 is 4 and 4.
4. $a = 3$, $b = -10$, $c = 8$, $ac = 24$
   List all factor pairs for 24: 1×24, 2×12, 3×8, 4×6, -1×-24, -2×-12, -3×-8, -4×-6. The factor pair whose sum is -10 is -4 and -6.
5. $20x + 6x$
6. $30x + 4x$
7. $40x + 3x$

8. $12y + 10y$
9. $-8y - 15y$
10. $-120x - x$
11. $-5v - 24v$
12. $-2v - 60v$
13. $(a + 3)(2a + 1)$
14. $(n + 2)(3n + 1)$
15. $(k + 1)(3k + 2)$
16. $(a + 1)(5a + 2)$
17. $(h + 2)(3h + 2)$
18. $(h + 1)(7h + 1)$
19. $(z + 3)(5z + 2)$
20. $(v + 2)(3v + 5)$
21. $(v + 1)(4v + 3)$
22. $(4p - 5)(2p - 1)$
23. $(7x - 1)(x - 2)$
24. $(2c - 1)(4c - 1)$
25. $(8c - 3)(c - 1)$
26. $(6a - 1)(2a - 1)$
★ 27. $-(3m - 1)(3m + 2)$
★ 28. $-(6x - 1)(x + 2)$
★ 29. $-(2x - 2)(x + 3)$
★ 30. $-(8b - 1)(b + 1)$
★ 31. $(2xy + 3)(2xy + 5)$

★ 32. $(2ab + 3mn)(2ab + 5mn)$
33. $\left(\frac{3}{7}, \frac{15}{14}\right)$
34. $(1, 3)$
35. $\left(-\frac{2}{11}, -\frac{7}{22}\right)$
36. $(0, 0)$
37. $\left(\frac{6}{13}, -\frac{30}{13}\right)$
38. $\left(\frac{16}{3}, \frac{10}{3}\right)$

★ **Advanced Exercises**

Factoring $ax^2 + bx + c$ when c is negative

Focus: *Students find the binomial factors of a quadratic trinomial whose constant term is negative and whose leading coefficient is not 1.*

EXAMPLES 1 and 2

PHASE 1: Presenting the Examples

Display $3x^2 + \square x + 8$ and ask students how to find numbers that could be written in the box to make the polynomial factorable. [**Sample: Multiply 3 by 8 to get 24. List all factor pairs of 24; the sum of each pair can be written in the box.**] Then ask students to list all such numbers. [**25, −25, 14, −14, 11, −11, 10, −10**]

Now change the equation to $3x^2 + \square x - 8$ and ask students how they would solve the problem now. Students should perceive that now they must find factor pairs of −24. (If they do not immediately see this, rewrite the equation above as $3x^2 + \square x + (-8)$.) Have students list all factor pairs of −24. [**1 and −24; −1 and 24; 2 and −12; −2 and 12; 3 and −8; −3 and 8; 4 and −6; −4 and 6**] Then ask them to find all numbers that can be written in the box to make the polynomial factorable. [**−23, 23, −10, 10, −5, 5, −2, 2**] Answer any questions and discuss the Examples.

PHASE 2: Providing Guidance for TRY THIS

Watch for students who confuse the signs. For instance, they may give the factorization for the Example 1 Try This as $(3k + 2)(k - 2)$ rather than $(3k - 2)(k + 2)$. This is a common error made in factoring trinomials of this type. Note that Skill B Exercises 1 through 4 on the facing page are designed to focus students' attention solely on identifying the correct signs of the factors.

EXAMPLE 3

PHASE 1: Presenting the Example

Ask students to recall the definition of prime number. [**A prime number has no factors besides 1 and itself.**] Then tell them that a polynomial can also be prime, meaning that it has no factors besides 1 and itself. When this happens, we say that the polynomial cannot be factored. Then discuss Example 3.

PHASE 2: Providing Guidance for TRY THIS

When splitting the middle term, students may sometimes make an algebra mistake and incorrectly conclude that the polynomial cannot be factored, as shown at right. This is especially common when the polynomial contains subtraction. Stress that the way to determine that a polynomial cannot be factored is to find that there are no factors of ac that add to b. If there are factors of ac that add to b, the polynomial is *always* factorable.

$$3x^2 + 5x - 2$$
$$= 3x^2 + 6x - x - 2$$
$$= (3x^2 + 6x) - (x - 2) \; \textsf{X}$$
$$= 3x(x + 2) - (x - 2)$$
cannot be factored

SKILL B — Factoring $ax^2 + bx + c$ when c is negative

To factor $ax^2 + bx + c$ when c is negative, follow the same procedure that you learned in Skill A. However, be careful when grouping terms that involve subtraction.

EXAMPLE 1 Factor $5b^2 + 4b - 1$.

> **Solution**
> 1. $a = 5, b = 4, c = -1$
> 2. $ac = (5)(-1) = -5$
> 3. Factors of -5 that add to 4 are 5 and -1.
> 4. Rewrite the middle term: $\qquad 5b^2 + 5b - 1b - 1$
> 5. Make two groups: $\qquad (5b^2 + 5b) - (1b + 1)$
> 6. Factor each group: $\qquad 5b(b + 1) - 1(b + 1)$
> 7. Apply the Distributive Property: $(5b - 1)(b + 1)$

TRY THIS   Factor $3k^2 + 4k - 4$.

EXAMPLE 2 Factor $10n^2 - 11n - 6$.

> **Solution**
> 1. $a = 10, b = -11, c = -6$
> 2. $ac = (10)(-6) = -60$
> 3. Factors of -60 whose sum is -11 are 4 and -15.
> 4. Rewrite the middle term: $\qquad 10n^2 + 4n - 15n - 6$
> 5. Make two groups: $\qquad (10n^2 + 4n) - (15n + 6)$
> 6. Factor each group: $\qquad 2n(5n + 2) - 3(5n + 2)$
> 7. Apply the Distributive Property: $(2n - 3)(5n + 2)$

TRY THIS   Factor $6m^2 - 7m - 49$.

Not all expressions of the form $ax^2 + bx - c$ can be factored.

EXAMPLE 3 Factor $-10n^2 + 21n - 5$.

> **Solution**
> 1. $a = -10, b = 21, c = -5$
> 2. $ac = (-10)(-5) = 50$
> 3. There are no factors of 50 whose sum is 21. Therefore, $-10n^2 + 21n - 5$ cannot be factored.

TRY THIS   Factor $-3x^2 + 15x - 2$.

EXERCISES

KEY SKILLS

Complete each grouping by filling in each blank with (+) or (−).

1. $8z^2 - 12z + 2z - 3 = (8z^2 - 12z) + (2z \underline{\ ?\ } 3)$
2. $15a^2 - 20a + 9a - 12 = (15a^2 - 20a) + (9a \underline{\ ?\ } 12)$
3. $3n^2 + 6n - 2n - 4 = (3n^2 + 6n) - (2n \underline{\ ?\ } 4)$
4. $20m^2 + 4m - 10m - 7 = (20m^2 + 4m) - (10m \underline{\ ?\ } 7)$

PRACTICE

Factor.

5. $3v^2 + 20v - 7$
6. $2b^2 + 3b - 5$
7. $3d^2 + 2d - 1$
8. $7p^2 + 12p - 4$
9. $11x^2 + 9x - 2$
10. $7y^2 + 4y - 3$
11. $3h^2 + h - 2$
12. $3g^2 - g - 2$
13. $3c^2 - 7c - 6$
14. $2z^2 - 7z - 15$
15. $2m^2 - 5m - 7$
16. $10a^2 + 21a - 10$
17. $2x^2 + 11x - 90$
18. $6x^2 - x - 5$
19. $6x^2 - 17x - 3$
20. $8x^2 - 4x - 4$

Show that the given polynomial cannot be factored.

21. $10x^2 + 21x + 6$
22. $6a^2 + a - 3$
23. $3x^2 + 3x - 4$
24. $9y^2 - y + 2$

25. **Critical Thinking** Suppose that a is a positive number. Find the value(s) of a so that $3x^2 + ax + 6$ can be factored as the product of two binomials. How does your answer change if a is a negative number?

MIXED REVIEW

Identify each system as consistent or inconsistent. Justify your response. (5.4 Skill A)

26. $\begin{cases} 4x + 2y = 0 \\ -6x - 5y = 1 \end{cases}$
27. $\begin{cases} 4m - 2n = 5 \\ -12m + 6n = -2 \end{cases}$
28. $\begin{cases} -3s - 5t = -7 \\ -2s + 6t = -7 \end{cases}$

Classify each system as specifically as possible. (5.4 Skill B)

29. $\begin{cases} 4c + 2d = 0 \\ -6c - 3d = 1 \end{cases}$
30. $\begin{cases} 7u + 2v = 1 \\ 14u + 4v = 2 \end{cases}$
31. $\begin{cases} -7m + 5n = 1 \\ 7m - 5n = 5 \end{cases}$

Answers for Lesson 7.4 Skill B

Try This Example 1
$(3k - 2)(k + 2)$

Try This Example 2
$(3m + 7)(2m - 7)$

Try This Example 3
$a = -3, b = 15, c = -2$, $ac = 6$; there are no factors of 6 that add to 15, so $-3x^2 + 15x - 2$ cannot be factored.

1. −
2. −
3. +
4. +
5. $(3v - 1)(v + 7)$
6. $(2b + 5)(b - 1)$
7. $(3d - 1)(d + 1)$
8. $(7p - 2)(p + 2)$
9. $(11x - 2)(x + 1)$
10. $(7y - 3)(y + 1)$
11. $(3h - 2)(h + 1)$
12. $(3g + 2)(g - 1)$
13. $(3c + 2)(c - 3)$
14. $(2z + 3)(z - 5)$
15. $(2m - 7)(m + 1)$
16. $(5a - 2)(2a + 5)$
17. $(2x - 9)(x + 10)$
18. $(6x + 5)(x - 1)$
19. $(6x + 1)(x - 3)$
20. $(8x + 4)(x - 1)$
21. No factors of 60 add to 21.
22. No factors of -18 add to 1.
23. No factors of -12 add to 3.
24. No factors of 18 add to -1.
25. $a = 19, a = 11,$ or $a = 9$. If $a < 0, a = -19, a = -11,$ or $a = -9$.
26. consistent
27. inconsistent
28. consistent
29. inconsistent
30. consistent, dependent
31. inconsistent

Focus: *Students perform factorizations that require two or more steps.*

EXAMPLE 1

PHASE 1: Presenting the Example

Display the multiplications $(6h + 2)(h + 5)$ and $(3h + 1)(2h + 10)$. Have students find each product. [$6h^2 + 32h + 10$; $6h^2 + 32h + 10$] Point out that, although the factors are not the same, the products are identical.

Tell students to examine each multiplication. Ask if any of the factors can be factored further. Lead them to the observations shown at right. Point out that each factorization contains a binomial factor whose terms have a common monomial factor, 2. When the binomials are factored, you arrive at a single factorization for $6h^2 + 32h + 10$, which is $2(3h + 1)(h + 5)$. Tell students that a polynomial is *factored completely* when you cannot factor any further, and that a polynomial's complete factorization is unique.

Draw a parallel to the factorization of an integer. For instance, point out that 12 can be factored as 1×12, 2×6, or 3×4. However, its prime factorization, $2^2 \times 3$, is unique. Now discuss Example 1.

$(6h + 2) \cdot (h + 5)$

$\downarrow$

$[2(3h + 1)] \cdot (h + 5)$
$2(3h + 1)(h + 5)$

$(3h + 1) \cdot (2h + 10)$

$\downarrow$

$(3h + 1) \cdot [2(h + 5)]$

$2(3h + 1)(h + 5)$

EXAMPLES 2 and 3

PHASE 2: Providing Guidance for TRY THIS

If your students are capable, challenge them to find one or more sets of steps for finding a factorization. Try This exercise b in Example 2 lends itself particularly well to this task, as shown below.

$4d^3 + 16d^2 + 16d$	$4d^3 + 16d^2 + 16d$	$4d^3 + 16d^2 + 16d$
$4d(d^2 + 4d + 4)$	$2d(2d^2 + 8d + 8)$	$d(4d^2 + 16d + 16)$
$4d(d + 2)^2$	$2d(2d + 4)(d + 2)$	$d(2d + 4)(2d + 4)$
	$2d[2(d + 2)](d + 2)$	$d[2(d + 2)][2(d + 2)]$
	$4d(d + 2)^2$	$4d(d + 2)^2$

If you display the alternative methods as shown, it should quickly become clear to the students why they are encouraged to begin by looking for the GCF of all terms.

PHASE 3: ASSESSMENT AND CLOSURE for Lesson 7.4

- Lead a discussion in which students summarize the types of factorization techniques that they have learned. Have each student create two original examples of each type. [**Answers will vary.**]

- This lesson concludes the sequence of instruction in factorization. Students will further use these skills in two major contexts: solving polynomial equations (Lesson 7.5) and simplifying rational expressions (Chapter 9).

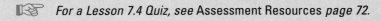 *For a Lesson 7.4 Quiz, see* Assessment Resources *page 72.*

The first step in factoring a polynomial is to check whether there is a GCF of all the terms. If so, factor it out before trying other factoring strategies.

Sometimes the GCF will be a number.

EXAMPLE 1 Factor each polynomial.   **a.** $-t^2 - 5t + 14$ **b.** $6x^2 + 10x + 4$

▶ Solution

a. $-t^2 - 5t + 14 = -(t^2 + 5t - 14)$ ⟵ Factor -1 from each term.
    $= -(t + 7)(t - 2)$ ⟵ Factor $t^2 + 5t - 14$.
b. $6x^2 + 10x + 4 = 2(3x^2 + 5x + 2)$ ⟵ Factor 2 from each term.
    $= 2(3x + 2)(x + 1)$ ⟵ Factor $3x^2 + 5x + 2$.

TRY THIS   Factor each polynomial.   **a.** $-a^2 + 7a - 12$ **b.** $6n^2 + 15n + 9$

In the Example above, you were able to continue factoring after factoring out the GCF. A polynomial is **factored completely** when it is written as a product that cannot be factored any further. Whenever you are factoring a polynomial, be sure to factor completely.

The GCF can also be a monomial.

EXAMPLE 2 Factor each polynomial.   **a.** $12c^3 - 75c$ **b.** $5d^3 + 50d^2 + 125d$

▶ Solution

a. $12c^3 - 75c = 3c(4c^2 - 25)$ ⟵ Factor $3c$ from each term.
    $= 3c(2c + 5)(2c - 5)$ ⟵ Factor the difference of two squares.
b. $5d^3 + 50d^2 + 125d$
    $= 5d(d^2 + 10d + 25)$ ⟵ Factor $5d$ from each term.
    $= 5d(d + 5)^2$ ⟵ Factor a perfect-square trinomial.

TRY THIS   Factor each polynomial.   **a.** $27g^3 - 48g$ **b.** $4d^3 + 16d^2 + 16d$

EXAMPLE 3 Factor $2z^3 + 6z^2 - 8z$.

▶ Solution

$2z^3 + 6z^2 - 8z = 2z(z^2 + 3z - 4)$ ⟵ Factor $2z$ from each term.
    $= 2z(z + 4)(z - 1)$ ⟵ Factor $z^2 + 3z - 4$.

TRY THIS   Factor $9r^3 - 24r^2 + 12r$.

EXERCISES

KEY SKILLS

Is each factorization complete? Justify your response.

1. $5x^3 - 5x^2 = 5x(x^2 - x)$
2. $28a^2 + 29a - 35 = (4a + 7)(7a - 5)$
3. $3t^3 - 6t^2 - 45t = 3t(t - 5)(t + 3)$
4. $-2x^3 - 14x^2 - 20x = -2x(x^2 + 7x + 10)$

PRACTICE

Factor completely.

5. $-x^2 + x + 6$
6. $-n^2 + 4n + 5$
7. $-u^2 + 7u - 12$
8. $-c^2 - c + 20$
9. $-z^2 - 4z + 21$
10. $-a^2 + 7a - 10$
11. $3v^2 + 18v + 15$
12. $3x^2 + 15x - 42$
13. $7x^2 + 42x + 35$
14. $2x^2 + 2x - 40$
15. $-3t^2 + 33t - 90$
16. $-4x^2 + 10x + 6$
17. $-30b^2 - 35b + 25$
18. $-42g^2 - 49g - 14$
19. $63m^2 - 42m + 7$
20. $45q^2 + 30q + 5$
21. $14ab + 7b + 21$
22. $9c + 6b - 15$
23. $125q^2 - 45$
24. $7y^3 - 63y$
25. $3a^3 + 30a^2 - 33a$
26. $20n^3 + 20n^2 - 15n$
27. $2x^3 + 12x^2 - 110x$
28. $16x^2y - 8xy^2 + 12xy$

29. Find n such that $12x^3 + 6x^2 - 4nx^2 - 2nx = 2x(3x - 4)(2x + 1)$.

MID-CHAPTER REVIEW

Use the GCF to factor each expression. (7.1 Skill C)

30. $5y^4 + 20y^2$
31. $15x^3v^2 + 20xv^2$
32. $15a^2b^2 + 20a^2b - 3ab$

Factor. (7.2 Skill A)

33. $25x^2y^2 - 49$
34. $m^2n^2 + 14mn + 49$
35. $a^2b^2 - x^2y^2$

Factor. (7.3 Skills A and B)

36. $x^2 - 35x + 34$
37. $a^2 - 14a - 15$
38. $y^2 - 17y - 60$

Factor. (7.4 Skills A and B)

39. $3x^2 + 5x + 2$
40. $2n^2 + 9n + 10$
41. $5k^2 + 13k - 6$

Answers for Lesson 7.4 Skill C

Try This Example 1
a. $-(a - 3)(a - 4)$
b. $3(2n + 3)(n + 1)$

Try This Example 2
a. $3g(3g + 4)(3g - 4)$
b. $4d(d + 2)(d + 2)$

Try This Example 3
$3r(3r - 2)(r - 2)$

1. No; x can be factored from $x^2 - x$.
2. Yes
3. Yes
4. No; $x^2 + 7x + 10$ can be factored as $(x + 5)(x + 2)$.
5. $-(x - 3)(x + 2)$
6. $-(n - 5)(n + 1)$
7. $-(u - 4)(u - 3)$
8. $-(c - 4)(c + 5)$
9. $-(z - 3)(z + 7)$
10. $-(a - 5)(a - 2)$

11. $3(v + 1)(v + 5)$
12. $3(x - 2)(x + 7)$
13. $7(x + 1)(x + 5)$
14. $2(x - 4)(x + 5)$
15. $-3(t - 6)(t - 5)$
16. $-2(x - 3)(2x + 1)$
17. $-5(2b - 1)(3b + 5)$
18. $-7(2g + 1)(3g + 2)$
19. $7(3m - 1)(3m - 1)$, or $7(3m - 1)^2$
20. $5(3q + 1)(3q + 1)$, or $5(3q + 1)^2$
21. $7(2ab + b + 3)$
22. $3(3c + 2b - 5)$
23. $5(5q + 3)(5q - 3)$
24. $7y(y - 3)(y + 3)$
★25. $3a(a + 11)(a - 1)$
★26. $5n(2n - 1)(2n + 3)$
★27. $2x(x + 11)(x - 5)$
★28. $4xy(4x - 2y + 3)$
★29. $n = 4$
30. $5y^2(y^2 + 4)$
31. $5xv^2(3x^2 + 4)$

32. $ab(15ab + 20a - 3)$
33. $(5xy - 7)(5xy + 7)$
34. $(mn + 7)^2$
35. $(ab + xy)(ab - xy)$
36. $(x - 1)(x - 34)$
37. $(a - 15)(a + 1)$
38. $(y - 20)(y + 3)$
39. $(3x + 2)(x + 1)$
40. $(n + 2)(2n + 5)$
41. $(5k - 2)(k + 3)$

★ **Advanced Exercises**

Solving Polynomial Equations by Factoring

TEACHING 7.5 LESSON

SKILL A *Solving quadratic equations by factoring*

Focus: *Students solve quadratic equations by applying the Zero-Product Property to equations of the form $(px + r)(qx + s) = 0$ or $ax^2 + bx + c = 0$.*

EXAMPLES 1 and 2

PHASE 1: Presenting the Examples

Review with students the Multiplication Property of Zero and the Zero-Product Property as stated at the top of the textbook page. Point out that each is given in the form of a conditional (*if-then*) statement. Note how the *if* and *then* clauses are interchanged, as shown at right. When an interchanging like this occurs, each statement is said to be the *converse* of the other.

For real numbers a and b,
If $\underline{a = 0}$ or $\underline{b = 0}$, then $\underline{ab = 0}$.

If $\underline{ab = 0}$, then $\underline{a = 0}$ or $\underline{b = 0}$.

The Multiplication Property of Zero is an *axiom* that is assumed to be true. However, the truth of a statement does not guarantee the truth of its converse. In this case, though, the Zero-Product Property can be *proved* true. You may want to develop the following proof with your students.

There are two cases to examine: either $a = 0$ or $a \neq 0$.

Case 1: If $a = 0$, it follows that the property is true.
Case 2: If $a \neq 0$, reason as follows:

1 $ab = 0$ Given

2 $\frac{1}{a}(ab) = \frac{1}{a}(0)$ Multiplication Property of Equality

3 $\frac{1}{a}(ab) = 0$ Multiplication Property of Zero

4 $\left(\frac{1}{a} \cdot a\right)b = 0$ Associative Property of Multiplication

5 $(1)b = 0$ Inverse Property of Multiplication

6 $b = 0$ Identity Property of Multiplication

Therefore, if $ab = 0$, then $a = 0$ or $b = 0$.

Now discuss the examples, which show how to use the Zero-Product Property to solve quadratic equations.

PHASE 2: Providing Guidance for **TRY THIS**

Some students simply inspect the factors and hastily write solutions "on sight." Watch for errors such the following:

Example 1

$(3d + 5)(d - 2) = 0$

$d = \frac{5}{3}, d = -2$ ✗

Example 2

$6t^2 + t - 15 = 0 \rightarrow (3t + 5)(2t - 3) = 0$

$t = -\frac{3}{5}, t = \frac{2}{3}$ ✗

Stress the importance of checking possible solutions in the original equation.

7.5 LESSON — Solving Polynomial Equations by Factoring

SKILL A *Solving quadratic equations by factoring*

A **quadratic equation** in x is any equation that can be written in the form $ax^2 + bx + c = 0$, where $a \neq 0$. This is called the **standard form of a quadratic equation**. The value of the variable in a standard form equation is called the *solution*, or the **root**, or the equation.

The **Multiplication Property of 0** states that if $a = 0$ or $b = 0$, then $ab = 0$. The *Zero-Product Property* enables you to conclude that if a product equals 0, then at least one of the factors of the product must equal 0.

Zero-Product Property
If a and b are real numbers and $ab = 0$, then $a = 0$ or $b = 0$.

Example 1 shows how to solve a quadratic equation once the quadratic expression has been factored into a product of two linear factors.

EXAMPLE 1 Solve $(4x + 5)(3x - 2) = 0$.

▶ **Solution**

Apply the Zero-Product Property.

$(4x + 5)(3x - 2) = 0$ ⟵ The quadratic expression has already been factored.
$4x + 5 = 0$ or $3x - 2 = 0$ ⟵ If $(4x + 5)(3x - 2) = 0$, then $(4x + 5) = 0$ or $(3x - 2) = 0$.
$4x = -5$ $3x = 2$
$x = -\dfrac{5}{4}$ $x = \dfrac{2}{3}$ ⟵ Solve each equation for x.

TRY THIS Solve $(3d + 5)(d - 2) = 0$.

Example 2 shows how to solve a quadratic equation when the quadratic expression is not given to you in factored form.

EXAMPLE 2 Solve $3t^2 - 8t + 5 = 0$ by factoring.

▶ **Solution**

$3t^2 - 8t + 5 = 0$
$(3t - 5)(t - 1) = 0$ ⟵ Factor $3t^2 - 8t + 5$.
$3t - 5 = 0$ or $t - 1 = 0$ ⟵ Apply the Zero-Product Property.
$t = \dfrac{5}{3}$ $t = 1$ ⟵ Solve each equation for t.

TRY THIS Solve $6t^2 + t - 15 = 0$ by factoring.

EXERCISES

KEY SKILLS

Solve each equation.

1. $x + 3 = 0$ or $x - 5 = 0$
2. $3z + 1 = 0$ or $2z - 3 = 0$
3. $5n + 3 = 0$ or $2n + 7 = 0$
4. $-2b + 5 = 0$ or $2b - 9 = 0$

PRACTICE

Solve each quadratic equation.

5. $(x + 5)(x + 2) = 0$
6. $(x - 7)(x + 3) = 0$
7. $(x - 7)(x - 5) = 0$
8. $(3d + 2)(7d - 1) = 0$
9. $(-5n - 1)(3n - 9) = 0$
10. $(6p + 5)(7p + 3) = 0$
11. $t^2 + 2t - 15 = 0$
12. $a^2 - 10a + 21 = 0$
13. $y^2 + 13y + 40 = 0$
14. $v^2 - 14v + 45 = 0$
15. $2s^2 + s - 15 = 0$
16. $3k^2 - 4k + 1 = 0$
17. $x^2 + 7x - 44 = 0$
18. $y^2 + 7y + 10 = 0$
19. $5y^2 - 46y + 9 = 0$
20. $2g^2 - 15g + 7 = 0$
21. $9x^2 - 12x + 4 = 0$
22. $4a^2 + 23a - 6 = 0$
23. $2x^2 - 5x + 3 = 0$
24. $3m^2 + 13m - 10 = 0$
25. $2a^2 - 7a - 4 = 0$
26. $3k^2 - 7k - 6 = 0$

27. Let m and n be real numbers and let $6x^2 - 2mx - 3nx + mn = 0$. Solve for x.

28. **Critical Thinking** A quadratic equation has 3 and -11 as its solutions. Work backwards to find the equation.

29. Solve $(x^2 - 16)(x^2 + 10x + 25) = 0$.

MIXED REVIEW

Find each product. (6.6 Skill A)

30. $2x^2(x^2 - 2x + 1)$
31. $3a^2(2a^2 - 2a - 2)$
32. $n^2(n^3 - 2n^2 - 5)$
33. $-5m^3(m^3 + m^2 + m)$
34. $k(k^3 + k^2 + k + 1)$
35. $y^3(y^3 - 2y^2 + 2y - 1)$
36. $-3b^5(-b^3 + 2b^2 + 3)$
37. $8d(d^7 + 9d^5 + d^3)$

Answers for Lesson 7.5 Skill A

Try This Example 1
$d = -\dfrac{5}{3}$ or $d = 2$

Try This Example 2
$t = -\dfrac{5}{3}$ or $t = \dfrac{3}{2}$

1. $x = -3$ or $x = 5$
2. $z = -\dfrac{1}{3}$ or $z = \dfrac{3}{2}$
3. $n = -\dfrac{3}{5}$ or $n = -\dfrac{7}{2}$
4. $b = \dfrac{5}{2}$ or $b = \dfrac{9}{2}$
5. $x = -2$ or $x = -5$
6. $x = 7$ or $x = -3$
7. $x = 5$ or $x = 7$
8. $d = -\dfrac{2}{3}$ or $d = \dfrac{1}{7}$
9. $n = -\dfrac{1}{5}$ or $n = 3$
10. $p = -\dfrac{5}{6}$ or $p = -\dfrac{3}{7}$

11. $t = -5$ or $t = 3$
12. $a = 3$ or $a = 7$
13. $y = -8$ or $y = -5$
14. $v = 5$ or $v = 9$
15. $s = -3$ or $s = \dfrac{5}{2}$
16. $k = 1$ or $k = \dfrac{1}{3}$
17. $x = 4$ or $x = -11$
18. $y = -5$ or $y = -2$
19. $y = \dfrac{1}{5}$ or $y = 9$
20. $g = \dfrac{1}{2}$ or $g = 7$
21. $x = \dfrac{2}{3}$
22. $a = \dfrac{1}{4}$ or $a = -6$
23. $x = \dfrac{3}{2}$ or $x = 1$
24. $m = \dfrac{2}{3}$ or $m = -5$
25. $a = -\dfrac{1}{2}$ or $a = 4$

26. $k = -\dfrac{2}{3}$ or $k = 3$
27. $x = \dfrac{m}{3}$ or $x = \dfrac{n}{2}$
28. $x^2 + 8x - 33 = 0$
29. $x = -4,\ x = 4,$ or $x = -5$
30. $2x^4 - 4x^3 + 2x^2$
31. $6a^4 - 6a^3 - 6a^2$
32. $n^5 - 2n^4 - 5n^2$
33. $-5m^6 - 5m^5 - 5m^4$
34. $k^4 + k^3 + k^2 + k$
35. $y^6 - 2y^5 + 2y^4 - y^3$
36. $3b^8 - 6b^7 - 9b^5$
37. $8d^8 + 72d^6 + 8d^4$

Focus: *Students solve quadratic equations by applying the Zero-Product Property to equations given in forms other than (px + r)(qx + s) = 0 or px² + bx + c = 0.*

EXAMPLE 1

PHASE 1: Presenting the Example

Display equation **1** at right and tell students to solve. $[v = -\frac{1}{4}$ or $v = \frac{3}{2}]$ Discuss the solution, taking care to emphasize the use of the Zero-Product Property.

1. $8v^2 - 10v - 3 = 0$
2. $8v^2 = 10v + 3$

Now display equation **2**. Ask students how it is different from equation **1**. They should readily see that equation **1** is in standard form, while equation **2** is not. They might also note that the Zero-Product Property cannot be used to solve equation **2** in its present form. Lead a discussion in which students brainstorm possible ways to remedy the situation. Then discuss the solution that is outlined on the textbook page.

PHASE 2: Providing Guidance for TRY THIS

Watch for students who factor immediately, writing $t(6t + 1) = 15$. Remind them that they have two goals, which they can remember from the term *zero-product*: one side of the equation must be *zero*, and the other must be a *product*. Point out that, while the left side of $t(6t + 1) = 15$ is indeed a product, the right side is not zero.

EXAMPLE 2

PHASE 2: Providing Guidance for TRY THIS

Given an equation in this form, it is a common error for students to quickly write the solution as $b = 3$ or $b = 2$. That is, they overlook the fact that the product $(b - 3)(b - 2)$ is not equal to zero. Continue to stress the significance of the term *zero-product* as discussed above.

EXAMPLE 3

PHASE 1: Presenting the Example

Some students may feel uneasy with the method taught in this example because the divisions on the right side of the equations involve the number zero. Remind them that it is the operation of division *by zero* that is undefined. They may feel more comfortable with the process if you insert steps in which the divisions are made explicit and the nonzero divisors are plainly visible, as shown at right.

$$2r^2 - 18 = 0 \qquad\qquad -3s^2 + 12s - 12 = 0$$

$$\frac{2r^2 - 18}{2} = \frac{0}{2} \qquad\qquad \frac{-3s^2 + 12s - 12}{-3} = \frac{0}{-3}$$

$$\frac{2r^2}{2} - \frac{18}{2} = \frac{0}{2} \qquad\qquad \frac{-3s^2}{-3} + \frac{12s}{-3} - \frac{12}{-3} = \frac{0}{-3}$$

$$r^2 - 9 = 0 \qquad\qquad s^2 - 4s + 4 = 0$$

$$(r + 3)(r - 3) = 0 \qquad\qquad (s - 2)(s - 2) = 0$$

Solving quadratic equations in multiple steps by factoring

In Skill A, all of the quadratic equations were written in standard form, $ax^2 + bx + c = 0$. If an equation is not given in standard form, rewrite the equation in standard form. Then solve it by factoring.

EXAMPLE 1

Solve by $8v^2 = 10v + 3$ by factoring.

▸ Solution

$$8v^2 = 10v + 3$$
$$8v^2 - 10v - 3 = 0 \quad \longleftarrow \text{First write in standard form.}$$
$$(4v + 1)(2v - 3) = 0 \quad \longleftarrow \text{Factor.}$$
$$4v + 1 = 0 \quad \text{or} \quad 2v - 3 = 0 \quad \longleftarrow \text{Apply the Zero-Product Property.}$$
$$v = -\frac{1}{4} \qquad v = \frac{3}{2}$$

TRY THIS   Solve $6t^2 + t = 15$ by factoring.

EXAMPLE 2

Solve $(z + 1)(z + 2) = 30$ by factoring.

▸ Solution

To apply the Zero-Product-Property, the equation must be in the form $ab = 0$.
$$(z + 1)(z + 2) = 30$$
$$z^2 + 3z + 2 = 30 \quad \longleftarrow \text{Multiply.}$$
$$z^2 + 3z - 28 = 0 \quad \longleftarrow \text{Write in standard form.}$$
$$(z - 4)(z + 7) = 0 \quad \longleftarrow \text{Factor.}$$
$$z - 4 = 0 \quad \text{or} \quad z + 7 = 0 \quad \longleftarrow \text{Apply the Zero-Product Property.}$$
$$z = 4 \qquad z = -7$$

TRY THIS   Solve $(b - 3)(b - 2) = 42$ by factoring.

Example 3 shows how you can make solving a quadratic equation easier if you divide both sides by the numerical GCF of all terms before factoring.

EXAMPLE 3

Solve by factoring.   **a.** $2r^2 - 18 = 0$ **b.** $-3s^2 + 12s - 12 = 0$

▸ Solution

a.
$$2r^2 - 18 = 0$$
$$r^2 - 9 = 0 \quad \longleftarrow \text{Divide each side by 2, the GCF of } 2r^2 \text{ and 18.}$$
$$(r + 3)(r - 3) = 0 \quad \longleftarrow \text{Factor.}$$
$$r = -3 \quad \text{or} \quad r = 3 \quad \longleftarrow \text{The solutions are 3 and } -3.$$

Because $(s - 2)$ appears twice as a factor, 2 is called a **double root.**

b.
$$-3s^2 + 12s - 12 = 0$$
$$s^2 - 4s + 4 = 0 \quad \longleftarrow \text{Divide each side by } -3.$$
$$(s - 2)(s - 2) = 0 \quad \longleftarrow \text{Factor.}$$
$$s = 2$$

TRY THIS   Solve by factoring.   **a.** $-9c^2 + 36 = 0$ **b.** $10s^2 + 20s + 10 = 0$

EXERCISES

KEY SKILLS

Write each quadratic equation in standard form.

1. $6x^2 = 7x + 5$
2. $r^2 + 5r + 6 = 42$
3. $4n^2 + 9 = 12n$
4. $6 = 22d - 12d^2$
5. $15t^2 - 1 = 2t$
6. $-12w^2 + 12 = 24w$

PRACTICE

Use factoring to solve.

7. $x^2 + 4x = 21$
8. $t^2 - 3t = 40$
9. $q^2 = -14q - 33$
10. $12a + a^2 = -35$
11. $w + w^2 = 42$
12. $39 + v^2 = 16v$
13. $4b + 15b^2 = 3$
14. $-d + 21d^2 = 2$
15. $9 - 15y = 14y^2$
16. $(x + 5)(x - 2) = 18$
17. $(z - 3)(z - 7) = 32$
18. $(a - 4)(a - 7) = -2$
19. $(2x - 6)(x - 7) = 24$
20. $(c + 1)(3c + 2) = 44$
21. $(3x + 5)(2x - 2) = 22$
22. $(4x - 1)(2x - 3) = 3$
23. $(7x + 2)(2x + 3) = -5$
24. $(7x - 3)(3x - 7) = -11$
25. $12x^2 + 12x + 3 = 0$
26. $12a^2 + 36a + 27 = 0$
27. $12x^2 - 3 = 0$
28. $16t^2 - 36 = 0$
29. $-16v^2 - 80v - 100 = 0$
30. $28c^2 - 84c + 63 = 0$
31. $100h^2 = 160h - 64$
32. $-144 + 144y = 36y^2$
33. $20 - 60x = -45x^2$
34. $30a^2 = 18 + 33a$

35. Solve $(x - 5)^2 = 36$ by writing the equation as the difference of two squares and then factoring.

36. **Critical Thinking**   Let $a = 0, 1, 2, 3, 4, 5$, or 6. For which values of a will $(x - 1)(x - 2) = a$ have integer solutions? What are the solutions?

MIXED REVIEW

Let r represent any real number. Is the given statement always, sometimes, or never true? Justify your response. (2.9 Skill A)

37. $r^2 = 7r$
38. $2(r - 5) + r = 2r - 2(r + 5)$
39. $2(r - 5) + 3(r - 5) = 5(r - 5) + 1$
40. $(r - 1)^2 = 2(r - 1)$

Answers for Lesson 7.5 Skill B

Try This Example 1
$t = -\dfrac{5}{3}$ or $t = \dfrac{3}{2}$

Try This Example 2
$b = -4$ or $b = 9$

Try This Example 3
a. $c = -2$ or $c = 2$
b. $s = -1$

1. $6x^2 - 7x - 5 = 0$
2. $r^2 + 5r - 36 = 0$
3. $4n^2 - 12n + 9 = 0$
4. $12d^2 - 22d + 6 = 0$
5. $15t^2 - 2t - 1 = 0$
6. $12w^2 + 24w - 12 = 0$
7. $x = -7$ or $x = 3$
8. $t = -5$ or $t = 8$
9. $q = -11$ or $q = -3$
10. $a = -7$ or $a = -5$
11. $w = -7$ or $w = 6$
12. $v = 3$ or $v = 13$
13. $b = -\dfrac{3}{5}$ or $b = \dfrac{1}{3}$

14. $d = -\dfrac{2}{7}$ or $d = \dfrac{1}{3}$
15. $y = -\dfrac{3}{2}$ or $y = \dfrac{3}{7}$
16. $x = -7$ or $x = 4$
17. $z = -1$ or $z = 11$
18. $a = 5$ or $a = 6$
19. $x = 1$ or $x = 9$
20. $c = 3$ or $c = -\dfrac{14}{3}$
21. $x = 2$ or $x = -\dfrac{8}{3}$
22. $x = 0$ or $x = \dfrac{7}{4}$
23. $x = -1$ or $x = -\dfrac{11}{14}$
24. $x = 2$ or $x = \dfrac{16}{21}$
25. $x = -\dfrac{1}{2}$
26. $a = -\dfrac{3}{2}$
27. $x = -\dfrac{1}{2}$ or $x = \dfrac{1}{2}$

28. $t = -\dfrac{3}{2}$ or $t = \dfrac{3}{2}$
29. $v = -\dfrac{5}{2}$
30. $c = \dfrac{3}{2}$
31. $h = \dfrac{4}{5}$
32. $y = 2$
33. $x = \dfrac{2}{3}$
34. $a = -\dfrac{2}{5}$ or $a = \dfrac{3}{2}$
35. $x = -1$ or $x = 11$
36. $a = 0$: 1 and 2
    $a = 2$: 0 and 3
    $a = 6$: -1 and 4
37. sometimes true, when $r = 0$ or $r = 7$
38. sometimes true, when $r = 0$
39. If $2(r - 5) + 3(r - 5) = 5(r - 5) + 1$, then $5r - 25 = 5r - 24$. This implies that $-25 = -24$. Since this is impossible, the equation is never true.
40. sometimes true, when $r = 1$ or $r = 3$

Focus: *Students use methods for solving quadratic equations in one variable to find the solutions of cubic equations in one variable.*

EXAMPLES 1 and 2

PHASE 1: Presenting the Example

Display equations **1** and **2** at right. Point out that **1** is a linear equation and **2** is a quadratic equation. Tell students to solve each. [**1** $x = \frac{5}{2}$ **2** $x = 0$ **or** $x = \frac{5}{2}$] Display equation **3** and ask students how it differs from **1** and **2**. Discuss their perceptions. Then review the definition of *cubic equation* given on the textbook page, and discuss.

1 $2x - 5 = 0$

2 $2x^2 - 5x = 0$

3 $2x^3 - 5x^2 = 0$

Point out the relationship between the degree of a polynomial and the number of solutions. A linear equation (degree 1) cannot have more than one solution; a quadratic equation (degree 2) cannot have more than two solutions; and a cubic equation (degree 3) cannot have more than 3 solutions.

PHASE 2: Providing Guidance for **TRY THIS**

In Skill B, students learned how to reduce some work by dividing each side of an equation by the same number. Consequently, they may try to solve the equations in this skill by dividing each side by the same *variable*. For example, some students may divide both sides of $10k^3 - 13k^2 + 4k = 0$ by k, as shown at right. Stress that the Division Property of Equality allows you to divide both sides of an equation by the same *nonzero* real number. However, we do not know that the variable k is always nonzero. Tell students that when they divide by k, they are assuming that k is not equal to zero, and they "lose" the solution $k = 0$. Warn students that they should not divide each side of an equation by any expression that contains a variable.

$$10k^3 - 13k^2 + 4k = 0$$
$$k(10k^2 - 13k + 4) = 0$$
$$\frac{k(10k^2 - 13k + 4)}{k} = \frac{0}{k} \quad \text{✗}$$
$$10k^2 - 13k + 4 = 0$$
$$(5k - 4)(2k - 1) = 0$$
$$k = \frac{4}{5} \text{ or } k = \frac{1}{2}$$

PHASE 3: ASSESSMENT AND CLOSURE for Lesson 7.5

- Have students write two quadratic equations with solutions $x = \frac{1}{2}$ or $x = -2$, using the form $ax^2 + bx + c$, where a, b, and c are integers. [**samples: $2x^2 + 3x - 2 = 0$, $4x^2 + 6x - 4 = 0$**] Then have them write two cubic equations with solutions $x = 0$, $x = 3$, or $x = -3$. [**samples: $x^3 - 9x = 0$, $2x^3 - 18x = 0$**]

- Not all quadratic equations can be solved by factoring over the integers. In Chapter 8, students will revisit quadratic equations and learn additional solution methods, including the quadratic formula. In the remainder of this chapter, students will begin their study of *quadratic functions*.

☞ *For a Lesson 7.5 Quiz, see* Assessment Resources *page 73.*

A **cubic equation** in x is any equation that can be written in the form $ax^3 + bx^2 + cx + d = 0$, where $a \neq 0$. This is called the **standard form of a cubic equation.** Using an extension of the Zero-Product Property, you can solve many cubic equations.

If a, b, and c represent real numbers and $abc = 0$, then $a = 0$, $b = 0$, or $c = 0$.

For example, if $x(x - 2)(3x + 4) = 0$, you can write the following.

$$x = 0 \quad \text{or} \quad x - 2 = 0 \quad \text{or} \quad 3x + 4 = 0$$

Thus, the solutions are 0, 2, and $-\frac{4}{3}$.

EXAMPLE 1 Solve $2n^3 + 8n^2 - 42n = 0$.

▸ Solution

$2n^3 + 8n^2 - 42n = 0$
$2n(n^2 + 4n - 21) = 0$ ⟵ Factor the GCF, $2n$, from each term.
$2n(n + 7)(n - 3) = 0$ ⟵ Factor $n^2 + 4n - 21$.
$2n = 0$ or $(n + 7) = 0$ or $(n - 3) = 0$ ⟵ Apply the Zero-Product Property.
$n = 0$ $n = -7$ $n = 3$ ⟵ Solve each equation for n.

The solutions are 0, -7, and 3.

TRY THIS Solve $10k^3 - 13k^2 + 4k = 0$.

EXAMPLE 2 Solve $m^3 + 22m^2 + 121m = 0$.

▸ Solution

$m^3 + 22m^2 + 121m = 0$
$m(m^2 + 22m + 121) = 0$ ⟵ Factor the GCF, m, from each term.
$m(m + 11)(m + 11) = 0$ ⟵ Factor $m^2 + 22m + 121$.
$m = 0$ or $m + 11 = 0$ or $m + 11 = 0$ ⟵ Apply the Zero-Product Property.
$m = 0$ $m = -11$ $m = -11$ ⟵ Solve each equation for m.

The solutions are 0 and -11.

TRY THIS Solve $z^3 - 14z^2 + 49z = 0$.

The steps for solving a polynomial equation by factoring are:
Step 1. Write the equation in standard form.
Step 2. Factor the GCF, if one exists, from each term in the equation.
Step 3. Factor the polynomial.
Step 4. Apply the Zero-Product Property and set each factor equal to zero.
Step 5. Solve for the variable.
Step 6. Check your solution(s) in the original equation.

EXERCISES

KEY SKILLS

Use the Zero-Product Property to write the solutions to each equation.

1. $(x - 1)(x - 2)(x - 3) = 0$
2. $a(a + 2)(a - 2) = 0$
3. $n(2n + 1)(3n - 2) = 0$
4. $x^2(x - 1) = 0$
5. $y(y - 3)(y - 3) = 0$
6. $(x - a)(x - b)(x - c) = 0$

PRACTICE

Use factoring to solve. Check your solutions.

7. $4x^3 - 3x^2 = 0$
8. $z^3 + 4z^2 = 0$
9. $7v^3 - 14v^2 = 0$
10. $2x^3 + 7x^2 = 0$
11. $k^3 - 4k^2 = 0$
12. $5b^3 + 10b^2 = 0$
13. $3y^3 + 9y^2 = 0$
14. $12g^3 - 144g^2 = 0$
15. $7s^3 - 5s^2 = 0$
16. $b^3 - 4b^2 + 3b = 0$
17. $t^3 + 11t^2 + 18t = 0$
18. $s^3 - 4s^2 + 3s = 0$
19. $x^3 - 2x^2 + x = 0$
20. $v^3 - 5v^2 + 4v = 0$
21. $7c^3 - 21c^2 = 0$
22. $6b^3 - 19b^2 + 15b = 0$
23. $4m^3 - 4m^2 - 35m = 0$
24. $6a^3 + 23a^2 + 21a = 0$
25. $2x^3 - 8x = 0$
26. $3h^3 - 27h = 0$
27. $25n^3 - 40n^2 + 16n = 0$
28. $36z^3 + 60z^2 + 25z = 0$
29. $2d^3 - 98d = 0$
30. $49y^3 + 14y^2 + y = 0$

31. Solve $(b^2 - 16)(b^2 - 9)(b^2 - 1) = 0$.
32. Solve $(n^2 - 25)(n^2 - 16)(n^2 + 2n + 1) = 0$.
33. Solve $(z^2 - 1)(z^2 - 4)(z^2 - 9)(z^2 - 16) = 0$.
34. Solve $(x^2 - 2xy + y^2)(x^2 + 2xy + y^2) = 0$.

35. **Critical Thinking** Justify each step in the following proof.
   1. $abc = a(bc)$
   2. If $abc = 0$, then $a(bc) = 0$.
   3. If $a(bc) = 0$, then $a = 0$ or $bc = 0$.
   4. If $a = 0$ or $bc = 0$, then $a = 0$ or $b = 0$ or $c = 0$.

MIXED REVIEW

Let r represent any real number. Is the given inequality sometimes, always, or never true? Justify your response. (4.8 Skill A)

36. $r \geq 20r$
37. $2(r - 3) - (r - 3) \geq r - 3$
38. $2(r + 5) - 3r \leq 9 - r$
39. $(r - 1)^2 \leq r^2$

Answers for Lesson 7.5 Skill C

Try This Example 1
$k = 0$, $k = \frac{1}{2}$, or $k = \frac{4}{5}$

Try This Example 2
$z = 0$ or $z = 7$

1. $x = 1$, $x = 2$, or $x = 3$
2. $a = -2$, $a = 0$, or $a = 2$
3. $n = 0$, $n = -\frac{1}{2}$, or $n = \frac{2}{3}$
4. $x = 0$ or $x = 1$
5. $y = 0$ or $y = 3$
6. $x = a$, $x = b$, or $x = c$
7. $x = 0$ or $x = \frac{3}{4}$
8. $z = -4$ or $z = 0$
9. $v = 0$ or $v = 2$
10. $x = 0$ or $x = -\frac{7}{2}$
11. $k = 0$ or $k = 4$
12. $b = -2$ or $b = 0$
13. $y = 0$ or $y = -3$
14. $g = 0$ or $g = 12$

15. $s = 0$ or $s = \frac{5}{7}$
16. $b = 0$, $b = 1$, or $b = 3$
17. $t = -9$, $t = -2$, or $t = 0$
18. $s = 0$, $s = 1$, or $s = 3$
19. $x = 0$ or $x = 1$
20. $v = 0$, $v = 1$, or $v = 4$
21. $c = 0$ or $c = 3$
22. $b = 0$, $b = \frac{3}{2}$, or $b = \frac{5}{3}$
23. $m = -\frac{5}{2}$, $m = 0$, or $m = \frac{7}{2}$
24. $a = -\frac{7}{3}$, $a = -\frac{3}{2}$, or $a = 0$
25. $x = -2$, $x = 0$, or $x = 2$
26. $h = -3$, $h = 0$, or $h = 3$
27. $n = 0$ or $n = \frac{4}{5}$
28. $z = -\frac{5}{6}$ or $z = 0$
29. $d = -7$, $d = 0$, or $d = 7$
30. $y = -\frac{1}{7}$ or $y = 0$

★ 31. $b = -4$, -3, -1, 1, 3, or 4
★ 32. $n = -5$, -4, -1, 4, or 5
★ 33. $z = \pm 1$, ± 2, ± 3, ± 4
★ 34. $x = \pm y$
35. 1. Associative Property of Multiplication
   2. Substitution Principle
   3. Zero-Product Property
   4. Zero-Product Property
36. sometimes true; if $r = 1$, the inequality is false. If $r = -1$, the inequality is true.
37. Because $2(r - 3) - (r - 3) = r - 3$, $2(r - 3) - (r - 3) \geq r - 3$ is always true.
38. If $2(r + 5) - 3r \leq 9 - r$, then $-r + 10 \leq 9 - r$. Therefore, $10 \leq 9$. Since this is impossible, the inequality is never true.
39. sometimes true; if $r = -1$, the inequality is false. If $r = 2$, the inequality is true.

★ **Advanced Exercises**

TEACHING
7.6
LESSON
Quadratic Functions and Their Graphs

SKILL A *Graphing a quadratic function*

Focus: *Given the equation of a quadratic function, students identify the axis of symmetry and vertex of its graph. They use this information to generate a table of ordered pairs and then graph the function.*

EXAMPLE

PHASE 1: Presenting the Example

Review with students the following key points about functions:

- A relation is a correspondence between two sets of numbers called the domain and range. If each member of the domain is assigned exactly one member of the range, then the relation is a function.

- A function can be represented as a list or a table of ordered pairs, a graph in the coordinate plane, or an equation in two variables.

Display the equation $y = x - 3$ and have students graph it. [**See the first graph at right.**] Point out that the right side of the equation is a polynomial. Ask them to classify it by degree. [**linear**] Tell them that a function defined by an equation of this form ($y = ax + b$) is called a *linear function*. Note that the graph is a line.

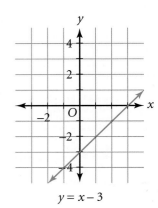

$y = x - 3$

Display the equation $y = x^2 + 2x - 3$. Ask students to classify the polynomial on the right side by degree. [**quadratic**] Tell them that a function defined by an equation of this form ($y = ax^2 + bx + c$) is called a *quadratic function*. Now they will investigate the form of the graph.

Have students make a table of ordered pairs for $y = x^2 + 2x - 3$ given $x = -4, -3, -2, -1, 0, 1, 2$. [$(-4, 5), (-3, 0), (-2, -3), (-1, -4), (0, -3), (1, 0), (2, 5)$] Tell them to graph the ordered pairs and connect them with a smooth curve. [**See the second graph at right.**] Tell students that the curve is called a *parabola*. Use their graphs to illustrate the terms *axis of symmetry*, *vertex*, and *minimum point* as described in the box on the textbook page. [**An equation of the axis of symmetry is $x = -1$. The vertex is $(-1, -4)$, which is a minimum point.**] Then discuss the Example.

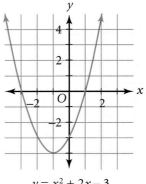

$y = x^2 + 2x - 3$

PHASE 2: Providing Guidance for TRY THIS

Watch for students who use only positive values of x in their tables. As a result, some may draw the graphs as partial curves or as straight lines. Stress the significance of first calculating $-\dfrac{b}{2a}$, which is the x-value of the maximum or minimum point. The table should be planned to include a few x-values less than this number and a few x-values greater than this number.

---

286 Chapter 7 Factoring Polynomials

7.6
LESSON

Quadratic Functions and Their Graphs

SKILL A *Graphing a quadratic function*

A **quadratic function** is any function of the form $y = ax^2 + bx + c$, where $a \neq 0$.

The Graph of a Quadratic Function

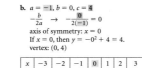

- The graph of $y = ax^2 + bx + c$ $(a \neq 0)$ has a U shape called a **parabola**. The lowest or highest point is called the **vertex**.
- The U shape is symmetric. It has an **axis of symmetry** whose equation is $x = -\frac{b}{2a}$. The value of $-\frac{b}{2a}$ is also the x-coordinate of the vertex.
- If $a > 0$, the graph opens upward and has a minimum point.
- If $a < 0$, the graph opens downward and has a maximum point.

EXAMPLE Find the coordinates of the vertex and an equation for the axis of symmetry. Then graph the function.

a. $y = x^2 - 4x + 3$ b. $y = -x^2 + 4$

Solution

a. $a = 1, b = -4, c = 3$
$-\frac{b}{2a} \rightarrow -\frac{-4}{2(1)} = 2$
axis of symmetry: $x = 2$
If $x = 2$, then $y = 2^2 - 4(2) + 3 = -1$.
vertex: $(2, -1)$

b. $a = -1, b = 0, c = 4$
$-\frac{b}{2a} \rightarrow -\frac{0}{2(-1)} = 0$
axis of symmetry: $x = 0$
If $x = 0$, then $y = -0^2 + 4 = 4$.
vertex: $(0, 4)$

Make a table.

x	-1	0	1	2	3	4	5
y	8	3	0	-1	0	3	8

x	-3	-2	-1	0	1	2	3
y	-5	0	3	4	3	0	-5

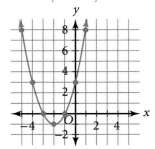

The graph opens down, and the vertex is the maximum point.

The graph opens up, and the vertex is the minimum point.

TRY THIS Rework the Example using the following equations:
a. $y = x^2 + 4x + 3$ b. $y = -x^2 + 2$

Find the coordinates of the vertex and write an equation for the axis of symmetry.

1.
2.
3.

Using $y = ax^2 + bx + c$, identify a, b, and c in each of the functions below.

4. $y = 2x^2 + 3x - 6$ 5. $y = -x^2 + 2x + 8$ 6. $y = x^2 - 5$

Each equation represents a quadratic function. Multiply and write each product in the form $ax^2 + bx + c$. Then identify a, b, and c.

7. $y = x(x - 3)$ 8. $y = (2x - 1)(x - 2)$ 9. $y = (x + 3)(x - 3)$
10. $y = (x - 1)(x + 1)$ 11. $y = (x + 2)^2$ 12. $y = -(x - 1)^2$

Does the graph of each function open upward or downward?

13. $y = x^2 + 3x - 2$ 14. $y = -x^2 + 4x - 3$ 15. $y = x^2 - 2x$

For Exercises 16–24:
a. Find the coordinates of the vertex and write an equation for the axis of symmetry.
b. Identify the vertex as the maximum or the minimum point.
c. Then make a table of ordered pairs and use it to graph each function.

16. $y = -x^2 - 2x + 3$ 17. $y = x^2 - 9$ 18. $y = -x^2 - 1$
19. $y = x^2 + x - 6$ 20. $y = x^2 - 3x + 2$ 21. $y = x^2 - 1$
22. $y = x^2 + 3x - 2$ 23. $y = -x^2 + 4x - 3$ 24. $y = x^2 - 2x$

Find the intercepts of the graph of each equation. Graph the line. (3.5 Skill A)

25. $y = -3.5$ 26. $x = 3.5$ 27. $x + y = 6$ 28. $x - y = 6$
29. $2x + 3y = 9$ 30. $2x - y = 10$ 31. $y = -\frac{2}{7}x + 1$ 32. $y = \frac{2}{5}x + 3$

Answers for Lesson 7.6 Skill A

Try This Example

a. vertex: $(-2, -1)$;
   axis of symmetry: $x = -2$

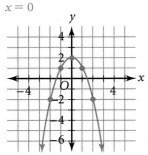

b. vertex: $(0, 2)$; axis of symmetry: $x = 0$

1. vertex: $(-3, 2)$; axis of symmetry: $x = -3$
2. vertex: $(0, 7)$; axis of symmetry: $x = 0$
3. vertex: $(4, 0)$; axis of symmetry: $x = 4$
4. $a = 2, b = 3, c = -6$
5. $a = -1, b = 2, c = 8$
6. $a = 1, b = 0, c = -5$
7. $y = x^2 - 3x; a = 1, b = -3, c = 0$
8. $y = 2x^2 - 5x + 2; a = 2, b = -5, c = 2$
9. $y = x^2 - 9; a = 1, b = 0, c = -9$
10. $y = x^2 - 1; a = 1, b = 0, c = -1$
11. $y = x^2 + 4x + 4; a = 1, b = 4, c = 4$
12. $y = -x^2 + 2x - 1; a = -1, b = 2, c = -1$
13. upward
14. downward
15. upward

16. a. vertex: $(-1, 4)$; axis of symmetry: $x = -1$
    b. maximum
    c.

x	-4	-3	-2	-1	0	1	2
y	-5	0	3	4	3	0	-5

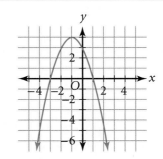

For answers to Exercises 17–32, see Additional Answers.

Focus: *Students identify the x-intercepts of a parabola by inspection and by applying the Zero-Product Property to the equation of a parabola.*

EXAMPLE 1

PHASE 1: Presenting the Example

Discuss Example 1 and its solution. Then use this reasoning to reinforce the conclusion that there are no *x*-intercepts: The vertex of a parabola that opens up is the minimum point of the parabola, and the *y*-coordinate of the vertex is the smallest *y*-value of the function. The coordinates of the vertex of this parabola are $(0, 2)$. So the smallest *y*-value of $y = x^2 + 2$ is 2, which is greater than 0. However, an *x*-intercept identifies a point where the *y*-value of a function is 0. Consequently, this graph will never have an *x*-intercept.

EXAMPLE 2

PHASE 1: Presenting the Example

Display the equation $y = x^2 - 2x - 8$. Have students make a table of ordered pairs and graph it. [**The graph is given at right.**] Ask them to identify the numbers that *appear* to be the *x*-intercepts. [**−2 and 4**]

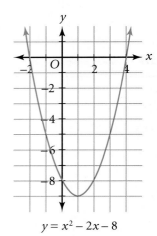

$y = x^2 - 2x - 8$

Return to the equation. Tell students to rewrite it by factoring the right side. [$y = (x - 4)(x + 2)$] Point out that the apparent *x*-intercepts are the solutions of the equation $(x - 4)(x + 2) = 0$. In fact, solving this equation confirms that the apparent *x*-intercepts are the *actual* *x*-intercepts.

Now display the equation in Example 2, $y = 2x^2 - 3x - 9$. Ask students to propose a method for finding the *x*-intercepts of its graph without graphing. Lead them to the solution that appears on the textbook page.

PHASE 2: Providing Guidance for TRY THIS

Watch for students who correctly factor $5x^2 - 26x + 5$ as $(5x - 1)(x - 5)$, and then hastily identify the *x*-intercepts as $-\frac{1}{5}$ and -5. Suggest that they graph the equation as a check on their work.

PHASE 3: ASSESSMENT AND CLOSURE for Lesson 7.6

- Ask students to state as many facts as they can about the graph of $y = -x^2 + 2x + 3$. [**It is a parabola that opens down, with axis of symmetry $x = 1$; vertex $(1, 4)$; maximum point $(1, 4)$; and x-intercepts -1 and 3.**]

- In Lesson 7.7, students will further investigate the graphs of quadratic functions, and they will relate quadratic functions to real-world situations.

☞ *For a Lesson 7.6 Quiz, see* Assessment Resources *page 74.*

Recall that an *x*-intercept of a graph is the *x*-coordinate of any point where the graph crosses the *x*-axis. The graph of a quadratic equation, a parabola, may have one of three possibilities: no *x*-intercepts, exactly one *x*-intercept, or two distinct *x*-intercepts. The values of *x* at the *x*-intercepts are the solutions, or roots, of the equation.

One way to find the *x*-intercepts of a quadratic function is to graph the function and see where it crosses the *x*-axis.

EXAMPLE 1 | **Use a graph to count the x-intercepts of the graph of $y = x^2 + 2$.**

▶ Solution

Use the coordinates of the vertex and a table of ordered pairs to graph $y = x^2 + 2$.

x-coordinate of vertex: $\frac{-b}{2a} = \frac{-0}{2(1)} = 0$

Make a table.

x	−2	−1	0	1	2
y	6	3	2	3	6

The graph never crosses the *x*-axis, so it has no *x*-intercepts.

TRY THIS | Use a graph to count the *x*-intercepts of $y = x^2 + x + 1$.

You can also use an algebraic approach to find the *x*-intercepts. To find the *x*-intercepts of the graph of $y = ax^2 + bx + c$, set $ax^2 + bx + c$ equal to 0, because $y = 0$ at the *x*-axis. Then solve for *x*. The solution gives the *x*-intercept.

EXAMPLE 2 | **Find and count the x-intercepts of the graph of $y = 2x^2 - 3x - 9$.**

▶ Solution

$2x^2 - 3x - 9 = 0$ ⟵ Set $2x^2 - 3x - 9$ equal to 0.
$(2x + 3)(x - 3) = 0$ ⟵ Factor.
$2x + 3 = 0$ or $x - 3 = 0$ ⟵ Apply the Zero-Product Property.
$x = -\frac{3}{2}$ or $x = 3$ ⟵ Solve each equation for x.

There are two *x*-intercepts of the graph, $-\frac{3}{2}$ and 3.

TRY THIS | Find and count the *x*-intercepts of the graph of $y = 5x^2 - 26x + 5$.

If you graph $y = x^2 + 2x + 1$ or use an algebraic approach to find the *x*-intercepts, you will find exactly one *x*-intercept, −1. If the graph of a quadratic function has exactly one *x*-intercept, then the *x*-intercept is a *double root* of the equation.

$x^2 + 2x + 1 = 0$
$(x + 1)^2 = 0$
$x = -1$

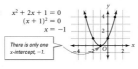

There is only one *x*-intercept, −1.

EXERCISES

How many *x*-intercepts does each graph have?

1.

2.

3.

PRACTICE

Use a graph to count the *x*-intercepts.

4. $y = x^2 + 4x + 4$
5. $y = -x^2 + 1$
6. $y = x^2 + x + 2$

Find and count the *x*-intercepts of the graph of each equation.

7. $y = x^2 + 7x + 6$
8. $y = -x^2 + 4x - 4$
9. $y = x^2 + 3x + 2$
10. $y = 4x^2 + 4x - 3$
11. $y = -x^2 + 9$
12. $y = x^2 + 6x + 9$

Let *r* be a real number. Find the values of *r* such that the graph has the specified number of *x*-intercepts.

13. $y = x^2 + r$
no *x*-intercepts
14. $y = x^2 + 2x + r$
one *x*-intercept
15. $y = x^2 + rx$
two *x*-intercepts

16. **Critical Thinking** Suppose that the vertex of the graph of $y = ax^2 + bx + c$ is above the *x*-axis and $a > 0$. How many *x*-intercepts does the graph have? Justify your response.

MIXED REVIEW

Find and simplify each sum. (6.4 Skill A)

17. $-3v^3 - 2v^2 + 1$ and $5v^3 - v^2 - 2$
18. $5.2c^2 - 2c$ and $-5.2c^2 - 3c - 3$
19. $2t^3 - 2t^2 - 5$ and $5t^2 - 2t - 3$
20. $-5n^2 - 2n + 1$ and $5n^2 - n - 1$

Find and simplify each product. (6.6 Skill B, C)

21. $2(a - 4)(a + 5)$
22. $(2r + 3)(3r + 7)$
23. $-(3d + 2)(5d + 1)$
24. $3(4g + 1)(4g - 1)$
25. $(7y - 2)(7y - 2)$
26. $-2(3w + 5)(2w - 3)$

Answers for Lesson 7.6 Skill B

Try This Example 1

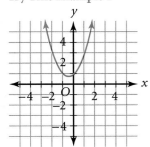

There are no *x*-intercepts.

Try This Example 2
$5x^2 - 26x + 5 = (5x - 1)(x - 5)$;
the equation has two real solutions.
Therefore, there are two *x*-intercepts,
5 and $\frac{1}{5}$.

1. two
2. none
3. one

4. one

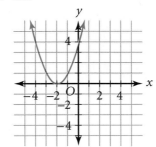

5. two

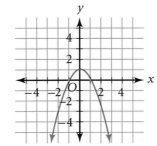

6. none

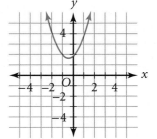

7. −1 and −6; two
8. 2; one
9. −1 and −2; two
10. $-\frac{3}{2}$ and $\frac{1}{2}$; two
11. 3 and −3; two
12. −3; one
★13. $r > 0$

For answers to Exercises 14–26, see Additional Answers.

★ Advanced Exercises

Analyzing the Graph of a Quadratic Function

SKILL A ► *Finding the range of a quadratic function*

Focus: *Given an equation for a quadratic function, students locate the vertex of its graph, identify the vertex as a maximum or minimum point, and use this information to state the range of the function.*

EXAMPLES 1 and 2

PHASE 1: Presenting the Examples

Display graph **1** at right. Remind students of the vertical-line test: No vertical line in the plane intersects the graph at more than one point, so the graph represents a function. Ask students to write the domain and range of this function. [**domain:** $\{-3, -2, -1, 0, 1, 2, 3\}$; **range:** $\{-6, -5, -2, 3\}$] Discuss their responses, and elicit the following observation: The domain is the set of all possible x-coordinates of the ordered pairs, and the range is the set of all possible y-coordinates.

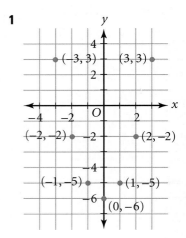

Display graph **2**. Tell students that the curve is a parabola. Ask if this graph, too, represents a function. [**yes**] Have students describe how it differs from graph **1**. They should observe that graph **1** is a discrete graph, consisting of separate points; in contrast, graph **2** is a continuous, or unbroken, graph.

Point out the arrowheads on graph **2**. Remind students that these show the graph continuing infinitely to the left and right. Ask what this indicates about the domain of the function. They should conclude that the domain is the set of all real numbers.

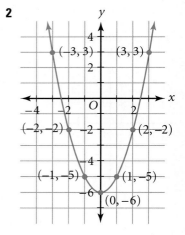

Note that the arrowheads also show graph **2** continuing infinitely upward. Ask if this indicates that the range, too, is the set of all real numbers. Students should observe that, in contrast to the domain, the vertex of the parabola places a lower bound on the range: The range is the set of all real numbers *greater than or equal to* -6. This can be denoted by the inequality $y \geq -6$.

Now discuss the Examples. Note that Example 1 shows another instance of a lower bound on the range. Example 2, on the other hand, illustrates a case in which the range has an upper bound.

PHASE 2: Providing Guidance for TRY THIS

Students may wonder why they are not expected to make tables of values to graph the functions in these exercises. Emphasize the fact that, to find the range of a quadratic function, they need only identify two characteristics of the parabola: the coordinates of the vertex and the direction in which it opens. So it is only necessary to make a quick sketch, as shown in the textbook solutions, rather than a detailed graph. Point out that the textbook sketches do not even include scales on the axes.

7.7 LESSON — Analyzing the Graph of a Quadratic Function

SKILL A *Finding the range of a quadratic function*

The domain of a quadratic function is the set of all real numbers. The range of the function can be represented by an inequality. When you find the range of a quadratic function, draw a quick sketch of the graph to help you.

Recall that if $a > 0$, the graph opens upward and the y-coordinate of the vertex is the minimum value of y.

EXAMPLE 1 **Find the range of the function represented by $y = 3x^2 - 6x - 1$.**

▶ Solution

Use $\frac{-b}{2a}$ to find the x-coordinate.

Find the coordinates of the vertex, V.

x-coordinate: $-\frac{-6}{2(3)} = 1$ ⟵ $a = 3$ and $b = -6$

y-coordinate: $3(1)^2 - 6(1) - 1 = -4$

Because the graph opens upward, all values of y are *at least* equal to the y-coordinate of the vertex.

Draw a quick sketch. a is positive, so the graph opens upward.

$V (1, -4)$

Therefore, the range is $y \geq -4$.

TRY THIS Find the range of the function represented by $y = 2x^2 + 5x$.

Recall that if $a < 0$, the graph opens downward and the y-coordinate of the vertex is the maximum value of y.

EXAMPLE 2 **Find the range of the function represented by $y = -5x^2 - 10x + 3$.**

▶ Solution

Use $\frac{-b}{2a}$ to find the x-coordinate.

Find the coordinates of the vertex, V.

x-coordinate: $-\frac{-10}{2(-5)} = -1$ ⟵ $a = -5$ and $b = -10$

y-coordinate: $-5(-1)^2 - 10(-1) + 3 = 8$

Draw a quick sketch. a is negative, so the graph opens downward.

$V (-1, 8)$

Because the graph opens downward, all values of y are *at most* equal to the y-coordinate of the vertex.

Therefore, the range is $y \leq 8$.

TRY THIS Find the range of the function represented by $y = -x^2 + 7x - 1$.

EXERCISES

KEY SKILLS

Find the range of each quadratic function graphed below.

1. 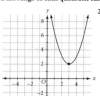   2. 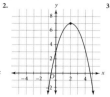   3.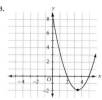

PRACTICE

Find the range of each quadratic function.

4. $y = x^2 - 4x + 3$
5. $y = -x^2 - 6x + 2$
6. $y = -x^2 - 5x + 2$
7. $y = x^2 - 3x + 1$
8. $y = -2x^2 - 5x + 3$
9. $y = 2x^2 + 7x + 3$
10. $y = -3x^2 - 9x + 7$
11. $y = 5x^2 - 10x + 4$
12. $y = \frac{1}{2}x^2 - 3x - 7$
13. $y = \frac{1}{2}x^2 + 5x - 1$
14. $y = \frac{3}{4}x^2 + x + 3$
15. $y = -\frac{3}{4}x^2 - 3x$
16. $y = x(x - 1)$
17. $y = x(x + 2)$
18. $y = (x + 2)(x - 1)$
19. $y = -(x - 5)(x - 3)$
20. $y = (2x + 5)(x + 1)$
21. $y = (2x + 7)(3x - 5)$

In Exercises 22–24, use the given range to determine if the graph of the quadratic function opens up or down.

22. The range is $y \geq -2$.
23. The range is $y \leq 3$.
24. The range is $y \leq 0$.

25. Given the domain $-2 \leq x \leq 2$, what is the range of $y = x^2$?

MIXED REVIEW

Simplify each expression. (6.5 Skill A)

26. $(2n^2)(3n^4)$
27. $(3n^3)(2n^2)(-2n^2)$
28. $(z^3)(z^2)^2$
29. $(k^2)(k^2)^2$

Simplify each expression. (6.5 Skill B)

30. $\frac{2(x^2y^2)^2}{(xy)^2}$
31. $\frac{(3a^2b)^2}{(9a^2b)^2}$
32. $\frac{(2c^2d^2)^2}{(c^4d^3)}$
33. $\frac{(2gh^2)^2}{(4g^2h^2)^2}$

Answers for Lesson 7.7 Skill A

Try This Example 1

$y \geq -\frac{25}{8}$

Try This Example 2

$y \leq \frac{45}{4}$

1. $y \geq 2$
2. $y \leq 7$
3. $y \geq -2$
4. $y \geq -1$
5. $y \leq 11$
6. $y \leq \frac{33}{4}$
7. $y \geq -\frac{5}{4}$
8. $y \leq \frac{49}{8}$
9. $y \geq -\frac{25}{8}$
10. $y \leq \frac{55}{4}$
11. $y \geq -1$

12. $y \geq -11.5$
13. $y \geq -13.5$
14. $y \geq \frac{8}{3}$
15. $y \leq 3$
16. $y \geq -\frac{1}{4}$
17. $y \geq -1$
18. $y \geq -\frac{9}{4}$
19. $y \leq 1$
20. $y \geq -\frac{9}{8}$
21. $y \geq -\frac{961}{24}$
22. up
23. down
24. down
25. $0 \leq y \leq 4$
26. $6n^6$
27. $-12n^7$
28. z^7
29. k^6

30. $2x^2y^2$
31. $\frac{1}{9}$
32. $4d$
33. $\frac{1}{4g^2}$

Focus: *Students write an equation for a quadratic function given the x-intercepts and the coordinates of a third point on its graph.*

EXAMPLE 1

PHASE 1: **Presenting the Example**

Remind students that a line is determined by two points. Tell them that they are going to explore whether the same is true of a parabola.

Display the equations $y = x^2 - 4$, $y = -x^2 + 4$, and $y = 2x^2 - 8$. Have students graph them on the same coordinate plane. [**The graphs are shown at right.**] Ask if the graphs have any points in common. [$(-2, 0)$ **and** $(2, 0)$] Emphasize the fact that these three distinct parabolas pass through two points, $(-2, 0)$ and $(2, 0)$. Therefore, given just these points, you would not be able to identify a unique parabola passing through them. So, unlike a line, a parabola is *not* determined by just two points.

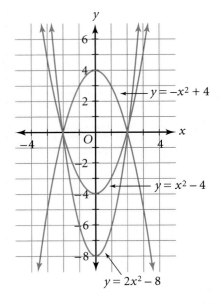

Have students factor the right sides of the three equations. [$y = (x + 2)(x - 2)$; $y = -(x + 2)(x - 2)$; $y = 2(x + 2)(x - 2)$] Use the rewritten equations to illustrate the text in the box near the top of the page. That is, note that each equation is of the form $y = a(x + 2)(x - 2)$. For the first equation, $a = 1$; for the second, $a = -1$; for the third, $a = 2$. Remind students that *any* equation of the form $y = a(x + 2)(x - 2)$ will have a graph with x-intercepts -2 and 2. So, given the points $(-2, 0)$ and $(2, 0)$, there actually are infinitely many parabolas that pass through them.

Therefore, the question is this: What differentiates one parabola with equation $y = a(x + 2)(x - 2)$ from another? One distinguishing characteristic is the y-intercept. That is, given the two x-intercepts and the y-intercept, you can write the equation of a unique parabola. Example 1 shows how to do this.

EXAMPLE 2

PHASE 2: **Providing Guidance for TRY THIS**

Be sure students are aware that the method used in this example is a generalization of the method used in Example 1. That is, here they find an equation of a parabola given x-intercepts r and s and coordinates (m, n) of any point on the parabola. In Example 1, they worked with x-intercepts r and s and coordinates $(0, p)$ of a specific point, the y-intercept. To underscore this connection, use the following as an extension of the Try This exercise:

> *The graph of a quadratic function has x-intercepts −4 and 4. The point (0, −32) is on the graph. Write an equation for the function.*

Students should arrive at the same equation, namely $y = 2x^2 - 32$.

Just as you can write an equation for a line given sufficient information, you can write an equation for a quadratic function given sufficient information about its graph.

Using Intercepts to Write a Quadratic Function
If a parabola has x-intercepts r and s, then for some nonzero real number a, $y = a(x - r)(x - s)$ represents the parabola.

EXAMPLE 1
The graph of a quadratic function has x-intercepts -2 and 3. The y-intercept of the graph is -12. Write an equation for the function.

▶ **Solution**
1. Because the x-intercepts are -2 and 3, the function has the form $y = a[x - (-2)](x - 3)$ for some nonzero real number a.

2. To find a, use the coordinates of the y-intercept, $(0, -12)$.
Substitute 0 for x and -12 for y and then solve the equation for a.
$a(x + 2)(x - 3) = y$
$a(0 + 2)(0 - 3) = -12$
$a(-6) = -12$
$a = 2$

3. Write the function and simplify it.
$y = 2(x + 2)(x - 3) = 2x^2 - 2x - 12$

An equation for the function is $y = 2x^2 - 2x - 12$.

TRY THIS
The graph of a quadratic function has x-intercepts 3 and 4. The y-intercept of the graph is 12. Write an equation for the function.

EXAMPLE 2
The graph of a quadratic function has x-intercepts -4 and 5. The point $(2, -36)$ is on the graph. Write an equation for the function.

▶ **Solution**
1. Since the x-intercepts are -4 and 5, the function has the form $y = a[x - (-4)](x - 5)$ for some nonzero real number a.

2. To find a, use $(2, -36)$. Substitute 2 for x and -36 for y and then solve the equation for a.
$a(x + 4)(x - 5) = y$
$a(2 + 4)(2 - 5) = -36$
$a(-18) = -36$
$a = 2$

3. Write the function and simplify it.
$y = 2(x + 4)(x - 5) = 2x^2 - 2x - 40$

An equation for the function is $y = 2x^2 - 2x - 40$.

TRY THIS
The graph of a quadratic function has x-intercepts -4 and 4. The point $(3, -14)$ is on the graph. Write an equation for the function.

EXERCISES

KEY SKILLS

Use the given x-intercept(s) to write a quadratic function in the form $y = a(x - r)(x - s)$.

1. x-intercepts: 2 and -2
2. x-intercepts: -5 and 2
3. x-intercepts: -1 and 3
4. x-intercepts: 0 and 4
5. x-intercept: 1
6. x-intercept: -3

PRACTICE

Use the given information to write a quadratic function in the form $y = ax^2 + bx + c$.

7. x-intercepts: -2 and 8; y-intercept 32
8. x-intercepts: 1 and 7; y-intercept 5
9. x-intercepts: -3 and 3; vertex $(0, 5)$
10. x-intercepts: -5 and 5; vertex $(0, -75)$
11. x-intercepts: -8 and 1
The point $(3, 44)$ is on the graph.
12. x-intercepts: 3 and 4
The point $(-1, -5)$ is on the graph.

Write an equation in the form $y = ax^2 + bx + c$ to represent each parabola.

13. 14. 15.

16. Let r be a nonzero real number. Show that if a parabola has x-intercepts r and $-r$, then the parabola is represented by $y = ax^2 - ar^2$ for some nonzero real number a.

17. Show that if a parabola has exactly one x-intercept r, then the parabola is represented by $y = a(x - r)^2$.

18. **Critical Thinking** Explain why the graph of any function represented by $y = ax^2 + b$, where $a > 0$ and $b > 0$, has no x-intercepts.

MIXED REVIEW

Use factoring to solve each equation. (7.5 Skill A)

19. $3a^2 - 75 = 0$
20. $6x^2 + 3x - 9 = 0$
21. $4x^2 - 8x + 3 = 0$

Use factoring to solve each equation. (7.5 Skill B)

22. $3a^2 = 4a$
23. $6c^2 = 132 - 61c$
24. $-35n^2 = 74n + 35$
25. $(2n + 1)(2n + 3) = 99$
26. $(3p - 5)(5p - 3) = 7$
27. $(4a + 3)(7a + 11) = 33$

Answers for Lesson 7.7 Skill B

Try This Example 1
$y = x^2 - 7x + 12$

Try This Example 2
$y = 2x^2 - 32$

1. $y = a(x - 2)(x + 2)$
2. $y = a(x + 5)(x - 2)$
3. $y = a(x + 1)(x - 3)$
4. $y = ax(x - 4)$
5. $y = a(x - 1)(x - 1)$
6. $y = a(x + 3)(x + 3)$
7. $y = -2x^2 + 12x + 32$
8. $y = \frac{5}{7}x^2 - \frac{40}{7}x + 5$
9. $y = -\frac{5}{9}x^2 + 5$
10. $y = 3x^2 - 75$
11. $y = 2x^2 + 14x - 16$
12. $y = -\frac{1}{4}x^2 + \frac{7}{4}x - 3$
13. $y = x^2 - 2x - 3$
14. $y = -2x^2 - 4x + 6$
15. $y = x^2 - 6x + 8$

★ 16. If r and $-r$ are the x-intercepts of the graph, then, for some nonzero real number a, $y = a(x + r)(x - r)$. Therefore, $y = a(x^2 - r^2) = ax^2 - ar^2$.

★ 17. If r is the only x-intercept of the graph, then for some nonzero real number a, $y = a(x - r)(x - r) = a(x - r)^2$.

18. If $y = ax^2 + b$ has an x-intercept, then $ax^2 + b = 0$ for some value of x. Then $ax^2 = -b$ and $x^2 = -\frac{b}{a}$. Since $a > 0$ and $b > 0$, $-\frac{b}{a}$ is negative. But no number squared equals a negative number. So there is no value of x such that $ax^2 + b = 0$, and this means that $y = ax^2 + b$ has no x-intercepts.

19. $a = 5$, or $a = -5$
20. $x = -\frac{3}{2}$ or $x = 1$
21. $x = \frac{1}{2}$ or $x = \frac{3}{2}$
22. $a = 0$ or $a = \frac{4}{3}$
23. $c = -12$ or $c = \frac{11}{6}$
24. $n = -\frac{5}{7}$ or $n = -\frac{7}{5}$
25. $n = 4$ or $n = -6$
26. $p = 2$ or $p = \frac{4}{15}$
27. $a = -\frac{65}{28}$ or $a = 0$

★ **Advanced Exercises**

SKILL C APPLICATIONS *Using the characteristics of a quadratic function to solve real-world problems*

Focus: *Students use quadratic functions to model real-world problems.*

EXAMPLE 1

PHASE 1: Presenting the Example

Tell students that many real-world relationships cannot be represented linearly but are instead modeled by quadratic functions. For example, the height of an object affected by gravity is a quadratic function of time. Then discuss Example 1.

Be sure students understand that the graph in Example 1 is not the actual path of the ball. Instead, the graph shows how the ball's height changes over time. To reinforce this concept, have students consider a ball that is thrown straight up into the air, reaches a height of 16 feet after one second, and returns to the thrower's hand after two seconds. (Consider the thrower's hand to be a height of 0.) The graph of this ball's height over time is the same as the graph for the soccer ball on the textbook page; however, the two balls have different paths.

EXAMPLE 2

PHASE 1: Presenting the Example

Students tend to believe that profit is linear; that is, the more you sell, the more you earn. So they may be surprised to see this quadratic model, which indicates a decrease in profits beyond a certain volume of sales. Point out that, as a company produces more items, it incurs additional expenses that "eat away" at profits. Ask students what some of these expenses might be. [**samples: wages of additional factory workers or salespersons; rent or purchase of a larger warehouse; rent or purchase of more retail space**]

PHASE 3: ASSESSMENT AND CLOSURE for Lesson 7.7

- Have students give equations for four different parabolas with x-intercepts -4 and 2. Then have them identify the range of the function that each parabola represents. [**samples:** $y = x^2 + 2x - 8$, **with range** $y \geq -9$; $y = -x^2 - 2x + 8$, **with range** $y \leq -9$; $y = 2x^2 + 4x - 16$, **with range** $y \geq -18$; $y = 3x^2 + 6x - 24$, **with range** $y \geq -27$]
- Quadratic functions are important because they describe many real-life phenomena. This chapter provided a strong visual image of a quadratic function—the parabola—and used the factoring of quadratic trinomials as a means of analyzing it. In Chapter 8, students will learn new techniques for solving quadratic equations, and then they will revisit quadratic functions.

☞ *For a Lesson 7.7 Quiz, see* Assessment Resources *page 75.*

☞ *For a Chapter 7 Assessment, see* Assessment Resources *pages 76–77.*

You can apply your knowledge of quadratic functions to real-world problems.

EXAMPLE 1

A soccer ball is kicked into the air. The function $y = -16x^2 + 32x$ describes its height in feet, y, after x seconds. Draw a graph of the function and use it to answer the questions below.
a. What is the maximum height reached by the ball?
b. How long does it take for the ball to reach its maximum height?
c. How long does it take for the ball to return to the ground?

▷ Solution

a. Because a, -16, is negative, the graph is a parabola that opens downward, and the y-coordinate of the vertex represents the maximum height, 16 feet.
b. The x-coordinate of the vertex represents the time (in seconds) when the ball reaches its maximum height. The vertex is (1, 16), which means that after 1 second, the ball reaches its maximum height of 16 feet.
c. The y-coordinate, which represents the height, increases until it reaches the vertex and then decreases until it reaches 0 at $x = 2$. Therefore, the ball returns to the ground after 2 seconds.

TRY THIS   Rework Example 1 for the function $y = -16x^2 + 64x$.

EXAMPLE 2

The profit p a company can make on the sale of x calendars is given by the function $p = 0.05x(1200 - x)$. How many calendars need to be sold to achieve the maximum profit? What is the maximum profit?

▷ Solution

1. Rewrite the function in standard form.
   $p = 0.05x(1200 - x) \rightarrow p = -0.05x^2 + 60x$
2. Because a, -0.05, is negative, the graph opens downward and the vertex contains the maximum value of p, the profit.
3. Find the coordinates of the vertex.
   x-coordinate: $-\dfrac{b}{2a} \rightarrow -\dfrac{60}{2(-0.05)} = 600$
   p-coordinate: $p = -0.05(600)^2 + 60(600) = 18,000$
   The maximum profit of $18,000 is reached when 600 calendars are sold.

TRY THIS   Rework Example 2 for the profit function $p = 0.04x(1500 - x)$.

EXERCISES

KEY SKILLS

Determine whether the graph of the function has a maximum or a minimum value.

1. $y = -5x^2 - 9x + 1$
2. $y = 2x^2 + 3x - 1$
3. $y = 9x^2 - 3x - 7$
4. $y = -x^2 + x$

PRACTICE

Refer to Example 1.

5. What is the axis of symmetry of the graph?
6. Does it make sense for the graph to extend beyond the first quadrant? Why or why not?
7. What do the x-intercepts of the graph represent?

In Exercises 8–15, each function represents the profit p made from the sale of x units of a product. How many units need to be sold to achieve the maximum profit? What is the maximum profit?

8. $p = 0.02x(4800 - x)$
9. $p = 0.08x(8000 - x)$
10. $p = 0.03x(4900 - x)$
11. $p = 0.05x(6300 - 9x)$
12. $p = 0.01x(2700 - 6x)$
13. $p = 0.04x(9600 - 12x)$
14. $p = 0.03x(2700 - 9x)$
15. $p = 0.06x(3600 - 4x)$

16. **Critical Thinking** Suppose that x units are sold and that profit p is given by $p = 0.04x(1500 - x)$. Show that maximum profit is reached when x is the average of the x-intercepts of the graph of the function.

MIXED REVIEW APPLICATIONS

Write an expression for the total volume of 36 boxes with each box having length, width, and height as specified. Do not simplify. (6.5 Skill C)

17. length $2.5a$ units, width $2a$ units, and height $4a$ units
18. length $3.5n$ units, width $2.5n$ units, and height $6n$ units
19. length $1.5z$ units, width $1.5z$ units, and height $2.5z$ units
20. length $\frac{4m}{5}$ units, width $\frac{5m}{16}$ units, and height $\frac{16m}{25}$ units

---

Answers for Lesson 7.7 Skill C

Try This Example 1
a. 64 feet
b. 2 seconds
c. 4 seconds

Try This Example 2
750 calendars; $22,500

1. maximum
2. minimum
3. minimum
4. maximum
5. $x = 1$
6. No; both time and height are nonnegative quantities, so they are both graphed in Quadrant I only.
7. The x-intercepts represent the times when the ball's height is 0.
8. 2400 units; $115,200
9. 4000 units; $1,280,000
10. 2450 units; $180,075
11. 350 units; $55,125
12. 225 units; $3037.50
13. 400 units; $76,800
14. 150 units; $6075

15. 450 units; $48,600
16. The x-intercepts are 0 and 1500. The average of these is 750. The value of $-\dfrac{b}{2a}$ is $-\dfrac{0.04 \times 1500}{2(-0.04)} = \dfrac{0.04 \times 1500}{2(0.04)} = \dfrac{1500}{2} = 750$.
    Since these agree, the average of the x-intercepts is the x-coordinate of the vertex. Since the graph opens downward, the x-coordinate of the vertex gives the maximum value of p, profit.
17. $(36)(2.5a)(2a)(4a)$
18. $(36)(3.5n)(2.5n)(6n)$
19. $(36)(1.5z)(1.5z)(2.5z)$
20. $(36)\left(\dfrac{4m}{5}\right)\left(\dfrac{5m}{16}\right)\left(\dfrac{16m}{25}\right)$

Write the prime factorization of each number.

1. 500 2. 1080 3. 31 4. 2401

Find the GCF of each set of monomials.

5. $5x^3$ and $35x^4$ 6. $6n^2d$ and $32n^4d^2$ 7. $5y^2z^2$, $18y^2$, and $60y^4z^2$

Use the GCF to factor each polynomial.

8. $3s^2 - 4s^3$ 9. $14z^4 + 4z^2$
10. $7a^3 + 14a^2 - 7a$ 11. $10n^4 - 14n^3 - 7n^2$
12. $3qp^4 - 12q^2p^2$ 13. $a^5b^2 + 10ab^3 - ab^2 - 15b^2$

Factor each polynomial.

14. $25d^2 - 120d + 144$ 15. $144z^2 - 25$
16. $4b^2 - 44bd + 121d^2$ 17. $16z^2 - 81y^2$
18. $4a^2b^2 + 4abc + c^2$ 19. $49m^2n^2 - 16p^2$
20. $4a^4 + 12a^2 + 9$ 21. $9hj + 6h - 12kj - 8k$
22. $k^2 + 16k + 39$ 23. $y^2 - 18y + 56$
24. $a^2 + 19a + 84$

Factor. If the expression cannot be factored using integers, write *not factorable* and justify your response.

25. $m^2 + 5m - 84$ 26. $c^2 + 13c - 8$ 27. $d^2 + 13d - 30$

Factor.

28. $15z^2 + 52z + 32$ 29. $15a^2 + 67a + 44$
30. $14m^2 + 57m + 55$ 31. $18b^2 + 49b - 49$
32. $33p^2 - 37p + 10$ 33. $12w^2 - 53w + 55$
34. $-u^2 + 12u - 35$ 35. $3u^2 - 36u + 81$
36. $5g^2 - 10g - 120$ 37. $25x^3 - 121x$
38. $49k^3 - 140k^2 + 100k$ 39. $42a^3 + 81a^2 + 30a$

Use factoring to solve each equation.

40. $14b^2 - 39b + 27 = 0$ 41. $36n^2 - 36n + 5 = 0$
42. $25q^2 - 75q + 26 = 0$ 43. $21d^2 - 50d - 99 = 0$
44. $8z^2 = 63z + 81$ 45. $(3s - 10)(s - 1) = 40$
46. $7r^2 - 63 = 0$ 47. $5v^2 - 10v + 5 = 0$
48. $3s^3 - 7s^2 = 0$ 49. $n^3 - 7n^2 + 6n = 0$
50. $p^3 + 18p^2 + 81p = 0$ 51. $15n^3 + 57n^2 - 12n = 0$

Find the coordinates of the vertex and an equation for the axis of symmetry. Is the vertex the maximum or minimum point on the graph?

52. $y = 3x^2 + 4x + 4$ 53. $y = 2x^2 - 4x + 1$
54. $y = 0.5x^2 - 2x + 0.2$ 55. $y = -1.5x^2 + 5x - 3$

Use a graph to determine if the graph of the function has zero, one, or two *x*-intercepts.

56. $y = -x^2 - 3$ 57. $y = x^2 - 6x + 9$

Find and count the *x*-intercepts of the graph of each function.

58. $y = 4x^2 - 20x + 21$ 59. $y = -x^2 - 2x - 1$

Find the range of each quadratic function.

60. $y = 2.5x^2 - 2x - 3$ 61. $y = -4x^2 - 7x + 5$
62. $y = -x^2 - 4$ 63. $y = 3x^2 - 9$

64. The graph of a quadratic function has *x*-intercepts −5 and 6. The point (1, 60) is on the graph. Write an equation for the function.

65. The graph of a quadratic function has *x*-intercepts −6 and −2. The point (0, 24) is on the graph. Write an equation for the function.

66. The profit a company can make on the same of *x* units of goods is represented by the function $p = 0.04x(1600 - 2x)$. How many units need to be sold to achieve the maximum profit possible? What is the maximum profit?

Answers for Chapter 7 Assessment

1. $2^2 \cdot 5^3$
2. $2^3 \cdot 3^3 \cdot 5$
3. prime
4. 7^4
5. $5x^3$
6. $2n^2d$
7. y^2
8. $s^2(3 - 4s)$
9. $2z^2(7z^2 + 2)$
10. $7a(a^2 + 2a - 1)$
11. $n^2(10n^2 - 14n - 7)$
12. $3qp^2(p^2 - 4q)$
13. $b^2(a^5 + 10ab - a - 15)$
14. $(5d - 12)^2$
15. $(12z + 5)(12z - 5)$
16. $(2b - 11d)^2$
17. $(4z - 9y)(4z + 9y)$
18. $(2ab + c)^2$
19. $(7mn - 4p)(7mn + 4p)$
20. $(2a^2 + 3)^2$
21. $(3h - 4k)(3j + 2)$

22. $(k + 3)(k + 13)$
23. $(y - 4)(y - 14)$
24. $(a + 7)(a + 12)$
25. $(m - 7)(m + 12)$
26. The factors of −8 are 1, 2, 4, 8, −1, −2, −4, and −8. The sums of factors are shown in the table below.

$-1 + 8$	sum: 7
$-2 + 4$	sum: 2
$-4 + 2$	sum: -2
$-8 + 1$	sum: -7

Since no sum equals 13, $c^2 + 13c - 8$ cannot be factored.

27. $(d - 2)(d + 15)$
28. $(3z + 8)(5z + 4)$
29. $(3a + 11)(5a + 4)$

30. $(7m + 11)(2m + 5)$
31. $(9b - 7)(2b + 7)$
32. $(3p - 2)(11p - 5)$
33. $(4w - 11)(3w - 5)$
34. $-(u - 7)(u - 5)$
35. $3(u - 3)(u - 9)$
36. $5(g + 4)(g - 6)$
37. $x(5x + 11)(5x - 11)$
38. $k(7k - 10)^2$
39. $3a(7a + 10)(2a + 1)$
40. $b = \dfrac{9}{7}$ or $b = \dfrac{3}{2}$
41. $n = \dfrac{5}{6}$ or $n = \dfrac{1}{6}$
42. $q = \dfrac{13}{5}$ or $q = \dfrac{2}{5}$
43. $d = \dfrac{11}{3}$ or $d = -\dfrac{9}{7}$
44. $z = 9$ or $z = -\dfrac{9}{8}$
45. $s = 6$ or $s = -\dfrac{5}{3}$
46. $r = 3$ or $r = -3$
47. $v = 1$

48. $s = 0$ or $s = \dfrac{7}{3}$
49. $n = 0$, $n = 1$, or $n = 6$
50. $p = 0$ or $p = -9$
51. $n = 0$, $n = -4$, or $n = \dfrac{1}{5}$
52. vertex: $\left(-\dfrac{2}{3}, \dfrac{8}{3}\right)$; axis of symmetry: $x = -\dfrac{2}{3}$; minimum
53. vertex: $(1, -1)$; axis of symmetry: $x = 1$; minimum
54. vertex: $(2, -1.8)$; axis of symmetry: $x = 2$; minimum
55. vertex: $\left(\dfrac{5}{3}, \dfrac{7}{6}\right)$; axis of symmetry: $x = \dfrac{5}{3}$; maximum

For answers to Exercises 56–66, see Additional Answers.

296 **Chapter 7** **Factoring Polynomials**

Chapter 8 Quadratic Functions and Equations	State or Local Standards	Corresponding Lessons in *Algebra 1*, Schultz et al.
8.1 Square Roots and the Equation $x^2 = k$		10.2
Skill A: Finding and approximating square roots		
Skill B: Simplifying square roots of numbers using properties of square roots		
Skill C: Using the exponent $\frac{1}{2}$ in proofs		
8.2 Solving Equations of the Form $ax^2 + c = 0$		10.2
Skill A: Finding the real roots of a quadratic equation		
Skill B: APPLICATIONS Using simple quadratic equations to solve geometry problems		
Skill C: Finding the imaginary roots of a quadratic equation		
8.3 Solving Quadratic Equations by Completing the Square		10.3, 10.4
Skill A: Solving equations of the form $a(x - r)^2 = s$		
Skill B: Solving $ax^2 + bx + c = 0$ by completing the square		
Skill C: APPLICATIONS Solving real-world problems using completing the square		
8.4 Solving Quadratic Equations by Using the Quadratic Formula		10.5
Skill A: Using the quadratic formula to solve quadratic equations		
Skill B: Using the discriminant $b^2 - 4ac$ to determine the number of real roots		
Skill C: Understanding the proof of the quadratic formula		
8.5 Solving Quadratic Equations and Logical Reasoning		11.6
Skill A: Choosing a method for solving a quadratic equation		
Skill B: Writing a solution process as a deductive argument		
8.6 Quadratic Patterns		11.6
Skill A: Recognizing and continuing a quadratic pattern		
Skill B: Recognizing quadratic patterns in visual representations		
Skill C: APPLICATIONS Using a system of equations in finding a quadratic pattern		
8.7 Quadratic Functions and Acceleration		9.8, 10.1, 10.2, 10.3, 10.4
Skill A: APPLICATIONS Using quadratic functions to solve problems involving falling objects that have no initial velocity		
Skill B: APPLICATIONS Using quadratic functions to solve problems involving falling objects that have an initial velocity		

CHAPTER 8

Chapter 8   **297**

Quadratic Functions and Equations

▶ What You Already Know

By now, you have gained much experience in solving linear equations. In particular, the four arithmetic operations of addition, subtraction, multiplication, and division of real numbers is key to your success here. Now it is time to extend those skills and learn new ones that will help you solve quadratic equations, equations of the form $ax^2 + bx + c$, where a, b, and c are real numbers and $a \neq 0$.

▶ What You Will Learn

First, you will learn about taking the square root of a number. This operation is quite different from the four operations mentioned above. Then you will see how to solve certain quadratic equations by using the four arithmetic operations along with taking the square root of a number.

Special and new equation-solving skills presented to you include

- factoring and applying the Zero-Product Property,

- completing the square, and

- applying the quadratic formula.

Your study of quadratic equations will lead you to

- write a solution as a deductive argument,

- use inductive reasoning to represent patterns as quadratic functions, and

- solve real-world problems involving the accelerated motion of an object under the force of gravity.

VOCABULARY	
completing the square	Pythagorean Theorem
discriminant	quadratic formula
imaginary unit	quadratic pattern
perfect square	Quotient Property of Square Roots
principal square root	second differences
Product Property of Square Roots	simplest radical form
pure imaginary number	square root

The diagram below shows how mathematical skills and mathematical reasoning are interrelated with the skills and concepts in Chapter 8. Notice that this chapter involves solving quadratic equations using different methods.

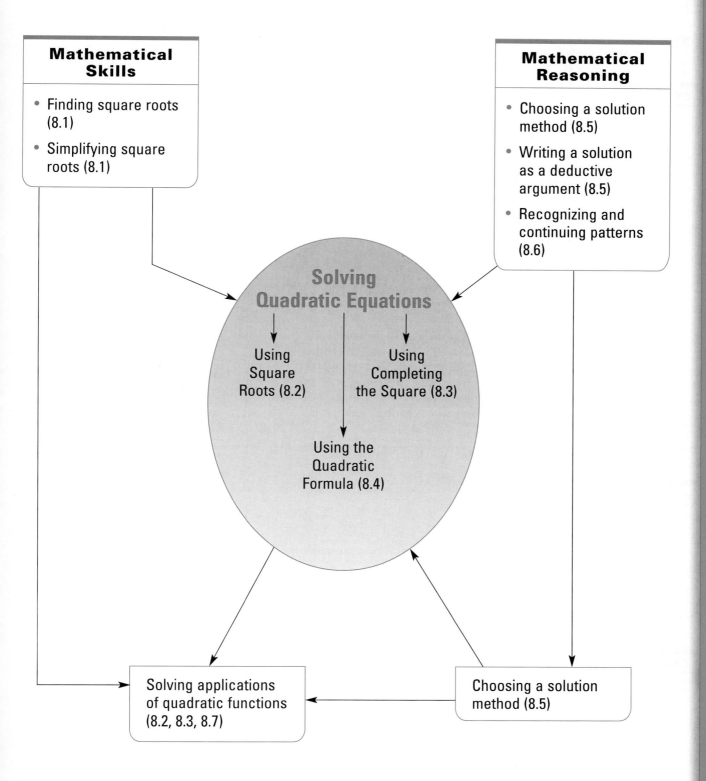

Mathematical Skills

- Finding square roots (8.1)
- Simplifying square roots (8.1)

Mathematical Reasoning

- Choosing a solution method (8.5)
- Writing a solution as a deductive argument (8.5)
- Recognizing and continuing patterns (8.6)

Solving Quadratic Equations

Using Square Roots (8.2)

Using Completing the Square (8.3)

Using the Quadratic Formula (8.4)

Solving applications of quadratic functions (8.2, 8.3, 8.7)

Choosing a solution method (8.5)

TEACHING
8.1 LESSON
Square Roots and the Equation $x^2 = k$

SKILL A ▶ *Finding and approximating square roots*

Focus: *Students calculate rational square roots, approximate irrational square roots, and solve simple quadratic equations.*

EXAMPLE 1

PHASE 1: Presenting the Example

Display the two statements shown at right. Have students complete each. [**9; 4**] Discuss their responses. Lead them to see that the second problem is, in effect, the reverse of the first. You may want to draw the squares at right to stress the following two points:

- For the first problem, they needed to calculate $3 \cdot 3$. Because of this geometric interpretation, multiplying a number by itself is called finding the *square* of the number.

- For the second problem, they needed to find which number gives the result 16 when multiplied by itself. This operation is called finding a *square root* of the number.

Now turn from the geometric model to an algebraic discussion. Ask if there is only one square of 3. [**Yes, because the product $3 \cdot 3$ is unique.**] Ask if there is only one square root of 16. [**No, because $4 \cdot 4$ and $(-4)(-4)$ each equal 16.**] Note that 4 is called the *positive square root* of 16, and -4 is the *negative square root* of 16. Point out that the negative square root did not arise in the geometric problem because a negative length has no meaning. Then discuss Example 1.

If the length of one side of a square is 3, then its area is ____?____.

If the area of a square is 16, then the length of one side is ____?____.

given: length of side = 3
area = $3 \cdot 3 = 9$

given: area = 16
length of side = $\sqrt{16} = 4$

EXAMPLE 2

PHASE 1: Presenting the Example

Display $\sqrt{64}$ and ask students to find its value. [**8**] Repeat the question with $\sqrt{58}$. In this case, they most likely will be unsure how to proceed, or some will try finding products such as $(7.5)(7.5)$. Use this activity as a lead-in to Example 2. Stress the statement: *If a whole number is not a perfect square, its square roots are irrational numbers.*

EXAMPLE 3

PHASE 2: Providing Guidance for **TRY THIS**

Watch for students whose response to part b is "-7.94 and 7.94" rather than "about -7.94 and 7.94." Suggest that they calculate $(-7.94)^2$ and $(7.94)^2$. Note that each result is exactly 63.0436. To further emphasize the distinction, display the statements at right.

$$-\sqrt{63.0436} = -7.94$$
$$-\sqrt{63} \approx -7.94$$
$$\sqrt{63.0436} = 7.94$$
$$\sqrt{63} \approx 7.94$$

---

300   Chapter 8   Quadratic Functions and Equations

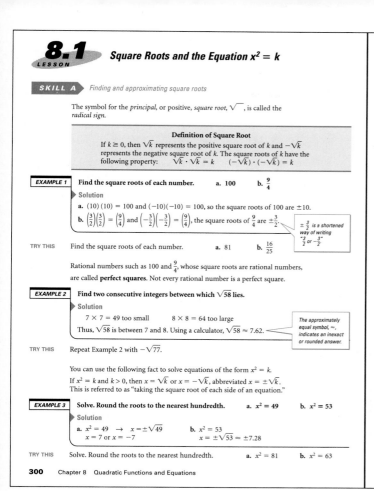

8.1 Square Roots and the Equation $x^2 = k$

LESSON

SKILL A *Finding and approximating square roots*

The symbol for the *principal*, or positive, *square root*, $\sqrt{}$, is called the *radical sign*.

> **Definition of Square Root**
> If $k \geq 0$, then $\sqrt{k}$ represents the positive square root of k and $-\sqrt{k}$ represents the negative square root of k. The square roots of k have the following property: $\sqrt{k} \cdot \sqrt{k} = k$ $\quad (-\sqrt{k}) \cdot (-\sqrt{k}) = k$

EXAMPLE 1 Find the square roots of each number. **a.** 100 **b.** $\frac{9}{4}$

▶ **Solution**
a. $(10)(10) = 100$ and $(-10)(-10) = 100$, so the square roots of 100 are ± 10.
b. $\left(\frac{3}{2}\right)\left(\frac{3}{2}\right) = \left(\frac{9}{4}\right)$ and $\left(-\frac{3}{2}\right)\left(-\frac{3}{2}\right) = \left(\frac{9}{4}\right)$, the square roots of $\frac{9}{4}$ are $\pm\frac{3}{2}$.

$\pm\frac{3}{2}$ is a shortened way of writing $\frac{-3}{2}$ or $-\frac{3}{2}$.

TRY THIS Find the square roots of each number. **a.** 81 **b.** $\frac{16}{25}$

Rational numbers such as 100 and $\frac{9}{4}$, whose square roots are rational numbers, are called **perfect squares**. Not every rational number is a perfect square.

EXAMPLE 2 Find two consecutive integers between which $\sqrt{58}$ lies.

▶ **Solution**
$7 \times 7 = 49$ too small $\qquad 8 \times 8 = 64$ too large
Thus, $\sqrt{58}$ is between 7 and 8. Using a calculator, $\sqrt{58} \approx 7.62$.

The approximately equal symbol, $\approx$, indicates an inexact or rounded answer.

TRY THIS Repeat Example 2 with $-\sqrt{77}$.

You can use the following fact to solve equations of the form $x^2 = k$.
If $x^2 = k$ and $k > 0$, then $x = \sqrt{k}$ or $x = -\sqrt{k}$, abbreviated $x = \pm\sqrt{k}$. This is referred to as "taking the square root of each side of an equation."

EXAMPLE 3 Solve. Round the roots to the nearest hundredth. **a.** $x^2 = 49$ **b.** $x^2 = 53$

▶ **Solution**
a. $x^2 = 49 \rightarrow x = \pm\sqrt{49}$
$\quad x = 7$ or $x = -7$
b. $x^2 = 53$
$\quad x = \pm\sqrt{53} \approx \pm 7.28$

TRY THIS Solve. Round the roots to the nearest hundredth. **a.** $x^2 = 81$ **b.** $x^2 = 63$

EXERCISES

KEY SKILLS

Find the square roots of each number.
1. 121 **2.** 4 **3.** 900 **4.** 169

PRACTICE

Find the square roots of each number. If the square roots are not rational numbers, then approximate the roots to the nearest hundredth.
5. 144 **6.** 50 **7.** 1 **8.** 56
9. 0.25 **10.** 2 **11.** $\frac{121}{64}$ **12.** 5

Find two consecutive integers between which each square root lies. Use a calculator to approximate the root to the nearest hundredth.
13. $\sqrt{68}$ **14.** $-\sqrt{160}$ **15.** $\sqrt{33}$ **16.** $-\sqrt{601}$

List all integers that are perfect squares that lie between each given pair of numbers.
17. 1 and 9, inclusive **18.** 10 and 20, inclusive **19.** 40 and 70, inclusive
20. 80 and 100, inclusive **21.** 200 and 400, inclusive **22.** 600 and 1000, inclusive

Solve each equation. Round irrational roots to the nearest hundredth.
23. $x^2 = 144$ **24.** $x^2 = 14$ **25.** $x^2 = 42$ **26.** $x^2 = 400$

27. Critical Thinking Show that there are no perfect squares between 961 and 1024.

28. Critical Thinking Let a be a real number. Show that the solutions to $x^2 = a^2$ are $-a$ and a by factoring $x^2 - a^2 = 0$.

29. Critical Thinking Show that the solutions to $x^2 - 5x + 4 = 0$ are perfect squares.

MIXED REVIEW

Write the prime factorization of each number. (7.1 Skill A)
30. 196 **31.** 725 **32.** 375 **33.** 2000

Simplify. (6.1 Skill A)
34. $1^4 \cdot 2^3$ **35.** $7^2 \cdot 7^2$ **36.** $3^5 \cdot 3$ **37.** $3^2 \cdot 5^0$

Answers for Lesson 8.1 Skill A

Try This Example 1
a. ± 9 **b.** $\pm\frac{4}{5}$

Try This Example 2
-8 and -9; -8.77

Try This Example 3
a. ± 9 **b.** ± 7.94

1. ± 11
2. ± 2
3. ± 30
4. ± 13
5. ± 12
6. ± 7.07
7. ± 1
8. ± 7.48
9. ± 0.5
10. ± 1.41
11. $\pm\frac{11}{8}$
12. ± 2.24
13. 8 and 9; 8.25
14. -12 and -13; -12.65

15. 5 and 6; 5.74
16. -24 and -25; -24.52
17. 1, 4, 9
18. 16
19. 49, 64
20. 81, 100
21. 225, 256, 289, 324, 361, 400
22. 625, 676, 729, 784, 841, 900, 961
23. $x = \pm 12$
24. $x = \pm 3.74$
25. $x = \pm 6.48$
26. $x = \pm 20$
27. Since $31^2 = 961$ and $32^2 = 1024$, there are no perfect squares between 961 and 1024.
28. If $x^2 = a^2$, then $x^2 - a^2 = 0$. Thus, $(x + a)(x - a) = 0$. By the Zero-Product Property, the solutions are $-a$ and a.
29. If $x^2 - 5x + 4 = 0$, then $(x - 1)(x - 4) = 0$. By the Zero-Product Property, the solutions are 1 and 4. Since $1 \times 1 = 1$ and $2 \times 2 = 4$, the solutions are perfect squares.

30. $2^2 \cdot 7^2$
31. $5^2 \cdot 29$
32. $3 \cdot 5^3$
33. $2^4 \cdot 5^3$
34. 8
35. 2401
36. 729
37. 9

Simplifying square roots of numbers using properties of square roots

Focus: *Students use the Product Property of Square Roots and the Quotient Property of Square Roots to write expressions in simplest radical form.*

EXAMPLE 1

PHASE 1: Presenting the Example

Display the expressions given in Column I at right. Have students copy them, filling in the blanks with the correct values. Be sure they multiply first and then take the square root. [**Answers are shown in red at right.**]

Now display Column II to the right of Column I, as shown. Have students copy it and fill in the blanks. This time, be sure that they take the square roots first and then multiply. [**Answers are shown in red at right.**]

Column I		Column II	
$\sqrt{4 \cdot 9} =$	6	$\sqrt{4} \cdot \sqrt{9} =$	6
$\sqrt{16 \cdot 4} =$	8	$\sqrt{16} \cdot \sqrt{4} =$	8
$\sqrt{9 \cdot 16} =$	12	$\sqrt{9} \cdot \sqrt{16} =$	12
$\sqrt{4 \cdot 25} =$	10	$\sqrt{4} \cdot \sqrt{25} =$	10
$\sqrt{25 \cdot 9} =$	15	$\sqrt{25} \cdot \sqrt{9} =$	15

Note that the expressions in each row have the same value. Lead students to the following observation: *The square root of the product of two positive numbers is equal to the product of the square roots.* Discuss the algebraic statement of this rule, the Product Property of Square Roots, which appears near the top of the textbook page.

Ask students how they might apply the rule to find the value of $\sqrt{324}$. Lead them to see that, by rewriting $\sqrt{324}$ as $\sqrt{4 \cdot 81}$, you can find the square root quickly by calculating $\sqrt{4} \cdot \sqrt{81} = 2 \cdot 9 = 18$.

Now discuss Example 1. Point out that, in this case, it is possible to find a rational value for $\sqrt{4}$ but not for $\sqrt{3}$. Therefore, you can only rewrite the expression in its simplest radical form, which is $2\sqrt{3}$.

PHASE 2: Providing Guidance for TRY THIS

Some students may calculate $\sqrt{9} \cdot \sqrt{8}$ and give the response $3\sqrt{8}$. Point out that $8 = 4 \cdot 2$, and so $\sqrt{8}$ can be further simplified as $2\sqrt{2}$. Therefore, $3\sqrt{8}$ is not the simplest radical form of the expression.

EXAMPLE 2

PHASE 1: Presenting the Example

Using the expressions given at right, parallel the presentation suggested for Example 1. Students should arrive at the following conclusion: *The square root of a quotient of two positive numbers is equal to the quotient of the square roots.* The algebraic statement of this rule is the Quotient Property of Square Roots, which appears in the box near the middle of the page.

Column I		Column II	
$\sqrt{64 \div 4} =$	4	$\sqrt{64} \div \sqrt{4} =$	4
$\sqrt{100 \div 25} =$	2	$\sqrt{100} \div \sqrt{25} =$	2
$\sqrt{144 \div 4} =$	6	$\sqrt{144} \div \sqrt{4} =$	6
$\sqrt{144 \div 9} =$	4	$\sqrt{144} \div \sqrt{9} =$	4
$\sqrt{225 \div 25} =$	3	$\sqrt{225} \div \sqrt{25} =$	3

Simplifying square roots of numbers using properties of square roots

Using properties of square roots, you can simplify a radical expression. A square-root expression is in **simplest radical form** when all of the following are true.

1. There are no perfect squares under the radical sign.
2. There are no fractions under the radical sign.
3. There are no radical expressions in the denominator.

One property of square roots that can help simplify expressions involving square roots is the Product Property of Square Roots.

Product Property of Square Roots
If a and b represent positive real numbers, then $\sqrt{ab} = \sqrt{a} \cdot \sqrt{b}$.

When you simplify a square root, you may need to take the square root of a perfect square. Use the following fact.

If x is a real number, then $\sqrt{x^2} = |x|$. If $x > 0$, then $\sqrt{x^2} = x$.

For example, $\sqrt{4^2} = 4$ and $\sqrt{(-3)^2} = |-3| = 3$.

EXAMPLE 1 Write $\sqrt{12}$ in simplest radical form.

▶ **Solution**
$\sqrt{12} = \sqrt{4 \cdot 3} = \sqrt{2^2 \cdot 3} = \sqrt{2^2} \cdot \sqrt{3} = 2\sqrt{3}$ ◀—— Apply the Product Property of Square Roots.
In simplest radical form, $\sqrt{12} = 2\sqrt{3}$.

TRY THIS   Write $\sqrt{72}$ in simplest radical form.

Another property of square roots that you can use to simplify a square root is the Quotient Property of Square Roots.

Quotient Property of Square Roots
If a and b represent positive real numbers, then $\sqrt{\dfrac{a}{b}} = \dfrac{\sqrt{a}}{\sqrt{b}}$.

EXAMPLE 2 Write $\sqrt{\dfrac{5}{16}}$ in simplest radical form.

▶ **Solution**
$\sqrt{\dfrac{5}{16}} = \dfrac{\sqrt{5}}{\sqrt{16}} = \dfrac{\sqrt{5}}{\sqrt{4^2}} = \dfrac{\sqrt{5}}{4}$ ◀—— Apply the Quotient Property of Square Roots.

In simplest radical form, $\sqrt{\dfrac{5}{16}} = \dfrac{\sqrt{5}}{4}$.

TRY THIS   Write $\sqrt{\dfrac{20}{49}}$ in simplest radical form.

EXERCISES

KEY SKILLS

Complete each simplification.

1. $\sqrt{288} = \sqrt{12^2 \cdot 2}$

2. $\sqrt{75} = \sqrt{5^2 \cdot 3}$

3. $\sqrt{98} = \sqrt{7^2 \cdot 2}$

4. $\sqrt{\dfrac{5}{36}} = \dfrac{\sqrt{5}}{\sqrt{6^2}}$

5. $\sqrt{\dfrac{7}{64}} = \dfrac{\sqrt{7}}{\sqrt{8^2}}$

6. $\sqrt{\dfrac{1}{25}} = \dfrac{\sqrt{1}}{\sqrt{5^2}}$

PRACTICE

Write each expression in simplest radical form.

7. $\sqrt{108}$
8. $\sqrt{245}$
9. $\sqrt{252}$
10. $\sqrt{192}$

11. $\sqrt{\dfrac{3}{25}}$
12. $\sqrt{\dfrac{7}{144}}$
13. $\sqrt{\dfrac{2}{49}}$
14. $\sqrt{\dfrac{5}{81}}$

15. $\sqrt{405}$
16. $\sqrt{363}$
17. $\sqrt{1200}$
18. $\sqrt{675}$

19. $\sqrt{\dfrac{72}{49}}$
20. $\sqrt{\dfrac{125}{81}}$
21. $\sqrt{468}$
22. $\sqrt{828}$

23. $\sqrt{\dfrac{343}{25}}$
24. $\sqrt{\dfrac{567}{16}}$
25. $\sqrt{3564}$
26. $\sqrt{2352}$

Find and simplify each product.

27. $\sqrt{\dfrac{15}{16}} \cdot \sqrt{\dfrac{32}{25}}$

28. $\sqrt{\dfrac{7}{9}} \cdot \sqrt{\dfrac{27}{49}}$

29. $\sqrt{\dfrac{2}{25}} \cdot \sqrt{\dfrac{2}{9}} \cdot \sqrt{\dfrac{1}{16}}$

30. $\sqrt{\dfrac{3}{16}} \cdot \sqrt{\dfrac{4}{25}} \cdot \sqrt{\dfrac{5}{36}}$

31. **Critical Thinking** Let a and b represent positive real numbers. Prove that $\sqrt{\dfrac{a}{b}} \cdot \sqrt{\dfrac{a}{b}} = \dfrac{a}{b}$.

MIXED REVIEW

Simplify. (6.5 Skill A)

32. $(-3r)(-4r)$
33. $s^3(-5s^3)$
34. $2g^2(3g^2)(-3g)$
35. $-h^2(h^2)(-h^2)$
36. $(2r^2t^3)^2$
37. $(-2a^2b^4)^3$
38. $(4ns^3)(6n^2s^3)$
39. $(-a^2z^4)(-a^2z^2)$

Simplify. (6.5 Skill B)

40. $\dfrac{24n^3}{n^2}$
41. $\dfrac{26z^5}{39z^2}$
42. $\dfrac{(4d^3m^2)^5}{(16d^5m^3)^3}$
43. $\dfrac{(5a^3b^3)^3}{(25a^3b^4)^3}$

Answers for Lesson 8.1 Skill B

Try This Example 1
$6\sqrt{2}$

Try This Example 2
$\dfrac{\sqrt{10}}{7}$

1. $12\sqrt{2}$
2. $5\sqrt{3}$
3. $7\sqrt{2}$
4. $\dfrac{\sqrt{5}}{6}$
5. $\dfrac{\sqrt{7}}{8}$
6. $\dfrac{1}{5}$
7. $6\sqrt{3}$
8. $7\sqrt{5}$
9. $6\sqrt{7}$
10. $8\sqrt{3}$
11. $\dfrac{\sqrt{3}}{5}$

12. $\dfrac{\sqrt{7}}{12}$
13. $\dfrac{\sqrt{2}}{7}$
14. $\dfrac{\sqrt{5}}{9}$
15. $9\sqrt{5}$
16. $11\sqrt{3}$
17. $20\sqrt{3}$
18. $15\sqrt{3}$
19. $\dfrac{6\sqrt{2}}{7}$
20. $\dfrac{5\sqrt{5}}{9}$
★21. $6\sqrt{13}$
★22. $6\sqrt{23}$
★23. $\dfrac{7\sqrt{7}}{5}$
★24. $\dfrac{9\sqrt{7}}{4}$
★25. $18\sqrt{11}$

★26. $28\sqrt{3}$
27. $\dfrac{\sqrt{30}}{5}$
28. $\dfrac{\sqrt{21}}{7}$
29. $\dfrac{1}{30}$
30. $\dfrac{\sqrt{15}}{60}$
31. $\sqrt{\dfrac{a}{b}} \cdot \sqrt{\dfrac{a}{b}} = \sqrt{\dfrac{a}{b} \cdot \dfrac{a}{b}}$
$= \sqrt{\dfrac{a^2}{b^2}} = \dfrac{\sqrt{a^2}}{\sqrt{b^2}} = \dfrac{a}{b}$
32. $12r^2$
33. $-5s^6$
34. $-18g^5$
35. h^6
36. $4r^4t^6$
37. $-8a^6b^{12}$
38. $24n^3s^6$
39. a^4z^6
40. $24n$

41. $\dfrac{2z^3}{3}$
42. $\dfrac{m}{4}$
43. $\dfrac{1}{125b^3}$

★ **Advanced Exercises**

Focus: *Students learn the meaning of the exponent $\frac{1}{2}$ and then use it in combination with properties of exponents to prove statements about square roots.*

EXAMPLE

PHASE 1: Presenting the Example

Display statement **1** at right. Ask students to copy it, filling in the blank with the correct value of the expression. [5] Discuss their responses, stressing the following reasoning: The value of $\sqrt{5} \cdot \sqrt{5}$ must be 5 because of the definition of a square root.

Now display statement **2** directly beneath statement **1**, as shown. Although this second statement will not be familiar to students, ask them to copy it and fill in the blank with what they believe to be the value of the expression. If they seem perplexed, suggest that they consider the Product Property of Exponents: $a^m \cdot a^n = a^{m+n}$. They should quickly arrive at the conclusion that the value must be $5^{\frac{1}{2}+\frac{1}{2}} = 5^1 = 5$.

Tell students to examine the two statements. Ask them to make a conjecture about the meaning of $5^{\frac{1}{2}}$. Lead them to the conclusion that $5^{\frac{1}{2}}$ must be equal to $\sqrt{5}$. Have students read the definition of $a^{\frac{1}{2}}$ that appears near the top of the textbook page, and answer any questions that they may have. Then discuss the Example.

1. $\sqrt{5} \cdot \sqrt{5} = \underline{\quad ? \quad}$

2. $5^{\frac{1}{2}} \cdot 5^{\frac{1}{2}} = \underline{\quad ? \quad}$

PHASE 2: Providing Guidance for TRY THIS

Suggest that, before writing their proofs, students write a brief summary of the properties of exponents so that it is available for easy reference.

PHASE 3: ASSESSMENT AND CLOSURE for Lesson 8.1

- Give students the list at right. Ask them to explain how the items in the list are related to each other. [**sample: $\sqrt{48}$ and $-\sqrt{48}$ are the two solutions to the equation $x^2 = 48$. The simplest radical forms of these numbers are $4\sqrt{3}$ and $-4\sqrt{3}$, respectively. Rational approximations of these numbers are 6.93 and -6.93, respectively.**]

$\sqrt{48}$	$-\sqrt{48}$
$4\sqrt{3}$	$-4\sqrt{3}$
6.93	-6.93
$x^2 = 48$	

- Students have relatively little difficulty with natural-number exponents because they are able to employ visual images. That is, to understand a^n when n is a natural number, they need only visualize a product in which a appears as a factor n times. In contrast, the meaning of a rational exponent is somewhat abstract. To help students become comfortable with the concept, this initial lesson in rational exponents is confined to the meaning of $a^{\frac{1}{2}}$. Students will work with other rational exponents in Chapter 10.

 For a Lesson 8.1 Quiz, see Assessment Resources page 78.

Radical expressions can be represented in an equivalent form using fractional exponents. The following definition shows the form for $\sqrt{a}$.

Definition of the Exponent $\frac{1}{2}$

If a is positive, then $a^{\frac{1}{2}}$ is defined as $\sqrt{a}$.

The following is an explanation of why this definition works mathematically.

By the definition of square root, any number whose square is a is represented as $\sqrt{a}$. And, by the Product Property of Exponents, $\left(a^{\frac{1}{2}}\right)\left(a^{\frac{1}{2}}\right) = a^1 = a$.

Therefore, because $\left(a^{\frac{1}{2}}\right)^2 = a$, $a^{\frac{1}{2}} = \sqrt{a}$. Some numerical examples are:

$$4^{\frac{1}{2}} = \sqrt{4} = 2 \qquad 9^{\frac{1}{2}} = \sqrt{9} = 3 \qquad \left(\frac{9}{4}\right)^{\frac{1}{2}} = \sqrt{\frac{9}{4}} = \frac{3}{2}$$

All of the laws of exponents you learned in Chapter 6 are true when the exponent is $\frac{1}{2}$ and the bases are positive.

EXAMPLE

Suppose $a > 0$. Using laws of exponents, prove that $\sqrt{4a} = 2\sqrt{a}$.

Solution

$$\sqrt{4a} = (4a)^{\frac{1}{2}} \qquad \longleftarrow \text{ Definition of the exponent } \frac{1}{2}$$
$$= (4)^{\frac{1}{2}}a^{\frac{1}{2}} \qquad \longleftarrow \text{ Power-of-a-Product Law of Exponents}$$
$$= (2^2)^{\frac{1}{2}}a^{\frac{1}{2}} \qquad \longleftarrow \text{ Definition of positive integer exponent}$$
$$= 2^{\left(2 \times \frac{1}{2}\right)}a^{\frac{1}{2}} \qquad \longleftarrow \text{ Power-of-a-Power Law of Exponents}$$
$$= 2^1 a^{\frac{1}{2}} \qquad \longleftarrow \text{ Multiplicative inverse}$$
$$= 2\sqrt{a} \qquad \longleftarrow \text{ Definition of the exponents 1 and } \frac{1}{2}$$

Thus, $\sqrt{4a} = 2\sqrt{a}$.

TRY THIS Suppose $a > 0$. Using laws of exponents, prove that $\sqrt{9a} = 3\sqrt{a}$.

You prove the Product Property of Square Roots as shown below.
For $a \geq 0$ and $b \geq 0$,

$$\sqrt{ab} = (ab)^{\frac{1}{2}} \qquad \longleftarrow \text{ Definition of the exponent } \frac{1}{2}$$
$$= a^{\frac{1}{2}}b^{\frac{1}{2}} \qquad \longleftarrow \text{ Power-of-a-Product Law of Exponents}$$
$$= \sqrt{a}\sqrt{b} \qquad \longleftarrow \text{ Definition of the exponent } \frac{1}{2}$$

EXERCISES

KEY SKILLS

Evaluate each expression.

1. $49^{\frac{1}{2}}$ 2. $121^{\frac{1}{2}}$ 3. $144^{\frac{1}{2}}$ 4. $400^{\frac{1}{2}}$

PRACTICE

The proof below states that if a and b are positive numbers, then $\left(\sqrt{ab}\right)^2 = ab$. In Exercises 5–8, give a reason for each step in the proof.

Step 1   Step 2   Step 3   Step 4

$$\left(\sqrt{ab}\right)^2 = \left((ab)^{\frac{1}{2}}\right)^2 = (ab)^{\frac{1}{2} \cdot 2} = (ab)^1 = ab$$

5. Step 1   6. Step 2   7. Step 3   8. Step 4

The proof below states that if a is positive and n is an integer, then $\sqrt{a^n} = \left(\sqrt{a}\right)^n$. In Exercises 9–13, give a reason for each step in the proof.

Step 1   Step 2   Step 3   Step 4   Step 5

$$\sqrt{a^n} = (a^n)^{\frac{1}{2}} = a^{n \cdot \frac{1}{2}} = a^{\frac{1}{2} \cdot n} = \left(a^{\frac{1}{2}}\right)^n = \left(\sqrt{a}\right)^n$$

9. Step 1   10. Step 2   11. Step 3   12. Step 4   13. Step 5

Use the properties of exponents with the exponent $\frac{1}{2}$ to prove each statement.

14. If a is a positive real number, then $\sqrt{25a} = 5\sqrt{a}$.

15. If a and b are positive real numbers, then $\sqrt{\frac{a}{b}} = \frac{\sqrt{a}}{\sqrt{b}}$.

16. If a is a positive real number, then $\frac{\sqrt{4a^2}}{\sqrt{9a^2}} = \frac{2}{3}$.

17. If a and b are positive real numbers, then $\frac{\sqrt{a^4b^4}}{\sqrt{a^2b^2}} = ab$.

MIXED REVIEW

Solve each system. (5.3 Skills A and B)

18. $\begin{cases} 2x + 3y = 6 \\ 2x - 3y = 6 \end{cases}$

19. $\begin{cases} 2x + y = 5 \\ 4x + y = 6 \end{cases}$

20. $\begin{cases} 5x + 2y = -7 \\ x + 3y = 9 \end{cases}$

21. $\begin{cases} -3a + 11b = 0 \\ 4a - 11b = 3 \end{cases}$

22. $\begin{cases} 5a + 6b = 4 \\ 4a + 5b = 0 \end{cases}$

23. $\begin{cases} -7x - 6y = 14 \\ 8x + 7y = 5 \end{cases}$

Answers for Lesson 8.1 Skill C

Try This Example

$$\sqrt{9a} = (9a)^{\frac{1}{2}} = (9)^{\frac{1}{2}}a^{\frac{1}{2}}$$
$$= (3^2)^{\frac{1}{2}}a^{\frac{1}{2}}$$
$$= 3^{\left(2 \times \frac{1}{2}\right)}a^{\frac{1}{2}} = 3a^{\frac{1}{2}} = 3\sqrt{a}$$

1. 7
2. 11
3. 12
4. 20
5. definition of the exponent $\frac{1}{2}$
6. Power-of-a-Power Property of Exponents
7. Multiplicative Inverse
8. definition of the exponent 1
9. definition of the exponent $\frac{1}{2}$
10. Power-of-a-Power Property of Exponents
11. Commutative Property of Multiplication
12. Power-of-a-Power Property of Exponents
13. definition of the exponent $\frac{1}{2}$

14. $\sqrt{25a} = (25a)^{\frac{1}{2}} = (25)^{\frac{1}{2}}a^{\frac{1}{2}}$
$$= (5^2)^{\frac{1}{2}}a^{\frac{1}{2}}$$
$$= 5^{\left(2 \times \frac{1}{2}\right)}a^{\frac{1}{2}} = 5a^{\frac{1}{2}} = 5\sqrt{a}$$

15. $\sqrt{\frac{a}{b}} = \sqrt{a \cdot \frac{1}{b}} = \sqrt{a} \cdot \sqrt{\frac{1}{b}} = \sqrt{a} \cdot \frac{1}{\sqrt{b}} = \frac{\sqrt{a}}{\sqrt{b}}$

★16. $\frac{\sqrt{4a^2}}{\sqrt{9a^2}} = \sqrt{\frac{4a^2}{9a^2}} = \sqrt{\frac{4}{9}} \cdot \sqrt{\frac{a^2}{a^2}} = \frac{2}{3}$

★17. $\frac{\sqrt{a^4b^4}}{\sqrt{a^2b^2}} = \sqrt{\frac{a^4b^4}{a^2b^2}}$
$$= \sqrt{a^2b^2} = \sqrt{a^2}\sqrt{b^2} = a \cdot b = ab$$

18. $(3, 0)$
19. $\left(\frac{1}{2}, 4\right)$
20. $(-3, 4)$
21. $\left(3, \frac{9}{11}\right)$
22. $(20, -16)$
23. $(-128, 147)$

★ **Advanced Exercises**

TEACHING
8.2
LESSON

Solving Equations of the Form $ax^2 + c = 0$

SKILL A *Finding the real roots of a quadratic equation*

Focus: *Students find real-number solutions to equations of the form $ax^2 + c = 0$ by applying properties of equality and using the definition of square root.*

EXAMPLE 1

PHASE 1: Presenting the Example

Display the equation $2d - 32 = 0$. Tell students to solve it using properties of equality. [$d = 16$] Have a volunteer write the steps of the solution on the chalkboard or on a transparency. [**A sample is shown at right.**]

$$2d - 32 = 0$$
$$2d - 32 + 32 = 0 + 32$$
$$2d = 32$$
$$\frac{2d}{2} = \frac{32}{2}$$
$$d = 16$$
$$\downarrow$$

Keep the volunteer's work displayed for all to see. As the class watches, insert the exponent 2 to the right of d throughout, as shown at right. Ask students how the new display is different. Lead them to the following conclusions: First, the initial equation is now $2d^2 - 32 = 0$, which is quadratic rather than linear. Second, the "bottom line" no longer shows the solution to the equation. Elicit the fact that another step is needed, namely, to identify the solutions as $d = -4$ or $d = 4$.

$$2d^2 - 32 = 0$$
$$2d^2 - 32 + 32 = 0 + 32$$
$$2d^2 = 32$$
$$\frac{2d^2}{2} = \frac{32}{2}$$
$$d^2 = 16$$

PHASE 2: Providing Guidance for **TRY THIS**

Some students might solve the equation by factoring, as shown below.

$$10d^2 - 1000 = 0 \rightarrow 10(d^2 - 100) = 0 \rightarrow 10(d + 10)(d - 10) = 0 \rightarrow d = -10 \text{ or } d = 10$$

This is a valid method and should be acknowledged as such. However, note that the solutions are integers. Students may have difficulty applying this method in Example 2, where the solutions are irrational numbers.

EXAMPLE 2

PHASE 2: Providing Guidance for **TRY THIS**

Students who try to solve by factoring will obtain $4(t^2 - 24) = 0$ and be faced with a problem, since $t^2 - 24$ is not factorable over the integers. If you wish, you can show them how to proceed by factoring over the real numbers:

$$4(t^2 - 24) = 0 \rightarrow 4(t + \sqrt{24})(t - \sqrt{24}) = 0 \rightarrow t = -\sqrt{24} \text{ or } t = \sqrt{24} \rightarrow t = -2\sqrt{6} \text{ or } t = 2\sqrt{6}$$

EXAMPLE 3

PHASE 1: Presenting the Example

Display $4z^2 - 109 = 91$. Ask students how this equation differs from the equations in Examples 1 and 2. They should observe that, in this case, the right side is not zero. Elicit their ideas about how to proceed with the solution process. Then discuss the steps that are given on the textbook page.

8.2 Solving Equations of the Form $ax^2 + c = 0$

LESSON

SKILL A *Finding the real roots of a quadratic equation*

In Lesson 8.1, you learned that the solutions to an equation of the form $x^2 = k$ are $x = \pm\sqrt{k}$. In this lesson, you will learn how to solve equations of the form $ax^2 + c = 0$ by first isolating x^2 on one side of the equation and *then* taking the square root of each side.

EXAMPLE 1 Solve $2d^2 - 32 = 0$.

▶ **Solution**

$$2d^2 - 32 = 0$$
$$2d^2 = 32$$
$$d^2 = 16$$
$$d = \pm\sqrt{16} \quad \longleftarrow \text{Take the square root of each side.}$$
$$d = \pm 4$$

The solutions to $2d^2 - 32 = 0$ are 4 and −4.

TRY THIS Solve $10d^2 - 1000 = 0$.

EXAMPLE 2 Solve $3t^2 - 54 = 0$. Write the solutions in simplest radical form.

▶ **Solution**

$$3t^2 - 54 = 0$$
$$3t^2 = 54$$
$$t^2 = 18$$
$$t = \pm\sqrt{18} \quad \longleftarrow \text{Take the square root of each side.}$$
$$t = \pm\sqrt{9 \cdot 2} = \pm 3\sqrt{2} \quad \longleftarrow \text{Simplify.}$$

The solutions to $3t^2 - 54 = 0$ are $3\sqrt{2}$ and $-3\sqrt{2}$.

TRY THIS Solve $4t^2 - 96 = 0$. Write the solutions in simplest radical form.

EXAMPLE 3 Solve $4z^2 - 109 = 91$. Write the solutions in simplest radical form.

▶ **Solution**

$$4z^2 - 109 = 91$$
$$4z^2 = 200$$
$$z^2 = 50$$
$$z = \pm\sqrt{50} \quad \longleftarrow \text{Take the square root of each side.}$$
$$z = \pm\sqrt{25 \cdot 2} = \pm 5\sqrt{2} \quad \longleftarrow \text{Simplify.}$$

The solutions to $4z^2 - 109 = 91$ are $5\sqrt{2}$ and $-5\sqrt{2}$.

TRY THIS Solve $3z^2 - 83 = -2$. Write the solutions in simplest radical form.

EXERCISES

KEY SKILLS

Write each equation in the form $x^2 = k$.

1. $7t^2 - 28 = 0$ 2. $-2x^2 + 50 = 0$ 3. $25z^2 - 96 = 4$ 4. $\frac{1}{2}h^2 - 2 = 0$

PRACTICE

Solve. Write the solutions in simplest radical form.

5. $x^2 - 1 = 0$ 6. $y^2 - 49 = 0$ 7. $2n^2 - 32 = 0$ 8. $3m^2 - 27 = 0$

9. $-11s^2 + 44 = 0$ 10. $-3v^2 + 75 = 0$ 11. $-\frac{1}{10}z^2 + 10 = 0$ 12. $-\frac{1}{2}m^2 + 2 = 0$

13. $7d^2 - 21 = 0$ 14. $-2y^2 + 4 = 0$ 15. $5q^2 - 35 = 0$ 16. $3k^2 - 6 = 0$

17. $6t^2 - 72 = 0$ 18. $-3p^2 + 72 = 0$ 19. $8v^2 - 96 = 0$ 20. $3h^2 - 60 = 0$

21. $3u^2 - 2 = 4$ 22. $3u^2 - 14 = 4$ 23. $-4u^2 + 36 = 4$ 24. $-2s^2 + 113 = 13$

25. $3n^2 - 104 = -8$ 26. $5z^2 - 21 = 39$ 27. $-7r^2 + 63 = 7$ 28. $-8w^2 + 107 = 11$

29. For what values of a will $av^2 - 144 = 0$ have one solution between 3 and 4 and the other solution between −4 and −3?

30. For what values of a will $av^2 - 144 = 0$ have one solution between 1 and 2 and the other solution between −2 and −1?

31. For what values of c will $3u^2 + c = 0$ have one solution between 3 and 4 and the other solution between −4 and −3?

32. **Critical Thinking** Let $ax^2 + c = 0$ and let a and c represent real numbers with $a \neq 0$.
   a. Write a formula that gives x in terms of a and c.
   b. Under what conditions will your formula give two solutions for x? one solution for x? no real solution for x?

MIXED REVIEW

Solve each equation for the specified variable. All variables represent real numbers. Assume that there is no division by 0. (2.7 Skill B)

33. $\frac{m}{n}x + a = y$; m 34. $y(x - c) = z$; x 35. $yz + 4 = xz - 4$; z

36. $x(y + b) = 2xy$; y 37. $a(x + 3) = b(x - 2)$; x 38. $x(y - n) = 2ay$; y

---

Answers for Lesson 8.2 Skill A

Try This Example 1
$d = \pm 10$

Try This Example 2
$t = \pm 2\sqrt{6}$

Try This Example 3
$z = \pm 3\sqrt{3}$

1. $t^2 = 4$
2. $x^2 = 25$
3. $z^2 = 4$
4. $h^2 = 4$
5. $x = \pm 1$
6. $y = \pm 7$
7. $n = \pm 4$
8. $m = \pm 3$
9. $s = \pm 2$
10. $v = \pm 5$
11. $z = \pm 10$
12. $m = \pm 2$
13. $d = \pm\sqrt{3}$

14. $y = \pm\sqrt{2}$
15. $q = \pm\sqrt{7}$
16. $k = \pm\sqrt{2}$
17. $t = \pm 2\sqrt{3}$
18. $p = \pm 2\sqrt{6}$
19. $v = \pm 2\sqrt{3}$
20. $h = \pm 2\sqrt{5}$
21. $u = \pm\sqrt{2}$
22. $u = \pm\sqrt{6}$
23. $u = \pm 2\sqrt{2}$
24. $s = \pm 5\sqrt{2}$
25. $n = \pm 4\sqrt{2}$
26. $z = \pm 2\sqrt{3}$
27. $r = \pm 2\sqrt{2}$
28. $w = \pm 2\sqrt{3}$
29. $9 < a < 16$
30. $36 < a < 144$
★31. $-48 < c < -27$

32. **a.** $x = \pm\sqrt{-\dfrac{c}{a}}$

   b. The solutions will be two distinct real numbers if a and c have different signs and $c \neq 0$. There will be one solution if $c = 0$. There will be no solution if a and c have the same sign.

33. $m = \dfrac{n(y - a)}{x}$

34. $x = \dfrac{z}{y} + c$

35. $z = \dfrac{-8}{y - x}$

36. $y = b$

37. $x = \dfrac{3a + 2b}{b - a}$

38. $y = \dfrac{xn}{x - 2a}$

★ **Advanced Exercises**

Focus: *Students use a quadratic equation to find the radius of a circle when the area is known. They also find the length of a leg of a right triangle when the length of the other leg and the length of the hypotenuse are known.*

EXAMPLE 1

PHASE 1: Presenting the Example

Review with students the circle area formula: The area, A, of a circle with radius r is given by $A = \pi r^2$. Remind them that a radius is a segment whose endpoints are the center of the circle and a point on the circle.

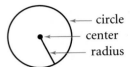

Have students use the formula to find the area of a circle with radius 5 meters. This is a straightforward calculation of $\pi \times 5^2$. If students have scientific or graphics calculators, check that they know how to use the π key. If they do not have a π key, tell them to use 3.14 to approximate π. With either method, they should find the area to be about 78.5 square meters. Tell them that they will now learn the reverse procedure. That is, given the area of a circle, they will find its radius. Then discuss Example 1.

PHASE 2: Providing Guidance for **TRY THIS**

Watch for students who approximate the radius as 4.8 meters. Most likely they used a calculator key sequence that did not evaluate $225 \div \pi$ before finding the square root. Work with them to identify a suitable key sequence.

EXAMPLE 2

PHASE 1: Presenting the Example

Review with students the definitions of right angle and right triangle: A *right angle* is an angle whose measure is 90°. A *right triangle* is a triangle with one right angle. Students most likely will be familiar with these terms from their previous study of geometry. Then discuss the terms shown on the textbook page, which may be new to them: In a right triangle, the side opposite the right angle is called the *hypotenuse*, and the other two sides are called the *legs*. Be sure students are aware that the hypotenuse of a right triangle is always the longest side and is represented by c in the Pythagorean Theorem.

Find x:

$$a^2 + b^2 = c^2$$
$$5^2 + 12^2 = x^2$$
$$25 + 144 = x^2$$
$$169 = x^2$$
$$x^2 = 169$$
$$x = -13 \text{ or } x = 13$$

The length of the hypotenuse is 13.

Example 2 requires students to apply the Pythagorean Theorem in the context of a real-world problem. If your students generally become anxious about "word problems" like this, you may want to first present the theorem in a purely geometric context. At right are two examples that you can use for this purpose. Stress the fact that, although the equation used in each situation has two solutions, only a positive solution has meaning as a length.

Find y:

$$a^2 + b^2 = c^2$$
$$4^2 + y^2 = 8^2$$
$$16 + y^2 = 64$$
$$y^2 = 48$$
$$y = -4\sqrt{3} \text{ or } y = 4\sqrt{3}$$

The length of the leg is $4\sqrt{3}$.

Some geometry problems can be solved using quadratic equations.

EXAMPLE 1

Use the formula area = $\pi \cdot$ (radius)^2 to find the radius of a circle whose area is 1000 square meters. Use a calculator to approximate the radius to the nearest tenth of a meter.

▶ Solution

$$\pi r^2 = 1000$$
$$r^2 = \frac{1000}{\pi}$$
$$r = \pm\sqrt{\frac{1000}{\pi}} \approx \pm 17.8$$

Reject −17.8 because radius cannot be negative.

The radius is about 17.8 meters.

TRY THIS   Find the radius of a circle whose area is 225 square meters.

You can often use the *Pythagorean Theorem* to solve problems involving right triangles. When two side lengths of a right triangle are known, the Pythagorean Theorem allows you to write a quadratic equation.

The Pythagorean Theorem

In any right triangle, the square of the length of the hypotenuse, c, equals the sum of the squares of the lengths, a and b, of the legs.

$$a^2 + b^2 = c^2$$

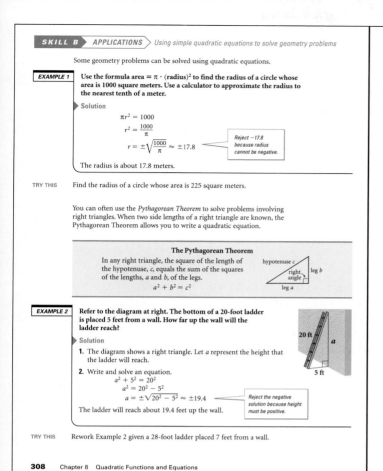

EXAMPLE 2

Refer to the diagram at right. The bottom of a 20-foot ladder is placed 5 feet from a wall. How far up the wall will the ladder reach?

▶ Solution

1. The diagram shows a right triangle. Let a represent the height that the ladder will reach.

2. Write and solve an equation.
$$a^2 + 5^2 = 20^2$$
$$a^2 = 20^2 - 5^2$$
$$a = \pm\sqrt{20^2 - 5^2} \approx \pm 19.4$$

Reject the negative solution because height must be positive.

The ladder will reach about 19.4 feet up the wall.

TRY THIS   Rework Example 2 given a 28-foot ladder placed 7 feet from a wall.

EXERCISES

Write an equation that you can use to solve each problem. Identify the variable. Do not solve.

1. What is the radius of a circle whose area is 1200 square feet?

2. If the hypotenuse of a triangle is 55 units and the length of one leg is 35 units, what is the length of the other leg?

Use the given area to find the radius of each circle. Round answers to the nearest tenth of a unit.

3. area: 500 square feet

4. area: 620 square meters

5. area: 314 square inches

6. area: 1240 square meters

Find the length of the other leg of each right triangle. Round answers to the nearest tenth of a unit.

7. hypotenuse: 24 units
   one leg: 18 units

8. hypotenuse: 35 feet
   one leg: 30 feet

9. hypotenuse: 500 inches
   one leg: 250 inches

10. hypotenuse: 625 yards
   one leg: 62.5 yards

Refer to rectangle *ABCD* at right. (The notation *AD* represents the length of the line segment with endpoints *A* and *D*.)

11. a. Write an expression for *AD*.
   b. Approximate *AD* to the nearest hundredth of a unit.

Write a simplified expression for the total perimeter of each figure. Lightly shaded rectangles are congruent and black rectangles are congruent. (6.4 Skill C)

12.

13.

14.

Answers for Lesson 8.2 Skill B

Try This Example 1
8.46 meters

Try This Example 2
27.1 feet

1. $\pi r^2 = 1200$; r represents the radius of the circle.

2. $35^2 + x^2 = 55^2$; x represents the unknown length of the other leg.

3. 12.6 feet

4. 14.0 meters

5. 10.0 inches

6. 19.9 meters

7. 15.9 units

8. 18.0 feet

9. 433.0 inches

10. 621.9 yards

11. a. $AD = \sqrt{7.16^2 - 6.32^2}$
   b. $AD \approx 3.36$

12. $20.4x$

13. $32.8x$

14. $47.6x$

Focus: *Students find imaginary-number solutions to equations of the form $ax^2 + c = 0$ by applying properties of equality and the definition of i.*

EXAMPLE 1

PHASE 1: Presenting the Example

Display statements **1, 2**, and **3** that are shown in Column I at right. Ask students to write a value of *a* that makes each statement true. [**9, 4, 1**] Discuss their responses. Then display statement **4** from Column I directly beneath the others, as shown. Ask students what value of *a* must make this statement true. They should conclude that $a = -1$.

Write Column II to the right of Column I as shown. Ask students to fill in the blanks to make each statement true. They should have no difficulty giving the first three responses as 3, 2, and 1, respectively. However, they probably will be baffled by row **4**. Discuss the cause of the confusion. Lead them to observe that there is no real number *a* for which $a \cdot a$ is negative.

Tell students that $\sqrt{-1}$ has no real-number value because people originally could find no real-world meaning for it. So $\sqrt{-1}$ was called *imaginary* and was given the special name *i*. Point out that many real-world applications of $\sqrt{-1}$ have since been discovered, and it is now known that this "imaginary" number does indeed exist. Use this discussion as a lead-in to Example 1.

Column I	Column II
1. $\sqrt{a} \cdot \sqrt{a} = 9$	$\sqrt{9} = $ _____
2. $\sqrt{a} \cdot \sqrt{a} = 4$	$\sqrt{4} = $ _____
3. $\sqrt{a} \cdot \sqrt{a} = 1$	$\sqrt{1} = $ _____
4. $\sqrt{a} \cdot \sqrt{a} = -1$	$\sqrt{-1} = $ _____

EXAMPLES 2 and 3

PHASE 2: Providing Guidance for TRY THIS

Students should continue the practice of checking possible solutions in the original equation. However, they may be uncomfortable with checking these solutions because the process involves arithmetic with imaginary numbers. You may want to do at least one solution check with them. A check for the Example 2 Try This exercise is shown at right.

Is $5i$ a solution?

$$a^2 + 25 = 0$$
$$(5i)^2 + 25 \overset{?}{=} 0$$
$$5^2 \cdot i^2 + 25 \overset{?}{=} 0$$
$$(25)(-1) + 25 \overset{?}{=} 0$$
$$-25 + 25 \overset{?}{=} 0$$
$$0 = 0 ✔$$

Is $-5i$ a solution?

$$a^2 + 25 = 0$$
$$(-5i)^2 + 25 \overset{?}{=} 0$$
$$(-5)^2 \cdot i^2 + 25 \overset{?}{=} 0$$
$$(25)(-1) + 25 \overset{?}{=} 0$$
$$-25 + 25 \overset{?}{=} 0$$
$$0 = 0 ✔$$

PHASE 3: ASSESSMENT AND CLOSURE for Lesson 8.2

- Display the equation at right. Tell students to fill in the boxes with numbers greater than 1 so that the solutions are -4 and 4. [**sample: $2d^2 + 3 = 35$**] Repeat for the solutions $-4i$ and $4i$. [**sample: $2d^2 + 35 = 3$**]

$$\Box d^2 + \Box = \Box$$

- In Lesson 8.3, students will see how they can extend the square-root method to solve equations of the form $a(x - r)^2 = s$. They will then see how this technique allows them to solve any quadratic equation with a linear term.

☞ *For a Lesson 8.2 Quiz, see* Assessment Resources *page 79.*

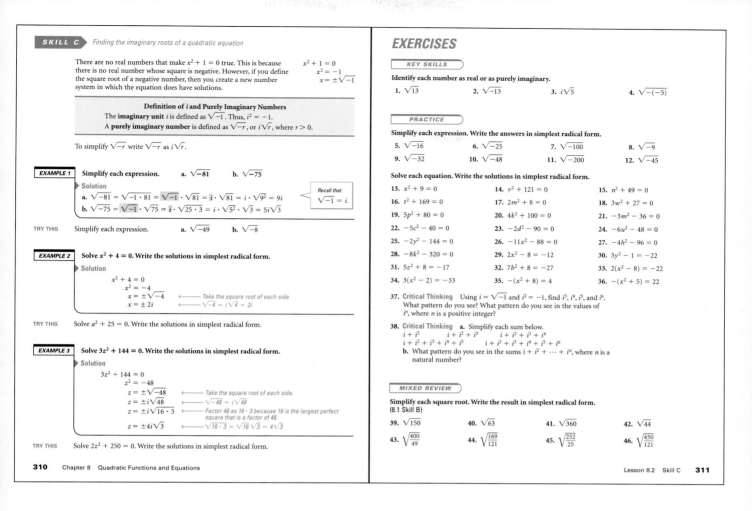

There are no real numbers that make $x^2 + 1 = 0$ true. This is because there is no real number whose square is negative. However, if you define the square root of a negative number, then you create a new number system in which the equation does have solutions.

$$x^2 + 1 = 0$$
$$x^2 = -1$$
$$x = \pm\sqrt{-1}$$

Definition of i and Purely Imaginary Numbers
The **imaginary unit** i is defined as $\sqrt{-1}$. Thus, $i^2 = -1$.
A **purely imaginary number** is defined as $\sqrt{-r}$, or $i\sqrt{r}$, where $r > 0$.

To simplify $\sqrt{-r}$ write $\sqrt{-r}$ as $i\sqrt{r}$.

EXAMPLE 1 Simplify each expression. a. $\sqrt{-81}$ b. $\sqrt{-75}$

▶ Solution
a. $\sqrt{-81} = \sqrt{-1 \cdot 81} = \sqrt{-1} \cdot \sqrt{81} = i \cdot \sqrt{81} = i \cdot \sqrt{9^2} = 9i$
b. $\sqrt{-75} = \sqrt{-1} \cdot \sqrt{75} = i \cdot \sqrt{25 \cdot 3} = i \cdot \sqrt{5^2} \cdot \sqrt{3} = 5i\sqrt{3}$

Recall that $\sqrt{-1} = i$.

TRY THIS Simplify each expression. a. $\sqrt{-49}$ b. $\sqrt{-8}$

EXAMPLE 2 Solve $x^2 + 4 = 0$. Write the solutions in simplest radical form.

▶ Solution
$$x^2 + 4 = 0$$
$$x^2 = -4$$
$$x = \pm\sqrt{-4} \quad \longleftarrow \text{Take the square root of each side.}$$
$$x = \pm 2i \quad \longleftarrow \sqrt{-4} = i\sqrt{4} = 2i$$

TRY THIS Solve $a^2 + 25 = 0$. Write the solutions in simplest radical form.

EXAMPLE 3 Solve $3z^2 + 144 = 0$. Write the solutions in simplest radical form.

▶ Solution
$$3z^2 + 144 = 0$$
$$z^2 = -48$$
$$z = \pm\sqrt{-48} \quad \longleftarrow \text{Take the square root of each side.}$$
$$z = \pm i\sqrt{48}$$
$$z = \pm i\sqrt{16 \cdot 3} \quad \longleftarrow \text{Factor 48 as } 16 \cdot 3 \text{ because 16 is the largest perfect square that is a factor of 48.}$$
$$z = \pm 4i\sqrt{3} \quad \longleftarrow \sqrt{16 \cdot 3} = \sqrt{16}\sqrt{3} = 4\sqrt{3}$$

TRY THIS Solve $2z^2 + 250 = 0$. Write the solutions in simplest radical form.

310 Chapter 8 Quadratic Functions and Equations

EXERCISES

Identify each number as real or as purely imaginary.

1. $\sqrt{13}$ 2. $\sqrt{-13}$ 3. $i\sqrt{5}$ 4. $\sqrt{-(-5)}$

Simplify each expression. Write the answers in simplest radical form.

5. $\sqrt{-16}$ 6. $\sqrt{-25}$ 7. $\sqrt{-100}$ 8. $\sqrt{-9}$
9. $\sqrt{-32}$ 10. $\sqrt{-48}$ 11. $\sqrt{-200}$ 12. $\sqrt{-45}$

Solve each equation. Write the solutions in simplest radical form.

13. $x^2 + 9 = 0$ 14. $v^2 + 121 = 0$ 15. $n^2 + 49 = 0$
16. $t^2 + 169 = 0$ 17. $2m^2 + 8 = 0$ 18. $3w^2 + 27 = 0$
19. $5p^2 + 80 = 0$ 20. $4k^2 + 100 = 0$ 21. $-3m^2 - 36 = 0$
22. $-5c^2 - 40 = 0$ 23. $-2d^2 - 90 = 0$ 24. $-6u^2 - 48 = 0$
25. $-2y^2 - 144 = 0$ 26. $-11x^2 - 88 = 0$ 27. $-4h^2 - 96 = 0$
28. $-8k^2 - 320 = 0$ 29. $2x^2 - 8 = -12$ 30. $3y^2 - 1 = -22$
31. $5z^2 + 8 = -17$ 32. $7b^2 + 8 = -27$ 33. $2(x^2 - 8) = -22$
34. $3(x^2 - 2) = -33$ 35. $-(x^2 + 8) = 4$ 36. $-(x^2 + 5) = 22$

37. **Critical Thinking** Using $i = \sqrt{-1}$ and $i^2 = -1$, find i^3, i^4, i^5, and i^6. What pattern do you see? What pattern do you see in the values of i^n, where n is a positive integer?

38. **Critical Thinking** a. Simplify each sum below.
$i + i^2$ $i + i^2 + i^3$ $i + i^2 + i^3 + i^4$
$i + i^2 + i^3 + i^4 + i^5$ $i + i^2 + i^3 + i^4 + i^5 + i^6$
b. What pattern do you see in the sums $i + i^2 + \cdots + i^n$, where n is a natural number?

Simplify each square root. Write the result in simplest radical form.
(8.1 Skill B)

39. $\sqrt{150}$ 40. $\sqrt{63}$ 41. $\sqrt{360}$ 42. $\sqrt{44}$
43. $\sqrt{\frac{400}{49}}$ 44. $\sqrt{\frac{169}{121}}$ 45. $\sqrt{\frac{252}{25}}$ 46. $\sqrt{\frac{450}{121}}$

Answers for Lesson 8.2 Skill C

Try This Example 1
a. $7i$ b. $2i\sqrt{2}$

Try This Example 2
$a = \pm 5i$

Try This Example 3
$z = \pm 5i\sqrt{5}$

1. real
2. purely imaginary
3. purely imaginary
4. real
5. $4i$
6. $5i$
7. $10i$
8. $3i$
9. $4i\sqrt{2}$
10. $4i\sqrt{3}$
11. $10i\sqrt{2}$
12. $3i\sqrt{5}$
13. $x = \pm 3i$
14. $v = \pm 11i$
15. $n = \pm 7i$

16. $t = \pm 13i$
17. $m = \pm 2i$
18. $w = \pm 3i$
19. $p = \pm 4i$
20. $k = \pm 5i$
21. $m = \pm 2i\sqrt{3}$
22. $c = \pm 2i\sqrt{2}$
23. $d = \pm 3i\sqrt{5}$
24. $u = \pm 2i\sqrt{2}$
25. $y = \pm 6i\sqrt{2}$
26. $x = \pm 2i\sqrt{2}$
27. $h = \pm 12i\sqrt{6}$
28. $k = \pm 2i\sqrt{10}$
29. $x = \pm i\sqrt{2}$
30. $y = \pm i\sqrt{7}$
31. $z = \pm i\sqrt{5}$
32. $b = \pm i\sqrt{5}$
33. $x = \pm i\sqrt{3}$
34. $x = \pm 3i$
35. $x = \pm 2i\sqrt{3}$
36. $x = \pm 3i\sqrt{3}$

37. $i^3 = -i$, $i^4 = 1$, $i^5 = i$, and $i^6 = -1$.

$$i^n = \begin{cases} 1 \text{ if } n = 4k \\ i \text{ if } n = 4k + 1 \\ -1 \text{ if } n = 4k + 2 \\ -i \text{ if } n = 4k + 3 \end{cases}$$

where k is an integer

38. a. $i + i^2 = -1 + i$
$i + i^2 + i^3 = -1$
$i + i^2 + i^3 + i^4 = 0$
$i + i^2 + i^3 + i^4 + i^5 = i$
$i + i^2 + i^3 + i^4 + i^5 + i^6 = -1 + i$

b. $i + i^2 + \cdots + i^n = \begin{cases} 0 \text{ if } n = 4k \\ i \text{ if } n = 4k + 1 \\ -1 + i \text{ if } n = 4k + 2 \\ -1 \text{ if } n = 4k + 3 \end{cases}$

where k is an integer
39. $5\sqrt{6}$
40. $3\sqrt{7}$
41. $6\sqrt{10}$

For answers to Exercises 42–46, see Additional Answers.

Solving Quadratic Equations by Completing the Square

SKILL A *Solving equations of the form $a(x - r)^2 = s$*

Focus: *Students solve equations of the form $a(x - r)^2 = s$ by applying properties of equality and using the definition of square root.*

EXAMPLE 1

PHASE 1: Presenting the Example

Display the equation $7\boxed{}^2 = 28$. Tell students to assume that the box is holding the place of any variable. Have them write a solution to the equation. Discuss their responses. Display all the steps of the solution as shown at right.

Ask students what the conclusion would be if the variable in the box were d. [$d = -2$ or $d = 2$] What if the variable were y? [$y = -2$ or $y = 2$] What if it were j? [$j = -2$ or $j = 2$]

Now have students go back to their solutions and write "$(x - 1)$" inside each box, as shown at right. Ask if they have identified the solutions to the equation $7(x - 1)^2 = 28$. Lead them to see that they have not, because another step is required. Specifically, they must now add 1 to each side of $x - 1 = -2$ and to each side of $x - 1 = 2$. Have them complete the solution. They should arrive at $x = -1$ or $x = 3$.

$$7\boxed{}^2 = 28$$

$$\frac{7\boxed{}^2}{7} = \frac{28}{7}$$

$$\boxed{}^2 = 4$$

$$\boxed{} = -2 \text{ or } \boxed{} = 2$$

$$7\boxed{(x-1)}^2 = 28$$

$$\frac{7\boxed{(x-1)}^2}{7} = \frac{28}{7}$$

$$\boxed{(x-1)}^2 = 4$$

$$\boxed{(x-1)} = -2 \text{ or } \boxed{(x-1)} = 2$$

PHASE 2: Providing Guidance for TRY THIS

Watch for students who arrive at $(z + 5)^2 = 64$ and carelessly identify the solutions as -8 and 8. Urge them to write detailed solutions, without skipping any steps, and use a colored marker to highlight the expression "$z + 5$" throughout. This should draw their attention to the fact that the equations in the statement "$z + 5 = -8$ or $z + 5 = 8$" are not yet solved for z.

EXAMPLE 2

PHASE 1: Presenting the Example

Students may wonder why you cannot simplify $2 + 2\sqrt{3}$ as $4\sqrt{3}$ and $2 - 2\sqrt{3}$ as $0\sqrt{3}$. Note that, just as $2x$ means $2 \cdot x$, $2\sqrt{3}$ means $2 \cdot \sqrt{3}$. So, just as you cannot simplify $2 + 2x$ and $2 - 2x$, you cannot simplify $2 + 2\sqrt{3}$ and $2 - 2\sqrt{3}$.

PHASE 2: Providing Guidance for TRY THIS

Some students may feel uncomfortable checking these solutions because the form is unfamiliar. You may want to do the first check together with them. The check for $-9 + 5\sqrt{3}$ is given at right.

$$10(n + 9)^2 = 750$$

$$10(-9 + 5\sqrt{3} + 9)^2 \overset{?}{=} 750$$

$$10(5\sqrt{3})^2 \overset{?}{=} 750$$

$$10[5^2 \cdot (\sqrt{3})^2] \overset{?}{=} 750$$

$$10 \cdot 25 \cdot 3 = 750 \ ✔$$

8.3 LESSON
Solving Quadratic Equations by Completing the Square

SKILL A Solving equations of the form $a(x - r)^2 = s$

In Lessons 8.1 and 8.2, you learned how to use properties of equality and square roots to solve different types of quadratic equations, such as equations of the forms $x^2 = a$ and $ax^2 + c = 0$. In this lesson, you will learn how to solve an equation of the form $a(x - r)^2 = s$, where $a \neq 0$.

EXAMPLE 1 Solve $7(x - 1)^2 = 28$.

▶ **Solution**

$$7(x - 1)^2 = 28$$
$$(x - 1)^2 = 4 \qquad \longleftarrow \text{ Apply the Division Property of Equality.}$$
$$\sqrt{(x - 1)^2} = \pm\sqrt{4} \qquad \longleftarrow \text{ Take the square root of each side.}$$
$$x - 1 = \pm 2$$
$$x = 1 \pm 2$$
$$x = 3 \text{ or } x = -1$$

TRY THIS Solve $-2(z + 5)^2 = -128$.

Example 2 shows solutions that are not rational.

EXAMPLE 2 Solve $\frac{1}{2}(v - 2)^2 = 6$.

▶ **Solution**

$$\frac{1}{2}(v - 2)^2 = 6$$
$$(v - 2)^2 = 12 \qquad \longleftarrow \text{ Apply the Multiplication Property of Equality.}$$
$$v - 2 = \pm\sqrt{12} \qquad \longleftarrow \text{ Take the square root of each side.}$$
$$v = 2 \pm \sqrt{12}$$
$$v = 2 \pm 2\sqrt{3} \qquad \longleftarrow \text{ Simplify } \sqrt{12} \text{ to } 2\sqrt{3}.$$
$$v = 2 + 2\sqrt{3} \text{ or } v = 2 - 2\sqrt{3}$$

TRY THIS Solve $10(n + 9)^2 = 750$.

The following is a summary of how to solve an equation of the form $a(x - r)^2 = s$.

Summary of the Method for Solving $a(x - r)^2 = s$
1. Isolate the squared expression on one side of the equation.
2. Take the square root of each side of the equation.
3. Solve for the variable.
4. Simplify the solution(s), if necessary.
---

312   Chapter 8   Quadratic Functions and Equations

EXERCISES

KEY SKILLS

Explain how the method summarized on the previous page can be applied to solve each equation. Do not solve.

1. $(x - 4)^2 = 49$ **2.** $(x + 1)^2 = 4$ **3.** $\frac{1}{3}(x - 3)^2 = 12$

PRACTICE

Solve each equation. Write the solutions in simplest radical form.

4. $(x - 4)^2 = 1$ **5.** $(x + 2)^2 = 9$ **6.** $(x + 3)^2 = 64$

7. $(x - 5)^2 = 100$ **8.** $(x + 7)^2 = 16$ **9.** $(x - 2)^2 = 81$

10. $(x + 1)^2 = 4$ **11.** $(x + 7)^2 = 9$ **12.** $2(d - 3)^2 = 48$

13. $-(m + 5)^2 = -75$ **14.** $-2(s - 7)^2 = -144$ **15.** $\frac{1}{3}(t + 4)^2 = 9$

16. $\frac{1}{2}(x - 8)^2 = 40$ **17.** $-3(n + 2)^2 = -60$ **18.** $0.5(k - 1)^2 = 49$

19. $\frac{3}{2}(p + 10)^2 = 42$ **20.** $(2y + 1)^2 = 36$ **21.** $(3n - 1)^2 = 49$

22. $(5t + 3)^2 = 25$ **23.** $(5c - 4)^2 = 1$ **24.** $2(2w - 3)^2 = 100$

Solve each equation by completing the square.

25. $x^2 + 4x + 4 = 36$ **26.** $n^2 - 10n + 25 = 100$ **27.** $9z^2 + 12z + 4 = 25$

28. $4s^2 + 12s + 9 = 1$ **29.** $25k^2 + 10k + 1 = 4$ **30.** $16c^2 + 56c + 49 = 81$

31. Let $a(x - r)^2 = s$, where $a > 0$ and $s > 0$. Give a justification for each step in the reasoning at right.
  a. Step 1   **b.** Step 2
  c. Step 3   **d.** Step 4

  Step 1: $a(x - r)^2 = s$
  Step 2: $(x - r)^2 = \frac{s}{a}$
  Step 3: $(x - r) = \pm\sqrt{\frac{s}{a}}$
  Step 4: $x = r \pm \sqrt{\frac{s}{a}}$

32. Write a formula for solving $a(bx - r)^2 = s$, where a and b are nonzero real numbers and $a \cdot s > 0$.

33. Critical Thinking For what real-number value(s) of n will $(x - 4)^2 = n - 5$ have two distinct real solutions?

MIXED REVIEW

Multiply. (6.5 Skill A)

34. $(2d^2n^2)^2(3dn^3)^2$ **35.** $(3a^2b^4)^3(-3a^2b^3)^2$ **36.** $(-4x^3y^3)^3(x^3y^3)^3$

Lesson 8.3   Skill A   **313**

Answers for Lesson 8.3 Skill A

Try This Example 1
$z = -13$ or $z = 3$

Try This Example 2
$n = -9 \pm 5\sqrt{3}$

1. Take the square root of each side of the equation. Then add 4 to each side of the equation.
2. Take the square root of each side of the equation. Then subtract 1 from each side of the equation.
3. Multiply each side of the equation by 3. Take the square root of each side of the equation. Then add 3 to each side of the equation.
4. $x = 3$ or $x = 5$
5. $x = -5$ or $x = 1$
6. $x = -11$ or $x = 5$
7. $x = -5$ or $x = 15$
8. $x = -11$ or $x = -3$
9. $x = -7$ or $x = 11$
10. $x = -3$ or $x = 1$

11. $x = -10$ or $x = -4$
12. $d = 3 \pm 2\sqrt{6}$
13. $m = -5 \pm 5\sqrt{3}$
14. $s = 7 \pm 6\sqrt{2}$
15. $t = -4 \pm 3\sqrt{3}$
16. $x = 8 \pm 4\sqrt{5}$
17. $n = -2 \pm 2\sqrt{5}$
18. $k = 1 \pm 7\sqrt{2}$
19. $p = -10 \pm 2\sqrt{7}$
20. $y = -\frac{7}{2}$ or $y = \frac{5}{2}$
21. $n = -2$ or $n = \frac{8}{3}$
22. $t = -\frac{8}{5}$ or $t = \frac{2}{5}$
23. $c = 1$ or $c = \frac{3}{5}$
★24. $w = \frac{3}{2} \pm \frac{5\sqrt{2}}{2}$
25. $x = -8$ or $x = 4$
26. $n = -5$ or $n = 15$

27. $z = -\frac{7}{3}$ or $z = 1$
28. $s = -2$ or $s = -1$
29. $k = -\frac{3}{5}$ or $k = \frac{1}{5}$
30. $c = -4$ or $c = \frac{1}{2}$
31. **a.** Given
   b. Division Property of Equality
   c. Definition of square root
   d. Addition Property of Equality
32. $x = \frac{1}{b}\left(\pm\sqrt{\frac{s}{a}} + r\right)$
33. $n > 5$
34. $36d^6n^{10}$
35. $243a^{10}b^{18}$
36. $-64x^{18}y^{18}$

★ **Advanced Exercises**

Lesson 8.3   Skill A   **313**

Focus: *Students solve a quadratic equation by rewriting it so that one side is a perfect-square trinomial.*

EXAMPLE 1

PHASE 1: Presenting the Example

Display equation **1** that is shown at right. Tell students to solve it. [$w + 1 = -4$ or $w + 1 = 4 \rightarrow w = -5$ or $w = 3$] Display equation **2** directly beneath equation **1**, as shown. Tell students that their goal is to solve equation **2** using the same method as was used for equation **1**. Ask them how this might be accomplished. Lead them to the conclusion that they must rewrite the left side as $(r + 5)^2$. Have them complete the solution. [$r + 5 = -6$ or $r + 5 = 6 \rightarrow r = -11$ or $r = 1$]

1. $(w + 1)^2 = 16$

2. $r^2 + 10r + 25 = 36$

3. $c^2 + 6c = 40$

Now display equation **3**. Tell students that their goal is to solve it using the same method used for equations **1** and **2**. Lead a discussion in which you elicit the following key observations: Unlike the situation in equations **1** and **2**, the left side of equation **3** is not a perfect-square trinomial. However, if the left side were $c^2 + 6c + 9$, it could be rewritten as $(c + 3)^2$.

Remind students that the Addition Property of Equality allows you to add the same quantity to each side of an equation. Tell them to write the result when 9 is added to each side of equation **3**. [$c^2 + 6c + 9 = 49$] They should recognize that this new equation can be solved using the desired method. Have them solve it. [$c + 3 = -7$ or $c + 3 = 7 \rightarrow c = -10$ or $c = 4$]

Tell students that the method used to solve equation **3** is called *completing the square*. Ask them why it is given this name. [**The equation is rewritten so that one side is a perfect-square trinomial.**] Then discuss Example 1.

PHASE 2: Providing Guidance for **TRY THIS**

Before they are able to solve an equation by completing the square, some students need practice in simply identifying the constant required to create a perfect-square trinomial. You can use Skill B Exercises 1 through 4 on the facing page for this purpose.

EXAMPLE 2

PHASE 1: Presenting the Example

PHASE 2: Providing Guidance for **TRY THIS**

Students can check solutions of this form by using a calculator with a memory function. For instance, to check that $-1 + \dfrac{\sqrt{7}}{2}$ is a solution to $4v^2 + 8v - 3 = 0$, have them first calculate $-1 + \sqrt{7} \div 2$ and enter that value into memory. (Caution them against using a key sequence that calculates $-1 + \sqrt{7 \div 2}$.) Then, using an appropriate key sequence, they can verify that $4 \times [\text{memory value}]\wedge 2 + 8 \times [\text{memory value}] - 3$ has a value of 0.

Completing the square is often a good method to use for solving a quadratic equation when the equation is not factorable using integers.

> **Completing the Square to Solve $ax^2 + bx + c = 0$**
> 1. Write the equation so that the constant term is isolated on the right side. Divide each side of the equation by a.
> 2. Find the square of one-half the coefficient of x. Add that number to each side of the equation.
> 3. Factor the left side of the equation. The result should have the form $(x + r)^2$.
> 4. Take the square root of each side of the equation. Then solve for x and simplify the solutions.

EXAMPLE 1   Solve $x^2 + 6x + 5 = 0$ by completing the square.

▸ **Solution**

1. $x^2 + 6x = -5$ ⟵ *Write variable terms on the left and the constant on the right.*
2. $x^2 + 6x + 3^2 = -5 + 9$ ⟵ *Add the square of one-half the coefficient of x to each side.*
3. $(x + 3)^2 = 4$ ⟵ *Write the left side of the equation as a squared binomial.*
4. $x + 3 = \pm\sqrt{4}$ ⟵ *Take the square root of each side of the equation.*
     $x + 3 = \pm 2$

$x = -3 + 2 = -1$ or $x = -3 - 2 = -5$

The solutions to $x^2 + 6x + 5 = 0$ are -1 and -5.

> *The coefficient of x is 6.*
> $\left(\frac{1}{2} \cdot 6\right)^2 = 3^2 = 9$

TRY THIS   Solve $z^2 - 2z - 15 = 0$ by completing the square.

EXAMPLE 2   Solve $4v^2 + 8v - 3 = 0$ by completing the square.

▸ **Solution**

1. $4v^2 + 8v = 3$ ⟵ *Write variable terms on the left and the constant on the right.*
     $v^2 + 2v = \frac{3}{4}$ ⟵ *Divide each side by the coefficient of v^2.*
2. $v^2 + 2v + 1^2 = \frac{3}{4} + 1$ ⟵ *Add the square of one-half the coefficient of v to each side.*
3. $(v + 1)^2 = \frac{7}{4}$ ⟵ *Write the left side as a squared binomial.*
4. $v + 1 = \pm\frac{\sqrt{7}}{2}$ ⟵ *Take the square root of each side of the equation.*

The solutions to $4v^2 + 8v - 3 = 0$ are $-1 + \frac{\sqrt{7}}{2}$ and $-1 - \frac{\sqrt{7}}{2}$.

> *The coefficient of v is 2.*
> $\left(\frac{1}{2} \cdot 2\right)^2 = 1^2 = 1$

TRY THIS   Solve $9y^2 + 18y + 4 = 0$ by completing the square.

EXERCISES

> **KEY SKILLS**

Write the number that makes the given expression a perfect-square trinomial.

1. $s^2 + 2s + \underline{\ ?\ }$
2. $y^2 + 10y + \underline{\ ?\ }$
3. $n^2 - 12n + \underline{\ ?\ }$
4. $v^2 - 16v + \underline{\ ?\ }$

Complete the solution to each equation.

5. $s^2 + 2s + 1 = 3 + 1$
    $(s + 1)^2 = 4$
6. $n^2 + 6n + 9 = 91 + 9$
    $(n + 3)^2 = 100$
7. $n^2 - 12n + 36 = 13 + 36$
    $(n - 6)^2 = 49$

> **PRACTICE**

Solve each equation by completing the square. Write the solutions in simplest radical form where necessary.

8. $x^2 + 2x = 3$
9. $m^2 + 4m = 5$
10. $t^2 - 10t = 11$
11. $q^2 - 6q = 7$
12. $u^2 + 4u = 77$
13. $h^2 + 8h = 9$
14. $x^2 - 12x = 28$
15. $r^2 - 18r = -17$
16. $d^2 - 4d - 12 = 0$
17. $f^2 - 10f - 11 = 0$
18. $h^2 + 8h - 20 = 0$
19. $m^2 - 16m - 36 = 0$
20. $4y^2 - 8y - 3 = 0$
21. $9z^2 + 18z - 11 = 0$
22. $4r^2 + 16r + 5 = 0$
23. $3t^2 - 18t + 21 = 0$
24. $4z^2 - 20z + 1 = 0$
25. $4m^2 + 12m - 11 = 0$
26. $12q^2 + 12q - 9 = 0$
27. $9q^2 - 36q + 1 = 0$

28. Use the method of completing the square to find n such that $x^2 + 2x = n$ has two distinct real solutions.

29. Use the method of completing the square to find m such that $x^2 + 6x = m$ has no real solutions.

30. **Critical Thinking** Find and graph the ordered pairs (a, b) that make $x^2 + 2x = a + b$ have two distinct real solutions. (*Hint:* Complete the square.)

31. **Critical Thinking** Solve the equation $x^2 + 2x - y^2 - 2y = 0$ for y.

> **MIXED REVIEW**

Evaluate each expression given $a = -2$, $b = 5$, and $c = 12$. (1.2 Skill C)

32. $\frac{3}{2}(a^2 + b^2 - c^2)$
33. $\frac{5}{2}(a^2 + b^2)^2 - c^2$
34. $\frac{1}{4}(a^2 + b^2 - c^2)^2$
35. $(2a + 4b - 5c)^2$
36. $b^2 - 4ac$
37. $a(ab^2 - 4ac)$

Answers for Lesson 8.3   Skill B

Try This Example 1
$z = -3$ or $z = 5$

Try This Example 2
$y = -1 \pm \dfrac{\sqrt{5}}{3}$

1. 1
2. 25
3. 36
4. 64
5. $(s + 1)^2 = 4$
    $s + 1 = \pm 2$
    $s = -1 \pm 2$
    $s = -3$ or $s = 1$
6. $(n + 3)^2 = 100$
    $n + 3 = \pm 10$
    $n = -3 \pm 10$
    $n = -13$ or $n = 7$
7. $(n - 6)^2 = 49$
    $n - 6 = \pm 7$
    $n = 6 \pm 7$
    $n = -1$ or $n = 13$

8. $x = -3$ or $x = 1$
9. $m = -5$ or $m = 1$
10. $t = -1$ and $t = 11$
11. $q = -1$ or $q = 7$
12. $u = -11$ or $u = 7$
13. $h = -9$ or $h = 1$
14. $x = -2$ or $x = 14$
15. $r = 1$ and $r = 17$
16. $d = -2$ or $d = 6$
17. $f = -1$ or $f = 11$
18. $h = 2$ or $h = -10$
19. $m = -2$ or $m = 18$
20. $y = 1 \pm \dfrac{\sqrt{7}}{2}$
21. $z = -1 \pm \dfrac{2\sqrt{5}}{3}$
22. $r = -2 \pm \dfrac{\sqrt{11}}{2}$
23. $t = 3 \pm \sqrt{2}$
24. $z = \dfrac{5}{2} \pm \sqrt{6}$

25. $m = -\dfrac{3}{2} \pm \sqrt{5}$
26. $q = -\dfrac{3}{2}$ or $q = \dfrac{1}{2}$
27. $q = 2 \pm \dfrac{\sqrt{35}}{3}$
28. $n > -1$
29. $m < -9$
30. $b > -a - 1$

31. $y = x$ or $y = -x - 2$
32. $-\dfrac{345}{2}$

33. $\dfrac{3917}{2}$
34. $\dfrac{13,225}{4}$
35. 1936
36. 121
37. -92

Focus: *Students use the Pythagorean Theorem to model a real-world problem, and then solve the resulting quadratic equation by completing the square.*

EXAMPLE

PHASE 1: Presenting the Example

Some students may be puzzled by the entries in the table at the beginning of the solution. Remind them of the familiar formula $d = rt$. That is, the distance, d, traveled by an object moving at a constant speed is the product of its rate of motion, r, and its time of travel, t. Then display the following:

Let r = speed of Boat A: Then $r + 7$ = speed of Boat B:

| d | = | r | $\times$ | t | | d | = | r | $\times$ | t |

| Boat A's distance | = | Boat A's rate | $\times$ | Boat A's time | | Boat B's distance | = | Boat B's rate | $\times$ | Boat B's time |

| d | = | r | $\times$ | 2 | | d | = | $(r + 7)$ | $\times$ | 2 |

For Boat A, $d = 2r$.   For Boat B, $d = 2(r + 7)$.

(Note that the textbook uses x to represent rate. The descriptive variables d, r, and t may be more helpful to students who are confused.)

Students should be aware that the essence of this problem lies in the fact that the boats begin traveling due east and due north from exactly the same point at exactly the same time. This means that their paths trace two legs of a right triangle, and the distances that they travel in two hours are the lengths of the legs. So the expressions $2r$ and $2(r + 7)$, which were derived above, can be translated to the diagram at the middle of the textbook page.

PHASE 2: Providing Guidance for **TRY THIS**

Students should have little difficulty generating $(2r)^2 + [2(r + 1)]^2 = 20^2$ as the required equation, since it is easily modeled on the solution of the Example. If they arrive at an incorrect solution, look for errors in applying the Power-of-a-Product Property of Exponents and the Distributive Property when simplifying $(2r)^2$ and $[2(r + 1)]^2$.

PHASE 3: ASSESSMENT AND CLOSURE for Lesson 8.3

- Give students the set of equations at right. Tell them these are the first five steps of a solution to a quadratic equation, but they are in the incorrect order. Have students arrange them in the correct order and then complete the solution. [$4t^2 - 24t - 9 = 0$; $4t^2 - 24t = 9$; $t^2 - 6t = \dfrac{9}{4}$; $t^2 - 6t + 9 = \dfrac{45}{4}$; $(t - 3)^2 = \dfrac{45}{4}$; $t - 3 = \dfrac{3\sqrt{5}}{2}$ or $t - 3 = -\dfrac{3\sqrt{5}}{2}$; $t = 3 + \dfrac{3\sqrt{5}}{2}$ or $t = 3 - \dfrac{3\sqrt{5}}{2}$]

    $$t^2 - 6t = \frac{9}{4}$$
    $$t^2 - 6t + 9 = \frac{45}{4}$$
    $$4t^2 - 24t - 9 = 0$$
    $$(t - 3)^2 = \frac{45}{4}$$
    $$4t^2 - 24t = 9$$

- In the next lesson, students will see how the method of completing the square is used to generate the *quadratic formula*.

☞ *For a Lesson 8.3 Quiz, see* Assessment Resources *page 80.*

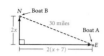

SKILL C ▷ APPLICATIONS ▷ *Solving real-world problems using completing the square*

You can use the *completing the square* method to solve quadratic equations which model real-world problems.

EXAMPLE

Boats A and B leave the dock at the same time and each sails at a constant speed. Boat A sails due north and Boat B sails due east. After 2 hours, they are 30 miles apart. Boat B sails 5 miles per hour faster than Boat A. Find the speed of each boat to the nearest tenth of a mile per hour. Use a calculator to approximate the value of the radical expression.

▶ **Solution**

Make a table.

1. Organize the information in a table. Let x be the speed of Boat A.

	speed (miles per hour)	distance (miles)
Boat A	x	$2x$
Boat B	$x + 7$	$2(x + 7)$

Draw a sketch. Because the paths of the boats form a right triangle, you can use the Pythagorean Theorem to help you solve the problem.

Make a diagram.

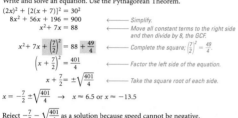

$$(\text{Boat A distance})^2 + (\text{Boat B distance})^2 = 30^2$$

Write an equation.

2. Write and solve an equation. Use the Pythagorean Theorem.

$$(2x)^2 + [2(x + 7)]^2 = 30^2$$
$$8x^2 + 56x + 196 = 900 \quad \longleftarrow \text{Simplify.}$$
$$x^2 + 7x = 88 \quad \longleftarrow \begin{array}{l}\text{Move all constant terms to the right side} \\ \text{and then divide by 8, the GCF.}\end{array}$$
$$x^2 + 7x + \left(\frac{7}{2}\right)^2 = 88 + \frac{49}{4} \quad \longleftarrow \text{Complete the square; } \left(\frac{7}{2}\right)^2 = \frac{49}{4}.$$
$$\left(x + \frac{7}{2}\right)^2 = \frac{401}{4} \quad \longleftarrow \text{Factor the left side of the equation.}$$
$$x + \frac{7}{2} = \pm\sqrt{\frac{401}{4}} \quad \longleftarrow \text{Take the square root of each side.}$$
$$x = -\frac{7}{2} \pm \sqrt{\frac{401}{4}} \quad \rightarrow \quad x \approx 6.5 \text{ or } x \approx -13.5$$

Reject $-\frac{7}{2} - \sqrt{\frac{401}{4}}$ as a solution because speed cannot be negative. The speed of Boat A is about 6.5 miles per hour. The speed of Boat B then is about $6.5 + 7$, or 13.5 miles per hour.

TRY THIS Rework the Example given that the speed of Boat B is 1 mile per hour more than Boat A and the distance between them is 20 miles.

EXERCISES

KEY SKILLS

Refer to the Example on the previous page.

1. Make a table to represent the information below. Boat B sails 5 miles per hour faster than Boat A. After 2 hours, the boats are 32 miles apart.

2. Use the Pythagorean Theorem to write an equation to represent the data in Exercise 1.

PRACTICE

Rework the Example on the previous page using the given modifications of the problem. Assume all other information is the same as in the Example.

3. The boats are 25 miles apart after 1 hour.

4. The boats are 45 miles apart after 2 hours.

5. The boats are 65 miles apart after 3 hours.

Boats A and B leave the dock at the same time and sail at a constant speed for two hours. Find the speed, s, of each boat in miles per hour.

6.

7.

8. **Critical Thinking** How far apart should points X and Z be so that $\triangle XYZ$ is a right triangle, $XZ = 2a - 4$, $YZ = a$, and $XY = 10$?

MIXED REVIEW APPLICATIONS

Represent the shaded region as a system of linear inequalities. (5.6 Skill C)

9.

10.

Answers for Lesson 8.3 Skill C

Try This Example
Boat A: about 6.6 miles per hour
Boat B: about 7.6 miles per hour

1.

	speed (miles per hour)	distance (miles)
Boat A	r	$2r$
Boat B	$r + 5$	$2(r + 5)$

2. $(2r)^2 + [2(r + 5)]^2 = 32^2$

3. Boat A: $\approx$ 13.8 miles per hour
   Boat B: $\approx$ 20.8 miles per hour

4. Boat A: $\approx$ 12.0 miles per hour
   Boat B: $\approx$ 19.0 miles per hour

5. Boat A: $\approx$ 11.4 miles per hour
   Boat B: $\approx$ 18.4 miles per hour

6. Boat A: 6 miles per hour
   Boat B: 8 miles per hour

7. Boat A: 8.4 miles per hour
   Boat B: 5.4 miles per hour

8. The distance between X and Z should be 8 units.

9. $\begin{cases} x \geq -2 \\ y \geq \dfrac{1}{2}x + 3 \\ y \leq -\dfrac{5}{6}x + 7 \end{cases}$

10. $\begin{cases} x \geq -3 \\ x \leq 5 \\ y \geq \dfrac{1}{2}x + 3 \\ y \leq \dfrac{1}{2}x + 6 \end{cases}$

Solving Quadratic Equations by Using the Quadratic Formula

> **SKILL A** *Using the quadratic formula to solve quadratic equations*

Focus: *Students use the quadratic formula to find rational and irrational solutions to quadratic equations.*

EXAMPLE 1

PHASE 1: Presenting the Example

Display the table below. Instruct students to copy and complete it. Then display equations **1** through **4** to the right of the table, as shown. Have students solve each equation. [**Answers are to the right of each equation.**]

a	b	c	$\dfrac{-b + \sqrt{b^2 - 4ac}}{2a}$	$\dfrac{-b - \sqrt{b^2 - 4ac}}{2a}$
1	4	3	-1	-3
1	-6	8	4	2
1	4	-5	1	-5
2	-5	-3	3	$-\dfrac{1}{2}$

1. $x^2 + 4x + 3 = 0$ $[x = -1 \text{ or } x = -3]$

2. $x^2 - 6x + 8 = 0$ $[x = 4 \text{ or } x = 2]$

3. $x^2 + 4x - 5 = 0$ $[x = 1 \text{ or } x = -5]$

4. $2x^2 - 5x - 3 = 0$ $[x = 3 \text{ or } x = -\dfrac{1}{2}]$

Ask students to describe any patterns that they see. They should make two important observations: First, the values of a, b, and c in each row are the values of a, b, and c in the quadratic equation $ax^2 + bx + c = 0$ to the right. Second, the calculated values of $\dfrac{-b + \sqrt{b^2 - 4ac}}{2a}$ and $\dfrac{-b - \sqrt{b^2 - 4ac}}{2a}$ in each row are equal to the solutions of the corresponding equations.

Have students read the statement of the quadratic formula that appears on the textbook page. Point out that, with this statement, the observations they made from their work with the table are generalized to any quadratic equation. They will see how to prove this generalization later in the lesson. Then discuss how the formula is applied to the equation in Example 1.

PHASE 2: Providing Guidance for TRY THIS

The symbol $\pm$ may be unfamiliar to many students. Tell them that it is a "shorthand" way to combine the two expressions that appear in the table.

EXAMPLE 2

PHASE 2: Providing Guidance for TRY THIS

Watch for students who identify 5 as the value of c. Stress that the formula only applies to a quadratic equation that is written in standard form.

8.4 LESSON
Solving Quadratic Equations by Using the Quadratic Formula

SKILL A *Using the quadratic formula to solve quadratic equations*

Recall from Lesson 7.5 that $ax^2 + bx + c = 0$ is the standard form of a quadratic equation. When a quadratic equation is written in standard form, you can apply the *quadratic formula* to find its solutions.

The Quadratic Formula

If $ax^2 + bx + c = 0$, where $a \neq 0$, then $x = \dfrac{-b \pm \sqrt{b^2 - 4ac}}{2a}$.

The proof of the quadratic formula is given in Skill C of this lesson.

EXAMPLE 1 Solve $x^2 + 7x + 6 = 0$ using the quadratic formula.

▶ **Solution**

$x = \dfrac{-b \pm \sqrt{b^2 - 4ac}}{2a}$

$x = \dfrac{-7 \pm \sqrt{7^2 - 4(1)(6)}}{2(1)}$ ⟵ Substitute $a = 1$, $b = 7$, and $c = 6$.

$x = \dfrac{-7 \pm \sqrt{25}}{2} = \dfrac{-7 \pm 5}{2}$

$x = -1$ or $x = -6$

TRY THIS Solve $w^2 - 3w - 4 = 0$ using the quadratic formula.

EXAMPLE 2 Solve $t^2 + 2t = 5$ using the quadratic formula.

▶ **Solution**

$t^2 + 2t = 5$

$t^2 + 2t - 5 = 0$ ⟵ Write the equation in standard form.

$t = \dfrac{-2 \pm \sqrt{2^2 - 4(1)(-5)}}{2(1)}$ ⟵ Apply the quadratic formula with $a = 1$, $b = 2$, and $c = -5$.

$t = \dfrac{-2 \pm \sqrt{24}}{2} = \dfrac{-2 \pm 2\sqrt{6}}{2}$ ⟵ $\sqrt{24} = 2\sqrt{6}$

$t = -1 + \sqrt{6}$ or $t = -1 - \sqrt{6}$

TRY THIS Solve $a^2 - 3a = 5$ using the quadratic formula.

Another name for a solution to a polynomial equation is *root*. In Example 2, $-1 + \sqrt{6}$ and $-1 - \sqrt{6}$ are the roots of the equation $t^2 + 2t = 5$.

EXERCISES

KEY SKILLS

Write each quadratic equation in standard form. Then identify a, b, and c.

1. $x^2 = 5x + 6$
2. $x^2 - 9x = -18$
3. $3x + 2x^2 = 20$
4. $-2 - 5x = 3x^2$

PRACTICE

Use the quadratic formula to solve each equation. Write the solutions in simplest radical form where necessary.

5. $x^2 + 11x + 30 = 0$
6. $y^2 - 15y + 56 = 0$
7. $w^2 + 15w + 56 = 0$
8. $v^2 + 13v + 42 = 0$
9. $z^2 - 8z = -7$
10. $m^2 + 27 = 12m$
11. $33 + 8x = x^2$
12. $45 - 4y = y^2$
13. $2k^2 = 11k - 15$
14. $6g^2 + 14 = 25g$
15. $6x^2 + 5 = 13x$
16. $7 + 10h^2 = 37h$
17. $y^2 = 6y - 2$
18. $m^2 = 2m + 6$
19. $p^2 = 8p - 11$
20. $44 + q^2 = 14q$
21. $x^2 = 4x - 2$
22. $z^2 = 4z - 1$
23. $4n^2 = 8n + 1$
24. $2t^2 - 6t = -1$
25. $3y^2 - 8y = 2$

Solve each equation for x in terms of a, b, and c. Then verify that you would arrive at the same expression for x if you solve using the quadratic formula.

26. $ax^2 + c = 0$
27. $ax^2 + bx = 0$
28. $x^2 + bx + c = 0$

29. For what value of b will $4x^2 + bx + 9 = 0$ have exactly one root? Use the quadratic formula to justify your response.

30. Use the quadratic formula to solve $2(v + 3)^2 - 4(v + 3) - 7 = 0$.

31. Use the quadratic formula to solve $6n^4 - 13n^2 + 5 = 0$.

32. Use the quadratic formula to show that $x^2 + 1 = 0$ has no real roots.

MIXED REVIEW

Find the coordinates of the vertex and an equation for the axis of symmetry. Identify the vertex as the maximum or the minimum point. Then graph the function. (7.6 Skill A)

33. $y = x^2 + 10x + 25$
34. $y = x^2 - 10x + 25$
35. $y = -x^2 + 4$
36. $y = -x^2 + 2x$
37. $y = x^2 - 4x$
38. $y = -\frac{1}{2}x^2 + x$

Answers for Lesson 8.4 Skill A

Try This Example 1
$w = -1$ or $w = 4$

Try This Example 2
$a = \dfrac{3}{2} \pm \dfrac{\sqrt{29}}{2}$

1. $x^2 - 5x - 6 = 0$; $a = 1, b = -5, c = -6$
2. $x^2 - 9x + 18 = 0$; $a = 1, b = -9, c = 18$
3. $2x^2 + 3x - 20 = 0$; $a = 2, b = 3, c = -20$
4. $-3x^2 - 5x - 2 = 0$; $a = -3, b = -5, c = -2$
5. $x = -6$ or $x = -5$
6. $y = 7$ or $y = 8$
7. $w = -7$ or $w = -8$
8. $v = -7$ or $v = -6$
9. $z = 1$ or $z = 7$
10. $m = 3$ or $m = 9$

11. $x = -3$ or $x = 11$
12. $y = -9$ or $y = 5$
13. $k = \dfrac{5}{2}$ or $k = 3$
14. $g = \dfrac{2}{3}$ or $g = \dfrac{7}{2}$
15. $x = \dfrac{1}{2}$ or $x = \dfrac{5}{3}$
16. $h = \dfrac{1}{5}$ or $h = \dfrac{7}{2}$
17. $y = 3 \pm \sqrt{7}$
18. $m = 1 \pm \sqrt{7}$
19. $p = 4 \pm \sqrt{5}$
20. $q = 7 \pm \sqrt{5}$
21. $x = 2 \pm \sqrt{2}$
22. $z = 2 \pm \sqrt{3}$
23. $n = 1 \pm \dfrac{\sqrt{5}}{2}$
24. $t = \dfrac{3}{2} \pm \dfrac{\sqrt{7}}{2}$

25. $y = \dfrac{4}{3} \pm \dfrac{\sqrt{22}}{3}$
26. $x = \pm\sqrt{-\dfrac{c}{a}}$
27. $x = 0$ or $x = -\dfrac{b}{a}$
28. $x = \dfrac{-b \pm \sqrt{b^2 - 4c}}{2}$
29. $b = \pm 12$
30. $v = -2 \pm \dfrac{3\sqrt{2}}{2}$
★31. $n = -\dfrac{\sqrt{15}}{3}, \dfrac{\sqrt{15}}{3},$ $-\dfrac{\sqrt{2}}{2},$ or $\dfrac{\sqrt{2}}{2}$

32. $x = \dfrac{-0 \pm \sqrt{0^2 - 4(1)(1)}}{2}$; Since $0^2 - 4(1)(1) < 0$, the formula gives no real numbers for x.

33. vertex: $(-5, 0)$; axis of symmetry: $x = -5$; minimum

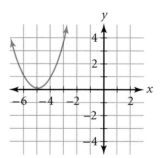

For answers to Exercises 34–38, see Additional Answers.

★ **Advanced Exercises**

Focus: *Students use the discriminant to determine whether a quadratic equation has two real roots, one real root, or no real roots.*

EXAMPLE 1

PHASE 1: Presenting the Example

Display equation **1** shown at right. Students should recognize that it represents the solutions to a quadratic equation. Ask them to simplify the solutions. They should arrive at $x = -2$ or $x = -4$.

1. $x = \dfrac{-6 \pm \sqrt{6^2 - 4(1)(8)}}{2(1)}$

Display equation **2** and repeat the instruction to simplify. Students will find exactly one solution, $x = -3$. Point out that there is only one solution because the value of the expression under the radical sign is 0. Consequently, $\dfrac{-6 + 0}{2}$ has the same value as $\dfrac{-6 - 0}{2}$.

2. $x = \dfrac{-6 \pm \sqrt{6^2 - 4(1)(9)}}{2(1)}$

3. $x = \dfrac{-6 \pm \sqrt{6^2 - 4(1)(10)}}{2(1)}$

Repeat the activity with equation **3**. This time, students most likely will be puzzled. Discuss the source of the confusion, which is the presence of a negative number under the radical sign. Note that there are two solutions, but each is imaginary. That is, there are no real solutions.

Equations **1**, **2**, and **3** represent three categories of solutions to quadratic equations: *two real solutions*, *one real solution*, and *no real solutions*. What distinguishes one category from the others is whether the expression under the radical sign is positive, zero, or negative. For this reason, $b^2 - 4ac$ is called the *discriminant* of the quadratic equation $ax^2 + bx + c = 0$.

PHASE 2: Providing Guidance for TRY THIS

If students remember that the discriminant is the expression under the square root sign, they will be able to remember what the value of the discriminant tells them about the roots. If $b^2 - 4ac < 0$, the square root has no real-number value, so the equation has no real roots. If $b^2 - 4ac = 0$, the square root has one value (0), so the equation has one real root. Finally, if $b^2 - 4ac > 0$, the square root has two real-number values, so the equation has two real roots.

Students sometimes become confused and think that, by calculating the discriminant, they are actually solving the equation. Stress the fact that solving the equation requires the use of the entire quadratic formula.

EXAMPLE 2

PHASE 1: Presenting the Example

In Lesson 7.6, students determined the number of x-intercepts of a parabola by locating its vertex and deciding if it opens up or down. Here they will make the same determination by the simple calculation of a discriminant.

In the quadratic formula, the expression under the radical sign, $b^2 - 4ac$, is called the **discriminant** of $ax^2 + bx + c = 0$.

The Quadratic Formula and the Discriminant

If $ax^2 + bx + c = 0$, where $a \neq 0$, then the equation has
- two real and distinct roots if $b^2 - 4ac > 0$,
- one real root if $b^2 - 4ac = 0$, and
- no real roots if $b^2 - 4ac < 0$.

EXAMPLE 1 Find the number of real roots of $3x^2 - 5x - 12 = 0$.

▶ **Solution**

Identify a, b, and c: $a = 3$, $b = -5$, and $c = -12$.

Evaluate the discriminant, $b^2 - 4ac$.

$$(-5)^2 - 4(3)(-12) = 169$$

Because the discriminant, 169, is positive, the equation has 2 real roots.

TRY THIS   Find the number of real roots of $2x^2 - 5x + 10 = 0$.

Recall from Lesson 7.6 that a parabola has either no x-intercept, one x-intercept, or two x-intercepts. The x-intercepts of the graph of $y = ax^2 + bx + c$ are the real roots of $ax^2 + bx + c = 0$.

EXAMPLE 2 Find the number of x-intercepts of the graph of $y = 3x^2 - 5x - 12$.

▶ **Solution**

The graph of $y = 3x^2 - 5x - 12$ is a parabola. Evaluate the discriminant of $3x^2 - 5x - 12 = 0$. From Example 1, we know that the discriminant, 169, is positive.

Because the discriminant is positive, the equation has two real, distinct roots and thus, two x-intercepts.

TRY THIS   Find the number of x-intercepts of the graph of $y = x^2 + 14x + 49$.

A quadratic equation has no real roots if $b^2 - 4ac < 0$. For example, the discriminant of $x^2 + 7x + 20 = 0$ is $7^2 - 4(1)(20) = -31$. The graph of such an equation has no x-intercepts. It does not touch or cross the x-axis at any point.

EXERCISES

KEY SKILLS

Each expression is the discriminant of a quadratic equation. How many real roots does the equation have?

1. $(-6)^2 - 4(2)(5)$ 2. $(-6)^2 - 4(2)(-5)$ 3. $(8)^2 - 4(4)(4)$ 4. $(-1)^2 - 4(1)(-3)$

PRACTICE

Use the discriminant to find the number of real roots of each quadratic equation.

5. $x^2 + 5x - 24 = 0$ 6. $2z^2 - z - 3 = 0$ 7. $n^2 + 6n + 11 = 0$
8. $k^2 + 10k + 28 = 0$ 9. $4y^2 - 12y + 9 = 0$ 10. $9d^2 - 42d + 49 = 0$
11. $25t^2 - 49 = 0$ 12. $25s^2 + 49 = 0$ 13. $2w^2 = -10w - 15$
14. $3h^2 + 2h = -2$ 15. $9v^2 + 16 = 24v$ 16. $p^2 + 1 = p$

Use the discriminant to find the number of x-intercepts of each quadratic function.

17. $y = x^2 - 12x + 36$ 18. $y = 4x^2 - 16x + 19$ 19. $y = 3x^2 - 8x + 7$
20. $y = 9x^2 - 42x + 49$ 21. $y = 3x^2 - 6x + 2$ 22. $y = 2x^2 + 6x + 1$
23. $y = 7x^2 + 1$ 24. $y = -x^2 - 11$ 25. $y = 6x^2 - 5x + 1$
26. $y = 6x^2 - 5x$ 27. $y = x^2 + 18x + 70$ 28. $y = x^2 + 18x + 68$

Write a quadratic equation that has the given number of real roots.

29. no real roots   30. one real root   31. two real roots

32. For what values of a will $ax^2 - 5x + 1 = 0$ have two distinct real roots?

33. For what values of c will $5x^2 - 7x + c = 0$ have no real roots?

34. For what values of b will $5x^2 - bx + 6 = 0$ have exactly one real root?

MIXED REVIEW

Solve each equation. Write the solution as a logical argument. Summarize the problem and its solution as a conditional statement. (2.9 Skill B)

35. $\frac{1}{2}(a + 3) = 5$ 36. $3x = -5x + 4$ 37. $-4t + 5 + 5t = 9$
38. $\frac{2}{3}(n + 3) = \frac{1}{3}n$ 39. $4m + 5m = m + 2$ 40. $\frac{2y - 2}{3} = 4$

Answers for Lesson 8.4 Skill B

Try This Example 1
none

Try This Example 2
one

1. none
2. two
3. one
4. two
5. two
6. two
7. none
8. none
9. one
10. one
11. two
12. none
13. none
14. none
15. one
16. none

17. one
18. none
19. none
20. one
21. two
22. two
23. none
24. none
25. two
26. two
27. two
28. two
29. Answers may vary. Sample answer: $x^2 + r = 0$, where r is any positive real number
30. Answers may vary. Sample answer: $(x - r)^2 = 0$, where r is any real number

31. Answers may vary. Sample answer: $x^2 + r = 0$, where r is any negative real number
32. $a < \dfrac{25}{4}$
33. $c > \dfrac{49}{20}$
34. $b = \pm 2\sqrt{30}$
35. If $\frac{1}{2}(a + 3) = 5$, then by the Multiplication Property of Equality, $a + 3 = 10$. If $a + 3 = 10$, then by the Subtraction Property of Equality, $a = 7$. Therefore, if $\frac{1}{2}(a + 3) = 5$, then $a = 7$.
36. If $3x = -5x + 4$, then by the Addition Property of Equality, $8x = 4$. If $8x = 4$, then by the Division Property of Equality, $x = \frac{1}{2}$. Therefore, if $3x = -5x + 4$, then $x = \frac{1}{2}$.

For answers to Exercises 37–40, see Additional Answers.

Focus: *Given a proof of the quadratic formula, students justify each step.*

PHASE 1: Presenting the Proof

PHASE 2: Providing Guidance for the Exercises

At this point students have had considerable experience in using the quadratic formula, and they should have memorized it. Now they will be asked to consider how the formula is derived. Note that students are not expected to memorize the derivation, but only to justify the steps.

Because of the complexity of the proof, the presentation of this skill varies slightly from the presentation of other skills in this textbook. That is, the steps of the proof are given on the skill page, and on the Exercises page students are asked to provide justifications for some of the steps.

Before students work with the proof, you may need to reacquaint them with literal equations. You can use Skill C Exercises 1 and 2 on the facing page for this purpose.

Point out that these exercises derive the simple formula $x = -\dfrac{b}{a}$ as a solution of the general *linear* equation $ax + b = 0$.

After students have completed the exercises and their questions have been answered, consider the following additional activity: Write the proof of the quadratic formula on a sheet of paper, but omit the number in front of each step. Make several copies of this proof. Cut each copy into strips, mix up the strips, and place them in an envelope. Give one envelope to each student or to each of several small groups of students, as you prefer. The students' task is then to reassemble the strips into a properly sequenced proof.

PHASE 3: ASSESSMENT AND CLOSURE for Lesson 8.4

- Give students the equations at right, one at a time. Tell them that each represents a student's work in applying the quadratic formula to a given quadratic equation. Ask them to identify whether each will result in exactly one real solution, two real solutions, or no real solutions. Then have them work backward to identify the given quadratic equation. [**1. one real solution;** $25x^2 + 10x + 1 = 0$ **2. two real solutions;** $25x^2 + 10x - 1 = 0$ **3. one real solution;** $25x^2 - 10x + 1 = 0$ **4. two real solutions;** $25x^2 - 10x - 1 = 0$ **5. no real solutions;** $25x^2 - 10x + 3 = 0$]

- Students have now learned four methods for solving quadratic equations: factoring, using square roots, completing the square, and applying the quadratic formula. As each method was taught, students were provided with several equations and were instructed to solve by applying that method. The next lesson will focus on choice. That is, given a quadratic equation, students must decide which is the most efficient method of solution.

1. $x = \dfrac{-10 \pm \sqrt{100 - 4(25)(1)}}{50}$

2. $x = \dfrac{-10 \pm \sqrt{100 - 4(25)(-1)}}{50}$

3. $x = \dfrac{10 \pm \sqrt{100 - 4(25)(1)}}{50}$

4. $x = \dfrac{10 \pm \sqrt{100 - 4(25)(-1)}}{50}$

5. $x = \dfrac{-10 \pm \sqrt{100 - 4(25)(3)}}{50}$

☞ *For a Lesson 8.4 Quiz, see* Assessment Resources *page 81.*

Recall from Lesson 2.9 that you can write the solution to an equation as a logical argument. For example, given the linear equation $ax + b = 0$, you can solve for x to write a formula for the solution. Examine the logical argument of this solution below.

$$ax + b = 0, a \neq 0 \quad \text{Given}$$
$$ax = -b \quad \text{Subtraction Property of Equality}$$
$$x = -\frac{b}{a} \quad \text{Division Property of Equality}$$

From the proof above, if $ax + b = 0$ and $a \neq 0$, then a formula for the solution is $x = \frac{-b}{a}$.

Similarly, given the quadratic equation $ax^2 + bx + c = 0$ and $a \neq 0$, you can solve for x to write a formula for the solution, $x = \frac{-b \pm \sqrt{b^2 - 4ac}}{2a}$.

The statements for the proof of this formula are shown at right.

The proof begins with the hypothesis "$ax^2 + bx + c = 0$ and $a \neq 0$" and ends with the conclusion "$x = \frac{-b \pm \sqrt{b^2 - 4ac}}{2a}$." The reasoning and the sequence of statements from the hypothesis to the conclusion can be justified with properties of real numbers, properties of equality, and other facts.

You will be asked to use the properties to justify some of the steps as part of the exercises on the next page.

1. $ax^2 + bx + c = 0, a \neq 0$
2. $ax^2 + bx = -c$
3. $x^2 + \frac{b}{a}x = -\frac{c}{a}$
4. $x^2 + \frac{b}{a}x + \left(\frac{b}{2a}\right)^2 = -\frac{c}{a} + \left(\frac{b}{2a}\right)^2$
5. $\left(x + \frac{b}{2a}\right)^2 = -\frac{c}{a} + \frac{b^2}{4a^2}$
6. $x + \frac{b}{2a} = \pm\sqrt{\frac{-c}{a} + \frac{b^2}{4a^2}}$
7. $x = -\frac{b}{2a} \pm \sqrt{\frac{-c}{a} + \frac{b^2}{4a^2}}$
8. $x = -\frac{b}{2a} \pm \sqrt{\frac{-c}{a} \cdot \frac{4a}{4a} + \frac{b^2}{4a^2}}$
9. $x = -\frac{b}{2a} \pm \sqrt{\frac{-4ac}{4a^2} + \frac{b^2}{4a^2}}$
10. $x = -\frac{b}{2a} \pm \sqrt{\frac{b^2 - 4ac}{4a^2}}$
11. $x = -\frac{b}{2a} \pm \sqrt{\frac{b^2 - 4ac}{(2a)^2}}$
12. $x = -\frac{b}{2a} \pm \frac{\sqrt{b^2 - 4ac}}{\sqrt{(2a)^2}}$
13. $x = -\frac{b}{2a} \pm \frac{\sqrt{b^2 - 4ac}}{2a}$
14. $x = \frac{-b \pm \sqrt{b^2 - 4ac}}{2a}$

EXERCISES

KEY SKILLS

Give a justification for each step at right.

Step 1: If $ax + b = 0$, then $ax = -b$.
Step 2: If $ax = -b$ and $a \neq 0$, then $x = -\frac{b}{a}$.

1. Step 1
2. Step 2

PRACTICE

Refer to the proof of the quadratic formula on the previous page. Give a justification for each specified step of the proof.

3. Step 2   4. Step 3   5. Step 4
6. Step 5   7. Step 6   8. Step 7
9. Step 8   10. Step 9   11. Step 12

The reasoning at right provides a derivation for a formula that can be used to solve equations of the form $ax^2 + b = rx^2 + t$, where $a \neq r$. Give a justification for each step.

Step 1: $ax^2 + b = rx^2 + t$, $a \neq r$
Step 2: $ax^2 = rx^2 + t - b$
Step 3: $ax^2 - rx^2 = t - b$
Step 4: $(a - r)x^2 = t - b$
Step 5: $x^2 = \frac{t - b}{a - r}$
Step 6: $x = \pm\sqrt{\frac{t - b}{a - r}}$

12. Step 1   13. Step 2
14. Step 3   15. Step 4
16. Step 5   17. Step 6

18. Derive a formula to find the solutions of $ax^2 + bx = rx^2 + tx$, where $a \neq r$. (*Hint:* Use factoring.)

MID-CHAPTER REVIEW

Solve each equation.

19. $x^2 = 25$ (8.1 Skill A)
20. $3a^2 + 46 = 100$ (8.2 Skill A)
21. $2z^2 + 48 = 0$ (8.2 Skill C)
22. $(2n + 3)^2 = 50$ (8.3 Skill A)
23. $4v^2 + 8v = 41$ (8.3 Skill B)
24. $4t^2 - 16t + 13 = 0$ (8.4 Skill A)

Solve. Round the answers to the nearest tenth. (8.2 Skill B)

25. The bottom of a 32-foot ladder is placed 8 feet from a wall. How far up the wall will the ladder reach?

26. The bottom of a 16-foot ladder is placed 5 feet from a wall. How far up the wall will the ladder reach?

Answers for Lesson 8.4   Skill C

1. Subtraction Property of Equality
2. Division Property of Equality
3. Subtraction Property of Equality
4. Distributive Property (factoring)
5. Addition Property of Equality
6. Addition Property of Equality to complete the square
7. Take the square root of each side of the equation.
8. Subtraction Property of Equality
9. Multiplicative Identity
10. Multiplying fractions
11. Quotient Property of Square Roots
12. given
13. Subtraction Property of Equality
14. Subtraction Property of Equality
15. Distributive Property (factoring)
16. Division Property of Equality
17. Take the square root of each side of the equation.
18. $x = 0$ or $x = \frac{t - b}{a - r}$
19. $x = \pm 5$
20. $a = \pm 3\sqrt{2}$
21. $z = \pm 2i\sqrt{6}$
22. $n = -\frac{3}{2} \pm \frac{5\sqrt{2}}{2}$
23. $v = -1 \pm \frac{3\sqrt{5}}{2}$
24. $t = 2 \pm \frac{\sqrt{3}}{2}$
25. 31.0 feet
26. 15.2 feet

Solving Quadratic Equations and Logical Reasoning

SKILL A *Choosing a method for solving a quadratic equation*

Focus: *Students analyze a given quadratic equation to determine the most appropriate solution method.*

EXAMPLE 1

PHASE 1: Presenting the Example

Display the equation $2n^2 - 18 = 0$. Divide the class into three groups. Assign one group to solve the equation by factoring, one by taking square roots, and one by using the quadratic formula. After they have completed their work, ask one student from each group to write the steps of the solution for all to see. Their work should essentially appear as follows:

Factoring	*Square Roots*	*Quadratic Formula*

$$
\begin{aligned}
2n^2 - 18 &= 0 \\
2(n^2 - 9) &= 0 \\
2(n + 3)(n - 3) &= 0 \\
n + 3 = 0 \quad &\text{or} \quad n - 3 = 0 \\
n = -3 \quad &\text{or} \quad n = 3
\end{aligned}
$$

$$
\begin{aligned}
2n^2 - 18 &= 0 \\
2n^2 &= 18 \\
n^2 &= 9 \\
n = -3 \quad &\text{or} \quad n = 3
\end{aligned}
$$

$$
2n^2 - 18 = 0
$$
$$
n = \frac{-0 \pm \sqrt{0^2 - 4(2)(-18)}}{2(2)}
$$
$$
n = \frac{\pm\sqrt{144}}{4}
$$
$$
n = 3 \quad \text{or} \quad n = -3
$$

Ask which solution method seems most appropriate. Students most likely will choose the square-root method, since it is most direct. Point out that the quadratic formula is the least efficient, because it involves much more work. Ask why no one was told to solve by completing the square. Lead students to see that the expression on the left side has no linear term, so it cannot be factored as a perfect-square trinomial. Then discuss Example 1.

EXAMPLE 2

PHASE 2: Providing Guidance for TRY THIS

After students have answered the question, ask how they would proceed in solving the equation. [**Simplify the left side to obtain $6x^2 - 11x - 7 = 3$; subtract 3 from each side to obtain $6x^2 - 11x - 10 = 0$; factor the left side or use the quadratic formula to obtain the solutions $\frac{5}{2}$ and $-\frac{2}{3}$.**]

EXAMPLE 3

PHASE 2: Providing Guidance for TRY THIS

Be sure students understand that the significance of the quadratic formula lies in its universality. That is, it applies to *any* quadratic equation. For equations such as $7x^2 = 175$, however, it is not the most efficient method of solution.

8.5 LESSON
Solving Quadratic Equations and Logical Reasoning

SKILL A *Choosing a method for solving a quadratic equation*

You have learned many different methods for solving a quadratic equation:

- Factoring. Use the Zero-Product Property.
- Taking a square root.
- Completing the square.
- Using the quadratic formula.

The method you choose should depend on the equation itself.

EXAMPLE 1 Identify the method you might choose to solve $7x^2 = 175$. Justify your choice.

▸ Solution

By dividing each side of the equation by 7 (Division Property of Equality), the result will be an equation that can be solved in one step by taking a square root.

TRY THIS   Explain why the method in Example 1 is also a good choice for solving $-3x^2 = -48$.

EXAMPLE 2 Identify the method you might choose to solve $(2x + 1)(3x - 7) = 0$. Justify your choice.

▸ Solution

Because the left side of the equation is given in factored form and the right side is 0, apply the Zero-Product Property. Then solve for x in the two linear equations that result.

TRY THIS   Explain why the choice in Example 2 will not immediately apply to solving $(2x + 1)(3x - 7) = 3$.

If the coefficients a, b, and c of a quadratic equation are not integers, it is often easiest to use the quadratic formula to solve the equation.

EXAMPLE 3 Identify the method you might choose to solve $2.5a^2 - 5.4a + 1 = 0$. Justify your choice.

▸ Solution

Because the coefficients are decimals, both factoring and completing the square would be very difficult.

Therefore, identify a, b, and c, and then apply the quadratic formula.

TRY THIS   Explain why using the quadratic formula to solve $7x^2 = 175$ is not the quickest method.

EXERCISES

KEY SKILLS

What solution method is used in each exercise?

1. $x^2 + 5x + 6 = 0$
    $(x + 2)(x + 3) = 0$
    $x + 2 = 0$ or $x + 3 = 0$
    $x = -2$ or $x = -3$

2. $3s^2 - 5s + 1 = 0$
    $$s = \frac{-(-5) \pm \sqrt{(-5)^2 - 4(3)(1)}}{2(3)}$$
    $$s = \frac{5 \pm \sqrt{13}}{6}$$

PRACTICE

Is the stated solution method suitable? Justify your response.

3. $2a^2 = 32$
    Take a square root.

4. $z^2 - 2.4z - 1.1 = 0$
    Use the quadratic formula.

5. $(v - 2)(v - 1.5) = 0$
    Apply the Zero-Product Property.

6. $3(n - 2)^2 = 27$
    Take a square root.

Identify the method you might choose to solve each quadratic equation. Justify your choice, but do not solve.

7. $0.5n^2 + 2.4n - 1 = 0$
8. $(2t + 3)(3t + 2) = 0$
9. $2h^2 - 98 = 0$
10. $-3(h + 3)^2 + 30 = 0$
11. $n^2 + 2n + 1 = 4$
12. $x^2 + 6x + 9 = 11$
13. $(-p + 5)(7p + 11) = 0$
14. $11y^2 = 13y - 37$
15. $0.5n^2 + 2.4n - 1 = 0$

Summarize a method for finding the number of x-intercepts.

16. $0 = x^2 + 2$
17. $0 = (2x + 3)(x + 11)$
18. $0 = 2.7x^2 - 2.4x - 1$
19. $0 = x^2 - 2$
20. $0 = (2x - 1)^2$
21. $0 = 1.3x^2 + 2x + 1.8$

22. How would you find b such that $y = 0.5x^2 + bx + 1$ has exactly one x-intercept?

23. **Critical Thinking** How would you find the real values of n such that $y = x^2 + nx + n + 1$ has one x-intercept?

MIXED REVIEW

Express the pattern shown in each table as a function with x as the independent variable. Predict the 60th term that would appear in the table. (3.8 Skill A)

24.
x	1	2	3	4	5
y	-0.5	1.5	3.5	5.5	7.5

25.
x	1	2	3	4	5
y	4	1	-2	-5	-8

Answers for Lesson 8.5 Skill A

Try This Example 1
Answers may vary. Sample answer: The equation $-3x^2 = -48$ has exactly the same form as $7x^2 = 175$. The steps in the solution would be the same but the numbers involved would be different.

Try This Example 2
Answers may vary. Sample answer: The Zero-Product Property requires that 0 be one of the sides of the given equation. That is not the case for $(2x + 1)(3x - 7) = 3$.

Try This Example 3
Answers may vary. Sample answer: Application of the quadratic formula would involve more steps than the solution given in Example 1. For example, it is not necessary to subtract 175 from each side of the equation in order to put it in standard form.

In Exercises 1–22, answers may vary.

1. factoring and the Zero-Product Property
2. quadratic formula
3. Yes; before taking square roots, divide each side of the equation by 2.
4. Yes; because the coefficients involve decimals.
5. Yes; the given equation is ready for the application of the Zero-Product Property.
6. Yes; before the square root is taken, divide each side of the equation by 3. After taking the square root, it is then necessary to add 2 to each side of the equation.
7. quadratic formula because the coefficients involve decimals.
8. Zero-Product Property because the equation is given in the form $ab = 0$
9. taking a square root after applying the Addition Property of Equality followed by the Division Property of Equality
10. taking a square root after applying the Addition Property of Equality followed by the Division Property of Equality; after taking square roots, add 3 to each side.
11. completing the square because $n^2 + 2n + 1 = 4$ is equivalent to $(n + 1)^2 = 4$
12. completing the square because $x^2 + 6x + 9 = 11$ is equivalent to $(x + 3)^2 = 11$

For answers to Exercises 13–25, see Additional Answers.

Focus: *Students organize the process of solving a quadratic equation into a logical argument that progresses from a given hypothesis to a justified conclusion.*

EXAMPLES 1 and 2

PHASE 1: Presenting the Examples

Begin by reviewing the terms *conditional statement, hypothesis, conclusion,* and *Transitive Property of Deductive Reasoning,* which were defined in Lesson 2.9. Then display each of the statements at right, one at a time.

$x^2 - 15x + 56 = (x - 7)(x - 8)$
If $x^2 - 15x + 56 = 0$, then $(x - 7)(x - 8) = 0$.
If $(x - 7)(x - 8) = 0$, then $x - 7 = 0$ or $x - 8 = 0$.
If $x - 7 = 0$ or $x - 8 = 0$, then $x = 7$ or $x = 8$.

After displaying each statement, tell students to write a justification for it. [**top to bottom: Factoring (Distributive Property), Substitution Principle, Zero-Product Property, Addition Property of Equality**] Discuss their responses, statement by statement, and answer any questions they may have.

Next display $x^2 - 15x + 56 = 0$, which is the equation in Example 1. Tell students that, when presented with an equation like this to solve, they are in fact being told to assume that it is a true statement. That is, they are to consider the equation as a given fact, or, more simply, as a *given*.

Now discuss the solution of Example 1 outlined in the text, where this given equation and the statements already justified are organized into a logically sequenced argument. Be sure students understand that, although the form of this argument omits the *if . . . then . . .* structure within individual statements, the Transitive Property of Deductive Reasoning remains the primary link in the chain of reasoning from hypothesis to conclusion.

PHASE 2: Providing Guidance for TRY THIS

Before they write an entire argument independently, some students may need practice in simply justifying the steps of a given argument. You can use Skill C Exercises 1 and 2 on the facing page for this purpose.

PHASE 3: ASSESSMENT AND CLOSURE for Lesson 8.5

- Have students create four quadratic equations: one that is best solved by factoring, one that is best solved by taking square roots, one that is best solved by completing the square, and one that is best solved by using the quadratic formula. Tell them to choose one of their equations and write the solution process for it as a deductive argument. [**Answers will vary.**]

- This lesson concludes the sequence of instruction in methods of solving quadratic equations. The two lessons that follow revisit the topic of quadratic functions, and students will see how the solutions of a quadratic equation can be used in analyzing the related quadratic function.

☞ *For a Lesson 8.5 Quiz, see* Assessment Resources *page 82.*

After deciding on a solution strategy, you can use *deductive reasoning* to write the steps in a solution as a logical argument. Recall from Lesson 2.9 that deductive reasoning is the process of starting with a statement or a hypothesis and following a sequence of logical steps to reach a conclusion.

In Lesson 2.9, you wrote solutions to linear equations as deductive arguments. You can also write solutions to quadratic equations as deductive arguments. The examples below show a two-column format for such deductive arguments.

EXAMPLE 1 Write the process for solving $x^2 - 15x + 56 = 0$ as a deductive argument in a two-column format.

▶ Solution

Statement	Reason
1. $x^2 - 15x + 56 = 0$	1. Given
2. $(x - 7)(x - 8) = 0$	2. Distributive Property (factoring)
3. $x - 7 = 0$ or $x - 8 = 0$	3. Zero-Product Property
4. $x = 7$ or $x = 8$	4. Addition Property of Equality

Thus, if $x^2 - 15x + 56 = 0$, then $x = 7$ or $x = 8$.

TRY THIS Write the process for solving $x^2 + 10x + 21 = 0$ as a deductive argument in a two-column format.

In many solutions, you can use the following fact from Lesson 8.1 as a justification in a logical argument.

If $x^2 = a$ and $a > 0$, then $x = \sqrt{a}$ or $x = -\sqrt{a}$.

It is the basis for taking the square root of each side of an equation.

EXAMPLE 2 Write the process for solving $x^2 + 6x = 7$ as a deductive argument in a two-column format.

▶ Solution

Statement	Reason
1. $x^2 + 6x = 7$	1. Given
2. $x^2 + 6x + 9 = 7 + 9$	2. Addition Property of Equality
3. $(x + 3)^2 = 16$	3. Distributive Property (factoring)
4. $x + 3 = \pm 4$	4. Take the square root of each side.
5. $x = -3 \pm 4$	5. Subtraction Property of Equality
6. $x = -7$ or $x = 1$	6. Simplify.

Thus, if $x^2 + 6x = 7$, then $x = -7$ or $x = 1$.

TRY THIS Write the process for solving $x^2 + 8x = 20$ as a deductive argument in a two-column format.

EXERCISES

KEY SKILLS

Give a reason for each step in the solution.

1.
Statements	Reasons
$x^2 - 6x = 0$	Given
$x(x - 6) = 0$	a. _?_
$x = 0$ or $x - 6 = 0$	b. _?_
$x = 0$ or $x = 6$	c. _?_

2.
Statements	Reasons
$-x^2 + 8x = -9$	Given
$x^2 - 8x = 9$	a. _?_
$x^2 - 8x - 9 = 0$	b. _?_
$(x - 9)(x + 1) = 0$	c. _?_
$x - 9 = 0$ or $x + 1 = 0$	d. _?_
$x = 9$ or $x = -1$	e. _?_

PRACTICE

The steps below are not listed in logical order. Write the steps in the solution of the given equation in logical order. Also give a reason for each step.

3. $x^2 - 6x + 5 = 32$ Given
$x^2 - 6x + 9 = 36$
$x - 3 = \pm 6$
$x^2 - 6x + 5 + 4 = 32 + 4$
$x = 9$ or $x = -3$
$(x - 3)^2 = 36$

4. $6m^2 = 7m + 5$ Given
$(2m + 1)(3m - 5) = 0$
$m = -\frac{1}{2}$ or $m = \frac{5}{3}$
$6m^2 - 7m - 5 = 0$
$2m = -1$ or $3m = 5$
$2m + 1 = 0$ or $3m - 5 = 0$

Write the process for solving each equation as a deductive argument in a two-column format.

5. $n^2 - 5n = 0$ 6. $4(t + 5)^2 = 32$ 7. $8b^2 + 2b - 3 = 0$

8. $d^2 - 12d = -36$ 9. $4r^2 - 9^2 = 0$ 10. $4(z - 5)^2 + 7 = 32$

Use a deductive argument to show that each statement is true.

11. The x-intercepts of $y = x^2 - 7x + 12$ are positive.

12. The x-intercepts of $y = x^2 - 3x - 18$ have opposite signs.

MIXED REVIEW

Solve each system of equations. (5.3 Skill B)

13. $\begin{cases} 3a - 2b = 1 \\ -2a + 7 = 0 \end{cases}$
14. $\begin{cases} -5r - 3s = 7 \\ -7r + 7s = 1 \end{cases}$
15. $\begin{cases} 2.5m - 3n = 0 \\ 3m + 2n = 2 \end{cases}$

16. $\begin{cases} 5x - 3y = 16 \\ 3x + 2y = 2 \end{cases}$
17. $\begin{cases} -7b - 3c = 5 \\ -2b + 2c = 30 \end{cases}$
18. $\begin{cases} 10u - 5v = 0 \\ 3u + 2v = 10 \end{cases}$

Answers for Lesson 8.5 Skill B

Try This Example 1

1 $x^2 + 10x + 21 = 0$
given

2 $x^2 + 10x + 21 = (x + 7)(x + 3)$
Distributive Property (factoring)

3 $(x + 7)(x + 3) = 0$
Substitution Principle with **1** and **2**

4 $x + 7 = 0$ or $x + 3 = 0$
Zero-Product Property

5 $x = -7$ or $x = -3$
Addition Property of Equality
Therefore, if $x^2 + 10x + 21 = 0$, then $x = -7$ or $x = -3$.

Try This Example 2

1 $x^2 + 8x = 20$
given

2 $x^2 + 8x + 16 = 20 + 16$
Addition Property of Equality

3 $(x + 4)^2 = 36$
Distributive Property (factoring)

4 $x + 4 = \pm 6$
Take the square root of each side.

5 $x = -10$ or $x = 2$
Subtraction Property of Equality
Therefore, if $x^2 + 8x = 20$, then $x = -10$ or $x = 2$.

1. a. Distributive Property (factoring)
   b. Zero-Product Property
   c. Addition Property of Equality
2. a. Multiplication Property of Equality
   b. Subtraction Property of Equality
   c. Distributive Property (factoring)
   d. Zero-Product Property
   e. Addition and Subtraction Properties of Equality
3. $x^2 - 6x + 5 = 32$
   given
   $x^2 - 6x + 5 + 4 = 32 + 4$
   Addition Property of Equality
   $x^2 - 6x + 9 = 36$
   addition of numbers
   $(x - 3)^2 = 36$
   Distributive Property (factoring)
   $x - 3 = \pm 6$
   Take the square root of each side.
   $x = 9$ or $x = -3$
   Addition Property of Equality

For answers to Exercises 4–18, see Additional Answers.

Quadratic Patterns

SKILL A   *Recognizing and continuing a quadratic pattern*

Focus: *Students use differences between successive terms to identify a given pattern as linear or quadratic and to extend a given quadratic pattern.*

EXAMPLE

PHASE 1: Presenting the Example

Display the first table shown at right. Ask students to describe the pattern among the numbers in the table. Discuss all their observations, ultimately eliciting the fact that the difference between successive x-terms is constantly 1 and the difference between successive y-terms is constantly 3. Remind them that a pattern like this is called a *linear pattern*.

x	-3	-2	-1	0	1	2	3
y	17	20	23	26	29	32	35

x	-3	-2	-1	0	1	2	3
y	17						

Now display the second table, which is like the first except that all numbers but 17 have been omitted from the second row. Tell students to find a different way to complete the table so that it shows a linear pattern. [**Answers will vary. Two possible responses are 22, 27, 32, 37, 42, 47 and 15, 13, 11, 9, 7, 5.**] Ask several volunteers to write their patterns on the chalkboard for all to see. Point out that infinitely many linear patterns are possible, but all will have one common characteristic: The difference between successive x-terms will be a constant and the difference between successive y-terms will be a constant. Then discuss the Example.

PHASE 2: Providing Guidance for TRY THIS

Students who are oriented toward a more visual style of learning may have greater success in continuing the pattern if they copy the table and extend it as shown below. This annotated table clearly shows that the *first differences* between successive y-values in the table are not constant. However, the differences between the first differences, called the *second differences*, are constant. Students can now work backward from the second differences to find that the next three first differences must be -7, -9, and -11, so the next three table entries must be -14, -23, and -34.

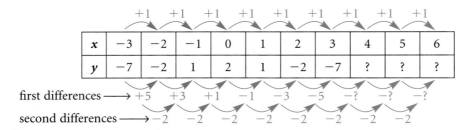

8.6 LESSON — Quadratic Patterns

SKILL A ▸ *Recognizing and continuing a quadratic pattern*

Recall from Lesson 3.9 that a table of ordered pairs (x, y) shows a linear pattern if the differences between successive x-terms are constant and the differences between successive y-terms are also constant. The table below represents a linear pattern.

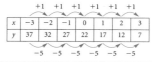

	$+1$	$+1$	$+1$	$+1$	$+1$	$+1$	
x	-3	-2	-1	0	1	2	3
y	37	32	27	22	17	12	7
	-5	-5	-5	-5	-5	-5	

Test for a Quadratic Pattern in a Table of Ordered Pairs

A table of ordered pairs (x, y) shows a quadratic pattern if the differences between successive x-values are constant and the differences between successive y-values form a linear pattern.

EXAMPLE

Does the table at right represent a quadratic pattern? Continue the table for $x = 4$, 5, and 6.

x	-3	-2	-1	0	1	2	3
y	17	7	1	-1	1	7	17

▸ **Solution**

First find the differences between successive x- and y-values.

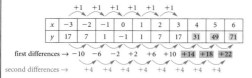

	$+1$	$+1$	$+1$	$+1$	$+1$	$+1$				
x	-3	-2	-1	0	1	2	3	4	5	6
y	17	7	1	-1	1	7	17	31	49	71

first differences → -10 -6 -2 $+2$ $+6$ $+10$ $+14$ $+18$ $+22$

second differences → $+4$ $+4$ $+4$ $+4$ $+4$ $+4$ $+4$ $+4$

Examine the differences of the y-values for a linear pattern by finding *their* differences, as shown above in red. Because these *second differences* are a constant 4, the table represents a quadratic pattern. You can use the patterns of differences to find the next three y-values. For example, $10 + 4 = 14$ and $17 + 14 = 31$. The y-values are 31, 49 and 71.

TRY THIS   Refer to the table in the Example. If the y-values are $-7, -2, 1, 2, 1, -2$, and -7, does the table represent a quadratic pattern? Continue the table for $x = 4$, 5, and 6.

328   Chapter 8   Quadratic Functions and Equations

EXERCISES

KEY SKILLS

Write the successive differences of the y-values. Are the differences constant?

1.

x	-3	-2	-1	0	1	2	3
y	-7	-2	3	8	13	18	23

2.

x	-3	-2	-1	0	1	2	3
y	27	14	5	0	-1	2	9

PRACTICE

Describe the pattern in each table as linear, quadratic, or neither.

3.

x	-3	-2	-1	0	1	2	3
y	4	1	0	1	4	9	16

4.

x	-3	-2	-1	0	1	2	3
y	4	0	-2	-2	0	4	10

5.

x	-3	-2	-1	0	1	2	3
y	48	22	6	0	4	18	42

6.

x	-3	-2	-1	0	1	2	3
y	24	22	20	18	16	14	12

7.

x	-3	-2	-1	0	1	2	3
y	-15	-10	-5	0	5	10	15

8.

x	-3	-2	-1	0	1	2	3
y	-23	-17	-11	-5	1	7	13

Describe the pattern in each list as linear or quadratic. Then give the next three ordered pairs for each list.

9. $(1, 1), (2, 4), (3, 9), (4, 16), (5, 25)$

10. $(-3, -1), (-2, -6), (-1, -9), (0, -10)$

11. $(-2, -14), (-1, -5), (0, 0), (1, 1), (2, -2)$

12. $(-2, -18), (-1, -8), (0, 0), (1, 6), (2, 10)$

13. Critical Thinking   Consider $y = ax^2 + bx + c$, where a, b, and c are fixed real numbers and $a \neq 0$.
 a. Write an expression for y when x is replaced by n.
 b. Show that when x is replaced by $n + 1$, then $y = an^2 + 2an + bn + a + b + c$.
 c. Show that the expression in part **b** minus the expression in part **a** equals $(2a)n + (a + b)$, which is a linear expression in n.

MIXED REVIEW

Write a function to predict the number of objects, t, in the nth diagram. Then predict the number of objects in the 40th diagram. (3.9 Skill B)

14.   Diagram 1   Diagram 2   Diagram 3   Diagram 4

15.   Diagram 1   Diagram 2   Diagram 3   Diagram 4

Answers for Lesson 8.6 Skill A

Try This Example
Yes, the pattern is quadratic.

4	5	6
-14	-23	-34

1. $5, 5, 5, 5, 5, 5$; yes
2. $-13, -9, -5, -1, 3, 7$; no
3. quadratic
4. quadratic
5. quadratic
6. linear
7. linear
8. linear
9. quadratic; $(6, 36), (7, 49), (8, 64)$
10. quadratic; $(1, -9), (2, -6), (3, -1)$
11. quadratic; $(3, -9), (4, -20), (5, -35)$
12. quadratic; $(3, 12), (4, 12), (5, 10)$

13. **a.** $y = an^2 + bn + c$
 b. $a(n + 1)^2 + b(n + 1) + c$
  $= an^2 + 2an + a + bn + b + c$
  $= an^2 + 2an + bn + a + b + c$
 c. $an^2 + 2an + bn + a + b + c - (an^2 + bn + c)$
  $= an^2 + 2an + bn + a + b + c - an^2 - bn - c$
  $= 2an + a + b = (2a)n + (a + b)$

14. $t = 6n - 3; 237$
15. $t = 7n - 5; 275$

Focus: *Students write equations of the form $y = ax^2 + bx + c$ to model geometric patterns.*

EXAMPLE

PHASE 1: Presenting the Example

Display the figure below.

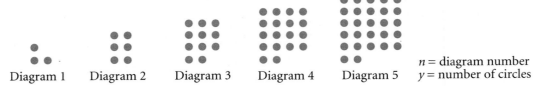

Remind students that they worked with figures such as this when they studied linear patterns in Chapter 3. Ask them to describe how the number of circles in each diagram is related to the diagram number. [**The number of circles is two more than the diagram number.**] Have them write a linear equation that represents this geometric pattern, with y as the dependent variable. [$y = n + 2$] Then display the following figure:

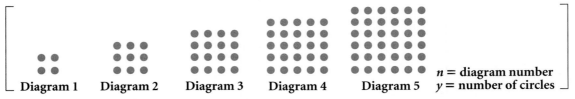

Tell students that this geometric pattern is not linear, but rather is quadratic. Ask them to describe the pattern. [**The number of circles is two more than the square of the diagram number.**] Have them write an equation that represents the pattern, again using y as the dependent variable. [$y = n^2 + 2$]

Now display the equation $y = (n + 1)^2$. Ask students to draw a set of diagrams that this equation might represent.

Point out to students that they have now seen how an equation of the form $y = ax^2 + bx + c$ can model a geometric pattern. Then discuss the Example.

PHASE 2: Providing Guidance for **TRY THIS**

If you had students draw the diagrams for $(n + 1)^2$ as suggested above, have them compare that set of diagrams to their diagrams for the Try This exercise. When placed side-by-side, they provide students with a graphic illustration of the fact that the expression $(n + 1)^2$ is not equivalent to the expression $n^2 + 1$.

In Lesson 3.9, you learned how to represent a linear pattern shown geometrically with an equation. You can also write a quadratic function to represent a quadratic pattern shown geometrically.

EXAMPLE

a. Represent the dot pattern below with an equation.
b. How many dots will be in the 50th group of dots?

Look for a pattern.

▸ **Solution**

a. Rearrange the dots so that a pattern may be more visible.

$$2^2 - 1 \quad 3^2 - 1 \quad 4^2 - 1 \quad 5^2 - 1$$
$$(1+1)^2 - 1 \quad (2+1)^2 - 1 \quad (3+1)^2 - 1 \quad (4+1)^2 - 1$$

This rearrangement suggests the following:

If n is the group number and y is the number of dots in each group, then $y = (n+1)^2 - 1$.

Substituting $(1, 3)$, $(2, 8)$, $(3, 15)$, and $(4, 24)$ in the equation verifies that the equation $y = (n+1)^2 - 1$ represents the pattern.

b. Substitute 50 for n in $y = (n+1)^2 - 1$ and solve for y.

$$y = (50+1)^2 - 1 = 2600$$

The 50th group will have 2600 dots.

You can also solve the example above by counting the dots in each group and then representing the counts in a table.

n	1	2	3	4
y	3	8	15	24
y	$4-1$	$9-1$	$16-1$	$25-1$
y	2^2-1	3^2-1	4^2-1	5^2-1

From the table, $y = (n+1)^2 - 1$.

TRY THIS Represent the dot pattern shown below with an equation. How many dots will be in the 50th group?

EXERCISES

KEY SKILLS

Refer to the pattern below.

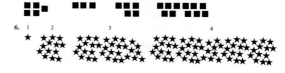

1. Make a table that shows the group number, n, and the number of squares in each group, y, for the pattern above.

2. Use guess-and-check to find a and b such that $y = 2(n-a)^2 + b$ represents the table you made in Exercise 1.

PRACTICE

Represent each pattern below with an equation. Then predict how many objects will appear in the 60th group of objects.

3.

4.

5.

6.

MIXED REVIEW

Solve each equation. (7.5 Skill A)

7. $(a+4)(a+7) = 0$
8. $(2n-1)(n+5) = 0$
9. $(7z+1)(3z-4) = 0$
10. $4t^2 - 12t + 9 = 0$
11. $4y^2 + 4y + 1 = 0$
12. $16d^2 - 56d + 49 = 0$
13. $9r^2 + 9r - 28 = 0$
14. $3z^2 - 10z - 77 = 0$
15. $16c^2 + 56c + 49 = 0$
16. $2h^2 + 11h - 90 = 0$
17. $3m^2 - 19m - 110 = 0$
18. $25g^2 - 5g - 2 = 0$

Answers for Lesson 8.6 Skill B

Try This Example
$y = n^2 + 1; 2501$

1.
n	1	2	3	4
y	3	5	11	21

2. $a = 1$ and $b = 3$; $y = 2(n-1)^2 + 3$
3. $y = (n+1)^2 + 1; 3722$
4. $y = (n+2)^2 + 1; 3845$
5. $y = 2(n-2)^2 + 3; 6731$
6. $y = 4n^2 - 3; 14{,}397$
7. $a = -4$ or $a = -7$
8. $n = \dfrac{1}{2}$ or $n = -5$
9. $z = -\dfrac{1}{7}$ or $z = \dfrac{4}{3}$
10. $t = \dfrac{3}{2}$
11. $y = -\dfrac{1}{2}$
12. $d = \dfrac{7}{4}$

13. $r = -\dfrac{7}{3}$ or $r = \dfrac{4}{3}$
14. $z = 7$ or $z = -\dfrac{11}{3}$
15. $c = -\dfrac{7}{4}$
16. $h = -10$ or $h = \dfrac{9}{2}$
17. $m = 10$ or $m = -\dfrac{11}{3}$
18. $g = -\dfrac{1}{5}$ or $g = \dfrac{2}{5}$

Focus: *Students work from a table of values to write an equation for a quadratic function and to verify that a set of points lie on the same parabola.*

EXAMPLE 1

PHASE 1: Presenting the Example

Display the function table and equations shown at right. Tell students that one of the equations represents the function. Have them identify which it is. [**equation 3**] Discuss how they found the correct equation. If no one observes it, point out that the table itself gives a clue: When $x = 0$, $y = -5$. Since each equation is of the form $y = ax^2 + bx + c$, you only need to look for the equation in which the value of c is -5. Tell students that they will now see how an observation like this is used to write an equation for a given quadratic function. Then discuss Example 1.

x	-3	-2	-1	0	1	2	3
y	-14	-13	-10	-5	2	11	22

1. $y = 3x^2 + 10x - 11$

2. $y = 2x^2 + 6x - 14$

3. $y = x^2 + 6x - 5$

4. $y = -x^2 + 9x + 22$

PHASE 2: Providing Guidance for TRY THIS

Be sure students understand that, to create the system of equations, they may choose any two ordered pairs from the table *except the pair with 0 as its x-coordinate*. However, it is wise to look for ordered pairs that will result in fairly simply calculations. In this case, $(-1, 1)$ and $(1, 1)$ are good choices.

EXAMPLE 2

PHASE 2: Providing Guidance for TRY THIS

To find the equation $y = 2x^2 - 3$, students will use three ordered pairs from the table. They probably will choose $(0, -3)$, $(-1, -1)$ and $(1, -1)$. Be sure they are aware that, geometrically, any three points determine a parabola. So, at this juncture, they have only found an equation of the parabola that contains the three points they used. To answer the question, they must show that all other ordered pairs in the table satisfy the equation.

PHASE 3: ASSESSMENT AND CLOSURE for Lesson 8.6

- Tell students to write an original equation for a quadratic function and use it to complete the table at right. Have them show how to work backward from the table to derive the equation. [**Answers will vary.**]

x	-3	-2	-1	0	1	2	3
y							

- In Lesson 8.7, students further investigate quadratic functions. Specifically, they will learn about functions that model the motion of a falling object.

☞ *For a Lesson 8.6 Quiz, see* Assessment Resources *page 83.*

If a table of ordered pairs can be represented by the equation $y = ax^2 + bx + c$, then you may be able to use a system of equations to find a, b, and c.

EXAMPLE 1 The table at right shows a quadratic pattern. Write a quadratic equation to represent the pattern.

x	-3	-2	-1	0	1	2	3
y	-7	-2	1	2	1	-2	-7

▶ Solution

Let $y = ax^2 + bx + c$ for some numbers a, b, and c.
From the table, we see that if $x = 0$, then $y = 2$.

$$2 = a(0)^2 + b(0) + c \rightarrow 2 = c \rightarrow y = ax^2 + bx + 2$$

Write and solve a system of equations. Use (1, 1) and (−1, 1).

$$x = 1: \quad a(1)^2 + b(1) + 2 = 1 \atop x = -1: \quad a(-1)^2 + b(-1) + 2 = 1 \rightarrow \begin{cases} a + b = -1 \\ a - b = -1 \end{cases} \rightarrow \begin{cases} a = -1 \\ b = 0 \end{cases}$$

Substitute the values for a and b into $y = ax^2 + bx + 2$. An equation that represents the table of ordered pairs is $y = -x^2 + 2$.

TRY THIS The table at right shows a quadratic pattern. Write a quadratic equation to represent the pattern.

x	-3	-2	-1	0	1	2	3
y	10	3	0	1	6	15	28

Recall that the graph of a quadratic function is a parabola.

EXAMPLE 2 The graphs of the ordered pairs at right lie along a parabola. Find an equation for the parabola. Verify that the ordered pairs satisfy the equation.

x	-3	-2	-1	0	1	2	3
y	29	14	5	2	5	14	29

▶ Solution

Assume that the ordered pairs satisfy $y = ax^2 + bx + c$ for some numbers a, b, and c. Because $y = 2$ when $x = 0$, $c = 2$ and $y = ax^2 + bx + 2$.

Write and solve a system of equations. Use (1, 5) and (2, 14).

$$x = 1: \quad a(1)^2 + b(1) + 2 = 5 \atop x = 2: \quad a(2)^2 + b(2) + 2 = 14 \rightarrow \begin{cases} a + b = 3 \\ 4a + 2b = 12 \end{cases} \rightarrow \begin{cases} a = 3 \\ b = 0 \end{cases}$$

Therefore, $y = 3x^2 + 2$.
Verify that $(-3, 29)$, $(-2, 14)$, $(-1, 5)$, and $(3, 29)$ satisfy $y = 3x^2 + 2$.

$$x = -3: \ 3(-3)^2 + 2 = 29 ✔ \qquad x = -1: \ 3(-1)^2 + 2 = 5 ✔$$
$$x = -2: \ 3(-2)^2 + 2 = 14 ✔ \qquad x = 3: \quad 3(3)^2 + 2 = 29 ✔$$

TRY THIS The graphs of the ordered pairs at right lie along a parabola. Find an equation for the parabola. Verify that the ordered pairs satisfy the equation.

x	-3	-2	-1	0	1	2	3
y	15	5	-1	-3	-1	5	15

EXERCISES

Verify that each table represents a quadratic pattern.

1.

x	-4	-3	-2	-1	0	1	2
y	80	45	20	5	0	5	20

2.

x	-3	-2	-1	0	1	2	3
y	16	6	0	-2	0	6	16

3.

x	-2	-1	0	1	2	3	4
y	-6	0	2	0	-6	-16	-30

4.

x	-4	-3	-2	-1	0	1	2
y	-47	-26	-11	-2	1	-2	-11

PRACTICE

Each table in Exercises 1–4 represents a quadratic pattern. Write a quadratic equation to represent each table.

5. Exercise 1   6. Exercise 2   7. Exercise 3   8. Exercise 4

The graphs of each set of ordered pairs lie along a parabola. Find an equation for the parabola. Verify that the ordered pairs satisfy the equation.

9. $y = x^2 - bx + 4$ for some real b

x	1	2	3	4	5	6
y	3	4	7	12	19	28

10. $y = -3x^2 + 7x + c$ for some real c

x	1	2	3	4	5	6
y	-1	-3	-11	-25	-45	-71

11. $y = ax^2 + x + c$ for some real a and c

x	1	2	3	4	5	6
y	-7	-12	-21	-34	-51	-72

12. $y = ax^2 + c$ for some real a and c

x	1	2	3	4	5	6
y	-7	2	17	38	65	98

13. Critical Thinking Find an equation of the form $y = ax^2 + c$, for some real numbers a and c, that represents the graph at right. Then find y for each value of x in the table below.

x	10	11	12	13	14	15	16
y	?	?	?	?	?	?	?

MIXED REVIEW

Solve each equation. Write the solutions in simplest radical form. (8.2 Skill A)

14. $4b^2 - 49 = 0$ 15. $81v^2 - 4 = 0$ 16. $9t^2 - 125 = 0$ 17. $-5u^2 + 120 = 0$

18. $-9r^2 + 44 = -5$ 19. $-3m^2 - 7 = -67$ 20. $-\frac{1}{3}h^2 + 10 = -1$ 21. $-\frac{1}{2}p^2 + 2 = -110$

Answers for Lesson 8.6 Skill C

Try This Example 1
$y = 2x^2 + 3x + 1$

Try This Example 2
$y = 2x^2 - 3$

1. Because the second differences are constant (-10), the table represents a quadratic pattern.
2. Because the second differences are constant (-4), the table represents a quadratic pattern.
3. Because the second differences are constant (4), the table represents a quadratic pattern.
4. Because the second differences are constant (6), the table represents a quadratic pattern.
5. $y = 5x^2$
6. $y = 2x^2 - 2$
7. $y = -2x^2 + 2$
8. $y = -3x^2 + 1$
9. $y = x^2 - 2x + 4$
10. $y = -3x^2 + 7x - 5$
11. $y = -2x^2 + x - 6$
12. $y = 3x^2 - 10$
13. $y = -\frac{1}{2}x^2;\ -50, -60.5, -72, -84.5, -98, -112.5, -128$

14. $b = \pm\frac{7}{2}$

15. $v = \pm\frac{2}{9}$

16. $t = \pm\frac{5\sqrt{5}}{3}$

17. $u = \pm2\sqrt{6}$

18. $r = \pm\frac{7}{3}$

19. $m = \pm2\sqrt{5}$
20. $h = \pm\sqrt{33}$
21. $p = \pm4\sqrt{14}$

Quadratic Functions and Acceleration

> **SKILL A** **APPLICATIONS** Using quadratic functions to solve
> problems involving falling objects
> that have no initial velocity

Focus: *Students find the height above ground of a free-falling object at a given time.*
They also find the time needed for a free-falling object to achieve a given height.

EXAMPLE 1

PHASE 1: Presenting the Example

A *projectile* is any object that is set in motion by an external action. Some of these actions
might be throwing the object, firing it, or simply dropping it. Whatever the action, the
origin of motion is not within the object itself.

Have students read the problem. Note that the skydiver initiates
motion by jumping. After this, however, and until a parachute opens,
the fall is determined by gravity. So, during this period of *free fall*, the
skydiver is a projectile.

Some students may be unaware that the rate of fall is not constant.
Point out that any object travels faster and faster, or *accelerates*, as it
falls. The rate of acceleration due to gravity is approximately 32 feet
per second *per second*. This means that, during every second of the fall,
the skydiver's speed increases by 32 feet per second. Ask students why
the familiar formula *distance* = *rate* × *time* would not apply to the fall.
[**To use this formula, the rate of motion must be constant.**]

Tell students that the distance traveled by an object in free fall is
given by the formula $d = 16t^2$, where d is distance in feet and t is
time in seconds. Ask them to use this fact to create a formula for the
skydiver's height above the ground, h, at any given time, t.
[$h = 5200 - 16t^2$, **or** $h = -16t^2 + 5200$] Now discuss the solution
of Example 1 that appears on the textbook page.

PHASE 2: Providing Guidance for TRY THIS

To give students a visual representation of the concept of acceleration,
extend the problem. Have them make a table like the one begun at
right, using equally-spaced 16-foot increments throughout the
Distance Fallen column.

Progress of a Skydiver from 5400 Feet (Without Parachute)			
Time Falling	Speed of Fall	Distance Fallen	Altitude
0 s	0 ft/s	0 ft	5400 ft
1 s	32 ft/s	16 ft	5384 ft
2 s	64 ft/s	64 ft	5336 ft
3 s	96 ft/s	144 ft	5256 ft
4 s	128 ft/s	256 ft	5144 ft
⋮	⋮	⋮	⋮

EXAMPLE 2

PHASE 1: Presenting the Example

After students read the problem, ask them how this situation differs from Example 1.
Lead them to see that, in this case, a height above the ground is given and they must find
the time required to reach this height. Then discuss the solution.

8.7 LESSON — Quadratic Functions and Acceleration

SKILL A ▶ APPLICATIONS Using quadratic functions to solve problems involving falling objects that have no initial velocity

Because of acceleration due to gravity, the speed of a falling object is not constant. The relationship between the distance that an object falls and the time that the object spends falling cannot be modeled by a linear equation.

Free-Falling Objects
If an object is dropped from an initial height of h_0 feet and is allowed to fall freely, then after t seconds it will fall $16t^2$ feet and will have height h given by $h = -16t^2 + h_0$.

EXAMPLE 1 A skydiver jumps from a plane at a height of 5200 feet above the ground. How far does the diver descend and what is the diver's height above the ground after 3 seconds, 4 seconds, and 5 seconds of free fall?

▶ Solution

Substitute 3, 4, and 5 for t into $16t^2$ and $-16t^2 + 5200$.

Make a table.

time	$t = 3$	$t = 4$	$t = 5$
distance	$16(3)^2 = 144$	$16(4)^2 = 256$	$16(5)^2 = 400$
height	$5200 - 144 = 5056$	$5200 - 256 = 4944$	$5200 - 400 = 4800$

After 3, 4, and 5 seconds, the diver descends 144 feet, 256 feet, and 400 feet; his height is 5056 feet, 4944 feet, and 4800 feet, respectively.

TRY THIS Rework Example 1 given an initial height of 5400 feet and elapsed times of 2 seconds, 4 seconds, and 6 seconds.

EXAMPLE 2 After how many seconds of free fall will an object dropped from 1200 feet have a height of 120 feet above the ground? Use a calculator.

▶ Solution

$$h = -16t^2 + h_0$$
$$120 = -16t^2 + 1200 \quad \longleftarrow \text{ Replace } h \text{ with 120 and } h_0 \text{ with 1200.}$$
$$-1080 = -16t^2$$
$$t = \pm\sqrt{\frac{1080}{16}} \quad \longleftarrow \text{ Take the square root of each side.}$$
$$t \approx 8.2 \quad \longleftarrow \text{ Reject the negative solution because time cannot be negative.}$$

The height will be 120 feet after about 8.2 seconds.

TRY THIS After how many seconds of free fall will an object dropped from 2500 feet have a height of 500 feet above the ground?

EXERCISES

KEY SKILLS

Find the exact positive value of t.

1. $125 = -16t^2 + 1300$
2. $140 = -16t^2 + 1500$
3. $160 = -16t^2 + 2800$

PRACTICE

A skydiver jumps from a plane at the given height and falls freely. Find how far the diver descends and the diver's height for each elapsed time.

4. 5000 feet; after 4 seconds
5. 5600 feet; after 4 seconds
6. 5200 feet; after 3 seconds
7. 4800 feet; after 5 seconds
8. 6500 feet; after 4.5 seconds
9. 6300 feet; after 5.5 seconds

10. A skydiver jumps from a plane at a height of 6800 feet. How far does the diver descend and what is the diver's height above the ground after 2 seconds, 3 seconds, and 6 seconds of free fall?

To the nearest tenth of a second, find how long it takes for an object in free fall to have the specified height.

11. initial height: 5300 feet
    specified height: 4000 feet
12. initial height: 6000 feet
    specified height: 3000 feet
13. initial height: 4200 feet
    specified height: 100 feet
14. initial height: 4800 feet
    specified height: 1000 feet
15. initial height: 7500 feet
    specified height: 0 feet
16. initial height: 100 feet
    specified height: 10 feet

17. **Critical Thinking** How long will it take an object in free fall to have a height that is one-half its initial height of h feet?

MIXED REVIEW APPLICATIONS

To the nearest tenth of a unit, approximate the unknown length of the third side in each right triangle. (8.2 Skill B)

18. hypotenuse: 100 meters
    length of leg: 70 meters
19. length of leg: 60 feet
    length of leg: 60 feet
20. hypotenuse: 2.5 inches
    length of leg: 1.0 inches
21. length of leg: 12 yards
    length of leg: 15 yards
22. length of leg: 120 feet
    length of leg: 200 feet
23. hypotenuse: 300 meters
    length of leg: 150 meters

Answers for Lesson 8.7 Skill A

Try This Example 1
2 seconds: 64 feet and 5336 feet
4 seconds: 256 feet and 5144 feet
6 seconds: 576 feet and 4824 feet

Try This Example 2
11.1 seconds

1. $t = \dfrac{5\sqrt{47}}{4}$
2. $t = \sqrt{85}$
3. $t = \sqrt{165}$
4. 256 feet and 4744 feet
5. 256 feet and 5344 feet
6. 144 feet and 5056 feet
7. 400 feet and 4400 feet
8. 324 feet and 6176 feet
9. 484 feet and 5816 feet
10. 2 seconds: 64 feet and 6736 feet
    3 seconds: 144 feet and 6656 feet
    6 seconds: 576 feet and 6224 feet
11. 9.0 seconds
12. 13.7 seconds

13. 16.0 seconds
14. 15.4 seconds
15. 21.7 seconds
16. 2.4 seconds
17. $\dfrac{\sqrt{2h}}{8}$ seconds
18. 71.4 meters
19. 84.9 feet
20. 2.3 inches
21. 19.2 yards
22. 233.2 feet
23. 259.8 meters

Focus: *Students find the height above ground of an object after it is propelled upward. They also find the time needed for such an object to achieve a given height.*

EXAMPLE 1

PHASE 1: Presenting the Example

Tell students that a ball is thrown directly up *from ground level* at a speed of 96 feet per second. Then lead them through the following problems:

Suppose there is no gravity pulling the ball back toward Earth. Copy and complete the table.

There still is no gravity, but now the ball is thrown from an altitude of 48 feet. Copy and complete the table.

The ball is again thrown from an altitude of 48 feet, but this time the pull of gravity is introduced. Copy and complete the table.

Time (s)	Altitude (ft)
1	96
2	192
3	288
t	$96t$

Time (s)	Altitude (ft)
1	144
2	240
3	336
t	$96t + 48$

Time (s)	Altitude (ft)
1	128
2	176
3	192
t	$96t + 48 - 16t^2$

Discuss the final expression. Point out that $-16t^2$ represents gravity, $96t$ represents initial upward velocity, and 48 represents initial height above sea level. Use this expression to develop the general formula $h = -16t^2 + v_0 t + h_0$.

EXAMPLE 2

PHASE 2: Providing Guidance for TRY THIS

Students may wonder why there are two answers. Show them the graph of $y = -16t^2 + 96t + 48$. (A graphing calculator display is shown at right.) The parabolic shape shows that the ball will achieve a height of 170 feet at two times, once on its way up and once on its way down. To illustrate a one-answer question, have students find the only time when the altitude is 32 feet. [**about 6.2 seconds**]

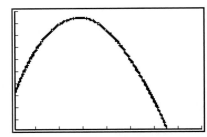

Xmin = 0	Ymin = 0
Xmax = 8	Ymax = 200
Xscl = 1	Yscl = 20

PHASE 3: ASSESSMENT AND CLOSURE for Lesson 8.7

- Have students describe the situation represented by $h = -16t^2 + 48t + 32$.
  [**A projectile is thrown upward from an altitude of 32 feet with an initial velocity of 48 feet per second. It achieves a height of *h* feet after *t* seconds.**]

- This lesson focused only on the *vertical component* of projectile motion. Students will analyze the horizontal component if they take an advanced algebra course.

☞ *For a Lesson 8.7 Quiz, see* Assessment Resources *page 84.*

☞ *For a Chapter 8 Assessment, see* Assessment Resources *pages 85–86.*

Using quadratic functions to solve problems involving falling objects that have an initial velocity

In Skill A of this lesson, you learned how to solve problems involving objects that were dropped from a still position, or that had no *initial velocity*. Now you will learn how to solve problems involving objects that do have an initial velocity.

Accelerated Motion with Initial Velocity

After t seconds, an object propelled upward from an initial height of h_0 feet and that has an initial velocity of v_0 feet per second will have height h feet given by $h = -16t^2 + v_0 t + h_0$.

EXAMPLE 1
A ball thrown directly upward from the edge of a cliff 48 feet above sea level is thrown at an initial velocity of 96 feet per second. Find the ball's height above sea level after 2 seconds and after 3 seconds.

▶ Solution

The height function is $h = -16t^2 + 96t + 48$.

Make a table.

t	$t = 2$	$t = 3$
h	$-16(2)^2 + 96(2) + 48 = 176$	$-16(3)^2 + 96(3) + 48 = 192$

After 2 and 3 seconds, the ball's height above sea level is 176 feet and 192 feet, respectively.

TRY THIS
Rework Example 1 given an initial height of 50 feet and an initial velocity of 128 feet per second.

EXAMPLE 2
Refer to the problem in Example 1. After how many seconds will the ball have a height of 180 feet? Round answer(s) to the nearest hundredth.

▶ Solution

Use a formula.

$$180 = -16t^2 + 96t + 48$$
$$0 = -16t^2 + 96t - 132 \quad \longleftarrow \text{Write in standard form.}$$
$$t = \frac{-96 \pm \sqrt{96^2 - 4(-16)(-132)}}{2(-16)} \quad \longleftarrow \begin{array}{l}\text{Apply the quadratic formula} \\ \text{with } a = -16, b = 96, \text{ and} \\ c = -132.\end{array}$$
$$t \approx 2.13 \text{ or } 3.87$$

The height is 180 feet after about 2.13 and 3.87 seconds.

TRY THIS
Rework Example 2 given a desired height of 170 feet.

You can use the following fact to find a projectile's maximum height.

If $h = -16t^2 + v_0 t + h_0$ models height as a function of time t, then maximum height is achieved after $\frac{v_0}{32}$ seconds.

EXERCISES

Write each equation in standard form. Then write the equation for t that results from applying the quadratic formula. Do not evaluate.

1. $-16t^2 + 96t + 60 = 180$

2. $-16t^2 + 96t + 100 = 200$

3. $-16t^2 + 96t = 180$

4. $-16t^2 + 96t + 500 = 1000$

A ball thrown directly upward from the edge of a cliff. The height of the cliff is given in feet above sea level and the specified initial velocity is given in feet per second. Find the height of the ball after 3 seconds and after 5 seconds.

5. height 50 ft; velocity 96 ft/sec

6. height 60 ft; velocity 96 ft/sec

7. height 96 ft; velocity 128 ft/sec

8. height 10 ft; velocity 128 ft/sec

9. height 160 ft; velocity 48 ft/sec

10. height 250 ft; velocity 48 ft/sec

A ball thrown directly upward from the edge of a cliff. The cliff is 64 feet above sea level and the ball has an initial velocity of 96 feet per second. Find the amount of time it takes the ball to have the specified height above sea level.

11. 170 feet

12. 200 feet

13. 64 feet

14. 128 feet

15. 0 feet

16. 100 feet

17. A ball thrown directly upward from the edge of a cliff 48 feet above sea level has an initial velocity of 96 feet per second.
  a. How long will it take the ball to reach its maximum height?
  b. Find the maximum height.

Simplify. Write answers in scientific notation. (6.1 Skill C)

18. $(2.3 \times 10^2)(3.1 \times 10^3)$

19. $(5.0 \times 10^1)(2.5 \times 10^4)$

20. $(6.3 \times 10^2)(4.0 \times 10^2)$

21. $(9.0 \times 10^3)(9.0 \times 10^2)$

22. $\frac{18.6 \times 10^5}{6.0 \times 10^4}$

23. $\frac{2.25 \times 10^2}{1.5 \times 10^3}$

24. $\frac{1.69 \times 10^6}{1.3 \times 10^2}$

25. $\frac{2.4 \times 10^3}{4.8 \times 10^5}$

Answers for Lesson 8.7 Skill B

Try This Example 1
after 2 seconds: 242 feet
after 3 seconds: 290 feet

Try This Example 2
after 1.8 seconds and after 4.2 seconds

1. $-16t^2 + 96t - 120 = 0$
$$t = \frac{-96 \pm \sqrt{96^2 - 4(-16)(-120)}}{2(-16)}$$

2. $-16t^2 + 96t - 100 = 0$
$$t = \frac{-96 \pm \sqrt{96^2 - 4(-16)(-100)}}{2(-16)}$$

3. $-16t^2 + 96t - 180 = 0$
$$t = \frac{-96 \pm \sqrt{96^2 - 4(-16)(-180)}}{2(-16)}$$

4. $-16t^2 + 96t - 500 = 0$
$$t = \frac{-96 \pm \sqrt{96^2 - 4(-16)(-500)}}{2(-16)}$$

5. after 3 seconds: 194 feet
  after 5 seconds: 130 feet

6. after 3 seconds: 204 feet
  after 5 seconds: 140 feet

7. after 3 seconds: 336 feet
  after 5 seconds: 336 feet

8. after 3 seconds: 250 feet
  after 5 seconds: 250 feet

9. after 3 seconds: 160 feet
  after 5 seconds: 0 feet

10. after 3 seconds: 250 feet
  after 5 seconds: 90 feet

11. 1.5 seconds and 4.5 seconds

12. 2.3 seconds and 3.7 seconds

13. 0 seconds and 6 seconds

14. 0.8 second and 5.2 seconds

15. 6.6 seconds

16. 0.4 second and 5.6 seconds

17. a. 3 seconds
  b. 192 feet

18. 7.13×10^5

19. 1.25×10^6

20. 2.52×10^5

21. 8.1×10^6

22. 3.1×10^1

23. 1.5×10^{-1}

24. 1.3×10^4

25. 5.0×10^{-3}

CHAPTER 8 ASSESSMENT

Find the square roots of each number. If the number is not a perfect square, approximate the roots to the nearest hundredth.

1. 225   2. 60   3. 0.49   4. 72

Write each square root in simplest radical form.

5. $\sqrt{98}$ 6. $\sqrt{300}$ 7. $\sqrt{\frac{2}{25}}$ 8. $\sqrt{\frac{50}{81}}$

9. Let $a > 0$. Using laws of exponents, prove that $\sqrt{36a^2} = 6a$.

Solve. Write irrational solutions in simplest radical form.

10. $9n^2 - 1 = 0$ 11. $4d^2 - 1 = 8$ 12. $16y^2 - 5 = 15$

13. In a right triangle, the length of the hypotenuse is 110 feet and the length of one leg is 65 feet. To the nearest hundredth of a foot, find the length of the other leg.

Solve. Write the solutions in simplest radical form where necessary.

14. $p^2 + 5 = 0$ 15. $2x^2 + 4 = 0$ 16. $2z^2 - 3 = -11$

17. $(x + 2)^2 = 4$ 18. $(k - 5)^2 = 27$ 19. $3(z - 1)^2 = 36$

Solve each equation by completing the square. Write the solutions in simplest radical form where necessary.

20. $4g^2 + 4g - 3 = 0$ 21. $z^2 - 4z - 46 = 0$

22. Boats A and B leave the dock at the same time and sail at a constant speed. Boat A sails due north and Boat B sails due east. After 2 hours, they are 26 miles apart. Boat B sails 7 miles per hour faster than Boat A. Find their speeds using completing the square.

Solve each equation using the quadratic formula. Write the solutions in simplest radical form where necessary.

23. $h^2 - 2h - 26 = 0$ 24. $s^2 - 14s + 31 = 0$ 25. $p^2 - 2p - 3 = 0$

26. $7z^2 + 24z - 55 = 0$ 27. $2d^2 = 19d - 45$ 28. $w^2 + 2w = 12$

Use the discriminant to find the number of real roots of each equation.

29. $3q^2 - q + 3 = 0$ 30. $4n^2 = 8n - 4$

Use the discriminant to find the number of x-intercepts of the graph of each quadratic function.

31. $y = 3t^2 - t - 3$ 32. $y = -9x^2 - 18x - 9$

33. $y = 3t^2 - t + 3$ 34. $y = 16x^2 - 49$

35. Let a, b, and c represent real numbers with $a \neq 0$.
 Prove that if $ax^2 + bx = -c$, then $a\left(x^2 + \frac{b}{a}x\right) = -c$.

36. Identify the solution method you might choose to solve $v^2 = 14v - 45$. Justify your choice but do not solve the equation.

37. Write the solution process to $2u^2 = -3u + 5$ as a deductive argument in a two-column format.

38. Continue the pattern in the table below for $x = 4, 5,$ and 6.

x	-3	-2	-1	0	1	2	3
y	30	16	6	0	-2	0	6

39. a. Represent the dot pattern below with an equation.
 b. How many dots will be in the 50th group of dots?

40. Write a quadratic function to represent the pattern in the table below.

x	-1	0	1	2	3	4	5
y	5	0	-1	2	9	20	35

41. A object is dropped from a height of 1375 feet. Approximate to the nearest tenth of a second how long it will take for the object to have a height above ground of 1000 feet. Use a calculator to approximate the value of the radical expressions.

42. A ball thrown directly upward from the edge of a cliff. The cliff is 100 feet above sea level and the ball has an initial velocity of 96 feet per second. After how many seconds will the ball have a height of 200 feet? Use a calculator to approximate the value of the radical expressions.

Answers for Chapter 8 Assessment

1. ± 15
2. ± 7.75
3. ± 0.7
4. ± 8.49
5. $7\sqrt{2}$
6. $10\sqrt{3}$
7. $\dfrac{\sqrt{2}}{5}$
8. $\dfrac{5\sqrt{2}}{9}$
9. $\sqrt{36a^2} =$
 $\sqrt{36}\sqrt{a^2} =$
 $\sqrt{6^2}\sqrt{a^2} = 6a$
10. $n = \pm\dfrac{1}{3}$
11. $d = \pm\dfrac{3}{2}$
12. $y = \pm\dfrac{\sqrt{5}}{2}$
13. 88.74 feet
14. $p = \pm i\sqrt{5}$

15. $x = \pm i\sqrt{2}$
16. $z = \pm 2i$
17. $x = 0$ or $x = -4$
18. $k = 5 \pm 3\sqrt{3}$
19. $z = 1 \pm 2\sqrt{3}$
20. $g = \dfrac{1}{2}$ or $g = -\dfrac{3}{2}$
21. $z = 2 \pm 5\sqrt{2}$
22. Boat A: 6.4 miles per hour; Boat B: 12.4 miles per hour
23. $h = 1 \pm 3\sqrt{3}$
24. $s = 7 \pm 3\sqrt{2}$
25. $p = -1$ or $p = 3$
26. $z = -5$ or $z = \dfrac{11}{7}$
27. $d = 5$ or $d = \dfrac{9}{2}$
28. $w = -1 \pm \sqrt{13}$
29. none
30. one
31. two

32. one
33. none
34. two
35. $ax^2 + bx = -c$ Given
 $a\left(x^2 + \dfrac{b}{a}x\right) = -c$ Distributive Property (factoring)
36. Write the equation in standard form. Then factor and use the Zero-Product Property since the coefficient of v^2 is 1.
37. $2u^2 = -3u + 5$ given
 $2u^2 + 3u - 5 = 0$ Addition Property of Equality
 $(2u + 5)(u - 1) = 0$ Distributive Property (factoring)
 $2u + 5 = 0$ or $u - 1 = 0$ Zero-Product Property
 $2u = -5$ or $u = 1$ Subtraction and Addition Properties of Equality
 $u = -\dfrac{5}{2}$ or $u = 1$ Division Property of Equality
38. $16, 30, 48$
39. a. $y = 3x^2 - 2x$
 b. 7400
40. $y = 2x^2 - 3x$
41. 4.8 seconds
42. 1.3 seconds and 4.7 seconds

Chapter 9 Rational Expressions, Equations, and Functions	State or Local Standards	Corresponding Lessons in *Algebra 1*, Schultz et al.
9.1 Rational Expressions and Functions		11.2
Skill A: Finding the domain of a rational expression or function		
9.2 Simplifying Rational Expressions		11.3
Skill A: Using monomial factoring to simplify rational expressions		
Skill B: Using polynomial factoring to simplify rational expressions		
Skill C: Simplifying rational expressions with more than one variable		
9.3 Multiplying and Dividing Rational Expressions		11.4
Skill A: Multiplying rational expressions		
Skill B: Dividing rational expressions		
Skill C: Combining multiplication and division with rational expressions		
9.4 Adding Rational Expressions		11.4
Skill A: Adding rational expressions with like denominators		
Skill B: Adding rational expressions with unlike monomial denominators		
Skill C: Adding rational expressions with unlike polynomial denominators		
9.5 Subtracting Rational Expressions		11.4
Skill A: Subtracting rational expressions with like denominators		
Skill B: Subtracting rational expressions with unlike denominators		
Skill C: APPLICATIONS Using operations with rational expressions to solve geometry problems		
9.6 Simplifying and Using Complex Fractions		11.4
Skill A: Simplifying complex fractions		
Skill B: APPLICATIONS Using complex fractions to solve real-world problems		
9.7 Solving Rational Equations		11.5
Skill A: Solving proportions involving rational expressions		
Skill B: Solving rational equations		
Skill C: APPLICATIONS Using work and rate relationships to solve real-world problems		
9.8 Inverse Variation		11.1
Skill A: Solving inverse-variation problems		
Skill B: Using a graph to represent an inverse-variation relationship		
Skill C: APPLICATIONS Using $y = \dfrac{k}{x^2}$ to solve inverse-variation problems		
9.9 Proportions and Deductive Reasoning		4.1, 11.6
Skill A: Using deductive reasoning to derive formulas and prove statements about proportions		

Chapter 9   **339**

CHAPTER

9

Rational Expressions, Equations, and Functions

▶ **What You Already Know**

In earlier mathematics courses, you learned how to add, subtract, multiply, and divide rational numbers. In Chapter 6, you learned how to add, subtract, multiply, and divide polynomials in one variable. In Chapter 9, you can bring these concepts and skills together to help you work with rational expressions and functions.

▶ **What You Will Learn**

First, you will learn what a rational expression is. In particular, you will learn how to find those real numbers for which the rational expression has meaning and those for which it does not.

A great deal of your time will be directed toward adding, subtracting, multiplying, and dividing rational expressions. Simplifying a rational expression will also be part of your study. You will also learn how to

• solve and use rational equations to solve problems involving electricity, ratios, and work,

• solve problems involving inverse variation and inverse variation as the square, problems such as those dealing with area and illumination, and

• use deductive reasoning to prove mathematical statements about proportions.

VOCABULARY

complex fraction	least common denominator (LCD)
constant of variation	least common multiple (LCM)
converse	rational equation
Cross-Product Property of Proportions	rational expression
domain (of a rational expression)	rational function
excluded value	reciprocal function
extraneous solution	similar (geometric figures)
inverse variation	work rate
inverse-square variation	

The diagram below shows how mathematical skills and mathematical reasoning are interrelated with the skills and concepts in Chapter 9. Notice that the focus of this chapter involves operations with rational expressions such as multiplication and addition.

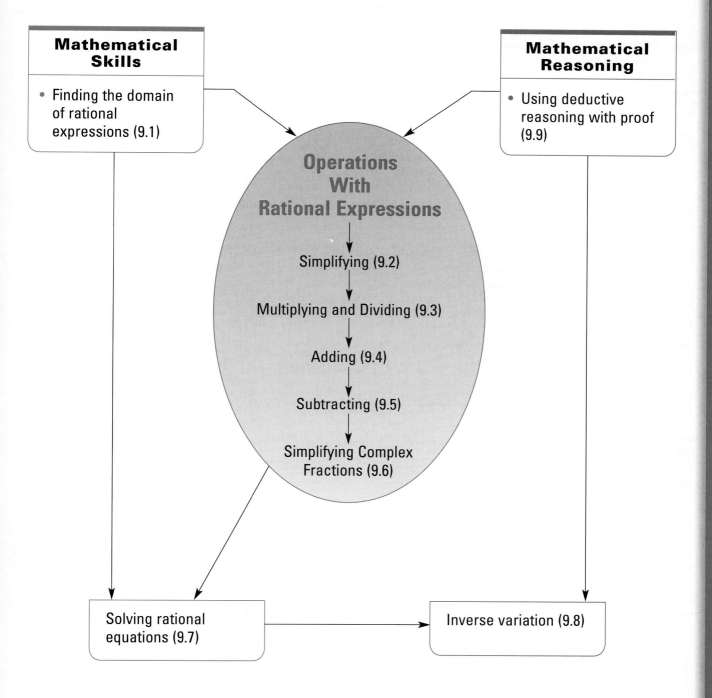

SKILL A ▶ *Finding the domain of a rational expression or function*

Focus: *Students identify those values that are excluded from the domain of a rational expression or function.*

EXAMPLE 1

PHASE 1: Presenting the Example

Have students write a definition of *rational number*. [**any number that can be represented by the expression $\frac{p}{q}$, where p and q are integers and $q \neq 0$.**] Discuss their responses. Ask them to explain why q cannot be zero. [**The expression represents $p \div q$, and division by zero is undefined.**]

Discuss the definition of *rational expression*. Point out to students that all rational numbers are rational expressions, and so rational expressions are not totally new to them. Rather, they will now extend their experience to encompass rational expressions that involve variables. Similarly, their work with *rational functions* will extend previous work with rational functions. Introduce the definition of *excluded value*. Then discuss Example 1.

PHASE 2: Providing Guidance for TRY THIS

Watch for students who write $2t + 3 = 0$. Stress the fact that only division *by 0* is undefined, so they need only solve $5t + 4 = 0$.

EXAMPLE 2

PHASE 2: Providing Guidance for TRY THIS

Students often think that a quadratic expression in a denominator implies that there must be two excluded values of the variable. To correct this error, have them name the excluded values for the series of functions below.

$$y = \frac{1}{x^2} \qquad y = \frac{1}{x^2 - 9} \qquad y = \frac{1}{x^2 + 8x - 9} \qquad y = \frac{1}{x^2 + 6x - 9} \qquad y = \frac{1}{x^2 + 9}$$

$$[x \neq 0] \qquad [x \neq -3, 3] \qquad [x \neq -9, 1] \qquad [x \neq -3] \qquad [\text{no exclusions}]$$

PHASE 3: ASSESSMENT AND CLOSURE for Lesson 9.1

- Have students write two different rational expressions with domain *all real numbers except 3*. Repeat the question for domain *all real numbers except 3 and −1*. [**See samples at right.**]

- Tell students that Chapter 9 will focus on performing operations with rational expressions and analyzing simple rational functions.

$$\begin{bmatrix} x \neq 3: \dfrac{x + 2}{x - 3}, \dfrac{2x + 1}{x - 3} \\[2mm] x \neq 3, -1: \\[2mm] \dfrac{1}{x^2 - 2x - 3}, \dfrac{x + 2}{x^2 - 2x - 3} \end{bmatrix}$$

☞ *For a Lesson 9.1 Quiz, see* Assessment Resources *page 91.*

9.1 LESSON Rational Expressions and Functions

SKILL A ▶ Finding the domain of a rational expression or function

A quotient of two polynomials is called a **rational expression.** Every polynomial is a rational expression since every polynomial can be written as a quotient whose denominator is 1. A **rational function** in x is a function defined by a rational expression in x.

> rational expression: $\dfrac{3x - 1}{2x - 7}$ rational function: $y = \dfrac{3x - 1}{2x - 7}$

The **domain** of a rational expression or a rational function in x, unless otherwise stated, is the set of all real values of x for which the denominator is not equal to 0. A value of the variable for which the denominator equals 0 is an **excluded value** of the domain.

EXAMPLE 1 Find the domain of the expression $\dfrac{3x - 1}{2x - 7}$.

▶ **Solution**

Solve $2x - 7 = 0$. $2x - 7 = 0$ → $x = 3.5$

domain: all real numbers except 3.5

> The denominator equals 0 when $x = 3.5$.

TRY THIS Find the domain of the expression $\dfrac{2t + 3}{5t + 4}$.

If the denominator of a rational expression or function is a quadratic polynomial, you will need to solve a quadratic equation to find the excluded values of the domain.

EXAMPLE 2 Find the domain of the function $y = \dfrac{2x - 1}{x^2 - 13x + 12}$.

▶ **Solution**

Set the denominator equal to 0 and then solve for x.

$x^2 - 13x + 12 = 0$

$(x - 12)(x - 1) = 0$ ←——— Factor.

$x - 12 = 0$ or $x - 1 = 0$ ←——— Apply the Zero-Product Property.

$x = 12$ or $x = 1$ ←——— Solve for x.

domain: all real numbers except 1 and 12

TRY THIS Find the domain of each function.

a. $y = \dfrac{3x^2 - 2x - 1}{6x^2 + x - 1}$ **b.** $y = \dfrac{3x^2}{x^2 + 4}$

KEY SKILLS

Write the equation you would solve to find the excluded values of the domain of each rational expression or function. Do not solve.

1. $\dfrac{6}{y - 2}$

2. $\dfrac{3}{x} - 5$

3. $y = \dfrac{3n + 7}{6n^2 - 11n - 35}$

PRACTICE

Find the domain of each rational expression or function.

4. $\dfrac{1}{x - 5}$

5. $\dfrac{4r - 1}{r + 3}$

6. $\dfrac{2a - 1}{5a + 1}$

7. $\dfrac{5x + 7x^2}{x^2}$

8. $\dfrac{3x^2 + 2x}{x^2}$

9. $\dfrac{t^2 - 5t + 6}{(3t + 2)(2t - 3)}$

10. $\dfrac{2x^2 - 5x + 2}{(3x - 1)(2x + 5)}$

11. $\dfrac{4n - 7}{n^2 - 5n + 4}$

12. $\dfrac{x^2 + 2x - 3}{x^2 + 4x - 5}$

13. $y = \dfrac{x^2 + 3x - 2}{(4x + 5)(2x + 5)}$

14. $y = \dfrac{x^2 + 5x - 4}{(4x + 3)(4x + 3)}$

15. $y = \dfrac{x^2 - x - 3}{16x^2 - 1}$

16. $y = \dfrac{x^2 - 3}{x^2 - 4}$

17. $y = \dfrac{x^2 + 1}{9x^2 + 25}$

18. $y = \dfrac{x^2 - 1}{4x^2 + x + 9}$

19. $y = \dfrac{x^2 + 7x - 1}{14x^2 + 17x + 5}$

20. $y = \dfrac{5x^2 + x - 2}{10x^2 - 29x + 21}$

21. $y = \dfrac{7x^2 + x - 1}{6x^2 - 35x + 49}$

22. Show that the domain of $y = \dfrac{2m - 1}{7m^2 + 10m + 5}$ is the set of all real numbers.

23. **Critical Thinking** **a.** Use a number line to find the distance between the excluded values of $y = \dfrac{4x - 9}{36x^2 - 100}$.

b. Let a and b represent real numbers with $a \neq 0$. Use a number line to find the distance between the excluded values of $y = \dfrac{1}{a^2x^2 - b^2}$.

MIXED REVIEW

Factor. (7.3 Skill A)

24. $x^2 + 15x + 50$

25. $c^2 + 16c + 63$

26. $z^2 + 18x + 77$

Factor. (7.3 Skill B)

27. $a^2 + 12a - 13$

28. $n^2 - 10n - 39$

29. $p^2 - 21p - 100$

Answers for Lesson 9.1 Skill A

Try This Example 1

all real numbers except $-\dfrac{4}{5}$

Try This Example 2

a. all real numbers except $-\dfrac{1}{2}$ and $\dfrac{1}{3}$

b. all real numbers

1. $y - 2 = 0$
2. $x = 0$
3. $6n^2 - 11n - 35 = 0$
4. all real numbers except 5
5. all real numbers except -3
6. all real numbers except $-\dfrac{1}{5}$
7. all real numbers except 0
8. all real numbers except 0
9. all real numbers except $\dfrac{3}{2}$ and $-\dfrac{2}{3}$
10. all real numbers except $\dfrac{1}{3}$ and $-\dfrac{5}{2}$

11. all real numbers except 1 and 4
12. all real numbers except -5 and 1
13. all real numbers except $-\dfrac{5}{4}$ and $-\dfrac{5}{2}$
14. all real numbers except $-\dfrac{3}{4}$
15. all real numbers except $-\dfrac{1}{4}$ and $\dfrac{1}{4}$
16. all real numbers except 2 and -2
17. all real numbers
18. all real numbers
★ 19. all real numbers except $-\dfrac{5}{7}$ and $-\dfrac{1}{2}$
★ 20. all real numbers except $\dfrac{7}{5}$ and $\dfrac{3}{2}$
★ 21. all real numbers except $\dfrac{7}{3}$ and $\dfrac{7}{2}$

★ 22. The discriminant of $7m^2 + 10m + 5 = 0$ is $10^2 - 4(7)(5) = -40$. Since the discriminant is negative, $7m^2 + 10m + 5 = 0$ has no real solutions. Thus, there are no excluded values. The domain is the set of all real numbers.

23. **a.** $\dfrac{10}{3}$ units

b. $\dfrac{2b}{a}$ units

24. $(x + 5)(x + 10)$
25. $(c + 7)(c + 9)$
26. $(z + 7)(z + 11)$
27. $(a - 1)(a + 13)$
28. $(n - 13)(n + 3)$
29. $(p - 25)(p + 4)$

★ **Advanced Exercises**

TEACHING 9.2 LESSON

Simplifying Rational Expressions

SKILL A ▶ *Using monomial factoring to simplify rational expressions*

Focus: *Students simplify rational expressions in one variable by factoring linear polynomials.*

EXAMPLE 1

PHASE 1: Presenting the Example

Display $\dfrac{3}{3 \cdot \boxed{}}$. Tell students to assume that the box is holding the place of any number or variable. Have them simplify the expression. [**See answer at right.**]

Ask students the following questions, one at a time: What would the result be if the box contained the number 8? the variable x? the expression x^2? the expression $(x + 4)$? Display the answers as shown at right.

Now display $\dfrac{3}{3x - 15}$, which is the expression in Example 1. Ask students what they must do in order to simplify this expression in the same manner as the others. Lead them to see that they must first factor $3x - 15$ as $3(x - 5)$. Then discuss the solution that appears on the textbook page.

$$\frac{3}{3 \cdot \boxed{}} = \frac{1}{\boxed{}}$$

$$\frac{3}{3 \cdot \boxed{8}} = \frac{1}{\boxed{8}}$$

$$\frac{3}{3 \cdot \boxed{x}} = \frac{1}{\boxed{x}}$$

$$\frac{3}{3 \cdot \boxed{x^2}} = \frac{1}{\boxed{x^2}}$$

$$\frac{3}{3 \cdot \boxed{(x + 4)}} = \frac{1}{\boxed{x + 4}}$$

PHASE 2: Providing Guidance for TRY THIS

Watch for students who "cancel" as shown at right. Stress the fact that they must divide by a factor of the *entire* numerator and the *entire* denominator. To explain why this is necessary, work with them to develop the following expanded simplification.

$$\frac{\overset{1}{\cancel{7}}}{\underset{3}{\cancel{21}}r - 49} = \frac{1}{3r - 49} \enspace \textbf{X}$$

$$\frac{7}{21r - 49} = \frac{7}{7(3r - 7)} = \frac{7 \cdot 1}{7(3r - 7)} = \frac{7}{7} \cdot \frac{1}{3r - 7} = 1 \cdot \frac{1}{3r - 7} = \frac{1}{3r - 7}$$

EXAMPLE 2

PHASE 1: Presenting the Example

Display the expression $\dfrac{4w + 8}{4w - 8}$. Ask students how it differs from the expressions in Example 1. Lead them to see that, in this case, both the numerator and denominator are binomials that must be factored.

EXAMPLE 3

PHASE 1: Presenting the Example

Students may think there are no excluded values because the variable has "disappeared" in the simplified expression. Stress that, if $x = 3$, the *original* expression has no meaning. Therefore, the number 3 must be excluded.

9.2 Simplifying Rational Expressions

LESSON

SKILL A ▸ Using monomial factoring to simplify rational expressions

Recall that when you simplify a fraction, you first factor the numerator and the denominator. Then you divide both the numerator and the denominator by all common factors.

$$\frac{60}{75} = \frac{3 \cdot 4 \cdot 5}{3 \cdot 5 \cdot 5} = \frac{\cancel{3} \cdot 4 \cdot \cancel{5}}{\cancel{3} \cdot \cancel{5} \cdot 5} = \frac{4}{5}$$

Simplify a rational expression in the same way. A common factor can be a number, a variable, or an expression.

EXAMPLE 1 Simplify $\frac{3}{3x - 15}$. Identify any excluded values.

▸ **Solution**

$$\frac{3}{3x - 15} = \frac{\cancel{3}^1}{\cancel{3}(x - 5)} \qquad \longleftarrow \text{Factor the denominator.}$$
$$= \frac{1}{x - 5} \qquad \longleftarrow \text{Divide both the numerator and the denominator by 3.}$$

The excluded value is 5 because the denominator, $x - 5$, equals 0 when $x = 5$.

TRY THIS Simplify $\frac{7}{21r - 49}$. Identify any excluded values.

EXAMPLE 2 Simplify $\frac{4w + 8}{4w - 8}$. Identify any excluded values.

▸ **Solution**

$$\frac{4w + 8}{4w - 8} = \frac{4(w + 2)}{4(w - 2)} \qquad \longleftarrow \text{Factor 4 from the numerator and the denominator.}$$
$$= \frac{w + 2}{w - 2} \quad w \neq 2 \qquad \longleftarrow \text{Divide both the numerator and denominator by 4.}$$

TRY THIS Simplify. Identify any excluded values. **a.** $\frac{6c + 2}{2c + 6}$ **b.** $\frac{8n - 8}{2n + 6}$

EXAMPLE 3 Simplify $\frac{3x - 9}{5x - 15}$. Identify any excluded values.

▸ **Solution**

$$\frac{3x - 9}{5x - 15} = \frac{3(x - 3)}{5(x - 3)} \qquad \longleftarrow \text{Factor the numerator and the denominator.}$$
$$= \frac{3}{5} \qquad \longleftarrow \text{Divide the numerator and the denominator by } (x - 3).$$

Although the expression simplifies to $\frac{3}{5}$, the factor $(x - 3)$ was originally in the denominator, so 3 is an excluded value.

TRY THIS Simplify. Identify any excluded values. **a.** $\frac{-9m + 108}{3m - 36}$ **b.** $\frac{8n^2 + 2n}{2n}$

KEY SKILLS

Simplify each rational expression. Identify any excluded values.

1. $\frac{4(x + 4)}{4(x - 4)}$ 2. $\frac{4(x + 5)}{5(x + 5)}$ 3. $\frac{2 \cdot 5}{5(x - 1)}$

PRACTICE

Simplify. Identify any excluded values.

4. $\frac{21}{7y + 35}$ 5. $\frac{-40}{10b + 50}$ 6. $\frac{2d + 4}{d + 2}$

7. $\frac{3h - 15}{h - 5}$ 8. $\frac{3z - 18}{3z + 12}$ 9. $\frac{10a + 20}{5a + 15}$

10. $\frac{4b - 12}{2b + 14}$ 11. $\frac{11c - 33}{11c + 22}$ 12. $\frac{2v + 10}{5v + 25}$

13. $\frac{-7v + 21}{3v - 9}$ 14. $\frac{-3z + 9}{10z - 30}$ 15. $\frac{z + 8}{3z + 24}$

Find a such that each equation is true for all values of n.

16. $\frac{n - a}{2n - 6} = \frac{1}{2}$ 17. $\frac{3n + 9}{5n - a} = \frac{3}{5}$ 18. $\frac{8n + a}{3n + 15} = \frac{8}{3}$

Find the unknown numerator or denominator.

19. $\frac{\text{numerator}}{2x - 6} = \frac{1}{2}$ 20. $\frac{2x - 6}{\text{denominator}} = \frac{2}{5}$

21. **Critical Thinking** Consider $\frac{ax + b}{cx + d}$, where a, b, c, and d are nonzero real numbers.
   a. What must be true of a, b, c, and d if the value of the expression is 1?
   b. What must be true of a, b, c, and d if the expression can be simplified to a constant?

MIXED REVIEW

Factor. (7.4 Skill A)

22. $10a^2 + 25a + 15$ 23. $4a^2 + 8a + 4$ 24. $24 + 10x - x^2$

25. $30 + x - x^2$ 26. $64 + 16x + x^2$ 27. $9 - 6x + x^2$

Answers for Lesson 9.2 Skill A

Try This Example 1

$\frac{1}{3r - 7}; r \neq \frac{7}{3}$

Try This Example 2

a. $\frac{3c + 1}{c + 3}; c \neq -3$

b. $\frac{4n - 4}{n + 3}; n \neq -3$

Try This Example 3

a. $-3; m \neq 12$

b. $4n + 1; n \neq 0$

1. $\frac{x + 4}{x - 4}; x \neq 4$

2. $\frac{4}{5}; x \neq -5$

3. $\frac{2}{x - 1}; x \neq 1$

4. $\frac{3}{y + 5}; y \neq 5$

5. $\frac{-4}{b + 5}; b \neq -5$

6. $2; d \neq -2$

7. $3; h \neq 5$

8. $\frac{z - 6}{z + 4}; z \neq -4$

9. $\frac{2a + 4}{a + 3}; a \neq -3$

10. $\frac{2b - 6}{b + 7}; b \neq -7$

11. $\frac{c - 3}{c + 2}; c \neq -2$

12. $\frac{2}{5}; v \neq -5$

13. $-\frac{7}{3}; v \neq 3$

14. $\frac{-3}{10}; z \neq 3$

15. $\frac{1}{3}; z \neq -8$

★ 16. $a = 3$

★ 17. $a = -15$

★ 18. $a = 40$

★ 19. $x - 3$

★ 20. $5x - 15$

21. **a.** $a = c$ and $b = d$
   b. $a = kc$ and $b = kd$ for some nonzero real number k

22. $5(2a + 3)(a + 1)$

23. $4(a + 1)^2$

24. $(12 - x)(x + 2)$

25. $(6 - x)(x + 5)$

26. $(x + 8)^2$

27. $(x - 3)^2$

★ **Advanced Exercises**

Focus: *Students simplify rational expressions in one variable by factoring quadratic or cubic polynomials.*

EXAMPLE 1

PHASE 1: Presenting the Example

Display the expression at right. Point out to students that the numerator and the denominator are each expressed as a product of two binomial factors. Ask them to rewrite the expression with the numerator and denominator in non-factored form.

$$\dfrac{(r + 1)(r + 2)}{(r + 1)(r + 5)}$$

$\left[\dfrac{r^2 + 3r + 2}{r^2 + 6r + 5}\right]$ Then have them return to the factored form and simplify the

expression. Remind them to identify any values of the variable that must be excluded.

$\left[\dfrac{r + 2}{r + 5}; r \neq -5, -1\right]$

Now display $\dfrac{a^2 + 5a + 6}{a^2 - 4}$, which is the expression in Example 1. Ask students what must

be done in order to simplify it. Elicit the response that the numerator and denominator must each be rewritten as a product of binomial factors. Tell students to perform the factorizations, and then simplify the expression. Have them compare their results to the solution outlined on the textbook page. Discuss why 2 and -2 must be excluded as values of a.

EXAMPLE 2

PHASE 1: Presenting the Example

Display the expression $\dfrac{c^3 - 10c^2 + 25c}{c^3 - 6c^2 + 5c}$. Ask students how it differs from the expressions

in Example 1. Elicit responses such as the following: Both the numerator and denominator of this expression are cubic polynomials rather than quadratic polynomials. All the terms of the numerator have a common factor of c, as do all the terms of the denominator.

PHASE 2: Providing Guidance for TRY THIS

Some students may find it less overwhelming to simplify in two distinct stages, as shown below.

$$\dfrac{p^3 + 2p^2 + p}{p^3 + 8p^2 + 7p} = \dfrac{p(p^2 + 2p + 1)}{p(p^2 + 8p + 7)} \qquad\qquad \dfrac{p^2 + 2p + 1}{p^2 + 8p + 7} = \dfrac{(p + 1)(p + 1)}{(p + 1)(p + 7)}$$

$$= \dfrac{p^2 + 2p + 1}{p^2 + 8p + 7} \qquad\qquad\qquad\qquad = \dfrac{p + 1}{p + 7}$$

Watch for students who correctly factor $p^2 + 2p + 1$ as $(p + 1)^2$, and then simplify as shown at right. Encourage them to factor using the expanded form $(p + 1)(p + 1)$ instead.

$$\dfrac{p(p + 1)^2}{p(p + 1)(p + 7)} = \dfrac{2}{p + 7} \quad \text{✗}$$

SKILL B — Using polynomial factoring to simplify rational expressions

In Chapter 7, you learned how to factor a quadratic expression into a product of linear factors. This skill is often needed to simplify a rational expression.

EXAMPLE 1 Simplify $\dfrac{a^2 + 5a + 6}{a^2 - 4}$. Identify any excluded values.

▶ **Solution**

$$\dfrac{a^2 + 5a + 6}{a^2 - 4} = \dfrac{(a + 2)(a + 3)}{(a + 2)(a - 2)} \longleftarrow \begin{array}{l}\text{Factor } a^2 + 5a + 6 \text{ as } (a + 2)(a + 3).\\ \text{Factor } a^2 - 4 \text{ as } (a + 2)(a - 2).\end{array}$$

$$= \dfrac{\cancel{(a + 2)}(a + 3)}{\cancel{(a + 2)}(a - 2)} \longleftarrow \begin{array}{l}\text{Divide both the numerator and the}\\ \text{denominator by } (a + 2).\end{array}$$

$$= \dfrac{a + 3}{a - 2}$$

The expression is undefined when $a^2 - 4 = 0$.

If $a^2 - 4 = 0$, then $(a + 2)(a - 2) = 0$. By the Zero-Product Property, $a + 2 = 0$ or $a - 2 = 0$. Thus, $a = -2$ or $a = 2$.

The excluded values of the given expression are 2 and -2.

TRY THIS Simplify $\dfrac{t^2 - 9}{t^2 - t - 6}$. Identify any excluded values.

In Lesson 7.5, you learned how to factor some types of cubic polynomials. This skill is often useful in simplifying rational expressions involving cubic numerators or denominators.

EXAMPLE 2 Simplify $\dfrac{c^3 - 10c^2 + 25c}{c^3 - 6c^2 + 5c}$. Identify any excluded values.

▶ **Solution**

$$\dfrac{c^3 - 10c^2 + 25c}{c^3 - 6c^2 + 5c} = \dfrac{c(c^2 - 10c + 25)}{c(c^2 - 6c + 5)} \longleftarrow \begin{array}{l}\text{Factor } c \text{ from both the numerator and}\\ \text{the denominator.}\end{array}$$

$$= \dfrac{c(c - 5)(c - 5)}{c(c - 5)(c - 1)} \longleftarrow \begin{array}{l}\text{Factor } c^2 - 10c + 25 \text{ as } (c - 5)(c - 5).\\ \text{Factor } c^2 - 6c + 5 \text{ as } (c - 5)(c - 1).\end{array}$$

$$= \dfrac{\cancel{c}\cancel{(c - 5)}(c - 5)}{\cancel{c}\cancel{(c - 5)}(c - 1)} \longleftarrow \text{Divide by the common factors, } c \text{ and } (c - 5).$$

$$= \dfrac{c - 5}{c - 1}$$

The expression is undefined when $c^3 - 6c^2 + 5c = 0$.

If $c^3 - 6c^2 + 5c = 0$, then $c(c - 5)(c - 1) = 0$.

By the Zero-Product Property, $c = 0$, $c - 5 = 0$, or $c - 1 = 0$.
The excluded values of the given expression are 0, 1, and 5.

TRY THIS Simplify $\dfrac{p^3 + 2p^2 + p}{p^3 + 8p^2 + 7p}$. Identify any excluded values.

EXERCISES

KEY SKILLS

Write the numerator and the denominator in factored form. Do not simplify.

1. $\dfrac{z^2 + 2z + 1}{z^2 - 1}$

2. $\dfrac{y^2 - 25}{y^2 - 6y + 5}$

3. $\dfrac{n^2 + 8n + 15}{n^2 + 6n + 5}$

PRACTICE

Simplify. Identify any excluded values.

4. $\dfrac{z - 3}{z^2 - 9}$

5. $\dfrac{z + 4}{z^2 - 16}$

6. $\dfrac{4v^2 - 25}{2v + 5}$

7. $\dfrac{9u^2 - 49}{3u - 7}$

8. $\dfrac{a^2 + 14a + 49}{a^2 - 49}$

9. $\dfrac{c^2 - 100}{c^2 - 20c + 100}$

10. $\dfrac{c^2 + 2c - 3}{c^2 - 11c + 10}$

11. $\dfrac{m^2 + 3m - 18}{m^2 - 11m + 24}$

12. $\dfrac{p^2 + p - 56}{p^2 + 13p + 40}$

13. $\dfrac{s^3 - s}{s^2 - 1}$

14. $\dfrac{t^2 - 1}{t^3 - t}$

15. $\dfrac{k^2 - 1}{k^3 - k^2}$

16. $\dfrac{x^3 - 2x^2}{x^3 - 2x}$

17. $\dfrac{y^3 - 2y^2 + y}{y^3 - 2y^2 + 3y}$

18. $\dfrac{y^3 + 4y^2 - 12y}{y^3 - 4y^2 + 4y}$

19. $\dfrac{h^3 - 2h^2 - 35h}{h^3 + 6h^2 + 5h}$

20. $\dfrac{t^3 + 11t^2 + 28t}{t^3 + 9t^2 + 14t}$

21. $\dfrac{a^3 - 20a^2 + 64a}{a^3 - 3a^2 - 4a}$

22. **Critical Thinking** Let x be any real number in the domain of $\dfrac{x^4 + 2ax^3 + a^2x^2}{x^4 - a^2x^2}$ and let a be a fixed nonzero real number. Simplify $\dfrac{x^4 + 2ax^3 + a^2x^2}{x^4 - a^2x^2}$. Identify the excluded values of the expression.

MIXED REVIEW

Simplify. Write the answers with positive exponents only. Assume that all variables are nonzero. (6.5 Skill B)

23. $\dfrac{4a^3b^4}{16ab^2}$

24. $\dfrac{25d^2f^2}{10d^3f^3}$

25. $\dfrac{24r^9s^2}{12r^9s^2}$

26. $\dfrac{7x^4y^2}{3.5x^3y^3}$

27. $\dfrac{(2xy^2)^3}{8x^2y^4}$

28. $\dfrac{9a^2b^4}{(3a^2b)^3}$

29. $\dfrac{(4m^2n^3)^3}{(4m^2n^3)^2}$

30. $\dfrac{(-3r^3t^3)^3}{(-3r^3t^3)^4}$

Answers for Lesson 9.2 Skill B

Try This Example 1

$\dfrac{t + 3}{t + 2}$; $t \neq 3, -2$

Try This Example 2

$\dfrac{p + 1}{p + 7}$; $p \neq 0, -1, -7$

1. $\dfrac{(z + 1)(z + 1)}{(z + 1)(z - 1)}$

2. $\dfrac{(y + 5)(y - 5)}{(y - 1)(y - 5)}$

3. $\dfrac{(n + 5)(n + 3)}{(n + 1)(n + 5)}$

4. $\dfrac{1}{z + 3}$; $z \neq 3, -3$

5. $\dfrac{1}{z - 4}$; $z \neq 4, -4$

6. $2v - 5$; $v \neq -\dfrac{5}{2}$

7. $3u + 7$; $u \neq \dfrac{7}{3}$

8. $\dfrac{a + 7}{a - 7}$; $a \neq 7, -7$

9. $\dfrac{c + 10}{c - 10}$; $c \neq 10, -10$

10. $\dfrac{c + 3}{c - 10}$; $c \neq 10, 1$

11. $\dfrac{m + 6}{m - 8}$; $m \neq 8, 3$

12. $\dfrac{p - 7}{p + 5}$; $p \neq -5, -8$

13. s; $s \neq 1, -1$

14. $\dfrac{1}{t}$; $t \neq 0, 1, -1$

15. $\dfrac{k + 1}{k^2}$; $k \neq 0, 1$

16. $\dfrac{x^2 - 2x}{x^2 - 2}$; $x \neq 0, \sqrt{2}, -\sqrt{2}$

17. $\dfrac{y^2 - 2y + 1}{y^2 - 2y + 3}$; $y \neq 0$

18. $\dfrac{y + 6}{y - 2}$; $y \neq 0, 2$

19. $\dfrac{h - 7}{h + 1}$; $h \neq 0, -1, -5$

20. $\dfrac{t + 4}{t + 2}$; $t \neq 0, -2, -7$

21. $\dfrac{a - 16}{a + 1}$; $a \neq 0, -1, 4$

22. $\dfrac{x + a}{x - a}$; $x \neq 0, a, -a$

23. $\dfrac{a^2b^2}{4}$

24. $\dfrac{5}{2df}$

25. 2

26. $\dfrac{2x}{y}$

27. xy^2

28. $\dfrac{b}{3a^4}$

29. $4m^2n^3$

30. $-\dfrac{1}{3r^3t^3}$

Focus: *Students simplify rational expressions in two variables.*

EXAMPLE 1

PHASE 1: Presenting the Example

Display the expression $\dfrac{ab}{3ab^2 + a^2b}$. Ask students how it differs from the rational
expressions that they simplified previously. Elicit the response that this expression
involves two variables rather than just one. Lead a discussion in which students
brainstorm their ideas for simplifying it. Then discuss the solution outlined on the
textbook page.

EXAMPLE 2

PHASE 2: Providing Guidance for **TRY THIS**

Watch for students who give responses such as $\dfrac{cd^2 + d^3}{cd^2 - d^3}$ and $\dfrac{cd + d^2}{cd - d^2}$.

Although these expressions are equivalent to the given expression, each can still be
simplified further. Encourage students to always begin their work by factoring both the
numerator and the denominator using the GCF of its terms. In this case, they could use
either $-5cd^3$ or $5cd^3$.

EXAMPLE 3

PHASE 1: Presenting the Example

Display the expression $a - b$. Have students evaluate it given $a = 7$ and $b = 3$. **[4]** Now
display $b - a$ and have them evaluate it given the same a and b. **[−4]** Ask them to make a
conjecture about the relationship between $a - b$ and $b - a$. **[They are opposites.]** Lead
students through the steps at right to demonstrate why $a - b$ and $b - a$ must be
opposites for all real numbers a and b. Tell them that they will now see how this
relationship is used in simplifying rational expressions. Then discuss Example 3.

$$-(a - b) = -a + b$$
$$= b + (-a)$$
$$= b - a$$

$$-(b - a) = -b + a$$
$$= a + (-b)$$
$$= a - b$$

PHASE 3: ASSESSMENT AND CLOSURE for Lesson 9.2

- Give students the two statements shown at right. Have them fill in the boxes with
positive integers to make the two statements true. $\left[\,\textbf{samples: } \dfrac{2c + 10}{2c + 12} = \dfrac{c + 5}{c + 6};\right.$
$\left.\dfrac{c^2 + 6c + 5}{c^2 + 7c + 6} = \dfrac{c + 5}{c + 6}\right]$

$$\dfrac{\square c + \square}{\square c + \square} = \dfrac{c + 5}{c + 6}$$

$$\dfrac{c^2 + \square c + \square}{c^2 + \square c + \square} = \dfrac{c + 5}{c + 6}$$

- In the lessons that follow, students will multiply, divide, add, and subtract rational
expressions. Success with each of the operations requires mastery of the
simplification skills taught in this lesson.

☞ *For a Lesson 9.2 Quiz, see* Assessment Resources *page 92.*

A rational expression may contain more than one variable. To simplify such a rational expression, look for common factors of the numerator and the denominator. Then divide the numerator and the denominator by their greatest common factor (GCF).

When simplifying expressions in the remainder of this chapter, assume that variables do not represent any excluded values. That is, assume the denominator does not equal 0.

EXAMPLE 1 Simplify $\dfrac{ab}{3ab^2 + a^2b}$.

▷ Solution

$\dfrac{ab}{3ab^2 + a^2b} = \dfrac{ab}{ab(3b + a)}$ ← Factor ab from each term in the denominator.

$= \dfrac{\overset{1}{ab}}{\underset{1}{ab}(3b + a)}$ ← Divide the numerator and the denominator by their GCF, ab.

$= \dfrac{1}{3b + a}$

TRY THIS Simplify $\dfrac{2a^2b^3}{2a^2b^3 + 4a^2b^2}$.

EXAMPLE 2 Simplify $\dfrac{8r^3s - 4r^2s^2}{8r^3s - 20r^2s}$.

▷ Solution

$\dfrac{8r^3s - 4r^2s^2}{8r^3s - 20r^2s} = \dfrac{4r^2s(2r - s)}{4r^2s(2r - 5)}$ ← Factor $4r^2s$ from each term in the numerator and the denominator.

$= \dfrac{2r - s}{2r - 5}$ ← Divide the numerator and the denominator by their GCF, $4r^2s$.

TRY THIS Simplify $\dfrac{-5c^2d^3 - 5cd^4}{-5c^2d^3 + 5cd^4}$.

In Example 3, you will use the Opposite of a Difference (or Multiplicative Property of -1) to factor: $(a - b) = -(b - a)$.

EXAMPLE 3 Simplify $\dfrac{g^2 - y^2}{4y^2 - 4g^2}$.

▷ Solution

$\dfrac{g^2 - y^2}{4y^2 - 4g^2} = \dfrac{g^2 - y^2}{4(y^2 - g^2)}$

$= \dfrac{g^2 - y^2}{-4(g^2 - y^2)}$ ← Recognize that $(y^2 - g^2) = -(g^2 - y^2)$ and rewrite.

$= -\dfrac{1}{4}$ ← Divide by the common factor, $(g^2 - y^2)$.

TRY THIS Simplify $\dfrac{4u^2 - a^2}{5a^2 - 20u^2}$.

EXERCISES

Identify the common factors in each rational expression.

1. $\dfrac{a^2b(2a - b)}{a^2b(2a + b)}$

2. $\dfrac{(3r - z)(4r - z)}{(2r - z)(3r - z)}$

3. $\dfrac{3(4q^2 - 81)}{-5(4q^2 - 81)}$

PRACTICE

Simplify.

4. $\dfrac{bc^2 - bc}{b}$

5. $\dfrac{s}{s^2 + sc}$

6. $\dfrac{x^2y^2 - xy}{xy}$

7. $\dfrac{w^2z^2}{w^2z^2 + wz}$

8. $\dfrac{a - b}{b - a}$

9. $\dfrac{ab - b}{ba - b}$

10. $\dfrac{pq^2 - q}{q^2 - q}$

11. $\dfrac{nb - b}{b^2 - b}$

12. $\dfrac{mn^2 - mn}{n^2 - 1}$

13. $\dfrac{p^2q^2 - pq}{pq^2 - q}$

14. $\dfrac{q^3a^2 - q^3a}{aq^3 - aq^2}$

15. $\dfrac{r^2s^2 - r^3s^2}{s^3r^2 - s^2r^2}$

16. $\dfrac{r^2s^2 - 4}{4 - r^2s^2}$

17. $\dfrac{4a^2c^2 - 9a^2}{8a^2c^2 - 18a^2}$

18. $\dfrac{6c^2 - 7ac - 5a^2}{9c^2 - 25a^2}$

19. $\dfrac{9y^2 + 30yz + 25z^2}{9y^2 - 25z^2}$

20. $\dfrac{3m^2 - mz - 2z^2}{6m^2 - 5mz - 6z^2}$

21. $\dfrac{6a^2 - 7ac - 20c^2}{3a^2 + 13ac + 12c^2}$

22. **Critical Thinking** What must be true of x and y so that $\dfrac{4x^2 - 81y^2}{4x^2 + 36xy + 81y^2} = -\dfrac{7}{11}$?

23. **Critical Thinking** Find the excluded values of $\dfrac{ab}{3ab^2 + a^2b}$ in terms of a and b.

MIXED REVIEW

Simplify. (previous courses)

24. $\dfrac{21}{25} \cdot \dfrac{10}{14}$

25. $\dfrac{18}{35} \cdot \dfrac{49}{6}$

26. $\dfrac{100}{121} \cdot \dfrac{33}{10}$

27. $\dfrac{45}{64} \cdot \dfrac{16}{90}$

28. $\left(\dfrac{4}{5}\right)^2$

29. $\left(\dfrac{2}{3}\right)^3$

30. $\dfrac{1}{5} \cdot \dfrac{5}{18} \cdot \dfrac{6}{7}$

31. $\dfrac{14}{48} \cdot \dfrac{4}{7} \cdot \dfrac{21}{16}$

Answers for Lesson 9.2 Skill C

Try This Example 1
$\dfrac{1}{b + 2}$

Try This Example 2
$\dfrac{c + d}{c - d}$

Try This Example 3
$-\dfrac{1}{5}$

1. a^2b
2. $3r - z$
3. $4q^2 - 81$
4. $c(c - 1)$
5. $\dfrac{1}{s + c}$
6. $xy - 1$
7. $\dfrac{wz}{wz + 1}$
8. -1
9. 1
10. $\dfrac{pq - 1}{q - 1}$

11. $\dfrac{n - 1}{b - 1}$
12. $\dfrac{mn}{n + 1}$
13. p
14. $\dfrac{q - aq}{q - 1}$
15. $\dfrac{1 - r}{s - 1}$
16. -1
17. $\dfrac{1}{2}$
18. $\dfrac{2c + a}{3c + 5a}$
19. $\dfrac{3y + 5z}{3y - 5z}$
20. $\dfrac{m - z}{2m - 3z}$
21. $\dfrac{2a - 5c}{a + 3c}$
22. $x \neq 0, y \neq 0,$ and $x = y$
23. $a \neq 0, -3b$ and $b \neq 0$
24. $\dfrac{3}{5}$

25. $\dfrac{21}{5}$
26. $\dfrac{30}{11}$
27. $\dfrac{1}{8}$
28. $\dfrac{16}{25}$
29. $\dfrac{8}{27}$
30. $\dfrac{1}{21}$
31. $\dfrac{7}{32}$

TEACHING

9.3 LESSON Multiplying and Dividing Rational Expressions

SKILL A *Multiplying rational expressions*

Focus: *Students use the rule for multiplying fractions to find the product of two or more rational expressions.*

EXAMPLE 1

PHASE 1: Presenting the Example

Display expression **1** at right. Point out to students that it shows a multiplication involving two rational expressions. Tell them to evaluate the two expressions *individually* given $n = 4$. $\left[\frac{n^2}{3} = \frac{4^2}{3} = \frac{16}{3}; \frac{15}{n} = \frac{15}{4}\right]$ Then have them find and simplify the product of $\frac{16}{3}$ and $\frac{15}{4}$. [**20**]

1 $\frac{n^2}{3} \cdot \frac{15}{n}$

2 $\frac{15n^2}{3n}$

Now display expression **2**. Remind students that this expression shows a division of a monomial by a monomial, and point out that the numerator is the product of the numerators in **1**, and the denominator is the product of the denominators in **1**. Tell them to find the quotient. [$\mathbf{5n}$] Then have them evaluate this quotient given $n = 4$. [**20**]

Tell students to compare the results of evaluating expressions **1** and **2** when $n = 4$. [**They are the same.**] Ask them to make a conjecture about finding a product of two rational expressions. [**sample: To multiply two rational expressions, multiply the numerators, multiply the denominators, and simplify the resulting expression.**] Then discuss Example 1.

PHASE 2: Providing Guidance for TRY THIS

Because the expression involves numbers that are bases as well as numbers that are exponents, some students may mistakenly subtract bases and give the response $\frac{35z}{72a}$. Encourage these students to work with the bases in prime-factored form, as shown at right.

$$\frac{49z^7}{81a^5} \cdot \frac{9a^4}{14z^6} = \frac{7^2z^7}{3^4a^5} \cdot \frac{3^2a^4}{2^17^1z^6}$$
$$= \frac{1}{2^1} \cdot \frac{3^2}{3^4} \cdot \frac{7^2}{7^1} \cdot \frac{a^4}{a^5} \cdot \frac{z^7}{z^6}$$
$$= \frac{1}{2^1} \cdot \frac{1}{3^2} \cdot \frac{7^1}{1} \cdot \frac{1}{a^1} \cdot \frac{z^1}{1} = \frac{7^1z^1}{2^13^2a^1} = \frac{7z}{18a}$$

EXAMPLE 2

PHASE 1: Presenting the Example

Display the expression that appears at right. Tell students to write the result when they apply the rule for multiplying rational expressions. $\left[\frac{(2x - 10)(4x + 12)}{(x^2 + 6x + 9)(x^2 - 25)}\right]$ Ask if they think this is the simplest form of the product. Lead them to see that each of the expressions within parentheses can be factored, and they can then look for common factors. Then continue with the solution outlined on the textbook page.

$$\frac{2x - 10}{x^2 + 6x + 9} \cdot \frac{4x + 12}{x^2 - 25}$$

9.3
LESSON

Multiplying and Dividing Rational Expressions

Multiplying rational expressions

Use the following rule to multiply two fractions or two rational expressions.

Multiplying Fractions

If a, b, c, and d represent real numbers with $b \neq 0$ and $d \neq 0$, then

$\frac{a}{b} \cdot \frac{c}{d} = \frac{ac}{bd}$.

For example, $\frac{6}{21} \cdot \frac{15}{14} = \frac{6 \cdot 15}{21 \cdot 14}$. To simplify the product of two fractions

such as $\frac{6 \cdot 15}{21 \cdot 14}$, first factor the numerator and the denominator so that

they only contain prime numbers. Then divide both the numerator and
denominator by their common factors.

$\frac{6}{21} \cdot \frac{15}{14} = \frac{6 \cdot 15}{21 \cdot 14} = \frac{\overset{2}{\cancel{2}} \cdot \overset{1}{\cancel{3}} \cdot 3 \cdot 5}{\underset{3}{\cancel{3}} \cdot 7 \cdot \underset{2}{\cancel{2}} \cdot 7} = \frac{15}{49}$

EXAMPLE 1 Simplify $\frac{3x^2}{4y^3} \cdot \frac{16y^4}{9x^3}$. Write the answer with positive exponents.

▸ **Solution**

$\frac{3x^2}{4y^3} \cdot \frac{16y^4}{9x^3} = \left(\frac{3}{4} \cdot \frac{16}{9}\right)\left(\frac{x^2}{x^3}\right)\left(\frac{y^4}{y^3}\right)$ ⟵ Separate like terms.

$= \frac{4}{3} \cdot \frac{1}{x} \cdot \frac{y}{1}$ ⟵ Apply the Quotient Property of Exponents.

$= \frac{4y}{3x}$ ⟵ Write a single fraction.

TRY THIS Simplify $\frac{49z^7}{81a^3} \cdot \frac{9a^4}{14z^6}$. Write the answer with positive exponents.

EXAMPLE 2 Simplify $\frac{2x - 10}{x^2 + 6x + 9} \cdot \frac{4x + 12}{x^2 - 25}$.

▸ **Solution**

$\frac{2x - 10}{x^2 + 6x + 9} \cdot \frac{4x + 12}{x^2 - 25} = \frac{2(x - 5)}{(x + 3)(x + 3)} \cdot \frac{4(x + 3)}{(x - 5)(x + 5)}$

$= \frac{2\cancel{(x - 5)}}{\cancel{(x + 3)}(x + 3)} \cdot \frac{4\cancel{(x + 3)}}{\cancel{(x - 5)}(x + 5)}$

$= \frac{8}{(x + 3)(x + 5)}, \text{ or } \frac{8}{x^2 + 8x + 15}$

Note that the excluded values of the answer include all the excluded values of the original expressions: -3, -5, and $+5$.

TRY THIS Simplify $\frac{-3y - 21}{y^2 - 10y + 25} \cdot \frac{4y - 20}{y^2 - 49}$.

KEY SKILLS

Simplify. Write the answers with positive exponents only.

1. $\left(\frac{5}{7} \cdot \frac{14}{15}\right)\left(\frac{a^3}{a^2}\right)\left(\frac{b}{b^3}\right)$

2. $\left(\frac{3}{8} \cdot \frac{24}{1}\right)\left(\frac{r}{r^2}\right)\left(\frac{s}{s^2}\right)$

3. $\left(\frac{1}{15} \cdot \frac{25}{1}\right)\left(\frac{n^3}{n^3}\right)\left(\frac{m^4}{m^2}\right)$

4. $\frac{3x(x - 5)}{(x + 1)(x - 5)} \cdot \frac{4(x + 1)}{6x(x - 1)}$

5. $\frac{(x + 1)(x + 1)}{(x - 1)(x + 1)} \cdot \frac{4(x - 1)}{6x(x + 1)}$

6. $\frac{3}{(x + 3)(x - 3)} \cdot \frac{4(x + 2)}{5x(x + 2)}$

PRACTICE

Simplify. Write the answers with positive exponents only.

7. $\frac{4n^3}{5m^2} \cdot \frac{55m}{12n^2}$

8. $\frac{11p^3}{7q^2} \cdot \frac{21p^3}{33q^2}$

9. $\frac{13a^3}{7k^2} \cdot \frac{21k^3}{26a^4}$

10. $\frac{25z^4}{34x^2} \cdot \frac{17x^3}{50z^4}$

11. $\frac{y^4}{32t^2} \cdot \frac{16t^2}{5y^5}$

12. $\frac{6w^4}{25t^2} \cdot \frac{50t^4}{3w^4}$

13. $\frac{3(x - 1)}{x^2 - 2x + 1} \cdot \frac{2x - 2}{x - 3}$

14. $\frac{3d^2 - 27}{d^2 - 6d + 5} \cdot \frac{d - 5}{d^2 - 9}$

15. $\frac{t + 1}{t^2 + 6t + 5} \cdot \frac{t - 5}{t^2 - 25}$

16. $\frac{(c - 6)(c - 5)}{c^2 - 36} \cdot \frac{(c + 5)(c + 6)}{c^2 - 25}$

17. $\frac{(x - 1)(x - 7)}{x^2 - 49} \cdot \frac{(x + 7)(x + 1)}{x^2 - 25}$

18. $\frac{n^2 + 3n + 2}{n^2 - 49} \cdot \frac{n^2 - 6n - 7}{n^2 - 3n - 4}$

19. $\frac{m^2 + 8m - 9}{m^2 + 9m + 8} \cdot \frac{m^2 + 2m + 1}{m^2 + 10m + 9}$

20. $\frac{z^2 + 11z + 10}{z^2 + 10z + 9} \cdot \frac{z^2 + 11z + 18}{z^2 + 12z + 20}$

21. $\frac{s^2 - 3s - 28}{s^2 + s - 2} \cdot \frac{s^2 + 2s - 3}{s^2 - 5s - 14}$

22. $\frac{y^2 - 9}{y^2 - 1} \cdot \frac{y^2 + 2y + 1}{y^2 + 6y + 9} \cdot \frac{y^2 + 4y - 5}{y^2 - 2y - 3}$

23. $\frac{v^2 + 14v + 49}{v^2 - 12 + 36} \cdot \frac{v^2 - 13v + 42}{v^2 + 7v} \cdot \frac{v^2 - 36}{v^2 - 49}$

MIXED REVIEW

Simplify. (previous courses)

24. $\frac{18}{11} \div \frac{54}{22}$

25. $\frac{1}{3} \div \frac{3}{14}$

26. $\frac{4}{7} \div \frac{8}{14}$

27. $\frac{13}{21} \div \frac{52}{7}$

Simplify. (2.6 Skill A)

28. $\frac{1}{7}(7x + 14)$

29. $-\frac{1}{2}(4x - 6)$

30. $\frac{-1}{8}(-4w + 16)$

31. $-\frac{1}{3}(-21t - 18)$

Answers for Lesson 9.3 Skill A

Try This Example 1
$\frac{7z}{18a}$

Try This Example 2
$\frac{-12}{y^2 - 12y + 35}$

1. $\frac{2a^3}{3b^2}$

2. $\frac{9}{rs}$

3. $\frac{5m^2}{3}$

4. $\frac{2}{x - 1}$

5. $\frac{2}{3x}$

6. $\frac{12}{5x(x + 3)(x - 3)}$

7. $\frac{11n}{3m}$

8. $\frac{p^6}{q^4}$

9. $\frac{3k}{2a}$

10. $\frac{x}{4}$

11. $\frac{1}{10y}$

12. $4t^2$

13. $\frac{6}{x - 3}$

14. $\frac{3}{d - 1}$

15. $\frac{1}{t^2 + 10t + 25}$

16. 1

17. $\frac{x^2 - 1}{x^2 - 25}$

18. $\frac{n^2 + 3n + 2}{n^2 + 3n - 28}$

19. $\frac{m - 1}{m + 8}$

20. 1

21. $\frac{s^2 + 7s + 12}{s^2 + 4s + 4}$

22. $\frac{y + 5}{y + 3}$

23. $\frac{v + 6}{v}$

24. $\frac{2}{3}$

25. $\frac{14}{9}$

26. 1

27. $\frac{1}{12}$

28. $x + 2$

29. $-2x + 3$

30. $\frac{1}{2}w - 2$

31. $7t + 6$

Focus: *Students find the quotient of two rational expressions by multiplying by the reciprocal of the divisor.*

EXAMPLE 1

PHASE 1: Presenting the Example

Display the division $\frac{3}{4} \div \frac{3}{8}$. Have students find the quotient. [2] Check their responses.

Review the term *reciprocal* and discuss the rule for dividing fractions given at the top of the textbook page. Then lead them through the steps of the division that are shown below. Repeat the activity for $\frac{11}{4} \div \frac{11}{8}$, displaying the steps directly beneath the steps for $\frac{3}{4} \div \frac{3}{8}$.

Now display $\frac{t}{4} \div \frac{t}{8}$ directly beneath $\frac{11}{4} \div \frac{11}{8}$. Ask students what they think the simplified product should be. [2] Have a volunteer come to the chalkboard and show how that product would be obtained. The steps should parallel the steps for $\frac{3}{4} \div \frac{3}{8}$, as shown at right.

$$\frac{3}{4} \div \frac{3}{8} = \frac{3}{4} \cdot \frac{8}{3} = \frac{3 \cdot 8}{4 \cdot 3} = \frac{24}{12} = 2$$

$$\frac{11}{4} \div \frac{11}{8} = \frac{11}{4} \cdot \frac{8}{11} = \frac{11 \cdot 8}{4 \cdot 11} = \frac{88}{44} = 2$$

$$\frac{t}{4} \div \frac{t}{8} = \frac{t}{4} \cdot \frac{8}{t} = \frac{t \cdot 8}{4 \cdot t} = \frac{8t}{4t} = 2$$

In summary, point out that the procedure for dividing rational expressions is the same as that for dividing fractions. Then discuss Example 1.

PHASE 2: Providing Guidance for TRY THIS

Remind students that division and multiplication are inverse operations. So, to check whether a quotient is correct, they can multiply it by the divisor; the result should be the original dividend. A sample check for the Try This exercise is shown at right.

$$\left(\frac{5t}{8s^3}\right) \div \frac{5t^3}{18s^2} \overset{?}{=} \frac{9}{4st^2}$$

$$\frac{5t^3}{18s^2} \cdot \frac{9}{4st^2} = \frac{5t^3 \cdot 9}{18s^2 4st^2} = \frac{45t^3}{72s^3t^2} = \left(\frac{5t}{8s^3}\right) ✔$$

EXAMPLE 2

PHASE 1: Presenting the Example

Display the expression at right. Tell students to write a set of steps that they would follow in order to simplify. Stress that they should *not* simplify at this time.

$$\frac{n^2 + 2n - 3}{n^2 - 16} \div \frac{n^2 - 9}{n^2 - 8n + 16}$$

Discuss the students' responses, consolidating them into a "master procedure" that all agree upon. [sample: **1. Rewrite the division as multiplication by the reciprocal of the divisor. 2. Factor all the numerators and denominators. 3. Divide by any factors that are common to a numerator and a denominator. 4. Multiply numerators and multiply denominators. 5. If necessary, simplify the resulting expression.**] Then discuss the simplification shown on the textbook page.

To divide one rational number or expression by another, multiply by the reciprocal of the divisor. Recall that the reciprocal of a fraction $\frac{x}{y}$ is $\frac{y}{x}$.

Dividing Fractions

If a, b, c, and d are real numbers with $b \neq 0$, $c \neq 0$, and $d \neq 0$, then
$$\frac{a}{b} \div \frac{c}{d} = \frac{a}{b} \times \frac{d}{c}.$$

For example, $\frac{12}{25} \div \frac{4}{15} = \frac{12}{25} \times \frac{15}{4}$.

$$\frac{12}{25} \div \frac{4}{15} = \frac{12}{25} \cdot \frac{15}{4} = \frac{\overset{3}{12}}{\underset{5}{25}} \cdot \frac{\overset{3}{15}}{\underset{1}{4}} = \frac{3}{5} \cdot \frac{3}{1} = \frac{9}{5}$$

EXAMPLE 1 Simplify $\frac{4x^3}{15y^3} \div \frac{12x^4}{35y^4}$. Write the answer with positive exponents only.

▶ Solution

$$\frac{4x^3}{15y^3} \div \frac{12x^4}{35y^4} = \frac{4x^3}{15y^3} \cdot \frac{35y^4}{12x^4}$$ ⟵ Multiply by $\frac{35y^4}{12x^4}$, the reciprocal of $\frac{12x^4}{35y^4}$.

$$= \left(\frac{4 \cdot 35}{15 \cdot 12}\right)\left(\frac{x^3}{x^4}\right)\left(\frac{y^4}{y^3}\right)$$ ⟵ Separate like terms.

$$= \frac{7}{9} \cdot \frac{1}{x} \cdot \frac{y}{1}$$ ⟵ Apply the Quotient Property of Exponents.

$$= \frac{7y}{9x}$$

TRY THIS Simplify $\frac{5t}{8s^3} \div \frac{5t^3}{18s^2}$. Write the answer with positive exponents only.

EXAMPLE 2 Simplify $\frac{n^2 + 2n - 3}{n^2 - 16} \div \frac{n^2 - 9}{n^2 - 8n + 16}$.

▶ Solution

$$\frac{n^2 + 2n - 3}{n^2 - 16} \div \frac{n^2 - 9}{n^2 - 8n + 16} = \frac{n^2 + 2n - 3}{n^2 - 16} \cdot \frac{n^2 - 8n + 16}{n^2 - 9}$$

$$= \frac{(n+3)(n-1)}{(n-4)(n+4)} \cdot \frac{(n-4)(n-4)}{(n+3)(n-3)}$$

$$= \frac{(n-1)(n-4)}{(n+4)(n-3)}, \text{ or } \frac{n^2 - 5n + 4}{n^2 + n - 12}$$

TRY THIS Simplify $\frac{m^2 + 2m + 1}{m^2 - 36} \div \frac{m^2 - 1}{m^2 - 12m + 36}$.

EXERCISES

KEY SKILLS

Write the reciprocal of each rational expression.

1. $\frac{2x + 3}{3x + 1}$ 2. $\frac{x^2 - 1}{x^2 - 25}$ 3. $\frac{x^2 + 7x + 6}{x^2 - 5x + 6}$ 4. $\frac{x^2 + 2x + 1}{x^2 - 9x - 10}$

PRACTICE

Simplify. Write the answers with positive exponents only.

5. $\frac{a}{3} \div \frac{a}{4}$ 6. $\frac{b}{5} \div \frac{b}{6}$ 7. $\frac{x^2}{2} \div \frac{x^3}{6}$

8. $\frac{y^3}{2} \div \frac{y^2}{3}$ 9. $\frac{4a^3}{5b} \div \frac{6a^4}{25b^2}$ 10. $\frac{12x^3}{5n} \div \frac{18x^4}{35n^2}$

11. $\frac{4y^3}{7p^6} \div \frac{18y^4}{49p^2}$ 12. $\frac{22z^2}{4a^4} \div \frac{33z^6}{16a^5}$ 13. $\frac{2(x + 3)}{(x + 4)} \div \frac{3(x + 3)}{5(x + 4)}$

14. $\frac{5(y - 1)}{(y + 1)} \div \frac{3(y - 1)}{7(y + 1)}$ 15. $\frac{5a - 15}{3a + 3} \div \frac{a - 3}{2a + 2}$ 16. $\frac{3n - 30}{n - 1} \div \frac{n - 10}{n - 1}$

17. $\frac{5k + 15}{5k + 25} \div \frac{k + 3}{2k^2 + 10k}$ 18. $\frac{5p - 5}{5p - 15} \div \frac{p - 1}{3p^2 - 9p}$ 19. $\frac{9d + 9}{2d - 20} \div \frac{d + 1}{5d^2 - 50d}$

20. $\frac{5b + 15}{3b + 33} \div \frac{b + 3}{3b^2 + 33b}$ 21. $\frac{c - 6}{c^2 + 7c} \div \frac{3c - 18}{3c + 21}$ 22. $\frac{w - 8}{7w^2 - 56w} \div \frac{w - 1}{3w - 24}$

23. $\frac{5g^2 - 15g}{g - 2} \div \frac{3g - 9}{2g - 4}$ 24. $\frac{v^2 + v}{v - 8} \div \frac{v + 1}{2v - 16}$ 25. $\frac{r^2 + 5r}{r - 1} \div \frac{r + 5}{r - 1}$

26. $\frac{x^2 + 2x - 3}{x^2 - 49} \div \frac{x^2 - 2x + 1}{x^2 + 6x - 7}$ 27. $\frac{x^2 + 3x - 10}{x^2 - 4x - 21} \div \frac{x^2 - 3x + 2}{x^2 - 2x - 15}$

28. $\frac{y^2 + 10y + 25}{y^2 - 6y - 7} \div \frac{y^2 - 25}{y^2 - 4y - 5}$ 29. $\frac{u^2 - 8u + 15}{u^2 - 6u + 5} \div \frac{u^2 + 2u - 15}{u^2 + 4u - 5}$

30. $\frac{h^2 - 6h + 9}{h^2 + 2h - 15} \div \frac{h^2 - 9}{h^2 + 4h - 5}$ 31. $\frac{n^2 - 9}{n^2 + 2n + 1} \div \frac{n^2 - 9}{n^2 - 1}$

32. **Critical Thinking** Find a such that $\frac{x^2 - a}{x^2 + x - 6} \times \frac{x^2 + 4x + 3}{x^2 + 3x + 2} = 1$.

MIXED REVIEW

Simplify. (2.6 Skill A)

33. $\frac{3}{4}(8a + 3) + a$ 34. $3 + \frac{1}{6}(24a - 3)$ 35. $\frac{1}{2}(10t - 50) + \frac{1}{3}(3t)$

36. $\frac{3}{8}(r - 8) + \frac{4r}{8}$ 37. $\frac{1}{3}(a + 2) + \frac{1}{2}(a + 2)$ 38. $\frac{5}{7}(2r - 2) + \frac{1}{3}(2r - 2)$

Answers for Lesson 9.3 Skill B

Try This Example 1
$\frac{9}{4st^2}$

Try This Example 2
$\frac{m^2 - 5m - 6}{m^2 + 5m - 6}$

1. $\frac{3x + 1}{2x + 3}$

2. $\frac{x^2 - 25}{x^2 - 1}$

3. $\frac{x^2 - 5x + 6}{x^2 + 7x + 6}$

4. $\frac{x^2 - 9x - 10}{x^2 + 2x + 1}$

5. $\frac{4}{3}$

6. $\frac{6}{5}$

7. $\frac{3}{x}$

8. $\frac{3y}{2}$

9. $\frac{10b}{3a}$

10. $\frac{14n}{3x}$

11. $\frac{14}{9p^4 y}$

12. $\frac{8a}{3z^4}$

13. $\frac{10}{3}$

14. $\frac{35}{3}$

15. $\frac{10}{3}$

16. 3

17. $2k$

18. $3p$

19. $\frac{45d}{2}$

20. $5b$

21. $\frac{1}{c}$

22. $\frac{3w - 24}{7w^2 - 7w}$

23. $\frac{10g}{3}$

24. $2v$

25. r

26. $\frac{x + 3}{x - 7}$

27. $\frac{x^2 - 25}{x^2 - 8x + 7}$

28. $\frac{y + 5}{y - 7}$

29. 1

30. $\frac{h - 1}{h + 3}$

31. $\frac{n - 1}{n + 1}$

32. $a = 4$

33. $7a + \frac{9}{4}$

34. $4a + \frac{5}{2}$

35. $6t - 25$

36. $\frac{7r}{8} - 3$

37. $\frac{5a}{6} + \frac{5}{3}$

38. $\frac{44r}{21} - \frac{44}{21}$

Focus: *Students simplify expressions that involve both multiplication and division of rational expressions.*

EXAMPLE 1

PHASE 1: Presenting the Example

Display expression **1** at right. Tell students to simplify. [16] Point out that both multiplication and division were involved. Ask them to describe how they applied the order of operations. [**Perform multiplication and division in order from left to right:** $4 \cdot 12 = 48; 48 \div 3 = 16.$]

1. $4 \cdot 12 \div 3$

2. $\dfrac{1}{4} \cdot \dfrac{1}{6} \div \dfrac{1}{2}$

Display expression **2**. Note that the operations are the same, but the numbers are now fractions. Have students simplify the expression using the order of operations. $\left[\dfrac{1}{4} \cdot \dfrac{1}{6} = \dfrac{1}{24}; \dfrac{1}{24} \div \dfrac{1}{2} = \dfrac{1}{24} \cdot \dfrac{2}{1} = \dfrac{2}{24} = \dfrac{1}{12}\right]$ Point out that an alternative method is to immediately apply the definition of division, rewriting the expression as $\dfrac{1}{4} \cdot \dfrac{1}{6} \cdot \dfrac{2}{1}$. Then it can be simplified using just multiplication: $\dfrac{1}{4} \cdot \dfrac{1}{6} \cdot \dfrac{2}{1} = \dfrac{1 \cdot 1 \cdot 2}{4 \cdot 6 \cdot 1} = \dfrac{2}{24} = \dfrac{1}{12}$. Now discuss Example 1.

PHASE 2: Providing Guidance for TRY THIS

Some students may wonder how to find the reciprocal of $r^4 s^4$. Ask them to name the reciprocal of a whole number such as 5. $\left[\dfrac{1}{5}\right]$ They should quickly realize that the desired reciprocal in this case is $\dfrac{1}{r^4 s^4}$.

EXAMPLE 2

PHASE 2: Providing Guidance for TRY THIS

Watch for students who attempt to replace $(x + 2) \div (x + 2)$ with 1. Remind them that they must perform divisions in order from left to right.

PHASE 3: ASSESSMENT AND CLOSURE for Lesson 9.3

- Give students the two statements shown at right. Have them fill in the boxes with positive integers *greater than 1* to make the statements true.

  $\left[\text{samples: } \dfrac{4x^2}{15y^2} \cdot \dfrac{3y^5}{2x^4} = \dfrac{2y^3}{5x^2}; \dfrac{4x^2}{15y^2} \div \dfrac{2x^4}{3y^5} = \dfrac{2y^3}{5x^2}\right]$

- This lesson begins a sequence in which students learn about operations with rational expressions. As students encounter each new topic, have them continue to look for parallels to operations with arithmetic fractions.

$\dfrac{\square x^{\square}}{\square y^{\square}} \cdot \dfrac{\square y^{\square}}{\square x^{\square}} = \dfrac{2y^3}{5x^2}$

$\dfrac{\square x^{\square}}{\square y^{\square}} \div \dfrac{\square y^{\square}}{\square x^{\square}} = \dfrac{2y^3}{5x^2}$

☞ *For a Lesson 9.3 Quiz, see* Assessment Resources *page 93.*

SKILL C ▸ Combining multiplication and division with rational expressions

In Skill B of this lesson, you learned to simplify a quotient of two rational expressions. You rewrote the division as multiplication by the reciprocal of the divisor. This technique still applies when you are simplifying rational expressions that involve both multiplication and division.

EXAMPLE 1 Simplify $\dfrac{2ab}{6a^2b} \cdot \dfrac{18ab^2}{12ab} \div \dfrac{a^3b}{2ab^3}$.

▸ **Solution**

$\dfrac{2ab}{6a^2b} \cdot \dfrac{18ab^2}{12ab} \div \dfrac{a^3b}{2ab^3} = \dfrac{2ab}{6a^2b} \cdot \dfrac{18ab^2}{12ab} \cdot \dfrac{2ab^3}{a^3b}$ ⟵ Change division to multiplication by the reciprocal.

$= \dfrac{2 \cdot 18 \cdot 2}{6 \cdot 12 \cdot 1} \cdot \dfrac{a \cdot a \cdot a}{a^2 \cdot a \cdot a^3} \cdot \dfrac{b \cdot b^2 \cdot b^3}{b \cdot b \cdot b}$ ⟵ Separate like terms.

$= 1 \cdot \dfrac{a^3}{a^6} \cdot \dfrac{b^6}{b^3}$

$= 1 \cdot \dfrac{1}{a^3} \cdot \dfrac{b^3}{1}$ ⟵ Apply the Quotient Property of Exponents.

$= \dfrac{b^3}{a^3}$

TRY THIS Simplify $\dfrac{1}{6rs} \cdot \dfrac{6r^2s^2}{18rs^2} \div (r^4s^4)$.

Recall that the reciprocal of 4 is $\frac{1}{4}$. Similarly, the reciprocal of $x - 1$ is $\dfrac{1}{x-1}$. This fact is important when you divide by a polynomial, as in the example below.

EXAMPLE 2 Simplify $\dfrac{x^2 - 3x + 2}{x^2 - 9} \div \dfrac{x^2 - 4x + 4}{x^2 - 6x + 9} \div (x - 1)$.

▸ **Solution**

$\dfrac{x^2 - 3x + 2}{x^2 - 9} \div \dfrac{x^2 - 4x + 4}{x^2 - 6x + 9} \div (x - 1)$

$\dfrac{x^2 - 3x + 2}{x^2 - 9} \cdot \dfrac{x^2 - 6x + 9}{x^2 - 4x + 4} \cdot \dfrac{1}{x - 1}$ ⟵ Change division to multiplication by the reciprocal.

$\dfrac{(x-1)(x-2)}{(x-3)(x+3)} \cdot \dfrac{(x-3)(x-3)}{(x-2)(x-2)} \cdot \dfrac{1}{x-1}$ ⟵ Factor and divide by common factors.

$\dfrac{x - 3}{(x + 3)(x - 2)}$, or $\dfrac{x - 3}{x^2 + x - 6}$

TRY THIS Simplify $\dfrac{x^2 + 4x + 4}{x^2 - 16} \div (x + 2) \div (x + 2)$.

EXERCISES

KEY SKILLS

Write each expression as a product.

1. $\dfrac{2m^2n}{5mn^2} \cdot \dfrac{15mn^2}{12m^2n^3} \div \dfrac{5m^3n}{2mn}$

2. $\dfrac{3x^2y}{6xy^2} \div \dfrac{15xy^2}{12x^2y^3} \div \dfrac{x^3y}{2xy}$

3. $\dfrac{x^2 - 5x + 6}{x^2 - 9} \div \dfrac{x^2 - 4x + 4}{x^2 - 6x + 9} \div (x - 2)$

4. $\dfrac{x^2 - 5x + 6}{x^2 - 9} \cdot \dfrac{x^2 - 6x + 5}{x^2 - 25} \div \dfrac{x - 2}{x - 5}$

PRACTICE

Simplify. Write the answers with positive exponents only.

5. $\dfrac{a}{4} \div \dfrac{a}{8} \cdot \dfrac{a^2}{2}$

6. $\dfrac{4n}{n} \div \dfrac{n}{4} \cdot \dfrac{n^2}{4}$

7. $\dfrac{5}{w} \cdot \dfrac{2}{w} \div \dfrac{15}{w}$

8. $\dfrac{c^2}{-3} \div \dfrac{c^3}{9} \div \dfrac{c}{12}$

9. $\dfrac{x^3y}{x^2y^2} \cdot \dfrac{5x^2y}{2x^2y^3} \div \dfrac{5x^3y^2}{2}$

10. $\dfrac{ab}{a^2b^2} \cdot \dfrac{5a^2b^2}{2a^3b^3} \div \dfrac{5ab}{10a^3b^3}$

11. $\dfrac{d^2c^2}{dc} \div \dfrac{7d^2c^2}{2dc} \div \dfrac{7dc}{10d^3c^3}$

12. $\dfrac{dc^3}{dc} \div \dfrac{d^2}{2c} \div \dfrac{dc}{10d^2c^3}$

13. $\dfrac{x^3y^3}{x^2y} \div \dfrac{xy^2}{2xy} \cdot \dfrac{xy}{12x^2y^3}$

14. $\dfrac{r^2s^3}{r^3s} \div \dfrac{r^2s^3}{r^3s} \cdot \dfrac{rs}{r^2s^3}$

15. $\dfrac{(ab)^2}{3b^2} \cdot \dfrac{a^2b^3}{(ab)^3} \cdot \dfrac{ab}{b}$

16. $\dfrac{(mn)^5}{m^2n^3} \cdot \dfrac{m^3n^4}{(mn)^4} \cdot \dfrac{(mn)^2}{mn}$

17. $\dfrac{(pq)^2}{p^2q^3} \div \dfrac{p^3q^4}{(pq)^3} \div (p^4q^4)$

18. $\dfrac{x^2 - 2x - 15}{x^2 - 9} \div \dfrac{x^2 - 25}{x^2 - 10x + 21} \div (x - 7)$

19. $\dfrac{x^2 - x - 2}{x^2 - 1} \div \dfrac{x^2 - 7x + 10}{x^2 - 2x + 1} \div (x - 1)$

20. $\dfrac{v^2 - 3v + 2}{v^2 - v - 2} \cdot \dfrac{v^2 - 2v + 1}{v^2 - 1} \div \dfrac{v + 1}{v - 1}$

21. $\dfrac{p^2 - 5p - 14}{p^2 - 3p - 10} \cdot \dfrac{p^2 - 6p + 5}{p^2 - 2p + 1} \div \dfrac{p - 7}{p - 1}$

22. $\dfrac{c^2 - 9}{c^2 + 3c - 10} \div \dfrac{c^2 + c - 6}{c^2 - 7c + 10} \cdot \dfrac{c - 2}{c - 3}$

23. $\dfrac{a^2 - 7a - 30}{a^2 - 7a + 10} \cdot \dfrac{a^2 + a - 6}{a^2 + 3a - 10} \cdot \dfrac{a - 2}{a - 10}$

24. $\dfrac{(a + b)^3}{(r + s)^3} \cdot \dfrac{(r + s)^2}{(a + b)^2} \div \dfrac{a + b}{r + s}$

25. $\dfrac{(m + n)^2}{(x + y)^2} \cdot \dfrac{(x + y)^3}{(m + n)^3} \cdot \dfrac{m + n}{x + y}$

MIXED REVIEW

Simplify. (previous courses)

26. $\dfrac{7}{11} + \dfrac{19}{22}$

27. $\dfrac{1}{35} + \dfrac{6}{7}$

28. $\dfrac{5}{18} + \dfrac{5}{24}$

29. $\dfrac{5}{14} + \dfrac{6}{49}$

30. $\dfrac{18}{25} + \dfrac{7}{15}$

31. $\dfrac{17}{24} + \dfrac{5}{6}$

32. $\dfrac{9}{14} + \dfrac{12}{21}$

33. $\dfrac{17}{20} + \dfrac{7}{10}$

Answers for Lesson 9.3 Skill C

Try This Example 1

$\dfrac{1}{18r^4s^5}$

Try This Example 2

$\dfrac{1}{x^2 - 16}$

1. $\dfrac{2m^2n}{5mn^2} \cdot \dfrac{15mn^2}{12m^2n^3} \cdot \dfrac{2mn}{5m^3n}$

2. $\dfrac{3x^2y}{6xy^2} \cdot \dfrac{12x^2y^3}{15xy^2} \cdot \dfrac{2xy}{x^3y}$

3. $\dfrac{x^2 - 5x + 6}{x^2 - 9} \cdot$ $\dfrac{x^2 - 6x + 9}{x^2 - 4x + 4} \cdot$ $\dfrac{1}{x - 2}$

4. $\dfrac{x^2 - 5x + 6}{x^2 - 9} \cdot$ $\dfrac{x^2 - 6x + 5}{x^2 - 25} \cdot$ $\dfrac{x - 5}{x - 2}$

5. a^2

6. $\dfrac{64}{n^3}$

7. $\dfrac{2}{3w}$

8. $-\dfrac{36}{c^2}$

9. $\dfrac{1}{x^2y^4}$

10. 5

11. $\dfrac{20d^2c^2}{49}$

12. $\dfrac{20c^5}{d}$

13. $\dfrac{1}{6xy}$

14. $\dfrac{1}{rs^2}$

15. $\dfrac{1}{a}$

16. m^3n^3

17. $\dfrac{1}{p^4q^6}$

18. $\dfrac{1}{x + 5}$

19. $\dfrac{1}{x - 5}$

20. $\dfrac{v^3 - 3v^2 + 3v - 1}{v^3 + 3v^2 + 3v + 1}$

21. 1

22. $\dfrac{c - 5}{c + 5}$

23. $\dfrac{a + 5}{a - 5}$

24. 1

25. $\dfrac{(x + y)^2}{(m + n)^2}$

26. $\dfrac{3}{2}$

27. $\dfrac{31}{35}$

28. $\dfrac{35}{72}$

29. $\dfrac{47}{98}$

30. $\dfrac{89}{75}$

31. $\dfrac{37}{24}$

32. $\dfrac{17}{14}$

33. $\dfrac{31}{20}$

TEACHING 9.4 LESSON

Adding Rational Expressions

Adding rational expressions with like denominators

Focus: *Students add two rational expressions with like denominators and simplify the sum as necessary.*

EXAMPLE 1

PHASE 1: Presenting the Example

Display the expression $\dfrac{\boxed{}}{9} + \dfrac{\overline{}}{9}$. Tell

students that the box and oval each hold the place of a real number. Have them write an expression for the sum. Display the result as shown on the first line at right.

Ask students what the sum would be if the box contained 5 and the oval contained 2. What if the box contained r and the oval contained $4r$? What if the box contained $r + 5$ and the oval contained $4r + 2$? One by one, display the results as shown at right.

$$\frac{\boxed{}}{9} + \frac{\overline{}}{9} = \frac{\boxed{} + \overline{}}{9}$$

$$\frac{\boxed{5}}{9} + \frac{\overline{2}}{9} = \frac{\boxed{5} + \overline{2}}{9} = \frac{7}{9}$$

$$\frac{\boxed{r}}{9} + \frac{\overline{4r}}{9} = \frac{\boxed{r} + \overline{4r}}{9} = \frac{5r}{9}$$

$$\frac{\boxed{r + 5}}{9} + \frac{\overline{4r + 2}}{9} = \frac{\boxed{r + 5} + \overline{4r + 2}}{9} = \frac{5r + 7}{9}$$

After completing this series of examples, students should understand that the procedure for adding rational expressions with like denominators is the same as the familiar procedure for adding fractions that have like denominators. That is, you add the numerators and write the resulting sum over the *common denominator.*

Remind students that sometimes a sum must be further simplified. To illustrate this point, repeat the activity above with the following values: box = 1 and oval = 2; box = r and oval = $5r$; box = $r + 1$ and oval = $5r + 2$. $\Big[$ **The resulting simplified sums**

will be, respectively: $\dfrac{1}{9} + \dfrac{2}{9} = \dfrac{3}{9} = \dfrac{1}{3}$; $\dfrac{r}{9} + \dfrac{5r}{9} = \dfrac{6r}{9} = \dfrac{2r}{3}$; and $\dfrac{r + 1}{9} + \dfrac{5r + 2}{9} =$

$\dfrac{6r + 3}{9} = \dfrac{3(2r + 1)}{9} = \dfrac{2r + 1}{3}.\Big]$ Then discuss Example 1.

EXAMPLE 2

PHASE 1: Providing Guidance for **TRY THIS**

When adding rational expressions, students commonly add the numerators *and* add the denominators. As an aid in determining the correct procedure, suggest that they model a simple fraction addition from arithmetic. The diagram at right, for instance, shows that $\dfrac{1}{4} + \dfrac{2}{4} = \dfrac{3}{4}$. That is, the sum is $\dfrac{1 + 2}{4}$, *not* $\dfrac{1 + 2}{4 + 4}$. Therefore, the model indicates that adding the denominators is the incorrect procedure.

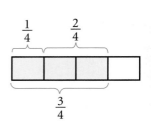

9.4 Adding Rational Expressions

SKILL A Adding rational expressions with like denominators

To add two fractions, first determine whether the denominators are like or unlike. If the two fractions have the same denominator, then you can use the rule below to add the fractions.

Adding Fractions with Like Denominators

Let a, b, and c be real numbers with $c \neq 0$. Then $\dfrac{a}{c} + \dfrac{b}{c} = \dfrac{a + b}{c}$.

EXAMPLE 1 Simplify.   **a.** $\dfrac{2}{9x} + \dfrac{4}{9x}$ **b.** $\dfrac{z - 5}{2} + \dfrac{z + 3}{2}$

▶ **Solution**

Write the sum of the numerators over the like denominator. Then simplify further, if necessary.

a. $\dfrac{2}{9x} + \dfrac{4}{9x} = \dfrac{6}{9x}$

    $= \dfrac{2}{3x}$

b. $\dfrac{z - 5}{2} + \dfrac{z + 3}{2} = \dfrac{(z - 5) + (z + 3)}{2}$

    $= \dfrac{2z - 2}{2}$

    $= \dfrac{2(z - 1)}{2}$

    $= z - 1$

TRY THIS Simplify.   **a.** $\dfrac{2}{15n} + \dfrac{-7}{15n}$ **b.** $\dfrac{n - 18}{-3} + \dfrac{2n + 3}{-3}$

The rule for addition applies whether the denominators are numbers or polynomials.

EXAMPLE 2 Simplify $\dfrac{x^2 - 3x}{x - 5} + \dfrac{x - 15}{x - 5}$.

▶ **Solution**

$\dfrac{x^2 - 3x}{x - 5} + \dfrac{x - 15}{x - 5} = \dfrac{(x^2 - 3x) + (x - 15)}{x - 5}$ ⟵ Add the numerators.

    $= \dfrac{x^2 - 2x - 15}{x - 5}$ ⟵ Combine like terms.

    $= \dfrac{(x - 5)(x + 3)}{x - 5}$ ⟵ Factor the numerator.

    $= \dfrac{(x - 5)(x + 3)}{x - 5}$ ⟵ Divide both the numerator and the denominator by $(x - 5)$.

    $= x + 3$

TRY THIS Simplify $\dfrac{a^2}{a + 9} + \dfrac{-81}{a + 9}$.

EXERCISES

KEY SKILLS

Write the numerator of each sum in simplified form. Do not simplify further.

1. $\dfrac{3n - 5}{3n^2} + \dfrac{2n - 1}{3n^2}$ **2.** $\dfrac{5z^2}{5(z + 1)} + \dfrac{-5z^2 + 3z}{5(z + 1)}$ **3.** $\dfrac{5a^2 - 3a}{-3(a - 4)} + \dfrac{a^2 + 3a}{-3(a - 4)}$

PRACTICE

Simplify.

4. $\dfrac{1}{2a} + \dfrac{5}{2a}$ **5.** $\dfrac{1}{9n} + \dfrac{2}{9n}$ **6.** $\dfrac{-5}{14k} + \dfrac{-3}{14k}$

7. $\dfrac{-11}{27k} + \dfrac{2}{27k}$ **8.** $\dfrac{13}{22b^2} + \dfrac{-2}{22b^2}$ **9.** $\dfrac{72}{25c^2} + \dfrac{-22}{25c^2}$

10. $\dfrac{w - 5}{3} + \dfrac{5w - 4}{3}$ **11.** $\dfrac{3n - 12}{-5} + \dfrac{7n - 3}{-5}$ **12.** $\dfrac{z - 12}{3z} + \dfrac{5z + 12}{3z}$

13. $\dfrac{t + 2}{6t} + \dfrac{11t - 2}{6t}$ **14.** $\dfrac{v + 2}{2v^2} + \dfrac{v - 2}{2v^2}$ **15.** $\dfrac{4u - 5}{-3u^2} + \dfrac{8u + 5}{-3u^2}$

16. $\dfrac{5y + 5}{y + 3} + \dfrac{-3y + 1}{y + 3}$ **17.** $\dfrac{5x + 53}{x + 6} + \dfrac{4x + 1}{x + 6}$ **18.** $\dfrac{5m + 19}{2m + 5} + \dfrac{5m + 6}{2m + 5}$

19. $\dfrac{q - 9}{2q - 3} + \dfrac{5q}{2q - 3}$ **20.** $\dfrac{a}{2a - 3} + \dfrac{9a - 15}{2a - 3}$ **21.** $\dfrac{b + 3}{-2b + 7} + \dfrac{-3b + 4}{-2b + 7}$

22. $\dfrac{2x + 3}{x^2 - 25} + \dfrac{-x + 2}{x^2 - 25}$ **23.** $\dfrac{3y - 3}{4y^2 - 36} + \dfrac{-y - 3}{4y^2 - 36}$ **24.** $\dfrac{2y - 1}{y^2 + 4y + 4} + \dfrac{-y + 3}{y^2 + 4y + 4}$

25. $\dfrac{5y + 1}{y^2 - 6y + 9} + \dfrac{-4y - 4}{y^2 - 6y + 9}$ **26.** $\dfrac{x^2 + 5x}{x + 3} + \dfrac{5x + 21}{x + 3}$

27. $\dfrac{x^2 + x}{x - 4} + \dfrac{x - 24}{x - 4}$ **28.** $\dfrac{5x^2 - 5x - 2}{2x - 3} + \dfrac{x^2 - 2x - 1}{2x - 3}$

29. $\dfrac{7a^2 + 10a - 2}{3a - 1} + \dfrac{2a^2 - a - 2}{3a - 1}$ **30.** $\dfrac{10v^2 - 5v - 6}{4v - 3} + \dfrac{10v^2 - 2v}{4v - 3}$

31. Critical Thinking Find m and n such that $\dfrac{mr + ms}{2x - 1} + \dfrac{nr - ns}{2x - 1} = \dfrac{r - 2s}{2x - 1}$.

MIXED REVIEW

Find each difference. (6.4 Skill B)

32. $(x^2 - 3x + 1) - (4x + 1)$ **33.** $(a^2 - 3a) - (-a^2 - 4a + 1)$ **34.** $(n^2 - 5n) - (4n^2 - n)$

35. $(b^2 - 4b) - (4b^3 + 2b)$ **36.** $(x^4 - x^2) - (x^3 - x)$ **37.** $n^4 - (n^3 - n^2 - n + 1)$

Answers for Lesson 9.4 Skill A

Try This Example 1

a. $-\dfrac{1}{3n}$

b. $-n + 5$

Try This Example 2

$a - 9$

1. $5n - 6$

2. $3z$

3. $6a^2$

4. $\dfrac{3}{a}$

5. $\dfrac{1}{3n}$

6. $\dfrac{-4}{7k}$

7. $-\dfrac{1}{3k}$

8. $\dfrac{1}{2b^2}$

9. $\dfrac{2}{c^2}$

10. $2w - 3$

11. $-2n + 3$

12. 2

13. 2

14. $\dfrac{1}{v}$

15. $\dfrac{-4}{u}$

16. 2

17. 9

18. 5

19. 3

20. 5

21. 1

22. $\dfrac{1}{x - 5}$

23. $\dfrac{1}{2y + 6}$

24. $\dfrac{1}{y + 2}$

25. $\dfrac{1}{y - 3}$

26. $x + 7$

27. $x + 6$

28. $3x + 1$

29. $3a + 4$

30. $5v + 2$

31. $m = \dfrac{1}{2}$ and $n = \dfrac{3}{2}$

32. $x^2 - 7x$

33. $2a^2 + a - 1$

34. $-3n^2 - 4n$

35. $-4b^3 + b^2 - 6b$

36. $x^4 - x^3 - x^2 + x$

37. $n^4 - n^3 + n^2 + n - 1$

SKILL B Adding rational expressions with unlike monomial denominators

Focus: *Students add rational expressions with unlike monomial denominators, rewriting one or more of the expressions to create a common denominator.*

EXAMPLE 1

PHASE 1: Presenting the Example

Display the expression $\frac{\Box}{6} + \frac{\bigcirc}{3}$. Again tell students that the box and oval each hold the place of a real number. Have them write an expression for the sum. Point out that, in this case, they must consider how to obtain a common denominator. Display the result as shown below.

$$\frac{\Box}{6} + \frac{\bigcirc}{3} = \frac{\Box}{6} + \frac{2\bigcirc}{2 \cdot 3} = \frac{\Box + 2\bigcirc}{6}$$

Ask students what the sum would be if the box contained 1 and the oval contained 4. What if the box contained d and the oval contained $5d$? What if the box contained $d + 1$ and the oval contained $5d + 4$? One by one, display the results as shown below. As you discuss these, check that all students understand common denominators and know how to rewrite a fraction so that it has a desired denominator.

$$\frac{\boxed{1}}{6} + \frac{\left(4\right)}{3} = \frac{\boxed{1}}{6} + \frac{2\left(4\right)}{2 \cdot 3} = \frac{\boxed{1} + 2\left(4\right)}{6} = \frac{1 + 8}{6} = \frac{9}{6} = \frac{3}{2}$$

$$\frac{\boxed{d}}{6} + \frac{\left(5d\right)}{3} = \frac{\boxed{d}}{6} + \frac{2\left(5d\right)}{2 \cdot 3} = \frac{\boxed{d} + 2\left(5d\right)}{6} = \frac{d + 10d}{6} = \frac{11d}{6}$$

$$\frac{\boxed{d+1}}{6} + \frac{\left(5d + 4\right)}{3} = \frac{\boxed{d+1}}{6} + \frac{2\left(5d + 4\right)}{2 \cdot 3} = \frac{\boxed{d+1} + 2\left(5d + 4\right)}{6} = \frac{d + 1 + 10d + 8}{6} = \frac{11d + 9}{6}$$

Now discuss Example 1. Point out that, in some additions, only one rational expression must be rewritten. This is the situation in part a. In other cases it will be necessary to rewrite all the expressions involved, as in part b.

PHASE 2: Providing Guidance for **TRY THIS**

Before they can rewrite expressions and add, some students need practice in just finding the least common denominator of a set of expressions. You can use Skill B Exercises 1 through 3 on the facing page for this purpose.

EXAMPLE 2

PHASE 2: Providing Guidance for **TRY THIS**

Some students may find it easier to rewrite fractions correctly if they work with the LCD in the expanded form of $2 \cdot 3 \cdot p \cdot p$, as shown at right.

$$\frac{5p + 1}{3p} + \frac{2}{p} + \frac{1}{2p^2}$$

$$= \frac{(5p + 1)}{3 \cdot p} \cdot \frac{2 \cdot p}{2 \cdot p} + \frac{2}{p} \cdot \frac{2 \cdot 3 \cdot p}{2 \cdot 3 \cdot p} + \frac{1}{2 \cdot p \cdot p} \cdot \frac{3}{3}$$

To add two fractions or rational expressions with unlike denominators, write new equivalent fractions or rational expressions that have the same denominator, and then add.

The **least common multiple** (LCM) of two numbers is the smallest number that is a multiple of both numbers. The **least common denominator** (LCD) of two fractions is the least common multiple of their denominators.

$$\frac{1}{3} + \frac{4}{5} = \frac{1}{3}\left(\frac{5}{5}\right) + \frac{4}{5}\left(\frac{3}{3}\right) = \frac{5}{15} + \frac{12}{15} = \frac{17}{15}, \text{ or } 1\frac{2}{15}$$

In the example above, 15 is the least common multiple of 3 and 5, and thus 15 is the least common denominator of $\frac{1}{3}$ and $\frac{4}{5}$.

EXAMPLE 1

Simplify.   **a.** $\frac{3}{a} + \frac{4}{a^2}$ **b.** $\frac{x-1}{2} + \frac{3}{x}$

▶ Solution

a. $\frac{3}{a} + \frac{4}{a^2} = \frac{3}{a} \cdot \frac{a}{a} + \frac{4}{a^2}$ **b.** $\frac{x-1}{2} + \frac{3}{x} = \frac{x-1}{2} \cdot \frac{x}{x} + \frac{3}{x} \cdot \frac{2}{2}$

$= \frac{3a}{a^2} + \frac{4}{a^2}$ $= \frac{x(x-1) + 3 \cdot 2}{2x}$

$= \frac{3a+4}{a^2}$ $= \frac{x^2 - x + 6}{2x}$

TRY THIS   Simplify.   **a.** $\frac{2}{b^2} + \frac{-5}{b^3}$ **b.** $\frac{2n-1}{3} + \frac{3}{2n}$

EXAMPLE 2

Simplify $\frac{5y+1}{3y} + \frac{2}{y^2} + \frac{1}{2y^2}$.

▶ Solution

$\frac{5y+1}{3y} + \frac{2}{y^2} + \frac{1}{2y^2} = \frac{5y+1}{3y} \cdot \frac{2y}{2y} + \frac{2}{y^2} \cdot \frac{6}{6} + \frac{1}{2y^2} \cdot \frac{3}{3}$ ◀— *The LCD of 3y, y², and 2y² is 6y².*

$= \frac{2y(5y+1) + 2(6) + 1(3)}{6y^2}$

$= \frac{10y^2 + 2y + 12 + 3}{6y^2}$

$= \frac{10y^2 + 2y + 15}{6y^2}$

TRY THIS   Simplify $\frac{5p+1}{3p} + \frac{2}{p} + \frac{1}{2p^2}$.

EXERCISES

KEY SKILLS

Write the least common denominator for each group of rational expressions.

1. $\frac{2z+7}{4z}$ and $\frac{z+3}{5z^2}$ **2.** $\frac{3a+1}{3a}$, $\frac{2a}{4}$, and $\frac{a-1}{a^2}$ **3.** $\frac{3n-5}{3n^2}$, $\frac{2n-1}{2n}$, and $\frac{2n-1}{5n^2}$

PRACTICE

Simplify.

4. $\frac{3}{2a} + \frac{5}{4a}$ **5.** $\frac{5}{2z} + \frac{-7}{8z}$ **6.** $\frac{-1}{3z} + \frac{3}{7z}$

7. $\frac{-2}{3b} + \frac{-5}{4b}$ **8.** $\frac{4}{5d^2} + \frac{1}{6d^2}$ **9.** $\frac{7}{6w^2} + \frac{5}{7w^2}$

10. $\frac{2}{3} + \frac{3a-2}{a}$ **11.** $\frac{2}{5} + \frac{2t+1}{t}$ **12.** $\frac{4}{5z} + \frac{2z-3}{z}$

13. $\frac{1}{2n} + \frac{n+2}{3n}$ **14.** $\frac{s}{2} + \frac{5s+1}{5s}$ **15.** $\frac{b}{3} + \frac{b+1}{5b}$

16. $\frac{d+1}{2} + \frac{d+1}{d}$ **17.** $\frac{m+1}{3} + \frac{m-1}{5m}$ **18.** $\frac{w+3}{3w} + \frac{w-2}{w^2}$

19. $\frac{3v+2}{3v^2} + \frac{v-2}{v}$ **20.** $\frac{p+2}{2p} + \frac{p-1}{p} + \frac{1}{p^2}$ **21.** $\frac{r+2}{5r^2} + \frac{2r-3}{r} + \frac{1}{r}$

22. $\frac{b}{2} + \frac{b+1}{b} + \frac{b+2}{b^2}$ **23.** $\frac{2}{a^3} + \frac{2a+1}{a} + \frac{2a+3}{a^2}$ **24.** $3 + \frac{2a+1}{a} + \frac{2a+3}{a^2}$

25. $7 + \frac{4z-5}{z} + \frac{4z-3}{z^2}$ **26.** $\frac{1}{x} + \frac{4x-5}{x^2} + \frac{x+7}{x^3}$ **27.** $\frac{u-1}{u} + \frac{u+1}{u^2} + \frac{u+3}{u^3}$

28. $\frac{1}{a} + \frac{1}{a^2} + \frac{1}{a^3} + \frac{1}{a^4}$ **29.** $\frac{1}{(3s)^2} + \frac{2}{s^2} + \frac{3}{(3s)^3}$ **30.** $\frac{5}{(-2r)^2} + \frac{3}{r^2} + \frac{1}{(-2r)^3}$

31. **Critical Thinking**   Let n represent a natural number. Find a formula for $\frac{1}{x} + \frac{1}{x^2} + \frac{1}{x^3} + \cdots + \frac{1}{x^n}$.

MIXED REVIEW

Solve each inequality. (4.5 Skill B)

32. $-4(r+5) - 6 < 0$ **33.** $6 - 4(s-1) \geq 3(s+1)$ **34.** $2(t-5) - 6(t-5) < 0$

35. $\frac{1}{3}x - 2x + 1 \leq \frac{1}{3}(3-x)$ **36.** $5 - 3(d+3) - 3 \geq \frac{2}{5}d$ **37.** $\frac{3}{7}(y-4) - \frac{4}{7}(y-4) > 2$

Answers for Lesson 9.4   Skill B

Try This Example 1

a. $\frac{2b-5}{b^3}$

b. $\frac{4n^2 - 2n + 9}{6n}$

Try This Example 2

$\frac{10p^2 + 14p + 3}{6p^2}$

1. $20z^2$

2. $12a^2$

3. $30n^2$

4. $\frac{11}{4a}$

5. $\frac{13}{8z}$

6. $\frac{2}{21z}$

7. $\frac{-23}{12b}$

8. $\frac{29}{30d^2}$

9. $\frac{79}{42w^2}$

10. $\frac{11a-6}{3a}$

11. $\frac{12t+5}{5t}$

12. $\frac{10z-11}{5z}$

13. $\frac{2n+7}{6n}$

14. $\frac{5s^2 + 10s + 2}{10s}$

15. $\frac{5b^2 + 3b + 3}{15b}$

16. $\frac{d^2 + 3d + 2}{2d}$

17. $\frac{5m^2 + 8m - 3}{15m}$

18. $\frac{w^2 + 6w - 6}{3w^2}$

19. $\frac{3v^2 - 3v + 2}{3v^2}$

20. $\frac{3p^2 + 2}{2p^2}$

21. $\frac{10r^2 - 9r + 2}{5r^2}$

22. $\frac{b^3 + 2b^2 + 4b + 4}{b^2}$

23. $\frac{2a^3 + 3a^2 + 3a + 2}{a^3}$

24. $\frac{5a^2 + 3a + 3}{a^2}$

25. $\frac{11z^2 - z - 3}{z^2}$

26. $\frac{5x^2 - 4x + 7}{x^3}$

27. $\frac{u^3 + 2u + 3}{u^3}$

28. $\frac{a^3 + a^2 + a + 1}{a^4}$

29. $\frac{19s + 1}{9s^3}$

30. $\frac{34r - 1}{8r^3}$

31. $\frac{x^{n-1} + x^{n-2} + \cdots + x^1 + x^0}{x^n}$

For answers to Exercises 32–37, see Additional Answers.

Focus: *Students add rational expressions with unlike polynomial denominators,*
rewriting one or more of the expressions to create a common denominator.

EXAMPLE 1

PHASE 1: Presenting the Example

Display the expression $\dfrac{2}{\boxed{}} + \dfrac{5}{\big(\big)}$. Once again tell students that the box and

oval each hold the place of a real number. Have them write an expression for the sum.
Display the result as shown below.

$$\frac{2}{\boxed{}} + \frac{5}{\big(\big)} = \frac{2}{\boxed{}} \cdot \frac{\big(\big)}{\big(\big)} + \frac{5}{\big(\big)} \cdot \frac{\boxed{}}{\boxed{}} = \frac{2\big(\big) + 5\boxed{}}{\boxed{} \cdot \big(\big)}$$

Ask students what the sum would be if the box contained 7 and the oval contained 4.
What if the box contained t and the oval contained 4? What if the box contained $t + 7$
and the oval contained 4? What if the box contained $t + 7$ and the oval contained $t + 4$?

One by one, display the results. $\left[\vphantom{\dfrac{A}{B}}\right.$ **The resulting simplified sums will be,**

respectively: $\dfrac{2 \cdot 4}{7 \cdot 4} + \dfrac{5 \cdot 7}{4 \cdot 7} = \dfrac{43}{28}; \dfrac{2 \cdot 4}{t \cdot 4} + \dfrac{5 \cdot t}{4 \cdot t} = \dfrac{5t + 8}{4t}; \dfrac{2 \cdot 4}{(t + 7) \cdot 4} + \dfrac{5 \cdot (t + 7)}{4 \cdot (t + 7)} =$

$\dfrac{5t + 43}{4t + 28};$ and $\dfrac{2 \cdot (t + 4)}{(t + 7) \cdot (t + 4)} + \dfrac{5 \cdot (t + 7)}{(t + 4) \cdot (t + 7)} = \dfrac{7t + 43}{t^2 + 11t + 28}.\left.\vphantom{\dfrac{A}{B}}\right]$ Then discuss

Example 1.

EXAMPLE 2

PHASE 2: Providing Guidance for **TRY THIS**

After students have successfully completed the Try This exercise, lead a discussion
in which they list all the previously learned math concepts and skills that were
needed to complete the task. Ask them to identify the step at which each concept
or skill was used. A partial list of concepts and skills is given at right.

- common monomial factoring
- least common denominator
- multiplicative identity
- fraction multiplication
- fraction addition
- multiplication of binomials

PHASE 3: ASSESSMENT AND CLOSURE for Lesson 9.4

- Give students the expression at right. Tell them that it shows the first step in the
  process of adding two rational expressions. Have them identify the original addition,
  and then continue the work to find the simplified sum. [**See answers at the right.**]

- In the lesson that follows, students will extend their addition skills to encompass the
  process of subtracting rational expressions.

$$\frac{3}{k + 2} \cdot \frac{k}{k} + \frac{5}{k} \cdot \frac{k + 2}{k + 2}$$

$\left[\text{addition: } \dfrac{3}{k + 2} + \dfrac{5}{k};\right.$

$\left.\text{sum: } \dfrac{8k + 10}{k^2 + 2k}\right]$

☞ *For a Lesson 9.4 Quiz, see* Assessment Resources *page 94.*

To add $\frac{2}{m}$ and $\frac{1}{m+3}$, you need to find a common denominator for the pair of rational expressions.

EXAMPLE 1   Simplify $\frac{2}{m} + \frac{1}{m+3}$.

▶ Solution

The least common denominator (LCD) of $\frac{2}{m}$ and $\frac{1}{m+3}$ is $m(m+3)$.

Multiply each expression by the equivalent of 1 that makes the two denominators the same.

$$\frac{2}{m} + \frac{1}{m+3} = \frac{2}{m} \cdot \frac{(m+3)}{(m+3)} + \frac{1}{(m+3)} \cdot \frac{m}{m}$$

$$= \frac{2(m+3)}{m(m+3)} + \frac{m}{m(m+3)} \qquad \longleftarrow \text{The denominators are now the same.}$$

$$= \frac{2(m+3)+m}{m(m+3)} \qquad \longleftarrow \text{Add the numerators.}$$

$$= \frac{3m+6}{m^2+3m} \qquad \longleftarrow \text{Simplify.}$$

TRY THIS   Simplify $\frac{1}{b-1} + \frac{5}{b+1}$.

You may need to factor each denominator before looking for the LCD. In Example 2, factor $b^2 + b$ as $b(b+1)$ and $b^2 - b$ as $b(b-1)$.

EXAMPLE 2   Simplify $\frac{1}{b^2+b} + \frac{1}{b^2-b}$.

▶ Solution

$$\frac{1}{b^2+b} + \frac{1}{b^2-b}$$

$$\frac{1}{b(b+1)} + \frac{1}{b(b-1)} \qquad \longleftarrow \text{Factor each denominator.}$$

$$\frac{1}{b(b+1)} \cdot \frac{b-1}{b-1} + \frac{1}{b(b-1)} \cdot \frac{b+1}{b+1} \qquad \longleftarrow \frac{b-1}{b-1} = 1 \text{ and } \frac{b+1}{b+1} = 1$$

$$\frac{(b-1)+(b+1)}{b(b-1)(b+1)} \qquad \longleftarrow \text{Write one denominator: } b(b-1)(b+1)$$

$$\frac{2b}{b(b-1)(b+1)} \qquad \longleftarrow \text{Add the numerators: } (b-1)+(b+1)$$

$$\frac{2}{(b-1)(b+1)}, \text{ or } \frac{2}{b^2-1}$$

TRY THIS   Simplify $\frac{2}{2c^2+c} + \frac{2}{2c^2-c}$.

EXERCISES

KEY SKILLS

Write the denominators in factored form. Then find the LCD. Do not simplify further.

1. $\frac{1}{z^2+3z} + \frac{1}{z+3}$

2. $\frac{6a}{a^2-25} + \frac{5a}{2a-10}$

3. $\frac{6k+1}{k^2-1} + \frac{5k+3}{k^2-k}$

PRACTICE

Simplify.

4. $\frac{1}{n} + \frac{1}{n+1}$

5. $\frac{2}{z} + \frac{1}{z-1}$

6. $\frac{3}{2k+1} + \frac{5}{k}$

7. $\frac{5}{3y-5} + \frac{3}{y}$

8. $\frac{2}{x^3} + \frac{1}{x-2}$

9. $\frac{5}{r-5} + \frac{5}{r^2}$

10. $\frac{1}{b+1} + \frac{1}{b-1}$

11. $\frac{2}{d+1} + \frac{3}{d+2}$

12. $\frac{5}{d-1} + \frac{2}{d-3}$

13. $\frac{5}{q+2} + \frac{2}{q+5}$

14. $\frac{5s+1}{s+1} + \frac{2s-3}{s+2}$

15. $\frac{2t-1}{t+3} + \frac{2t+1}{t+2}$

16. $\frac{3v-1}{v-3} + \frac{3v+1}{v+3}$

17. $\frac{2u+1}{u-1} + \frac{2u+1}{u+1}$

18. $\frac{2a}{a^2-1} + \frac{2a}{a+1}$

19. $\frac{3g}{g^2-4} + \frac{g}{g-2}$

20. $\frac{c}{c^2-1} + \frac{c-2}{c^2-c}$

21. $\frac{z}{z^2+2z} + \frac{5z}{z^2-2z}$

22. $\frac{3}{h^2+4h+4} + \frac{5}{h^2+2h}$

23. $\frac{3}{b^2+4b+4} + \frac{2}{b^2-4}$

24. $\frac{1}{b^2+2b+1} + \frac{1}{b^2+b}$

25. $\frac{1}{a+1} + \frac{1}{(a+1)^2} + \frac{1}{(a+1)^3}$

26. $\frac{1}{n-1} + \frac{2}{(n-1)^2} + \frac{3}{(n-1)^3}$

27. $\frac{3}{m} + \frac{2}{m+1} + \frac{1}{(m+1)^2}$

28. **Critical Thinking**   Find A such that $\frac{7x-19}{x^2-5x+4} = \frac{A}{x-4} + \frac{4}{x-1}$.

MIXED REVIEW

Solve each system of equations. (5.3 Skill A)

29. $\begin{cases} 2x+7y=0 \\ -2x+6y=0 \end{cases}$

30. $\begin{cases} 2a+7b=3 \\ 2a-7b=4 \end{cases}$

31. $\begin{cases} -3m+2n=6 \\ 3m+2n=4 \end{cases}$

32. $\begin{cases} 3u-5v=2 \\ 3u+5v=3 \end{cases}$

33. $\begin{cases} -3b-4d=-1 \\ 3b+5d=-1 \end{cases}$

34. $\begin{cases} 11w-6y=35 \\ 10w+6y=-14 \end{cases}$

Answers for Lesson 9.4 Skill C

Try This Example 1

$\dfrac{6b-4}{(b-1)(b+1)}$, or $\dfrac{6b-4}{b^2-1}$

Try This Example 2

$\dfrac{8}{(2c+1)(2c-1)}$, or $\dfrac{8}{4c^2-1}$

1. $z^2 + 3z = z(z+3)$; LCD $= z(z+3)$

2. $a^2 - 25 = (a+5)(a-5)$; $2a - 10 = 2(a-5)$; LCD $= 2(a+5)(a-5)$

3. $k^2 - 1 = (k+1)(k-1)$; $k^2 - k = k(k-1)$; LCD $= k(k+1)(k-1)$

4. $\dfrac{2n+1}{n(n+1)}$, or $\dfrac{2n+1}{n^2+n}$

5. $\dfrac{3z-2}{z(z+1)}$, or $\dfrac{3z-2}{z^2-z}$

6. $\dfrac{13k+5}{k(2k+1)}$, or $\dfrac{13k+5}{2k^2+k}$

7. $\dfrac{14y-15}{y(3y-5)}$, or $\dfrac{14y-15}{3y^2-5y}$

8. $\dfrac{x^2+2x-4}{x^2(x-2)}$, or $\dfrac{x^2+2x-4}{x^3-2x^2}$

9. $\dfrac{5r^2+5r-25}{r^2(r-5)}$, or $\dfrac{5r^2+5r-25}{r^3-5r^2}$

10. $\dfrac{2b}{(b+1)(b-1)}$, or $\dfrac{2b}{b^2-1}$

11. $\dfrac{5d+7}{(d+1)(d+2)}$, or $\dfrac{5d+7}{d^2+3d+2}$

12. $\dfrac{7d-17}{(d-1)(d-3)}$, or $\dfrac{7d-17}{d^2-4d+3}$

13. $\dfrac{7q+29}{(q+2)(q+5)}$, or $\dfrac{7q+29}{q^2+7q+10}$

14. $\dfrac{7s^2+10s-1}{(s+1)(s+2)}$, or $\dfrac{7s^2+10s-1}{s^2+3s+2}$

15. $\dfrac{4t^2+10t+1}{(t+3)(t+2)}$, or $\dfrac{4t^2+10t+1}{t^2+5t+6}$

16. $\dfrac{6v^2-6}{(v-3)(v+3)}$, or $\dfrac{6v^2-6}{v^2-9}$

17. $\dfrac{4u^2+2u}{(u-1)(u+1)}$, or $\dfrac{4u^2+2u}{u^2-1}$

18. $\dfrac{2a^2}{(a-1)(a+1)}$, or $\dfrac{2a^2}{a^2-1}$

19. $\dfrac{g^2+5g}{(g-2)(g+2)}$, or $\dfrac{g^2+5g}{g^2-4}$

20. $\dfrac{2c^2-c-2}{c(c+1)(c-1)}$, or $\dfrac{2c^2-c-2}{c^3-c}$

21. $\dfrac{6z+8}{(z+2)(z-2)}$, or $\dfrac{6z+8}{z^2-4}$

22. $\dfrac{7h+10}{h(h+2)(h+2)}$, or $\dfrac{7h+10}{h^3+4h^2+4h}$

23. $\dfrac{5b-2}{(b-2)(b+2)(b+2)}$, or $\dfrac{5b-2}{b^3+2b^2-4b-8}$

24. $\dfrac{2b+1}{b(b+1)(b+1)}$, or $\dfrac{2b+1}{b^3+2b^2+b}$

25. $\dfrac{a^2+3a+3}{(a+1)^3}$, or $\dfrac{a^2+3a+3}{a^3+3a^2+3a+1}$

For answers to Exercises 26–34, see Additional Answers.

TEACHING
9.5
LESSON

Subtracting Rational Expressions

Focus: *Students subtract two rational expressions with like denominators and simplify*
the difference as necessary.

EXAMPLE 1

PHASE 1: Presenting the Example

To present part a of Example 1, first display the addition $\dfrac{4}{p} + \dfrac{7}{p}$. Have students simplify it.

$\left[\dfrac{11}{p}\right]$ Then, as students watch, erase the "+" symbol and insert the "−" symbol. Have

them write what they believe to be the simplified difference. They most likely will

respond $\dfrac{4-7}{p} = \dfrac{-3}{p}$. Point out that this is usually rewritten as $-\dfrac{3}{p}$.

For part b, first have students simplify $\dfrac{3t}{t-3} + \dfrac{9}{t-3}$. $\left[\dfrac{3t+9}{t-3}\right]$ Be sure they note

that this cannot be further simplified. In contrast, when you ask them to simplify

$\dfrac{3t}{t-3} - \dfrac{9}{t-3}$, they can rewrite the result $\dfrac{3t-9}{t-3}$ as $\dfrac{3(t-3)}{t-3} = 3$.

PHASE 2: Providing Guidance for **TRY THIS**

After students have successfully completed the Try This exercises, ask
them to rewrite each of the given subtractions as an addition of
rational expressions. [**The additions are shown at right.**]

a. $\dfrac{2}{n} - \dfrac{-3}{n} = \dfrac{2}{n} + \dfrac{3}{n}$

b. $\dfrac{4z}{2z-3} - \dfrac{6}{2z-3} = \dfrac{4z}{2z-3} + \dfrac{-6}{2z-3}$

EXAMPLES 2 and 3

PHASE 1: Presenting the Examples

Continue the process of first having students perform an addition, and then changing the
addition to a "look-alike" subtraction.

For instance, in Example 2 have them simplify $\dfrac{3a-8}{a-5} + \dfrac{17-2a}{a-5}$. $\left[\dfrac{a+9}{a-5}\right]$ When you

change this to $\dfrac{3a-8}{a-5} - \dfrac{17-2a}{a-5}$, several may write $\dfrac{5a+9}{a-5}$, forgetting that they must

subtract 17 *and* subtract −2a. Have them carefully examine the use of parentheses in the
simplification on the textbook page.

In Example 3, first have students simplify $\dfrac{4d^2}{2d+5} + \dfrac{25}{2d+5}$. $\left[\dfrac{4d^2+25}{2d+5}\right]$ Once again,

point out that this result cannot be further simplified. However, the result of the

subtraction $\dfrac{4d^2}{2d+5} - \dfrac{25}{2d+5}$, which is $\dfrac{4d^2-25}{2d+5}$, can be simplified after recognizing

$4d^2 - 25$ as a difference of two squares.

9.5 LESSON — Subtracting Rational Expressions

When you subtract $\frac{b}{c}$ from $\frac{a}{c}$, you subtract b from a and write the difference over c.

$$\frac{a}{c} - \frac{b}{c} = \frac{a}{c} + \frac{-b}{c} = \frac{a + (-b)}{c} = \frac{a - b}{c}$$

EXAMPLE 1 **Simplify.** **a.** $\frac{4}{p} - \frac{7}{p}$ **b.** $\frac{3t}{t-3} - \frac{9}{t-3}$

▶ Solution

a. $\frac{4}{p} - \frac{7}{p} = \frac{4-7}{p}$

$= \frac{-3}{p}$, or $-\frac{3}{p}$

b. $\frac{3t}{t-3} - \frac{9}{t-3} = \frac{3t-9}{t-3}$

$= \frac{3(t-3)}{t-3} = 3$

Divide both the numerator and the denominator by $(t-3)$.

TRY THIS Simplify. **a.** $\frac{2}{n} - \frac{-3}{n}$ **b.** $\frac{4z}{2z-3} - \frac{6}{2z-3}$

EXAMPLE 2 **Simplify** $\frac{3a-8}{a-5} - \frac{17-2a}{a-5}$.

▶ Solution

$\frac{3a-8}{a-5} - \frac{17-2a}{a-5} = \frac{(3a-8) + [-(17-2a)]}{a-5}$ ← *Write subtraction as addition of the opposite.*

$= \frac{5a-25}{a-5}$ ← *Simplify.*

$= \frac{5(a-5)}{a-5}$ ← *Factor 5 from the numerator.*

$= 5$ ← *Divide both the numerator and the denominator by $(a-5)$.*

TRY THIS Simplify $\frac{11z+10}{2z-3} - \frac{5z+19}{2z-3}$.

EXAMPLE 3 **Simplify** $\frac{4d^2}{2d+5} - \frac{25}{2d+5}$.

▶ Solution

$\frac{4d^2}{2d+5} - \frac{25}{2d+5} = \frac{4d^2-25}{2d+5}$

$= \frac{(2d+5)(2d-5)}{2d+5}$ ← *Factor $4d^2 - 25$ as $(2d+5)(2d-5)$.*

$= 2d-5$

TRY THIS Simplify $\frac{9t^2+49}{3t-7} - \frac{42t}{3t-7}$.

EXERCISES

KEY SKILLS

Write each subtraction as addition of the opposite. Do not simplify further.

1. $\frac{3t-1}{t} - \frac{2t-2}{t}$

2. $\frac{3y^2-1}{y+5} - \frac{3y^2+2}{y+5}$

3. $\frac{3v^2-3}{v-7} - \frac{2v^2+3}{v-7}$

PRACTICE

Simplify.

4. $\frac{5}{s} - \frac{7}{s}$

5. $\frac{5}{2v} - \frac{3}{2v}$

6. $\frac{2t}{t-7} - \frac{14}{t-7}$

7. $\frac{a}{7a+7} - \frac{-1}{7a+7}$

8. $\frac{4r+1}{2r+1} - \frac{-1}{2r+1}$

9. $\frac{9w+1}{3w-1} - \frac{4}{3w-1}$

10. $\frac{5b-3}{b+2} - \frac{b-11}{b+2}$

11. $\frac{6n-18}{n-2} - \frac{n-8}{n-2}$

12. $\frac{11x-2}{3x-1} - \frac{2x+1}{3x-1}$

13. $\frac{11r+2}{5r+2} - \frac{6r}{5r+2}$

14. $\frac{6n^2+3n}{2n-1} - \frac{6n^2+2n}{2n-1}$

15. $\frac{2s^2}{3s-2} - \frac{2s^2-4s+1}{3s-2}$

16. $\frac{10d^2-6}{d+1} - \frac{2d^2+2}{d+1}$

17. $\frac{14p^2-200}{2p+7} - \frac{2p^2+94}{2p+7}$

18. $\frac{5q^2-2q+1}{q^2+6q+9} - \frac{5q^2-q+4}{q^2+6q+9}$

19. $\frac{4z^2+6z+1}{4z^2+4z+1} - \frac{4z^2-2z-3}{4z^2+4z+1}$

20. $\frac{6v^2-v}{9v^2-25} - \frac{15}{9v^2-25}$

21. $\frac{5n^2+n}{4n^2-9} - \frac{n^2-7n-3}{4n^2-9}$

22. **Critical Thinking** **a.** Let a and b represent nonzero real numbers. Assume that $a \neq b$. Prove that $\frac{b}{a-b} + 1 = \frac{a}{a-b}$. Justify each step.

b. Simplify $\frac{rx+s}{ax+b} - \frac{ux+v}{ax+b}$. Assume all of the variables represent positive real numbers.

c. Find r and s such that $\frac{rx+s}{3x-5} - \frac{2x+7}{3x-5} = 1$ for all values of x.

d. Find a and b such that $\frac{ax-b}{2x+1} - \frac{(b-1)x+2a}{2x+1} = \frac{2x-5}{2x+1}$ for all values of x.

MIXED REVIEW

Simplify. (2.3 Skill A)

23. $\frac{5}{6} - \frac{1}{3}$

24. $\frac{9}{14} - \frac{1}{2}$

25. $\frac{5}{7} - \frac{3}{5}$

26. $\frac{5}{11} - \frac{13}{33}$

27. $\frac{1}{3} - \frac{11}{18}$

28. $\frac{2}{5} - \frac{6}{7}$

29. $\frac{3}{11} - \frac{15}{33}$

30. $\frac{2}{7} - \frac{5}{9}$

Answers for Lesson 9.5 Skill A

Try This Example 1

a. $\frac{5}{n}$ **b.** 2

Try This Example 2

3

Try This Example 3

$3t - 7$

1. $\frac{3t-1}{t} + \frac{-2t+2}{t}$

2. $\frac{3y^2-1}{y+5} + \frac{-3y^2-2}{y+5}$

3. $\frac{3v^2-3}{v-7} + \frac{-2v^2-3}{v-7}$

4. $-\frac{2}{s}$

5. $\frac{1}{v}$

6. 2

7. $\frac{1}{7}$

8. 2

9. 3

10. 4

11. 5

12. 3

13. 1

14. $\frac{n}{2n-1}$

15. $\frac{4s-1}{3s-2}$

16. $8(d-1)$

17. $\frac{12p^2-294}{2p+7}$

18. $\frac{-1}{q+3}$

19. $\frac{4}{2z+1}$

20. $\frac{2v+3}{3v-5}$

21. $\frac{2n+1}{2n-3}$

★ **22.** **a.** By the Inverse Property of Multiplication, $\frac{b}{a-b} + 1$

$= \frac{b}{a-b} + \frac{a-b}{a-b}$. By the definition of addition of

fractions, $\frac{b}{a-b} + \frac{a-b}{a-b} = \frac{b+(a-b)}{a-b}$. By the

Associative and Commutative Properties of Addition,

$\frac{b+(a-b)}{a-b} = \frac{b+(-b)+a}{a-b}$. By the Inverse

Property of Addition, $\frac{b+(-b)+a}{a-b} = \frac{0+a}{a-b}$.

By the Identity Property of Addition, $\frac{0+a}{a-b} = \frac{a}{a-b}$.

b. $\frac{(r-u)x + (s-v)}{ax+b}$

c. $r = 5$ and $s = 2$

d. $a = 2$ and $b = 1$

23. $\frac{1}{2}$

24. $\frac{1}{7}$

For answers to Exercises 25–30, see Additional Answers.

★ **Advanced Exercise**

Focus: *Students subtract two rational expressions with unlike denominators and simplify the difference as necessary.*

EXAMPLE 1

PHASE 1: Presenting the Example

To introduce part a, display the addition $\frac{5}{b} + \frac{4}{b^2}$. Have students simplify. $\left[\frac{5b+4}{b^2}\right]$ Ask a student volunteer to write the steps of the simplification on the chalkboard. [**See the sample at right.**] Ask students how these steps would differ if the given expression were $\frac{5}{b} - \frac{4}{b^2}$. They should readily see that the steps would be very similar. The only difference is that each "+" symbol would be replaced by a "−" symbol.

Repeat the activity for part b. In this case, begin by having students simplify $\frac{6}{n} + \frac{2n}{n+1}$.

$\left[\dfrac{2n^2 + 6n + 6}{n^2 + n}\right]$

$$\frac{5}{b} + \frac{4}{b^2}$$

$$\frac{5 \cdot b}{b \cdot b} + \frac{4}{b \cdot b}$$

$$\frac{5 \cdot b + 4}{b \cdot b}$$

$$\frac{5b + 4}{b^2}$$

EXAMPLES 2 and 3

PHASE 1: Presenting the Examples

Repeat the activity suggested for Example 1, continuing to stress the connections between addition and subtraction of rational expressions.

In Example 2, first have students simplify $\frac{5x-3}{3x-6} + \frac{4}{x-2}$. $\left[\dfrac{5x+9}{3x-6}\right]$ In Example 3, have them simplify $\frac{c-7}{c^2-3c} + \frac{c-11}{c^2-9}$. $\left[\dfrac{2c^2 - 15c - 21}{c^3 - 9c}\right]$

PHASE 2: Providing Guidance for TRY THIS

Students sometimes overlook the fact that one of the denominators of a given expression can be factored. Consequently, they might work with a common denominator that is not the *least* common denominator. This is not an error in itself. However, they should be aware that this type of oversight generally makes the simplification much more complicated and introduces several opportunities for error.

For example, the Try This exercise of Example 2 is simplified at right. Notice that $5y - 5$ was not factored as $5(y-1)$. Instead, $(y-1)(5y-5)$ was used as the common denominator. Have students compare the large amount of work involved in this process to the work involved in the process below, which uses the LCD $5(y-1)$.

$$\frac{y}{y-1} - \frac{2}{5y-5} = \frac{5}{5} \cdot \frac{y}{(y-1)} - \frac{2}{5(y-1)} = \frac{5y-2}{5(y-1)} = \frac{5y-2}{5y-5}$$

$$\frac{y}{y-1} - \frac{2}{5y-5}$$

$$\frac{y}{y-1} \cdot \frac{5y-5}{5y-5} - \frac{2}{5y-5} \cdot \frac{y-1}{y-1}$$

$$\frac{y(5y-5) - 2(y-1)}{(y-1)(5y-5)}$$

$$\frac{5y^2 - 5y - 2y + 2}{(y-1)(5y-5)}$$

$$\frac{5y^2 - 7y + 2}{(y-1)(5y-5)}$$

$$\frac{(5y-2)(y-1)}{(y-1)(5y-5)}$$

$$\frac{5y-2}{5y-5}$$

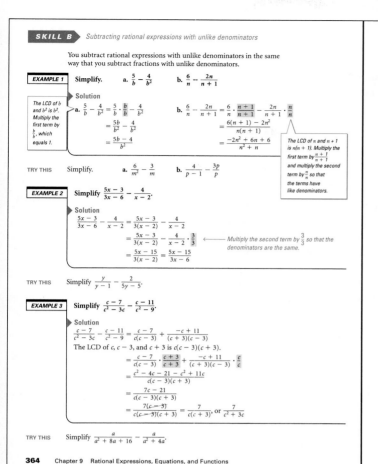

SKILL B Subtracting rational expressions with unlike denominators

You subtract rational expressions with unlike denominators in the same way that you subtract fractions with unlike denominators.

EXAMPLE 1 Simplify. **a.** $\frac{5}{b} - \frac{4}{b^2}$ **b.** $\frac{6}{n} - \frac{2n}{n+1}$

Solution

The LCD of b and b^2 is b^2. Multiply the first term by $\frac{b}{b}$, which equals 1.

a. $\frac{5}{b} - \frac{4}{b^2} = \frac{5}{b} \cdot \frac{b}{b} - \frac{4}{b^2}$

$= \frac{5b}{b^2} - \frac{4}{b^2}$

$= \frac{5b - 4}{b^2}$

b. $\frac{6}{n} - \frac{2n}{n+1} = \frac{6}{n} \cdot \frac{n+1}{n+1} - \frac{2n}{n+1} \cdot \frac{n}{n}$

$= \frac{6(n+1) - 2n^2}{n(n+1)}$

$= \frac{-2n^2 + 6n + 6}{n^2 + n}$

The LCD of n and $n+1$ is $n(n+1)$. Multiply the first term by $\frac{n+1}{n+1}$ and multiply the second term by $\frac{n}{n}$ so that the terms have like denominators.

TRY THIS Simplify. **a.** $\frac{6}{m^2} - \frac{3}{m}$ **b.** $\frac{4}{p-1} - \frac{3p}{p}$

EXAMPLE 2 Simplify $\frac{5x - 3}{3x - 6} - \frac{4}{x - 2}$.

Solution

$\frac{5x - 3}{3x - 6} - \frac{4}{x - 2} = \frac{5x - 3}{3(x - 2)} - \frac{4}{x - 2}$

$= \frac{5x - 3}{3(x - 2)} - \frac{4}{x - 2} \cdot \frac{3}{3}$ ← Multiply the second term by $\frac{3}{3}$ so that the denominators are the same.

$= \frac{5x - 15}{3(x - 2)} = \frac{5x - 15}{3x - 6}$

TRY THIS Simplify $\frac{y}{y - 1} - \frac{2}{5y - 5}$.

EXAMPLE 3 Simplify $\frac{c - 7}{c^2 - 3c} - \frac{c - 11}{c^2 - 9}$.

Solution

$\frac{c - 7}{c^2 - 3c} - \frac{c - 11}{c^2 - 9} = \frac{c - 7}{c(c - 3)} + \frac{-c + 11}{(c + 3)(c - 3)}$

The LCD of c, $c - 3$, and $c + 3$ is $c(c - 3)(c + 3)$.

$= \frac{c - 7}{c(c - 3)} \cdot \frac{c + 3}{c + 3} + \frac{-c + 11}{(c + 3)(c - 3)} \cdot \frac{c}{c}$

$= \frac{c^2 - 4c - 21 - c^2 + 11c}{c(c - 3)(c + 3)}$

$= \frac{7c - 21}{c(c - 3)(c + 3)}$

$= \frac{7(c - 3)}{c(c - 3)(c + 3)} = \frac{7}{c(c + 3)}$, or $\frac{7}{c^2 + 3c}$

TRY THIS Simplify $\frac{a}{a^2 + 8a + 16} - \frac{a}{a^2 + 4a}$.

EXERCISES

KEY SKILLS

Tell whether the fractions have like or unlike denominators. If the fractions have unlike denominators, find the LCD. Do not perform the subtraction.

1. $\frac{5}{3x + 6} - \frac{4}{3(x + 2)}$ 2. $\frac{4}{3x - 6} - \frac{1}{x - 2}$ 3. $\frac{10}{5x - 15} - \frac{1}{x + 3}$ 4. $\frac{7}{5x - 15} - \frac{4}{3x - 9}$

PRACTICE

Simplify.

5. $\frac{2}{x} - \frac{1}{x}$ 6. $\frac{7}{a} - \frac{4}{5a}$ 7. $\frac{4}{3z} - \frac{2}{5z}$ 8. $\frac{3}{2v} - \frac{6}{7v}$

9. $\frac{1}{x} - \frac{1}{x - 1}$ 10. $\frac{1}{n} - \frac{1}{n + 2}$ 11. $\frac{3z}{z - 2} - \frac{5}{2z}$ 12. $\frac{7w}{w - 1} - \frac{5}{2w}$

13. $\frac{2x - 3}{5x - 15} - \frac{1}{x - 3}$ 14. $\frac{m + 5}{m + 2} - \frac{3}{2m + 4}$ 15. $\frac{z - 2}{z + 2} - \frac{7}{5z + 10}$

16. $\frac{3v - 2}{2v - 10} - \frac{v}{v - 5}$ 17. $\frac{3u - 2}{2u - 6} - \frac{u + 2}{3u + 9}$ 18. $\frac{3u + 4}{2u + 6} - \frac{u + 1}{4u + 12}$

19. $\frac{x - 1}{x + 1} - \frac{x + 1}{x}$ 20. $\frac{c + 1}{c - 1} - \frac{c + 1}{c}$ 21. $\frac{a - 1}{a + 1} - \frac{a + 1}{a - 1}$

22. $\frac{3s + 1}{s - 1} - \frac{s - 3}{s + 2}$ 23. $\frac{2b}{b^2 - 1} - \frac{b}{b^2 - b}$ 24. $\frac{2d}{d^2 + 2d} - \frac{d + 1}{d^2 - 4}$

25. $\frac{y - 1}{y^2 + 2y + 1} - \frac{y + 1}{y^2 - 1}$ 26. $\frac{t - 2}{t^2 - 9} - \frac{t + 2}{t^2 + 6t + 9}$ 27. $\frac{g - 1}{4 - g^2} - \frac{g + 1}{g^2 + 4g + 4}$

28. Let a and b represent nonzero real numbers. Simplify $\frac{a}{b} - \frac{b}{a}$.

29. Let a and b represent nonzero real numbers such that $a \neq \pm b$. Simplify $\frac{a}{a - b} - \frac{b}{a + b}$.

30. **Critical Thinking** Show that if n is any positive real number, then $\frac{1}{n} - \frac{1}{n + 1}$ is positive.

MIXED REVIEW

Simplify. Write answers with positive exponents only. Assume that all variables are nonzero. (9.3 Skill A)

31. $\frac{4a^2}{-2b^3} \cdot \frac{4b}{12a}$ 32. $\frac{15m^2}{-21n^3} \cdot \frac{-7n^2}{15m^2}$ 33. $\frac{24m^2}{-11n^3} \cdot \frac{11n^3}{24m^2}$ 34. $\frac{3u^3}{21v^4} \cdot \frac{7v^3}{u^3}$

35. $\frac{3y^4}{-z} \cdot \frac{-z^3}{-7y^3}$ 36. $\frac{3a^2}{b^3} \cdot \frac{5b}{-7a^4}$ 37. $\frac{-5d^5}{c^5} \cdot \frac{5c^4}{-5d^4}$ 38. $\frac{2.5s^4}{3.5r^5} \cdot \frac{7r^4}{5s^4}$

Answers for Lesson 9.5 Skill B

Try This Example 1

a. $\frac{6 - 3m}{m^2}$

b. $\frac{-3p + 7}{(p - 1)}$

Try This Example 2

$\frac{5y - 2}{5y - 5}$

Try This Example 3

$\frac{-4}{a^2 + 8a + 16}$

1. like
2. unlike; $3(x - 2)$, or $3x - 6$
3. unlike; $(x - 3)(x + 3)$, or $x^2 - 9$
4. unlike; $15(x - 3)$, or $15x - 45$
5. $-\frac{1}{3x}$
6. $\frac{31}{5a}$
7. $\frac{14}{15z}$
8. $\frac{9}{14v}$
9. $\frac{-1}{x(x - 1)}$
10. $\frac{2}{n(n + 2)}$
11. $\frac{6z^2 - 5z + 10}{2z(z - 2)}$
12. $\frac{14w^2 - 5w + 5}{2w(w - 1)}$
13. $\frac{2(x - 4)}{5(x - 3)}$
14. $\frac{2m + 7}{2(m + 2)}$
15. $\frac{5z - 17}{5(z + 2)}$
16. $\frac{v - 2}{2(v - 5)}$
17. $\frac{7u^2 + 23u - 6}{6(u^2 - 9)}$
18. $\frac{5(u + 7)}{4(u + 3)}$
19. $\frac{-(3x + 1)}{x(x + 1)}$
20. $\frac{c + 1}{c(c - 1)}$
21. $\frac{-4a}{a^2 - 1}$
22. $\frac{2s^2 + 11s - 1}{(s - 1)(s + 2)}$
23. $\frac{1}{b + 1}$
24. $\frac{d - 5}{(d - 2)(d + 2)}$
25. $\frac{-4y}{(y - 1)(y + 1)^2}$
26. $\frac{2t}{(t + 3)^2(t - 3)}$
27. $\frac{-2g^2 + g + 2}{(g + 2)^2(g - 2)}$
28. $\frac{a^2 - b^2}{ab}$
29. $\frac{a^2 + b^2}{a^2 - b^2}$

For answers to Exercises 30–38, see **Additional Answers.**

Focus: *Students add, subtract, and multiply rational expressions in order to write a simplified expression for the area of an irregular plane figure.*

EXAMPLE

PHASE 1: Presenting the Example

Students have probably studied similarity in previous mathematics courses, and so they should be somewhat familiar with the concept. Review the definition of similar figures given in the textbook. Some students may recall the following alternative form of the definition: *Two plane figures are similar if the lengths of their corresponding sides are in proportion.* Point out that similar figures have the same shape, but not necessarily the same size.

This example also requires knowledge of some basic geometry principles involving area. Depending upon the extent of your students' prior experience, you may need to review one or more of the following facts:

- The area, A, of a rectangle with length ℓ and width w is given by the formula $A = \ell w$.

- The area of a region of a plane is equal to the sum of the areas of all its nonoverlapping parts.

- When one region of a plane is contained within another, the area between them is equal to the area of the larger region minus the area of the smaller region.

PHASE 2: Providing Guidance for **TRY THIS**

Students will be asked to generalize the solution of this problem in Exercise 5 on the facing page. If time allows, prepare them for this task by having them find the shaded area when the dimensions of the inner T are one-fourth the corresponding dimensions of the outer T; then repeat the problem when the fraction is one-fifth. $\left[\dfrac{45x^2}{128} ; \dfrac{9x^2}{25} \right]$

The repetitiveness of the task should help students see the pattern that leads to the generalization.

PHASE 3: ASSESSMENT AND CLOSURE for Lesson 9.5

- Ask students to compare subtraction of rational expressions to addition of rational expressions. Have them write an addition and a subtraction that they think are very similar, and another addition and subtraction that they think are somewhat different. Ask them to describe how addition and subtraction differ from multiplication and division. [**Answers will vary.**]

- This lesson concludes the sequence of instruction in operations with rational expressions. In the lessons that follow, students will use these operations as they learn to simplify complex fractions and to solve rational equations.

☞ *For a Lesson 9.5 Quiz, see* Assessment Resources *page 95.*

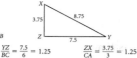

SKILL C **APPLICATIONS** Using operations with rational expressions to solve geometry problems

Two triangles are *similar* if the ratios of the lengths of the corresponding sides are equal. Triangles ABC and XYZ shown below are similar. The ratio of each pair of corresponding sides is 1.25.

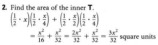

$\frac{XY}{AB} = \frac{8.75}{7} = 1.25$ $\frac{YZ}{BC} = \frac{7.5}{6} = 1.25$ $\frac{ZX}{CA} = \frac{3.75}{3} = 1.25$

You can use what you know about operations on rational expressions to solve problems involving other similar figures. For figures with four or more sides, the measures of corresponding angles must also be equal in order for the figures to be similar.

EXAMPLE

Refer to the diagram at right. All angles in the diagram are right angles. The lengths of the sides of the outer **T** are proportional to the corresponding sides of the inner **T**. The dimensions of the inner **T** are all $\frac{1}{2}$ of the corresponding dimensions in the outer **T**. Write an expression for the shaded area A.

▶ **Solution**

The area of the shaded region equals the area of the outer **T** minus the area of the inner **T**.

Use a formula.

1. Find the area of the outer **T**.
$$x\left(\frac{x}{4}\right) + \left(\frac{x}{2}\right)\left(\frac{x}{4}\right) = \frac{x^2}{4} + \frac{x^2}{8}$$
$$= \frac{2x^2}{8} + \frac{x^2}{8} = \frac{3x^2}{8} \text{ square units}$$

2. Find the area of the inner **T**.
$$\left(\frac{1}{2} \cdot x\right)\left(\frac{1}{2} \cdot \frac{x}{4}\right) + \left(\frac{1}{2} \cdot \frac{x}{2}\right)\left(\frac{1}{2} \cdot \frac{x}{4}\right)$$
$$= \frac{x^2}{16} + \frac{x^2}{32} = \frac{2x^2}{32} + \frac{x^2}{32} = \frac{3x^2}{32} \text{ square units}$$

3. Subtract the result of Step **2** from the result of Step **1** to find the shaded area A.
$$A = \frac{3x^2}{8} - \frac{3x^2}{32} = \frac{3x^2}{8} \cdot \frac{4}{4} - \frac{3x^2}{32} = \frac{12x^2 - 3x^2}{32} = \frac{9x^2}{32}$$

Therefore, the shaded area is $\frac{9x^2}{32}$ square units.

TRY THIS Rework the Example given that each of the dimensions of the inner **T** are one-third those of the corresponding dimensions in the outer **T**.

EXERCISES

KEY SKILLS

Refer to the diagram at right. Assume the four smaller squares are all the same size.

1. Write expressions for the area of the large square and an expression for the area of each of the smaller squares.

2. Write an expression for the area of all five squares together.

3. How would the answer to Exercise 2 change if each small square has sides one third as long as the large square?

PRACTICE

Refer to the Example on the previous page.

4. Show that the grey-shaded portion of the outer **T** is 75% of the entire area of the outer **T**.

5. Let n represent a natural number. Suppose that the dimensions of the inner **T** are $\frac{1}{n}$ times as long as the corresponding dimensions of the outer **T**. Find a formula for the shaded area.

6. Refer to Exercise 5. What percent of the area of the outer **T** is the shaded area? Justify your answer.

MID-CHAPTER REVIEW

7. Find the excluded values of $y = \frac{3x + 2}{x^2 + 10x + 21}$. (9.1 Skill A)

Simplify. (9.2–9.4)

8. $\frac{-2x - 10}{4x + 8}$

9. $\frac{4u^2 - 9}{2u^2 - 9u + 9}$

10. $\frac{6m^3n^2 + 3m^2n^3}{6m^3n^2 - 3m^2n^3}$

11. $\frac{x + 1}{x^2 - 4} \cdot \frac{x - 2}{x^2 + 2x + 1}$

12. $\frac{a - 3}{a^2 - 4} \div \frac{a^2 - 6a + 9}{a + 2}$

13. $\frac{6u^3v^2}{uv^3} + \frac{7u^2v^3}{2uv} \cdot \frac{14u^2v}{12u^3v^3}$

14. $\frac{n^2 - 6n}{n - 4} + \frac{8}{n - 4}$

15. $\frac{5}{3z^2} + \frac{1}{6z}$

16. $\frac{3}{t^2 + t} + \frac{t}{t^2 - 1}$

17. $\frac{k^2 + 4k}{k + 3} - \frac{-3}{k + 3}$

18. $\frac{z + 5}{2z - 14} - \frac{3z + 8}{3z - 21}$

19. The lengths of the sides of the outer and inner rectangles at right are proportional. The dimensions of the inner rectangle are two-thirds of the corresponding dimensions of the outer rectangle. Write an expression for the shaded area. (9.5 Skill C)

Answers for Lesson 9.5 Skill C

Try This Example

$A = \frac{x^2}{3}$

1. large square: x^2; small square: $\frac{x^2}{4}$

2. $2x^2$

3. $\frac{13x^2}{9}$

4. $\frac{9x^2}{32} \div \frac{3x^2}{8} = \frac{9x^2}{32} \cdot \frac{8}{3x^2} = \frac{3}{4}; \frac{3}{4} = 75\%$

★ **5.** $\frac{3x^2(n^2 - 1)}{8n^2}$

★ **6.** Ratio of the area of the shaded **T** to that of the outer **T** is given by $\frac{3x^2(n^2 - 1)}{8n^2} \div \frac{3x^2}{8}$.

$\frac{3x^2(n^2 - 1)}{8n^2} \div \frac{3x^2}{8} = \frac{3x^2(n^2 - 1)}{8n^2} \cdot \frac{8}{3x^2}$

$= \frac{n^2 - 1}{n^2} = 1 - \frac{1}{n^2}$

So, the percent is $\left[100\left(1 - \frac{1}{n^2}\right)\right]\%$.

7. $x \neq -3, -7$

8. $\frac{-x - 5}{2x + 4}$

9. $\frac{2u + 3}{u - 3}$

10. $\frac{2m + n}{2m - n}$

11. $\frac{1}{(x + 2)(x + 1)}$, or $\frac{1}{x^2 + 3x + 2}$

12. $\frac{1}{(a - 2)(a - 3)}$, or $\frac{1}{a^2 - 5a + 6}$

13. $\frac{2}{v^5}$

14. $n - 2$

15. $\frac{z + 10}{6z^2}$

16. $\frac{t^2 + 3t - 3}{t(t^2 - 1)}$, or $\frac{t^2 + 3t - 3}{t^3 - t}$

17. $k + 1$

18. $\frac{-3z - 1}{6(z - 7)}$, or $-\frac{3z + 1}{6z - 42}$

19. $\frac{5a^2}{12}$

★ **Advanced Exercises**

Lesson 9.5 Skill C **367**

TEACHING **9.6** LESSON

Simplifying and Using Complex Fractions

SKILL A *Simplifying complex fractions*

Focus: *Students use division to simplify complex fractions involving rational numbers and complex fractions involving rational expressions.*

EXAMPLE 1

PHASE 1: Presenting the Example

Display the expression $\square \div \bigcirc$. Remind students that another symbol for division is the fraction bar. Tell them to rewrite this expression as a fraction. Display the result as shown at right.

Ask students the following questions, one at a time: What would the quotient be if the box contained 24 and the oval contained 3? What if the box contained 96 and the oval contained 16? What if the box contained 8.4 and the oval contained 7? Display their answers one beneath the other, as shown.

Now write $\frac{3}{8}$ in the numerator box and $\frac{5}{7}$ in the denominator oval. Tell students that this is a new type of fraction called a *complex fraction.* Discuss the definition of complex fraction at the top of the textbook page. Point out that, although the complex fraction looks different, the fraction bars still represent division. Tell them to simplify the complex fraction by modeling their work on the previous divisions. Display the result as shown at right, answering any questions that students may have. Then discuss Example 1.

$$\frac{\square}{\bigcirc} = \square \div \bigcirc$$

$$\frac{\boxed{24}}{(3)} = \boxed{24} \div (3) = 8$$

$$\frac{\boxed{96}}{(16)} = \boxed{96} \div (16) = 6$$

$$\frac{\boxed{8.4}}{(7)} = \boxed{8.4} \div (7) = 1.2$$

$$\frac{\boxed{\frac{3}{8}}}{\left(\frac{5}{7}\right)} = \boxed{\frac{3}{8}} \div \left(\frac{5}{7}\right)$$

$$= \frac{3}{8} \times \frac{7}{5} = \frac{21}{40}$$

PHASE 2: Providing Guidance for TRY THIS

Some students may become confused and simply multiply the numerator of a complex fraction by its denominator. Suggest that they simplify using the steps shown at right. Although this method involves a bit more work, it clearly demonstrates why it is necessary to multiply by the reciprocal of the denominator.

$$\frac{\frac{10}{21}}{\frac{5}{28}} = \frac{\frac{10}{21}}{\frac{5}{28}} \cdot \frac{\frac{28}{5}}{\frac{28}{5}}$$

$$= \frac{\frac{10}{21} \cdot \frac{28}{5}}{\frac{5}{28} \cdot \frac{28}{5}} = \frac{\frac{10}{21} \cdot \frac{28}{5}}{1} = \frac{10}{21} \cdot \frac{28}{5}$$

EXAMPLE 2

PHASE 2: Providing Guidance for TRY THIS

Watch for students who use the reciprocal of $\frac{b}{a}$ and the reciprocal of $\frac{a}{b}$ individually. Remind them that a fraction bar is a grouping symbol, and so they must find a reciprocal for the entire expression $\frac{b}{a} - \frac{a}{b}$. It may be easier for students to see the structure of the complex fraction if they rewrite the numerator ab as a simple fraction, as shown at right.

$$\frac{ab}{\frac{b}{a} - \frac{a}{b}} = \frac{\frac{ab}{1}}{\frac{b}{a} - \frac{a}{b}}$$

Simplifying and Using Complex Fractions

SKILL A ▶ Simplifying complex fractions

KEY SKILLS

A *complex fraction* is a quotient of two rational expressions written as a fraction. The expressions at right are both complex fractions.

numerical complex fraction: $\dfrac{\frac{2}{3}}{\frac{4}{5}}$

algebraic complex fraction: $\dfrac{2x+5}{4x-1}{\Big/}\dfrac{x+7}{3x+9}$

To simplify a complex fraction, you may need to simplify parts of the expression one at a time.

Write each complex fraction as a product. Then write the numerators and the denominators in factored form. Do not simplify further.

1. $\dfrac{\frac{2}{15}}{\frac{12}{35}}$ 2. $\dfrac{\frac{3x}{49}}{\frac{12x}{21}}$ 3. $\dfrac{\frac{2(a-1)}{12}}{\frac{3a-3}{4}}$ 4. $\dfrac{\frac{5t+25}{3t-3}}{\frac{4t+20}{5t-5}}$

EXAMPLE 1 Simplify.   a. $\dfrac{\frac{18}{25}}{\frac{2}{15}}$ b. $\dfrac{\frac{3p}{20}}{\frac{4p}{15}}$ c. $\dfrac{\frac{21z^2}{25}}{\frac{7z}{50}}$

PRACTICE

Simplify.

▶ Solution

5. $\dfrac{\frac{3}{5}}{\frac{9}{25}}$ 6. $\dfrac{\frac{8}{21}}{\frac{16}{35}}$ 7. $\dfrac{\frac{3n^2}{7}}{\frac{5n}{28}}$

a. $\dfrac{\frac{18}{25}}{\frac{2}{15}} = \dfrac{18}{25} \div \dfrac{2}{15}$

b. $\dfrac{\frac{3p}{20}}{\frac{4p}{15}} = \dfrac{3p}{20} \div \dfrac{4p}{15}$

c. $\dfrac{\frac{21z^2}{25}}{\frac{7z}{50}} = \dfrac{21z^2}{25} \div \dfrac{7z}{50}$

8. $\dfrac{\frac{6r^2}{35}}{\frac{3r}{70}}$ 9. $3 + \dfrac{1}{2+\frac{3}{2}}$ 10. $3 - \dfrac{1}{3-\frac{1}{3}}$

$= \dfrac{18}{25} \cdot \dfrac{15}{2}$ $= \dfrac{3p}{20} \cdot \dfrac{15}{4p}$ $= \dfrac{21z^2}{25} \cdot \dfrac{50}{7z}$

$= \dfrac{27}{5}$, or $5\frac{2}{5}$ $= \dfrac{9}{16}$ $= 6z$

11. $\dfrac{\frac{2b+6}{5b-15}}{\frac{3b+9}{25b-75}}$ 12. $\dfrac{\frac{4r+4}{7r+14}}{\frac{7r+7}{2r+4}}$ 13. $\dfrac{\frac{3a+21}{7a-14}}{\frac{7a+49}{7a-14}}$

TRY THIS   Simplify.   a. $\dfrac{\frac{10}{21}}{\frac{5}{28}}$ b. $\dfrac{\frac{11a}{30}}{\frac{33a}{50}}$ c. $\dfrac{\frac{35k^2}{16}}{\frac{7k}{48}}$

14. $\dfrac{\frac{4z+6}{3z-36}}{\frac{4z+16}{3z-36}}$ 15. $\dfrac{\frac{1}{a+b}}{\frac{1}{a-b}}$ 16. $\dfrac{\frac{a}{a-b}}{\frac{b}{a+b}}$

EXAMPLE 2 Simplify $\dfrac{1}{\frac{b}{a}+\frac{a}{b}}$.

17. **Critical Thinking** Find an expression in simplified form for the sum of the reciprocals of two nonzero real numbers, r and s, divided by the difference of the reciprocals of r and s.

▶ Solution

$\dfrac{1}{\frac{b}{a}+\frac{a}{b}} = \dfrac{1}{\frac{b}{a}\cdot\frac{b}{b}+\frac{a}{b}\cdot\frac{a}{a}}$ ◀── The LCD of $\frac{b}{a}$ and $\frac{a}{b}$ is ab. Multiply $\frac{b}{a}$ by $\frac{b}{b}$ and multiply $\frac{a}{b}$ by $\frac{a}{a}$ so that their denominators are both ab.

$= \dfrac{1}{\frac{a^2+b^2}{ab}}$

$= \dfrac{ab}{a^2+b^2}$

MIXED REVIEW

Solve each equation. (2.8 Skill B)

18. $5 - 3(r-9) = 4(r+1)$ 19. $3(x-3) + 5(x-3) = 0$ 20. $4 - (3y+1) = 4y$

21. $3d + 5 - 4d = 4(d-1)$ 22. $4g - 5 = 4 - 7(g+1)$ 23. $4x + 5x + 6x = 12(x+7)$

TRY THIS   Simplify $\dfrac{ab}{\frac{b}{a}-\frac{a}{b}}$.

Solve each equation. (7.5 Skill C)

24. $9f^2 + 27f - 22 = 0$ 25. $3w^2 + 38w + 99 = 0$ 26. $27w^2 - 69w - 70 = 0$
27. $16h^2 - 8h + 1 = 0$ 28. $121z^2 - 49 = 0$ 29. $49a^2 + 140a + 100 = 0$

Answers for Lesson 9.6 Skill A

Try This Example 1

a. $\dfrac{8}{3}$

b. $\dfrac{5}{9}$

c. $15k$

Try This Example 2

$\dfrac{a^2b^2}{b^2-a^2}$

1. $\dfrac{2}{15} \cdot \dfrac{35}{12} = \dfrac{2}{3\cdot5} \cdot \dfrac{7\cdot5}{2\cdot2\cdot3}$

2. $\dfrac{3x}{49} \cdot \dfrac{21}{12x} = \dfrac{3\cdot x}{7\cdot7} \cdot \dfrac{3\cdot7}{2\cdot2\cdot3\cdot x}$

3. $\dfrac{2(a-1)}{12} \cdot \dfrac{4}{3a-3} = \dfrac{2(a-1)}{2\cdot2\cdot3} \cdot \dfrac{2\cdot2}{3(a-1)}$

4. $\dfrac{5t+25}{3t-3} \cdot \dfrac{5t-5}{4t+20} = \dfrac{5(t+5)}{3(t-1)} \cdot \dfrac{5(t-1)}{4(t+5)}$

5. $\dfrac{5}{3}$

6. $\dfrac{5}{6}$

7. $\dfrac{12n}{5}$

8. $4r$

9. $\dfrac{23}{7}$

10. $\dfrac{21}{8}$

11. $\dfrac{10}{3}$

12. $\dfrac{8}{49}$

13. $\dfrac{3}{7}$

14. $\dfrac{2z+3}{2(z+4)}$

15. $\dfrac{a-b}{a+b}$

16. $\dfrac{a(a+b)}{b(a-b)}$

17. $\dfrac{s+r}{s-r}$

18. $r = 4$

19. $x = 3$

20. $y = \dfrac{3}{7}$

21. $d = \dfrac{9}{5}$

22. $g = \dfrac{2}{11}$

23. $x = 28$

24. $f = \dfrac{2}{3}$ or $f = -\dfrac{11}{3}$

25. $w = -9$ or $w = -\dfrac{11}{3}$

26. $w = -\dfrac{7}{9}$ or $w = \dfrac{10}{3}$

27. $h = \dfrac{1}{4}$

28. $z = \pm\dfrac{7}{11}$

29. $a = -\dfrac{10}{7}$

Focus: *Students use complex fractions to find the total amount of resistance in an electrical circuit and to calculate an average speed for an automobile trip.*

EXAMPLE 1

PHASE 1: Presenting the Example

A *circuit* is the closed path traveled by an electric current as it flows. Anything that impedes the flow of current in a circuit is called a *resistance*. Students may already be familiar with the term *resistor*, which is a device that introduces a desired amount of resistance into a circuit.

In a *series connection*, all the electrical devices in a circuit are connected one after another. If any device is turned off, the circuit is broken and the current stops flowing. A familiar example of a series connection is a string of holiday lights; all of the lights go out when one bulb is removed. In a series connection, the total resistance is the sum of all the individual resistances.

In contrast, a *parallel connection* provides two or more paths between two points of a circuit. Electrical devices along these paths are independent of one another. For example, a typical house is wired in a parallel connection; if a lamp is turned off, other electrical devices in the house are not affected. In a parallel connection, the total resistance is the sum of the *reciprocals* of all the individual resistances. This is the situation addressed in Example 1.

EXAMPLE 2

PHASE 1: Presenting the Example

Display the familiar formula $d = rt$. Ask students to write a brief explanation of its meaning. [**The formula gives the distance, *d*, traveled when a constant rate, *r*, and time, *t*, are known.**] Have students solve the formula for *t*. $\left[t = \dfrac{d}{r} \right]$ Point out that now they have a formula for the amount of time spent traveling, *t*, when distance, *d*, and constant rate, *r*, are known. Then discuss Example 2.

PHASE 3: ASSESSMENT AND CLOSURE for Lesson 9.6

- Give students the fractions at right. Ask why all are complex fractions. [**The numerator or denominator contains one or more rational expressions.**] Discuss how the process of simplifying will be similar among the three fractions and how it will be different. Then have them simplify. [**Answers appear beneath the fractions.**]

$$1 \quad \frac{\left(\dfrac{a}{b}\right)}{\left(\dfrac{c}{d}\right)} \qquad 2 \quad \frac{x}{\left(\dfrac{y}{z}\right)} \qquad 3 \quad \frac{\dfrac{m}{n}}{\dfrac{p}{q} + \dfrac{r}{s}}$$

$$\left[\frac{ad}{bc}\right] \qquad \left[\frac{xz}{y}\right] \qquad \left[\frac{mqs}{nps + nqr}\right]$$

- Students will learn more about complex fractions in advanced algebra courses. In the next lesson, they will learn how to solve rational equations.

☞ *For a Lesson 9.6 Quiz, see* Assessment Resources *page 96.*

Light-bulb filaments are resistors in an electric circuit. The diagram at right shows two resistors in a parallel arrangement.

The total resistance R_T is given by the formula $R_T = \dfrac{1}{\dfrac{1}{A} + \dfrac{1}{B}}$.

EXAMPLE 1 A parallel circuit has two resistances A and B. Show that the total effective resistance R_T is given by $R_T = \dfrac{AB}{A + B}$.

▶ Solution

$$R_T = \frac{1}{\dfrac{1}{A} + \dfrac{1}{B}} = \frac{1}{\dfrac{1}{A} \cdot \dfrac{B}{B} + \dfrac{1}{B} \cdot \dfrac{A}{A}} = \frac{1}{\dfrac{B + A}{AB}} = \frac{AB}{A + B}$$

TRY THIS A parallel circuit has resistances A, B, and C. Show that the total resistance, $R_T = \dfrac{1}{\dfrac{1}{A} + \dfrac{1}{B} + \dfrac{1}{C}}$, can be written as $R_T = \dfrac{ABC}{BC + AC + AB}$.

EXAMPLE 2 Jack drove from his shop to a job site at an average speed of 55 miles per hour and returned along the same highway at an average speed of 45 miles per hour. Find his average speed for the entire trip.

r_1 : 55 mi/h

shop — d — job site

r_2 : 45 mi/h

▶ Solution

1. Let d represent the length of the trip one way and let t_1 and t_2 represent his travel time to and from the site, respectively. Use the relationship

   Use a formula.

   $\text{rate} = \dfrac{\text{distance}}{\text{time}}$.

   average speed over the entire trip: $\dfrac{\text{total distance (in miles)}}{\text{total time (in hours)}} \rightarrow \dfrac{d + d}{t_1 + t_2}$

   Because $d = 55t_1 = 45t_2$, then $t_1 = \dfrac{d}{55}$ and $t_2 = \dfrac{d}{45}$.

2. Simplify $\dfrac{2d}{\dfrac{d}{55} + \dfrac{d}{45}}$.

   $\dfrac{2d}{\dfrac{d}{55} + \dfrac{d}{45}} = \dfrac{2d}{\dfrac{9d + 11d}{495}} = 2d \times \dfrac{495}{20d} = 49.5 \text{ miles/hour}$

   Jack's average speed over the entire trip is 49.5 miles per hour.

TRY THIS Rework Example 2 using 44 miles per hour and 50 miles per hour.

EXERCISES

KEY SKILLS

In Exercises 1 and 2, use the formula distance (*d*) = rate (*r*) × time (*t*).

1. Which properties justify *If d = rt, then* $t = \dfrac{d}{r}$?

2. Which properties justify *If* $t_1 = \dfrac{d}{55}$, $t_2 = \dfrac{d}{45}$, *and* $s = \dfrac{d + d}{t_1 + t_2}$, *then* $s = \dfrac{2d}{\dfrac{d}{55} + \dfrac{d}{45}}$?

PRACTICE

3. A parallel circuit has resistors of 15 ohms and of 12 ohms. Simplify $\dfrac{1}{\dfrac{1}{12} + \dfrac{1}{15}}$, an expression for the total resistance.

Electrical resistance is measured in ohms (Ω). Find the total effective resistance to the nearest tenth for resistances A and B in a parallel circuit.

4. A: 12 Ω and B: 9 Ω
5. A: 10 Ω and B: 12 Ω
6. A: 8 Ω and B: 14 Ω
7. A: 10 Ω and B: 10 Ω

8. Show that if resistances A and B are equal, then the total resistance is one half of one of them.

Refer to Example 2. Find the average speed for the entire trip.

9. to the site: 40 miles per hour   from the site: 45 miles per hour
10. to the site: 40 miles per hour   from the site: 50 miles per hour

11. Show that if Jack drives to and from a job site at *r* miles per hour, then his average speed for the entire trip is *r* miles per hour.

12. **Critical Thinking** Write and simplify a formula for the average speed for a trip of *d* miles one way and *d* miles back if the driver travels at r_1 miles per hour to the destination and r_2 miles per hour from the destination.

MIXED REVIEW APPLICATIONS

An object propelled upward from an initial altitude h_0 in feet and given an initial velocity v_0 in feet per second will have altitude *h* in feet given by $h = -16t^2 + v_0 t + h_0$.

A ball thrown directly upward from the edge of a cliff 84 feet above sea level is given an initial velocity of 96 feet per second. Find the amount of time it takes the ball to have the specified altitude above sea level. Give answers to the nearest tenth of a second. (8.7 Skill B)

13. 130 feet
14. 140 feet
15. 200 feet
16. 180 feet

Answers for Lesson 9.6 Skill B

Try This Example 1

$$\frac{1}{\dfrac{1}{A} + \dfrac{1}{B} + \dfrac{1}{C}} = \frac{1}{\dfrac{1}{A} \cdot \dfrac{BC}{BC} + \dfrac{1}{B} \cdot \dfrac{AC}{AC} + \dfrac{1}{C} \cdot \dfrac{AB}{AB}} =$$

$$\frac{1}{\dfrac{BC}{ABC} + \dfrac{AC}{ABC} + \dfrac{AB}{ABC}} = \frac{1}{\dfrac{BC + AC + AB}{ABC}} =$$

$$\frac{ABC}{BC + AC + AB}$$

Try This Example 2
about 46.8 miles per hour

1. Division Property of Equality and Symmetric Property of Equality
2. definition of multiplication by 2 and the Substitution Principle
3. $\dfrac{60}{9}$, or $6\dfrac{2}{3}$ ohms
4. 5.1Ω
5. 5.5 Ω
6. 5.1 Ω
7. 5 Ω

8. If $A = B$, then $\dfrac{AB}{A + B}$ becomes $\dfrac{A^2}{2A}$ or $\dfrac{B^2}{2B}$. These expressions simplify to $\dfrac{A}{2}$ or $\dfrac{B}{2}$, which respresents one half of each of them.

9. about 42.4 miles per hour
10. about 44.4 miles per hour

11. $\dfrac{2d}{\dfrac{d}{r} + \dfrac{d}{r}} = \dfrac{2d}{\dfrac{2d}{r}} = \dfrac{2d}{1} \cdot \dfrac{r}{2d} = r$

12. $\dfrac{2r_1 r_2}{r_1 + r_2}$

13. 0.5 second and 5.5 seconds
14. 0.7 second and 5.3 seconds
15. 1.7 seconds and 4.3 seconds
16. 1.3 seconds and 4.7 seconds

Solving Rational Equations

SKILL A ▶ *Solving proportions involving rational expressions*

Focus: *Students use cross products to solve proportions.*

EXAMPLE 1

PHASE 1: Presenting the Example

Display equation **1** at right. Have students write whether it is true or false. [**true**] Discuss the strategies they used in arriving at their responses. Some probably noted that each fraction is equal to $\frac{2}{3}$. Most likely others found the cross products, $4 \cdot 9$ and $6 \cdot 6$, and noted that each is equal to 36. Use the second method as an introduction to the Cross Product Property of Proportions that is discussed at the top of the textbook page.

1 $\quad \dfrac{4}{6} = \dfrac{6}{9}$

2 $\quad \dfrac{4}{6} = \dfrac{6}{r}$

3 $\quad \dfrac{4}{r-3} = \dfrac{6}{r}$

Display equation **2**. Tell students to imagine they do not know the value of r. Ask how they could use the Cross Product Property to solve for r. Lead them to see that, when set equal, the cross products $4 \cdot r$ and $6 \cdot 6$ form a simple linear equation: $4r = 36$. Have them solve this equation. [$r = 9$] Point out that 9 is the value of the second denominator in equation **1**.

Proceed to display equation **3**. Lead students through the same reasoning as used with equation **2**. Now the cross products are $4 \cdot r$ and $(r - 3) \cdot 6$, and they result in a multi-step linear equation: $4r = 6r - 18$. Have students solve it. [$r = 9$] Note that $r = 9$ gives the value of the second denominator in equation **1**, and $r - 3 = 9 - 3 = 6$ gives the value of the first denominator. Then discuss how the cross product method is applied in Example 1.

$$\frac{a}{b} = \frac{c}{d}$$

$$b \cdot \frac{a}{b} = b \cdot \frac{c}{d}$$

$$a = b \cdot \frac{c}{d}$$

PHASE 2: Providing Guidance for **TRY THIS**

Using the Cross Product Property is a simple method for solving proportions. However, some students might perceive it as just one of many rules to be memorized, and they may become confused. Show them how they can quickly use the Multiplication Property of Equality to derive the Cross Product Property of Proportions, as shown at right.

$$a \cdot d = b \cdot \frac{c}{d} \cdot d$$

$$ad = bc$$

EXAMPLE 2

PHASE 1: Presenting the Example

Display $\dfrac{a}{2} = \dfrac{2}{a+3}$. Ask students to compare and contrast this proportion to the proportions in Example 1. Elicit observations such as these: Both here and in Example 1, two parts of the proportion are variable expressions. In Example 1, both variable expressions are denominators or both are numerators. Here one variable expression is a numerator and the other is a denominator.

Now discuss Example 2. Students should note that, when the Cross Product Property is applied, the positioning of the variable expressions results in a quadratic equation rather than the linear equations of Example 1.

9.7 Solving Rational Equations
LESSON

Solving proportions involving rational expressions

A **rational equation** is an equation containing at least one rational expression. A solution to a rational equation is any number that satisfies the equation.

Recall that a proportion is the equality of two ratios. Any proportion containing a variable is a rational equation. You can use *cross products* to solve proportions.

Cross Product Property of Proportions

If a, b, c, and d are real numbers with $b \neq 0$ and $d \neq 0$, and $\frac{a}{b} = \frac{c}{d}$, then $ad = bc$.

Cross products: ad and bc

The equation $\frac{3}{z} = \frac{4}{5}$ is a proportion. Using the property above, you can write $3 \cdot 5 = 4z$. Then divide both sides by 4 to solve for z.

EXAMPLE 1 Solve $\frac{3}{x-1} = \frac{8}{x}$.

▶ Solution

$$\frac{3}{x-1} = \frac{8}{x}$$
$$3x = 8(x-1) \quad \longleftarrow \text{Apply the Cross Product Property of Proportions.}$$
$$x = \frac{8}{5}, \text{ or } 1\tfrac{3}{5}$$

TRY THIS   Solve.   **a.** $\frac{y+2}{7} = \frac{y}{5}$ **b.** $\frac{-3}{n-2} = \frac{4}{n}$

EXAMPLE 2 Solve $\frac{a}{2} = \frac{2}{a+3}$.

▶ Solution

$$\frac{a}{2} = \frac{2}{a+3}$$
$$a(a+3) = 2(2) \quad \longleftarrow \text{Apply the Cross Product Property of Proportions.}$$
$$a^2 + 3a - 4 = 0$$
$$(a+4)(a-1) = 0$$
$$a = -4 \text{ or } a = 1 \quad \longleftarrow \text{Apply the Zero-Product Property and solve for } a.$$

The solutions are -4 and 1.

TRY THIS   Solve.   **a.** $\frac{n-6}{-2} = \frac{4}{n}$ **b.** $\frac{r-4}{-2} = \frac{2}{r}$

KEY SKILLS

Write the equation that results after you apply the Cross Product Property of Proportions. Do not solve.

1. $\frac{5}{z} = \frac{2}{7}$ **2.** $\frac{3}{x-2} = \frac{2}{x+2}$ **3.** $\frac{5}{a-1} = \frac{2}{a}$ **4.** $\frac{3}{x-1} = \frac{2x}{5}$

PRACTICE

Solve each proportion.

5. $\frac{5}{7} = \frac{w}{21}$ **6.** $\frac{s}{8} = \frac{8}{32}$ **7.** $\frac{5}{9} = \frac{t}{6}$ **8.** $\frac{z}{3} = \frac{13}{5}$

9. $\frac{6}{b} = \frac{11}{2}$ **10.** $\frac{6}{7} = \frac{4}{d}$ **11.** $\frac{5}{y} = \frac{12}{7}$ **12.** $\frac{1}{9} = \frac{13}{n}$

13. $\frac{1}{3n} = \frac{3}{n-1}$ **14.** $\frac{5}{3m} = \frac{2}{m-2}$ **15.** $\frac{1}{4} = \frac{3k+1}{k-2}$ **16.** $\frac{4p-1}{5} = \frac{3p+1}{2}$

17. $\frac{y-1}{y} = \frac{y}{4}$ **18.** $\frac{v-4}{v} = \frac{v}{-2}$ **19.** $\frac{x}{2} = \frac{3x-4}{x}$ **20.** $\frac{a}{10} = \frac{2}{a+1}$

Solve each equation for the specified variable.

21. $\frac{ax}{b} = \frac{c}{d}$ for x **22.** $\frac{z}{t} = \frac{r}{z}$ for z **23.** $\frac{s}{t} = \frac{rn}{an+b}$ for n

The *geometric mean* between positive numbers a and b is defined as the positive solution to the equation $\frac{a}{x} = \frac{x}{b}$. Find each geometric mean. Write the answers in simplest radical form.

24. 2 and 3   **25.** 10 and 20   **26.** 12 and 24   **27.** 7 and 70

28. Show that the geometric mean of a and b is $\sqrt{ab}$.

29. Critical Thinking

a. Show that if $\frac{a}{b} = \frac{c}{d}$, then $a = \frac{bc}{d}$.

b. Show that if $\frac{a}{b} = \frac{c}{d}$ is true, then $\frac{b}{a} = \frac{d}{c}$ is also true.

c. Describe how the information in parts **a** and **b** can make solving proportions easier.

MIXED REVIEW

Find the range of the function represented by each equation. (7.7 Skill A)

30. $y = 2x^2 - 12x + 7$ **31.** $y = -5x^2 + 10x + 1$ **32.** $y = 3.5x^2 - 7x$

33. $y = -3x^2 + 3x - 11$ **34.** $y = 4x^2 + 9x - 13$ **35.** $y = -5x^2 + 15x - 9$

Answers for Lesson 9.7 Skill A

Try This Example 1
a. $y = 5$
b. $n = \frac{8}{7}$

Try This Example 2
a. $n = 2$ or $n = 4$
b. $r = 2$

1. $(5)(7) = 2z$
2. $3(x+2) = 2(x-2)$
3. $5a = 2(a-1)$
4. $(3)(5) = 2x(x-1)$
5. $w = 15$
6. $s = 2$
7. $t = \frac{10}{3}$
8. $z = \frac{39}{5}$
9. $b = \frac{12}{11}$
10. $d = \frac{14}{3}$

11. $y = \frac{35}{12}$
12. $n = 117$
13. $n = -\frac{1}{8}$
14. $m = -10$
15. $k = -\frac{6}{11}$
16. $p = -1$
17. $y = 2$
18. $v = -4$ or $v = 2$
19. $x = 2$ or $x = 4$
20. $a = 4$ or $a = -5$
21. $x = \frac{bc}{ad}$
22. $z = \pm\sqrt{rt}$
★ **23.** $n = \frac{sb}{rt - as}$
★ **24.** $\sqrt{6}$
★ **25.** $10\sqrt{2}$
★ **26.** $12\sqrt{2}$
★ **27.** $7\sqrt{10}$

★ **28.** If $\frac{x}{a} = \frac{b}{x}$, then $x^2 = ab$. The solutions are $\pm\sqrt{ab}$. Thus, the geometric mean is $\sqrt{ab}$.

29. a. If $\frac{a}{b} = \frac{c}{d}$, then $ad = bc$ by the Cross Product Property. If $ad = bc$, then $a = \frac{bc}{d}$ by the Division Property of Equality. Thus, if $\frac{a}{b} = \frac{c}{d}$, then $a = \frac{bc}{d}$.

b. If $\frac{a}{b} = \frac{c}{d}$, then $ad = bc$ by the Cross Product Property. If $ad = bc$, then $d = \frac{bc}{a}$ by the Division Property of Equality. If $d = \frac{bc}{a}$, then $\frac{d}{c} = \frac{b}{a}$ by the Division Property of Equality. If $\frac{d}{c} = \frac{b}{a}$, then $\frac{b}{a} = \frac{d}{c}$ by the Symmetric Property of Equality.

Thus, if $\frac{a}{b} = \frac{c}{d}$, then $\frac{b}{a} = \frac{d}{c}$.

c. The information in parts a and b allows you to solve a proportion in one step, rather than the two steps required by the Cross Product Property.

★ Advanced Exercises

For answers to Exercises 30–35, see Additional Answers.

Focus: *Students solve rational equations by transforming them into proportions, and then using cross products to solve. They also learn to evaluate proposed solutions to determine whether any are extraneous.*

EXAMPLE 1

PHASE 1: Presenting the Example

Display equation **1** at right. Have students explain why it is *not* a proportion. [**If it were a proportion, there would be only one rational expression on each side of the equal sign.**] Ask if they can think of a way to transform it into a proportion. They should observe that the sum of the two expressions on the left side is $\dfrac{10}{c}$, resulting in the proportion $\dfrac{10}{c} = \dfrac{5}{4}$. Have them solve this proportion. [$c = 8$]

Now display equation **2**. Ask if this equation can be transformed into a proportion in the same way as equation **1**. This time, students should note that an additional step is needed. That is, $\dfrac{1}{c}$ must be rewritten as $\dfrac{2}{2c}$ before the expressions on the left side can be added. Have students add the expressions, and then write and solve the resulting proportion. $\left[\dfrac{11}{2c} = \dfrac{5}{4}; c = 4\dfrac{2}{5}, \text{ or } c = 4.4\right]$ Then discuss Example 1.

Note that an alternative method for solving a rational equation is to multiply each side by the LCD of all the denominators that appear in the equation. You may want to use part b of Example 1 to demonstrate this method, as shown at right.

1 $\dfrac{9}{c} + \dfrac{1}{c} = \dfrac{5}{4}$

2 $\dfrac{9}{2c} + \dfrac{1}{c} = \dfrac{5}{4}$

$$\dfrac{a}{2} - \dfrac{1}{3a} = \dfrac{1}{6}$$

$$\left(\dfrac{a}{2} - \dfrac{1}{3a}\right) \cdot 6a = \dfrac{1}{6} \cdot 6a$$

$$\dfrac{a \cdot 6a}{2} - \dfrac{1 \cdot 6a}{3a} = \dfrac{1 \cdot 6a}{6}$$

$$3a^2 - 2 = a$$

$$3a^2 - a - 2 = 0$$

$$(3a + 2)(a - 1) = 0$$

$$a = -\dfrac{2}{3} \text{ or } a = 1$$

PHASE 2: Providing Guidance for TRY THIS

Have students work in pairs. For part a, one student in each pair should solve using the proportion method shown on the textbook page while the other solves using the LCD method shown above. Have the partners switch methods for part b. After each exercise, the partners should compare results. Students can then decide which method they prefer.

EXAMPLE 2

PHASE 2: Providing Guidance for TRY THIS

The idea of extraneous solutions might trouble some students. In particular, they may wonder how they can get a "wrong answer" when all their steps are justified. Stress the fact that extraneous solutions are not mistakes. When solving rational equations, extraneous solutions often arise because the rational expressions in the original equation are not defined for all real numbers. You might suggest that they begin their work by listing any excluded values, and then check the possible solutions against the list.

SKILL B ▸ *Solving rational equations*

EXAMPLE 1 Solve. a. $x + \frac{2}{x} = 3$ b. $\frac{a}{2} - \frac{1}{3a} = \frac{1}{6}$

▸ **Solution**

a.
$$x + \frac{2}{x} = 3$$
$$\frac{x^2}{x} + \frac{2}{x} = 3$$
$$\frac{x^2 + 2}{x} = 3$$
$$x^2 + 2 = 3x$$
$$x^2 - 3x + 2 = 0$$
$$(x - 2)(x - 1) = 0$$
$$x = 1 \quad \text{or} \quad x = 2$$

b.
$$\frac{a}{2} - \frac{1}{3a} = \frac{1}{6}$$
$$\frac{3a^2}{6a} - \frac{2}{6a} = \frac{1}{6}$$
$$\frac{3a^2 - 2}{6a} = \frac{1}{6}$$
$$3a^2 - 2 = a$$
$$3a^2 - a - 2 = 0$$
$$(3a + 2)(a - 1) = 0$$
$$a = -\frac{2}{3} \quad \text{or} \quad a = 1$$

TRY THIS Solve. a. $2d - \frac{6}{d} = 1$ b. $\frac{5s}{3} + \frac{6}{s} = 7$

An **extraneous solution** to a given equation is a number found as a solution to an equation that does not satisfy the given equation. An extraneous solution can be a number that causes the denominator to equal zero, which is not allowed.

EXAMPLE 2 Solve $\frac{2}{z^2 - 2z} - \frac{1}{z - 2} = 1$. Identify extraneous solutions.

▸ **Solution**

$$\frac{2}{z^2 - 2z} - \frac{1}{z - 2} = 1$$
$$\frac{2}{z(z - 2)} - \frac{z}{z(z - 2)} = 1 \quad \longleftarrow \text{Write using a common denominator.}$$
$$\frac{2 - z}{z(z - 2)} = 1 \quad \longleftarrow \text{Subtract.}$$
$$2 - z = z(z - 2)$$
$$2 - z = z^2 - 2z$$
$$z^2 - z - 2 = 0 \quad \longleftarrow \text{Write a quadratic equation in standard form.}$$
$$(z - 2)(z + 1) = 0 \quad \longleftarrow \text{Factor.}$$
$$z = 2 \quad \text{or} \quad z = -1 \quad \longleftarrow \text{Apply the Zero-Product Property and solve for } z.$$

Substitute 2 and −1 for z in the original equation to see if the numbers are solutions. Only −1 satisfies the original equation. The solution is −1.

Check: $\frac{2}{(2)^2 - 2(2)} - \frac{1}{(2) - 2} \overset{?}{=} 1 \longrightarrow \frac{2}{0} - \frac{1}{0} \neq 1$ ✗

$\frac{2}{(-1)^2 - 2(-1)} - \frac{1}{(-1) - 2} \overset{?}{=} 1 \longrightarrow \frac{2}{3} - \left(-\frac{1}{3}\right) = 1$ ✔

TRY THIS Solve $\frac{1}{a^2 - a} - \frac{1}{a - 1} = \frac{1}{2}$. Identify extraneous solutions.

EXERCISES

KEY SKILLS

Substitute the given numbers into the given equation. Which are actual solutions and which, if any, are extraneous?

1. n: -1 and $\frac{5}{2}$
$$n - \frac{3}{2n + 1} = 2$$

2. s: 0 and 3
$$\frac{3s}{s^2 - s} - \frac{1}{s - 1} = 1$$

3. p: 1
$$\frac{1}{p^2 - p} - \frac{1}{p - 1} = 0$$

PRACTICE

Solve each equation. Identify extraneous solutions.

4. $x + \frac{6}{x} = -5$
5. $x - \frac{10}{x} = 3$
6. $x + \frac{1}{6x} = \frac{5}{6}$

7. $9a - \frac{2}{a} = 3$
8. $2x + \frac{1}{x - 1} = -1$
9. $2y + \frac{3}{y + 2} = 1$

10. $2r - \frac{1}{3r + 1} = 1$
11. $3t - \frac{1}{2t - 1} = 4$
12. $\frac{k}{k^2 + k} + \frac{1}{k + 1} = \frac{-1}{2k}$

13. $\frac{2v}{v^2 + 3v} - \frac{2}{v + 3} = \frac{2}{v}$
14. $\frac{3y}{y^2 - 2y} + \frac{5}{y - 2} = \frac{2}{y - 2}$
15. $\frac{x}{x - 3} + \frac{2x}{x + 3} = \frac{18}{x^2 - 9}$

Solve each equation for x. For what value(s) of a will there be a solution?

16. $\frac{1}{x} + \frac{1}{a} = 1$
17. $\frac{1}{x} + \frac{1}{a + 1} = 1$
18. $\frac{1}{x(a + 1)} - \frac{a}{x} = 1$

19. Suppose that you choose a number n, take its reciprocal, and add it to the reciprocal of $n + 1$. How many values of n will give the sum $\frac{3}{2}$?

20. **Critical Thinking** For what value(s) of a will $\frac{1}{n} + \frac{1}{n + 1} = \frac{a}{n}$ always have a solution for n, except $n \neq 0$ and $n \neq -1$?

MIXED REVIEW

Suppose that $a > 0$. Use laws of exponents and $\sqrt{a} = a^{\frac{1}{2}}$ to prove each statement. (8.1 Skill C)

21. $\sqrt{81a} = 9\sqrt{a}$
22. $\sqrt{100a} = 10\sqrt{a}$
23. $\sqrt{\frac{a}{25}} = \frac{\sqrt{a}}{5}$
24. $\sqrt{\frac{a}{49}} = \frac{\sqrt{a}}{7}$
25. $\sqrt{\frac{64a}{121}} = \frac{8\sqrt{a}}{11}$
26. $\sqrt{\frac{16a}{9}} = \frac{4\sqrt{a}}{3}$

Answers for Lesson 9.7 Skill B

Try This Example 1

a. $d = 2$ or $d = -\frac{3}{2}$

b. $s = 3$ or $s = \frac{6}{5}$

Try This Example 2

$a = -2$; 1 is extraneous.

1. -1 and $\frac{5}{2}$ are both solutions.

2. 3 is a solution and 0 is extraneous.

3. 1 is extraneous.

4. $x = -3$ or $x = -2$

5. $x = 5$ or $x = -2$

6. $x = \frac{1}{3}$ or $x = \frac{1}{2}$

7. $a = \frac{2}{3}$ or $a = -\frac{1}{3}$

8. $x = 0$ or $x = \frac{1}{2}$

9. $y = -1$ or $y = -\frac{1}{2}$

10. $r = -\frac{1}{2}$ or $r = \frac{2}{3}$

11. $t = \frac{1}{3}$ or $t = \frac{3}{2}$

12. $k = -\frac{1}{5}$; 0 is extraneous.

13. no solution; 0 and -3 are extraneous.

14. no solution; 0 and 2 are extraneous.

15. $x = -2$; 3 is extraneous.

★16. $x = \frac{a}{a - 1}$; $a \neq 0, 1$

★17. $x = \frac{a + 1}{a}$; $a \neq 0, -1$

★18. $x = \frac{-1 - a^2}{a}$, $a \neq 0, -1$

★19. two

20. no values of a

21. $\sqrt{81a} = (81a)^{\frac{1}{2}}$
$$= 81^{\frac{1}{2}}a^{\frac{1}{2}}$$
$$= (9^2)^{\frac{1}{2}}a^{\frac{1}{2}}$$
$$= 9^{\left(2 \cdot \frac{1}{2}\right)}a^{\frac{1}{2}}$$
$$= 9a^{\frac{1}{2}} = 9\sqrt{a}$$

22. $\sqrt{100a} = (100a)^{\frac{1}{2}}$
$$= 100^{\frac{1}{2}}a^{\frac{1}{2}}$$
$$= (10^2)^{\frac{1}{2}}a^{\frac{1}{2}}$$
$$= 10^{\left(2 \cdot \frac{1}{2}\right)}a^{\frac{1}{2}}$$
$$= 10a^{\frac{1}{2}} = 10\sqrt{a}$$

For answers to Exercises 23–26, see Additional Answers.

★ **Advanced Exercises**

Focus: *Students use a rational equation to model a real-world problem involving rate of work.*

EXAMPLES 1 and 2

PHASE 1: Presenting the Examples

Remind students of the familiar relationship *distance* = *rate* × *time*. If they think of a distance traveled as "work done," it is easy to transform this relationship into the following formula, which is the backbone of this skill:

$$\text{work done} = \text{rate of work} \times \text{work time}$$

In Example 1, students should note that the number of days needed to complete the entire project is unknown, so it is called t. This translates to a work rate of $\dfrac{1 \text{ project}}{t \text{ days}}$, or $\dfrac{1}{t}$ project per day. Thus the above formula yields the following:

$$\text{work done in 5 days} = \frac{1 \text{ project}}{t \text{ days}} \times 5 \text{ days} = \frac{5}{t} \text{ project}$$

However, point out that the following fact is given in the problem:

$$\text{work done in 5 days} = \frac{3}{5} \text{ project}$$

Since both $\dfrac{3}{5}$ and $\dfrac{5}{t}$ represent the part of the project done in 5 days, students can then write the equation used in Example 1, which is $\dfrac{3}{5} = \dfrac{5}{t}$.

Using like reasoning, they can form the equation for Example 2 as follows:

Alan

work rate: $\dfrac{1 \text{ project}}{t \text{ hours}}$

$$\text{work done in 15 hours} = \frac{1 \text{ project}}{t \text{ hours}} \times 15 \text{ hours}$$
$$= \frac{15}{t} \text{ project}$$

Louis

work rate: $\dfrac{1 \text{ project}}{2t \text{ hours}}$

$$\text{work done in 15 hours} = \frac{1 \text{ project}}{2t \text{ hours}} \times 15 \text{ hours}$$
$$= \frac{15}{2t} \text{ project}$$

Alan and Louis together

$$\text{work done in 15 hours} = \frac{15}{t} \text{ project} + \frac{15}{2t} \text{ project} = \left(\frac{15}{t} + \frac{15}{2t}\right) \text{ project} \rightarrow \frac{15}{t} + \frac{15}{2t} = 1$$
$$\text{work done in 15 hours} = 1 \text{ project}$$

PHASE 3: ASSESSMENT AND CLOSURE for Lesson 9.7

- Have students write a general procedure for solving a rational equation. [**sample: 1. If necessary, add or subtract so there is exactly one rational expression on each side. 2. Cross multiply. 3. Solve the new equation. 4. Check each solution in the given equation. Omit extraneous solutions.**]

- In Chapter 3, students learned about direct variations. In the next lesson they will use rational equations in working with *inverse* variations.

☞ *For a Lesson 9.7 Quiz, see* Assessment Resources *page 97.*

If it takes 10 days to complete a task and the worker spends 1 day on it, then the worker completes $\frac{1}{10}$ of the work in one day. This unit rate is called the *work rate*. If the worker spends t days on the task and the work rate is constant, then the worker completes $\frac{t}{10}$ of the task. This can be expressed as a formula:

$$\text{portion of task completed} = \text{rate} \times \text{time}$$

EXAMPLE 1

After working 5 days at a constant rate, a student completed 60%, or $\frac{3}{5}$, of a project. How long would it take the student to complete the entire project?

▷ **Solution**

1. Let t represent the time it would take to complete the whole project.
2. Solve $\frac{3}{5} = \frac{5}{t}$. $t = \frac{25}{3}$, or $8\frac{1}{3}$

Write an equation.

The project would take $8\frac{1}{3}$ days to complete.

TRY THIS Rework Example 1 if 75% of a project is completed in 6 days.

If two workers, worker A and worker B, work together, then you can use the equation below to solve combined-work problems.

$$\frac{\text{work time}}{\text{time needed}_A} + \frac{\text{work time}}{\text{time needed}_B} = \text{portion of task completed}$$

EXAMPLE 2

Louis takes twice as long as Alan to complete a school project. It takes them 15 hours to complete the project together. How long would it take each student to complete the project if he works alone?

▷ **Solution**

1. Let t and $2t$ represent time needed for Alan and Louis to complete the project working alone, respectively.

2. Write an equation.

Write an equation.

Alan: $\frac{\text{work time}}{\text{time needed}_A} \rightarrow \frac{15}{t}$ Louis: $\frac{\text{work time}}{\text{time needed}_B} \rightarrow \frac{15}{2t}$

$$\frac{15}{t} + \frac{15}{2t} \rightarrow 100\%, \text{ or } 1$$

3. Solve $\frac{15}{t} + \frac{15}{2t} = 1$: $t = \frac{45}{2}$, or $22\frac{1}{2}$.

It would take Alan $22\frac{1}{2}$ hours to complete the project. It would take Louis 45 hours.

TRY THIS Rework Example 2 given that it takes Louis three times as long as Alan to complete the project.

EXERCISES

▷ **KEY SKILLS**

For Exercises 1 and 2, find each work rate. Then write an equation for work, w, in terms of time spent working, t.

1. 12 days to complete the work
2. 15 days to complete the work

3. Amit needs 1.5 times the number of hours it takes Juan to complete a project. Together, it takes them 18 hours to complete it. Write an equation you can use to find how long it would take each student working alone to complete the project. Do not solve.

▷ **PRACTICE**

For Exercises 4 and 5, calculate how long would it take the worker to complete the project. Assume that work rate is constant.

4. After 7 days, 35% of the work is completed.
5. After 4.5 days, 40% of the work is completed.

How long would it take each student to complete the project if he or she works alone? Assume that work rate is constant.

6. Sam takes 1.6 times as long as Nina to do the project working alone. Together, it takes them 15 days to complete the project.

7. Lee takes the same amount of time as Kim to do the project working alone. Together, it takes them 18 hours to complete the project.

Let t represent the amount of time it takes Students A and B working together to complete a task. Let a and b represent the respective amounts of time it takes them working alone.

8. Solve $\frac{t}{a} + \frac{t}{b} = 1$ for t.
9. Solve $\frac{t}{a} + \frac{t}{b} = 1$ for a.

▷ **MIXED REVIEW APPLICATIONS**

Solve each problem. (2.8 Skill C)

10. How much of $5000 should an investor put into accounts paying 5% and 7.5% simple interest to earn $300 interest in one year?

11. A motorist drove 420 miles in 8 hours. During part of the trip she drove 50 miles per hour and during the rest of it she drove 54 miles per hour. How far did she drive at each speed?

12. A chemist wants to mix a 5% salt solution and a 10% salt solution to make 500 milliliters of an 8% salt solution. How much of the 5% and 10% solutions are needed?

Answers for Lesson 9.7 Skill C

Try This Example 1
8 days

Try This Example 2
Alan: 20 hours
Louis: 60 hours

1. $\frac{1}{12}$; $w = \frac{t}{12}$

2. $\frac{1}{15}$; $w = \frac{t}{15}$

3. Let t represent the amount of time Juan needs to complete the project working alone; $\frac{18}{t} + \frac{18}{1.5t} = 1$

4. 20 days
5. 11.25 days
6. Nina: $24\frac{3}{8}$ days
   Sam: 39 days
7. Lee: 36 hours
   Kim: 36 hours
★8. $t = \dfrac{ab}{a + b}$

★ 9. $a = \dfrac{bt}{b - t}$
10. $3000 in the 5% account and $2000 in the 7.5% account
11. 3 hours at 50 miles per hour and 5 hours at 54 miles per hour
12. 200 milliliters of the 5% solution and 300 milliliters of the 10% solution

★ **Advanced Exercises**

Inverse Variation

SKILL A   APPLICATIONS   *Solving inverse-variation problems*

Focus: *Students use proportions to solve inverse-variation problems.*

EXAMPLES 1 and 2

PHASE 1: Presenting the Examples

Display the table at right. Tell students to find the product of the entries in each row. In each case, they should arrive at 360. Lead them to see that the table lists several rate/time combinations that result in a distance of 360 miles. Stress the following key aspect of the relationship: As the rate increases, the time needed to travel 360 miles decreases.

Review the concept of direct variation (Lesson 3.3). Then have students read the box entitled *Inverse Variation* that appears at the top of the textbook page. Ask if the relationship between rate and time is a direct or inverse variation. [**inverse**] Have students identify the constant of variation. [**360**] Then have them write an inverse-variation equation to represent the relationship. [$rt = 360$, $r = \dfrac{360}{t}$, or $t = \dfrac{360}{r}$,

where r represents rate in miles per hour and t represents time in hours] When discussing their responses, display all three forms of the equation. Point out that the relationship can be described in two ways: *Rate varies inversely as time* and *time varies inversely as rate*.

Rate (miles per hour)	Time spent traveling (hours)
10	36
20	18
30	12
40	9
50	7.2
60	6

Now have students read the box entitled *Alternative Form of Inverse Variation.* Ask them to pick any two rate/time pairs from the table. Tell them to use their rate/time pairs to illustrate the alternative form. Display several of the resulting proportions as shown below.

(10, 36)	(10, 36)	(20, 18)	(30, 12)	(60, 6)	(50, 7.2)
(20, 18)	(40, 9)	(50, 7.2)	(60, 6)	(40, 9)	(10, 36)

$$\frac{10}{20} = \frac{18}{36} \qquad \frac{10}{40} = \frac{9}{36} \qquad \frac{20}{50} = \frac{7.2}{18} \qquad \frac{30}{60} = \frac{6}{12} \qquad \frac{60}{40} = \frac{9}{6} \qquad \frac{50}{10} = \frac{36}{7.2}$$

Then discuss Examples 1 and 2, in which proportions are used to solve one numerical problem and one application problem related to inverse variation.

PHASE 2: Providing Guidance for TRY THIS

Inverse-variation problems can also be solved by first identifying the constant of variation, and then writing and solving a variation equation. After students have used a proportion to solve each Try This problem, you may wish to demonstrate the alternative method as shown below.

Example 1 Try This:

$$xy = k \qquad y = \frac{k}{x}$$

$$(5)(30) = k \quad \rightarrow \quad y = \frac{150}{12}$$

$$k = 150 \qquad y = 12.5$$

Example 2 Try This:

$$VP = k \qquad V = \frac{k}{P}$$

$$(45)(75) = k \quad \rightarrow \quad V = \frac{3375}{80}$$

$$k = 3375 \qquad V = 42.1875$$

9.8 Inverse Variation

SKILL A ▸ **APPLICATIONS** ▸ *Solving inverse-variation problems*

A relationship between y and x is an *inverse-variation* relationship if y varies with the reciprocal, or multiplicative inverse, of x. That is, as one variable gets larger, the other variable gets smaller.

Inverse Variation

If there is a fixed nonzero number k such that $xy = k$, then you can say that y *varies inversely as* x. The relationship between x and y is called an **inverse-variation** relationship and k is called the *constant of variation*.

The equation $xy = k$ can also be written as $y = \frac{k}{x}$.

Alternative Form of Inverse Variation

If (x_1, y_1) and (x_2, y_2) satisfy $xy = k$, then $x_1 y_1 = x_2 y_2$ and $\frac{x_1}{x_2} = \frac{y_2}{y_1}$.

EXAMPLE 1 Suppose that y varies inversely as x. If $y = 36$ when $x = 4$, find x when $y = 24$.

▸ Solution

Use $(4, 36)$ and $(x, 24)$ as (x_1, y_1) and (x_2, y_2).

$$\frac{x_1}{x_2} = \frac{y_2}{y_1} \;\rightarrow\; \frac{4}{x} = \frac{24}{36} \;\rightarrow\; 4 \cdot 36 = 24x \;\rightarrow\; 6 = x$$

When $y = 24$, $x = 6$.

TRY THIS Suppose that y varies inversely as x. If $y = 30$ when $x = 5$, find y when $x = 12$.

At a constant temperature, pressure P in pounds per square inch and volume V in cubic inches of an ideal gas are related by inverse variation. That is, $PV = k$, for some nonzero constant k.

EXAMPLE 2 A gas with volume 30 cubic inches is under a pressure that is 40 pounds per square inch. Find the pressure when the volume is 60 cubic inches.

▸ Solution

Write an equation. Use $(30, 40)$ and $(60, P_2)$ as (V_1, P_1) and (V_2, P_2).

$$\frac{V_1}{V_2} = \frac{P_2}{P_1} \;\rightarrow\; \frac{30}{60} = \frac{P_2}{40} \;\rightarrow\; 60P_2 = 30 \times 40 \;\rightarrow\; P_2 = 20$$

The pressure will be 20 pounds per square inch.

TRY THIS A gas with volume 45 cubic inches is under a pressure that is 75 pounds per square inch. Find the pressure when the volume is 80 cubic inches.

EXERCISES

KEY SKILLS

In Exercises 1 and 2, speed s varies inversely with elapsed time t.

1. Samantha and Derek plan to bicycle a distance of 8 miles. Copy and complete the time/speed table below.

t	$\frac{1}{4}$ hour	$\frac{1}{2}$ hour	1 hour	2 hours	3 hours	4 hours
s						

2. How would you find the bicycling speed they need if they want to travel 8 miles in $1\frac{1}{2}$ hours?

PRACTICE

In Exercises 3–6, y varies inversely as x. Find the value of the indicated variable.

3. If $y = 12$ when $x = 4$, find y when $x = 12$.

4. If $y = 10$ when $x = 2$, find y when $x = 6$.

5. If $y = 36$ when $x = 5$, find x when $y = 12$.

6. If $y = 9$ when $x = 1$, find x when $y = 10$.

7. Show that if (x_1, y_1) and (x_2, y_2) satisfy $xy = k$, then $\frac{x_1}{x_2} = \frac{y_2}{y_1}$.

Solve each problem.

8. Given a fixed number as the area of a rectangle, length varies inversely as width. What happens to the length of the rectangle if area is constant and width is doubled?

9. The volume of a gas is 45 cubic inches when the pressure is 42 pounds per square inch. Given constant temperature, find the pressure when the volume is cut by 50%.

10. Two boxes balance on a seesaw when box weight and distance from the fulcrum satisfy an inverse-variation relationship. How far from the fulcrum should Box B be placed to balance Box A if Box A weighs 45 pounds, Box B weighs 60 pounds and Box A is 4.5 feet from the fulcrum?

MIXED REVIEW APPLICATIONS

In Exercises 11 and 12, y varies directly as x. (3.3 Skill B)

11. If $y = 18$ when $x = 6$, find y when $x = 2.5$.

12. If $y = 100$ when $x = 25$, find y when $x = 5$.

Answers for Lesson 9.8 Skill A

Try This Example 1
$y = 12.5$

Try This Example 2
$42\frac{3}{16}$, or 42.1875 pounds per square inch

1. 32 miles per hour, 16 miles per hour, 8 miles per hour, 4 miles per hour, $2\frac{2}{3}$ miles per hour, 2 miles per hour

2. Solve $1\frac{1}{2}s = 8$ for s.

3. $y = 4$

4. $y = \dfrac{10}{3}$

5. $x = 15$

6. $x = \dfrac{9}{10}$

7. If (x_1, y_1) and (x_2, y_2) satisfy $xy = k$, then $x_1 y_1 = k$ and $x_2 y_2 = k$. By the Transitive Property of Equality, $x_1 y_1 = x_2 y_2$. Divide each side of the equation by $y_1 x_2$. The result is $\dfrac{x_1}{x_2} = \dfrac{y_2}{y_1}$.

8. Length is reduced by half when width is doubled.

9. 84 pounds per square inch

10. $3\dfrac{3}{8}$, or 3.375 feet

11. $y = 7.5$

12. $y = 20$

Focus: *Students examine basic properties of an inverse-variation function.*

EXAMPLE 1

PHASE 1: Presenting the Example

Display the equation $y = 4x$ and tell students to graph it. [**See the graph at right.**] Ask them to find the slope of the graph. [**4**] Remind them that any line through the origin is a "picture" of a direct variation. Ask them to name the constant for the direct variation pictured by this graph. [**4**]

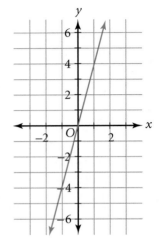

Display $xy = 4$. Remind students that this equation represents an inverse variation. Ask if they think its graph is a line. Lead them to observe the following: When you solve an equation of a line for y, you obtain a slope-intercept form, $y = mx + b$.

However, when you solve $xy = 4$ for y, you obtain $y = \dfrac{4}{x}$. This is not slope-intercept form, so its graph is not a line.

Now discuss Example 1, where students see that the graph is a curve with two branches. Tell them that this curve is called a *hyperbola*.

PHASE 2: Providing Guidance for **TRY THIS**

Some students may graph only in the first quadrant. Point out that a graph of an equation must represent *all* solutions. By excluding the third quadrant, they omit a large set of solutions. Ask them to describe the third-quadrant solutions. [**all pairs of negative numbers whose product is 4**]

EXAMPLE 2

PHASE 1: Presenting the Example

Display the graph at right. Have students explain why it represents a function. [**No vertical line crosses the graph in more than one point.**]

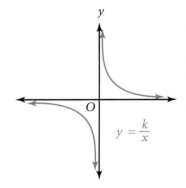

Tell students to write what they think is the domain of the function. Several may write *all real numbers*. Ask if any number must be excluded. Lead them to see that, because the equation is of the form $y = \dfrac{k}{x}$, x cannot take on the value 0. So the domain is all real numbers *except 0*.

Now have them write what they think is the range. Again, several may believe it to be all real numbers. Display the sequences at the right, which show some values for just one specific inverse-variation function, $y = \dfrac{4}{x}$, as x becomes infinitely large and infinitely small.

$$\dfrac{4}{10}, \dfrac{4}{100}, \dfrac{4}{1000}, \dfrac{4}{10,000}, \dfrac{4}{100,000}, \dfrac{4}{1,000,000}, \cdots$$

$$\dfrac{4}{-10}, \dfrac{4}{-100}, \dfrac{4}{-1000}, \dfrac{4}{-10,000}, \dfrac{4}{-100,000}, \dfrac{4}{-1,000,000}, \cdots$$

Lead students to see that the values of the function get closer and closer to 0, but never actually equal 0. Tell them that the function will behave in a similar manner no matter what the value of k. So the range of any function with an equation of the form $y = \dfrac{k}{x}$ is all real numbers *except 0*. Then discuss Example 2.

If y varies inversely as x and the constant of variation is k, then $y = \frac{k}{x}$. In particular, the function represented by $y = \frac{1}{x}$ is called the *reciprocal function*.

EXAMPLE 1 Graph $xy = 4$.

▶ Solution

Write y in terms of x: $y = \frac{4}{x}$

Make tables of ordered pairs.

Make a table.

x	-8	-4	-2	-1	-0.5
y	-0.5	-1	-2	-4	-8

x	0.5	1	2	4	8
y	8	4	2	1	0.5

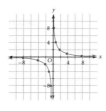

Sketch smooth curves. The graph gets closer and closer to each axis but does not touch or cross either axis.

TRY THIS Graph $xy = 16$.

EXAMPLE 2 The graph at right represents an equation of the form $xy = k$.
  a. Does the graph represent a function?
  b. What are the domain and range?
  c. Find y when $x = 250$.
  d. Is $(10, 0.25)$ on the graph? Justify your response.

▶ Solution

  a. Apply the vertical-line test. No vertical line intersects the graph in more than one point. So, the graph represents a function.
  b. domain: all real numbers except 0   range: all real numbers except 0
  c. Use the point given on the graph to find k: $k = (5)(5) = 25$.
      Substitute 250 for x to find y: $250y = 25 \rightarrow y = \frac{1}{10}$.
  d. No, $(10, 0.25)$ is not on the graph, because $10 \cdot 0.25 \neq 5 \cdot 5$.

TRY THIS The graph at right represents an equation of the form $xy = k$.
  a. Does the graph represent a function?
  b. What are the domain and range?
  c. Find y when $x = 120$.
  d. Is $(80, 0.25)$ on the graph? Justify your response.

EXERCISES

KEY SKILLS

Refer to the graph at right. The graph represents an equation of the form $y = \frac{k}{x}$. Point $P(7, 7)$ is on the graph.

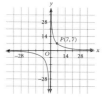

  1. What is the value of k?

  2. Explain why $Q(-7, -7)$ is also on the graph.

  3. Explain why there is no point on the graph that corresponds to $x = 0$.

PRACTICE

Graph each equation.

  4. $xy = 1$ 5. $xy = 9$ 6. $xy = 36$

Use each graph in Exercises 7–9 to answer the questions below.
  a. Does the graph represent a function?
  b. What is the domain? the range?
  c. What is the value of y when $x = 150$?
  d. Is the point $(100, 0.25)$ on the graph? Justify your response.

  7.   8.   9.

10. **Critical Thinking** Let $y = \frac{1}{n}$, where n is a natural number. Is $\frac{1}{n+1}$ greater than or less than $\frac{1}{n}$? Justify your response. What does your conclusion tell you about the graph of $y = \frac{1}{n}$ as n increases?

MIXED REVIEW

Give a counterexample to disprove each statement. (4.7 Skill C)

11. The product $(3n - 1)(2n + 3)$ is always less than -6.

12. The product $(-2n - 1)(2n - 3)$ is always greater than 5.

13. The product $(-3n)(n + 1)$ is always positive.

Answers for Lesson 9.8 Skill B

Try This Example 1

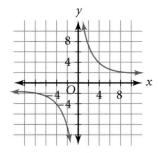

Try This Example 2
a. yes
b. domain: all real numbers except 0;
    range: all real numbers except 0
c. $y = 1.2$
d. If $xy = 144$ and $x = 80$, then
    $y = 1.8 \neq 0.25$. The point $(80, 0.25)$
    is not on the graph.

1. $k = 49$

2. If $P(7, 7)$ is on the graph, then
    $xy = 49$. Since $(-7)(-7) = 49$,
    then $(-7, -7)$ is a solution and
    $Q(-7, -7)$ is on the graph.

3. If $xy = 49$, then $y = \frac{49}{x}$. Since
    $\frac{49}{x}$ is not defined for $x = 0$, there
    is no point on the graph corre-
    sponding to $x = 0$.

4.

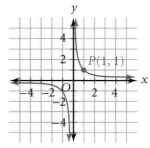

5.

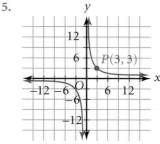

6.

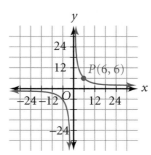

For answers to Exercises 7–13, see
Additional Answers.

Focus: *Students use proportions to solve problems in which one quantity varies inversely as the square of another.*

EXAMPLE 1

PHASE 2: Providing Guidance for **TRY THIS**

Some students may model their work on the solution for Example 1 and give the response $y = -\dfrac{1}{64}$ or $y = \dfrac{1}{64}$. Stress the difference between the two problems. That is, in Example 1 the unknown value was x, resulting in a quadratic equation ($x^2 = 9$) that had two solutions. Point out that here the unknown value is y, yielding a one-solution linear equation $\left(y = \dfrac{4}{256} \right)$.

EXAMPLE 2

PHASE 1: Presenting the Example

This example addresses a physical principle called the *Law of Illumination*. To help students visualize this principle, you can conduct an informal demonstration such as the following: Shine a flashlight on a wall from a small distance, perhaps about one foot. Have a student use a measuring tape to find the distance between the flashlight and the wall. Then have another student find a point twice as far from the wall, and move the flashlight there. Repeat for a point that is three times the original distance from the wall.

Discuss the demonstration with the class. Elicit the following observation: As the distance between the flashlight and the wall increases, the brightness, or *intensity*, of the light on the wall decreases. Point out that this language suggests an inverse-variation relationship. Then discuss Example 2, which defines the exact nature of the relationship.

PHASE 3: ASSESSMENT AND CLOSURE for Lesson 9.8

- Summarize the students' work with all types of variation by giving them the table at the right. Have them fill it out so that: y_1 varies directly as x; y_2 varies directly as the square of x; y_3 varies inversely as x; and y_4 varies inversely as the square of x. In each case, the constant of variation should be 36. [y_1: **36, 72, 108, 144, 180**; y_2: **36, 144, 324, 576, 900**; y_3: **36, 18, 12, 9, 7.2**; y_4: **36, 9, 4, 2.25, 1.44**]

x	y_1	y_2	y_3	y_4
1				
2				
3				
4				
5				

- In this course, students have learned about the two fundamental types of variation, direct and inverse. If they take an advanced algebra course, they will learn about *joint variation*, in which one quantity varies directly as two or more others; and *combined variation*, in which a quantity varies directly as one quantity and inversely as another.

☞ *For a Lesson 9.8 Quiz, see* Assessment Resources *page 98.*

Recall from Lesson 6.2 that y varies directly as the square of x if there is a nonzero number k such that $y = kx^2$.

Inverse Variation as the Square

If there is a fixed nonzero number k such that $x^2y = k$, then you can say that y *varies inversely as the square of x*. The relationship between x and y is called an **inverse-square-variation** relationship and k is called the *constant of variation.*

If y varies *inversely* as the square of x, you can say that y varies *directly* as the square of the *reciprocal* of x. This can be expressed symbolically as shown below.

$$y = k\left(\frac{1}{x}\right)^2, \ y = k \cdot \frac{1}{x^2}, \text{ or } y = \frac{k}{x^2}$$

EXAMPLE 1 If y varies inversely as the square of x and $y = \frac{3}{100}$ when $x = 10$, find x when $y = \frac{1}{3}$.

▶ Solution

If $y = \frac{k}{x^2}$, $x = 10$, and $y = \frac{3}{100}$, then $k = 3$. Solve $\frac{1}{3} = \frac{3}{x^2}$.

$$\frac{1}{3} = \frac{3}{x^2} \rightarrow x^2 = 9 \rightarrow x = \pm 3$$

If $y = \frac{1}{3}$, then $x = \pm 3$.

TRY THIS If y varies inversely as the square of x and $y = \frac{4}{100}$ when $x = 10$, find y when $x = 16$.

EXAMPLE 2 The illumination, I, of a point varies inversely as the square of d, the distance from the light source. How does the illumination of a point compare with that of a point twice the distance from the source?

▶ Solution

The variables I and d are related by $I = \frac{k}{d^2}$, where $k \neq 0$.

Point A: d units from source   Point B: $2d$ units from source

$$I_A = \frac{k}{d^2} \qquad\qquad I_B = \frac{k}{(2d)^2}$$

Simplify the ratio $\frac{I_B}{I_A}$. $\frac{I_B}{I_A} = \frac{\frac{k}{(2d)^2}}{\frac{k}{d^2}} = \frac{k}{(2d)^2} \cdot \frac{d^2}{k} = \frac{1}{4}$

If distance is doubled, illumination is one quarter of what it was.

TRY THIS Rework Example 2 for the case where distance is tripled.

EXERCISES

In Exercises 1 and 2, $y = \frac{1}{x^2}$.

1. Copy and complete the table below.

x	1	2	3	4	5
y	?	?	?	?	?

2. What can you say from the table in Exercise 1 about the values of y as x increases?

PRACTICE

In Exercises 3–6, y varies inversely as the square of x.

3. If $y = 0.05$ when $x = 10$, find y when $x = 15$.

4. If $y = 3$ when $x = 4$, find y when $x = 12$.

5. If $y = \frac{1}{7}$ when $x = 5$, find x when $y = \frac{1}{28}$.

6. If $y = 2$ when $x = 8$, find x when $y = \frac{2}{9}$.

In Exercises 7–9, refer to Example 2 on the previous page.

7. Fill in the steps that show that $\frac{I_B}{I_A} = \frac{1}{4}$.

8. Rework Example 2 for the case where distance is quadrupled.

9. Suppose that distance d is multiplied by r, where $r > 0$. How does illumination at distance rd compare with the illumination at distance d?

10. Points P and Q are situated so that the illumination at P is one-sixteenth that of the illumination at Q. How are the distances of P and Q from the light source related?

MIXED REVIEW APPLICATIONS

A car has a stopping distance, d, that varies directly as the square of the car's speed, s. Round to the nearest hundredth. (Lesson 6.2 Skill B)

If $d = 120$ ft when $s = 30$ mph, find d when:

11. $s = 15$ mph

12. $s = 40$ mph

13. $s = 50$ mph

14. $s = 60$ mph

Answers for Lesson 9.8 Skill C

Try This Example 1

$y = \frac{1}{64}$

Try This Example 2

$\frac{I_B}{I_A} = \frac{1}{9}$

1.

x	1	2	3	4	5
y	1	$\frac{1}{4}$	$\frac{1}{9}$	$\frac{1}{16}$	$\frac{1}{25}$

2. As x increases, y decreases. However, for a given value of x, the value of y for $y = \frac{1}{x^2}$ is less than the corresponding value of y for $\frac{1}{x}$. That is, y decreases faster when variation is with the inverse of the square than when it is with the inverse.

3. $y = \frac{1}{45}$

4. $y = \frac{1}{3}$

5. $x = \pm 10$

6. $x = \pm 24$

7. $\frac{I_B}{I_A} = \dfrac{\frac{k}{(2d)^2}}{\frac{k}{d^2}}$

$= \frac{k}{(2d)^2} \cdot \frac{d^2}{k}$

$= \frac{k}{4d^2} \cdot \frac{d^2}{k}$

$= \frac{k}{k} \cdot \frac{d^2}{d^2} \cdot \frac{1}{4}$

$= \frac{1}{4}$

8. $\frac{1}{16}$ times illumination at distance d

9. $\frac{1}{r^2}$ times illumination at distance d

10. The distance from P to the light source is 4 times the distance from Q to the light source.

11. 30 feet

12. 213.33 feet

13. 333.33 feet

14. 480 feet

Proportions and Deductive Reasoning

 Using deductive reasoning to derive formulas and prove statements about proportions

Focus: *Students write deductive arguments to solve literal equations involving proportions and to prove properties of proportions.*

EXAMPLE 1

PHASE 1: Presenting the Example

Give students the three equations shown at right and tell them to solve. [**1.** $x = -2$ **2.** $x = -5$ **3.** $x = -1.875$] They should note that each equation has the same form, generalized as $\frac{x - a}{x} = b$, and so each could be solved using the same set of steps. Discuss the proof in Example 1. Tell students this proof guarantees that any equation of the form $\frac{x - a}{x} = b$ can be solved by finding the value of $\frac{a}{1 - b}$, provided that $x \neq 0$ and $b \neq 1$. To demonstrate this fact, have them use the result to solve $\frac{x - 1.46}{x} = 3$. [$x = \frac{1.46}{1 - 3} = \frac{1.46}{-2} = -0.73$]

1 $\dfrac{x - 8}{x} = 5$

2 $\dfrac{x - 10}{x} = 3$

3 $\dfrac{x - 15}{x} = 9$

PHASE 2: Providing Guidance for TRY THIS

Watch for students who simply multiply each side by x and write $bx = x + a$. Point out that the work is not done until x is isolated on one side of the equal sign.

EXAMPLE 2

PHASE 2: Providing Guidance for TRY THIS

Given $\frac{a}{b} = \frac{c}{d}$, a and d are called the *extremes* of the proportion, while b and c are called the *means*. The property that students prove in the Try This exercise is sometimes called the *Interchange-Means Property*.

PHASE 3: ASSESSMENT AND CLOSURE for Lesson 9.9

- Give students the five equations at right. Tell them to rearrange the equations to form a deductive argument. Have them justify each step of the argument, and then write a conditional statement to summarize it. $\left[\frac{a}{b} = \frac{c}{x + d}; a(x + d) = bc; ax + ad = bc; \right.$ $ax = bc - ad; x = \frac{bc - ad}{a};$ If $\frac{a}{b} = \frac{c}{x + d}$, then $x = \frac{bc - ad}{a}.\left. \right]$

- This lesson concludes the study of rational expressions and rational equations in this course. If students proceed to a course in geometry, they will see how these concepts are applied in the analysis of similar figures.

$a(x + d) = bc$

$x = \dfrac{bc - ad}{a}$

$ax = bc - ad$

$\dfrac{a}{b} = \dfrac{c}{x + d}$

$ax + ad = bc$

☞ *For a Lesson 9.9 Quiz, see* Assessment Resources *page 99.*

☞ *For a Chapter 9 Assessment, see* Assessment Resources *pages 100–101.*

9.9 LESSON — Proportions and Deductive Reasoning

SKILL A — Using deductive reasoning to derive formulas and prove statements about proportions

EXAMPLE 1

Let $\frac{x-a}{x} = b$. Use deductive reasoning to write a formula for x in terms of a and b. State any restrictions on the values of a, b, and x.

▶ **Solution**

$$\frac{x-a}{x} = b$$
$$x - a = bx \quad \longleftarrow \quad \text{Multiplication Property of Equality}$$
$$x = bx + a \quad \longleftarrow \quad \text{Subtraction Property of Equality}$$
$$x - bx = a \quad \longleftarrow \quad \text{Subtraction Property of Equality}$$
$$x(1 - b) = a \quad \longleftarrow \quad \text{Distributive Property}$$
$$x = \frac{a}{1-b} \quad \longleftarrow \quad \text{Division Property of Equality}$$

Collect all the terms with x on one side.

If $\frac{x-a}{x} = b$, then $x = \frac{a}{1-b}$ for $x \neq 0$ and $b \neq 1$.

TRY THIS

Let $\frac{x+a}{x} = b$. Use deductive reasoning to write a formula for x in terms of a and b. State restrictions on variables.

Recall from Lesson 9.7 the Cross Product Property of Proportions:

If a, b, c, and d are real numbers, $b \neq 0$, $d \neq 0$, and $\frac{a}{b} = \frac{c}{d}$, then $ad = bc$.

You can prove the *converse* statement of this property:

If a, b, c, and d are real numbers, $b \neq 0$, $d \neq 0$, and $ad = bc$, then $\frac{a}{b} = \frac{c}{d}$.

EXAMPLE 2

Prove: If a, b, c, and d are real numbers, and $b \neq 0$, $d \neq 0$, and $ad = bc$, then $\frac{a}{b} = \frac{c}{d}$.

▶ **Solution**

$$ad = bc$$
$$\frac{ad}{bd} = \frac{bc}{bd} \quad \longleftarrow \quad \text{Division Property of Equality}$$
$$\frac{a}{b} \cdot \frac{d}{d} = \frac{b}{b} \cdot \frac{c}{d} \quad \longleftarrow \quad \text{Definition of multiplication of fractions}$$
$$\frac{a}{b} \cdot 1 = 1 \cdot \frac{c}{d} \quad \longleftarrow \quad \text{Multiplicative inverse}$$
$$\frac{a}{b} = \frac{c}{d} \quad \longleftarrow \quad \text{Multiplicative identity}$$

TRY THIS

Prove: If a, b, c, and d are nonzero real numbers and $\frac{a}{b} = \frac{c}{d}$, then $\frac{a}{c} = \frac{b}{d}$.

EXERCISES

KEY SKILLS

State any restrictions on the values of a, b, c, and x. Do not prove the statement.

1. If $\frac{x-a}{b} = c$, then $x = a + bc$.

2. If $\frac{x+a}{x-b} = c$, then $x = \frac{a+bc}{c-1}$.

PRACTICE

Use deductive reasoning to solve for x in terms of the other variables. State any restrictions on the values of a, b, c, d, and x.

3. $\frac{ax}{b} = \frac{c}{d}$

4. $\frac{a}{bx} = \frac{c}{d}$

5. $\frac{a}{bx+c} = 1$

6. $\frac{a}{bx-c} = 1$

7. $\frac{x+a}{x+b} = c$

8. $\frac{a}{bx+c} = d$

9. $\frac{ax-b}{ax+b} = c$

10. $\frac{a}{bx} = \frac{cx}{d}$

11. $\frac{1}{ax+b} = ax + b$

Give the justification for each step. Assume a, b, c, and d represent nonzero real numbers.

12. If $\frac{a}{b} = \frac{c}{d}$, then $bd \cdot \frac{a}{b} = bd \cdot \frac{c}{d}$.

13. $\frac{ad}{bd} = \frac{a}{b} \cdot \frac{d}{d}$

Prove each statement below. Assume a, b, c, and d represent nonzero real numbers.

14. If $\frac{1}{b} = \frac{c}{d}$, then $\frac{d}{b} = c$.

15. If $\frac{1}{b} = \frac{c}{d}$, then $\frac{b}{d} = \frac{1}{c}$.

16. If $\frac{a}{b} = \frac{c}{d}$, then $\frac{a+b}{b} = \frac{c+d}{d}$.

17. If $\frac{a}{b} = \frac{c}{d}$, then $\frac{a-b}{b} = \frac{c-d}{d}$.

18. Prove: If a, b, c, and d are real numbers, $b \neq 0$, $d \neq 0$, and $\frac{a}{b} = \frac{c}{d}$, then $\frac{a^2}{b^2} = \frac{c^2}{d^2}$.

19. **Critical Thinking** Is the statement below always, sometimes or never true? If the statement is sometimes true, when is it true?

If a, b, c, and d are real numbers, and $b \neq 0$, $d \neq 0$, and $\frac{a}{b} = \frac{c}{d}$, then $\frac{a}{d} = \frac{c}{b}$.

MIXED REVIEW

Find the domain of each rational expression or function. (9.1 Skill A)

20. $\frac{3x^2 - 5}{4x^2 - x}$

21. $\frac{7a^2 + 1}{-4a^2 + 1}$

22. $\frac{v + 7}{4v^2 + v - 5}$

23. $y = \frac{-7x + 3}{2x + 11}$

24. $y = \frac{1}{4x^2 + 28x + 49}$

25. $y = \frac{2x^2 - 12}{4x^2 - 16x}$

Answers for Lesson 9.9 Skill A

Try This Example 1

$x = \dfrac{a}{b-1}$; $x \neq 0$ and $b \neq 1$

Try This Example 2

If $\frac{a}{b} = \frac{c}{d}$, then by the Cross Product Property of Proportions, $ad = bc$. By the Multiplication Property of Equality, $\frac{bd}{cd} \cdot \frac{a}{b} = \frac{bd}{cd} \cdot \frac{c}{d}$. By definition of multiplication of fractions, $\frac{bda}{cdb} = \frac{bdc}{cdd}$. By the Commutative and Associative Properties of Multiplication, $\frac{bda}{bdc} = \frac{bdc}{ddc}$. By definition of multiplication of fractions, $\frac{bd}{bd} \cdot \frac{a}{c} = \frac{dc}{dc} \cdot \frac{b}{d}$. By the Inverse Property of Multiplication, $1 \cdot \frac{a}{c} = 1 \cdot \frac{b}{d}$. By the Identity Property of Multiplication, $\frac{a}{c} = \frac{b}{d}$.

1. $b \neq 0$

2. $x \neq b$ and $c \neq 1$

3. $x = \dfrac{bc}{ad}$; $b \neq 0$, $d \neq 0$, $a \neq 0$

4. $x = \dfrac{ad}{bc}$; $b \neq 0$, $d \neq 0$, $c \neq 0$

5. $x = \dfrac{a-c}{b}$; $x \neq -\dfrac{c}{b}$ and $b \neq 0$

6. $x = \dfrac{a+c}{b}$; $x \neq \dfrac{c}{b}$ and $b \neq 0$

7. $x = \dfrac{a-bc}{c-1}$; $x \neq -b$ and $c \neq 1$

8. $x = \dfrac{a-cd}{bd}$; $x \neq -\dfrac{c}{b}$; $b \neq 0$, $d \neq 0$

9. $x = \dfrac{b(1+c)}{a(1-c)}$; $x \neq -\dfrac{b}{a}$, $a \neq 0$, $c \neq 1$

★10. $x = \pm\sqrt{\dfrac{ad}{bc}}$; $b \neq 0$, $c \neq 0$, $d \neq 0$ and two or all four of a, b, c, and d must have the same sign.

★11. $x = \dfrac{-b \pm 1}{a}$; $a \neq 0$, $x \neq -\dfrac{b}{a}$

For answers to Exercises 12–25, see Additional Answers.

★ Advanced Exercises

Identify any excluded values of the domain.

1. $\dfrac{2.5x^2 - x + 1}{4x + 3}$

2. $\dfrac{a^2 - 1}{4a^2 - 9}$

3. $y = \dfrac{5a + 2}{5a + 2}$

Simplify. Identify any excluded values.

4. $\dfrac{11}{22n - 55}$

5. $\dfrac{-3d + 6}{9d - 18}$

6. $\dfrac{15v + 60}{15v + 30}$

7. $\dfrac{t^2 - 25}{t^2 - 2t - 15}$

8. $\dfrac{a^2 - 8a + 7}{a^2 - 2a + 1}$

9. $\dfrac{m^2 + 8m - 9}{m^2 + 7m - 18}$

Simplify.

10. $\dfrac{p^2 t^2 + 4pt}{8pt + pt^2}$

11. $\dfrac{r^2 s^2 - 9}{2rs - 6}$

12. $\dfrac{3a^2 b - 12a}{a^2 b^2 - 16}$

Write the answer with positive exponents only.

13. $\dfrac{30h^3}{17g^2} \cdot \dfrac{34g}{15h^2}$

14. $\dfrac{-3x + 9}{x^2 + 5x + 6} \cdot \dfrac{x^2 + 6x + 9}{9 - x^2}$

15. $\dfrac{16a^2}{45z^2} \div \dfrac{4a^4}{9z^3}$

16. $\dfrac{2x - 6}{x^2 - 5x + 6} \div \dfrac{5x - 15}{x^2 - 6x + 9}$

17. $\dfrac{3a - 6}{a^2 - 25} \div \dfrac{a^2 - 1}{a^2 - 4a - 5} \cdot \dfrac{a^2 - 4a + 3}{a^2 - 8a + 15}$

18. $\dfrac{5g + 2}{3g^2} + \dfrac{4g + 7}{3g^2}$

19. $\dfrac{15z^2}{3z + 1} + \dfrac{11z + 2}{3z + 1}$

20. $\dfrac{2h + 1}{3h} + \dfrac{5}{6h^2}$

21. $\dfrac{1}{2} + \dfrac{1}{3z} + \dfrac{4}{5z^2}$

22. $\dfrac{3c + 1}{3c^2 - c} + \dfrac{2}{c - 1}$

23. $\dfrac{1}{x^2 + 2x + 1} + \dfrac{1}{x^2 - 1}$

24. $\dfrac{4y^3 + 1}{3y^3} + \dfrac{y^3 + 4}{3y^3}$

25. $\dfrac{4a^2 + 2a}{2a + 1} - \dfrac{-1}{2a + 1}$

26. $\dfrac{u}{7u + 14} - \dfrac{u}{u + 2}$

27. $\dfrac{n + 3}{n^2 + 6n + 9} - \dfrac{n - 3}{n^2 - 9}$

28. The lengths of the sides of the outer and inner rectangles below are in proportion. The dimensions of the inner rectangle are all one quarter of the corresponding dimensions in the outer rectangle. Write an expression for the shaded area.

Simplify.

29. $\dfrac{\frac{24z^2}{49n^3}}{\frac{81z^4}{14n^4}}$

30. $\dfrac{\frac{4a^2 - 8ab + 4b^2}{a^2 - 2ab + b^2}}{\frac{2a - 2b}{a + b}}$

31. In a parallel electrical circuit, the total effective resistance of two resistances is given in ohms by $\dfrac{1}{\frac{1}{9} + \frac{1}{12}}$. Write this as a single fraction.

Solve.

32. $\dfrac{-4}{3z} = \dfrac{7}{5z - 2}$

33. $\dfrac{s}{s - 1} = \dfrac{-3}{s - 5}$

34. $c - \dfrac{4}{c} = 3$

35. $\dfrac{x}{x - 3} + \dfrac{2x}{x + 3} = \dfrac{18}{x^2 - 9}$

36. Ken needs twice as long as Maria to complete a project. Together, it takes them 24 hours to complete it. How long would it take each student working alone?

37. Suppose that y varies inversely as x. If $y = 15$ when $x = 3$, find x when $y = 10$.

38. The graph of an equation of the form $xy = k$ ($k \neq 0$) contains the point (4, 12). Does the graph represent a function? What are its domain and range? Find y when $x = 50$. Is (5, 50) on the graph? Justify your response.

39. Suppose that y varies inversely as the square of x. If $y = 16$ when $x = 1$, find x when $y = 25$.

40. Let $\dfrac{ax + b}{x} = 1$. Use deductive reasoning to write a formula for x in terms of a and b. State the restrictions on variables.

Answers for Chapter 9 Assessment

1. $x \neq -\dfrac{3}{4}$

2. $a \neq \pm\dfrac{3}{2}$

3. $a \neq -\dfrac{2}{5}$

4. $\dfrac{1}{2n - 5}; n \neq \dfrac{5}{2}$

5. $-\dfrac{1}{3}; d \neq 2$

6. $\dfrac{v + 4}{v + 2}; v \neq -2$

7. $\dfrac{t + 5}{t + 3}; t \neq 5, -3$

8. $\dfrac{a - 7}{a - 1}; a \neq 1$

9. $\dfrac{m - 1}{m - 2}; m \neq 2, -9$

10. $\dfrac{pt + 4}{8 + t}$

11. $\dfrac{rs + 3}{2}$

12. $\dfrac{3a}{ab + 4}$

13. $\dfrac{4h}{g}$

14. $\dfrac{3}{x + 2}$

15. $\dfrac{4}{5a^2}$

16. $\dfrac{2x - 6}{5x - 10}$

17. $\dfrac{3a - 6}{a^2 - 25}$

18. $\dfrac{3g + 3}{g^2}$

19. $5z + 2$

20. $\dfrac{4h^2 + 2h + 5}{6h^2}$

21. $\dfrac{15z^2 + 10z + 24}{30z^2}$

22. $\dfrac{9c^2 - 4c - 1}{3c^3 - 4c^2 + c}$

23. $\dfrac{2x}{(x + 1)(x - 1)(x + 1)}$, or $\dfrac{2x}{x^3 + x^2 - x - 1}$

24. $\dfrac{5y^3 + 5}{3y^3}$

25. $\dfrac{4a^2 + 2a + 1}{2a + 1}$

26. $\dfrac{-6u}{7u + 14}$

27. 0

28. $\dfrac{5a^2}{4}$

29. $\dfrac{16n}{189z}$

30. $\dfrac{2a + 2b}{a - b}$

31. $\dfrac{36}{7}$ ohms

32. $z = \dfrac{8}{41}$

33. $s = -1$ or $s = 3$

34. $c = -1$ or $c = 4$

35. $x = -2$

For answers to Exercises 36–40, see **Additional Answers.**

Chapter 10 *Radical Expressions, Equations, and Functions*	State or Local Standards	Corresponding Lessons in *Algebra 1,* Schultz et al.
10.1 Square-Root Expressions and Functions		12.1
Skill A: Finding the domain of a square-root expression or function		
10.2 Solving Square-Root Equations		12.2, 12.4
Skill A: Solving equations in which one side is a square-root expression		
Skill B: APPLICATIONS Using square-root equations to solve real-world problems		
Skill C: Solving equations in which both sides are square-root expressions		
10.3 Numbers of the Form $a + b\sqrt{2}$		12.1
Skill A: Adding and subtracting numbers of the form $a + b\sqrt{2}$		
Skill B: Multiplying and dividing numbers of the form $a + b\sqrt{2}$		
Skill C: Proving statements and solving equations that involve $a + b\sqrt{2}$		
10.4 Operations on Square-Root Expressions		12.1
Skill A: Using properties of square roots to simplify products and quotients		
Skill B: Adding and subtracting square-root expressions		
Skill C: Multiplying and dividing square-root expressions		
10.5 Powers, Roots, and Rational Exponents		12.1
Skill A: Simplifying expressions that involve exponents of the form $\frac{1}{n}$		
Skill B: Simplifying expressions that involve exponents of the form $\frac{m}{n}$		
Skill C: APPLICATIONS Using rational exponents to solve real-world problems		
10.6 Assessing and Justifying Statements About Square Roots		11.6, 12.1, 12.2
Skill A: Exploring and proving statements that involve square roots		
Skill B: APPLICATIONS Using square-root equations to solve real-world problems		
Skill C: Approximating irrational numbers		

CHAPTER

10

Radical Expressions, Equations, and Functions

▶ **What You Already Know**

By this stage in the course, you have learned concepts and skills that involve: square roots, functions of various kinds, integer exponents, solving linear and quadratic equations, and solving linear inequalities.

Now it will be time to bring all this experience together to learn more about square-root expressions and start to learn about rational exponents.

▶ **What You Will Learn**

In Chapter 10, you will use what you learned about solving linear inequalities in one variable to find the domain and range for a square-root expression. You will also use what you learned about solving linear and quadratic equations to help you solve equations involving square roots.

The set of real numbers of the form $a + b\sqrt{2}$, where a and b are rational numbers, is a special set of real numbers. In Lesson 10.3, you will see how to work with such numbers and discover that this set is closed under both addition and under multiplication. In Lesson 10.4, your study is extended to numbers of the form $a + b\sqrt{p}$, where a and b are rational numbers.

The last two lessons are somewhat independent of one another. You will see how to extend the concept of exponent so that rational numbers are allowed as exponents. Lastly, you will use your critical-thinking skills to consider the truth of statements about square roots.

VOCABULARY

compound interest	index
conjugate	Property of Equality of Squares
cube-root radical	radicand
Density Property of Real Numbers	rationalizing the denominator
Distance Formula	square-root equation
field	square-root expression
fourth-root radical	square-root function

The diagram below shows how mathematical skills and mathematical reasoning are interrelated with the skills and concepts in Chapter 10. Notice that this chapter involves radical expressions and equations with an emphasis on square-root expressions and equations.

Mathematical Skills

- Finding the domain of a square-root expression (10.1)
- Simplifying products and quotients involving square roots (10.4)

Mathematical Reasoning

- Justifying statements involving square roots (10.6)
- Understanding and writing proofs (10.3, 10.6)

Radical Expressions and Equations

Solving Square-Root Equations (10.2)

Operations on Square-Root Expressions (10.3, 10.4)

Simplifying Expressions Involving Fractional Exponents (10.5)

Solving application problems (10.2, 10.5, 10.6)

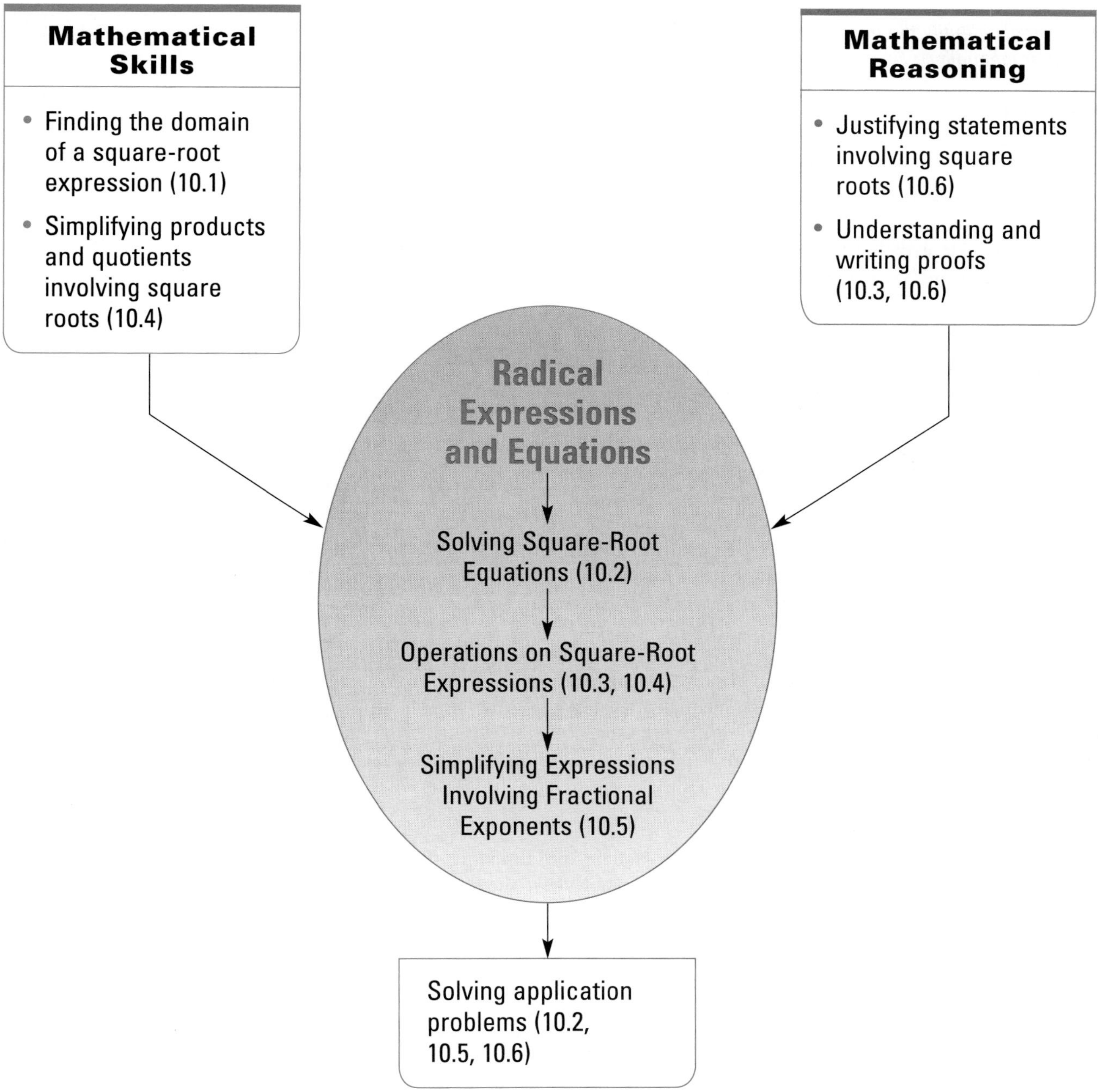

10.1
LESSON

Square-Root Expressions and Functions

Finding the domain of a square-root expression or function

Focus: *Students identify the domain of a square-root expression or function by solving a linear inequality.*

EXAMPLE

PHASE 1: Presenting the Example

Display the expressions shown at right. Have students find the value of each. [**1.** 5 **2.** –5 **3.** 5*i*] Ask them to explain how the values of **1** and **2** differ from the value of **3**. Elicit the response that 5 and –5 are real numbers, while 5*i* is an imaginary number.

1. $\sqrt{25}$

2. $-\sqrt{25}$

3. $\sqrt{-25}$

Display $\sqrt{\boxed{}}$. Tell students that it represents a real number. Ask what values must be excluded from the box. [**all negative real numbers**] Tell them this means that any nonnegative real number may be *included* in the box. Write $\boxed{} \geq 0$ next to the expression, as shown at right below.

Remind students that the set of all possible values of a variable is called the domain of the variable. Tell them to suppose that the box contains the expression $x + 15$. Ask them how they could identify the domain of x. Lead them to see that the domain is described by $x + 15 \geq 0$. Have them solve this inequality. [$x \geq -15$] Point out that the domain is *all real numbers greater than or equal to –15.*

$$\sqrt{\boxed{}} \rightarrow \boxed{} \geq 0$$

$$\sqrt{\boxed{x+15}} \rightarrow \boxed{x+15} \geq 0 \rightarrow x \geq -15$$

$$\sqrt{\boxed{2x}} \rightarrow \boxed{2x} \geq 0 \rightarrow x \geq 0$$

$$\sqrt{\boxed{-2x}} \rightarrow \boxed{-2x} \geq 0 \rightarrow x \leq 0$$

$$\sqrt{\boxed{-2x+15}} \rightarrow \boxed{-2x+15} \geq 0 \rightarrow x \leq 7.5$$

Repeat this activity with $2x$, $-2x$, and $-2x + 15$ in the box. One by one, have students solve the resulting inequality. Display the conclusions as shown at right. The final result is the solution of part a of the Example.

PHASE 2: Providing Guidance for TRY THIS

Be sure students understand that the range of any square-root function will not contain negative real numbers. If graphics calculators are available, have them graph several square-root functions to convince them of this fact.

PHASE 3: ASSESSMENT AND CLOSURE for Lesson 10.1

- For each set of numbers at right, have students write a square-root expression whose domain is that set. [**samples: 1.** $\sqrt{x}$ **2.** $\sqrt{-x}$ **3.** $\sqrt{x-3}$ **4.** $\sqrt{x+6}$

- The lessons that follow will focus on solving square-root equations and performing operations with square-root expressions.

☞ *For a Lesson 10.1 Quiz, see* Assessment Resources *page 102.*

1. all real numbers greater than or equal to 0
2. all real numbers less than or equal to 0
3. all real numbers greater than or equal to 3
4. all real numbers less than or equal to –6

10.1 LESSON — Square-Root Expressions and Functions

▶ *Finding the domain of a square-root expression or function*

A **square-root expression** is an expression that contains a square-root symbol. The expressions $\sqrt{20}$ and $\sqrt{2x-15}$ are examples of square-root expressions. The expression under the square-root symbol, or *radical*, is called the **radicand**. For example, in $\sqrt{2x-15}$, the radicand is $2x-15$.

A **square-root function** is a function defined by a square-root expression. For example, $y = \sqrt{x}$ and $y = 3\sqrt{x-7}$ are square-root functions.

The square root of a negative number is not defined in the set of real numbers. Thus, the domain of $\sqrt{x}$ is the set of all nonnegative real numbers.

The Domain and Range of $y = \sqrt{x}$
The domain of $y = \sqrt{x}$ is the set of all nonnegative real numbers. The range of $y = \sqrt{x}$ is the set of all nonnegative real numbers.

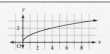

EXAMPLE
a. Find the domain of $\sqrt{-2x+15}$.
b. Find the domain and range of $y = \sqrt{3(x-5)}+1$.

▶ Solution

a. Because the square root of a negative number is not defined in the set of real numbers, the radicand must be greater than or equal to 0. Therefore, solve the inequality $-2x+15 \geq 0$.
$$-2x+15-15 \geq 0-15 \quad \longleftarrow \text{Subtraction Property of Inequality}$$
$$-2x \geq -15$$
$$x \leq \frac{15}{2}, \text{ or } 7.5 \quad \longleftarrow \text{Division Property of Inequality}$$

Reverse the inequality when you divide by a negative number.

The domain of $\sqrt{-2x+15}$ is the set of all real numbers 7.5 or less.

b. The radicand cannot be negative, so solve $3(x-5)+1 \geq 0$.
$$3(x-5)+1 \geq 0$$
$$3x-15+1 \geq 0 \quad \longleftarrow \text{Distributive Property}$$
$$3x \geq 14$$
$$x \geq \frac{14}{3}$$

The domain of $y = \sqrt{3(x-5)}+1$ is $x \geq \frac{14}{3}$. Because y can have any nonnegative value, the range is the set of all nonnegative real numbers.

TRY THIS
a. Find the domain of $\sqrt{3x+5}$.
b. Find the domain and range of $y = \sqrt{2(x-1)}+3$.

EXERCISES

KEY SKILLS

Write the inequality that you would solve to find the domain of each expression or function. Do not solve.

1. $\sqrt{3m}$
2. $\sqrt{2(3t+1)}$
3. $y = \sqrt{3(x-4)}-12$

PRACTICE

Find the domain of each square-root expression.

4. $\sqrt{4x}$
5. $\sqrt{15n}$
6. $\sqrt{-3r}$
7. $\sqrt{-7a}$
8. $\sqrt{2x-9}$
9. $\sqrt{2a+9}$
10. $\sqrt{5(d-3)}$
11. $\sqrt{4(m+5)}$
12. $\sqrt{-3(z+2)}+6$

Find the domain and range of each square-root function.

13. $y = \sqrt{11t}$
14. $y = \sqrt{-x}$
15. $y = \sqrt{3(g+3)}$
16. $y = \sqrt{-2(k-3)}+5$
17. $y = \sqrt{7(s-1)}-2$
18. $y = \sqrt{-(3t-2)}+4$
19. $y = \sqrt{2(x-3)}+2(x+1)$
20. $y = \sqrt{3(t+1)}-(t-1)$
21. $y = \sqrt{-(r+3)}-2(r-3)$

22. Let a and b represent positive real numbers. Find the domain and range of $y = \sqrt{ax+b}$.

23. Find a such that the domain of $y = \sqrt{ax-3}$ is the set of all real numbers greater than or equal to 5.

24. Find the domain and range of $y = \sqrt{2x-9}+3$. Justify your response.

25. **Critical Thinking** Show that all the numbers in the domain of $y = \sqrt{2x-7}$ are also in the domain of $y = \sqrt{2x-5}$. Then show that the domain of $y = \sqrt{2x-5}$ contains numbers that are not in the domain of $y = \sqrt{2x-7}$.

MIXED REVIEW

Solve. Write the solutions in simplest radical form. (8.2 Skill A)

26. $7x^2 = 28$
27. $-6n^2 = -150$
28. $4p^2+3 = 28$
29. $3x^2+5 = 14$
30. $-2a^2+11 = -61$
31. $7v^2-11 = 38$
32. $11 = 11d^2$
33. $24 = 3u^2-3$
34. $-7 = -9t^2+2$

Answers for Lesson 10.1 Skill A

Try This Example
a. $x \geq -\dfrac{5}{3}$

b. domain: $x \geq -\dfrac{1}{2}$
range: $y \geq 0$

1. $3m \geq 0$
2. $2(3t+1) \geq 0$
3. $3(x-4)-12 \geq 0$
4. $x \geq 0$
5. $n \geq 0$
6. $r \leq 0$
7. $a \leq 0$
8. $x \geq \dfrac{9}{2}$
9. $a \geq -\dfrac{9}{2}$
10. $d \geq 3$
11. $m \geq -5$
12. $z \leq 0$
13. domain: $t \geq 0$;
range: $y \geq 0$

14. domain: $x \leq 0$;
range: $y \geq 0$
15. domain: $g \geq -3$;
range: $y \geq 0$
16. domain: $k \leq \dfrac{11}{2}$;
range: $y \geq 0$
17. domain: $s \geq \dfrac{9}{7}$;
range: $y \geq 0$
18. domain: $t \leq 2$;
range: $y \geq 0$
19. domain: $x \geq 1$;
range: $y \geq 0$
20. domain: $t \geq -2$;
range: $y \geq 0$
21. domain: $r \leq 1$;
range: $y \geq 0$
★ 22. domain: $x \geq -\dfrac{b}{a}$;
range: $y \geq 0$
★ 23. $a = \dfrac{3}{5}$

★ 24. domain: $x \geq \dfrac{9}{2}$; range: $y \geq 3$. Since every square root is greater than or equal to 0, y must be at least 3.

25. The domain of $y = \sqrt{2x-7}$ is all real numbers such that $2x-7 \geq 0$, that is, all real numbers such that $x \geq \dfrac{7}{2}$. The domain of $y = \sqrt{2x-5}$ is all real numbers such that $2x-5 \geq 0$, that is, all real numbers such that $x \geq \dfrac{5}{2}$. Thus, the domain of $y = \sqrt{2x-7}$ is contained in the domain of $y = \sqrt{2x-5}$. The number 3 is in the domain of $y = \sqrt{2x-5}$ but not in the domain of $y = \sqrt{2x-7}$.

26. $x = \pm 2$
27. $n = \pm 5$
28. $p = \pm\dfrac{5}{2}$
29. $x = \pm\sqrt{3}$
30. $a = \pm 6$
31. $v = \pm\sqrt{7}$
32. $d = \pm 1$
33. $u = \pm 3$
34. $t = \pm 1$

★ **Advanced Exercises**

TEACHING
10.2
LESSON
Solving Square-Root Equations

SKILL A | *Solving equations in which one side is a square-root expression*

Focus: *Students solve an equation with a square-root expression on one side by squaring each side, and then solving the resulting equation.*

EXAMPLE 1

PHASE 1: Presenting the Example

Display the equations shown at right. Have students solve each one. Display the solutions as shown. Stress the fact that each is a simple one-operation equation that is solved by using the inverse operation. That is: Subtraction "undoes" the addition; addition "undoes" the subtraction; division "undoes" the multiplication; and multiplication "undoes" the division.

1 $k + 4 = 11 \;\rightarrow\; k + 4 - 4 = 11 - 4 \;\rightarrow\; k = 7$

2 $z - 3 = 18 \;\rightarrow\; z - 3 + 3 = 18 + 3 \;\rightarrow\; z = 21$

3 $5b = 85 \qquad\rightarrow\qquad \dfrac{5b}{5} = \dfrac{85}{5} \qquad\rightarrow\; b = 17$

4 $\dfrac{r}{3} = 24 \qquad\rightarrow\qquad 3 \cdot \dfrac{r}{3} = 3 \cdot 24 \qquad\rightarrow\; r = 72$

$\sqrt{x} = 4 \qquad\rightarrow\qquad (\sqrt{x})^2 = 4^2 \qquad\rightarrow\; x = 16$

Now display the equation $\sqrt{x} = 4$. Ask students what operation is involved. **[taking the square root]** Have them name the inverse operation. **[squaring]** Tell them that, just as with the first four equations, $\sqrt{x} = 4$ can be solved by using an inverse operation. That is, squaring each side "undoes" the square root. Display the solution directly beneath the first four equations as shown.

EXAMPLES 2 and 3

PHASE 1: Presenting the Examples

Display $\sqrt{\boxed{}} = 5$. Ask students what the value of the box must be. **[25]** Display this response as shown on the first line at right.

Now tell students to suppose that the box contains $m + 1$. Ask what the value of m must be. Repeat the question for $2m$ and $2m + 1$. Display the results as shown at right. Then discuss Example 2.

$$\sqrt{\boxed{}} = 5 \;\rightarrow\; \boxed{} = 25$$

$$\sqrt{\boxed{m+1}} = 5 \;\rightarrow\; \boxed{m+1} = 25 \;\rightarrow\; m = 24$$

$$\sqrt{\boxed{2m}} = 5 \;\rightarrow\; \boxed{2m} = 25 \;\rightarrow\; m = 12.5$$

$$\sqrt{\boxed{2m+1}} = 5 \;\rightarrow\; \boxed{2m+1} = 25 \;\rightarrow\; m = 12$$

To introduce Example 3, display the equation $\sqrt{x^2 + x - 2} = 2$. Ask students how the radicand differs from the radicands in Example 1. Lead them to see that the radicand is now a quadratic expression rather than a linear expression. Then discuss the solution outlined on the textbook page.

PHASE 2: Providing Guidance for TRY THIS

Watch for students who apply a property of equality to part of a radicand, as shown at right with the Example 2 Try This exercise. Stress the fact that the square-root symbol is a grouping symbol. So a property of equality must be applied to the entire square-root expression, not just part of it.

$$\sqrt{3(x-5)} - 2 = 7$$
$$\sqrt{3(x-5)} - 2 + 2 = 7 + 2$$
$$\sqrt{3(x-5)} = 9 \qquad X$$

---

392 Chapter 10 Radical Expressions, Equations, and Functions

10.2 Solving Square-Root Equations
LESSON

SKILL A   *Solving equations in which one side is a square-root expression*

Substitute the given values of x into the equation. Are the values of x solutions to the given square-root equation?

A **square-root equation** is an equation that contains a square-root expression. For example, $\sqrt{x} = 3$ and $\sqrt{x^2 + 2x} = 1$ are square-root equations. To solve such equations, apply the following property.

1. $x = 4$ or $x = -1$
$\sqrt{x^2 - 3x} = 2$

2. $x = -3$ or $x = 12$
$\sqrt{x^2 - 9x} = 6$

3. $x = 4$ or $x = 1$
$\sqrt{x^2 - 5x} = -2$

Property of Equality of Squares
If $r = s$, then $r^2 = s^2$. In words, if two expressions r and s are equal, then their squares, r^2 and s^2, are also equal.

Solve each square-root equation. Check your solution.

EXAMPLE 1   Solve $\sqrt{x} = 4$.

4. $\sqrt{t} = 6$
5. $\sqrt{c} = 9$
6. $10 = \sqrt{z}$

> **Solution**
> $$\sqrt{x} = 4$$
> $$\left(\sqrt{x}\right)^2 = 4^2 \longleftarrow \text{ Square each side of the equation.}$$
> $$x = 16 \longleftarrow \left(\sqrt{x}\right)^2 = x \text{ and } 4^2 = 16$$

7. $15 = \sqrt{d}$
8. $10 = 2\sqrt{d}$
9. $-6 = -3\sqrt{d}$

10. $\sqrt{3x - 1} = 2$
11. $\sqrt{5n - 7} = 2$
12. $\sqrt{5n} = 5$

TRY THIS   Solve each square-root equation.   **a.** $\sqrt{v} = \frac{1}{2}$ **b.** $2.5 = \sqrt{m}$

13. $\sqrt{-3p} = 2$
14. $\sqrt{-3(k + 1)} - 5 = 2$
15. $\sqrt{2(z + 1)} - z = 5$

EXAMPLE 2   Solve $\sqrt{2(c - 2)} + 3 = 5$. Check your solution.

16. $3\sqrt{5k + 1} = 27$
17. $-4\sqrt{4n - 3} = -12$
18. $1 = \sqrt{-3(s - 3)} + (s - 1)$

> **Solution**
> $$\sqrt{2(c - 2)} + 3 = 5$$
> $$2(c - 2) + 3 = 25 \longleftarrow \text{ Square each side.}$$
> $$c = 13 \longleftarrow \text{ Solve for } c.$$
>
> **Check:** Replace c with 13.
> $$\sqrt{2(13 - 2)} + 3 \overset{?}{=} 5$$
> $$5 \overset{?}{=} 5 ✔$$

19. $5 = \sqrt{4(b - 1) - 2(b - 2)}$ **20.** $\sqrt{x^2 + 3} = 2$
21. $\sqrt{x^2 - 9} = 4$

22. $\sqrt{(v - 1)^2 + 84} = 10$ **23.** $5 = \sqrt{(u + 3)^2 - 11}$
24. $\sqrt{g^2 - 2g + 1} = 2$

25. $\sqrt{z^2 - 3z - 3} = 5$ **26.** $6 = \sqrt{n^2 + n + 6}$
27. $4 = \sqrt{q^2 - 4q + 19}$

TRY THIS   Solve $\sqrt{3(x - 5)} - 2 = 7$. Check your solution.

28. Let a represent a real number. Solve $\sqrt{x^2 - a^2} = 1$ for x.

EXAMPLE 3   Solve $\sqrt{x^2 + x - 2} = 2$. Check your solution.

29. Let b represent a real number. Solve $\sqrt{x^2 + b^2} = 1$ for x.

> **Solution**
> $$\sqrt{x^2 + x - 2} = 2$$
> $$x^2 + x - 2 = 4 \longleftarrow \text{ Square each side.}$$
> $$x^2 + x - 6 = 0$$
> $$x = -3 \text{ or } x = 2 \longleftarrow \text{ Apply the Zero-Product Property and solve for } x.$$
>
> **Check:** Replace x with -3.
> $$\sqrt{(-3)^2 + (-3) - 2} \overset{?}{=} 2$$
> $$2 \overset{?}{=} 2 ✔$$
> Replace x with 2.
> $$\sqrt{2^2 + 2 - 2} \overset{?}{=} 2$$
> $$2 \overset{?}{=} 2 ✔$$

30. Critical Thinking   Let n represent a real number. For what value(s) of n will $\sqrt{x^2 - (x + n)^2} = 5$ not have a solution? Justify your response.

TRY THIS   Solve $\sqrt{x^2 - 11x + 19} = 3$. Check your solution.

Multiply. (6.6 Skill B)

31. $(x + 1)(2x - 5)$ **32.** $(3y - 2)(y + 7)$ **33.** $(2n - 4)(2n + 2)$
34. $(1 + bx)(1 + ax)$ **35.** $(1 + bx)(1 - ax)$ **36.** $(1 - bx)(1 + ax)$
37. $(a + bx)(a + bx)$ **38.** $(a - bx)(a + bx)$ **39.** $(a + bx)(a - bx)$

Answers for Lesson 10.2 Skill A

Try This Example 1
a. $v = \frac{1}{4}$ **b.** $m = 6.25$

Try This Example 2
$x = 22$

Try This Example 3
$x = 1$ or $x = 10$

1. yes
2. yes
3. no
4. $t = 36$
5. $c = 81$
6. $z = 100$
7. $d = 225$
8. $d = 25$
9. $d = 4$
10. $x = \frac{5}{3}$
11. $n = \frac{11}{5}$
12. $n = 5$

13. $p = -\frac{4}{3}$
14. $k = -4$
15. $z = 23$
16. $k = 16$
17. $n = 3$
18. $s = \frac{7}{2}$
19. $b = \frac{25}{2}$
20. $x = -1$ or $x = 1$
21. $x = -5$ or $x = 5$
22. $v = -3$ or $v = 5$
23. $u = -9$ or $u = 3$
24. $g = -1$ or $g = 3$
25. $z = -4$ or $z = 7$
26. $n = -6$ or $n = 5$
27. $q = 1$ or $q = 3$
28. $x = \pm\sqrt{a^2 + 1}$
29. $x = \pm\sqrt{1 - b^2}$

30. $x^2 - (x + n)^2 = -2xn - n^2$;
If $\sqrt{x^2 - (x + n)^2} = 5$, then
$-2xn - n^2 = 25$.
If $-2xn - n^2 = 25$, then
$x = \frac{n^2 + 25}{-2n}$. This equation has a
solution except when $n = 0$.
31. $2x^2 - 3x - 5$
32. $3y^2 + 19y - 14$
33. $4n^2 - 4n - 8$
34. $abx^2 + ax + bx + 1$
35. $-abx^2 - ax + bx + 1$
36. $-abx^2 + ax - bx + 1$
37. $b^2x^2 + 2abx + a^2$
38. $a^2 - b^2x^2$
39. $a^2 - b^2x^2$

Focus: *Students use square-root equations to solve problems involving distance to the horizon. They also learn the formula for the distance between two points in the coordinate plane and apply the formula to a real-world setting.*

EXAMPLE 1

PHASE 1: Presenting the Example

Many students are fascinated by the formula discussed in Example 1, and they are usually surprised to see how easily it is obtained. If time permits, you may wish to discuss the following derivation with your students.

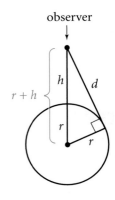

Display the figure at right. It depicts an observer at an altitude, h, above Earth's surface. The radius of Earth is r, so $r + h$ is the observer's distance above the center of Earth. The distance that the observer can see to the horizon is d. At the horizon, the observer's line of sight forms a right angle with Earth's radius. So a right triangle is formed, and you can apply the Pythagorean Theorem as shown below the figure.

Focus on the final equation below the figure, which is $d = \sqrt{2rh + h^2}$. Note that, in the problems presented in this lesson, the value of h is very, very small when compared to the radius of Earth. In cases like this it is permissible to omit the term h^2 in the radicand, thus obtaining the formula $d \approx \sqrt{2rh}$.

Point out that an approximation for the radius of Earth is 6400 kilometers. Substitute this value into $d \approx \sqrt{2rh}$ as follows:

$$r^2 + d^2 = (r + h)^2$$
$$r^2 + d^2 = r^2 + 2rh + h^2$$
$$d^2 = 2rh + h^2$$
$$d = \sqrt{2rh + h^2}$$

$$d \approx \sqrt{2 \cdot 6400 \cdot h} \quad \rightarrow \quad d \approx \sqrt{12{,}800 \cdot h} \quad \rightarrow \quad d \approx \sqrt{12{,}800} \cdot \sqrt{h} \quad \rightarrow \quad d \approx 113.14\sqrt{h}$$

This is the formula that is used in Example 1.

EXAMPLE 2

PHASE 1: Presenting the Example

Display the right triangle shown at right. Ask students to write an equation that describes the relationship among the lengths of the sides. [$a^2 + b^2 = c^2$] Remind them that this relationship is called the Pythagorean Theorem. Have them solve the equation for c. [$c = \sqrt{a^2 + b^2}$]

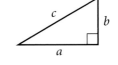

Now display the graph at right. It shows a segment with endpoints $P(x_1, y_1)$ and $Q(x_2, y_2)$. Point out that segment PQ together with the dashed segments forms a right triangle. The lengths of the dashed segments are $x_2 - x_1$ and $y_2 - y_1$, as labeled. Substituting these values into the formula derived for the triangle above, $PQ = \sqrt{(x_2 - x_1)^2 + (y_2 - y_1)^2}$. Tell students that this is called the *Distance Formula*. Then discuss Example 2, which applies this formula to a real-world situation.

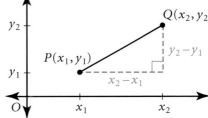

You can use square-root equations to solve distance problems.

EXAMPLE 1

From an altitude of h kilometers, the approximate distance d in kilometers you can see to the horizon is given by $d = 113.14\sqrt{h}$. To see 25 kilometers to the horizon, how high up will you need to be?

▸ **Solution**

$$d = 113.14\sqrt{h} \quad \rightarrow \quad 25 = 113.4\sqrt{h} \qquad \longleftarrow \text{Replace } d \text{ with 25.}$$
$$\frac{25}{113.14} = \sqrt{h}$$
$$h = \left(\frac{25}{113.14}\right)^2 \approx 0.05 \qquad \longleftarrow \begin{array}{l}\text{Square each side of the equation. Use a}\\\text{calculator to find the approximate value of } h.\end{array}$$

You will need to be about 0.05 kilometers, or 50 meters, above Earth.

TRY THIS Rework Example 1 if the desired distance is to be 150 kilometers.

If you apply the Pythagorean Theorem to find the distance between two points $P(x_1, y_1)$ and $Q(x_2, y_2)$ in the coordinate plane, you can derive what is called the *Distance Formula*.

The Distance Formula

If $P(x_1, y_1)$ and $Q(x_2, y_2)$ are two points in the coordinate plane, then the distance between them is given by the formula below.

$$PQ = \sqrt{(x_2 - x_1)^2 + (y_2 - y_1)^2}$$

EXAMPLE 2

Sylvia is 4 miles due east of City Hall. How far due north or due south must she travel so that her distance from City Hall will be 5 miles?

▸ **Solution**

Make a graph. Let City Hall be at the origin of a coordinate grid. Since Sylvia is 4 miles east of City Hall, let $P(4, 0)$ be her location. Let points Q and Q' be y miles due north and due south of Sylvia's location, $P(4, 0)$. Apply the Distance Formula to find the distance y that Sylvia must travel.

Use a formula.
$$5 = \sqrt{(4 - 0)^2 + (y - 0)^2}$$
$$5 = \sqrt{16 + y^2}$$
$$25 = 16 + y^2 \qquad \longleftarrow \begin{array}{l}\text{Because distance cannot}\\\text{be negative, reject } -3 \text{ as}\\\text{an answer.}\end{array}$$
$$y = 3 \quad \text{or} \quad y = -3$$

Sylvia must travel 3 miles due north or due south of her present location.

TRY THIS Rework Example 2 if the desired distance is to be 10 miles.

EXERCISES

〔 **KEY SKILLS** 〕

Use the distance formula to write an equation for the distance between points P and Q. Do not solve.

1. $P(0, 4)$, $Q(1, 7)$ **2.** $P(2, 0)$, $Q(3, 1)$

3. $P(-1, -1)$, $Q(3, 5)$ **4.** $P(6, 1)$, $Q(4, 4)$

〔 **PRACTICE** 〕

Refer to Example 1 on the previous page. Find each altitude needed to see the given distance to the horizon. Round each altitude to the nearest hundredth of a kilometer.

5. 30 kilometers   **6.** 24 kilometers

7. 36 kilometers   **8.** 154 kilometers

Find the coordinates of Q and Q' that meet the specified conditions.

9. 12 units from $O(0, 0)$ and directly above or below $P(4, 0)$

10. 15 units from $O(0, 0)$ and directly left or right of $P(0, 4)$

11. 8 units from $C(2, 2)$ and directly above or below $Z(7, 2)$, as shown in the graph

12. 8 units from $K(4, 1)$ and directly left or right of $L(4, 6)$, as shown in the graph

13. Solve the equation in Example 1 for altitude h.

〔 **MIXED REVIEW APPLICATIONS** 〕

If each student can complete a class project in the given amount of time, how long will it take the two students to complete the project if they work together? Round answers to the nearest whole unit. (9.7 Skill C)

14. John: 20 hours; Viola: 24 hours   **15.** Carol: 7 days; Carmen: 14 days

16. Ken: 34 hours; Reema: 48 hours   **17.** Jackie: 10 days; Bill: 12 days

---

Answers for Lesson 10.2 Skill B

Try This Example 1
about 1.76 kilometers

Try This Example 2
about 9.2 miles north or 9.2 miles south

1. $PQ = \sqrt{(1 - 0)^2 + (7 - 4)^2}$

2. $PQ = \sqrt{(3 - 2)^2 + (1 - 0)^2}$

3. $PQ = \sqrt{[3 - (-1)]^2 + [5 - (-1)]^2}$

4. $PQ = \sqrt{(4 - 6)^2 + (4 - 1)^2}$

5. about 0.07 kilometer

6. about 0.04 kilometer

7. about 0.10 kilometer

8. about 1.85 kilometers

9. about 11.3 units above or below $P(4, 0)$; $Q(4, 11.3)$ and $Q'(4, -11.3)$

10. about 14.5 units left or right of $P(0, 4)$; $Q(14.5, 4)$ and $Q'(-14.5, 4)$

11. about 6.2 units above or below $Z(7, 2)$; $Q(7, 8.2)$ and $Q'(7, -4.2)$

12. about 6.2 units left or right from $L(4, 6)$; $Q(10.2, 6)$ and $Q'(-2.2, 6)$

13. $h = \left(\dfrac{d}{113.14}\right)^2$

14. 11 hours

15. 5 days

16. 20 hours

17. 5 days

Focus: *Students solve equations with square-root expressions on both sides by squaring each side, and then solving the resulting equation.*

EXAMPLES 1 AND 2

PHASE 1: Presenting the Examples

Display the equation $\sqrt{\boxed{}} = \sqrt{\overline{\big(\big)}}$. Ask students how the form of this equation differs from the form of the equations that they have solved before. Lead them to see that this equation has a square-root expression on each side; in previous equations, there was a square root expression on just one side.

Ask students to describe the relationship between the value of the box and the value of the oval. Elicit the response that the values are equal. Display this response as shown on the first line at right below. Stress the fact that this conclusion is justified by the Property of Equality of Squares.

Now tell students to suppose that the box contains h and the oval contains 8. Ask what the value of h must be. Repeat the question with $4h$ in the box and 8 in the oval; $4h$ in the box and $8h$ in the oval; and with $4h + 10$ in the box and $8h$ in the oval. Display the results as shown at right. Then discuss Example 1.

$$\sqrt{\boxed{}} = \sqrt{\overline{\big(\big)}} \rightarrow \boxed{} = \overline{\big(\big)}$$

$$\sqrt{\boxed{h}} = \sqrt{\overline{\big(8\big)}} \rightarrow \boxed{h} = \overline{\big(8\big)} \rightarrow h = 8$$

$$\sqrt{\boxed{4h}} = \sqrt{\overline{\big(8\big)}} \rightarrow \boxed{4h} = \overline{\big(8\big)} \rightarrow h = 2$$

$$\sqrt{\boxed{4h}} = \sqrt{\overline{\big(8h\big)}} \rightarrow \boxed{4h} = \overline{\big(8h\big)} \rightarrow h = 0$$

$$\sqrt{\boxed{4h + 10}} = \sqrt{\overline{\big(8h\big)}} \rightarrow \boxed{4h + 10} = \overline{\big(8h\big)} \rightarrow h = 2.5$$

To introduce Example 2, repeat the activity with the following values: box $= s^2$ and oval $= 36$; box $= s^2 + 11$ and oval $= 36$; box $= s^2 + 9s$ and oval $= 36$. [**respectively: $s = -6$ or $s = 6$; $s = -5$ or $s = 5$; $s = -12$ or $s = 3$**] Then discuss Example 2.

PHASE 2: Providing Guidance for **TRY THIS**

Point out the discussion of extraneous solutions that appears at the bottom of the textbook page. Once again remind students that it is important to check all proposed solutions *in the original equation.*

PHASE 3: ASSESSMENT AND CLOSURE for Lesson 10.2

- Give students the two columns of equations at right. Have them match each equation in column I with an equation in column II that has *exactly* the same set of solutions. [**1.** C **2.** A **3.** B **4.** D]

- Students will learn to solve more complex square-root equations and other types of *radical equations* if they pursue an advanced algebra course.

I	**II**
1. $\sqrt{x^2} = 2$	**A.** $\sqrt{x^2} = \sqrt{2x}$
2. $\sqrt{2x} = x$	**B.** $\sqrt{x + 2} = 2$
3. $\sqrt{x + 2} = x$	**C.** $\sqrt{x^2} = \sqrt{4}$
4. $\sqrt{x + 2} = \sqrt{x^2}$	**D.** $\sqrt{x^2 - x} = \sqrt{2}$

☞ *For a Lesson 10.2 Quiz, see* Assessment Resources *page 103.*

A square-root equation may have a square-root expression on both sides of the equal sign. To solve such an equation, apply the Property of Equality of Squares; that is, square each side of the equation. Then solve the equation that results.

EXAMPLE 1 Solve $\sqrt{3x + 5} = \sqrt{7x - 31}$. Check your solution.

▶ Solution

$$\sqrt{3x + 5} = \sqrt{7x - 31}$$
$$3x + 5 = 7x - 31 \quad \longleftarrow \text{Square each side of the equation.}$$
$$x = 9 \quad \longleftarrow \text{Solve for } x.$$

Check: $\sqrt{3(9) + 5} \stackrel{?}{=} \sqrt{7(9) - 31}$

$$\sqrt{32} \stackrel{?}{=} \sqrt{32} \checkmark$$

TRY THIS   Solve $\sqrt{2(t - 2) + 1} = \sqrt{t + 5}$. Check your solution.

EXAMPLE 2 Solve $\sqrt{x^2 - 2x} = \sqrt{-x + 20}$. Check your solutions.

▶ Solution

$$\sqrt{x^2 - 2x} = \sqrt{-x + 20}$$
$$x^2 - 2x = -x + 20 \quad \longleftarrow \text{Square each side of the equation.}$$
$$x^2 - x - 20 = 0 \quad \longleftarrow \text{Solve a quadratic equation.}$$
$$(x - 5)(x + 4) = 0$$
$$x = 5 \text{ or } x = 4$$

Check: Replace x with 5.   Replace x with -4.

$\sqrt{5^2 - 2(5)} \stackrel{?}{=} \sqrt{-5 + 20}$ $\sqrt{(-4)^2 - 2(-4)} \stackrel{?}{=} \sqrt{-(-4) + 20}$

$\sqrt{15} \stackrel{?}{=} \sqrt{15} \checkmark$ $\sqrt{24} \stackrel{?}{=} \sqrt{24} \checkmark$

TRY THIS   Solve $\sqrt{x^2 + 2x} = \sqrt{7x - 4}$. Check your solution(s).

- A square-root equation may have no solution in the set of real numbers. The equation $\sqrt{x} + 1 = 0$, for example, has no real solution. (However, recall from Lesson 8.2 that $\sqrt{x} + 1 = 0$ *can* be solved using imaginary numbers.)
- A square-root equation may have an *extraneous solution*, a solution that does not satisfy the given equation. The equation $\sqrt{x} = x - 2$ has the solutions 1 and 4 when you solve it by squaring each side of the equation. However, when you substitute both values back into the original equation, you will find that 1 is *not* a solution to the given equation.

$\sqrt{4} \stackrel{?}{=} 4 - 2$ $\sqrt{1} \stackrel{?}{=} 1 - 2$
$2 \stackrel{?}{=} 2 \checkmark$ $1 \stackrel{?}{=} -1 ✗$

EXERCISES

KEY SKILLS

Write the equation that results from squaring each side of the given equation. Identify the equation as linear or quadratic. Do not solve.

1. $\sqrt{3(x - 2) + 5} = \sqrt{3x - 1}$ 2. $\sqrt{(s - 1)^2 + 1} = \sqrt{2s^2}$ 3. $\sqrt{r^2 - 2r} = \sqrt{3r}$

PRACTICE

Solve each square-root equation. Check your solution.

4. $\sqrt{3d} = \sqrt{-2d}$ 5. $\sqrt{3a} = \sqrt{a - 5}$
6. $\sqrt{x} = \sqrt{2x - 3}$ 7. $\sqrt{b + 2} = \sqrt{2 - b}$
8. $\sqrt{3n + 2} = \sqrt{4n}$ 9. $\sqrt{4(k - 1)} = \sqrt{5k}$
10. $\sqrt{8} = \sqrt{5(z + 2)}$ 11. $\sqrt{4(g + 1)} - 5 = \sqrt{6 - (g - 2)}$
12. $\sqrt{3} - 2(v - 2) = \sqrt{6v - 2}$ 13. $\sqrt{n^2} = \sqrt{4n - 4}$
14. $\sqrt{k^2} = \sqrt{6k - 9}$ 15. $\sqrt{z^2} = \sqrt{-8z - 16}$
16. $\sqrt{p^2 + 25} = \sqrt{-10p}$ 17. $\sqrt{a^2 + 12} = \sqrt{8a}$
18. $\sqrt{c^2 - 15} = \sqrt{-2c}$ 19. $\sqrt{x^2 - 10x + 10} = \sqrt{x - 14}$

Show that each square-root equation has no solution.

20. $\sqrt{x - 3} = \sqrt{x - 4}$ 21. $\sqrt{3t + 12} = \sqrt{3(4 + t) - 1}$

In Exercises 22–24, let a, b, c, and d represent real numbers. Solve each equation for x.

22. $\sqrt{ax} = \sqrt{b + c}$ 23. $\sqrt{ax + b} = \sqrt{cx + d}$ 24. $\sqrt{ax^2 + b} = \sqrt{bx^2 + a}$

25. **Critical Thinking** Triangle OAB has vertices $O(0, 0)$, $A(6, r)$, and $B(r, 6)$ for some nonzero real number r. Show that the triangle always has two sides that are equal in length.

26. Solve $\sqrt{n + 1} = 1 + \sqrt{n - 12}$. Check your solution(s).

MIXED REVIEW

Write each expression in simplest radical form. (8.1 Skill B)

27. $\sqrt{147}$ 28. $\sqrt{320}$ 29. $\sqrt{891}$ 30. $\sqrt{625}$
31. $\sqrt{\dfrac{3}{100}}$ 32. $\sqrt{\dfrac{7}{225}}$ 33. $\sqrt{\dfrac{98}{121}}$ 34. $\sqrt{\dfrac{75}{169}}$

Answers for Lesson 10.2 Skill C

Try This Example 1
$t = 8$

Try This Example 2
$x = 1$ or $x = 4$

1. $3(x - 2) + 5 = 3x - 1$; linear
2. $(s - 1)^2 + 1 = 2s^2$; quadratic
3. $r^2 - 2r = 3r$; quadratic
4. $d = 0$
5. no real solution
6. no real solution
7. $b = 0$
8. $n = 2$
9. no real solution
10. $z = -\dfrac{2}{5}$
11. $g = \dfrac{9}{5}$
12. $v = \dfrac{9}{8}$
13. $n = 2$
14. $k = 3$

15. $z = -4$
16. $p = -5$
17. $a = 2$ or $a = 6$
18. $c = -5$
19. no real solution
20. If $\sqrt{x - 3} = \sqrt{x - 4}$, then $x - 3 = x - 4$. This implies that $-3 = -4$. Since this is impossible, $\sqrt{x - 3} = \sqrt{x - 4}$ has no solution.
21. If $\sqrt{3t + 12} = \sqrt{3(4 + t) - 1}$, then $3t + 12 = 3(4 + t) - 1$. This implies that $3t + 12 = 3t + 11$, and that $12 = 11$. Since this is impossible, $\sqrt{3t + 12} = \sqrt{3(4 + t) - 1}$ has no solution.
★ 22. $x = \dfrac{b + c}{a}$; $a \neq 0$
★ 23. $x = \dfrac{d - b}{a - c}$; $a \neq c$
★ 24. $x = \pm 1$; $a \neq b$

25. $OA = \sqrt{(6 - 0)^2 + (r - 0)^2} = \sqrt{36 + r^2}$;
$OB = \sqrt{(r - 0)^2 + (6 - 0)^2} = \sqrt{r^2 + 36}$;
Since $36 + r^2 = r^2 + 36$ for all real numbers r, then $\sqrt{36 + r^2} = \sqrt{r^2 + 36}$ for all real numbers r. Thus, $OA = OB$ for all nonzero values of r. This implies that two sides of triangle OAB are equal in length.
★ 26. $n = 48$
27. $7\sqrt{3}$
28. $8\sqrt{5}$
29. $9\sqrt{11}$
30. 25
31. $\dfrac{\sqrt{3}}{10}$

For answers to Exercises 32–34, see Additional Answers.

★ **Advanced Exercises**

TEACHING

10.3 LESSON Numbers of the Form $a + b\sqrt{2}$

SKILL A *Adding and subtracting numbers of the form $a + b\sqrt{2}$*

Focus: *Students find sums and differences of real numbers of the form $a + b\sqrt{2}$ by combining like parts of these numbers.*

EXAMPLES 1 and 2

PHASE 1: Presenting the Examples

Ask students what it means for a set of numbers to be *closed* under an operation. [**When that operation is performed on any two numbers in the set, the result is also in the set.**] Display the set {–1, 0, 1}. Ask students to write an addition to show that this set is *not* closed under addition. [$\mathbf{1 + 1 = 2}$ **or** $\mathbf{-1 + (-1) = -2}$]

Display the set of integers as {…, −3, −2, −1, 0, 1, 2, 3, …}. Ask students to explain why this set *is* closed under addition. [**The sum of any two integers is always another integer.**]

Tell students that they are now going to examine another subset of the real numbers that is closed under addition. Have them read the description of the set $a + b\sqrt{2}$ at the top of the textbook page. Be sure they note that a and b must be rational numbers.

To teach students how to add two numbers of this form, draw a parallel to addition of like terms. Display the variable expression $(3 + 5z) + (-5 + 7z)$ and have students simplify. Display the simplification as shown below.

$$(3 + 5z) + (-5 + 7z) = [3 + (-5)] + (5z + 7z) = -2 + (5 + 7)z = -2 + 12z$$

Now display the square-root expression given in Example 1, which is $(3 + 5\sqrt{2}) + (-5 + 7\sqrt{2})$. Tell students that, just as $5z$ represents $5 \cdot z$ and $7z$ represents $7 \cdot z$, so too $5\sqrt{2}$ represents $5 \cdot \sqrt{2}$ and $7\sqrt{2}$ represents $7 \cdot \sqrt{2}$. Ask them to simplify the square-root expression. Display the simplification directly beneath the simplification of the variable expression, as shown.

$$\begin{aligned}(3 + 5z) + (-5 + 7z) &= [3 + (-5)] + (5z + 7z) = -2 + (5 + 7)z = -2 + 12z\\(3 + 5\sqrt{2}) + (-5 + 7\sqrt{2}) &= [3 + (-5)] + (5\sqrt{2} + 7\sqrt{2}) = -2 + (5 + 7)\sqrt{2} = -2 + 12\sqrt{2}\end{aligned}$$

To introduce Example 2, have students first simplify $(3 + 5t) - (-4 + 6t)$. [$\mathbf{7 - t}$] Then show a parallel simplification of $(3 + 5\sqrt{2}) - (-4 + 6\sqrt{2})$.

PHASE 2: Providing Guidance for TRY THIS

Some students may perceive these additions and subtractions simply as a maze of symbols. It may help them to visually rearrange the expressions using the vertical format shown at right. When rewriting a subtraction, however, caution them to write the opposite of each term in the expression being subtracted.

Example 1
Try This Exercise a:

$$\begin{array}{r}-2 + 5\sqrt{2}\\ \underline{2 - 3\sqrt{2}}\\ 2\sqrt{2}\end{array}$$

Example 2
Try This Exercise a:

$$\begin{array}{r}3 + 5\sqrt{2}\\ \underline{-7 + 3\sqrt{2}}\\ -4 + 8\sqrt{2}\end{array}$$

10.3

LESSON

Numbers of the form $a + b\sqrt{2}$

SKILL A   *Adding and subtracting numbers of the form $a + b\sqrt{2}$*

If a and b are rational numbers, then all numbers of the form $a + b\sqrt{2}$ are members of a subset of the real numbers. This subset of the real numbers follows all of the basic properties of real numbers for addition and multiplication that you learned in Lesson 1.2 Skill A: Closure, Associative, Identity, and Inverse.

To add two numbers of the form $a + b\sqrt{2}$, group like terms.

EXAMPLE 1   Simplify $\left(3 + 5\sqrt{2}\right) + \left(-5 + 7\sqrt{2}\right)$.

▶ Solution

$$\left(3 + 5\sqrt{2}\right) + \left(-5 + 7\sqrt{2}\right) = (3 + (-5)) + \left(5\sqrt{2} + 7\sqrt{2}\right)$$
$$= -2 + (5 + 7)\sqrt{2}$$
$$= -2 + 12\sqrt{2}$$

TRY THIS   Simplify.   **a.** $\left(-2 + 5\sqrt{2}\right) + \left(2 + (-3)\sqrt{2}\right)$ **b.** $\left(4 + 5\sqrt{2}\right) + \left(3 + (-7)\sqrt{2}\right)$

Every number of the form $a + b\sqrt{2}$, has an opposite, or additive inverse, $(-a) + (-b)\sqrt{2}$. To subtract one number of the form $a + b\sqrt{2}$ from another number of this form, add the opposite.

EXAMPLE 2   Simplify $\left(3 + 5\sqrt{2}\right) - \left(-4 + 6\sqrt{2}\right)$.

▶ Solution

$$\left(3 + 5\sqrt{2}\right) - \left(-4 + 6\sqrt{2}\right) = \left(3 + 5\sqrt{2}\right) + \left(4 - 6\sqrt{2}\right) \longleftarrow \text{Write subtraction as}$$
$$= (3 + 4) + \left(5\sqrt{2} - 6\sqrt{2}\right) \quad\text{addition of the opposite.}$$
$$= 7 + (5 - 6)\sqrt{2}$$
$$= 7 - \sqrt{2}$$

TRY THIS   Simplify.   **a.** $\left(3 + 5\sqrt{2}\right) - \left(7 - 3\sqrt{2}\right)$ **b.** $\left(7 + 5\sqrt{2}\right) - \left(2 + 5\sqrt{2}\right)$

The sum of two numbers of the form $a + b\sqrt{2}$ is another number of the form $a + b\sqrt{2}$. Thus, the set of numbers of the form $a + b\sqrt{2}$ is *closed* under addition. That is, the set follows the Closure Property of Addition. For example, the sum $\left(3 + 5\sqrt{2}\right) + \left(-5 + 7\sqrt{2}\right)$ equals $-2 + 12\sqrt{2}$, which is another number of the form $a + b\sqrt{2}$.

EXERCISES

KEY SKILLS

Simplify each number.

1. $\frac{3}{4} + 0\sqrt{2}$ 2. $0 + 5\sqrt{2}$ 3. $0 + 1\sqrt{2}$ 4. $0 - 1\sqrt{2}$

PRACTICE

Simplify. Write answers in the form $a + b\sqrt{2}$.

5. $3\sqrt{2} + 2\sqrt{2}$ 6. $3\sqrt{2} - 2\sqrt{2}$

7. $7\sqrt{2} + (-2)\sqrt{2}$ 8. $4\sqrt{2} - (-3)\sqrt{2}$

9. $-11\sqrt{2} + (-9)\sqrt{2}$ 10. $-\sqrt{2} - (-13)\sqrt{2}$

11. $\left(7 + 3\sqrt{2}\right) + \left(5 + \sqrt{2}\right)$ 12. $\left(4 - 4\sqrt{2}\right) + \left(-7 + 2\sqrt{2}\right)$

13. $\left(5 + 4\sqrt{2}\right) - \left(-3 + 4\sqrt{2}\right)$ 14. $\left(5 + 4\sqrt{2}\right) - \left(5 - 4\sqrt{2}\right)$

15. $\left(5 + 4\sqrt{2}\right) + \left(-5 - 4\sqrt{2}\right)$ 16. $\left(7 + 4\sqrt{2}\right) + \left(-7 - 3\sqrt{2}\right)$

17. $\left(-7 - 5\sqrt{2}\right) - \left(7 - 3\sqrt{2}\right)$ 18. $\left(-11 - 5\sqrt{2}\right) + \left(5 - 11\sqrt{2}\right)$

19. $\left(3 - 3\sqrt{2}\right) + \left(5 - 5\sqrt{2}\right) + \left(2 - 2\sqrt{2}\right)$ 20. $\left(3 - 5\sqrt{2}\right) + \left(6 - 3\sqrt{2}\right) - \left(7 - 3\sqrt{2}\right)$

Two expressions $a + b\sqrt{2}$ and $c + d\sqrt{2}$ are *equal* if $a = c$ and $b = d$. Find r and s such that each equation is true.

21. $\left(3 + 7\sqrt{2}\right) + \left(r + s\sqrt{2}\right) = -1 + 5\sqrt{2}$

22. $\left(6 - 5\sqrt{2}\right) - \left(r + s\sqrt{2}\right) = 7 + 2\sqrt{2}$

23. $\left[(r + s) - (r - s)\sqrt{2}\right] + (r + s)\sqrt{2} = 6 + 4\sqrt{2}$

24. $\left(r^2 + 4\right) + \left(s^2 + 9\right)\sqrt{2} = -4r + 6s\sqrt{2}$

Let a and b represent rational numbers. Prove each statement using properties of addition, multiplication, and equality.

25. $\left(a + b\sqrt{2}\right) + \left[(-a) + (-b)\sqrt{2}\right] = 0$ 26. $\left(a + b\sqrt{2}\right) + \left(a - b\sqrt{2}\right)$ is a rational number.

MIXED REVIEW

Multiply. (6.6 Skill B)

27. $(2 + 3a)(4 + 2a)$ 28. $(-2 + 3a)(-5 + 2a)$ 29. $(-2 - 2a)(-3 + 2a)$

30. $(5 + 3b)(5 + 3b)$ 31. $(1 + b)(-2 + b)$ 32. $(3 - 4b)(-1 + 3b)$

Answers for Lesson 10.3   Skill A

Try This Example 1
a. $2\sqrt{2}$ **b.** $7 - 2\sqrt{2}$

Try This Example 2
a. $-4 + 8\sqrt{2}$ **b.** 5

1. $\frac{3}{4}$

2. $5\sqrt{2}$

3. $\sqrt{2}$

4. $-\sqrt{2}$

5. $5\sqrt{2}$

6. $\sqrt{2}$

7. $5\sqrt{2}$

8. $7\sqrt{2}$

9. $-20\sqrt{2}$

10. $12\sqrt{2}$

11. $12 + 4\sqrt{2}$

12. $-3 - 2\sqrt{2}$

13. 8

14. $8\sqrt{2}$

15. 0

16. $\sqrt{2}$

17. $-14 - 2\sqrt{2}$

18. $-6 - 16\sqrt{2}$

19. $10 - 10\sqrt{2}$

20. $2 - 5\sqrt{2}$

21. $r = -4, s = -2$

22. $r = -1, s = -7$

23. $r = 4, s = 2$

24. $r = -2, s = 3$

25. $\left(a + b\sqrt{2}\right) + \left[(-a) + (-b)\sqrt{2}\right]$
$$= \left[a + b\sqrt{2} + (-a)\right] + (-b)\sqrt{2})$$
$$= a + (-a) + b\sqrt{2} + (-b)\sqrt{2}$$
$$= 0 + [b + (-b)]\sqrt{2}$$
$$= 0 + 0$$
$$= 0$$

26. $\left(a + b\sqrt{2}\right) + \left(a - b\sqrt{2}\right)$
$$= \left(a + b\sqrt{2} + a\right) - b\sqrt{2}$$
$$= a + a + b\sqrt{2} - b\sqrt{2}$$
$$= 2a + b\sqrt{2} + (-b)\sqrt{2}$$
$$= 2a + [b + (-b)]\sqrt{2}$$
$$= 2a$$
Since a is rational, $2a$ is rational. Thus, $\left(a + b\sqrt{2}\right) + \left(a - b\sqrt{2}\right)$ is rational.

27. $6a^2 + 16a + 8$
28. $6a^2 - 19a + 10$
29. $-4a^2 + 2a + 6$
30. $9b^2 + 30b + 25$
31. $b^2 - b - 2$
32. $-12b^2 + 13b - 3$

SKILL B Multiplying and dividing numbers of the form $a + b\sqrt{2}$

Focus: *Students find products and quotients involving real numbers of the form $a + b\sqrt{2}$ and eliminate $\sqrt{2}$ from the denominators of any fractions that result.*

EXAMPLE 1

PHASE 1: Presenting the Example

Once again display the set $\{-1, 0, 1\}$. Ask students to show that this set *is* closed under multiplication. [**$(-1)(-1) = 1$; $(-1)(0) = 0$; $(-1)(1) = -1$; $(0)(-1) = 0$; $(0)(0) = 0$; $(0)(1) = 0$; $(1)(-1) = -1$; $(1)(0) = 0$; $(1)(1) = 1$**] Then have students again consider the set of integers. Ask them whether it, too, is closed under multiplication. [**Yes; the product of any two integers is always another integer.**]

Tell students that the set of numbers $a + b\sqrt{2}$ is also closed under multiplication. To teach them how to perform the multiplications, draw a parallel to the multiplication of two binomials. Display the expression $(3 + 4r)(-5 + 2r)$ and have students simplify. Present the simplification as shown below. Next to it, develop the simplification of the expression in Example 1, as shown.

$$\begin{aligned}
&(3 + 4r)(-5 + 2r)\\
&= 3(-5 + 2r) + 4r(-5 + 2r)\\
&= (3)(-5) + (3)(2r) + (4r)(-5) + (4r)(2r)\\
&= -15 + 6r + (-20r) + 8r^2\\
&= -15 + (-14r) + 8r^2
\end{aligned}$$

$$\begin{aligned}
&(3 + 4\sqrt{2})(-5 + 2\sqrt{2})\\
&= 3(-5 + 2\sqrt{2}) + 4\sqrt{2}(-5 + 2\sqrt{2})\\
&= (3)(-5) + (3)(2\sqrt{2}) + (4\sqrt{2})(-5) + (4\sqrt{2})(2\sqrt{2})\\
&= -15 + 6\sqrt{2} + (-20\sqrt{2}) + [8(\sqrt{2})^2]\\
&= -15 + (-14\sqrt{2}) + [8(2)]\\
&= -15 + (-14\sqrt{2}) + 16\\
&= 1 - 14\sqrt{2}
\end{aligned}$$

EXAMPLE 2

PHASE 1: Presenting the Example

Display the expression $7\sqrt{2} + \frac{6}{\sqrt{2}}$. Ask students if it can be simplified. It is likely that most will believe the answer is no. Have them multiply $\frac{6}{\sqrt{2}}$ by $\frac{\sqrt{2}}{\sqrt{2}}$ and simplify the result. [**$3\sqrt{2}$; see the simplification at right.**] Since the value of $\frac{\sqrt{2}}{\sqrt{2}}$ is 1, this means that $\frac{6}{\sqrt{2}} = 3\sqrt{2}$. So $3\sqrt{2}$ can replace $\frac{6}{\sqrt{2}}$ in the original expression, resulting in $7\sqrt{2} + 3\sqrt{2}$, or $10\sqrt{2}$. Tell students that, because of situations like this, they should eliminate any square roots from the denominator when simplifying a fraction.

$$\begin{aligned}
\frac{6}{\sqrt{2}} \cdot \frac{\sqrt{2}}{\sqrt{2}} &= \frac{6 \cdot \sqrt{2}}{\sqrt{2} \cdot \sqrt{2}}\\
&= \frac{6\sqrt{2}}{2}\\
&= 3\sqrt{2}
\end{aligned}$$

Now discuss Example 2, where divisions involving numbers of the form $a + b\sqrt{2}$ yield fractions. In part b, note that eliminating the square root in the denominator $3 + \sqrt{2}$ requires multiplying by the *conjugate* $3 - \sqrt{2}$. Students should observe that the multiplication takes the form of a difference of two squares:
$(3 + \sqrt{2})(3 - \sqrt{2}) = 3^2 - (\sqrt{2})^2 = 9 - 2 = 7$.

When you multiply or divide numbers of the form $a + b\sqrt{2}$, the result is another number of the form $a + b\sqrt{2}$.

EXAMPLE 1 Simplify $(3 + 4\sqrt{2})(-5 + 2\sqrt{2})$.

▶ Solution

$(3 + 4\sqrt{2})(-5 + 2\sqrt{2}) = 3(-5 + 2\sqrt{2}) + 4\sqrt{2}(-5 + 2\sqrt{2})$ ← *Apply the Distributive Property.*
$= -15 + 6\sqrt{2} - 20\sqrt{2} + 8\sqrt{2} \cdot \sqrt{2}$
$= -15 + 6\sqrt{2} - 20\sqrt{2} + 8(2)$ ← $\sqrt{2} \cdot \sqrt{2} = 2$
$= (-15 + 16) + (6 - 20)\sqrt{2}$
$= 1 - 14\sqrt{2}$

TRY THIS Simplify. **a.** $6\sqrt{2}(5 - 3\sqrt{2})$ **b.** $(7 + 5\sqrt{2})(-1 + (-5)\sqrt{2})$

To simplify a quotient that has a square root in the denominator, multiply the numerator and the denominator by a number that will eliminate the square root sign from the denominator. To simplify $(a + b\sqrt{2}) \div (c + d\sqrt{2})$, multiply the numerator and the denominator by $c - d\sqrt{2}$, which is the **conjugate** of $c + d\sqrt{2}$. When you multiply a number by its conjugate, the result has no radical signs. For example, the conjugate of $3 + \sqrt{2}$ is $3 - \sqrt{2}$ and $(3 + \sqrt{2})(3 - \sqrt{2}) = 7$.

EXAMPLE 2 Simplify. **a.** $(2 + 9\sqrt{2}) \div (5\sqrt{2})$ **b.** $[5 + (-4)\sqrt{2}] \div (3 + \sqrt{2})$

▶ Solution

a. $\dfrac{2 + 9\sqrt{2}}{5\sqrt{2}} = \dfrac{2 + 9\sqrt{2}}{5\sqrt{2}}\left(\dfrac{\sqrt{2}}{\sqrt{2}}\right)$

$= \dfrac{(2 + 9\sqrt{2})(\sqrt{2})}{5 \cdot 2}$

$= \dfrac{2\sqrt{2} + 9 \cdot 2}{10}$

$= \dfrac{18 + 2\sqrt{2}}{10}$

$= \dfrac{9 + \sqrt{2}}{5}$

b. $\dfrac{5 + (-4)\sqrt{2}}{3 + \sqrt{2}} = \dfrac{5 + (-4)\sqrt{2}}{3 + \sqrt{2}}\left(\dfrac{3 - \sqrt{2}}{3 - \sqrt{2}}\right)$

$= \dfrac{(5 - 4\sqrt{2})(3 - \sqrt{2})}{(3 + \sqrt{2})(3 - \sqrt{2})}$

$= \dfrac{(5 - 4\sqrt{2})(3 - \sqrt{2})}{9 - 3\sqrt{2} + 3\sqrt{2} - 2}$

$= \dfrac{15 - 5\sqrt{2} - 12\sqrt{2} + 4 \cdot 2}{7}$

$= \dfrac{23 - 17\sqrt{2}}{7}$

TRY THIS Simplify. **a.** $\dfrac{5}{3\sqrt{2}}$ **b.** $\dfrac{5 + 7\sqrt{2}}{3\sqrt{2}}$ **c.** $\dfrac{2 + 5\sqrt{2}}{1 + 3\sqrt{2}}$

• The product of two numbers of the form $a + b\sqrt{2}$ is another number of the form $a + b\sqrt{2}$. Thus, the set of numbers of the form $a + b\sqrt{2}$ is *closed* under multiplication.
• Every nonzero number of the form $a + b\sqrt{2}$, where a and b are rational numbers, has a reciprocal, or multiplicative inverse, that is also in the set of numbers of the form $a + b\sqrt{2}$.

EXERCISES

Complete the simplification of each product.

1. $(2 - 4\sqrt{2})(3 + 5\sqrt{2}) = 2 \cdot 3 + 2 \cdot 5\sqrt{2} - 4\sqrt{2} \cdot (3) - 4\sqrt{2} \cdot 5\sqrt{2}$

2. $(2 + 5\sqrt{2})(3 - 2\sqrt{2}) = 2 \cdot 3 - 2 \cdot 2\sqrt{2} + 5\sqrt{2} \cdot (3) - 5\sqrt{2} \cdot 2\sqrt{2}$

PRACTICE

Simplify. Write the answers in the form $a + b\sqrt{2}$.

3. $3(2 + 5\sqrt{2})$ 4. $4(2 - 5\sqrt{2})$ 5. $-2(1 + 7\sqrt{2})$ 6. $-3(1 - 5\sqrt{2})$

7. $2\sqrt{2}(1 + \sqrt{2})$ 8. $2\sqrt{2}(3 - 2\sqrt{2})$ 9. $-5\sqrt{2}(1 - 3\sqrt{2})$ 10. $-4\sqrt{2}(-3 - 5\sqrt{2})$

11. $\dfrac{1}{\sqrt{2}}$ 12. $\dfrac{3}{5\sqrt{2}}$ 13. $\dfrac{-4}{\sqrt{2}}$ 14. $\dfrac{4}{-3\sqrt{2}}$

15. $(3 + 5\sqrt{2})(3 - \sqrt{2})$ 16. $(3 - \sqrt{2})(7 - \sqrt{2})$ 17. $(-2 + 4\sqrt{2})(7 - 3\sqrt{2})$

18. $(-2 - 3\sqrt{2})(5 - 6\sqrt{2})$ 19. $(2 - 5\sqrt{2})^2$ 20. $(3 + 7\sqrt{2})^2$

21. $\dfrac{3 + \sqrt{2}}{6 + \sqrt{2}}$ 22. $\dfrac{3 + 5\sqrt{2}}{6 - \sqrt{2}}$ 23. $\dfrac{-3 + 2\sqrt{2}}{3 - \sqrt{2}}$

24. $\dfrac{7 - 2\sqrt{2}}{3 + 2\sqrt{2}}$ 25. $\dfrac{5 - 2\sqrt{2}}{5 - 7\sqrt{2}}$ 26. $\dfrac{3 - 7\sqrt{2}}{1 - 7\sqrt{2}}$

Find $(r + s\sqrt{2})^3$ by using $(r + s\sqrt{2})^2(r + s\sqrt{2})$. Write the answers in the form $a + b\sqrt{2}$.

27. $(1 + \sqrt{2})^3$ 28. $(3 - \sqrt{2})^3$ 29. $(2 + 3\sqrt{2})^3$ 30. $(5 - 3\sqrt{2})^3$

Let a and b represent rational numbers. Prove each statement.

31. $(a + b\sqrt{2})(a - b\sqrt{2}) = a^2 - 2b^2$

32. $\dfrac{1}{a + b\sqrt{2}} = \dfrac{a}{a^2 - 2b^2} - \dfrac{b}{a^2 - 2b^2}(\sqrt{2})$ $a \neq 0$ and $b \neq 0$

MIXED REVIEW

Simplify. (previous courses)

33. $4\frac{2}{3} + 4\frac{4}{7}$ 34. $4\frac{2}{3} + 11\frac{1}{5}$ 35. $7\frac{2}{5} + 7\frac{2}{5}$ 36. $3\frac{3}{11} + 5\frac{1}{2}$

37. $7\frac{5}{9} - 5\frac{2}{3}$ 38. $10\frac{1}{10} - 6\frac{9}{10}$ 39. $11\frac{4}{9} - 11\frac{3}{11}$ 40. $1\frac{1}{3} - \frac{4}{13}$

Answers for Lesson 10.3 Skill B

Try This Example 1

a. $-36 + 30\sqrt{2}$ **b.** $-57 - 40\sqrt{2}$

Try This Example 2

a. $\dfrac{5\sqrt{2}}{6}$ **b.** $\dfrac{14 + 5\sqrt{2}}{6}$ **c.** $\dfrac{28 + \sqrt{2}}{17}$

1. $-34 - 2\sqrt{2}$
2. $-14 + 11\sqrt{2}$
3. $6 + 15\sqrt{2}$
4. $8 - 20\sqrt{2}$
5. $-2 - 14\sqrt{2}$
6. $-3 + 15\sqrt{2}$
7. $4 + 2\sqrt{2}$
8. $-8 + 6\sqrt{2}$
9. $30 - 5\sqrt{2}$
10. $40 + 12\sqrt{2}$
11. $\left(\dfrac{1}{2}\right)\sqrt{2}$
12. $\left(\dfrac{3}{10}\right)\sqrt{2}$

13. $\left(-\dfrac{2}{5}\right)\sqrt{2}$
14. $\left(-\dfrac{2}{3}\right)\sqrt{2}$
15. $-1 + 12\sqrt{2}$
16. $23 - 10\sqrt{2}$
17. $-38 + 34\sqrt{2}$
18. $26 - 3\sqrt{2}$
19. $54 - 20\sqrt{2}$
20. $107 + 42\sqrt{2}$
21. $\dfrac{8}{17} + \left(\dfrac{3}{34}\right)\sqrt{2}$
22. $\dfrac{14}{17} + \left(\dfrac{33}{34}\right)\sqrt{2}$
23. $-\dfrac{5}{7} + \left(\dfrac{3}{7}\right)\sqrt{2}$
24. $29 - 20\sqrt{2}$
25. $\dfrac{3}{73} - \left(\dfrac{25}{73}\right)\sqrt{2}$
26. $\dfrac{95}{97} - \left(\dfrac{14}{97}\right)\sqrt{2}$
27. $7 + 5\sqrt{2}$

28. $45 - 29\sqrt{2}$
29. $116 + 90\sqrt{2}$
30. $395 - 279\sqrt{2}$
31. $(a + b\sqrt{2})(a - b\sqrt{2})$
$= a(a - b\sqrt{2}) + b\sqrt{2}$
$(a - b\sqrt{2})$
$= a^2 - ab\sqrt{2} + ab\sqrt{2} -$
$(b\sqrt{2})(b\sqrt{2})$
$= a^2 - (b\sqrt{2})(b\sqrt{2})$
$= a^2 - 2b^2$

32. $\dfrac{1}{a + b\sqrt{2}}$

$= \dfrac{1}{a + b\sqrt{2}} \cdot \dfrac{a - b\sqrt{2}}{a - b\sqrt{2}}$

$= \dfrac{a - b\sqrt{2}}{(a + b\sqrt{2})(a - b\sqrt{2})}$

$= \dfrac{a - b\sqrt{2}}{a^2 - 2b^2}$

$= \dfrac{a}{a^2 - 2b^2} - \dfrac{b}{a^2 - 2b^2}(\sqrt{2})$

For answers to Exercises 33–40, see Additional Answers.

Focus: *Students prove that sums and products of numbers of the form a + b√2 are numbers of the same form. They also solve simple linear equations involving numbers of this form.*

EXAMPLE 1

PHASE 2: Providing Guidance for **TRY THIS**

Try This simply requires students to emulate the proof in Example 1 for a specific case. If you would like to give your students a more challenging problem, have them write a deductive proof that the set of all numbers of the form $a + b\sqrt{2}$ is closed under *multiplication*. A sample proof follows:

Let m, n, p, and q be rational numbers.

$$(m + n\sqrt{2})(p + q\sqrt{2})$$
$$= m(p + q\sqrt{2}) + n\sqrt{2}\,(p + q\sqrt{2}) \quad \leftarrow \textit{Distributive Property}$$
$$= mp + mq\sqrt{2} + np\sqrt{2} + 2nq \quad \leftarrow \textit{Distributive Property}$$
$$= (mp + 2nq) + (mq\sqrt{2} + np\sqrt{2}) \quad \leftarrow \textit{Commutative and Associative Properties of Addition}$$
$$= (mp + 2nq) + (mq + np)\sqrt{2} \quad \leftarrow \textit{Distributive Property}$$

Since the set of rational numbers is closed under addition and multiplication, $mp + 2nq$ and $mq + np$ are rational numbers. Let $a = mp + 2nq$ and $b = mq + np$. Then the product has the form $a + b\sqrt{2}$, where a and b are rational numbers. Therefore, the set of numbers of the form $a + b\sqrt{2}$ is closed under multiplication.

EXAMPLE 2

PHASE 2: Providing Guidance for **TRY THIS**

Some students find square-root symbols to be intimidating, so they may lose sight of the fact that the underlying equation is a simple linear equation in one variable. Suggest that they use a colored highlighting marker to identify the coefficient of the variable and the constants as shown at right.

$$(1 + 2\sqrt{2})x + (3 - 2\sqrt{2}) = 4 - 2\sqrt{2}$$
$$\underline{\qquad -(3 - 2\sqrt{2}) \quad -(3 - 2\sqrt{2})}$$
$$(1 + 2\sqrt{2})x = 1$$
$$x = \frac{1}{(1 + 2\sqrt{2})}$$

PHASE 3: ASSESSMENT AND CLOSURE for Lesson 10.3

- Have students compare the process of adding $4 + 5\sqrt{2}$ and $1 - \sqrt{2}$ to the process of multiplying the same two expressions. [**Comparisons will vary. The sum is $5 + 4\sqrt{2}$ and the product is $-6 + \sqrt{2}$.**]

- Students have now learned how to add, subtract, multiply, and divide real numbers of the form $a + b\sqrt{2}$. In the next lesson they will generalize these operations to a larger set of real numbers involving square roots.

☞ *For a Lesson 10.3 Quiz, see* Assessment Resources *page 104.*

Proving statements and solving equations that involve $a + b\sqrt{2}$

What follows is a summary of what you learned so far in this lesson.

> **Addition, Multiplication, and Numbers of the Form** $a + b\sqrt{2}$
> - Addition and multiplication of these numbers are closed.
> - Every number of this form has an additive inverse and every nonzero number of this form has a multiplicative inverse.
> - The Commutative, Associative, and Identity properties of addition and multiplication are true for these numbers.

Any set of numbers that satisfies these properties and the Distributive Property is called a **field.** The set of real numbers, the set of numbers of the form $a + b\sqrt{2}$, and the set of rational numbers are all *fields.*

EXAMPLE 1 Prove that addition of numbers of the form $a + b\sqrt{2}$, where a and b are rational numbers, is closed.

▶ Solution

To prove that addition of numbers of the form $a + b\sqrt{2}$ is closed, we must show that the sum of any two numbers of that form is another number of that form. Let r, s, t, and u be rational numbers.

$$(r + s\sqrt{2}) + (t + u\sqrt{2})$$
$$= (r + t) + (s\sqrt{2} + u\sqrt{2}) \quad \longleftarrow \text{Commutative and Associative Properties of Addition}$$
$$= (r + t) + (s + u)\sqrt{2} \quad \longleftarrow \text{Distributive Property}$$

Since addition and multiplication of rational numbers are closed, then $r + t$ and $s + u$ must also be rational numbers. Therefore, the sum has the form $a + b\sqrt{2}$, where $a = r + t$ and $b = s + u$.

TRY THIS   Prove that $(-7 + 3\sqrt{2}) + (1 + (-5)\sqrt{2})$ has the form $a + b\sqrt{2}$.

You can solve simple equations involving numbers of the form $a + b\sqrt{2}$.

EXAMPLE 2 Solve $(2 + 3\sqrt{2})x + (5 - 3\sqrt{2}) = 4 - 2\sqrt{2}$.

▶ Solution

$$(2 + 3\sqrt{2})x + (5 - 3\sqrt{2}) = 4 - 2\sqrt{2}$$
$$(2 + 3\sqrt{2})x = -1 + \sqrt{2} \quad \longleftarrow \text{Subtract } 5 - 3\sqrt{2} \text{ from each side.}$$
$$x = \frac{-1 + \sqrt{2}}{2 + 3\sqrt{2}} \quad \longleftarrow \text{Divide each side by } 2 + 3\sqrt{2}.$$
$$x = \frac{-1 + \sqrt{2}}{2 + 3\sqrt{2}} \left(\frac{2 - 3\sqrt{2}}{2 - 3\sqrt{2}}\right) \quad \longleftarrow \text{Multiply both the numerator and the denominator by } 2 - 3\sqrt{2}, \text{ the conjugate of } 2 + 3\sqrt{2}.$$
$$x = \frac{4}{7} + \frac{-5}{14}\sqrt{2}$$

TRY THIS   Solve $(1 + 2\sqrt{2})x + (3 - 2\sqrt{2}) = 4 - 2\sqrt{2}$.

EXERCISES

KEY SKILLS

Write the additive inverse of each number.

1. $-2.5 + 5\sqrt{2}$ 2. $-4 - 7\sqrt{2}$ 3. $4 - 1.4\sqrt{2}$

PRACTICE

Show that each sum or product has the form $a + b\sqrt{2}$.

4. $(2 + \sqrt{2}) + (3 - 7\sqrt{2})$ 5. $(2 - \sqrt{2}) + (5 + 4\sqrt{2})$ 6. $(4 - 3\sqrt{2})(3 - \sqrt{2})$

Solve each equation. Write the solutions in the form $a + b\sqrt{2}$.

7. $n + (2 - 4\sqrt{2}) = 1 - 3\sqrt{2}$ 8. $m - (2 - 3\sqrt{2}) = 6 - 5\sqrt{2}$

9. $(1 + 2\sqrt{2})k = 4 + 3\sqrt{2}$ 10. $\frac{d}{1 + 2\sqrt{2}} = 2 - \sqrt{2}$

11. $2n + 3 = 7 - 2\sqrt{2}$ 12. $2t + (5 + \sqrt{2}) = 4 - 3\sqrt{2}$

13. Find n such that $nx = 3 + 5\sqrt{2}$ has $2 + \sqrt{2}$ as the solution for x.

MID-CHAPTER REVIEW

Find the domain of each expression or function. (10.1 Skill A)

14. $\sqrt{3 - 4(a + 1)}$ 15. $y = \sqrt{4(x + 5) - 2(x + 6)}$

Solve each equation. (10.2 Skill A)

16. $\sqrt{6(t - 7)} + 5 = 3$ 17. $\sqrt{12v - 4v^2} = 3$

18. Let O represent the origin of a coordinate system. Find the coordinates of the points Q and Q' directly above and below $P(-3, 0)$ and such that $OQ = OQ' = 5$. (10.2 Skill B)

Solve each equation. (10.2 Skill C)

19. $\sqrt{3(r - 1)} - 4 = \sqrt{10 - 2(r + 5)}$ 20. $\sqrt{6m^2 + 21m} = \sqrt{10m + 35}$

Simplify. (10.3 Skills A and B)

21. $(4 - 6\sqrt{2}) - (11 + 11\sqrt{2})$ 22. $(-1 - \sqrt{2})(10 + 10\sqrt{2})$ 23. $\frac{5 - \sqrt{2}}{3\sqrt{2}}$

Solve each equation. (10.3 Skill C)

24. $(3 + 2\sqrt{2})q + (3 + \sqrt{2}) = 3 + 4\sqrt{2}$ 25. $(3 + 5\sqrt{2})v - (3 - \sqrt{2}) = -3 + 2\sqrt{2}$

Answers for Lesson 10.3 Skill C

Try This Example 1

By the Commutative and Associative Properties of Addition, $(-7 + 3\sqrt{2}) + [1 + (-5)\sqrt{2}] = (-7 + 1) + 3\sqrt{2} + (-5)\sqrt{2}$.

By the Distributive Property, $(-7 + 1) + 3\sqrt{2} + (-5)\sqrt{2} = (-7 + 1) + [3 + (-5)]\sqrt{2}$.

By addition, $(-7 + 1) + [3 + (-5)]\sqrt{2} = -6 + (-2)\sqrt{2}$. The sum has the form $a + b\sqrt{2}$, where $a = -6$ and $b = -2$, both rational numbers.

Try This Example 2

$x = -\frac{1}{7} + \frac{2}{7}\sqrt{2}$

1. $2.5 - 5\sqrt{2}$

2. $4 + 7\sqrt{2}$

3. $-4 + 1.4\sqrt{2}$

4. $(2 + \sqrt{2}) + (3 - 7\sqrt{2}) = (2 + 3) + \sqrt{2} - 7\sqrt{2} = (2 + 3) + (1 - 7)\sqrt{2} = 5 + (-6)\sqrt{2}$.
   The sum has the form $a + b\sqrt{2}$, where $a = 5$ and $b = -6$, both rational numbers.

5. $(2 - \sqrt{2}) + (5 + 4\sqrt{2}) = (2 + 5) + (-1)\sqrt{2} + 4\sqrt{2}$
   $= (2 + 5) + [(-1) + 4]\sqrt{2} = 7 + 3\sqrt{2}$. The sum has the form $a + b\sqrt{2}$, where $a = 7$ and $b = 3$, both rational numbers.

6. $(4 - 3\sqrt{2})(3 - \sqrt{2}) = 4(3) - 4\sqrt{2} + (-3\sqrt{2})$
   $(3) - (-3\sqrt{2})(\sqrt{2}) = 12 - 4\sqrt{2} - 9\sqrt{2} + 6 = 12 + 6 - 4\sqrt{2} - 9\sqrt{2} = 12 + 6 + (-4 - 9)\sqrt{2} = 18 - 13\sqrt{2}$. The sum has the form $a + b\sqrt{2}$, where $a = 18$ and $b = -13$, both rational numbers.

7. $n = -1 + 1\sqrt{2}$

8. $m = 8 - 8\sqrt{2}$

9. $k = \frac{8}{7} + \left(\frac{5}{7}\right)\sqrt{2}$

10. $d = -2 + 3\sqrt{2}$

11. $n = 2 - \sqrt{2}$

12. $t = -\frac{1}{2} - 2\sqrt{2}$

13. $n = -2 + \frac{7}{2}\sqrt{2}$

For answers to Exercises 14–25, see Additional Answers.

Operations on Square-Root Expressions

SKILL A *Using properties of square roots to simplify products and quotients*

Focus: *Students simplify expressions by using the Product Property of Square Roots and Quotient Property of Square Roots and by rationalizing denominators.*

EXAMPLES 1 and 2

PHASE 1: Presenting the Examples

Display the expression $\sqrt{18}$ and have students simplify. [$3\sqrt{2}$] Discuss their responses, displaying the simplification as shown at right. Focus their attention on the second line of the simplification. Ask what property justifies rewriting $\sqrt{9 \cdot 2}$ as $\sqrt{9}\sqrt{2}$. [**Product Property of Square Roots**] Then display the following statement of this property from Lesson 8.1:

$$\sqrt{18} = \sqrt{9 \cdot 2}$$
$$= \sqrt{9}\sqrt{2}$$
$$= 3\sqrt{2}$$

If a and b represent nonnegative real numbers, then $\sqrt{ab} = \sqrt{a}\sqrt{b}$.

Ask students to explain why the following statement must also be true:

If a and b represent nonnegative real numbers, then $\sqrt{a}\sqrt{b} = \sqrt{ab}$.

Lead them to see that the second statement is a direct result of the first when the Reflexive Property of Equality is applied to $\sqrt{ab} = \sqrt{a}\sqrt{b}$.

Now display the expression given in Example 1, $\sqrt{3}\sqrt{12}$. Ask students how they might use the second statement of the Product Property to simplify. Elicit the simplification that is shown on the textbook page.

Use a similar technique to present Example 2. This time, first have students simplify $\sqrt{\frac{3}{16}}$. $\left[\frac{\sqrt{3}}{4}\right]$ Display the simplification shown at right. Discuss the following two statements of the Quotient Property of Square Roots.

$$\sqrt{\frac{3}{16}} = \frac{\sqrt{3}}{\sqrt{16}}$$
$$= \frac{\sqrt{3}}{4}$$

If a and b represent positive real numbers, then $\sqrt{\frac{a}{b}} = \frac{\sqrt{a}}{\sqrt{b}}$.

If a and b represent positive real numbers, then $\frac{\sqrt{a}}{\sqrt{b}} = \sqrt{\frac{a}{b}}$.

Then discuss Example 2.

EXAMPLE 3

PHASE 2: Providing Guidance for **TRY THIS**

In Try This exercise b, some students may multiply both numerator and denominator by $\sqrt{24}$. This is valid but may be more complex than necessary. Ask for volunteers to show a simpler approach at the chalkboard. Two alternatives are shown at right.

$$\frac{\sqrt{75}}{\sqrt{24}} = \sqrt{\frac{75}{24}}$$
$$= \sqrt{\frac{25}{8}} = \frac{\sqrt{25}}{\sqrt{8}}$$
$$= \frac{5}{2\sqrt{2}} \cdot \frac{\sqrt{2}}{\sqrt{2}}$$
$$= \frac{5\sqrt{2}}{2\sqrt{4}} = \frac{5}{4}\sqrt{2}$$

$$\frac{\sqrt{75}}{\sqrt{24}} = \frac{5\sqrt{3}}{2\sqrt{6}}$$
$$= \frac{5\sqrt{3}}{2\sqrt{6}} \cdot \frac{\sqrt{6}}{\sqrt{6}}$$
$$= \frac{5\sqrt{18}}{2\sqrt{36}} = \frac{5 \cdot 3\sqrt{2}}{2 \cdot 6}$$
$$= \frac{15\sqrt{2}}{12} = \frac{5}{4}\sqrt{2}$$

10.4 Operations on Square-Root Expressions
LESSON

SKILL A Using properties of square roots to simplify products and quotients

Recall the Product Property of Square Roots from Lesson 8.1. It can be restated in the form given below.

If a and b are nonnegative real numbers, then $\sqrt{a} \cdot \sqrt{b} = \sqrt{ab}$.

EXAMPLE 1 Simplify $\sqrt{3} \cdot \sqrt{12}$. Write the answer in simplest radical form.

▶ Solution
$\sqrt{3} \cdot \sqrt{12} = \sqrt{3 \cdot 12} = \sqrt{36} = \sqrt{6^2} = 6$

Notice that an answer in "simplest radical form" does not have to contain a radical.

TRY THIS Simplify $\sqrt{54} \cdot \sqrt{6}$. Write the answer in simplest radical form.

Recall the Quotient Property of Square Roots from Lesson 8.1. It can be restated in the form given below.

If a and b are positive real numbers, then $\dfrac{\sqrt{a}}{\sqrt{b}} = \sqrt{\dfrac{a}{b}}$.

EXAMPLE 2 Simplify. Write the answers in simplest radical form.   a. $\dfrac{\sqrt{405}}{\sqrt{5}}$ b. $\dfrac{\sqrt{300}}{\sqrt{15}}$

▶ Solution
a. $\dfrac{\sqrt{405}}{\sqrt{5}} = \sqrt{\dfrac{405}{5}} = \sqrt{81} = 9$ b. $\dfrac{\sqrt{300}}{\sqrt{15}} = \sqrt{\dfrac{300}{15}} = \sqrt{20} = \sqrt{4 \cdot 5} = 2\sqrt{5}$

TRY THIS Simplify. Write the answers in simplest radical form.   a. $\dfrac{\sqrt{75}}{\sqrt{3}}$ b. $\dfrac{\sqrt{252}}{\sqrt{21}}$

To simplify the expressions in Example 3, multiply each expression by a clever choice for 1 to make the denominator an integer. This process of changing a fraction with a radical in the denominator to an *equivalent* fraction with a rational denominator is called **rationalizing the denominator**.

EXAMPLE 3 Simplify by rationalizing the denominator.   a. $\dfrac{\sqrt{25}}{\sqrt{3}}$ b. $\dfrac{\sqrt{27}}{\sqrt{8}}$

▶ Solution
a. $\dfrac{\sqrt{25}}{\sqrt{3}} = \dfrac{5}{\sqrt{3}} = \dfrac{5}{\sqrt{3}}\left(\dfrac{\sqrt{3}}{\sqrt{3}}\right) = \dfrac{5\sqrt{3}}{3}$ b. $\dfrac{\sqrt{27}}{\sqrt{8}} = \dfrac{3\sqrt{3}}{2\sqrt{2}}\left(\dfrac{\sqrt{2}}{\sqrt{2}}\right) = \dfrac{3\sqrt{6}}{4}$

TRY THIS Simplify by rationalizing the denominator.   a. $\dfrac{\sqrt{36}}{\sqrt{7}}$ b. $\dfrac{\sqrt{75}}{\sqrt{24}}$

EXERCISES

KEY SKILLS

Write each expression in simplest radical form.

1. $\sqrt{5^2} \cdot \sqrt{7}$ 2. $\sqrt{5^2} \cdot \sqrt{5}$ 3. $\sqrt{3^2} \cdot \sqrt{5^2}$ 4. $\sqrt{2^2} \cdot \sqrt{3^2} \cdot \sqrt{11}$

PRACTICE

Simplify each expression.

5. $\sqrt{49} \cdot \sqrt{2}$ 6. $\sqrt{144} \cdot \sqrt{3}$ 7. $\sqrt{121} \cdot \sqrt{3}$ 8. $\sqrt{64} \cdot \sqrt{5}$

9. $\sqrt{6} \cdot \sqrt{24}$ 10. $\sqrt{10} \cdot \sqrt{40}$ 11. $\sqrt{7} \cdot \sqrt{14}$ 12. $\sqrt{5} \cdot \sqrt{10}$

13. $\dfrac{\sqrt{200}}{\sqrt{2}}$ 14. $\dfrac{\sqrt{27}}{\sqrt{3}}$ 15. $\dfrac{\sqrt{343}}{\sqrt{7}}$ 16. $\dfrac{\sqrt{605}}{\sqrt{5}}$

17. $\dfrac{\sqrt{60}}{\sqrt{20}}$ 18. $\dfrac{\sqrt{112}}{\sqrt{14}}$ 19. $\dfrac{\sqrt{105}}{\sqrt{35}}$ 20. $\dfrac{\sqrt{96}}{\sqrt{12}}$

21. $\dfrac{\sqrt{5}}{\sqrt{3}}$ 22. $\dfrac{\sqrt{2}}{\sqrt{7}}$ 23. $\dfrac{\sqrt{5}}{\sqrt{8}}$ 24. $\dfrac{\sqrt{50}}{\sqrt{12}}$

25. $\dfrac{\sqrt{3}}{\sqrt{200}}$ 26. $\dfrac{\sqrt{2000}}{\sqrt{72}}$ 27. $\dfrac{\sqrt{343}}{\sqrt{75}}$ 28. $\dfrac{\sqrt{512}}{\sqrt{12}}$

Critical Thinking Write each expression as a single square-root expression in simplest form. State any restrictions on the variables. (*Hint*: Remember that an expression under a square-root sign must be nonnegative.)

29. $\dfrac{\sqrt{x^2 + 5x + 6}}{\sqrt{x^2 + 6x + 9}}$ 30. $\dfrac{\sqrt{a^2 + 10a + 25}}{\sqrt{a^2 - 25}}$ 31. $\dfrac{\sqrt{2z^2 + 19z - 33}}{\sqrt{z^2 + 22z + 121}}$

32. **Critical Thinking** If $4 \le x \le 16$, between what two numbers will the value of $\dfrac{\sqrt{x^3}}{\sqrt{2}}$ be found?

MIXED REVIEW

Evaluate each expression. (8.1 Skill C)

33. $1^{\frac{1}{2}}$ 34. $169^{\frac{1}{2}}$ 35. $225^{\frac{1}{2}}$ 36. $256^{\frac{1}{2}}$

37. $100^{\frac{1}{2}}$ 38. $\left(\dfrac{9}{16}\right)^{\frac{1}{2}}$ 39. $\left(\dfrac{81}{25}\right)^{\frac{1}{2}}$ 40. $\left(\dfrac{1}{49}\right)^{\frac{1}{2}}$

Answers for Lesson 10.4 Skill A

Try This Example 1
18

Try This Example 2
a. 5
b. $2\sqrt{3}$

Try This Example 3
a. $\dfrac{6\sqrt{7}}{7}$
b. $\dfrac{5\sqrt{2}}{4}$

1. $5\sqrt{7}$
2. $5\sqrt{5}$
3. 15
4. $6\sqrt{11}$
5. $7\sqrt{2}$
6. $12\sqrt{3}$
7. $11\sqrt{3}$
8. $8\sqrt{5}$
9. 12
10. 20

11. $7\sqrt{2}$
12. $5\sqrt{2}$
13. 10
14. 3
15. 7
16. 11
17. $\sqrt{3}$
18. $2\sqrt{2}$
19. $\sqrt{3}$
20. $2\sqrt{2}$
21. $\dfrac{\sqrt{15}}{3}$
22. $\dfrac{\sqrt{14}}{7}$
23. $\dfrac{\sqrt{10}}{4}$
24. $\dfrac{5\sqrt{6}}{9}$
25. $\dfrac{\sqrt{6}}{20}$
26. $\dfrac{5\sqrt{10}}{3}$

27. $\dfrac{7\sqrt{21}}{15}$
28. $\dfrac{8\sqrt{6}}{3}$
29. $\sqrt{\dfrac{x+2}{x+3}}; x < -3 \text{ or } x \ge -2$
30. $\sqrt{\dfrac{a+5}{a-5}}; a < -5 \text{ or } a > 5$
31. $\sqrt{\dfrac{2z-3}{z+11}}; z < -11 \text{ or } z \ge \dfrac{3}{2}$
32. between $4\sqrt{2}$ and $32\sqrt{2}$, inclusive
33. 1
34. 13
35. 15
36. 16
37. 10
38. $\dfrac{3}{4}$
39. $\dfrac{9}{5}$
40. $\dfrac{1}{7}$

SKILL B Adding and subtracting square-root expressions

Focus: *Students find sums and differences of numbers of the form $a + b\sqrt{n}$, where n is a positive integer, by combining like parts of the numbers.*

EXAMPLE 1

PHASE 1: Presenting the Example

Display the expression $-11\sqrt{2} + 8\sqrt{2}$ and have students simplify. [$\mathbf{-3\sqrt{2}}$] As they watch, erase the radicand "2" in each term and replace it with 7. This creates the expression from part a of the example, $-11\sqrt{7} + 8\sqrt{7}$. Ask students what they think will be the result of simplifying it. They probably will respond, correctly, that $-11\sqrt{7} + 8\sqrt{7}$ is equivalent to $-3\sqrt{7}$.

Before proceeding to part b, have students generalize part a for $-11\sqrt{n} + 8\sqrt{n}$, where n is a positive integer. They should conclude that, by the Distributive Property, $-11\sqrt{n} + 8\sqrt{n} = (-11 + 8)\sqrt{n} = -3\sqrt{n}$.

PHASE 2: Providing Guidance for TRY THIS

Watch for students who add radicands. For example, they may simplify the expression in Try This exercise a as $-5\sqrt{10}$. Demonstrate how simplifying this expression is similar to combining like terms involving variables, as shown at right.

$-2m + (-3)m$	$-2\sqrt{5} + (-3)\sqrt{5}$
$[-2 + (-3)]m$	$[-2 + (-3)]\sqrt{5}$
$-5m$	$-5\sqrt{5}$

EXAMPLES 2 and 3

PHASE 1: Presenting the Examples

Display expression **1**, shown at right. Have students simplify. [$\mathbf{2\sqrt{5}}$] Then display expression **2**. Point out that, like expression **1**, it contains two terms. In this case, however, the radicands are not the same. Ask students if this means that the expression cannot be simplified. Elicit the observation that $\sqrt{45}$ can be rewritten as $3\sqrt{5}$, and so $\sqrt{5} + \sqrt{45}$ can be rewritten as $\sqrt{5} + 3\sqrt{5}$. Have students simplify the rewritten expression. [$\mathbf{4\sqrt{5}}$] Then discuss how this technique is applied to the expression in Example 2.

1 $\sqrt{5} + \sqrt{5}$

2 $\sqrt{5} + \sqrt{45}$

Example 3 combines several skills that students have already learned. The challenge is to keep track of all the simplifications. To present the example, display the expression $(2 + 3\sqrt{12}) - (-5 + \sqrt{27})$ and have students work on it independently for a few minutes. Then discuss the results with the class.

PHASE 2: Providing Guidance for TRY THIS

When simplifying an expression of the form $b\sqrt{n}$, students may correctly take a square root, then mistakenly add it to the factor outside the radical sign instead of multiplying. In Example 2 Try This exercise a, for instance, they may rewrite $3\sqrt{75}$ as $(3 + 5)\sqrt{3} = 8\sqrt{3}$. Encourage them to insert multiplication symbols into their work as shown below.

$$3\sqrt{75} = 3 \times \sqrt{75} = 3 \times \sqrt{25 \times 3} = 3 \times \sqrt{25} \times \sqrt{3} = 3 \times 5 \times \sqrt{3} = 15 \times \sqrt{3} = 15\sqrt{3}$$

Adding and subtracting square-root expressions

If p is a prime number, then $\sqrt{p}$ is an irrational number. For example, $\sqrt{2}$, $\sqrt{3}$, and $\sqrt{5}$ are irrational numbers. Using properties of addition and muliplication and the Distributive Property, you can simplify expressions involving $\sqrt{p}$.

EXAMPLE 1   **Simplify.**   **a.** $-11\sqrt{7} + 8\sqrt{7}$ **b.** $(10 + 7\sqrt{3}) + (5 + 4\sqrt{3})$.

▶ **Solution**

Combine like terms by using the Distributive Property.
a. $-11\sqrt{7} + 8\sqrt{7} = (-11 + 8)\sqrt{7} = -3\sqrt{7}$
b. $(10 + 7\sqrt{3}) + (5 + 4\sqrt{3}) = (10 + 5) + (7 + 4)\sqrt{3} = 15 + 11\sqrt{3}$

TRY THIS   Simplify.   **a.** $-2\sqrt{5} + ((-3)\sqrt{5})$ **b.** $(-1 + 3\sqrt{5}) + (-4 - 3\sqrt{5})$

In Examples 2 and 3, you need to write the expressions in simplest radical form before you can combine like terms.

EXAMPLE 2   **Simplify** $\sqrt{75} - \sqrt{108}$.

▶ **Solution**

$\sqrt{75} - \sqrt{108} = \sqrt{25 \cdot 3} - \sqrt{36 \cdot 3}$
$= 5\sqrt{3} - 6\sqrt{3}$ ← Write in simplest radical form.
$= (5 - 6)\sqrt{3}$ ← Factor $\sqrt{3}$ from each term.
$= -\sqrt{3}$

TRY THIS   Simplify.   **a.** $\sqrt{3} + 3\sqrt{75}$ **b.** $-2\sqrt{54} + 3\sqrt{24}$

Recall from Lesson 2.3 that $-(a + b) = -a - b$.

EXAMPLE 3   **Simplify** $(2 + 3\sqrt{12}) - (-5 + \sqrt{27})$.

▶ **Solution**

$(2 + 3\sqrt{12}) - (-5 + \sqrt{27}) = (2 + 3\sqrt{12}) + (5 - \sqrt{27})$ ← $-(-5 + \sqrt{27}) = 5 - \sqrt{27}$
$= (2 + 6\sqrt{3}) + (5 - 3\sqrt{3})$ ← Write $3\sqrt{12}$ and $\sqrt{27}$ in
$= (2 + 5) + (6\sqrt{3} - 3\sqrt{3})$ simplest radical form.
$= 7 + (6 - 3)\sqrt{3}$
$= 7 + 3\sqrt{3}$

TRY THIS   Simplify.   **a.** $(10 + 7\sqrt{75}) - (5 + \sqrt{3})$ **b.** $(-3 + \sqrt{75}) - (4 - 3\sqrt{27})$

EXERCISES

KEY SKILLS

Simplify using mental math.

1. $(6 + \sqrt{11}) - 5\sqrt{11}$ 2. $(7 - 5\sqrt{2}) + 5\sqrt{2}$ 3. $(7 + \sqrt{3}) + (-7 + 5\sqrt{3})$
4. $(6 + \sqrt{5}) + (6 - 5\sqrt{5})$ 5. $(-3 + \sqrt{5}) + (7 + 4\sqrt{5})$ 6. $(-7 + \sqrt{7}) - (-7 - \sqrt{7})$

PRACTICE

Simplify. Write the answers in simplest radical form. (Recall that simplest radical form may consist of numbers with no radical sign.)

7. $(-1 + 2\sqrt{5}) + (-6 - 2\sqrt{5})$ 8. $(-7 - 2\sqrt{2}) + (1 - 2\sqrt{2})$
9. $(-17 - 7\sqrt{7}) + (2 + 7\sqrt{7})$ 10. $(-4 + 2\sqrt{11}) + (-8 - 2\sqrt{11})$
11. $(7 + 4\sqrt{5}) - (1 - \sqrt{5})$ 12. $(-7 + \sqrt{7}) - (3 + 4\sqrt{7})$
13. $(3 - 4\sqrt{3}) - (-3 - 4\sqrt{3})$ 14. $(2.5 - 7\sqrt{13}) - (4.5 - 7\sqrt{13})$
15. $\sqrt{27} + \sqrt{27}$ 16. $\sqrt{24} + \sqrt{54}$
17. $\sqrt{125} - \sqrt{5}$ 18. $\sqrt{98} - \sqrt{8}$
19. $(4 - 4\sqrt{27}) + (3 - \sqrt{147})$ 20. $(4 + 8\sqrt{125}) + (-7 + 3\sqrt{5})$
21. $(1 - 7\sqrt{32}) - (7 + 7\sqrt{128})$ 22. $(3 - 2\sqrt{32}) - (-3 + 2\sqrt{128})$
23. $3 - 2\sqrt{32} + 2\sqrt{128} - 3\sqrt{8}$ 24. $(-2 - 7\sqrt{5}) + 2\sqrt{125} - 3\sqrt{625}$

Let a and b represent rational numbers with $b > 0$. Simplify each expression. Write the answers in simplest radical form.

25. $(a + 8\sqrt{2b^2}) + (3a + 3\sqrt{8b^2})$ 26. $(a^2 + 8\sqrt{3b^2}) - (a^2 + 3\sqrt{27b^2})$

27. **Critical Thinking**   Suppose that $x > 0$. Simplify the expression.
$\sqrt{x} + \sqrt{x^3} + \sqrt{x^5} + \sqrt{x^7} + \sqrt{x^9} + \sqrt{x^{11}} + \sqrt{x^{13}} + \sqrt{x^{15}} + \sqrt{x^{17}}$.

MIXED REVIEW

Simplify. (10.3 Skill B)

28. $(3 - 2\sqrt{2})(-3 + 2\sqrt{2})$ 29. $(1 - \sqrt{2})(7 + 7\sqrt{2})$ 30. $(5 - 5\sqrt{2})(5 - 5\sqrt{2})$
31. $(-7 - \sqrt{2})(1 + 7\sqrt{2})$ 32. $(3.5 - \sqrt{2})(2 + \sqrt{2})$ 33. $(3 - 2\sqrt{2})(2.5 + 3\sqrt{2})$
34. $\dfrac{2}{3 - 5\sqrt{2}}$ 35. $\dfrac{1 + \sqrt{2}}{1 - \sqrt{2}}$ 36. $\dfrac{1 + 3\sqrt{2}}{\sqrt{2}}$

Answers for Lesson 10.4   Skill B

Try This Example 1
a. $-5\sqrt{5}$
b. -5

Try This Example 2
a. $16\sqrt{3}$
b. 0

Try This Example 3
a. $5 + 34\sqrt{3}$
b. $-7 + 14\sqrt{3}$

1. $6 - 4\sqrt{11}$
2. 7
3. $6\sqrt{3}$
4. $12 - 4\sqrt{5}$
5. $4 + 5\sqrt{5}$
6. $2\sqrt{7}$
7. -7
8. $-6 - 4\sqrt{2}$
9. -15
10. -12
11. $6 + 5\sqrt{5}$

12. $-10 - 3\sqrt{7}$
13. 6
14. -2
15. $6\sqrt{3}$
16. $5\sqrt{6}$
17. $4\sqrt{5}$
18. $5\sqrt{2}$
19. $7 - 19\sqrt{3}$
20. $-3 + 43\sqrt{5}$
21. $-6 - 84\sqrt{2}$
22. $6 - 24\sqrt{2}$
23. $3 + 2\sqrt{2}$
24. $-77 + 3\sqrt{5}$
★25. $4a + 14b\sqrt{2}$
★26. $-b\sqrt{3}$
27. $(1 + x + x^2 + x^3 + x^4 + x^5 + x^6 + x^7 + x^8)\sqrt{x}$
28. $-13 + 12\sqrt{2}$
29. -7
30. $75 - 50\sqrt{2}$

31. $-21 - 50\sqrt{2}$
32. $5 + \dfrac{3}{2}\sqrt{2}$
33. $-4.5 + 4\sqrt{2}$
34. $-\dfrac{6}{41} - \dfrac{10}{41}\sqrt{2}$
35. $-3 - 2\sqrt{2}$
36. $3 + \dfrac{\sqrt{2}}{2}$

★ **Advanced Exercises**

Focus: *Students find products and quotients involving numbers of the form* $a + b\sqrt{n}$
and rationalize a denominator when necessary.

EXAMPLE 1

PHASE 1: Presenting the Example

The simplifications required in Example 1 involve a combination of several skills that students have previously learned. Introduce the skills gradually by having students simplify the sequence of expressions at right, one at a time. [**1.** $3y + 9$ **2.** $3y + 3\sqrt{3}$ **3.** $y\sqrt{3} + 3$ **4.** $5\sqrt{3} + 3$ **5.** $5\sqrt{3} + 6$] As each expression is simplified, answer any questions that the students may have. Note that the simplification of expression **5** is outlined in Example 1.

1. $3(y + 3)$

2. $3(y + \sqrt{3})$

3. $\sqrt{3}(y + \sqrt{3})$

4. $\sqrt{3}(5 + \sqrt{3})$

5. $\sqrt{3}(5 + 2\sqrt{3})$

EXAMPLE 2

PHASE 2: Providing Guidance for TRY THIS

If students have not already made the observation, point out that the expressions in Example 2 are identical in form to the products of two binomials in Chapter 7. Those students who chose to multiply binomials using the FOIL method should be encouraged to do the same here. The FOIL multiplication for Try This exercise a is shown at right.

$$18\sqrt{3} \qquad -5\sqrt{3}$$
$$(6\sqrt{3} - 5)(3 + \sqrt{3})$$
$$-15$$
$$18$$

EXAMPLE 3

PHASE 1: Presenting the Example

Remind students that they simplified expressions of this type in Lesson 10.3. There the expressions were of the form $a + b\sqrt{2}$. Now they will generalize the process to $a + b\sqrt{n}$, where n is any positive integer.

PHASE 2: Providing Guidance for Try This

Students may need practice in choosing the correct multiplier. You can use Skill C Exercises 4 through 6 on the facing page for this purpose.

PHASE 3: ASSESSMENT AND CLOSURE for Lesson 10.4

- Have students simplify the four expressions shown at right. Then lead a discussion in which they compare the simplifications.

- This lesson served as a very brief introduction to operations with square-root expressions. Students will study the operations in greater detail if they pursue a further course in algebra.

👉 *For a Lesson 10.4 Quiz, see* Assessment Resources *page 105.*

1 $\sqrt{20} + \sqrt{45}$ $\quad [5\sqrt{5}]$

2 $\sqrt{20} - \sqrt{45}$ $\quad [-\sqrt{5}]$

3 $\sqrt{20}\sqrt{45}$ $\quad [30]$

4 $\dfrac{\sqrt{20}}{\sqrt{45}}$ $\quad \left[\dfrac{2}{3}\right]$

To multiply one radical expression by another, apply the Distributive Property, simplify the resulting products, and then combine like terms.

EXAMPLE 1 Simplify $\sqrt{3}(5 + 2\sqrt{3})$.

▶ Solution

$\sqrt{3}(5 + 2\sqrt{3}) = \sqrt{3} \cdot 5 + \sqrt{3} \cdot 2\sqrt{3}$ ←— Distribute $\sqrt{3}$ to each term.

$= 5\sqrt{3} + 2 \cdot 3$ ←— $\sqrt{3} \cdot \sqrt{3} = 3$

$= 6 + 5\sqrt{3}$

TRY THIS   Simplify each expression.   **a.** $\sqrt{7}(10\sqrt{7} + 2\sqrt{98})$ **b.** $\sqrt{2}(10\sqrt{2} - 5)$

EXAMPLE 2 Simplify $(2 - 5\sqrt{2})(3\sqrt{2} + 1)$.

▶ Solution

$(2 - 5\sqrt{2})(3\sqrt{2} + 1) = 2(3\sqrt{2} + 1) - 5\sqrt{2}(3\sqrt{2} + 1)$ ←— Apply the Distributive Property.

$= 6\sqrt{2} + 2 - 15 \cdot 2 - 5\sqrt{2}$

$= \sqrt{2} - 28$, or $-28 + \sqrt{2}$

TRY THIS   Simplify each expression.   **a.** $(6\sqrt{3} - 5)(3 + \sqrt{3})$ **b.** $(5 + 4\sqrt{3})(5 - 4\sqrt{3})$

Recall from Skill A that in order to simplify a quotient that has a radical in the denominator, you need to rationalize the denominator. In Example 3, the denominator of the quotient is $5 - 3\sqrt{2}$. To eliminate the square root from the denominator, multiply the quotient by $\frac{5 + 3\sqrt{2}}{5 + 3\sqrt{2}}$, which equals 1.

EXAMPLE 3 Simplify $\frac{4 + \sqrt{2}}{5 - 3\sqrt{2}}$.

▶ Solution

$\frac{4 + \sqrt{2}}{5 - 3\sqrt{2}} = \frac{4 + \sqrt{2}}{5 - 3\sqrt{2}}\left(\frac{5 + 3\sqrt{2}}{5 + 3\sqrt{2}}\right)$ ←— $\frac{5 + 3\sqrt{2}}{5 + 3\sqrt{2}} = 1$

$= \frac{20 + 12\sqrt{2} + 5\sqrt{2} + 3 \cdot 2}{25 + 15\sqrt{2} - 15\sqrt{2} - 9 \cdot 2}$

$= \frac{26 + 17\sqrt{2}}{7}$, or $\frac{26}{7} + \frac{17}{7}\sqrt{2}$

TRY THIS   Simplify each expression.   **a.** $\frac{5 + \sqrt{5}}{1 + 3\sqrt{5}}$ **b.** $\frac{1 + \sqrt{7}}{5 - 2\sqrt{7}}$

It is important to note that when you rationalize a denominator, you are not changing the value of the original expression. Because you are multiplying the numerator and the denominator by the same quantity, you are really just multiplying by 1: $\frac{a}{a} = 1$.

EXERCISES

KEY SKILLS

Apply the Distributive Property to each expression. Do not simplify further.

1. $\sqrt{5}(5 - 3\sqrt{5})$ **2.** $(1 + \sqrt{3})(1 - 3\sqrt{3})$ **3.** $(2 - \sqrt{7})(7 - 3\sqrt{7})$

What expression would you multiply the given quotient by in order to eliminate the radical from the denominator?

4. $\frac{1}{3 - 4\sqrt{5}}$ **5.** $\frac{-3}{2 + 5\sqrt{3}}$ **6.** $\frac{-3 + 4\sqrt{3}}{1 - 3\sqrt{3}}$

PRACTICE

Simplify. Write the answers in simplest radical form.

7. $\sqrt{2}(3 + 5\sqrt{2})$ **8.** $3\sqrt{11}(6 - \sqrt{11})$ **9.** $(1 - 3\sqrt{2})(1 - \sqrt{2})$

10. $(1 + 2\sqrt{13})(3 - 2\sqrt{13})$ **11.** $(4 + 3\sqrt{5})(4 - 3\sqrt{5})$ **12.** $(3 - 2\sqrt{7})(3 + 2\sqrt{7})$

13. $(3 + 3\sqrt{3})(5 - 3\sqrt{3})$ **14.** $(-5 + \sqrt{7})(-5 - 3\sqrt{7})$ **15.** $\sqrt{5}(-2 + \sqrt{5})(-5 + \sqrt{5})$

16. $\frac{1}{1 + \sqrt{2}}$ **17.** $\frac{3}{1 - \sqrt{11}}$ **18.** $\frac{\sqrt{3}}{2 + \sqrt{3}}$

19. $\frac{\sqrt{5}}{3 - \sqrt{5}}$ **20.** $\frac{1 - \sqrt{5}}{2 - \sqrt{5}}$ **21.** $\frac{4 + \sqrt{7}}{3 + \sqrt{7}}$

22. Critical Thinking Let a and b represent rational numbers with $ab \neq 0$. Simplify the expression $\frac{a + b\sqrt{p}}{a - b\sqrt{p}}$.

23. Critical Thinking   a. Is $\sqrt{a^2 + b^2} \leq a + b$ always true? Justify your response.
b. What does your answer indicate about squaring both sides as a method of solving an inequality?

MIXED REVIEW

Multiply. (6.5 Skill A)

24. $a^2 \cdot a^3$ **25.** $n^0 \cdot n^5$ **26.** $m^1 \cdot m^2 \cdot m^3$ **27.** $z^3 \cdot z^1 \cdot z^3$

28. $(2h^3)(3h^2)$ **29.** $(-5p^3)(-5p^4)$ **30.** $(4r^2)(r^3)(11r)$ **31.** $\left(\frac{3}{4}b^2\right)(b^3)\left(\frac{14}{33}b^4\right)$

Answers for Lesson 10.4 Skill C

Try This Example 1
a. $70 + 14\sqrt{14}$
b. $20 - 5\sqrt{2}$

Try This Example 2
a. $3 + 13\sqrt{3}$
b. -23

Try This Example 3
a. $\frac{5}{22} + \frac{7}{22}\sqrt{5}$
b. $-\frac{19}{3} - \frac{7}{3}\sqrt{7}$

1. $\sqrt{5}(5) - 3\sqrt{5}\sqrt{5}$
2. $(1)(1) - (1)3\sqrt{3} + \sqrt{3}(1) - 3\sqrt{3}\sqrt{3}$
3. $(2)(7) - (2)3\sqrt{7} - \sqrt{7}(7) + 3\sqrt{7}\sqrt{7}$
4. $\frac{3 + 4\sqrt{5}}{3 + 4\sqrt{5}}$
5. $\frac{2 - 5\sqrt{3}}{2 - 5\sqrt{3}}$

6. $\frac{1 + 3\sqrt{3}}{1 + 3\sqrt{3}}$
7. $10 + 3\sqrt{2}$
8. $-33 + 18\sqrt{11}$
9. $7 - 4\sqrt{2}$
10. $-49 + 4\sqrt{13}$
11. -29
12. -19
13. $-12 + 6\sqrt{3}$
14. $4 + 10\sqrt{7}$
15. $-35 + 15\sqrt{5}$
16. $-1 + \sqrt{2}$
17. $-\frac{3}{10} - \frac{3}{10}\sqrt{11}$
18. $-3 + 2\sqrt{3}$
19. $\frac{5}{4} + \frac{3}{4}\sqrt{5}$
20. $3 + \sqrt{5}$
21. $\frac{5}{2} - \frac{1}{2}\sqrt{7}$

22. $\frac{a^2 + pb^2}{a^2 - pb^2} + \frac{2ab}{a^2 - pb^2}\sqrt{p}$

23. a. No; for example, if $a = -1$ and $b = -1$, then $\sqrt{(-1)^2 + (-1)^2} \leq -1 + (-1)$ is false.
b. Squaring both sides of $\sqrt{a^2 + b^2} \leq a + b$ gives $a^2 + b^2 \leq a^2 + 2ab + b^2$. This inequality is true when $a = -1$ and $b = -1$. However, from part a, the original inequality is not true when $a = -1$ and $b = -1$. So, squaring both sides of an inequality does not always give an equivalent inequality. Therefore, squaring both sides is not a valid method for solving an inequality.

24. a^5
25. n^5
26. m^6
27. z^7
28. $6h^5$
29. $25p^7$
30. $44r^6$
31. $\frac{7}{22}b^9$

★ **Advanced Exercises**

10.5 Powers, Roots, and Rational Exponents

> **SKILL A** ▶ *Simplifying expressions that involve exponents of the form* $\frac{1}{n}$

Focus: *Students simplify the expressions* $\frac{1}{a^n}$ *and* $\sqrt[n]{a}$ *given positive integers a and n.*

EXAMPLES 1 and 2

PHASE 1: Presenting the Examples

Remind students that the formula for the volume, *V*, of a cube is $V = e \cdot e \cdot e$, or $V = e^3$, where *e* is the length of one edge. Then display the two statements shown at right. Have students complete each. [8; 3] Discuss their responses. Lead them to see that the second problem is the "reverse" of the first. Stress the following two points:

· For the first problem, they needed to calculate $2 \cdot 2 \cdot 2$. Because of this geometric interpretation, using a number as a factor three times is called finding the *cube* of the number.

· For the second problem, they needed to find which number gives the result 27 when used as a factor three times. This operation is called finding a *cube root* of the number.

Tell students that, just as there is a symbol to designate the square root of a number, there is also a symbol for the cube root: $\sqrt[3]{\ }$. Therefore, you can write the following for any real number *a*:

$$\sqrt[3]{a} \cdot \sqrt[3]{a} \cdot \sqrt[3]{a} = a$$

Now display the expression $a^{\frac{1}{3}} \cdot a^{\frac{1}{3}} \cdot a^{\frac{1}{3}}$. Have students simplify it. [*a*] Display the simplification as shown below.

$$a^{\frac{1}{3}} \cdot a^{\frac{1}{3}} \cdot a^{\frac{1}{3}} = a^{\frac{1}{3}+\frac{1}{3}+\frac{1}{3}} = a^1 = a$$

Note that, since both $\sqrt[3]{a} \cdot \sqrt[3]{a} \cdot \sqrt[3]{a}$ and $a^{\frac{1}{3}} \cdot a^{\frac{1}{3}} \cdot a^{\frac{1}{3}}$ are equal to *a*, the Transitive Property of Equality allows you to write the following:

$$\sqrt[3]{a} \cdot \sqrt[3]{a} \cdot \sqrt[3]{a} = a^{\frac{1}{3}} \cdot a^{\frac{1}{3}} \cdot a^{\frac{1}{3}}$$

Ask students to use this equation to make a conjecture about the meaning of $a^{\frac{1}{3}}$. Elicit the response that $a^{\frac{1}{3}}$ must represent the cube root of *a*. That is, $a^{\frac{1}{3}} = \sqrt[3]{a}$. Review the discussion at the top of the textbook page that leads to the generalization $a^{\frac{1}{n}} = \sqrt[n]{a}$. Then discuss the examples.

PHASE 2: Providing Guidance for **TRY THIS**

Students should be aware that some expressions can be simplified in different ways. For instance, at right are shown two slightly different methods for simplifying the expression in Try This exercise b of Example 2.

If the length of one side of a cube is 2, then its volume is ___?___. If the volume of a cube is 27, then the length of one side is ___?___.

given: length of edge = 2
volume = $2\cdot2\cdot2 = 2^3 = 8$

given: volume = 27
length of edge = $\sqrt[3]{27} = 3$

$$\sqrt[4]{256} = 256^{\frac{1}{4}} = \left(16^2\right)^{\frac{1}{4}} = 16^{2\times\frac{1}{4}} = 16^{\frac{1}{2}} = 4$$

$$\sqrt[4]{256} = 256^{\frac{1}{4}} = \left(2^8\right)^{\frac{1}{4}} = 2^{8\times\frac{1}{4}} = 2^2 = 4$$

10.5 Powers, Roots, and Rational Exponents

LESSON

Simplifying expressions that involve exponents of the form $\frac{1}{n}$

In Lesson 8.1, you learned that the equations below are all related.

$$5 \times 5 = 25 \qquad 5^2 = 25 \qquad 5 = \sqrt{25} \qquad 25^{\frac{1}{2}} = 5$$

Using the equations above, you can form patterns as follows.

$$5 \times 5 \times 5 = 125 \qquad 5^3 = 125 \qquad 5 = \sqrt[3]{125} \qquad 125^{\frac{1}{3}} = 5$$

$$5 \times 5 \times 5 \times 5 = 625 \qquad 5^4 = 625 \qquad 5 = \sqrt[4]{625} \qquad 625^{\frac{1}{4}} = 5$$

The symbols $\sqrt[3]{}$ and $\sqrt[4]{}$ are called *cube-root* and *fourth-root radicals*, respectively. The numbers 3 and 4 are each called the **index** of the radical. Notice the relationship in the equations above between the radical expression and the exponential expression with a rational-number exponent.

index →

$$\sqrt[3]{125} \qquad \sqrt[4]{625}$$

← radicand

$$\sqrt{25} = 25^{\frac{1}{2}} \qquad \sqrt[3]{125} = 125^{\frac{1}{3}} \qquad \sqrt[4]{625} = 625^{\frac{1}{4}}$$

Definition of the Exponent $\frac{1}{n}$

If $a > 0$ and n is a positive integer, $a^{\frac{1}{n}} = \sqrt[n]{a}$.

EXAMPLE 1 Write each radical expression as an exponential expression and write each exponential expression as a radical expression.

a. $\sqrt[3]{10}$ b. $\sqrt[4]{24}$ c. $78^{\frac{1}{2}}$ d. $75^{\frac{1}{4}}$

▶ Solution

a. $\sqrt[3]{10} = 10^{\frac{1}{3}}$ b. $\sqrt[4]{24} = 24^{\frac{1}{4}}$ c. $78^{\frac{1}{2}} = \sqrt{78}$ d. $75^{\frac{1}{4}} = \sqrt[4]{75}$

TRY THIS   Write each radical expression as an exponential expression and write each exponential expression as a radical expression.

a. $\sqrt[3]{3}$ b. $\sqrt[3]{30}$ c. $190^{\frac{1}{4}}$ d. $75^{\frac{1}{3}}$

EXAMPLE 2 Evaluate.   a. $\sqrt[3]{8}$ b. $\sqrt[4]{81}$ c. $16^{\frac{1}{4}}$ d. $64^{\frac{1}{3}}$

▶ Solution

a. $\sqrt[3]{8} = \sqrt[3]{2^3} = 2^{3 \cdot \frac{1}{3}} = 2^1 = 2$ b. $\sqrt[4]{81} = \sqrt[4]{3^4} = 3^{4 \cdot \frac{1}{4}} = 3^1 = 3$ $(a^m)^n = a^{mn}$

c. $16^{\frac{1}{4}} = (2^4)^{\frac{1}{4}} = 2^{4 \cdot \frac{1}{4}} = 2$ d. $64^{\frac{1}{3}} = (4^3)^{\frac{1}{3}} = 4^{3 \cdot \frac{1}{3}} = 4$

TRY THIS   Evaluate.   a. $\sqrt[3]{27}$ b. $\sqrt[4]{256}$ c. $16^{\frac{1}{4}}$ d. $125^{\frac{1}{3}}$

EXERCISES

Write each equation using a radical and using an exponential expression with a rational-number exponent.

1. $7^2 = 49$ 2. $2^6 = 64$ 3. $10^3 = 1000$ 4. $3^5 = 243$

PRACTICE

Write each radical expression as an exponential expression and write each radical expression as an exponential expression.

5. $\sqrt[3]{13}$ 6. $\sqrt{5}$ 7. $100^{\frac{1}{3}}$ 8. $101^{\frac{1}{2}}$

9. $\sqrt[3]{43}$ 10. $\sqrt[3]{5}$ 11. $90^{\frac{1}{3}}$ 12. $60^{\frac{1}{2}}$

13. $\sqrt[5]{55}$ 14. $\sqrt[6]{6}$ 15. $33^{\frac{1}{3}}$ 16. $222^{\frac{1}{2}}$

Evaluate each expression.

17. $(2^3)^{\frac{1}{3}}$ 18. $(5^4)^{\frac{1}{4}}$ 19. $(7^4)^{\frac{1}{4}}$ 20. $(7^6)^{\frac{1}{6}}$

21. $\sqrt[3]{1}$ 22. $\sqrt[4]{1}$ 23. $\sqrt[3]{512}$ 24. $\sqrt[4]{10,000}$

Simplify. Assume that the variables represent positive numbers.

25. $(a^2)^{\frac{1}{2}}$ 26. $(u^4)^{\frac{1}{4}}$ 27. $(n^3)^{\frac{1}{3}}$ 28. $(y^{10})^{\frac{1}{10}}$

29. $(a^3)^{\frac{1}{3}}(b^4)^{\frac{1}{4}}$ 30. $(y^5)^{\frac{1}{5}}(z^6)^{\frac{1}{6}}$ 31. $\sqrt[5]{r^5 s^5}$ 32. $\sqrt[3]{a^3 b^3 c^3}$

33. **Critical Thinking** How are real numbers r and s related if $(rs)^{\frac{1}{4}} = 1$?

34. **Critical Thinking** How are real numbers m and n related if $\sqrt[3]{3m + n} = 1$?

MIXED REVIEW

Write each expression in simplest radical form. (8.1 Skill B)

35. $\sqrt{275}$ 36. $\sqrt{117}$ 37. $\sqrt{484}$ 38. $\sqrt{363}$

39. $\sqrt{\frac{1}{100}}$ 40. $\sqrt{\frac{121}{100}}$ 41. $\sqrt{\frac{24}{49}}$ 42. $\sqrt{\frac{125}{64}}$

Answers for Lesson 10.5 Skill A

Try This Example 1

a. $3^{\frac{1}{4}}$ b. $30^{\frac{1}{3}}$ c. $\sqrt[4]{190}$ d. $\sqrt[3]{75}$

Try This Example 2

a. 3   b. 4   c. 2   d. 5

1. $7 = \sqrt{49}$ and $7 = 49^{\frac{1}{2}}$

2. $2 = \sqrt[6]{64}$ and $2 = 64^{\frac{1}{6}}$

3. $10 = \sqrt[3]{1000}$ and $10 = 1000^{\frac{1}{3}}$

4. $3 = \sqrt[5]{243}$ and $3 = 243^{\frac{1}{5}}$

5. $13^{\frac{1}{3}}$

6. $5^{\frac{1}{2}}$

7. $\sqrt[3]{100}$

8. $\sqrt{101}$

9. $43^{\frac{1}{4}}$

10. $5^{\frac{1}{3}}$

11. $\sqrt[3]{90}$

12. $\sqrt{60}$

13. $55^{\frac{1}{5}}$

14. $6^{\frac{1}{6}}$

15. $\sqrt[3]{33}$

16. $\sqrt{222}$

17. 2

18. 5

19. 7

20. 7

21. 1

22. 1

23. 8

24. 10

25. a

26. u

27. n

28. y

29. ab

30. yz

31. rs

32. abc

33. r and s are nonzero real numbers such that $r = \frac{1}{s}$.

34. $3m + n = 1$, $n = -3m + 1$, or $m = \frac{-n + 1}{3}$

35. $5\sqrt{11}$

36. $3\sqrt{13}$

37. 22

38. $11\sqrt{3}$

39. $\frac{1}{10}$

40. $\frac{11}{10}$

41. $\frac{2\sqrt{6}}{7}$

42. $\frac{5\sqrt{5}}{8}$

SKILL B Simplifying expressions that involve exponents of the form $\frac{m}{n}$

Focus: *Students simplify expressions involving $a^{\frac{m}{n}}$ and $\sqrt[n]{a^m}$ given positive integers a, m, and n.*

EXAMPLE 1

PHASE 1: Presenting the Example

Display the expression $(\sqrt[4]{16})^3$. Remind students that, according to the order of operations, any operation within parentheses must be performed first. Ask them to simplify the expression. **[8]** Display the simplification as shown at right.

$$(\sqrt[4]{16})^3 = (2)^3 = 8$$

Again display $(\sqrt[4]{16})^3$. Focus students' attention on the expression within parentheses, $\sqrt[4]{16}$. Ask them to rewrite it using a rational exponent. $\left[\mathbf{16^{\frac{1}{4}}}\right]$ Write $\left(16^{\frac{1}{4}}\right)^3$ beneath the original expression as shown at right. Tell students to apply the Power-of-a-Power Property of Exponents to this expression. $\left[\mathbf{16^{\frac{3}{4}}}\right]$ Display the result as shown.

$$(\sqrt[4]{16})^3 = \left(16^{\frac{1}{4}}\right)^3 = 16^{\frac{1}{4} \times 3} = 16^{\frac{3}{4}}$$

Point out to students that, according to the properties of equality, the logical conclusion is that $16^{\frac{3}{4}} = (\sqrt[4]{16})^3 = 8$. Generalize this specific example to the definition of the exponent $\frac{m}{n}$ that appears in the box on the textbook page. Then discuss how the definition is applied in the example.

$$16^{\frac{3}{4}} = (\sqrt[4]{16})^3 = 8$$

PHASE 2: Providing Guidance for TRY THIS

In Try This exercise b, recognizing that $5^{\frac{3}{2}}$ is equal to $5^{1 + \frac{1}{2}}$ is probably the most direct approach to simplifying the expression. However, some students may develop alternative strategies that are equally valid. Invite them to share these with the class. One possible alternative is shown at right.

$$5^{\frac{3}{2}} = (5^3)^{\frac{1}{2}} = \sqrt{5^3} = \sqrt{5^2 \cdot 5^1} = \sqrt{5^2} \cdot \sqrt{5^1} = 5\sqrt{5}$$

EXAMPLE 2

PHASE 1: Presenting the Example

Although the expression in Example 2 may look complex to some students, simplifying it merely involves recalling the familiar Product Property of Exponents and applying their newfound knowledge of rational exponents. To make this point, have students simplify the sequence of expressions at right, one at a time. In each case, be sure students understand that their goal is to write a single exponential expression with x as the base.

1. $x^a \cdot x^b$
2. $x^a \cdot x^4$
3. $x^2 \cdot x^4$
4. $x^2 \cdot x^{\frac{4}{3}}$
5. $x^2 \cdot \sqrt[3]{x^4}$

[**1.** x^{a+b} **2.** x^{a+4} **3.** $x^{2+4} = x^6$ **4.** $x^{2+\frac{4}{3}} = x^{\frac{10}{3}}$ **5.** $x^{2+\frac{4}{3}} = x^{\frac{10}{3}}$] As each expression is simplified, answer any questions that the students may have. Note that the simplification of expression **5** is outlined in Example 2.

Simplifying expressions that involve exponents of the form $\frac{m}{n}$

Suppose that m and n are integers and $n \neq 0$. You know that a^m represents the mth power of a. From Skill A of this lesson, you know that $a^{\frac{1}{n}}$ represents $\sqrt[n]{a}$, the nth root of a. You can now define the rational number, $\frac{m}{n}$, as an exponent as follows.

Definition of the Exponent $\frac{m}{n}$

If a is a positive number, m and n are integers, and $n \neq 0$, then
$$a^{\frac{m}{n}} = \sqrt[n]{a^m}, \text{ or } \left(\sqrt[n]{a}\right)^m.$$

If the bases of exponential expressions represent positive numbers and the exponents are rational numbers, then all of the properties of exponents you learned in Chapter 6 continue to be true.

EXAMPLE 1 Write each expression as a radical expression in simplest form.

  a. $27^{\frac{2}{3}}$ b. $2^{\frac{4}{3}}$

▶ Solution

  a. Method 1: Use the definition of the exponent $\frac{m}{n}$.
  $$27^{\frac{2}{3}} = \left(\sqrt[3]{27}\right)^2 = \left(\sqrt[3]{3^3}\right)^2 = 3^2 = 9$$
  Method 2: Use the Power-of-a-Power Property of Exponents.
  $$27^{\frac{2}{3}} = (3^3)^{\frac{2}{3}} = (3)^{3 \times \frac{2}{3}} = 3^2 = 9$$

  b. Method 1: Use the definition of the exponent $\frac{m}{n}$.
  $$2^{\frac{4}{3}} = \sqrt[3]{2^4} = \sqrt[3]{2^3 \cdot 2^1} = \sqrt[3]{2^3} \cdot \sqrt[3]{2^1} = 2 \cdot \sqrt[3]{2} = 2\sqrt[3]{2}$$

  Method 2: Write $\frac{4}{3}$ as $1 + \frac{1}{3}$. Then use the Product Property of Exponents.
  $$2^{\frac{4}{3}} = 2^{1+\frac{1}{3}} = 2^1 \cdot 2^{\frac{1}{3}} = 2 \cdot \sqrt[3]{2} = 2\sqrt[3]{2}$$

TRY THIS Simplify each expression.   a. $32^{\frac{4}{5}}$ b. $5^{\frac{3}{2}}$

Just as you can simplify exponential expressions whose bases are numbers, you can simplify exponential expressions whose bases are variables.

EXAMPLE 2 Write $(x^2)\sqrt[3]{x^4}$ as an expression with a rational exponent.

▶ Solution
$$(x^2)\sqrt[3]{x^4} = x^2 \cdot x^{\frac{4}{3}} = x^{2+\frac{4}{3}} \longleftarrow a^m \cdot a^n = a^{m+n}$$
$$= x^{\frac{10}{3}}$$

TRY THIS Write $(a^3)\sqrt{a^5}$ as an expression with a rational exponent.

EXERCISES

Write each radical expression as an exponential expression with a rational exponent.

1. $\sqrt[3]{5^2}$ 2. $\sqrt[4]{7^3}$ 3. $\left(\sqrt[4]{3}\right)^3$ 4. $\left(\sqrt[3]{10}\right)^2$

Write each exponential expression as a radical expression.

5. $11^{\frac{2}{3}}$ 6. $5^{\frac{3}{2}}$ 7. $3^{2\frac{1}{2}}$ 8. $7^{3\frac{2}{3}}$

PRACTICE

Write each expression as a radical expression in simplest form.

9. $125^{\frac{4}{3}}$ 10. $8^{\frac{5}{3}}$ 11. $81^{\frac{7}{4}}$ 12. $32^{\frac{8}{5}}$

13. $3^{\frac{3}{2}}$ 14. $7^{\frac{4}{3}}$ 15. $2^{\frac{5}{3}}$ 16. $5^{\frac{9}{4}}$

Write each expression as an expression with a rational exponent. Assume all of the variables represent positive real numbers.

17. $x\sqrt{x}$ 18. $a^3\sqrt{a}$ 19. $(t^2)\sqrt[3]{t}$ 20. $(v^2)\sqrt[3]{v^2}$

21. $n^{\frac{4}{3}} \cdot n^{\frac{1}{3}}$ 22. $k^{\frac{1}{2}} \cdot k^{\frac{1}{3}}$ 23. $b^{\frac{1}{2}} \cdot \sqrt[3]{b^2}$ 24. $c^{\frac{3}{2}} \cdot \sqrt{c^3}$

25. Let $t > 0$. Write the expression at right as an exponential expression with a single rational exponent. $\dfrac{(t^5)\sqrt[4]{t^3}}{(t^2)\sqrt[3]{t^2}}$

26. Let $a > 0$ and let n be a nonzero integer. Prove that $a^{\frac{n+1}{n}} = a\sqrt[n]{a}$.

27. Let $a > 0$ and let n be a nonzero integer. Prove that $a^{\frac{n-1}{n}} = a\left(\sqrt[n]{\frac{1}{a}}\right)$.

MIXED REVIEW

Evaluate each formula for the given values of the variables. (1.2 Skill C)

28. $A = P(1 + rt)$; $P = 1000$, $r = 0.05$, and $t = 4$

29. $A = P(1 + rt)$; $P = 2000$, $r = 0.05$, and $t = 3$

30. $A = P(1 + r)^t$; $P = 2000$, $r = 0.05$, and $t = 2$

31. $A = P(1 + r)^t$; $P = 4000$, $r = 0.04$, and $t = 4$

32. $A = P(1 + r)^t$; $P = 5000$, $r = 0.04$, and $t = 2$

33. $A = P(1 + r)^t$; $P = 8500$, $r = 0.06$, and $t = 3$

Answers for Lesson 10.5 Skill B

Try This Example 1

a. 16

b. $5\sqrt{5}$

Try This Example 2

$a^{\frac{11}{2}}$

1. $5^{\frac{2}{3}}$

2. $7^{\frac{3}{4}}$

3. $3^{\frac{3}{4}}$

4. $10^{\frac{2}{3}}$

5. $\sqrt[3]{11^2}$

6. $\sqrt{5^3}$

7. $\sqrt{3^5}$

8. $\sqrt[3]{7^{11}}$

9. 625

10. 32

11. 2187

12. 256

13. $3\sqrt{3}$

14. $7\sqrt[3]{7}$

15. $2\sqrt[3]{4}$

16. $25\sqrt[4]{5}$

17. $x^{\frac{3}{2}}$

18. $a^{\frac{7}{2}}$

19. $t^{\frac{7}{3}}$

20. $v^{\frac{8}{3}}$

21. $n^{\frac{5}{3}}$

22. $k^{\frac{5}{6}}$

23. $b^{\frac{7}{6}}$

24. $c^{\frac{5}{2}}$

25. $t^{\frac{37}{12}}$

26. $a^{\frac{n+1}{n}} = a^{1+\frac{1}{n}} = a^1 \cdot a^{\frac{1}{n}} = a \cdot \sqrt[n]{a}$

27. $a^{\frac{n-1}{n}} = a^{1-\frac{1}{n}}$
$$= a(a)^{-\frac{1}{n}}$$
$$= a(a)^{-1 \cdot \frac{1}{n}}$$
$$= a\left(\frac{1}{a}\right)^{\frac{1}{n}}$$
$$= a\left(\sqrt[n]{\frac{1}{a}}\right)$$

28. $1200

29. $2300

30. $2205

31. $4679.43

32. $5400

33. $10,123.64

Focus: *Students use rational exponents in a compound-interest formula.*

EXAMPLE

PHASE 1: Presenting the Example

Before introducing the example, discuss the following with students:

- An amount of money that you invest in an account at a financial institution is called *principal*. The institution uses your principal to make loans to other people and to make its own investments. *Interest* is the money that the institution pays you for this use of your money.

- *Simple interest* is paid only on the principal. *Compound interest* is paid both on the principal and on previously earned interest.

Tell students to consider a principal of $1000 that is invested in an account that earns 5% simple interest annually. Give them the simple-interest formula at right. Have them find the interest earned at the end of two years. [$I = prt = \$1000 \cdot 0.05 \cdot 2 = \$100$] So the amount in the account at the end of the second year would be $1000 + $100 = $1100.

$I = prt$

I: interest
p: principal
r: rate of interest
t: time

Now have them consider a principal of $1000 that is invested in an account that earns 5% *compound* interest annually. Have them use the simple interest formula to find the amount of interest earned at the end of the first year. [$\$1000 \cdot 0.05 \cdot 1 = \$50$] Stress that the amount in the account is now $1000 + $50 = $1050. Have them use that amount to find the second year's interest. [$\$1050 \cdot 0.05 \cdot 1 = \$52.50$] Point out that the amount in the account at the end of the second year would be $1050 + $52.50 = $1102.50. Compare this to the $1100 amount using simple interest. Then discuss the Example.

PHASE 2: Providing Guidance for TRY THIS

Be sure students are aware of the difference between the compound interest formula and the simple interest formula. The compound interest formula yields the entire amount in the account, not just the interest.

PHASE 3: ASSESSMENT AND CLOSURE for Lesson 10.5

- Give students the equations at right. Have them fill in each box with a rational exponent to make a true statement. [**1.** $\frac{1}{6}$ **2.** $\frac{1}{3}$ **3.** $\frac{1}{2}$ **4.** $\frac{2}{3}$ **5.** $\frac{5}{6}$] Then have them create a similar set of equations using 729 as the base. [$729^{\frac{1}{6}} = 3$; $729^{\frac{1}{3}} = 9$; $729^{\frac{1}{2}} = 27$; $729^{\frac{2}{3}} = 81$; $729^{\frac{5}{6}} = 243$]

  1. $64^{\square} = 2$
  2. $64^{\square} = 4$
  3. $64^{\square} = 8$
  4. $64^{\square} = 16$
  5. $64^{\square} = 32$

- This lesson provided a brief introduction to rational exponents. Students will study this topic in greater detail if they pursue a further course in algebra.

☞ *For a Lesson 10.5 Quiz, see* Assessment Resources *page 106.*

Suppose that you invest P dollars in a bank account that pays interest compounded annually. Then after t years, the amount A in the account (in dollars) is given by the *compound-interest formula* below, where r represents the annual interest rate in decimal form.

$$A = P(1 + r)^t$$

If t is a positive integer, then the money remains in the account for a whole number of years. How can you find the amount in the account if t is a period of time such as 1 year and 6 months? The example below illustrates how to answer that question.

EXAMPLE Suppose that an investor puts $5000 into an account that pays 4.5% interest compounded annually. Assuming that no money is withdrawn and no additional money is put into the account, how much money will be in the account after 1 year and 6 months?

▷ Solution

Use a formula.

1. Substitute the numbers into the compound-interest formula. Since time is 1 year and 6 months, use $1\frac{1}{2}$, or 1.5, for t. Rewrite 4.5% as .045.

$$A = 5000(1 + .045)^{1.5} \longleftarrow \text{Let } P = 5000, r = 45, \text{ and } t = 1.5.$$

2. Use a calculator and enter a key sequence like the one below.

exponent key

5000 $(\ $ 1 $+$ .045 $)$ $\wedge$ 1.5 ENTER

3. The calculator display will show 5341.268868.

The account will have $5341.27 in it after 1 year and 6 months.

TRY THIS Rework the Example given an initial deposit of $6000, an annual interest rate of 5.5% and a time of 2 years and 3 months.

Interest may be compounded every 6 months, every 3 months, or even monthly. In these situations, you can use the compound-interest formula given below to find the amount A in the account.

$$A = P\left(1 + \frac{r}{n}\right)^{nt}$$

n: number of compound periods in one year
t: number of years
r: annual rate of interest in decimal form

Refer to the Example. If the interest had been compounded monthly rather than annually, then the amount in the account could be found as shown below.

$$A = 5000\left(1 + \frac{.045}{12}\right)^{12 \cdot 1.5} = \$5348.48$$

EXERCISES

KEY SKILLS

Suppose money is placed into an account that pays 6% as the annual interest rate and the money is in the account for 3 years. Suppose also that compound interest is applied every 6 months.

1. How many times in one year will interest be paid?

2. What percent will be applied each six months?

3. How many times will the interest be applied over 3 years?

PRACTICE

Suppose that $1500 is deposited into an account that pays compound interest at 5% annually. To the nearest cent, how much money will be in the account after each amount of time?

4. 1 year and 9 months
5. 2 years and 6 months
6. 3 years and 3 months
7. 3 years and 8 months

Mr. Tyge invests $2400 into an account that pays 6% compound interest annually. He intends to keep the money there for 2.5 years.

8. a. How much will be in the account if interest is paid annually?

   b. How much will be in the account if interest is paid semi-annually?

   c. By how much will the amounts in parts **a** and **b** differ?

9. Suppose that interest is compounded annually. Will the amount Mr. Tyge have after 5 years be double the amount he will have after 2.5 years?

10. Suppose that interest is compounded annually and that Mr. Tyge puts the money into an account that pays 12% instead of 6%. Will the amount in the first account be double that of the second account after 2.5 years? Justify your response.

MIXED REVIEW APPLICATIONS

A ball thrown directly upward from the edge of a cliff 130 feet above sea level is given an initial velocity of 128 feet per second. To the nearest hundredth of a second, find the amount of time it takes the ball to have the specified altitude above sea level. (8.7 Skill B)

11. 180 feet
12. 210 feet
13. 300 feet
14. 240 feet
15. 130 feet
16. 90 feet

Answers for Lesson 10.5 Skill C

Try This Example
$6768.14

1. two
2. 3%
3. six
4. $1633.70
5. $1694.59
6. $1757.75
7. $1793.85
8. a. $2776.36
   b. $2782.26
   c. $5.90
9. no
10. No, doubling the interest rate will not result in doubling the money. At 6%, he will have $2776.36 in the account; at 12%, he will have $3186.08.
11. 0.41 second and 7.59 seconds
12. 0.68 second and 7.32 seconds
13. 1.68 seconds and 6.32 seconds
14. 0.98 second and 7.02 seconds
15. 0 seconds and 8 seconds
16. 8.30 seconds

Assessing and Justifying Statements about Square Roots

SKILL A *Exploring and proving statements that involve square roots*

Focus: *Students decide whether some statements about square roots are true or false, and whether others are always, sometimes, or never true. They also judge the validity of a logical argument that contains square roots.*

EXAMPLE 1

PHASE 1: Presenting the Example

Display the following statement:

$$\text{If } x \text{ is any real number, then } \sqrt{x^2} = x.$$

Ask students to write whether it is true or false. [**false**] Discuss their responses. Ask them to identify a case for which the statement is true and a case for which it is false. [**samples: It is true when $x = 1$; it is false when $x = -1$.**] Be sure students understand that the statement is false when x represents any negative real number. Remind them that a statement like this can only be identified as true if it is true for *every* real-number value of x. Then discuss Example 1.

EXAMPLE 2

PHASE 1: Presenting the Example

You might wish to present Example 2 as the logical chain of conditional statements shown at right. Remind students that a logical chain is only a valid argument if every statement can be justified. Using this form, they should readily see that the chain is broken at the very beginning because step **1** has no justification.

1. If $n = \sqrt{3^2 + 4^2}$, then $n = \sqrt{3^2} + \sqrt{4^2}$.
2. If $n = \sqrt{3^2} + \sqrt{4^2}$, then $n = 3 + 4$.
3. If $n = 3 + 4$, then $n = 7$.
4. If $n = \sqrt{3^2 + 4^2}$, then $n = 7$.

Have students make the necessary adjustments to create a valid argument. [**1**. If $n = \sqrt{3^2 + 4^2}$, then $n = \sqrt{9 + 16}$. **2**. If $n = \sqrt{9 + 16}$, then $n = \sqrt{25}$. **3**. If $n = \sqrt{25}$, then $n = 5$. **4**. If $n = \sqrt{3^2 + 4^2}$, then $n = 5$.]

EXAMPLE 3

PHASE 2: Providing Guidance for TRY THIS

Students may be tempted to answer *always true* because all negative numbers and 0 are excluded. Ask them if they can think of any positive number x for which $x = \sqrt{x}$. [**1**] If time permits, you might also ask them to identify the set of positive real numbers x for which $x < \sqrt{x}$. [**$0 < x < 1$**]

416   Chapter 10   Radical Expressions, Equations, and Functions

10.6 Assessing and Justifying Statements About Square Roots

SKILL A ▶ *Exploring and proving statements that involve square roots*

Using problem-solving strategies and properties of square roots, you can analyze statements and answer questions about them.

EXAMPLE 1 Prove or disprove: If x is any real number, then $\sqrt{2x} - \sqrt{x}$ is positive.

▶ **Solution**
The statement is false. The expression $\sqrt{2x} - \sqrt{x}$ is not defined in the set of real numbers unless $x \geq 0$. Thus, the statement cannot be true for all real numbers x. Furthermore, if $x = 0$, then $\sqrt{2(0)} - \sqrt{0} = 0$, which is not positive.

TRY THIS Prove or disprove: If x is any real number, then $\sqrt{x} - \sqrt{x-1}$ is positive.

EXAMPLE 2 Find and correct the error in the reasoning below.
$$\sqrt{3^2 + 4^2} = \sqrt{3^2} + \sqrt{4^2} = 3 + 4 = 7 \quad \text{✗}$$

▶ **Solution**
Writing $\sqrt{3^2} + \sqrt{4^2}$ in place of $\sqrt{3^2 + 4^2}$ is incorrect. It assumes that $\sqrt{a+b} = \sqrt{a} + \sqrt{b}$ for all real numbers a and b. However, $\sqrt{a+b} = \sqrt{a} + \sqrt{b}$ is not always true. The correct solution is shown below.
$$\sqrt{3^2 + 4^2} = \sqrt{9 + 16} = \sqrt{25} = 5 \quad \text{✓}$$

TRY THIS Find and correct the error in the reasoning below.
$$\sqrt{5^2 - 4^2} = \sqrt{5^2} - \sqrt{4^2} = 5 - 4 = 1 \quad \text{✗}$$

EXAMPLE 3 Is the statement below sometimes, always, or never true?
If a is any positive number, then $\sqrt{2a} = 2\sqrt{a}$.

▶ **Solution**
If $\sqrt{2a} = 2\sqrt{a}$, then, by squaring each side of the equation, $\left(\sqrt{2a}\right)^2 = \left(2\sqrt{a}\right)^2$. Therefore, $2a = 4a$. The only solution to this equation is 0. Since a represents a *positive* number, 0 is excluded from consideration. Thus, the statement is never true.

TRY THIS Is the statement below sometimes, always, or never true?
If x is any positive number, then $x > \sqrt{x}$.

KEY SKILLS

Let x represent any real number and let $\sqrt{x+5} = \sqrt{x} + \sqrt{5}$.

1. Find three values of x for which the equation is undefined.
2. Find three values of x for which the equation is defined but is false.
3. Find one value of x for which the equation is true.

PRACTICE

Prove or disprove.

4. If x is any real number, then $\sqrt{x+3} > \sqrt{x} + \sqrt{3}$.
5. Let p and q represent squares of positive integers, that is, $p = a^2$ and $q = b^2$ for some positive integers a and b. Then $\sqrt{pq}$ is a positive integer.

Find and correct any error.

6. $\dfrac{12\sqrt{3} - 18\sqrt{5}}{6} = \dfrac{12\sqrt{3}}{6} - 18\sqrt{5} = 2\sqrt{3} - 18\sqrt{5}$ ✗
7. $\dfrac{\sqrt{10(3+7)}}{5} = \sqrt{2(3+7)} = \sqrt{2(10)} = \sqrt{20}$ ✗

In Exercises 8 and 9, is the given statement sometimes, always, or never true? Justify your response.

8. If r represents any positive real number, then $r^2 \geq \sqrt{r}$.
9. If t represents a positive real number, then $\sqrt{\dfrac{t}{2}} = \dfrac{\sqrt{t}}{2}$.
10. For which positive integers n will $\sqrt{n}$ be an even integer?

MIXED REVIEW

In Exercises 11–16, y varies inversely as x. (9.8 Skill A)

11. If $y = 12.5$ when $x = 8$, find x when $y = 25$.
12. If $y = 49$ when $x = 7$, find x when $y = 7$.
13. If $y = 0.25$ when $x = 24$, find x when $y = 1$.
14. If $y = \dfrac{1}{3}$ when $x = 15$, find y when $x = 25$.
15. If $y = 0.2$ when $x = 2$, find y when $x = 2.5$.
16. If $y = 14$ when $x = 0.4$, find x when $x = 0.7$.

Answers for Lesson 10.6 Skill A

Try This Example 1
The statement is false. The expression $\sqrt{x} - \sqrt{x-1}$ is only defined when $x \geq 1$. For these real numbers, the statement is true.

Try This Example 2
Writing $\sqrt{5^2 - 4^2} = \sqrt{5^2} - \sqrt{4^2}$ in Step **1** is incorrect. It assumes that $\sqrt{a-b} = \sqrt{a} - \sqrt{b}$ for all real numbers a and b for which the expressions are defined. However, $\sqrt{a-b} = \sqrt{a} - \sqrt{b}$ is not always true.
$$\sqrt{5^2 - 4^2} = \sqrt{25 - 16} = \sqrt{9} = 3$$

Try This Example 3
If $x = \dfrac{1}{4}$, then $\dfrac{1}{4} > \sqrt{\dfrac{1}{4}}$, or $\dfrac{1}{2}$. This inequality is false. If $x = 4$, then $4 > \sqrt{4}$, or 2. This inequality is true. Therefore, the statement is sometimes true.

1. Answers may vary, but all values of x given should be negative.
2. Answers may vary, but all values of x given should be positive.
3. $x = 0$

4. Answers may vary. Sample answers: The statement is not even defined for negative values of x. Therefore, the statement is false.
   If $x = 0$, the two expressions are equal. Therefore, the statement is false. Counterexamples using positive values of x show that the statement is false.
5. $\sqrt{pq} = \sqrt{a^2 b^2} = \sqrt{a^2}\sqrt{b^2} = ab$
   Since a and b are positive integers, ab is a positive integer. Thus, $\sqrt{pq}$ is a positive integer.
6. The Distributive Property was not applied properly.
   $$\dfrac{12\sqrt{3} - 18\sqrt{5}}{6} = \dfrac{12\sqrt{3}}{6} - \dfrac{18\sqrt{5}}{6} = 2\sqrt{3} - 3\sqrt{5}$$
7. The order of operations was not followed. Operations under the square-root symbol take precedence over division by 5.
   $$\dfrac{\sqrt{10(3+7)}}{5} = \dfrac{\sqrt{10(10)}}{5} = \dfrac{\sqrt{100}}{5} = \dfrac{10}{5} = 2$$

For answers to Exercises 8–16, see Additional Answers.

Focus: *Students use a square-root equation to find the speed of a vehicle given its stopping distance after the brakes are applied.*

EXAMPLE 1

PHASE 1: Presenting the Example

Students previously investigated stopping distance in Lesson 6.2. There the topic was presented in the context of direct variation. Here they will study the same topic as it relates to square-root equations.

Begin by asking students to explain the meaning of the statement *y varies directly as the square of x.* Elicit the response that the relationship between *y* and *x* can be represented by $y = kx^2$, where *k* is a nonzero constant of variation. Then display the following situation:

> A car's braking distance in feet, *d*, varies directly as the square of its speed in miles per hour, *s*. A certain car that is traveling at 40 miles per hour requires 20 feet to stop.

Have students write a variation equation for this situation. $[d = \frac{1}{80}s^2]$ If necessary, review the process of determining the constant of variation, as shown at right. Point out that the equation $d = \frac{1}{80}s^2$ expresses stopping distance in terms of speed.

Now have students solve $d = \frac{1}{80}s^2$ for *s*. $[s = \sqrt{80d}]$ Note that the transformed equation expresses speed in terms of stopping distance. Ask students if they can think of a situation in which it may be more useful to express the relationship in this way. [**Sample: In a high-speed accident, a vehicle often leaves skid marks after the brakes are applied. After the skid marks are measured, the formula can be used to find the speed that the vehicle was traveling.**]

Have students read the paragraphs at the top of the textbook page. Point out that they have now been given a more general equation for speed in terms of stopping distance. Check that they understand how to read the table. Then discuss Examples 1 and 2.

Although students are now using stopping distance to find speed, be sure they understand that the cause-effect relationship between the two quantities has not changed. In other words, stopping distance still depends on speed; speed does not depend on stopping distance. The graph on page 419 has distance on the horizontal axis only to show students the graph of the square-root relationship.

$$d = ks^2$$
$$20 = k \cdot 40^2$$
$$20 = 1600k$$
$$\frac{20}{1600} = \frac{1600k}{1600}$$
$$\frac{1}{80} = k$$

$$d = \frac{1}{80}s^2$$
$$80 \cdot d = 80 \cdot \frac{1}{80}s^2$$
$$80d = s^2$$
$$\sqrt{80d} = \sqrt{s^2}$$
$$\sqrt{80d} = s$$

EXAMPLE 2

PHASE 2: Providing Guidance for **TRY THIS**

Be sure students understand that the response "$\sqrt{2}$" does not suffice. The order of the comparison must be specified. That is, they must respond as follows: "The speed on dry asphalt is $\sqrt{2}$ times the speed on wet asphalt." A second acceptable response would be, "The speed on wet asphalt is $\frac{\sqrt{2}}{2}$ times the speed on dry asphalt."

The relationship between a vehicle's speed s (in miles per hour) and the distance d (in feet) it takes the vehicle to stop once the brakes are applied can be modeled by the following square-root equation.

$$s = \sqrt{30fd}$$

In the equation above, the variable f represents the *coefficient of friction*. Its value depends on road surface, such as concrete or asphalt, and road conditions, such as wet or dry.

f	wet	dry
concrete	0.4	0.8
asphalt	0.5	1.0

EXAMPLE 1 Find the speed of a vehicle that was traveling on wet concrete and required 200 feet to stop after the brakes were applied. Approximate your answer to the nearest hundredth.

▸ **Solution**

Evaluate $s = \sqrt{30fd}$ with $f = 0.4$ and $d = 200$.

Use a formula.

$$s = \sqrt{30(0.4)(200)}$$
$$= \sqrt{30 \cdot \frac{4}{10} \cdot 200}$$
$$= \sqrt{2400} \qquad \longleftarrow \text{Simplify.}$$
$$s \approx 48.99 \qquad \longleftarrow \text{Use a calculator.}$$

The vehicle was traveling about 48.99 miles per hour.

TRY THIS   Rework Example 1 given that the vehicle was traveling on dry concrete.

In Example 2, you are asked to compare the speed of a vehicle given one set of road conditions with the speed of the same vehicle under different road conditions.

EXAMPLE 2 Two vehicles take the same distance to stop after the brakes are applied. One is traveling on wet concrete. The other is traveling on dry concrete. Compare the speeds of the two vehicles before braking.

▸ **Solution**

Use a ratio to compare $\sqrt{30(0.8)d}$ with $\sqrt{30(0.4)d}$.

$$\frac{\text{speed on dry concrete}}{\text{speed on wet concrete}} \;\rightarrow\; \frac{\sqrt{30(0.8)d}}{\sqrt{30(0.4)d}} = \sqrt{\frac{30(0.8)d}{30(0.4)d}} = \sqrt{2}$$

For the vehicle to stop in the same distance on dry concrete, the vehicle can travel at $\sqrt{2}$, or about 1.4, times the speed of the vehicle on wet concrete.

TRY THIS   Rework Example 2 given dry and wet asphalt.

EXERCISES

The speed that a vehicle travels before braking on dry asphalt is given by the formula $s = \sqrt{30d}$. The graph of this relationship is shown at right.

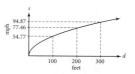

1. In general terms, what change in speed would increase the stopping distance from 100 feet to 200 feet to 300 feet?

2. What change in speed would double the stopping distance from 100 feet to 200 feet?

To the nearest hundredth of a mile per hour, approximate the speed a vehicle was traveling, given each stopping distance and road condition.

3. 50 feet, wet asphalt   4. 50 feet, dry concrete

5. 200 feet, wet concrete   6. 260 feet, dry asphalt

Compare the speeds of two vehicles. Assume each pair of vehicles travels under the same set of conditions.

7. The stopping distance of a vehicle is 1.2 times that of the other vehicle.

8. The stopping distance of a vehicle is 2.4 times that of the other vehicle.

9. **Critical Thinking** Let r represent a positive number. One vehicle takes d feet to stop. A second vehicle takes rd feet to stop under the same road conditions. Which motorist has the greater speed? Justify your response.

At a constant temperature, pressure P (in pounds per square inch) and volume V (in cubic inches) of an ideal gas are related by the inverse-variation equation $PV = k$, for some nonzero constant k. (9.8 Skill C)

10. A gas with volume 45 cubic inches is under a pressure that is 60 pounds per square inch. Find the volume when the pressure is 90 pounds per square inch.

11. A gas with volume 40 cubic inches is under a pressure that is 45 pounds per square inch. Find the volume when the pressure is 75 pounds per square inch.

Answers for Lesson 10.6 Skill B

Try This Example 1
69.28 miles per hour

Try This Example 2
On dry asphalt, the speed of the vehicle before braking would be $\sqrt{2}$, or about 1.4, times the speed on wet asphalt.

10. 30 cubic inches
11. 24 cubic inches

1. An increase in speed will cause an increase in stopping distance.

2. Increasing the speed from 54.77 miles per hour to 77.46 miles per hour will cause the stopping distance to double from 100 feet to 200 feet.

3. about 27.4 miles per hour

4. about 34.6 miles per hour

5. about 49.0 miles per hour

6. about 88.3 miles per hour

7. The second vehicle was traveling at about 110% of the speed of the first vehicle.

8. The second vehicle was traveling at about 155% of the speed of the first vehicle.

9. If $0 < r < 1$, the second vehicle will have the lesser speed. If $r = 1$, both vehicles will have the same speed. If $r > 1$, the first vehicle will have the lesser speed.

Focus: *Students approximate a square root by locating it within successively smaller intervals on a number line.*

EXAMPLE

PHASE 1: Presenting the Example

Most students should have little difficulty with this method of approximating square roots. However, they still might not appreciate the significance of the process. If time permits, you may wish to develop the following situation prior to discussing the Example.

If possible, have metric tape measures available in the classroom. Display the first figure at right. Tell students that it is a plan for a rectangular wooden gate with side lengths of 2 meters and 5 meters. The gate will also have a diagonal brace, indicated by the dashed line labeled c. In order to build the gate, a carpenter needs to find the length of the brace.

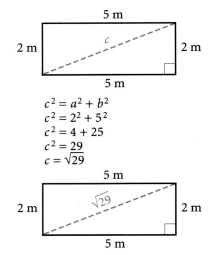

$$c^2 = a^2 + b^2$$
$$c^2 = 2^2 + 5^2$$
$$c^2 = 4 + 25$$
$$c^2 = 29$$
$$c = \sqrt{29}$$

Students should readily see that the length is calculated by applying the Pythagorean Theorem, as shown at right. The result is that the desired length is $\sqrt{29}$ meters. However, no tape measure has a marking for $\sqrt{29}$. So they need to find the best approximation for this length.

Now begin the approximation process on the textbook page. After each approximation, discuss its relevance to the problem of the brace.

1   This approximation places the length between 5 meters and 6 meters. Clearly, this is useless to the carpenter who must cut the brace.

2–3   This gives a length of 5.4 meters. Although this is much more reasonable, students should notice that the tape measure allows for greater precision than this.

4–5   This gives a length of 5.38 meters. This is probably enough precision for the carpenter; however, students may notice that the tape meaure allows for greater precision by showing ten millimeters within each centimeter. Tell students that they can repeat Steps 2 and 3 one more time to find the most precise measure available on the measuring tape.

PHASE 3: ASSESSMENT AND CLOSURE for Lesson 10.6

- Give students the following statement: If x is a positive real number, then $\sqrt{3x}$ ___?___. Have them fill in the blank to create statements that are sometimes, always, and never true. [**See samples at right.**]

  sometimes: $< 3x$;
  always: > 0;
  never: $= 3\sqrt{x}$

- Students will study square roots in greater depth if they pursue a further course in algebra.

  ☞ *For a Lesson 10.6 Quiz, see* Assessment Resources *page 107.*

  ☞ *For a Chapter 10 Assessment, see* Assessment Resources *pages 108–109.*

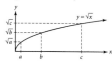

SKILL C *Approximating irrational numbers*

The graph of the square-root function $y = \sqrt{x}$ is shown below.

Notice that if $a < b < c$, then $\sqrt{a} < \sqrt{b} < \sqrt{c}$. For example, since $1 < 2 < 4$, you can say that $1 < \sqrt{2} < 2$. This means that a decimal approximation of $\sqrt{2}$ is a rational number between 1 and 2. Using a numerical method based on the reasoning above, you can approximate any square root to any degree of accuracy.

EXAMPLE Approximate $\sqrt{47}$ to the nearest hundredth without using a calculator.

▶ Solution

Step 1: Find the two perfect squares that lie on either side of 47 on the number line: 36 and 49. Therefore, the square root of 47 will lie between the square roots of 36 and 49, which are 6 and 7.

$$36 < 47 < 49$$
$$\sqrt{36} < \sqrt{47} < \sqrt{49}$$
$$6 < \sqrt{47} < 7$$

Step 2: 47 is closer to 49 than to 36, and so 7 is a closer approximation of $\sqrt{47}$. Divide 47 by 7. Stop when the quotient has one more decimal place than the divisor.

$$7\overline{)47.0} \quad 6.7$$

Step 3: Find the average of 7 and 6.7. The average should have the same number of decimal places as the quotient.

$$\frac{7 + 6.7}{2} = \frac{13.7}{2} \approx 6.8$$

Repeat Steps **2** and **3** using the average 6.8 as the divisor.

Step 4: Divide 47 by 6.8.

$$6.8\overline{)47.00} \quad 6.91$$

Step 5: Find the average of 6.8 and 6.91.

$$\frac{6.8 + 6.91}{2} = \frac{13.71}{2} \approx 6.85$$

Thus, $\sqrt{47} \approx 6.85$.

TRY THIS Approximate $\sqrt{75}$ to the nearest hundredth.

If you continue to divide and average as in Steps 4 and 5, you will get closer and closer approximations of the irrational number $\sqrt{47}$.

The process of finding new numbers within smaller and smaller intervals illustrates the **Density Property of Real Numbers**, which states that between any two real numbers there is another real number.

420   Chapter 10   Radical Expressions, Equations, and Functions

---

EXERCISES

Answer each question and justify your response.

1. Is $\sqrt{5}$ between 2.2 and 2.3 or between 2.3 and 2.4?

2. Suppose that you square 3.3 and find 10.89 while you are trying to approximate $\sqrt{11}$. Would your next estimate be greater than 3.3 or less than 3.3?

3. Given that $4^2 = 16$, $4.5^2 = 20.25$, and $5^2 = 25$, will you find $\sqrt{20}$ between 4 and 4.5 or between 4.5 and 5?

PRACTICE

Use the method in the Example to approximate each square root to the nearest hundredth. Then use a calculator to check your approximation.

4. $\sqrt{5}$ 5. $\sqrt{11}$ 6. $\sqrt{50}$ 7. $\sqrt{45}$

Given the extended inequality below, locate each square root between two consecutive numbers in the inequality.

$$11 < 12 < 13 < 14 < 15 < 16 < 17 < 18$$

8. $\sqrt{260}$ 9. $\sqrt{205}$ 10. $\sqrt{123}$ 11. $\sqrt{230}$

12. Order $\sqrt{73}, \sqrt{37}, \sqrt{30.7}, \sqrt{3.07}, \sqrt{7.3}$, and $\sqrt{3.7}$ from least to greatest.

13. Without doing any calculations, order $\frac{1}{\sqrt{73}}, \frac{1}{\sqrt{37}}, \frac{1}{\sqrt{30.7}}, \frac{1}{\sqrt{3.07}}, \frac{1}{\sqrt{7.3}}$, and $\frac{1}{\sqrt{3.7}}$ from least to greatest.

14. **Critical Thinking** Use guess-and-check to determine which of the statements below is true for all positive real numbers x.

$$\sqrt{x+2} \le \sqrt{x} + \sqrt{2} \qquad \sqrt{x+2} = \sqrt{x} + \sqrt{2} \qquad \sqrt{x+2} \ge \sqrt{x} + \sqrt{2}$$

MIXED REVIEW

Solve each equation. Write the solutions in simplest radical form. (Answers may contain imaginary terms.) (8.2 Skill C)

15. $5x^2 + 21 = -54$ 16. $3n^2 + 14 = -13$

17. $-2n^2 - 8 = 0$ 18. $11k^2 + 176 = 0$

19. $5a^2 + 1 = -9$ 20. $2d^2 + 31 = -9$

21. $3y^2 + 50 = 11$ 22. $-7v^2 + 114 = 16$

Answers for Lesson 10.6 Skill C

Try This Example
8.66

1. between 2.2 and 2.3
2. greater than 3.3
3. between 4 and 4.5
4. 2.23
5. 3.31
6. 7.07
7. 6.70
8. between 16 and 17
9. between 14 and 15
10. between 11 and 12
11. between 15 and 16
12. $\sqrt{3.07}, \sqrt{3.7}, \sqrt{7.3}, \sqrt{30.7}, \sqrt{37}$, and $\sqrt{73}$
13. $\frac{1}{\sqrt{73}}, \frac{1}{\sqrt{37}}, \frac{1}{\sqrt{30.7}}, \frac{1}{\sqrt{7.3}}, \frac{1}{\sqrt{3.7}}, \frac{1}{\sqrt{3.07}}$
14. $\sqrt{x+2} \le \sqrt{x} + \sqrt{2}$
15. $x = \pm 5i$
16. $n = \pm 3i$
17. $n = \pm 2i$
18. $k = \pm 4i$

19. $a = \pm i\sqrt{2}$
20. $d = \pm 2i\sqrt{5}$
21. $y = \pm i\sqrt{13}$
22. $v = \pm \sqrt{14}$

Find the domain of each square-root expression.

1. $\sqrt{-7(3-d)} + 2$

2. $\sqrt{2(n+3) - (3-n)}$

Find the domain and range of each square-root function.

3. $y = \sqrt{-3(4x - 3x + 1)} - 2$

4. $y = \sqrt{\frac{6(x+7) - 12}{3}}$

Solve each square-root equation. Check your solution.

5. $\sqrt{4 - 5(x+1)} - 3 = 1$

6. $\sqrt{5(p+3) - (-1)(2-p)} = 5$

7. $\sqrt{y^2 + 5y} = 6$

8. $\sqrt{\frac{a^2 + 8a}{3}} = 4$

9. Let $O(0, 0)$ represent the origin of the coordinate plane. Find the coordinates of Q and Q' that are directly above and below $P(5, 0)$, respectively, and that are 10 units from the origin.

Solve each square-root equation. Check your solution.

10. $\sqrt{4(g-5)} = \sqrt{3g+4}$

11. $\sqrt{\frac{4x+1}{3}} = \sqrt{\frac{3x-2}{5}}$

12. $\sqrt{12z^2 - 66z} = \sqrt{14z - 77}$

13. $\sqrt{15a^2} = \sqrt{34a - 15}$

Simplify. Write the answers in the form $a + b\sqrt{2}$, where a and b are rational numbers.

14. $(-2 - 5\sqrt{2}) + (-2 - 11\sqrt{2})$

15. $(-10 + 7\sqrt{2}) - (8 - 10\sqrt{2})$

16. $(-1 + 7\sqrt{2})(1 - \sqrt{2})$

17. $(4 - 7\sqrt{2})(4 - 2\sqrt{2})$

18. $\frac{10 - 2\sqrt{2}}{3\sqrt{2}}$

19. $\frac{-1 - 5\sqrt{2}}{1 - \sqrt{2}}$

Solve. Write the answers in the form $a + b\sqrt{2}$, where a and b are rational numbers.

20. $(1 - \sqrt{2})x + 5 = 3\sqrt{2}$

21. $\left(\frac{1}{2} + \sqrt{2}\right)n + (2 - \sqrt{2}) = -\sqrt{2}$

Simplify. Write the answers in simplest radical form.

22. $\sqrt{50}\sqrt{2}$

23. $\frac{\sqrt{192}}{\sqrt{3}}$

24. $\frac{\sqrt{500}}{\sqrt{50}}$

25. $\frac{\sqrt{3}}{\sqrt{18}}$

26. $7\sqrt{13} + (-5 + 13\sqrt{13})$

27. $(5 + 7\sqrt{7}) + (-5 + 7\sqrt{7})$

28. $-10\sqrt{17} - (-5 - \sqrt{17})$

29. $(3 - 3\sqrt{13}) - (13 + 13\sqrt{13})$

30. $-2\sqrt{5}(-1 + 11\sqrt{5})$

31. $(3 - \sqrt{5})(-2 + 10\sqrt{5})$

32. $\frac{-3 + 9\sqrt{7}}{\sqrt{7}}$

33. $\frac{-5 - 8\sqrt{11}}{5 + \sqrt{11}}$

Write each radical expression as an exponential expression and write each exponential expression as a radical expression.

34. $\sqrt[3]{7}$

35. $\sqrt[4]{29}$

36. $11^{\frac{1}{3}}$

37. $15^{\frac{1}{4}}$

Simplify each expression.

38. $\sqrt[5]{1}$

39. $\sqrt[3]{64}$

40. $27^{\frac{1}{3}}$

41. $625^{\frac{1}{4}}$

Write each expression in simplest radical form.

42. $4^{\frac{3}{2}}$

43. $1000^{\frac{1}{3}}$

44. $32^{\frac{2}{5}}$

45. $27^{\frac{2}{3}}$

Write each expression as an expression with a rational exponent. Assume all variables represent positive real numbers.

46. $a\sqrt[3]{a}$

47. $(b^2)\sqrt[3]{b}$

48. $(c^3)\sqrt[3]{c^2}$

49. $(z^5)\sqrt[5]{z}$

50. If $3500 is invested in an account paying 5% compound interest annually, how much will be in the account after 2 years and 3 months?

51. Is the statement below sometimes, always, or never true?
If a is any positive real number, then $\sqrt{6.25a^2} = 2.5a$.

52. Let $s = \sqrt{30(0.8)d}$, where s represents speed of a vehicle (in miles per hour) when the brakes are applied and d represents the stopping distance of the vehicle (in feet). To the nearest hundredth of a mile per hour, approximate the speed of a vehicle that takes 240 feet to stop.

53. Using inequalities, approximate $\sqrt{7}$ to the nearest hundredth.

Answers for Chapter 10 Assessment

1. $d \geq \frac{19}{7}$

2. $n \geq -1$

3. domain: $x \leq -\frac{5}{3}$; range: $y \geq 0$

4. domain: $x \geq -5$; range: $y \geq 0$

5. $x = -1$

6. $p = 2$

7. $y = -9$ or $y = 4$

8. $a = -12$ or $a = 4$

9. $Q(5, 8.7)$ and $Q'(5, -8.7)$

10. $g = 24$

11. no real solution

12. $z = \frac{11}{2}$

13. $a = \frac{3}{5}$ or $a = \frac{5}{3}$

14. $-4 - 16\sqrt{2}$

15. $-18 + 17\sqrt{2}$

16. $-15 + 8\sqrt{2}$

17. $44 - 36\sqrt{2}$

18. $-\frac{2}{3} + \frac{5}{3}\sqrt{2}$

19. $11 + 6\sqrt{2}$

20. $x = -1 + 2\sqrt{2}$

21. $n = \frac{4}{7} - \frac{8}{7}\sqrt{2}$

22. 10

23. 8

24. $\sqrt{10}$

25. $\frac{\sqrt{6}}{6}$

26. $-5 + 20\sqrt{13}$

27. $14\sqrt{7}$

28. $5 - 9\sqrt{17}$

29. $-10 - 16\sqrt{13}$

30. $-110 + 2\sqrt{5}$

31. $-56 + 32\sqrt{5}$

32. $9 - \frac{3}{7}\sqrt{7}$

33. $\frac{9}{2} - \frac{5}{2}\sqrt{11}$

34. $7^{\frac{1}{3}}$

35. $29^{\frac{1}{4}}$

36. $\sqrt[3]{11}$

37. $\sqrt[4]{15}$

38. 1

39. 4

40. 3

41. 5

42. 8

43. 10

44. 4

45. 9

46. $a^{\frac{4}{3}}$

47. $b^{\frac{7}{3}}$

48. $c^{\frac{11}{3}}$

49. $z^{\frac{26}{5}}$

50. $3906.11

51. The statement is always true.
Since $6.25 = 2.5^2$,
$\sqrt{6.25a^2} = $
$\sqrt{2.5^2 \cdot a^2}$
$= \sqrt{2.5^2}\sqrt{a^2} = $
$2.5 \cdot a = 2.5a$

52. 75.89 miles per hour

53. 2.64

Glossary

A

absolute value   The absolute value of any real number x, written $|x|$, is the distance from x to 0 on a number line. If x is greater than or equal to 0, then $|x| = x$. If x is less than 0, then $|x| = -x$. (44)

acceleration   The rate of change of velocity with respect to time. (334)

Addition Property of Equality   If a, b, and c are real numbers and $a = b$, then $a + c = b + c$. (20, 62)

Addition Property of Inequality   If a, b, and c are real numbers and $a < b$, then $a + c < b + c$. This statement is also true if $<$ is replaced by $>$, $\leq$ or $\geq$. (148)

additive identity   The number 0 is the additive identity because 0 added to any number equals that number. (8)

additive inverses   Two numbers are additive inverses if their sum is 0. For example, $3 + (-3) = 0$, so 3 and -3 are additive inverses, or opposites.

algebra   The branch of mathematics in which operations of arithmetic are generalized by the use of letters to represent unknown or varying quantities. (4)

algebraic expression   An expression that contains numbers, variables, and operation symbols. (12)

Associative Property of Addition   If a, b, and c are real numbers, then $(a + b) + c = a + (b + c)$. (8)

Associative Property of Multiplication   If a, b, and c are real numbers, then $(a \times b) \times c = a \times (b \times c)$. (8)

axis of symmetry (of a parabola)   The vertical line which divides a parabola into two parts that are mirror images of each other. (286)

B

boundary line   The graph of the related linear equation for a linear inequality. The line is the edge of the solution region of the inequality and is either solid or dashed. (206)

base (of an exponential expression)   The number that is raised to an exponent. In an expression of the form x^a, x is the base. (220)

binomial   A polynomial with exactly two terms. (230)

C

closure   A set of numbers is said to be closed under an operation if for every pair of numbers in the set, the result of the operation is also a member of the set. For example, the natural numbers are closed under addition and multiplication but not under subtraction and division; the difference of two natural numbers can be a negative number and the quotient can be a fraction. (8)

coefficient   The number used as a factor in a monomial, such as 6 in $6a$. If only a variable is written, the coefficient is understood to be 1. (66)

collinear   Points are collinear if they lie on the same line. (108)

common binomial factor   A binomial that appears as a factor in more than one term of a polynomial. (268)

Commutative Property of Addition   For all real numbers a and b, $a + b = b + a$. (8)

Commutative Property of Multiplication   For all real numbers a and b, $a \times b = b \times a$. (8)

completing the square   A technique for modifying a quadratic polynomial to obtain a perfect-square trinomial; it provides a general method for solving quadratic equations. (314)

complex fraction   A fraction whose numerator or denominator (or both) contains a fraction. (368)

composite number   A number that has factors other than 1 and itself. (260)

compound inequality   A pair of inequalities joined by *or* (disjunction) or *and* (conjunction). (146)

compound interest   Interest paid on earned interest. The formula for calculating compound interest is $A = P(1 + r)^t$, where A is the total amount of principal and interest, P is the original amount of principal, r is the rate of interest for the compounding period, and t is the number of compounding periods. (414)

conclusion   The consequence that follows the conditions in a conditional statement. If the statement is in *If-then* form, the conclusion is the part that follows *then*. (84)

conditional statement   A statement in which a specific conclusion follows a given hypothesis or hypotheses. The statement may be written in *If-then* form. The part following *If* is the hypothesis and part following *then* is the conclusion. (84)

congruent polygons   Polygons that can be made to coincide. They have the same size and shape and their corresponding sides and angles are equal. (238)

conjugates (radical)   A pair of expressions of the forms $a + b\sqrt{c}$ and $a - b\sqrt{c}$. The product of two conjugates contains no radical sign. (400)

conjunction   A pair of inequalities joined by *and*. A conjunction can be written in the form $a < x < b$. (170)

consistent system   A system of equations or inequalities that has at least one solution. (202)

constant   A term in an algebraic expression that does not contain any variables. (230)

constant of variation   The constant, k, in an equation of direct variation $\left(k = \dfrac{y}{x}\right)$ or of inverse variation ($k = xy$). (102, 378)

continuous graph   An unbroken graph. (116)

converse   A statement obtained by interchanging the parts of a conditional statement. Given *If p, then q,* the converse is *If q, then p.* The converse of a true statement is not necessarily true. (384)

coordinate axes   The vertical and horizontal lines in a coordinate plane that intersect at the origin, $(0, 0)$. Both axes represent number lines. The horizontal axis usually represents the values of the independent variable and the vertical axis usually represents values of the dependent variable. (92)

coordinate plane   A plane that is divided into four regions by a horizontal and a vertical number line.

counterexample   An example used to show that a statement is not always true. (16)

Cross-Product Property of Proportions   If a, b, c, and d are real numbers with $b \neq 0$ and $d \neq 0$, and $\dfrac{a}{b} = \dfrac{c}{d}$, then $ad = bc$. (372)

cross products   In a proportion $\dfrac{a}{b} = \dfrac{c}{d}$, the cross products are bc (the product of the means) and ad (the product of the extremes). (372)

cube-root radical   The symbol $\sqrt[3]{}$, which indicates to take the cube root of the expression under it. The cube root of a is the number that, when used as a factor 3 times, gives a as the product. (410)

cubic equation   An equation that can be written in the form $ax^3 + bx^2 + cx + d = 0$, where $a \neq 0$. (284)

cubic polynomial   A polynomial whose degree is three. (230)

D

data   (plural) Factual information, such as measurements or statistics. (116)

deductive reasoning   The process of reasoning logically from a hypothesis to a conclusion, using only certain clearly stated assumptions (axioms), defined terms, and statements (theorems) that have been previously proven. These elements are put in a sequence that follows the rules of formal logic. (84, 134, 326)

degree (of a polynomial)   The greatest exponent in all of the terms of a simplified polynomial. (230)

Density Property of Real Numbers   The Density Property states that between any two real numbers, there is another real number. (420)

dependent system   A system of equations that has infinitely many solutions. (204)

dependent variable   In a function of two variables, the variable of the range. (100)

descending order (of a polynomial)   A polynomial in one variable is said to be in descending order when the exponents of the terms decrease from left to right. (230)

difference of two squares   A binomial of the form $a^2 - b^2$. It can be factored as $(a - b)(a + b)$. (250, 266)

direct variation   If y varies directly as x, then $y = kx$, where k is a fixed nonzero number called the *constant of variation*. (102)

discriminant   The expression under the radical in the quadratic formula, $b^2 - 4ac$. (320)

discrete graph   A graph consisting of points that are not connected. (116)

disjunction   A pair of inequalities joined by *or*. (172)

Distance Formula   If $P(x_1, y_1)$ and $Q(x_2, y_2)$ are two points in the coordinate plane, then the distance between them is given by the formula
$PQ = \sqrt{(x_2 - x_1)^2 + (y_2 - y_1)^2}$. (394)

Distributive Property   For all real numbers a, b, and c, $a(b + c) = ab + ac$ and $(b + c)a = ba + ca$. (8)

Division Property of Equality   If a, b, and c are real numbers and $a = b$ and $c \neq 0$, then $\frac{a}{c} = \frac{b}{c}$. (30, 64)

Division Property of Inequality   Let a, b, and c be real numbers. If c is positive and $a < b$, then $\frac{a}{c} < \frac{b}{c}$. If c is negative and $a < b$, then $\frac{a}{c} > \frac{b}{c}$. Similar statements can be written using the other inequality symbols. (154)

domain   The set of all numbers which can be assigned to the independent variable in a function with two variables. The domain of a rational expression is restricted to values that do not cause the denominator to equal 0. (96, 342)

double root   If the polynomial member of an equation has two identical factors, then the equation has a double root. For example, in $y = (x - r)^2$, r is a double root. This double root will be represented in the graph of the equation as touching, but not crossing, the x-axis. (288)

elimination method   A method of solving a system of equations by multiplying and combining the equations in the system in order to eliminate a variable. (196)

equation   A mathematical statement which uses the equal symbol ($=$) to show that two expressions are equivalent. (14)

equivalent equations   Equations that have the same solution, such as $5x = 20$ and $2x = 8$. (70)

evaluate (an algebraic expression)   Substitute a given value for the variable(s) in the expression and simplify. (232)

excluded value   A value that is excluded from the domain of a function because no range value exists for it. The term is commonly used in reference to values in rational functions that cause the denominator to equal 0. (342)

exponent   The number that tells how many times the *base* is used as a factor. In an expression of the form x^a, a is the exponent. (10, 220)

exponential expression   An algebraic expression in which the exponent is a variable and the base is a fixed number. (220)

extraneous solution   A solution, found in the process of solving an equation, that is not a solution to the original equation. (374, 396)

F

factor (noun) A divisor of a number or of an expression. (verb) To write a number or an expression as the product of two or more numbers or expressions. (260)

factored completely The description of a polynomial which has been written as a product of factors that cannot be factored any further. It is the equivalent of prime factorization of numbers. (278)

field Any set that has the Associative, Closure, Commutative, Distributive, Identity, and Inverse Properties for any two operations. (402)

FOIL (method) A mnemonic for the procedure of multiplying two binomials. Add the products of the First terms, Outer terms, Inner terms, and Last terms. (248)

formula An equation that expresses a relationship between two or more quantities. (12)

fourth-root radical A radical with the number 4 as an *index*, $\sqrt[4]{}$. (410)

function A relation in which each member of the first set, the domain, is assigned exactly one member of the second set, the range. Functions can be expressed in the form of algebraic equations which give rules for assigning values in the range to values in the domain. For example, $y = 5x$ is a function that pairs each domain value, x, with a range value 5 times greater. (96)

G

greatest common factor (GCF) The largest number that is a divisor of two or more given numbers. Also, the monomial with all the integer and variable factors common to every term of a polynomial. (262)

H

horizontal-addition format (in the addition and subtraction of polynomials) The placement of expressions to be added side by side; the calculations are performed by grouping and combining like terms. (234)

horizontal line A line parallel to the x-axis. The equation of the horizontal line that contains the point $(0, s)$ is $y = s$. (112)

hypotenuse In a right triangle, the side opposite the right angle. (308)

hypothesis The condition(s) given in a conditional statement. If a statement is in *If-then* form, the hypothesis is the part that follows *If*. (84)

I

identity An equation that is true regardless of the value of the variable(s). (14)

Identity Property of Addition For all real numbers a, $a + 0 = a$ and $0 + a = a$; 0 is the identity element for addition. (8)

Identity Property of Multiplication For all real numbers a, $a \times 1 = a$ and $1 \times a = a$; 1 is the identity element for multiplication. (8)

imaginary unit The imaginary number i, which represents the quantity $\sqrt{-1}$. (310)

inconsistent system A system of equations that has no solution. (202)

independent system A consistent system of equations that has exactly one solution. (204)

independent variable In a function, the variable of the domain. (100)

index The number used with the radical sign to indicate the root. For example, in the expression $\sqrt[3]{5}$, the index 3 indicates that the expression represents the cube root of 5. (410)

indirect reasoning   The process by which a statement is proved to be true by showing that the negation of the statement is false. (86)

inductive reasoning   Reasoning that proceeds from individual or specific cases to general rules or theories. The scientific method is based upon inductive reasoning, and therefore scientific conclusions can never actually be proven. They can only appear to be supported more or less strongly by collected evidence. On the other hand, mathematical conclusions are based upon deductive reasoning and can be proved to be true or false. (134)

inequality   A statement that the value of one expression is not equal to the value of another. Strict inequality is shown by the symbols $>$ (is greater than) or $<$ (is less than). The symbols $\leq$ (is less than or equal to) and $\geq$ (is greater than or equal to) allow for the possibility of equality. (142)

infinite number of digits   An unlimited number of digits. (40, 420)

integers   The set of whole numbers and their opposites: ... $-4, -3, -2, -1, 0, 1, 2, 3, 4, ...$(40)

inverse operations   Operations that "undo" one another. For example, addition and subtraction are inverse operations, as are multiplication and division. (32)

Inverse Property of Addition   For every real number a, there is an opposite real number, $-a$, such that $a + (-a) = -a + a = 0$. (8)

Inverse Property of Multiplication   For every real number a, there is a multiplicative inverse $\frac{1}{a}$ such that $a \times \frac{1}{a} = \frac{1}{a} \times a = 1$. (8)

inverse-square variation   The relationship between x and y in the equation $x^2y = k$. y is said to vary inversely as the square of x. (382)

inverse variation   The relationship between x and y in $xy = k$, where k is a constant. The variation is inverse because when x increases, y decreases. An alternative form of inverse variation can be written as $x_1y_1 = x_2y_2$, when (x_1, y_1) and (x_2, y_2) both satisfy $xy = k$. (378)

irrational numbers   The set of all nonrepeating and non-terminating decimals. Irrational numbers cannot be written in the form $\frac{p}{q}$ where p and q are integers and $q \neq 0$. (42)

leading coefficient   The numerical part of the leading term of a polynomial. (230)

leading term   In a polynomial, the term with the greatest exponent. (230)

least common denominator (LCD)   The smallest number that is a multiple of the denominators of two or more fractions. (358)

least common multiple (LCM)   The smallest number that is a multiple of two or more given numbers. (76, 358)

leg   In a right triangle, one of the sides adjacent to the right angle. (308)

like terms   Monomials whose variable parts are the same, including the powers to which they are raised. (66)

linear equation in two variables   An equation that can be written in the form $Ax + By = C$, where A, B, and C are real numbers, and A and B are not both 0. The graph of a linear equation is a straight line in the coordinate plane. (112)

linear inequality in standard form   An inequality of the form $Ax + By < C$, where A, B, and C are real numbers, and A and B are not both 0. (The $<$ may be replaced by $>$, $\geq$, or $\leq$.) (212)

linear function   Any function that can be defined by a linear equation, and usually written in the form $y = mx + b$. (132)

linear inequality   An inequality formed when the equal sign ($=$) of a linear equation is replaced by an inequality symbol: $>$, $<$, $\geq$, or $\leq$. (206)

linear pattern   A pattern that can be modeled by a linear function. (132)

linear (polynomial) A polynomial whose degree is one. (230)

logic The study of the structure of statements and the formal laws of reasoning. (4)

mathematical expression A quantity made up of variables, numbers, operation symbols, and inclusion symbols. (4)

mathematical proof A logical argument, or sequence of statements, used to show that a statement is true. (18)

maximum The greatest value (of a function, or of a situation). (158)

minimum The least value (of a function, or of a situation). (158)

monomial An algebraic expression that is either a constant, a variable, or a product of a constant and one or more variables; one term in a polynomial. (66, 230)

Multiplication Property of Equality If a, b, and c are real numbers and $a = b$, then $ac = bc$. (28, 64)

Multiplication Property of Inequality Let a, b, and c be real numbers. If c is positive and $a < b$, then $ac < bc$. If c is negative and $a < b$, then $ac > bc$. Similar statements can be written using the other inequality symbols. (154)

Multiplication Property of 0 If $a = 0$ or $b = 0$, then $ab = 0$. (280)

multiplicative identity The number 1, because $a \times 1 = 1 \times a = a$. (8)

multiplicative inverse A reciprocal. If $a \neq 0$, then $\frac{1}{a}$ is the multiplicative inverse, or reciprocal, of a. (58)

Multiplicative Property of −1 For all real numbers a, $-1(a) = -a$. (52)

natural numbers The numbers that are used in counting: 1, 2, 3, ... (40)

negation Turning a statement into its opposite. The negation of "it is raining" is "it is not raining." In symbolic logic, the negation of statement p is written $\sim p$. (86)

negative exponent If a and n are real numbers and $a \neq 0$, then $a^{-n} = \frac{1}{a^n}$. (222)

numerical expression An expression that contains numbers and operation symbols, but no variables. (10)

open sentence An equation that contains at least one variable. (14)

opposite of a difference For all real numbers a and b, $-(a - b) = b - a$. (52)

opposite of a sum For all real numbers a and b, $-(a + b) = -a - b$. (52)

opposites Two numbers that lie on opposite sides of zero and are the same distance from zero on a number line, such as 3 and -3. Also called *additive inverses*. (8, 44)

order of operations When simplifying an expression, the order of operations is: 1) simplify expressions within grouping symbols, 2) simplify expressions involving exponents, 3) multiply and divide from left to right, and 4) add and subtract from left to right. (10)

ordered pair A pair of numbers, usually written in the form (x, y), which corresponds to a point in the coordinate plane. (92)

origin On the real number line, the origin is the number 0. In the coordinate plane, the origin is the point where the axes intersect, $(0, 0)$. (44, 92)

P

parabola   The U-shaped graph of a quadratic function. (286)

parallel lines   Lines in the same plane that never intersect. Two lines are parallel if they have the same slope. (126)

parallelogram   A quadrilateral in which both pairs of opposite sides are parallel. (131)

parameter   The variable in parametric equations upon which the x- and y-values are dependent. Frequently, the parameter represents time. (136)

parametric equations   Equations by which the coordinates of a graph may be described as two functions of a single variable. That variable is called the parameter, and the x- and y-values are both dependent upon it. (136)

pentagon   A 5-sided closed figure. (214)

perfect square   A rational number whose square roots are also rational numbers. (300)

perfect-square trinomial   The result of squaring a binomial; a trinomial of the form $a^2 + 2ab + b^2$ or $a^2 - 2ab + b^2$, which can be factored as $(a + b)^2$. (250, 266)

perpendicular lines   Lines that intersect at right angles. Two lines in a coordinate plane are perpendicular if the product of their slopes is -1. (128)

point-slope form   The point-slope form of a linear equation is $y - y_1 = m(x - x_1)$ where the coordinates x_1 and y_1 are taken from a given point (x_1, y_1), and m is the slope. (118)

polynomial   The sum of one or more monomials. (230)

power function   A function of the form $y = kx^n$, where $k \neq 0$ and n is a positive integer. (226)

Power-of-a-Fraction Property   If a, b, and n are real numbers and $b \neq 0$, then $\left(\dfrac{a}{b}\right)^n = \dfrac{a^n}{b^n}$. (222)

Power-of-a-Power Property   If a, m, and n are real numbers and $a \neq 0$, then $(a^m)^n = a^{mn}$. (220)

Power-of-a-Product Property   If a, b, and m are real numbers, then $(ab)^m = a^m b^m$. (220)

prime factorization   A product that contains only prime numbers or powers of prime numbers as factors. (260)

prime number   Any natural number greater than 1 whose only factors are 1 and itself. (260)

principal square root   The positive square root of a positive number. (300)

Product Property of Square Roots   If a and b are positive real numbers, then $\sqrt{ab} = \sqrt{a}\sqrt{b}$. (302)

profit   In business, the amount remaining after expenses are subtracted from income. (238)

Property of Equality of Squares   If two expressions are equal, then their squares are also equal: if $r = s$, then $r^2 = s^2$. (392)

proportion   A mathematical statement that two ratios are equal: $\dfrac{a}{b} = \dfrac{c}{d}$. (104)

pure imaginary number   A number of the form bi, where $b \neq 0$. (310)

Pythagorean Theorem   In any right triangle, the square of the length of the hypotenuse, c, equals the sum of the squares of the lengths, a and b, of the legs: $a^2 + b^2 = c^2$. (308)

quadrants   The four sections of the coordinate plane formed by the coordinate axes. (92)

quadratic equation   An equation that can be written in the form $ax^2 + bx + c = 0$, where $a \neq 0$. (280)

quadratic formula   A formula that can be used to solve quadratic equations: If $ax^2 + bx + c = 0$, then $x = \dfrac{-b \pm \sqrt{b^2 - 4ac}}{2a}$. (318)

quadratic function   Any function defined by a quadratic equation. (286)

quadratic pattern   A pattern that can be modeled by a quadratic equation. (332)

quadratic polynomial   A polynomial whose degree is two. (230)

quadrilateral   A four-sided closed figure. Squares, rectangles, trapezoids, and kites are all examples of quadrilaterals. (130)

Quotient Property of Exponents   If a, m, and n are real numbers and $a \neq 0$, then $\dfrac{a^m}{a^n} = a^{m-n}$. (222)

Quotient Property of Square Roots   If a and b represent positive real numbers, then $\sqrt{\dfrac{a}{b}} = \dfrac{\sqrt{a}}{\sqrt{b}}$. (302)

radical sign   The sign that shows a number is a radical. For example, $\sqrt{a}$ is used to denote the *principal square root* of a. (300, 390)

radicand   The expression under a radical sign. For example, in $\sqrt{x-1}$, $x-1$ is the radicand. (390)

range   The second numbers in a set of ordered pairs. In a function, there is only one number in the range assigned to a number in the domain. (96)

rate   A ratio that compares quantities of different kind of units, such as miles per hour or minutes per day. (102)

rate of change   In a linear function, the ratio of the change in the y-values to the change in the x-values. This relationship is represented by the slope of the graph of the function. (110)

ratio   A quotient that compares two quantities. A ratio of 5 boys to 2 girls can be written as $\dfrac{5}{2}$. (102)

rational equation   An equation that contains at least one rational expression. (372)

rational expression   An algebraic expression that is or can be expressed as the quotient of two polynomials. (342)

rational function   A function defined by a rational equation. (342)

rationalizing the denominator   The process of changing a fraction with an irrational denominator to an equivalent fraction with a rational denominator. (404)

rational numbers   The set of numbers that can be written in the form $\dfrac{p}{q}$, where p and q are integers and $q \neq 0$. (40)

real numbers   The set of numbers that consists of all the rational and irrational numbers. It does not include the set of imaginary numbers. (42)

reciprocal   For any nonzero number a, the reciprocal is $\dfrac{1}{a}$. A number multiplied by its reciprocal equals 1. Also called the *multiplicative inverse*. (8)

reciprocal function   The function in which the dependent variable is the reciprocal of the independent variable, $y = \dfrac{1}{x}$. This function is a form of the inverse-variation function in which $k = 1$. (380)

Reflexive Property of Equality   For any real number a, $a = a$ (a number is equal to itself). (18)

relation   A pairing of the elements in two sets. A relation can be represented by a set of ordered pairs, (x, y). (96)

relatively prime numbers   Two or more numbers that have no common factors other than 1. (261)

repeating decimal   A number in decimal form that has a string of one or more digits that repeat infinitely, such as 0.5555… or 4.123412341234… These repeating decimals can be indicated by a bar over the repeated numbers, as in $0.\overline{5}$ or $4.\overline{1234}$. (40)

replacement set   The set of all numbers that satisfy the conditions as possible replacements for the variable. (14)

right triangle   A triangle that has a 90° angle. (130)

root (of an equation)   A solution to an equation. (280, 318)

scientific notation   A number written in scientific notation is a product of the form $a \times 10^n$, where $1 \leq a < 10$ and n is an integer. (224)

second differences   In a table of ordered pairs, the numerical difference between one entry and the next of one of the variables is called the first difference. The difference between one first difference and the next is called the second difference. Second differences of the dependent variable can be used to determine whether the pattern is quadratic or not. (328)

sequence   An ordered set of real numbers. Each number in a sequence is called a term. (132)

similar geometric figures   Geometric shapes in which the corresponding sides have the same ratio and corresponding angles have the same measurement. If corresponding sides have the same ratio in a triangle, the corresponding angles will be equal. (366)

simplest radical form   The form of a square-root radical expression when there are no perfect squares or fractions under the radical sign and no radical expressions in the denominator of a fraction. (302)

simplify   To carry out all indicated operations in an expression. (10)

slope   A measurement of the steepness of a line in the coordinate plane. Given two points on a line, (x_1, y_1) and (x_2, y_2), the slope, m, of the line is given by $m = \dfrac{\text{rise}}{\text{run}} = \dfrac{y_2 - y_1}{x_2 - x_1}$.

slope-intercept form   A linear equation in the form $y = mx + b$, where m is the slope of the line and b is the y-intercept. (114)

solution (of an absolute-value equation)   If $|x| = a$ and $a \geq 0$, then $x = a$ or $x = -a$. (176)

solution (of an absolute-value inequality)   If $|x| < a$ and $a > 0$, then $x < a$ and $x > -a$ (which can be written $-a < x < a$). If $|x| > a$ and $a > 0$, then $x > a$ or $x < -a$. (178)

solution (of an open sentence)   Any value of the variable(s) that makes the equation true. (14)

solution (of an equation in two variables)   An ordered pair of numbers (x, y) that makes the equation true. (94, 112)

solution (of a system of linear equations)   Any ordered pair that satisfies each of the equations in the system. Graphically, the solution is the point of intersection of the graphs. (188)

solution region   The set of all possible solutions of a linear inequality. In a graph, the solution region is indicated by shading the region. (206)

splitting the middle term   The process of factoring an expression of the form $ax^2 + bx + c$ when $a \neq 1$. In this process, the middle term is rewritten as the sum of two terms and then factoring by grouping is applied. (274)

square root   The number c is a square root of a if $c^2 = a$. The principal, or positive, square root of a can be represented as $\sqrt{a}$ or as $a^{\frac{1}{2}}$. (300)

square-root equation   An equation in one variable that contains a square-root expression, such as $\sqrt{5x + 2} = 12$. (392)

square-root expression   An expression that contains a radical, $\sqrt{\quad}$. (343090)

square-root function   A function defined by a square-root expression, such as $y = \sqrt{3x}$ or $y = 4 - 2\sqrt{x - 5}$. (390)

standard form of a cubic equation   The form $ax^3 + bx^2 + cx + d = 0$, where $a \neq 0$. (284)

standard form of a quadratic equation   The form $ax^2 + bx + c = 0$, where $a \neq 0$. (280)

subset   A set that is contained in another set. A set that has fewer elements than the set that contains it is called a proper subset. (42)

substitution method   A procedure for solving a system of equations in which variables are replaced with known values or algebraic expressions. (190)

Substitution Principle   If $a = b$, then b may replace a in any true statement containing a and the resulting statement will be true. (12)

Subtraction Property of Equality   If a, b, and c are real numbers and $a = b$, then $a - c = b - c$. (22, 62)

Subtraction Property of Inequality   If a, b, and c are real numbers and $a < b$, then $a - c < b - c$. This statement is also true if $<$ is replaced by $>$, $\leq$ or $\geq$. (150)

Symmetric Property of Equality   For all real numbers a and b, if $a = b$, then $b = a$. (18)

system of linear equations (or inequalities)   A set of linear equations (or inequalities) with the same variables. (188, 210)

term (of a polynomial)   Any monomial in a polynomial. (230)

term (of a sequence)   Each number in a sequence of numbers. (132)

terminating decimal   A decimal number that has a finite number of nonzero digits to the right of the decimal point, such as 0.25 or 0.333. (40)

tolerance   The acceptable range that a measurement may differ from a fixed standard. (180)

Transitive Property of Deductive Reasoning   If p, q, and r represent statements and if $p \Rightarrow q$ and $q \Rightarrow r$, then $p \Rightarrow r$. (84)

Transitive Property of Equality   For all numbers a, b, and c, if $a = b$ and $b = c$, then $a = c$. (18)

Transitive Property of Inequality   For all numbers a, b, and c, if $a < b$ and $b < c$, then $a < c$. This is also true if $<$ is replaced by $>$, $\leq$, or $\geq$. (146)

trapezoid   A quadrilateral with exactly one pair of parallel sides. (130)

trinomial   A polynomial with exactly three terms. (230)

variable   A letter used to represent numbers in an algebraic expression. (12)

vertex (of a parabola)   The point of intersection of a parabola and its axis of symmetry; this point will represent either the minimum or maximum value of the function. (286)

vertical-addition format (in the addition or subtraction of polynomials)   The placement of the like terms above (or below) one another before calculation. (234)

vertical line   A line parallel to the y-axis. The equation of the vertical line that contains the point $(r, 0)$ is $x = r$. (112)

vertical-line test   A method of determining whether a graph represents a function: If no vertical line in the coordinate plane intersects the graph in more than one point, then the graph represents a function. (96)

whole numbers   The set of natural numbers and 0. (40)

work rate   The proportional amount of a task that is completed in one unit of time. For example, if a task takes a person 5 hours to complete, then the work rate is $\frac{1}{5}$ of the task per hour. (376)

x-axis   The horizontal axis in a coordinate plane, usually representing the values of the independent variable. (92)

x-coordinate   The first number in an ordered pair. It represents the horizontal distance from the origin of a point in the coordinate plane. (92)

x-intercept   The x-coordinate of the point where a graph crosses the x-axis. (112)

y-axis   The vertical axis in a coordinate plane, usually representing the values of the dependent variable. (92)

y-coordinate   The second number in an ordered pair. It represents the vertical distance from the origin of a point in the coordinate plane. (92)

y-intercept   The y-coordinate of the point where a graph crosses the y-axis. (112)

zero exponent   For any nonzero number a, $a^0 = 1$. (222)

Zero-Product Property   If a and b are real numbers and $ab = 0$, then $a = 0$ or $b = 0$. (280)

Additional Answers

CHAPTER 1

Lesson 1.3 Skill B (page 17)

19. 2.25; 225%   **20.** 2.4; 240%   **21.** 0.16; $\frac{4}{25}$

22. 0.62; $\frac{31}{50}$ **23.** 1.25; $\frac{5}{4}$ **24.** 3.00; $\frac{3}{1}$ **25.** 0.165; $\frac{33}{200}$

26. 0.001; $\frac{1}{1000}$

Lesson 1.3 Skill C (page 19)

★ **6.** $\frac{3}{12} = \frac{3}{3} \cdot \frac{1}{4}$ Multiplication of fractions

$\frac{3}{3} \cdot \frac{1}{4} = \frac{1}{4}$ Identity Property of Multiplication

$\frac{1}{4} = \frac{1}{4} \cdot \frac{25}{25}$ Identity Property of Multiplication

$\frac{1}{4} \cdot \frac{25}{25} = \frac{25}{100}$ Multiplication of fractions

$\frac{25}{100} = 25\%$ Definition of percent

$\frac{3}{12} = 25\%$ Transitive Property of Equality

7. $3(x + 3) + 4 = 3x + 9 + 4$ Distributive Property
$3x + 9 + 4 = 3x + 13$ Addition
$3(x + 3) + 4 = 3x + 13$
   Transitive Property of Equality

8. $2(x + 1) + 5 = 2x + 2 + 5$ Distributive Property
$2x + 2 + 5 = 2x + 7$ Addition
$2(x + 1) + 5 = 2x + 7$
   Transitive Property of Equality

9. $5a = a + a + a + a + a$ Multiplication by 5
$a + a + a + a + a = (a + a + a) + (a + a)$
   Associative Property of Addition
$(a + a + a) + (a + a) = 3a + 2a$
   Multiplication by 3 and Multiplication by 2
$5a = 3a + 2a$
   Transitive Property of Equality

10. $6a = a + a + a + a + a + a$ Multiplication by 6
$a + a + a + a + a + a = (a + a + a + a + a) + a$
   Associative Property of Addition
$(a + a + a + a + a) + a = 5a + a$
   Multiplication by 5
$6a = 5a + a$ Transitive Property of Equality

11. $3 + 4(y + 2) = 3 + 4y + 8$ Distributive Property
$3 + 4y + 8 = 4y + 3 + 8$
   Commutative Property of Addition
$4y + 3 + 8 = 4y + 11$ Addition
$3 + 4(y + 2) = 4y + 11$
   Transitive Property of Equality

12. $6 + 4(r + 3) = 6 + 4r + 12$ Distributive Property
$6 + 4r + 12 = 4r + 6 + 12$
   Commutative Property of Addition
$4r + 6 + 12 = 4r + 18$ Addition
$6 + 4(r + 3) = 4r + 18$
   Transitive Property of Equality

13. even, since $2n + 2n = 2(n + n)$ by the Distributive Property

14. odd, since $2n + 2n + 1 = 2(n + n) + 1$ by the Distributive Property

15. even, since $2n + 1 + 2n + 1 = 2(2n + 1)$ by the Distributive Property

16. even, since $2n + 1 + 2n + 3 = 2(2n + 2)$ by the Distributive Property

17. $3 + 5$ **18.** $2.5 + 8$ **19.** $1 + 2 \times 10 = 21$

20. $(3 + 4) - 2 = 5$, or $3 + (4 - 2) = 5$

21. $\frac{25}{3}$, or $8\frac{1}{3}$ **22.** 48   **23.** 82   **24.** 144   **25.** 3

26. none   **27.** none   **28.** 4

Lesson 1.9 Skill A (page 33)

26. $\frac{ax}{b} = 10$; $\frac{b}{a} \cdot \frac{ax}{b} = \frac{b}{a} \cdot 10$; $x = \frac{10b}{a}$

$\dfrac{x}{\left(\frac{b}{a}\right)} = 10$; $\dfrac{b}{a} \cdot \dfrac{x}{\left(\frac{b}{a}\right)} = \dfrac{b}{a} \cdot 10$; $\dfrac{b}{a} \cdot x \cdot \dfrac{a}{b} = \dfrac{b}{a} \cdot 10$;

$\frac{b}{a} \cdot \frac{a}{b} \cdot x = \frac{b}{a} \cdot 10$; $x = \frac{10b}{a}$

27. $x = 6$ **28.** $z = 10.4$ **29.** $n = 21$ **30.** $z = 9$

31. $c = 5.4$ **32.** $d = 9.8$ **33.** $x = 3.9$ **34.** $n = 2$

35. $t = 0.75$

CHAPTER 2

Lesson 2.1 Skill C (page 45)

37. 15 **38.** 3.75 **39.** 51 **40.** rational, real

41. rational, real **42.** integer, rational, real

Lesson 2.9 Skill B (page 85)

★ 8. $7h - (3h - 5) = 9 \Rightarrow 7h - 3h + 5 = 9$
(Opposite of a Difference)
$7h - 3h + 5 = 9 \Rightarrow 4h + 5 = 9$
(Combining like terms)
$4h + 5 = 9 \Rightarrow 4h = 4$
(Subtraction Property of Equality)
$4h = 4 \Rightarrow h = 1$
(Division Property of Equality)
$7h - (3h - 5) = 9 \Rightarrow h = 1$
(Transitive Property of Deductive Reasoning)

★ 9. $6(x - 3) + 4(x + 2) = 6 \Rightarrow 6x - 18 + 4x + 8 = 6$
(Distributive Property)
$6x - 18 + 4x + 8 = 6 \Rightarrow 10x - 10 = 6$
(Combining like terms)
$10x - 10 = 6 \Rightarrow 10x = 16$
(Addition Property of Equality)
$10x = 16 \Rightarrow x = 1.6$
(Division Property of Equality)
$6(x - 3) + 4(x + 2) = 6 \Rightarrow x = 1.6$
(Transitive Property of Deductive Reasoning)

10. If $3p + 3 - p + 1 = 6$, then $2p + 4 = 6$.
(Combining like terms)

11. $-6.1, -2, 0, 0.5, 9.2$ **12.** $-0.2, -0.1, 0, 0.1, 0.2, 5$

13. $3.6, 3.65, 3.7, 3.75, 3.754$ **14.** $=$ **15.** $<$ **16.** $>$

17. $<$

Lesson 2.9 Skill C (page 87)

15. The first part of the statement is always true because no triangle can have all sides equal in length and no sides equal in length at the same time. The second part of the statement is always true since, if all sides are equal in length, then at least two sides are equal in length. The full statement is always true.

16. $x = -\dfrac{1}{3}$ **17.** $x = -\dfrac{1}{2}$ **18.** $x = \dfrac{6}{11}$ **19.** $x = \dfrac{1}{2}$

20. $c = -\dfrac{3}{2}$, or $-1\dfrac{1}{2}$ **21.** $r = 3$ **22.** $n = 2$

23. $m = -\dfrac{1}{2}$

Chapter 2 Assessment (pages 88–89)

53. The Distributive Property was not applied correctly and the addition of integers was incorrect.

$$-5(2 - 3) + 4 = -10 + 15 + 4$$
$$= 5 + 4$$
$$= 9$$

54. If $4x = 2(x - 5)$, then $4x = 2x - 10$ by the Distributive Property.
If $4x = 2x - 10$, then $2x = -10$ by the Subtraction Property of Equality.
If $2x = -10$, then $x = -5$ by the Division Property of Equality. Therefore, if $4x = 2(x - 5)$, then $x = -5$.

55. always true
$5(h + 3) - 2 = 5h + 15 - 2$ Distributive Property
$\qquad\qquad\quad = 5h + (15 - 2)$ Associative Property of Addition
$\qquad\qquad\quad = 5h + 13$ Subtraction

CHAPTER 3

Lesson 3.1 Skill B (page 95)

7.

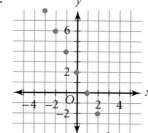

8.

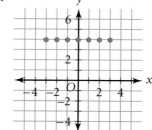

9.

x	−3	−2	−1	0	1	2	3
y	0	1	2	3	4	5	6

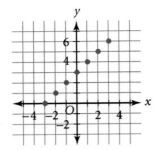

10.

x	−3	−2	−1	0	1	2	3
y	−5	−4	−3	−2	−1	0	1

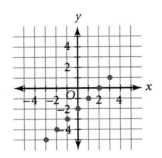

11.

x	0	1	2	3	4	5
y	0	2	4	6	8	10

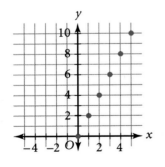

12.

x	−3	−2	−1	0	1	2	3
y	6	4	2	0	−2	−4	−6

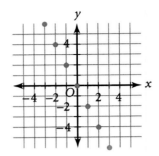

13.

x	−3	−2	−1	0	1	2	3
y	1	0	−1	−2	−3	−4	−5

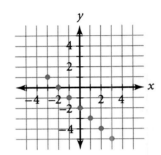

14.

x	−3	−2	−1	0	1	2	3
y	2	1	0	−1	−2	−3	−4

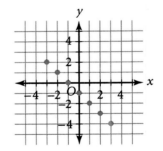

15.

x	−2	−1	0	1	2
y	7	1	−1	1	7

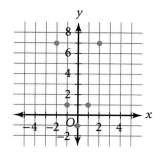

16.

x	0	1	2	3	4	5
y	0	−3	−4	−3	0	5

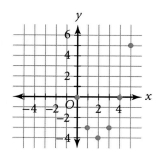

17.

x	0	1	2	3	4	5	6
y	−2	−2	−2	−2	−2	−2	−2

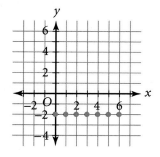

18.

x	−4	−2	0	2	4	6
y	3	3	3	3	3	3

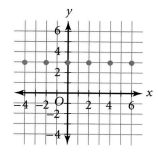

19.

x	−3	−2	−1	0	1	2	3
y	6	5	4	3	2	1	0

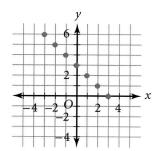

20.

x	−3	−2	−1	0	1	2	3
y	−7	−6	−5	−4	−3	−2	−1

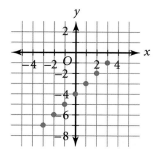

21.

x	−3	−2	−1	0	1	2	3
y	8	6	4	2	0	−2	−4

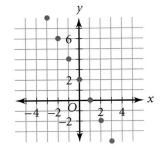

22.

x	−3	−2	−1	0	1	2	3
y	−4.5	−4	−3.5	−3	−2.5	−2	−1.5

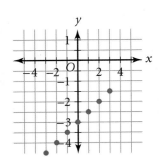

23. $(5)(9) - 2$ **24.** $3(12 - 5)$ **25.** $5 - (2)(5)$

Lesson 3.2 Skill B (page 101)

4.

t	0	1	2	3	4	5
d	0	0.1	0.2	0.3	0.4	0.5

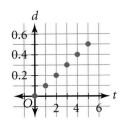

5.

n	1	2	3	4	5	6
c	24	48	72	96	120	144

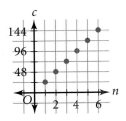

6. domain: all whole numbers

range: all nonnegative multiples of 10

7. domain: all whole numbers
range: all nonnegative multiples of 6

8. domain: all nonnegative real numbers
range: all nonnegative real numbers

9. $666.67 at 6% and $1333.33 at 12%

10. 235 milliliters and 165 milliliters

Lesson 3.4 Skill A (page 107)

24. If $3a - 2.6 = 4$, then $3a = 6.6$ by the Addition Property of Equality.
If $3a = 6.6$, then $a = 2.2$ by the Division Property of Equality.
Therefore, by the Transitive Property of Equality, if $3a - 2.6 = 4$, then $a = 2.2$.

25. If $-4 = 3x + 5x$, then $-4 = 8x$ by the Distributive Property.

If $-4 = 8x$, then $-\frac{1}{2} = x$ by the Division Property of Equality.

If $-\frac{1}{2} = x$, then $x = -\frac{1}{2}$ by the Symmetric Property of Equality. Therefore, by the Transitive Property of Equality, if $-4 = 3x + 5x$, then $x = -\frac{1}{2}$.

26. If $-3r - r + 5 = 6$, then $-4r + 5 = 6$ by the Distributive Property.
If $-4r + 5 = 6$, then $-4r = 1$ by the Subtraction Property of Equality.

If $-4r = 1$, then $r = -\frac{1}{4}$ by the Division Property of Equality. Therefore, by the Transitive Property of Equality, if $-3r - r + 5 = 6$, then $r = -\frac{1}{4}$.

27. If $-2(x - 1) + 3 = 7$, then $-2(x - 1) = 4$ by the Subtraction Property of Equality.
If $-2(x - 1) = 4$, then $x - 1 = -2$ by the Division Property of Equality.
If $x - 1 = -2$, then $x = -1$ by the Addition Property of Equality.
Therefore, by the Transitive Property of Equality, if $-2(x - 1) + 3 = 7$, then $x = -1$.

28. If $4b - 1 = 2b + 8$, then $2b - 1 = 8$ by the Subtraction Property of Equality.
If $2b - 1 = 8$, then $2b = 9$ by the Addition Property of Equality.
If $2b = 9$, then $b = \dfrac{9}{2}$ by the Division Property of Equality.
Therefore, by the Transitive Property of Equality, if $4b - 1 = 2b + 8$, then $b = \dfrac{9}{2}$.

Lesson 3.4 Skill B (page 109)

4. a.

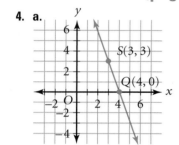

b. Answers may vary. Sample answer: $Q(4, 0)$

5. a.

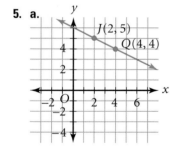

b. Answers may vary. Sample answer: $Q(4, 4)$

6. a.

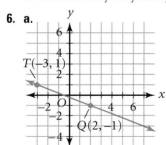

b. Answers may vary. Sample answer: $Q(2, -1)$

7. collinear

slope of $\overleftrightarrow{AB}$: $\dfrac{2 - 0}{2 - 0} = \dfrac{2}{2} = 1$

slope of $\overleftrightarrow{BC}$: $\dfrac{4 - 2}{4 - 2} = \dfrac{2}{2} = 1$

8. not collinear

slope of $\overleftrightarrow{DE}$: $\dfrac{5 - 1}{6 - 3} = \dfrac{4}{3}$;

slope of $\overleftrightarrow{EF}$: $\dfrac{8 - 5}{9 - 6} = \dfrac{3}{3} = 1$

9. not collinear

slope of $\overleftrightarrow{PQ}$: $\dfrac{2 - 5}{2 - 6} = \dfrac{3}{4}$;

slope of $\overleftrightarrow{QR}$: $\dfrac{-2 - 2}{0 - 2} = \dfrac{4}{2} = 2$

10. collinear

slope of $\overleftrightarrow{XY}$: $\dfrac{11 - 9}{5 - 4} = \dfrac{2}{1} = 2$;

slope of $\overleftrightarrow{YZ}$: $\dfrac{11 - 9}{5 - 4} = \dfrac{2}{1} = 2$

11. parallel   **12.** not parallel   **13.** If PQ has a slope of $\dfrac{m}{n}$, then it is the same as line l and Q is on l.

slope of $\overleftrightarrow{PQ}$: $\dfrac{(b + m) - b}{(a + n) - a} = \dfrac{b - b + m}{a - a + n} = \dfrac{m}{n}$

Since $\overleftrightarrow{PQ}$ contains $P(a, b)$ and has slope $\dfrac{m}{n}$, then it is line l.

Thus, $Q(a + n, b + m)$ is on line l.

14.

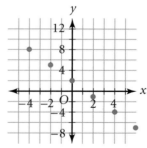

15. Yes; every member of the domain is assigned exactly one member of the range.

16. domain: all real numbers
range: all real numbers

17. Let s represent length and P represent perimeter. Then $P = 4s$; domain: all nonnegative real numbers; range: all nonnegative real numbers

18. $\dfrac{6}{7}$ **19.** $y = 77$

Lesson 3.5 Skill A (page 113)

14. x-intercept: 6
y-intercept: 4

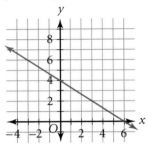

15. x-intercept: none
y-intercept: -1

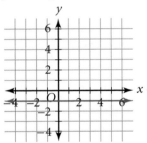

16. x-intercept: -1
y-intercept: none

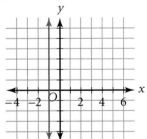

17. x-intercept: none
y-intercept: $\dfrac{5}{2}$

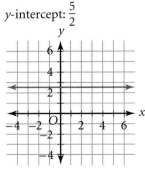

18. x-intercept: $\dfrac{5}{2}$
y-intercept: none

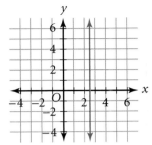

19. x-intercept: 4
y-intercept: 4

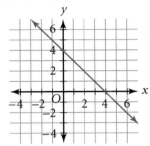

20. x-intercept: 4
y-intercept: -4

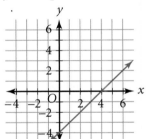

21. x-intercept: 2
y-intercept: 2

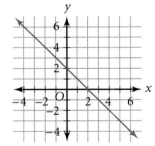

22. *x*-intercept: 6
y-intercept: 9

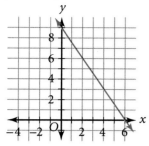

23. *x*-intercept: 5
y-intercept: 3

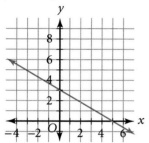

24. *x*-intercept: 5
y-intercept: $-\dfrac{5}{2}$

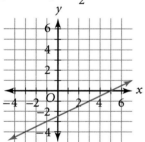

25. *x*-intercept: $-\dfrac{4}{3}$
y-intercept: 4

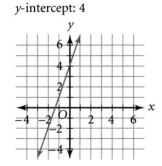

26. *x*-intercept: $-\dfrac{3}{2}$
y-intercept: -3

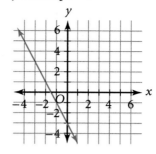

27. *x*-intercept: $-\dfrac{5}{3}$
y-intercept: 5

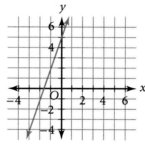

28. *x*-intercept: $\dfrac{1}{2}$
y-intercept: 1

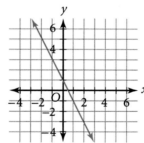

29. neither   ★**30.** $2x + y = 4$

31. a. Answers may vary. Sample answer:
$y = x - 3$
b. $x = 3$

32. *x*-intercept: *a*   *y*-intercept: *b*   **33.** $\dfrac{7}{2}$ **34.** $-\dfrac{3}{2}$

35. $\dfrac{2}{5}$ **36.** $-\dfrac{3}{5}$ **37.** $\dfrac{3}{2}$ **38.** -2 **39.** 0   **40.** $-\dfrac{1}{8}$

Lesson 3.5 Skill B (page 115)

12.

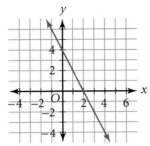

13.

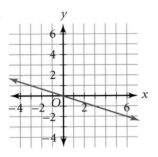

14.

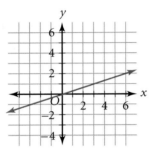

15.

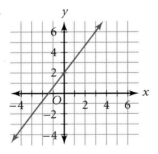

16.

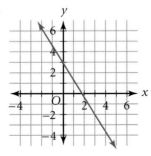

17.

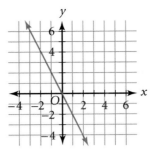

18.

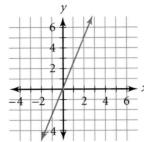

19.

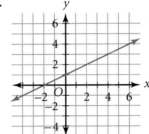

20.

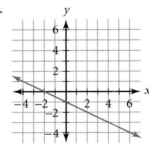

21. $y = \frac{1}{2}x + 4$

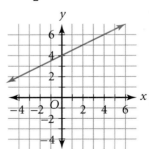

22. $y = -\frac{2}{3}x + 2$

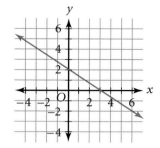

23. $y = \frac{2}{3}x - 2$

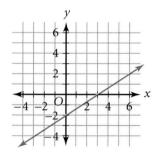

24. $y = \frac{3}{5}x + 3$

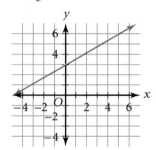

25.

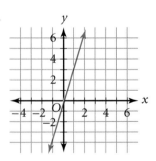

$y = 3x$

26.

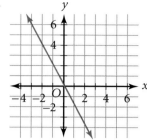

$y = -2x$

27.

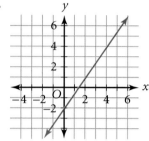

$y = 1.5x - 2$

28.

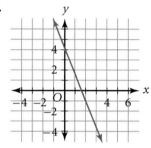

$y = -2.5x + 4$

29.

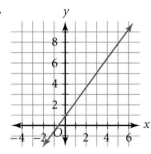

$y = \frac{7}{5}x + 1$

30.

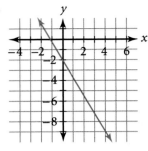

$$y = -\frac{7}{4}x - 2$$

31. $\frac{7}{3}$; $2.\overline{3}$ **32.** $-\frac{18}{5}$; -3.6 **33.** $\frac{21}{2}$; 10.5

34. $-\frac{28}{3}$; $-9.\overline{3}$ **35.** $x = 2$ **36.** $x = -3$ **37.** $x = -\frac{19}{4}$

Lesson 3.5 Skill C (page 117)

7. Let v represent volume remaining and let t represent elapsed time; $v = 200 - 40t$.

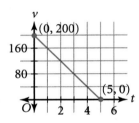

The graph is continuous.

8. Let h represent the number of hours worked and w represent the amount of money earned; $w = 12.5h$.

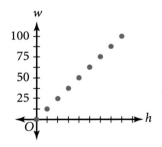

The graph is discrete.

9. Let p represent the original price and let s represent the sale price; $s = 0.75p$.

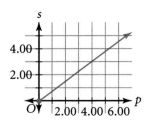

The graph is continuous.

10. When t is between 0 and 5, inclusive, $v = 3t$. When t is between 5 and 10, inclusive, $v = 15$. When t is between 10 and 17.5, inclusive, $v = 15 - 2t$.

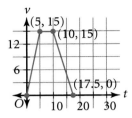

The graph is continuous.

11. Let v represent volume after t minutes; $v = 10t$; 55 cubic feet

12. Let v represent volume after t minutes; $v = 125t$; 687.5 gallons

Lesson 3.6 Skill A (page 119)

19. If the line contains $P(a, b)$ and has slope 0, then $y - b = 0(x - a)$. Thus, $y - b = 0$ and the equation has the form $y = b$.

20. If the line contains $P(0, 0)$ and has slope m, then $y - 0 = m(x - 0)$. Thus, $y = mx$.

21. -1 **22.** -4 **23.** 10

24.

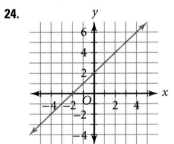

25.

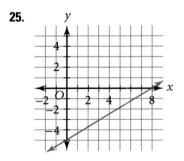

444   Additional Answers

26.

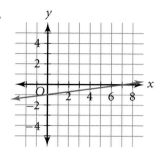

27. vertical; $(r, 0)$; undefined

Lesson 3.7 Skill A (page 123)

23. $(0, 4), (6, 0)$

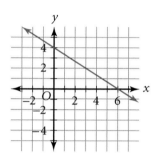

24.

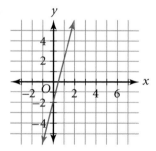

25.

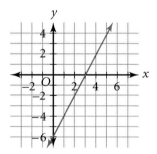

26.

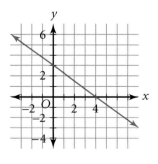

27. $y = -\dfrac{3}{2}x + \dfrac{5}{2}$ **28.** $y = \dfrac{9}{20}x + \dfrac{111}{10}$

Lesson 3.8 Skill B (page 129)

25.

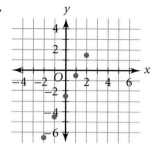

26.

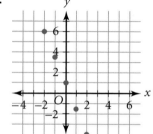

Lesson 3.8 Skill C (page 131)

13. Figure *ABCD* is a parallelogram because slope of *AB* = 1, slope of *BC* = −1, slope of *CD* = 1, and slope of *DA* = −1. *AB* is parallel to *CD* and *BC* is parallel to *AD*.

14. Figure *KLMN* is a parallelogram because slope of *KL* = 1, slope of *LM* = −2, slope of *MN* = 1, and slope of *NK* = −2. *KL* is parallel to *MN* and *LM* is parallel to *NK*.

15. Figure *PQRS* is a parallelogram and a rectangle.

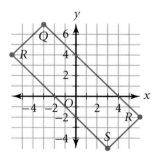

slope of $\overleftrightarrow{PQ}$: 1; slope of $\overleftrightarrow{QR}$: −1;
slope of $\overleftrightarrow{RS}$: 1; slope of $\overleftrightarrow{SR}$: −1

16. Let n represent the number of nickels and d represent the number of dimes; $5n + 10d = 30$

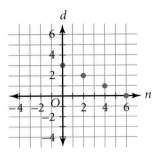

17. Let t represent time in minutes and T represent temperature in degrees Celsius; $T = 5t + 25$

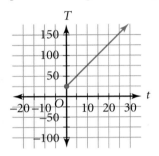

Chapter 3 Assessment (pages 138–139)

10.

x	-3	-2	-1	0	1	2	3
y	16	9	4	1	0	1	4

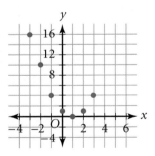

11. Yes; no member of the domain is assigned more than one member of the range.

12. domain: $-6, -4, -2, 0, 2, 4, 6$; range: $-4, -2, -1, 1$

13.

n	1	2	3	4	5	6
c	1.5	3	4.5	6	7.5	9

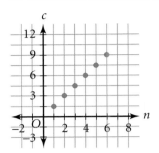

14. a. \$27.80   **b.** about 18.0 gallons   **15.** $y = 144$

16. 400 feet per minute (downward)   **17.** $\dfrac{3}{11}$ **18.** $-\dfrac{13}{2}$

19. Coordinates of the second point may vary.
Sample answer: $L(1, 4)$

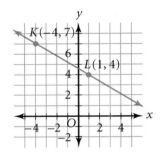

20. x-intercept: -6
y-intercept: 8

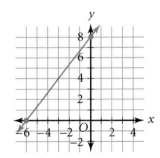

21. no x-intercept
y-intercept: 3.6

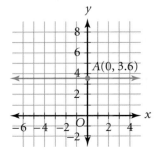

22.

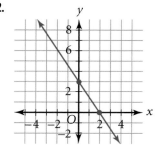

23.

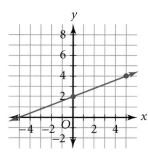

24. Let n represent the number of nickels and d represent the number of dimes; $5n + 10d = 40$.

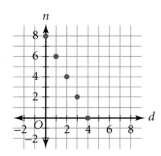

25. $y - (-5) = \frac{3}{7}[x - (-3)]$; $y = \frac{3}{7}x - \frac{26}{7}$

26. $y = -3$

27. Let v represent the value after n years; $v = -400n + 5500$; \$3900

28. $y = 3x - 11$ **29.** $y = \frac{10}{9}x + \frac{65}{9}$

30. $y = 19x + 30$; \$163; the slope gives the cost per day, \$19; the y-intercept gives the fixed fee, \$30.

31. $y = \frac{3}{5}x - 1$ **32.** $y = \frac{2}{3}x + 3$

33. No; no two slopes are negative reciprocals of one another; slope $\overleftrightarrow{KL}$: $-\frac{4}{3}$; slope $\overleftrightarrow{LM}$: $\frac{2}{5}$; slope $\overleftrightarrow{KM}$: $-\frac{1}{4}$

34. $y = -7x + 20$; -330 **35.** $t = -2.5n + 12.5$; -125

36. $x = 3n - 8$; $y = 2n - 5$; $(172, 115)$

CHAPTER 4

Lesson 4.1 Skill B (page 145)

30. $q = -1.25$ **31.** $x = 0.25$ **32.** $t = -3.25$

Lesson 4.1 Skill C (page 147)

26. $d = -\frac{3}{5}$ **27.** $y = 6.3$ **28.** $g = -15\frac{3}{4}$, or $-\frac{63}{4}$

29. $w = -36$ **30.** $p = -4$ **31.** $q = 0$ **32.** $x = -7$

Lesson 4.2 Skill A (page 149)

22. positive numbers   **23.** positive and negative numbers

24. $8, 9$ **25.** $1, 2, 3, 4, 5, 6, 7, 8$ **26.** $5, 6, 7$ **27.** 7

★ **28.** $8 < y < 9$

★ **29.** $x > 8$ or $x < 6$

★ **30.** $h \geq 5$ or $h < 3$

★ **31.** $0 < n \leq 8$

32. $x > b - a$; the solutions are positive if $b > a$.

33. -11 **34.** -7 **35.** -7.6

36. -5 **37.** -3.4 **38.** $-\frac{27}{5}$, or $-5\frac{2}{5}$

Lesson 4.2 Skill B (page 151)

20. negative numbers   **21.** negative numbers

22. negative and positive numbers

23. negative and positive numbers

24. $-5, -4, -3, -2, -1, 0, 1, 2, 3, 4$ **25.** $6, 7, 8$

26. $-6, -5, -4, -3, -2, -1, 0$

27. $-7, -6, -5, -4, -3, -2, -1, 0, 1, 2$

28. $x = 4$

29. 33 **30.** -56 **31.** -7 **32.** 16 **33.** -2 **34.** -9

35. $-\dfrac{1}{5}$ **36.** -1 **37.** $\dfrac{1}{7}$ **38.** $-\dfrac{5}{3}$ **39.** $-\dfrac{5}{2}$ **40.** $-\dfrac{5}{2}$

Lesson 4.3 Skill A (page 155)

19. $w > 9$

20. $k > -1.5$

21. $k < 2$

22. $z \geq -\dfrac{5}{3}$

23. $s \leq \dfrac{10}{9}$

24. $a \leq 2$

25. $s \leq -\dfrac{2}{3}$

26. If $\dfrac{x}{-3} \leq \dfrac{2}{5}$, then $-3 \cdot \dfrac{x}{-3} \geq -3 \cdot \dfrac{2}{5}$ by the Multiplication Property of Inequality. Therefore, $x \geq -\dfrac{6}{5}$. If $\dfrac{5}{2}x \geq -3$, then $\dfrac{2}{5} \cdot \dfrac{5}{2}x \geq \dfrac{2}{5} \cdot (-3)$ by the Multiplication Property of Inequality. Therefore, $x \geq -\dfrac{6}{5}$.

27. If $\dfrac{7}{3} > \dfrac{x}{4}$, then $\dfrac{7}{3} \cdot 4 > \dfrac{x}{4} \cdot 4$ by the Multiplication Property of Inequality. Therefore, $\dfrac{28}{3} > x$ and $x < \dfrac{28}{3}$. If $\dfrac{3}{7}x < 4$, then $\dfrac{7}{3} \cdot \dfrac{3}{7}x < \dfrac{7}{3} \cdot 4$ by the Multiplication Property of Inequality. Therefore, $x < \dfrac{28}{3}$.

28. If $a > 0$, then $x > ab$. If $a < 0$, then $x < ab$.

29. -1 **30.** 60 **31.** $-\dfrac{1}{4}$ **32.** -2 **33.** $-\dfrac{1}{14}$

34. 1 **35.** $\dfrac{9}{16}$ **36.** -1 **37.** $-\dfrac{3}{16}$

Lesson 4.3 Skill B (page 157)

23. $0 < x < 3$

24. $-2 < y < 4$

25. $0 \leq x \leq 3$

26. $6 \leq t \leq 8$

27. between -3 and 5 **28.** $-7n + 6.3$ **29.** $-3j$

30. $1.4c + 1$ **31.** $-21x - 35$ **32.** $-\dfrac{1}{2}y + 2$ **33.** $x - 9$

Lesson 4.4 Skill A (page 161)

18. a. Multiply each side by $-\dfrac{2}{13}$.
 b. yes
 c. $f > -\dfrac{4}{13}$

19. a. Add h to each side.
 b. no
 c. $h < 13$

20. a. Multiply each side by -1.
 b. yes
 c. $d < -4$

21. a. Add z to each side.
   b. no
   c. $z < 5$

22. a. Multiply each side by -1.
   b. yes
   c. $g \geq 2$

23. $6 < x < 7$

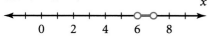

24. $-1 \leq w \leq 7$

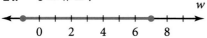

25. a. $x > 4$
   b.

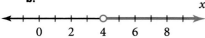

26. $x = -6$ **27.** $a = 28$ **28.** $x = 4$ **29.** $x = -12$

30. $z = 50$ **31.** $x = 12$ **32.** $k = -1$ **33.** $n = 9$

34. $x = -10$

Lesson 4.5 Skill A (page 165)

32.

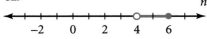

33.

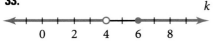

34.

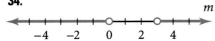

Lesson 4.6 Skill A (page 171)

20. $1 < k \leq 7$

21. $-2 < q \leq 2$

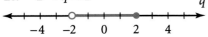

★22. If $2(p - 3) + 3 < 3 - 2(p + 3)$, then $p < 0$.
   If $-4 + 2(p + 2) > 8 - (p + 1)$, then $p > \dfrac{7}{3}$.
   There are no real numbers that are less than 0 and greater than $\dfrac{7}{3}$ at the same time. Therefore, there is no solution.

★23. If $-3 - 4(s + 3) > 0$, then $s < -\dfrac{15}{4}$.
   If $4 - 3(s - 3) < 10$, then $s > 1$. There are no real numbers that are less than $-\dfrac{15}{4}$ and greater than 1 at the same time. Therefore, there is no solution.

24. $a = -\dfrac{1}{2}$ **25.** $y = 11x - 41$; 509

26. $y = -12n + 36$; -564

27.

28.

29.

30.

Lesson 4.6 Skill B (page 173)

18. $r > 3 \ or \ r > -4$

19. The solution is $x < -a \ or \ x < a$. If $a \geq 0$, the solution is $x < a$. If $a < 0$, the solution is $x < -a$.

20. The solution is $x < -a \ or \ x > -a$. This includes all real numbers except $-a$.

21. 12.9; 12.9   **22.** -111; 111

23. -2.7; 2.7   **24.** 31.3; 31.3

25. $x > 4$

26. $d < 5$

27. $a \le 2.05$

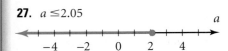

28. $d < \dfrac{3}{4}$

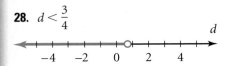

Lesson 4.7 Skill A (page 177)

39. $y < -1$ **40.** $t \ge \dfrac{12}{5}$

Lesson 4.7 Skill B (page 179)

19. $f < -\dfrac{1}{3}$ or $f > -\dfrac{1}{3}$

20. no solution

21. $-\dfrac{8}{3} \le g \le 0$

22. $-4 \le n \le -2$

23. $-2 < m < 4$

24. $t \le 2$ or $t \ge 3$

25. any negative real number

26. If $|x + 4| > r$, then $x < -4 - r$ or $x \ge r - 4$. In order for the endpoints to be opposites of one another, r must be 0. Since 0 is not allowed, there is no nonzero r such that the endpoints are opposites of one another.

★**27.** $x < 1$ and $x > \dfrac{1}{3}$, or $\dfrac{1}{3} < x < 1$

28. $5 < x \le 6$

29. The distance between x and 0 is more than 3.

30. The distance between x and -2 is less than 3.

31. If $|ax + b| \le c$, $-c \le ax + b \le c$. Thus,
$-c - b \le ax \le c - b$, or $-b - c \le ax \le c - b$.

If $a > 0$, then $\dfrac{-b - c}{a} \le x \le \dfrac{c - b}{a}$.

32. If $-x + 4x - 1 = 3x - 1$, then $-3x - 1 = -3x - 1$. The equation is always true.

33. If $-(x - 4) = 3x - 1$, then $-x + 4 = 3x - 1$.

Thus, $x = \dfrac{5}{4}$. The equation is sometimes true, when

$x = \dfrac{5}{4}$.

34. If $-(r - 4r) = 2r - 1$, then $3r = 3r - 1$. This implies that $0 = -1$. Thus, the given equation is never true.

Lesson 4.8 Skill A (page 183)

18. If $x < 0$, then $x^2 > 0$. Thus, $x^2 > x$ is true when $x < 0$.
If $x = 0$, then $x^2 = 0$. Thus, $x^2 > x$ is false when $x = 0$.
If $x > 0$, then $x^2 > x$ implies that $x > 1$. Thus, $x^2 > x$ is sometimes true when $x > 0$.
Therefore, the given statement is sometimes true.

19. If $x(-x) \le 0$, then $-x^2 \le 0$. This implies that $x^2 \ge 0$.
If $x < 0$, then $x^2 = x \cdot x > 0$ because the product of two negative numbers is positive. Thus, $x^2 \ge 0$ is true when $x < 0$.
If $x = 0$, then $x^2 = 0^2 = 0$. Thus, $x^2 \ge 0$ is true when $x = 0$.
If $x > 0$, then $x^2 = x \cdot x > 0$ because the product of two positive numbers is positive. Thus, $x^2 \ge 0$ is true when $x > 0$.
Therefore, the given statement is always true.

20. If $x^2 + 1 \ge 1$, then $x^2 \ge 0$. By the same reasoning used in Exercise 19, the given statement is always true.

21. If $x < 0$, then $x^2 > 0$ and $2x < 0$. Thus, $x^2 \ge 2x$ is true when $x < 0$.
If $x = 0$, then $x^2 = 0$ and $2x = 0$. Thus, $x^2 \ge 2x$ is true when $x = 0$.
If $x > 0$, then $x^2 \ge 2x$ implies that $x \ge 2$. Thus, $x^2 \ge 2x$ is sometimes true when $x > 0$.
Therefore, the given statement is sometimes true.

22. If $x < 0$, then $\dfrac{x}{5} \ge \dfrac{5}{x}$ implies that $x^2 \le 25$, or

$-5 \le x < 0$. Thus, $\dfrac{x}{5} \ge \dfrac{5}{x}$ is sometimes true

when $x < 0$.

If $x = 0$, then $\dfrac{5}{x}$ is undefined. Thus, $\dfrac{x}{5} \ge \dfrac{5}{x}$ is false

when $x = 0$.

If $x > 0$, then $\dfrac{x}{5} \ge \dfrac{5}{x}$ implies that $x^2 \ge 25$, or

$x \ge 5$. Thus, $\dfrac{x}{5} \ge \dfrac{5}{x}$ is sometimes true when $x > 0$.

Therefore, the given statement is sometimes true.

23. If $x < 0$, then $\frac{x}{3} \le \frac{3}{x}$ implies that $x^2 \ge 9$, or $x \le -3$.

Thus, $\frac{x}{3} \le \frac{3}{x}$ is sometimes true when $x < 0$.

If $x = 0$, then $\frac{3}{x}$ is undefined. Thus, $\frac{x}{3} \le \frac{3}{x}$ is false when $x = 0$.

If $x > 0$, then $\frac{x}{3} \le \frac{3}{x}$ implies that $x^2 \le 9$, or $0 < x \le 3$.

Thus, $\frac{x}{3} \le \frac{3}{x}$ is sometimes true when $x > 0$.
Therefore, the given statement is sometimes true.

24. Let r represent a real number. Then $\frac{1}{r}$ represents its reciprocal and $-r$ represents its opposite.
If $r < 0$, then $\frac{1}{r} > -r$ implies that $-r^2 > 1$, or $r^2 < -1$. But the product of two negative numbers is positive, so $r^2 = r \cdot r < -1$ is never true. Thus, $\frac{1}{r} > -r$ is never true when $r < 0$.

If $r = 0$, then $\frac{1}{r}$ is undefined. Thus, $\frac{1}{r} > -r$ is false when $r = 0$.
If $r > 0$, then $\frac{1}{r} > -r$ implies that $r^2 > 1$, or $r > 1$.

Thus, $\frac{1}{r} > -r$ is sometimes true when $r > 0$.
Therefore, the given statement is sometimes true.

25.
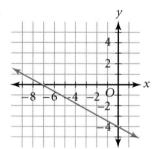

26.
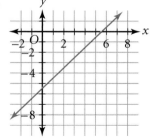

27.
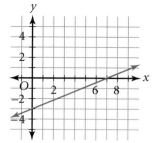

28.
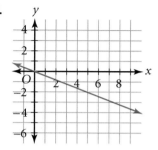

29.
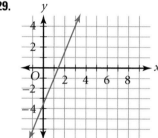

30.
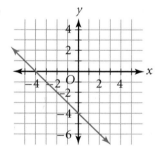

Chapter 4 Assessment (pages 184–185)

26. $w < 7$

27. $x \le 8$

ADDITIONAL ANSWERS

28. $p > 2$

$$\text{-----|----|----|--○----|---- } p$$
$$\quad -4 \quad -2 \quad 0 \quad 2 \quad 4$$

29. $r < \dfrac{13}{3}$

$$\text{◄----|----|--○----|----|----} r$$
$$\quad 0 \quad 2 \quad 4 \quad 6 \quad 8 \quad 10$$

30. $d < 10$

$$\text{◄----|----|----|----|--○} d$$
$$\quad 0 \quad 2 \quad 4 \quad 6 \quad 8 \quad 10$$

31. $w \leq -2.5$

$$\text{◄----|--●--|----|----|----} w$$
$$\quad -4 \quad -2 \quad 0 \quad 2 \quad 4$$

32. at least $214.29

33. $w < -\dfrac{5}{2}$

$$\text{◄----|-○--|----|----|----} w$$
$$\quad -4 \quad -2 \quad 0 \quad 2 \quad 4$$

34. $6 \leq a \leq 9$

$$\text{-----|----|--●--●--|----} a$$
$$\quad 0 \quad 2 \quad 4 \quad 6 \quad 8 \quad 10$$

35. $x \geq -8 \; or \; x \leq -2$

$$\text{◄----|----|----|----|----►} x$$
$$\quad -8 \quad -6 \quad -4 \quad -2 \quad 0$$

36. $w \leq \dfrac{5}{4}$

$$\text{◄----|----|----●--|----} w$$
$$\quad -4 \quad -2 \quad 0 \quad 2 \quad 4$$

37. all real numbers

$$\text{◄----|----|----|----|----►} a$$
$$\quad 0 \quad 2 \quad 4 \quad 6 \quad 8 \quad 10$$

38. no solution

$$\text{-----|----|----|----|----} x$$
$$\quad -4 \quad -2 \quad 0 \quad 2 \quad 4$$

39. between 2.2 pounds and 2.6 pounds, inclusive

40. $x = -\dfrac{16}{3} \; or \; x = \dfrac{8}{3}$ **41.** $d = -\dfrac{17}{2} \; or \; d = \dfrac{19}{2}$

42. $x = -3 \; or \; x = 3$

43. $x \leq -\dfrac{4}{3} \; or \; x \geq -\dfrac{2}{3}$

$$\text{◄----|--●●-|----|----|----} x$$
$$\quad -4 \quad -2 \quad 0 \quad 2 \quad 4$$

44. $-4 < x < 2$

$$\text{-----|--○--|----|--○--|----} x$$
$$\quad -6 \quad -4 \quad -2 \quad 0 \quad 2$$

45. $-1 < x < 1$

$$\text{-----|----○--○----|----} x$$
$$\quad -4 \quad -2 \quad 0 \quad 2 \quad 4$$

46. acceptable: between 31.98 ounces and 32.02 ounces, inclusive; unacceptable: less than 31.98 ounces or greater than 32.02 ounces

47. If $2r > r$, then $r > 0$. The statement is sometimes true.

CHAPTER 5

Lesson 5.1 Skill A (page 189)

10. none

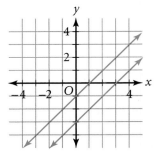

11. none

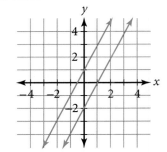

12. none

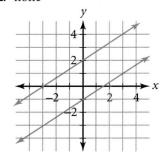

13. $(3, 4)$

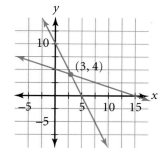

14. $(-3, -4)$

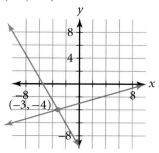

15. $\left(0, -\dfrac{1}{2}\right)$

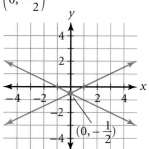

16. $(8, -2)$

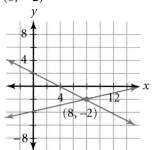

17. none

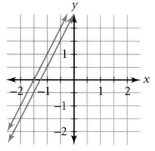

18. $(5, -1)$

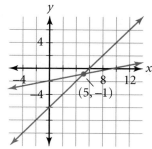

19. $y = \dfrac{4}{3}x - 2$: slope $\dfrac{4}{3}$ and y-intercept -2;
$y = 2$: slope 0 and y-intercept 2. The lines intersect because they have different slopes.

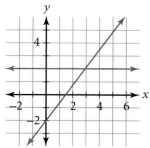

20. $y = -\dfrac{3}{5}x + 3$: slope $-\dfrac{3}{5}$ and y-intercept 3;
$y = 2x + 1$: slope 2 and y-intercept 1. The lines intersect because they have different slopes.

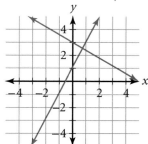

21. $y = \dfrac{1}{5}x + 1$: slope $\dfrac{1}{5}$ and y-intercept 1;
$y = -2x + 4$: slope -2 and y-intercept 4. The lines intersect because they have different slopes.

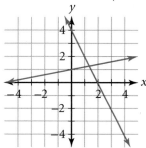

22. $2x - 6$ **23.** $-\frac{4}{5}x - 2$ **24.** $-8n - 14$

25. $-9y + 4$ **26.** $2.5k + 3.5$ **27.** $5x + 2.5$

Lesson 5.4 Skill A (page 203)

21.
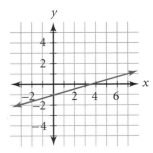

22.
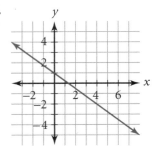

23.
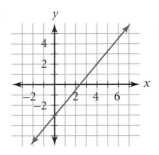

24.
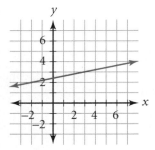

Lesson 5.5 Skill A (page 207)

10.
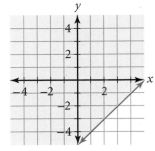

11.
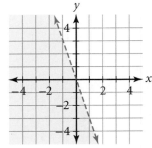

12.
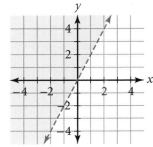

13.
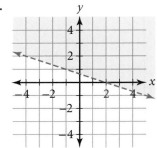

14.
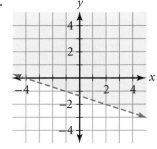

15.

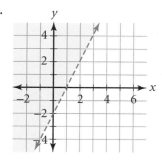

16.

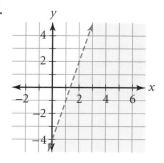

17.

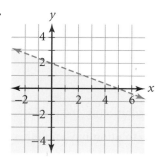

18.

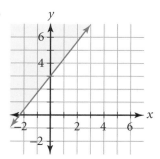

19.

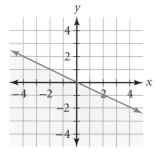

20.

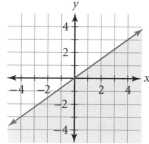

21.

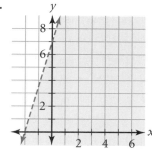

22.

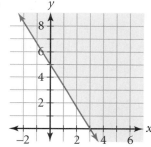

★ **23.** $y \geq \frac{4}{3}x + \frac{1}{3}$ ★ **24.** $y < \frac{3}{4}x + \frac{9}{2}$

★ **25.** $b = 0$ and $m < 5$ ★ **26.** $b = 0$ and $m > -\frac{7}{2}$

27. $1 < x < 3$

28. $3 < a \leq 6$

29. $y < -2$ or $y > 4$

30. $y \leq -2$ or $y \geq 2$

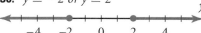

Lesson 5.5 Skill B (page 209)

5.

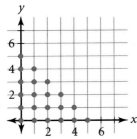

6.

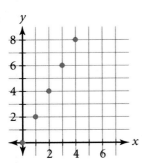

7. Let *r* represent the number of red chips and *b* represent the number of blue chips.

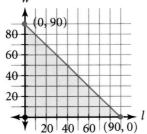

8. (1, 1), (2, 2), (3, 3), . . . , (88, 88), (89, 89), and (90, 90)

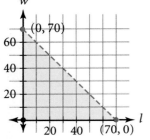

9. Let *l* represent length and *w* represent width. The ordered pairs (*w*, 2*w*), where 0 < *w* < 35, give rectangles in which the length is twice the width.

10.

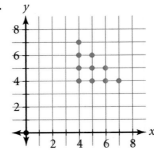

11. The restriction $x \geq 5$ means that all solutions must be to the right of the line $x = 5$. This is line *p*. The restriction $y \geq 5$ means that all solutions must be above the line $y = 5$. This is line *q*. However, the restriction that the sum be less than 9 means that all solutions must be below the line $x + y = 9$. This is line *r*. There are no such points satisfying all these restrictions.

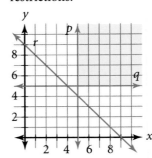

12. solution A: 40 milliliters; solution B: 160 milliliters

13. solution A: 210 milliliters; solution B: 90 milliliters

Lesson 5.6 Skill A (page 211)

8.

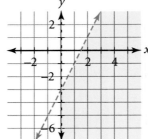

9.

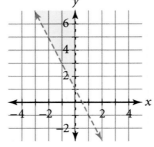

10.

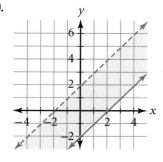

11.

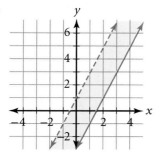

12.

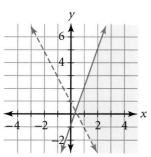

13.

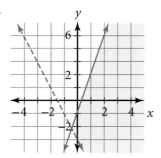

14.

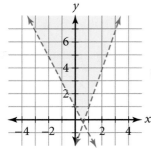

15.

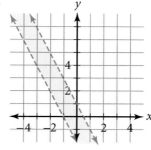

16.

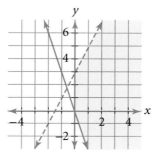

17.

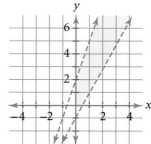

18.

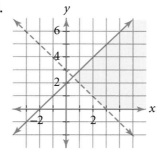

19. $m_1 = m_2$ and $b_2 > b_1$ **20.** inconsistent

21. inconsistent   **22.** consistent and dependent

Lesson 5.6 Skill B (page 213)

8.

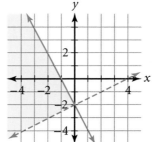

9.

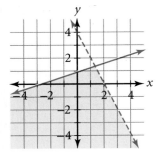

10.

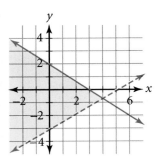

11.

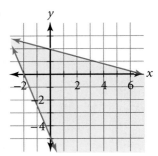

12.

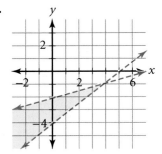

★**13.** If (x, y) is a solution to $\begin{cases} 2x + 3y \le -1 \\ 2x + 3y \ge -1 \end{cases}$, then (x, y)

must satisfy $2x + 3y = -1$. Thus, the solution to the system is all points on a line.

14. The solution to the first pair of equations,

$\begin{cases} y \ge x + 1 \\ y \le x + 1 \end{cases}$, is the line $y = x + 1$. The solution to the

second pair of equations, $\begin{cases} y \ge -x + 1 \\ y \le -x + 1 \end{cases}$, is the line

$y = -x + 1$. Since the graphs of $y = x + 1$ and $y = -x + 1$ have different slopes, the lines intersect. Therefore, the solution to the system is the solution to

$\begin{cases} y = x + 1 \\ y = -x + 1 \end{cases}$, or $(0, 1)$.

15. $a \le b$

16. Since $-2x + a < -2x + (a + 1)$ for all real numbers a, the system has a solution. Since the graphs of $y = -2x + a$ and $y = -2x + (a + 1)$ have the same slope, the boundaries of the solution are parallel lines. Since $a \ne a + 1$ for any real number a, the boundaries are distinct. Therefore, the solution is a region with parallel boundaries.

17. 8 **18.** 32 **19.** 9 **20.** 25 **21.** 16 **22.** 27

23. 1 **24.** 10,000 **25.** 25 **26.** 25 **27.** 8 **28.** 1

Chapter 5 Assessment (pages 216–217)

35.

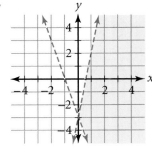

36.

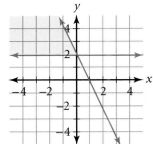

37.

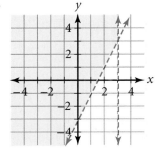

38.

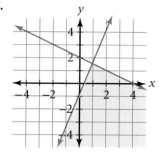

39.

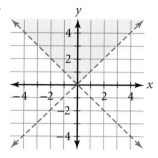

40.

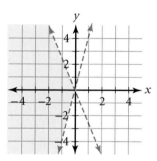

41.

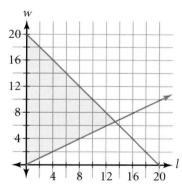

CHAPTER 6

Lesson 6.6 Skill A (page 247)

34.

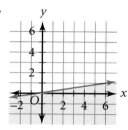

35.

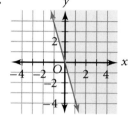

CHAPTER 7

Lesson 7.2 Skill B (page 269)

33.

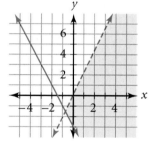

34.

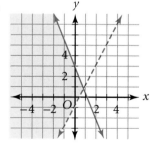

35.

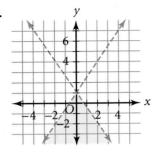

Lesson 7.3 Skill B (page 273)

30. $(x + r)(x + s) = x(x + s) + r(x + s)$
$= x^2 + xs + rx + rs$
$= x^2 + (s + r)x + rs$
$= x^2 + (r + s)x + rs$

If $x^2 + bx + c = (x + r)(x + s)$, then
$x^2 + bx + c = x^2 + (r + s)x + rs$.
Therefore, $b = r + s$ and $c = rs$.

31. $(n + 11)^2$ **32.** $(k - 13)^2$

33. $(2a + 5)^2$ **34.** $36(z - 1)(z + 1)$

35. $(3y - b)(3y + b)$ **36.** $(uv - 3)(uv + 3)$

37. $(2 + 3z)(z + 1)$ **38.** $(3 - 2a)(a + 2)$

39. $(-1 + 3n)(n - 1)$ **40.** $(x^2 - 3)(y + 1)$

41. $(x - y)(x - 20)$ **42.** $(w - 1)(4x^2 + 5)$

Lesson 7.6 Skill A (page 287)

17. a. vertex: $(0, -9)$; axis of
symmetry: $x = 0$

b. minimum

c.

x	-3	-2	-1	0	1	2	3
y	0	-5	-8	-9	-8	-5	0

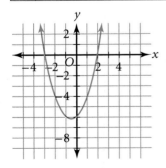

18. a. vertex: $(0, -1)$; axis of
symmetry: $x = 0$

b. maximum

c.

x	-3	-2	-1	0	1	2	3
y	-10	-5	-2	-1	-2	-5	-10

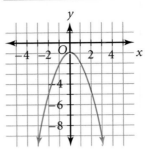

19. a. vertex: $\left(-\dfrac{1}{2}, -\dfrac{25}{4}\right)$; axis of

symmetry: $x = -\dfrac{1}{2}$

b. minimum

c.

x	-3	-2	-1	0	1	2	3
y	0	-4	-6	-6	-4	0	6

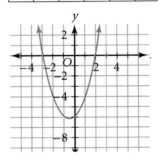

20. **a.** vertex: $\left(\dfrac{3}{2}, -\dfrac{1}{4}\right)$; axis of

symmetry: $x = \dfrac{3}{2}$

b. minimum

c.

x	−2	−1	0	1	2	3
y	12	6	2	0	0	2

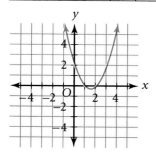

21. **a.** vertex: $(0, -1)$; axis of symmetry: $x = 0$

b. minimum

c.

x	−3	−2	−1	0	1	2	3
y	8	3	0	−1	0	3	8

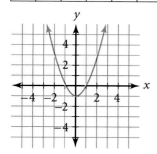

22. **a.** vertex: $\left(-\dfrac{3}{2}, -\dfrac{17}{4}\right)$; axis of

symmetry: $x = -\dfrac{3}{2}$

b. minimum

c.

x	−3	−2	−1	0	1	2
y	−2	−4	−4	−2	2	8

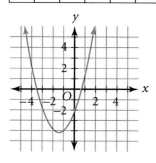

23. **a.** vertex: $(2, 1)$; axis of symmetry: $x = 2$

b. maximum

c.

x	−1	0	1	2	3
y	−8	−3	0	1	0

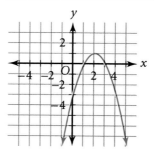

24. **a.** vertex: $(1, -1)$; axis of symmetry: $x = 1$

b. minimum

c.

x	−2	−1	0	1	2	3
y	8	3	0	−1	0	3

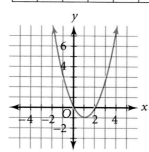

25. no x-intercept; y-intercept: -3.5

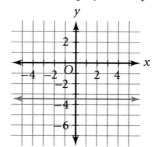

26. no *y*-intercept; *x*-intercept: 3.5

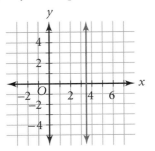

27. *x*-intercept: 6; *y*-intercept: 6

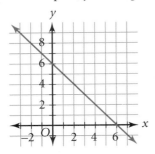

28. *x*-intercept: 6; *y*-intercept: −6

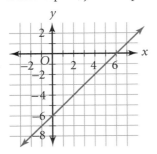

29. *x*-intercept: 4.5; *y*-intercept: 3

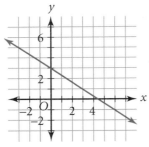

30. *x*-intercept: 5; *y*-intercept: −4

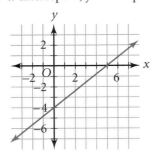

31. *x*-intercept: $\frac{7}{2}$; *y*-intercept: 1

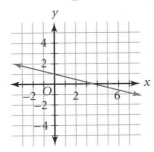

32. *x*-intercept: $-\frac{15}{2}$; *y*-intercept: 3

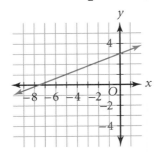

Lesson 7.6 Skill B (page 289)

★**14.** $r = 1$ ★**15.** $r \neq 0$

16. None; if the graph opens up, then the vertex contains the minimum value of *y*. Since the minimum value of *y* is positive, it can never be 0. Therefore, there are no values of *x* such that *y* equals 0.

17. $2v^3 - 3v^2 - 1$ **18.** $-5c - 3$

19. $2t^3 + 3t^2 - 2t - 8$ **20.** $-3n$

21. $2a^2 + 2a - 40$ **22.** $6r^2 + 23r + 21$

23. $-15d^2 - 13d - 2$ **24.** $48g^2 - 3$

25. $49y^2 - 28y + 4$ **26.** $-12w^2 - 2w + 30$

Chapter 7 Assessment (page 296)

56. none

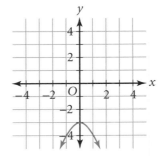

57. one

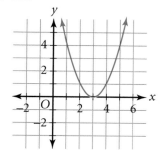

58. $\dfrac{3}{2}$ and $\dfrac{7}{2}$; two   **59.** -1; one   **60.** $y \geq -3\dfrac{2}{5}$

61. $y \leq 8.0625$ **62.** $y \geq -4$ **63.** $y \geq -9$

64. $y = -2x^2 + 2x + 60$ **65.** $y = 2x^2 + 16x + 24$

66. 400 units; \$12,800

CHAPTER 8

Lesson 8.2 Skill C (page 311)

42. $2\sqrt{11}$ **43.** $\dfrac{20}{7}$ **44.** $\dfrac{13}{11}$ **45.** $\dfrac{6\sqrt{7}}{5}$ **46.** $\dfrac{15\sqrt{2}}{11}$

Lesson 8.4 Skill A (page 319)

34. vertex: $(5, 0)$; axis of symmetry: $x = 5$; minimum

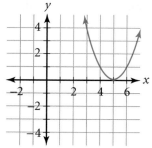

35. vertex: $(0, 4)$; axis of symmetry: $x = 0$; maximum

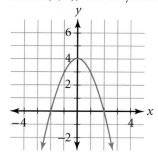

36. vertex: $(1, 1)$; axis of symmetry: $x = 1$; maximum

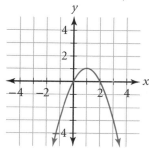

37. vertex: $(2, -4)$; axis of symmetry: $x = 2$; minimum

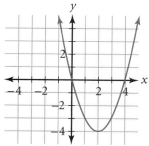

38. vertex: $\left(1, \dfrac{1}{2}\right)$; axis of

symmetry: $x = 1$; maximum

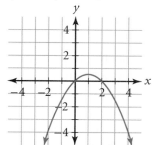

Lesson 8.4 Skill B (page 321)

37. If $-4t + 5 + 5t = 9$, then by the Commutative Property of Addition, $-4t + 5t + 5 = 9$. If $-4t + 5t + 5 = 9$, then by the Distributive Property, $t + 5 = 9$. If $t + 5 = 9$, then by the Subtraction Property of Equality, $t = 4$. Therefore, if $-4t + 5 + 5t = 9$, then $t = 4$.

38. If $\dfrac{2}{3}(n + 3) = \dfrac{1}{3}n$, then by the Distributive Property,

$\dfrac{2}{3}n + 2 = \dfrac{1}{3}n$. If $\dfrac{2}{3}n + 2 = \dfrac{1}{3}n$, then by the Subtraction

Property of Equality, $\dfrac{1}{3}n = -2$. If $\dfrac{1}{3}n = -2$, then by the

Multiplication Property of Equality, $n = -6$.

Therefore, if $\dfrac{2}{3}(n + 3) = \dfrac{1}{3}n$, then $n = -6$.

39. If $4m + 5m = m + 2$, then by the Distributive Property, $9m = m + 2$. If $9m = m + 2$, then by the Subtraction Property of Equality, $8m = 2$. If $8m = 2$, then by the Division Property of Equality, $m = \dfrac{1}{4}$. Therefore, if $4m + 5m = m + 2$, then $m = \dfrac{1}{4}$.

40. If $\dfrac{2y - 2}{3} = 4$, then by the Multiplication Property of Equality, $2y - 2 = 12$. If $2y - 2 = 12$, then by the Addition Property of Equality, $2y = 14$. If $2y = 14$, then by the Division Property of Equality, $y = 7$. Therefore, if $\dfrac{2y - 2}{3} = 4$, then $y = 7$.

Lesson 8.5 Skill A (page 325)

13. Zero-Product Property because the given equation is given in the form $ab = 0$

14. quadratic formula because the coefficients may not permit easy factoring. (First put the equation in standard form.)

15. quadratic formula because the coefficients involve decimals

16. Graph $y = x^2 + 2$. Observe that the graph does not cross the x-axis.

17. Observe that the Zero-Product Property will give two distinct solutions. Two distinct solutions indicate two distinct x-intercepts.

18. Apply the quadratic formula and observe that the two solutions are real and distinct. Alternatively, calculate the discriminant and observe that it is a positive number. Two distinct solutions indicate two distinct x-intercepts.

19. Graph $y = x^2 - 2$. Observe that the graph crosses the x-axis in two distinct points.

20. Observe that the Zero-Product Property will give one solution. One solution indicates one x-intercept.

21. Calculate the discriminant and observe that it is a negative number. A negative discriminant indicates that there are no x-intercepts.

22. Set the discriminant equal to 0 and solve for b. $b^2 - 4(0.5)(1) = 0$. There are two values of b that meet the specification. They are $\pm\sqrt{2}$. The graph of $y = 0.5x^2 + \sqrt{2}x + 1$ has one x-intercept. The graph of $y = 0.5x^2 - \sqrt{2}x + 1$ has one x-intercept.

23. Write the discriminant $n^2 - 4(n + 1)$, or $n^2 - 4n - 4$. Solve $n^2 - 4n - 4 = 0$. You will get $2 \pm \sqrt{2}$ as solutions. The graph of $y = x^2 + nx + n + 1$ will have one x-intercept when $n = 2 - 2\sqrt{2}$ or when $n = 2 + 2\sqrt{2}$.

24. $y = 2x - 2.5$; 117.5 **25.** $y = -3x + 7$; -173

Lesson 8.5 Skill B (page 327)

4. $6m^2 = 7m + 5$ given $6m^2 - 7m - 5 = 0$
Subtraction Property of Equality
$(2m + 1)(3m - 5) = 0$ Distributive Property
$2m + 1 = 0$ or $3m - 5 = 0$ Zero-Product Property
$2m = -1$ or $3m = 5$ Subtraction and Addition Properties of Equality
$m = -\dfrac{1}{2}$ or $m = \dfrac{5}{3}$
Division Property of Equality

In Exercises 5–10, arguments may vary. Sample arguments are given.

5. $n^2 - 5n = 0$ given $n(n - 5) = 0$
Distributive Property (factoring)
$n = 0$ or $n - 5 = 0$ Zero-Product Property
$n = 0$ or $n = 5$ Addition Property of Equality

6. $4(t + 5)^2 = 32$ given
$(t + 5)^2 = 8$ Division Property of Equality
$t + 5 = \pm\sqrt{8}$ Take the square root of each side.
$t = -5 \pm\sqrt{8}$ Subtraction Property of Equality
$t = -5 \pm 2\sqrt{2}$ Product Property of Square Roots

7. $8b^2 + 2b - 3 = 0$ given
$(4b + 3)(2b - 1) = 0$ Distributive Property (factoring)
$4b + 3$ or $2b - 1 = 0$ Zero Product Property
$4b = -3$ or $2b = 1$ Subtraction and Addition Properties of Equality
$b = -\dfrac{3}{4}$ or $b = \dfrac{1}{2}$ Division Property of Equality

8. $d^2 - 12d = -36$ given
$d^2 - 12d + 36 = 0$ Addition Property of Equality
$(d - 6)^2 = 0$ Distributive Property (factoring)
$d - 6 = 0$ Take the square root of each side.
$d = 6$ Subtraction Property of Equality

9. $4r^2 - 9^2 = 0$ given
$(2r - 9)(2r + 9) = 0$ Distributive Property (factoring)
$2r - 9 = 0$ or $2r + 9 = 0$ Zero-Product Property
$2r = 9$ or $2r = -9$ Addition and Subtraction Properties of Equality
$r = \pm\dfrac{9}{2}$ Division Property of Equality

10. $4(z-5)^2 + 7 = 32$ given

$4(z-5)^2 = 25$ Subtraction Property of Equality

$(z-5)^2 = \dfrac{25}{4}$ Division Property of Equality

$(z-5) = \pm\sqrt{\dfrac{25}{4}}$ Take the square root of each side.

$(z-5) = \pm\dfrac{5}{2}$ Simplify the square root.

$z = 5 \pm \dfrac{5}{2}$ $z = \dfrac{15}{2}$ or $z = \dfrac{5}{2}$

Add and subtract numbers.

11. By definition, the x-intercepts of the graph of $y = x^2 - 7x + 12$ are the real solutions to $x^2 - 7x + 12 = 0$. By the Distributive Property, the x-intercepts are the real solutions to $(x-4)(x-3) = 0$. By the Zero-Product Property, the solutions are 4 and 3. Since 4 and 3 are positive, the x-intercepts are positive.

12. By definition, the x-intercepts of the graph of $y = x^2 - 3x - 18$ are the real solutions to $x^2 - 3x - 18 = 0$. By the Distributive Property, the x-intercepts are the real solutions to $(x-6)(x+3) = 0$. By the Zero-Product Property, the solutions are 6 and -3. Since 6 is positive and -3 is negative, the x-intercepts have opposite signs.

13. $\left(\dfrac{7}{2}, \dfrac{19}{4}\right)$ **14.** $\left(-\dfrac{13}{14}, -\dfrac{11}{14}\right)$ **15.** $\left(\dfrac{3}{7}, \dfrac{5}{14}\right)$

16. $(2, -2)$ **17.** $(-5, 10)$ **18.** $\left(\dfrac{10}{7}, \dfrac{20}{7}\right)$

CHAPTER 9

Lesson 9.4 Skill B (page 359)

32. $r > -\dfrac{13}{2}$ **33.** $s \le 1$ **34.** $t > 5$

35. $x \ge 0$ **36.** $d \le -\dfrac{35}{17}$ **37.** $y < -10$

Lesson 9.4 Skill C (page 361)

26. $\dfrac{n^2+2}{(n-1)^3}$, or $\dfrac{n^2+2}{n^3-3n^2+3n-1}$

27. $\dfrac{5m^2+9m+3}{m(m+1)^2}$, or $\dfrac{5m^2+9m+3}{m^3+2m^2+m}$

28. $A = 3$ **29.** $(0, 0)$ **30.** $\left(\dfrac{7}{4}, -\dfrac{1}{14}\right)$ **31.** $\left(-\dfrac{1}{3}, \dfrac{5}{2}\right)$

32. $\left(\dfrac{5}{6}, \dfrac{1}{10}\right)$ **33.** $(3, -2)$ **34.** $(1, -4)$

Lesson 9.5 Skill A (page 363)

25. $\dfrac{4}{35}$ **26.** $\dfrac{2}{33}$ **27.** $-\dfrac{5}{18}$

28. $-\dfrac{16}{35}$ **29.** $-\dfrac{2}{11}$ **30.** $-\dfrac{17}{63}$

Lesson 9.5 Skill B (page 365)

30. $\dfrac{1}{n} - \dfrac{1}{n+1} = \dfrac{1}{n(n+1)}$; since $n > 0$, n, $n+1$, and $n(n+1)$ are all positive. Since $\dfrac{1}{n(n+1)}$ is the quotient of two positive numbers, $\dfrac{1}{n(n+1)}$ is positive.

31. $-\dfrac{2a}{3b^2}$ **32.** $\dfrac{1}{3n}$ **33.** -1 **34.** $\dfrac{1}{v}$

35. $-\dfrac{3yz^2}{7}$ **36.** $-\dfrac{15}{7a^2b^2}$ **37.** $\dfrac{5d}{c}$ **38.** $\dfrac{1}{r}$

Lesson 9.7 Skill A (page 373)

30. $y \ge -11$ **31.** $y \le 6$ **32.** $y \ge -3.5$

33. $y \le -\dfrac{41}{4}$ **34.** $y \ge -\dfrac{289}{16}$ **35.** $y \le \dfrac{9}{4}$

Lesson 9.7 Skill B (page 375)

23. $\sqrt{\dfrac{a}{25}} = \left(\dfrac{a}{25}\right)^{\frac{1}{2}} = \dfrac{a^{\frac{1}{2}}}{25^{\frac{1}{2}}} = \dfrac{a^{\frac{1}{2}}}{(5^2)^{\frac{1}{2}}}$

$= \dfrac{a^{\frac{1}{2}}}{5^{2 \cdot \frac{1}{2}}} = \dfrac{a^{\frac{1}{2}}}{5} = \dfrac{\sqrt{a}}{5}$

24. $\sqrt{\dfrac{a}{49}} = \left(\dfrac{a}{49}\right)^{\frac{1}{2}} = \dfrac{a^{\frac{1}{2}}}{49^{\frac{1}{2}}} = \dfrac{a^{\frac{1}{2}}}{(7^2)^{\frac{1}{2}}}$

$= \dfrac{a^{\frac{1}{2}}}{7^{2 \cdot \frac{1}{2}}} = \dfrac{a^{\frac{1}{2}}}{7} = \dfrac{\sqrt{a}}{7}$

25. $\sqrt{\dfrac{64a}{121}} = \left(\dfrac{64a}{121}\right)^{\frac{1}{2}} = \dfrac{(64a)^{\frac{1}{2}}}{121^{\frac{1}{2}}} = \dfrac{(8^2 \cdot a)^{\frac{1}{2}}}{(11^2)^{\frac{1}{2}}}$

$= \dfrac{(8^{2 \cdot \frac{1}{2}})a^{\frac{1}{2}}}{11^{2 \cdot \frac{1}{2}}} = \dfrac{8a^{\frac{1}{2}}}{11} = \dfrac{8\sqrt{a}}{11}$

26. $\sqrt{\dfrac{16a}{9}} = \left(\dfrac{16a}{9}\right)^{\frac{1}{2}} = \dfrac{(16a)^{\frac{1}{2}}}{9^{\frac{1}{2}}} = \dfrac{(4^2 \cdot a)^{\frac{1}{2}}}{(3^2)^{\frac{1}{2}}}$

$\qquad = \dfrac{(4^{2 \cdot \frac{1}{2}})a^{\frac{1}{2}}}{3^{2 \cdot \frac{1}{2}}} = \dfrac{4a^{\frac{1}{2}}}{3} = \dfrac{4\sqrt{a}}{3}$

Lesson 9.8 Skill B (page 381)

7. a. yes

   b. domain: all real numbers except 0; range: all real numbers except 0

   c. $y = \dfrac{8}{25}$

   d. No; if $x = 100$, then $y = 0.64 \neq 0.25$.

8. a. yes

   b. domain: all real numbers except 0; range: all real numbers except 0

   c. $y = \dfrac{3}{25}$

   d. No; if $x = 100$, then $y = 0.18 \neq 0.25$.

9. a. yes

   b. domain: all real numbers except 0; range: all real numbers except 0

   c. $y = \dfrac{8}{25}$

   d. No; if $x = 100$, then $y = 0.48 \neq 0.25$.

10. $\dfrac{1}{n + 1} < \dfrac{1}{n}$ for all natural numbers n. If n is a natural number, then $n + 1 > n$. Since $n(n + 1) > 0$, $\dfrac{n + 1}{n(n + 1)} > \dfrac{n}{n(n + 1)}$ by the Division Property of Inequality. Thus, $\dfrac{1}{n} > \dfrac{1}{n + 1}$, or $\dfrac{1}{n + 1} < \dfrac{1}{n}$. This means that as n increases, the y-coordinate is positive and decreases. Thus, the graph gets closer and closer to the x-axis while staying above it.

11. Sample counterexample: If $n = 0$, then $(3n - 1)(2n + 3) = (-1)(3) = -3 > -6$.

12. Sample counterexample: If $n = 0$, then $(-2n - 1)(2n - 3) = (-1)(-3) = 3 < 5$.

13. Sample counterexample: If $n = 0$, then $(-3n)(n + 1) = (0)(1) = 0$, which is not positive.

Lesson 9.9 Skill A (page 385)

12. Multiplication Property of Equality

13. definition of multiplication of fractions

★**14.** If $\dfrac{1}{b} = \dfrac{c}{d}$, then by the Multiplication Property of Equality, $d \cdot \dfrac{1}{b} = d \cdot \dfrac{c}{d}$. By definition of multiplication of fractions, $\dfrac{d}{b} = \dfrac{dc}{d}$ and $\dfrac{d}{b} = \dfrac{d}{d} \cdot c$. By the Inverse Property of Multiplication, $\dfrac{d}{b} = 1 \cdot c$. By the Identity Property of Multiplication, $\dfrac{d}{b} = c$.

★**15.** If $\dfrac{1}{b} = \dfrac{c}{d}$, then by the Cross Product Property of Proportions, $d = bc$. By the Division Property of Equality, $\dfrac{d}{dc} = \dfrac{bc}{dc}$. By the definition of multiplication of fractions, $\dfrac{d}{d} \cdot \dfrac{1}{c} = \dfrac{b}{d} \cdot \dfrac{c}{c}$. By the Inverse Property of Multiplication, $1 \cdot \dfrac{1}{c} = \dfrac{b}{d} \cdot 1$. By the Identity Property of Multiplication, $\dfrac{1}{c} = \dfrac{b}{d}$. By the Symmetric Property of Equality, $\dfrac{b}{d} = \dfrac{1}{c}$.

★**16.** If $\dfrac{a}{b} = \dfrac{c}{d}$, then $\dfrac{a}{b} + 1 = \dfrac{c}{d} + 1$ by the Addition Property of Equality. By the Substitution Principle, $\dfrac{a}{b} + \dfrac{b}{b} = \dfrac{c}{d} + \dfrac{d}{d}$. Thus, by the definition of addition of fractions, $\dfrac{a + b}{b} = \dfrac{c + d}{d}$.

★**17.** If $\dfrac{a}{b} = \dfrac{c}{d}$, then $\dfrac{a}{b} - 1 = \dfrac{c}{d} - 1$ by the Subtraction Property of Equality. By the Substitution Principle, $\dfrac{a}{b} - \dfrac{b}{b} = \dfrac{c}{d} - \dfrac{d}{d}$. Thus, by the definition of subtraction of fractions, $\dfrac{a - b}{b} = \dfrac{c - d}{d}$.

★**18.** If $\dfrac{a}{b} = \dfrac{c}{d}$, then by the Multiplication Property of Equality, $\dfrac{a}{b} \cdot \dfrac{a}{b} = \dfrac{c}{d} \cdot \dfrac{a}{b}$. By definition of multiplication of fractions and the exponent 2, $\dfrac{a^2}{b^2} = \dfrac{c}{d} \cdot \dfrac{a}{b}$. By the Substitution Principle, $\dfrac{c}{d} \cdot \dfrac{a}{b} = \dfrac{c}{d} \cdot \dfrac{c}{d}$. By definition of multiplication of fractions and the exponent 2, $\dfrac{c}{d} \cdot \dfrac{c}{d} = \dfrac{c^2}{d^2}$. Therefore, by the Transitive Property of Equality, $\dfrac{a^2}{b^2} = \dfrac{c^2}{d^2}$.

19. Suppose that $\frac{a}{b} = \frac{c}{d}$ and $\frac{a}{d} = \frac{c}{b}$. Then $ad = bc$ and $ab = dc$. Thus, $ad - ab = bc - dc$. So, $a(d - b) = -c(d - b)$. If $b \neq d$, then $a = -c$. The statement is true when $b = d$ or when $a = -c$.

20. $x \neq 0, \frac{1}{4}$ **21.** $a \neq \pm\frac{1}{2}$ **22.** $v \neq 1, -\frac{5}{4}$

23. $x \neq -\frac{11}{12}$ **24.** $x \neq -\frac{7}{2}$ **25.** $x \neq 0, 4$

Chapter 9 Assessment (page 386)

36. Maria: 36 hours Ken: 72 hours

37. $x = 4.5$

38. yes; the domain and range are both all real numbers except 0; $y = \frac{24}{25}$; no, because $5 \cdot 50 \neq 4 \cdot 12$

39. $x = \pm\frac{4}{5}$

40. If $\frac{ax + b}{x} = 1$, then by the Multiplication Property of Equality, $ax + b = x$. If $ax + b = x$, then by the Subtraction Property of Equality, $b = x - ax$. By the Distributive Property, $b = x(1 - a)$. By the Division Property of Equality, $\frac{b}{1 - a} = x$. By the Symmetric Property of Equality, $x = \frac{b}{1 - a}$. The restrictions are $x \neq 0$ and $a \neq 1$.

CHAPTER 10

Lesson 10.2 Skill C (page 397)

32. $\frac{\sqrt{7}}{15}$ **33.** $\frac{7\sqrt{2}}{11}$ **34.** $\frac{5\sqrt{3}}{13}$

Lesson 10.3 Skill B (page 401)

33. $9\frac{5}{21}$ **34.** $15\frac{13}{15}$ **35.** $14\frac{4}{5}$ **36.** $8\frac{17}{22}$

37. $1\frac{8}{9}$ **38.** $3\frac{1}{5}$ **39.** $\frac{17}{99}$ **40.** $1\frac{1}{39}$

Lesson 10.3 Skill C (page 403)

14. $a \leq -\frac{1}{4}$ **15.** $x \geq -4$ **16.** $t = \frac{23}{3}$ **17.** $v = \frac{3}{2}$

18. $Q(-3, 4)$ and $Q'(-3, -4)$

19. no real solution

20. $m = -\frac{7}{2}$ or $m = \frac{5}{3}$ **21.** $-7 - 17\sqrt{2}$

22. $-30 - 20\sqrt{2}$ **23.** $-\frac{1}{3} + \frac{5}{6}\sqrt{2}$

24. $q = -12 + 9\sqrt{2}$ **25.** $v = \frac{10}{41} - \frac{3}{41}\sqrt{2}$

Lesson 10.6 Skill A (page 417)

8. The statement is sometimes true. If $r = 4$, then $4^2 \geq \sqrt{4}$ or $16 \geq 2$. This inequality is true. If $r = \frac{1}{4}$, then $\left(\frac{1}{4}\right)^2 \leq \sqrt{\frac{1}{4}}$ or $\frac{1}{16} \leq \frac{1}{2}$. This inequality is false.

9. If $\sqrt{\frac{t}{2}} = \frac{\sqrt{t}}{2}$, then, by squaring each side of the equation, $\frac{t}{2} = \frac{t}{4}$. This equation is true only when $t = 0$. Since t is a positive real number, the statement is never true.

10. If n is the square of an even positive integer, then $\sqrt{n}$ will be an even integer.

11. $x = 4$ **12.** $x = 49$ **13.** $x = 6$

14. $y = \frac{1}{5}$ **15.** $y = \frac{4}{25}$ **16.** $y = 8$

Index

Definitions of boldface entries can be found in the glossary.

A

Absolute value, 44

Absolute-value equations, 176

Absolute-value inequalities, 178

Adding fractions with like denominators, 356

Addition
  Distributive Property, 66
  fractions, 356–361
  horizontal, 234
  inverse property, 8, 56
  polynomials, 234, 238
  properties of, 8
  rational expressions, 356–361
  of real numbers, 46, 48
  solving inequalities with, 146, 148, 152
  square-root expressions, 398, 402, 406
  vertical, 234

Addition Property of Equality, 20, 62, 72, 76

Addition Property of Inequality, 148

Additive Identity Property, 8, 20

Additive inverse, 50, 56, 402

Algebraic expressions, 12–13

"And," 170

Applications
  academics, 6, 162, 376
  agriculture, 214
  banking and finance, 34, 54, 72, 80, 168, 414
  chemistry, 168, 200
  electricity, 370
  engineering, 100, 104, 116
  entertainment, 34, 124

food preparation, 174

gardening, 74, 174, 208

geometry, 6, 130, 214, 232, 238, 244, 308, 366

inventory, 152

landscaping, 194

manufacturing, 120, 180

personal finance, 13, 100, 116, 152, 158

physics, 228, 294, 334, 336, 378, 382, 418–419

retail/sales, 6, 102, 104, 124, 158, 238, 294

sports/fitness, 152

temperature, 26, 54, 72

travel/transportation, 110, 120, 228, 316, 370, 394, 418–419

veterinary science, 194

voting, 162

Approximations of functions, 420

Assessments, 36–37, 88–89, 138–139, 184–185, 216–217, 256–257, 296–297, 338–339, 386–387, 422–423. *See also* Mid-chapter reviews

Associative Property of Addition, 8, 18, 48, 54, 402

Associative Property of Multiplication, 8, 402

Axis of symmetry, 286

B

Base (of an exponential expression), 220

Basic Properties of Equality, 18–19

Basic Properties of Numbers, 8

Binomials, 230, 254, 268

Boundary lines, 206–207, 210

Braces, 10, 14

Brackets, 10

Braking distance, 228, 418–419

C

Canceling, 60

Circles, 308

Closed sets, 234, 252, 398, 400, 402

Closure
  Property of Addition, 8, 398
  Property of Multiplication, 8
  of numbers of the form $a + b\sqrt{2}$, 398, 400, 402

Coefficient of friction, 418

Coefficients, 66

Collinear points, 108

Combining like terms, 66

Common binomial factor, 268

Commutative Property of Addition, 8, 14, 22, 54, 66, 402

Commutative Property of Multiplication, 8, 402

Completing the square, 312, 314, 316

Complex fractions, 368–371

Composite numbers, 260

Compound inequalities, 146, 170–175

Conclusions, 84

Conditional statements, 84

Congruent rectangles, 238

Conjectures, 16

Conjugates, 400

Conjunctions, 146, 170

Consistent systems, 202

Constant of variation, 102, 378, 382

Constant slope, 110